Modified Theories of Gravity and Cosmological Applications

Modified Theories of Gravity and Cosmological Applications

Collection Editors

Panayiotis Stavrinos
Emmanuel N. Saridakis

MDPI

Basel • Beijing • Wuhan • Barcelona • Belgrade • Novi Sad • Cluj • Manchester

Collection Editors
Panayiotis Stavrinos
Department of Mathematics
National and Kapodistrian
University of Athens
Athens
Greece

Emmanuel N. Saridakis
National Observatory
of Athens
Athens
Greece

Editorial Office
MDPI AG
Grosspeteranlage 5
4052 Basel, Switzerland

This is a reprint of the Topical Collection, published open access by the journal *Universe* (ISSN 2218-1997), freely accessible at: https://www.mdpi.com/journal/universe/topical_collections/MGCA.

For citation purposes, cite each article independently as indicated on the article page online and as indicated below:

Lastname, A.A.; Lastname, B.B. Article Title. *Journal Name* **Year**, *Volume Number*, Page Range.

ISBN 978-3-7258-5603-9 (Hbk)
ISBN 978-3-7258-5604-6 (PDF)
https://doi.org/10.3390/books978-3-7258-5604-6

Contents

About the Editors

Panayiotis Stavrinos

Panayiotis Stavrinos is an organizer of International Conferences and Seminars and an Editorial Board Member of international journals. He has published four monographs in the English language. He is a founding member of the Balkan Society of Geometers and served as its Vice President from 2000 to 2004. He is also a founding member of the Hellenic Relativity, Gravitation and Cosmology Society. He is a referee in many international journals and an invited speaker at Universities and Workshops worldwide. He contributed to the development of the research topic of Finslerian Gravitation and Finslerian Cosmology, with more than 100 research papers published in international journals and proceedings in international conferences since 1999. In particular, he has introduced a special type of Finsler Geometry and a cosmological model. In addition, he has introduced the Raychaudhuri equation in the Finslerian anisotropic framework. He is a member of the scientific and organizing committees of more than 20 conferences.

Emmanuel N. Saridakis

Emmanuel N. Saridakis is an author of 240 publications, with a total of 17,000 citations and an h-index of 75 (Google Scholar). He is a representative of Greece and a Working Group Leader at COST Action "Addressing observational tensions in cosmology with systematics and fundamental physics" (CosmoVerse), a member of the "Hellenic Relativity, Gravitation and Cosmology Society" (HSRGC), and a member of the management board. He is on the review boards of various journals and has performed more than 850 reviews of scientific works. He has given 70 talks at internationally established conferences, among which 50 were invited talks. He is also a member of the organizing committee of 15 conferences.

universe

MDPI

Editorial

Editorial: Modified Theories of Gravity and Cosmological Applications—Topical Collection

Panayiotis Stavrinos [1,*] and Emmanuel N. Saridakis [2,3,4]

1 Department of Mathematics, National and Kapodistrian University of Athens, Panepistimiopolis, 15784 Athens, Greece
2 National Observatory of Athens, Lofos Nymfon, 11852 Athens, Greece; msaridak@noa.gr
3 CAS Key Laboratory for Research in Galaxies and Cosmology, University of Science and Technology of China, Hefei 230026, China
4 Departamento de Matemáticas, Universidad Católica del Norte, Avda. Angamos 0610, Antofagasta 1249004, Chile
* Correspondence: pstavrin@math.uoa.gr

General relativity (GR) has been remarkably successful in describing the gravitational interaction through the curvature of spacetime. It has provided an unparalleled theoretical framework that accurately explains a wide range of astrophysical and cosmological phenomena. The theory's predictions, such as the perihelion precession of Mercury, the bending of light by massive bodies (gravitational lensing), and the redshift of light in a gravitational field (gravitational redshift), have been confirmed with high precision. Moreover, the detection of gravitational waves by LIGO and Virgo has provided direct observational evidence for the dynamic nature of spacetime predicted by GR, further solidifying its validity. In the cosmological context, GR underpins the standard ΛCDM model, successfully describing the large-scale structure of the universe, cosmic expansion, and the Cosmic Microwave Background (CMB) radiation. Despite these successes, GR is not without challenges [1,2]. It predicts singularities, such as those found in black holes and the Big Bang, where the theory itself breaks down and quantum effects are expected to become significant. Moreover, GR is non-renormalizable and hence it cannot be quantized according to the standard procedures. Additionally, the existence of dark matter and dark energy—introduced to reconcile observations with the theory—raises questions about whether GR remains valid on cosmic scales or requires modification. Finally, recently, there have been some tensions between theoretical predictions and observations, such as the H_0 and σ_8 ones, the possible effects of local anisotropy, etc. All these issues suggest that while GR is a powerful and elegant theory, it may not be the final description of gravitational phenomena, motivating ongoing research into alternative and extended theories of gravity.

Modified theories of gravity [3–5] extend the framework of general relativity through various methods, leading to different field equations and, consequently, distinct cosmological implications. Broadly, modified gravity theories can be constructed through several approaches. One of the most common methods is the extension of the Einstein–Hilbert action by introducing arbitrary functions of curvature invariants, leading to $f(R)$ gravity [6–9]. Another prominent approach is the formulation of gravity using torsion, such as in teleparallel gravity and its extensions like $f(T)$ gravity [10–27] and $f(T,B)$ gravity [20,28], as well as non-metricity, such as in $f(Q)$ gravity [29–35] and in $f(Q,C)$ gravity [36]. An alternative class of modifications involves scalar–tensor theories, where additional scalar fields are coupled to gravity, such as in Brans–Dicke theory and Horndeski theories, offering a framework to study cosmological inflation and dark energy dynamics [37]. Another widely explored modification involves extra-dimensional theories, such as Kaluza–Klein and brane-world models, which attempt to explain gravity behavior at different scales by considering higher-dimensional spacetimes [38]. Furthermore, anisotropic cosmological Finsler and Finsler-like models have been developed when the underlying geometry of spacetime adopts a more generalized metric structure than the traditional Riemannian one [39–50]. In particular, Finsler-based cosmological scenarios may provide solutions to the anisotropy tension, since the internal geometric anisotropy changes the luminosity distance toward varying directions for a fixed redshift [51,52].

Modified gravity does not exclude standard general relativity but instead highlights its limitations. These modifications play an essential role and contribute significantly to modern cosmology, providing a foundation for understanding the physical phenomena of the Universe. Experiments involving gravitational wave detectors, such as LIGO, VIRGO, and KAGRA, amplify the scope of modified gravity investigations, extending models beyond general relativity. In the present Special Issue "Modified Theories of Gravity and Cosmological Applications", the published articles contribute different approaches and consequences to the theoretical and observational treatment of fundamental open problems of cosmology, offering deeper understanding of gravitational phenomena beyond the bounds of GR. In these considerations, novel cosmological models play a significant role.

In [53], the authors present a review of energy–momentum squared gravity (EMSG). The inclusion of quadratic contributions from the energy–momentum components has intriguing cosmological implications, particularly during the Universe's early epochs. These effects dominate under high-energy conditions, enabling EMSG to potentially address unresolved issues in general relativity (GR). The exploration of EMSG presents a novel framework for extending of understanding of gravitational phenomena beyond the limits of GR. This review offers to shed light on some of the fundamental questions in modern cosmology and gravitational theory.

In [54], the authors elucidate the back-reaction within macroscopic gravity in terms of the non-metricity of average geometry. With a $1+3$ decomposition of the spacetime, they analyze how geometric flows are modified by deriving the Raychaudhuri and Sachs equations, which govern the expansion of the geodesic flows. Physically, these equations characterize the flow of the cosmic fluid and photon bundles, respectively. The work presents interesting results.

In [55], the authors propose a novel dark-energy equation-of-state parametrization, with a single parameter that quantifies the deviation from ΛCDM cosmology. Additionally, under this scenario, dark energy behaves like a cosmological constant at high redshifts, while the deviation becomes significant at low and recent redshifts, especially in the future. In light of this, the authors conduct a detailed analysis with observations, namely, with Hubble function (OHD), Pantheon, and baryon acoustic oscillations (BAO) data.

In [56], Palatini-like theories of gravity are studied in connection to models incorporating linear generalized uncertainty principles. In this consideration, the authors delve into the thermodynamics of systems comprising both Bose and Fermi gases. This analysis encompasses the equations of state for various systems and liquid helium. Additionally, the authors derive the energy and specific heat for photon and phonon gases.

In [57], the authors demonstrate reconstruction of the forms of $F(R)$ in the modified theory of gravity and the metric-compatible $F(T)$ together with the symmetric $F(Q)$ in alternative teleparallel theories of gravity, from different perspectives, primarily placing emphasis on a viable FLRW radiation era. Inflation has also been studied in this framework for a particular choice of scalar potential. The inflationary parameters are found to agree appreciably with the recently released observational data. This paper offers an interesting perspective of the corresponding research area.

In [58], the authors use the cosmography approach to study a particular self-tuning filter solution focused on a zero-curvature fixed point to study the H0 tension. In this approach, the equations restrict the universe's evolution to certain scenarios, including radiation-like expansion, matter-like expansion, and late-time acceleration. Furthermore, they build the cosmographic series of the Fab-Four theory to obtain the kinematic parameters as the Hubble constant H0 and the deceleration parameter for all the scenarios mentioned. These results represent an interesting approach to the open problem of the current discrepancy of cosmological tensions.

In [59], the authors use generalized type attractor models of single-field inflation in order to accommodate the production of primordial black holes (PBHs) by adding a near-inflection point to the inflaton scalar potential at smaller scales, in good agreement with measurements of cosmic microwave background (CMB) radiation. Fine-tuning of the parameters in these models was needed not only to match the CMB measurements (inflation is robust in the E-models) but also to generate PBH production at lower scales during the USR phase, which would lead to a sizeable portion of PBHs in the current dark matter.

In [60], Newtonian fractional-dimension gravity (NFDG) is applied in an alternative gravitational model to some notable cases of galaxies with little or no dark matter. In the case of the ultra-diffuse galaxy AGC 114905, the authors show that NFDG methods can effectively reproduce the observed rotation curve using a variable fractional dimension D(R).

In [61], the authors investigate a general class of pseudo-Finsler spaces with (α, β)-metrics. They establish necessary and sufficient conditions such that these exhibit a Finsler spacetime structure. This means that the fundamental tensor has a Lorentzian signature on a conic subbundle of the tangent bundle and thus the existence of a cone of future-pointing time-like vectors is ensured. The work offers an interesting consideration of applications of Finsler geometry to the gravitational theory.

In [62], Hořava–Lifshitz (HL) theory of gravity, in which equations governing cosmological evolution include a new term scaling similarly to the dark radiation term in the Friedmann equations, enabling a bounce of the universe instead of initial singularity. This review describes past works on the stability of such a bounce in different formulations of HL theory, an initial detailed balance scenario, and further projectable versions containing higher-than-quadratic terms to the original action.

In [63], the authors focus on the theoretical research and application of strong lenses and weak lenses. The authors begin by introducing the basic principles of geometric gravitational lensing, the representation of deflection angles, and the curvature relationship in different geometric spaces. In addition, they propose interesting views regarding the detection of dark matter and research on dark energies such as PLANCK and BIGBOSS.

In [64], the author reviews the construction of the deformed quaplectic group that is given by the semidirect product of $U(1,3)$ with the deformed (non-commutative) Weyl–Heisenberg group corresponding to non-commutative fiber coordinates and momenta. This construction leads to more general algebras being given by a two-parameter family of deformations of the quaplectic algebra. This analysis allows the author to find a direct link between non-commutative curved phase spaces in lower dimensions and commutative flat phase spaces in higher dimensions.

In [65], the authors explore a bouncing cosmology with modified generalized Chaplygin Gas in a bulk viscosity framework. In addition, the authors study the equation of state parameter under various cases and review the stability of the models through the sign of the squared speed of sound. Additionally, the authors propose a bouncing scale factor of a special form. The bouncing point appeared at $t = 0$. This model also realizes big bounce in the framework of Einstein gravity as well as $f(T)$-modified gravity, which violates the null energy condition.

In [66], the author provides a pedagogical introduction to the degenerate Hamiltonian systems, showing both very simple mechanical examples and general arguments about how such systems work. For the familiar field theory models, he explains why the gauge freedom there "hits twice" in the sense of producing twice as many first-class constraints as gauge symmetries, and why primary, and only primary, constraints should be put into the total Hamiltonian. The discussion of the author on the role of constraints and degrees of freedom in the Hamiltonian Formalism offers useful information.

In [67], the authors consider a Friedmann–Lemâitre–Robertson–Walker cosmological model dominated by bulk viscous cosmic fluid in $f(Q)$ gravity, where Q represents the non-metricity scalar. They find an interesting result: that the $f(Q)$ cosmological model efficiently describes the observational data. In addition, the authors present the evolution of their deceleration parameter with redshift, and show that it properly predicts a transition from decelerated to accelerated phases of the universe's expansion.

In [68], a simple cooling model of white dwarf stars is re-analyzed in Palatini $f(R)$ gravity. Modified gravity affects the white dwarf structures and, consequently, their ages. The author finds that the resulting super-Chandrasekhar white dwarfs need more time to cool down than sub-Chandrasekhar ones. In addition, he discusses the basic formalism of Palatini $f(R)$ gravity and briefly reviews how it modifies the hydrostatic balance equations which he employs to derive the modified temperature equation in this framework.

In [69], the authors study a perturbation theory for embedding gravity equations in a background for which corrections to the embedding function are linear with respect to corrections to the flat metric. It is shown that this function can be chosen in such a way that the gravitational potential is in a good agreement with the observed distribution of dark matter in a galactic halo. In this approach, the problem of constructing a perturbation theory for Regge–Teitelboim equation to provide background information is considered, and an additional energy–momentum tensor is proposed. It is induced by this embedding, with additional fictitious embedding matter simulated with dark matter. This is a novel and interesting approach.

In [70], the authors explore wormhole geometry in spiral galaxies under third-order Lovelock gravity. Using the cubic spline interpolation technique, they find the rotational velocity of test particles in the halo region of our spiral galaxy from observed values of radial distances and rotational velocities. Taking this value of the rotational velocity, they are able to show that it is possible to present a mathematical model regarding the viable existence of wormholes in the galactic halo region of the Milky Way under Lovelock gravity. An important result that they obtain from the present investigation is that the galactic wormhole in the halo region can exist with normal matter as well as exotic matter.

In [71], the authors present an analysis of a chiral cosmological scenario from the perspective of K-essence formalism. In this setup, several scalar fields interact within the kinetic and potential sectors. In particular, in this framework, the authors consider a flat Friedmann–Robertson–Lamaître–Walker universe coupled minimally to two quintom fields: one quintessence and one phantom. In addition, they present an interesting explanation of the "big-bang" singularity by means of a "big-bounce". Moreover, for the quantum counterpart, the Wheeler–DeWitt equation is analytically solved for various cases.

In [72], the authors argue that the Barrow entropy should run based on black hole thermodynamics (i.e., it should be energy-scale dependent), which is reasonable given that quantum gravitational corrections are expected to be important only in the high-energy regime. When applied to the Friedmann equations, the authors suggest that such a running Barrow entropy index could give rise to a dynamical effective dark energy, which is asymptotically positive and vanishing, but negative at the big bang. Such a sign switching dark energy could help to alleviate the Hubble tension. Other cosmological implications are also discussed.

In [73], two different Hamiltonian formulations of the metric gravity are discussed in this work and applied to describe a free gravitational field in d-dimensional Riemann spacetime. The theory of canonical transformations, which relates equivalent Hamiltonian formulations of metric gravity, is investigated in detail and the theory of integral invariants and its applications to the Hamiltonian metric gravity are also discussed. In this approach, Dirac's modifications of the classical Hamilton method contain several crucial advantages, which indicate the obvious superiority of this method in order to develop various non-contradictory Hamiltonian theories of many physical fields.

In [74], the authors review the history of, as well as recent progress in, the composite view of the gravitational constant. This a useful approach which could be applied in future research aimed at understanding and discussing Newton's gravitational constant. This work has provided an overview of much of the work that has been conducted in relation to the composite view of the gravitational constant. The implications of this work may open doors of insight into relationships between macroscopic gravity phenomena and the Planck scale.

In [75], the authors present a brief description of non-compactified higher-dimensional theories from the perspective of general relativity. More concretely, the space–time–matter theory, or induced matter theory, and the reduction procedure used to construct the modified Brans–Dicke theory and the modified Sáez–Ballester theory are briefly explained. Finally, the authors apply the latter to the Friedmann–Lemaître–Robertson–Walker (FLRW) cosmological models in arbitrary dimensions and analyze the corresponding solutions.

The remarkable success of general relativity in describing gravitational interactions has been firmly established through numerous experimental and observational confirmations. However, the challenges it faces at both theoretical and observational levels have spurred the exploration of modified gravity theories as viable extensions. These alternative theories, such as $f(R)$, $f(T)$, and $f(Q)$ gravity, scalar–tensor theories, extra-dimensional models, Finsler-based constructions, etc., offer new insights into possible cosmological problems, including the nature of dark matter and dark energy, the resolution of cosmic tensions like H_0 and σ_8, possible anisotropies in the CMB spectrum, and the quest for a consistent quantum theory of gravity.

The contributions in this Special Issue focus on the latest advancements in the field of modified gravity and cosmological applications. The diverse range of approaches presented highlights the growing interdisciplinary efforts to accumulate theoretical developments with observational data. The articles present novel approaches to some fundamental questions, using investigations which extend upon ΛCDM cosmology, proposing new relations between macroscopic gravity phenomena and the Planck scale, anisotropic extensions based on the Finsler geometry for studying of the gravitational field, and higher-dimensional theories from the perspective of general relativity to construct the modified Brans–Dicke theory. Palatini $f(R)$, Hořava–Lifshitz, and Lovelock gravity techniques are also applied to the cosmological framework. Furthermore, Barrow entropy is also examined in relation to the Friedmann equations. The investigation of Nature as a whole, and not only its specific parts, seeks to reveal the truth—namely the physical laws that govern objective reality—and this is a hard and arduous procedure. It is not a solitary or instant pursuit, but demands collective effort from the scientific community in many generations and different research directions. Nature is out there, it is knowable, and the physical laws, first principles and the highest causes, are waiting to to be discovered and explored further. We believe that the works published in this Issue will contribute proportionally to deepening our understanding of the fundamental nature of spacetime and the universe.

Author Contributions: The authors equivalently contributed in the review and editing of the papers. All authors have read and agreed to the published version of the manuscript.

Conflicts of Interest: The authors declare no conflict of interest.

References

1. Abdalla, E.; Franco Abellán, G.; Aboubrahim, A.; Agnello, A.; Akarsu, O.; Akrami, Y.; Alestas, G.; Aloni, D.; Amendola, L.; Anchordoqui, L.A.; et al. Cosmology intertwined: A review of the particle physics, astrophysics, and cosmology associated with the cosmological tensions and anomalies. *J. High Energy Astrophys.* **2022**, *34*, 49–211. [CrossRef]
2. Perivolaropoulos, L.; Skara, F. Challenges for ΛCDM: An update. *New Astron. Rev.* **2022**, *95*, 101659.
3. Saridakis, E.N. et al. [CANTATA]. *Modified Gravity and Cosmology. An Update by the CANTATA Network*; Springer: Cham, Switzerland, 2021; ISBN 978-3-030-83714-3/978-3-030-83717-4/978-3-030-83715-0. [CrossRef]
4. Capozziello, S.; De Laurentis, M. Extended Theories of Gravity. *Phys. Rep.* **2011**, *509*, 167. [CrossRef]
5. Nojiri, S.; Odintsov, S.D.; Oikonomou, V.K. Modified Gravity Theories on a Nutshell: Inflation, Bounce and Late-time Evolution. *Phys. Rep.* **2017**, *692*, 1–104. [CrossRef]
6. Starobinsky, A.A. A New Type of Isotropic Cosmological Models Without Singularity. *Phys. Lett. B* **1980**, *91*, 99. [CrossRef]
7. De Felice, A.; Tsujikawa, S. $f(R)$ theories. *Living Rev. Relativ.* **2010**, *13*, 3. [CrossRef]
8. Nojiri, S.; Odintsov, S.D. Unified cosmic history in modified gravity: From $F(R)$ theory to Lorentz non-invariant models. *Phys. Rep.* **2011**, *505*, 59–144. [CrossRef]
9. Capozziello, S. Curvature quintessence. *Int. J. Mod. Phys. D* **2002**, *11*, 483–492. [CrossRef]
10. Cai, Y.F.; Capozziello, S.; De Laurentis, M.; Saridakis, E.N. $f(T)$ teleparallel gravity and cosmology. *Rep. Prog. Phys.* **2016**, *79*, 106901. [CrossRef]
11. Linder, E.V. Einstein's Other Gravity and the Acceleration of the Universe. *Phys. Rev. D* **2010**, *81*, 127301; Erratum in *Phys. Rev. D* **2010**, *82*, 109902. [CrossRef]
12. Chen, S.H.; Dent, J.B.; Dutta, S.; Saridakis, E.N. Cosmological perturbations in $f(T)$ gravity. *Phys. Rev. D* **2011**, *83*, 023508. [CrossRef]
13. Wu, P.; Yu, H.W. Observational constraints on $f(T)$ theory. *Phys. Lett. B* **2010**, *693*, 415. [CrossRef]
14. Cai, Y.-F.; Chen, S.-H.; Dent, J.B.; Dutta, S.; Saridakis, E.N. Matter Bounce Cosmology with the $f(T)$ Gravity. *Class. Quant. Grav.* **2011**, *28*, 2150011. [CrossRef]
15. Li, M.; Miao, R.X.; Miao, Y.G. Degrees of freedom of $f(T)$ gravity. *J. High Energy Phys.* **2011**, *1107*, 108. [CrossRef]
16. Bamba, K.; Myrzakulov, R.; Nojiri, S.; Odintsov, S.D. Reconstruction of $f(T)$ gravity: Rip cosmology, finite-time future singularities and thermodynamics. *Phys. Rev. D* **2012**, *85*, 104036. [CrossRef]
17. Wu, Y.P.; Geng, C.Q. Primordial Fluctuations within Teleparallelism. *Phys. Rev. D* **2012**, *86*, 104058. [CrossRef]
18. Myrzakulov, R. Accelerating universe from $F(T)$ gravity. *Eur. Phys. J. C* **2011**, *71*, 1752.
19. Otalora, G. Scaling attractors in interacting teleparallel dark energy. *J. Cosmol. Astropart. Phys.* **2013**, *1307*, 044. [CrossRef]
20. Bahamonde, S.; Böhmer, C.G.; Wright, M. Modified teleparallel theories of gravity. *Phys. Rev. D* **2015**, *92*, 104042. [CrossRef]
21. Farrugia, G.; Said, J.L. Stability of the flat FLRW metric in $f(T)$ gravity. *Phys. Rev. D* **2016**, *94*, 124054. [CrossRef]
22. Awad, A.M.; Capozziello, S.; Nashed, G.G.L. *D*-dimensional charged Anti-de-Sitter black holes in $f(T)$ gravity. *J. High Energy Phys.* **2017**, *1707*, 136. [CrossRef]
23. Hohmann, M.; Jarv, L.; Ualikhanova, U. Dynamical systems approach and generic properties of $f(T)$ cosmology. *Phys. Rev. D* **2017**, *96*, 043508.
24. Bahamonde, S.; Böhmer, C.G.; Kršxsxák, M.M. New classes of modified teleparallel gravity models. *Phys. Lett. B* **2017**, *775*, 37. [CrossRef]
25. Awad, A.; El Hanafy, W.; Nashed, G.G.L.; Saridakis, E.N. Phase Portraits of general $f(T)$ Cosmology. *J. Cosmol. Astropart. Phys.* **2018**, *1802*, 052. [CrossRef]
26. Hohmann, M.; Järv, L.; Ualikhanova, U. Covariant formulation of scalar-torsion gravity. *Phys. Rev. D* **2018**, *97*, 104011. [CrossRef]
27. Abedi, H.; Capozziello, S.; D'Agostino, R.; Luongo, O. Effective gravitational coupling in modified teleparallel theories. *Phys. Rev. D* **2018**, *97*, 084008. [CrossRef]
28. Bahamonde, S.; Capozziello, S. Noether Symmetry Approach in $f(T,B)$ teleparallel cosmology. *Eur. Phys. J. C* **2017**, *77*, 107. [CrossRef]
29. Beltrán Jiménez, J.; Heisenberg, L.; Koivisto, T. Coincident General Relativity. *Phys. Rev. D* **2018**, *98*, 044048. [CrossRef]
30. Heisenberg, L. Review on $f(Q)$ gravity. *Phys. Rep.* **2024**, *1066*, 1–78. [CrossRef]

31. Capozziello, S.; Shokri, M. Slow-roll inflation in $f(Q)$ non-metric gravity. *Phys. Dark Univ.* **2022**, *37*, 101113. [CrossRef]
32. Capozziello, S.; D'Agostino, R. Model-independent reconstruction of $f(Q)$ non-metric gravity. *Phys. Lett. B* **2022**, *832*, 137229. [CrossRef]
33. Gadbail, G.N.; Mandal, S.; Sahoo, P.K. Reconstruction of ΛCDM universe in $f(Q)$ gravity. *Phys. Lett. B* **2022**, *835*, 137509. [CrossRef]
34. Wang, Q.; Ren, X.; Cai, Y.F.; Luo, W.; Saridakis, E.N. Observational Test of $f(Q)$ Gravity with Weak Gravitational Lensing. *Astrophys. J.* **2024**, *974*, 7. [CrossRef]
35. Hu, Y.M.; Li, B.C.; Yu, Y.; Kršxsxák, M.; Saridakis, E.N.; Cai, Y.F. Lowering the strong coupling mode of modified teleparallel gravity theories. *J. Cosmol. Astropart. Phys.* **2025**, *2025*, 88. [CrossRef]
36. Do, A.; Loo, T.H.; Saridakis, E.N. Non-metricity with boundary terms: f(Q,C) gravity and cosmology. *J. Cosmol. Astropart. Phys.* **2024**, *2024*, 50. [CrossRef]
37. Kobayashi, T. Horndeski theory and beyond: A review. *Rep. Prog. Phys.* **2019**, *82*, 086901. [CrossRef]
38. Maartens, R.; Koyama, K. Brane-World Gravity. *Living Rev. Relativ.* **2010**, *13*, 5. [CrossRef]
39. Stavrinos, P.C.; Vacaru, S.I. Cyclic and Ekpyrotic Universes in Modified Finsler Osculating Gravity on Tangent Lorentz Bundles. *Class. Quant. Grav.* **2013**, *30*, 055012. [CrossRef]
40. Basilakos, S.; Kouretsis, A.P.; Saridakis, E.N.; Stavrinos, P. Resembling dark energy and modified gravity with Finsler-Randers cosmology. *Phys. Rev. D* **2013**, *88*, 123510. [CrossRef]
41. Kapsabelis, E.; Triantafyllopoulos, A.; Basilakos, S.; Stavrinos, P.G. Applications of the Schwarzschild–Finsler–Randers model. *Eur. Phys. J. C* **2021**, *81*, 990. [CrossRef]
42. Hohmann, M.; Pfeifer, C. Geodesics and the magnitude-redshift relation on cosmologically symmetric Finsler spacetimes. *Phys. Rev. D* **2017**, *95*, 104021. [CrossRef]
43. Hohmann, M.; Pfeifer, C.; Voicu, N. Finsler gravity action from variational completion. *Phys. Rev. D* **2019**, *100*, 064035. [CrossRef]
44. Pfeifer, C. Finsler spacetime geometry in Physics. *arXiv* **2019**, arXiv:1903.10185. [CrossRef]
45. Bogoslovsky, G.Y.; Goenner, H.F. Finslerian spaces possessing local relativistic symmetry. *Gen. Rel. Grav.* **1999**, *31*, 1565. [CrossRef]
46. Chang, Z.; Li, X. Lorentz invariance violation and symmetry in Randers-Finsler spaces. *Phys. Lett. B* **2008**, *663*, 103. [CrossRef]
47. Vacaru, S.I. Principles of Einstein-Finsler Gravity and Perspectives in Modern Cosmology. *Int. J. Mod. Phys. D* **2012**, *21*, 1250072. [CrossRef]
48. Kostelecky, V.A.; Mewes, M. Astrophysical Tests of Lorentz and CPT Violation with Photons. *Astrophys. J.* **2008**, *689*, L1. [CrossRef]
49. Kostelecky, A. Riemann-Finsler geometry and Lorentz-violating kinematics. *Phys. Lett. B* **2011**, *701*, 137. [CrossRef]
50. Alan Kostelecky, V.; Russell, N.; Tso, R. Bipartite Riemann–Finsler geometry and Lorentz violation. *Phys. Lett. B* **2012**, *716*, 470. [CrossRef]
51. Aluri, P.K.; Cea, P.; Chingangbam, P.; Chu, M.C.; Clowes, R.G.; Hutsemékers, D.; Kochappan, J.P.; Lopez, A.M.; Liu, L.; Martens, N.C.M. Is the observable Universe consistent with the cosmological principle? *Class. Quant. Grav.* **2023**, *40*, 094001. [CrossRef]
52. Kapsabelis, E.; Kevrekidis, P.G.; Stavrinos, P.G.; Triantafyllopoulos, A. Schwarzschild–Finsler–Randers spacetime: Geodesics, dynamical analysis and deflection angle. *Eur. Phys. J. C* **2022**, *82*, 1098. [CrossRef]
53. Cipriano, R.A.C.; Ganiyeva, N.; Harko, T.; Lobo, F.S.N.; Pinto, M.A.S.; Rosa, J.L. Energy-Momentum Squared Gravity: A Brief Overview. *Universe* **2024**, *10*, 339. [CrossRef]
54. Agashe, A.; Modumudi, S.M. On the Effects of Non-Metricity in an Averaged Universe. *Universe* **2024**, *10*, 261. [CrossRef]
55. Singh, J.K.; Singh, P.; Saridakis, E.N.; Myrzakul, S.; Balhara, H. New Parametrization of the Dark-Energy Equation of State with a Single Parameter. *Universe* **2024**, *10*, 246. [CrossRef]
56. Wojnar, A.; Gomes, D.A. Bose and Fermi Gases in Metric-Affine Gravity and Linear Generalized Uncertainty Principle. *Universe* **2024**, *10*, 217. [CrossRef]
57. Saha, D.; Chakrabortty, M.; Sanyal, A.K. Reconstructing Modified and Alternative Theories of Gravity. *Universe* **2024**, *10*, 44. [CrossRef]
58. Escamilla-Rivera, C.; de Albornoz-Caratozzolo, J.M.; Nájera, S. Fab-Four Cosmography to Tackle the Hubble Tension. *Universe* **2023**, *9*, 311. [CrossRef]
59. Frolovsky, D.; Ketov, S.V. Production of Primordial Black Holes in Improved E-Models of Inflation. *Universe* **2023**, *9*, 294. [CrossRef]
60. Varieschi, G.U. Newtonian Fractional-Dimension Gravity and Galaxies without Dark Matter. *Universe* **2023**, *9*, 246. [CrossRef]
61. Voicu, N.; Friedl-Szász, A.; Popovici-Popescu, E.; Pfeifer, C. The Finsler Spacetime Condition for (α,β)-Metrics and Their Isometries. *Universe* **2023**, *9*, 198. [CrossRef]
62. Czuchry, E. Resolution of Cosmological Singularity in Hořava–Lifshitz Cosmology. *Universe* **2023**, *9*, 160. [CrossRef]
63. Shen, B.; Yu, M. Geometric Outlines of the Gravitational Lensing and Its Astronomic Applications. *Universe* **2023**, *9*, 153. [CrossRef]

64. Castro Perelman, C. On Born's Reciprocal Relativity, Algebraic Extensions of the Yang and Quaplectic Algebra, and Noncommutative Curved Phase Spaces. *Universe* **2023**, *9*, 144. [CrossRef]
65. Saha, S.; Chattopadhyay, S. Realization of Bounce in a Modified Gravity Framework and Information Theoretic Approach to the Bouncing Point. *Universe* **2023**, *9*, 136. [CrossRef]
66. Golovnev, A. On the Role of Constraints and Degrees of Freedom in the Hamiltonian Formalism. *Universe* **2023**, *9*, 101. [CrossRef]
67. Solanki, R.; Arora, S.; Sahoo, P.K.; Moraes, P.H.R.S. Bulk Viscous Fluid in Symmetric Teleparallel Cosmology: Theory versus Experiment. *Universe* **2023**, *9*, 12. [CrossRef]
68. Kalita, S.; Sarmah, L.; Wojnar, A. Cooling Process of White Dwarf Stars in Palatini $f(R)$ Gravity. *Universe* **2022**, *8*, 647. [CrossRef]
69. Kuptsov, S.; Ioffe, M.; Manida, S.; Paston, S. Weak Field Limit for Embedding Gravity. *Universe* **2022**, *8*, 635. [CrossRef]
70. Chakraborty, K.; Rahaman, F.; Ray, S.; Sen, B.; Deb, D. Galactic Wormhole under Lovelock Gravity. *Universe* **2022**, *8*, 581. [CrossRef]
71. Socorro, J.; Pérez-Payán, S.; Hernández-Jiménez, R.; Espinoza-García, A.; Díaz-Barrón, L.R. Quintom Fields from Chiral K-Essence Cosmology. *Universe* **2022**, *8*, 548. [CrossRef]
72. Di Gennaro, S.; Ong, Y.C. Sign Switching Dark Energy from a Running Barrow Entropy. *Universe* **2022**, *8*, 541. [CrossRef]
73. Frolov, A.M. Metric Gravity in the Hamiltonian Form—Canonical Transformations—Dirac's Modifications of the Hamilton Method and Integral Invariants of the Metric Gravity. *Universe* **2022**, *8*, 533. [CrossRef]
74. Haug, E.G. Progress in the Composite View of the Newton Gravitational Constant and Its Link to the Planck Scale. *Universe* **2022**, *8*, 454. [CrossRef]
75. Rasouli, S.M.M.; Jalalzadeh, S.; Moniz, P. Noncompactified Kaluza–Klein Gravity. *Universe* **2022**, *8*, 431. [CrossRef]

universe MDPI

Review

Noncompactified Kaluza–Klein Gravity

Seyed Meraj Mousavi Rasouli [1,2,*], Shahram Jalalzadeh [3] and Paulo Moniz [1]

1 Departamento de Física, Centro de Matemática e Aplicações (CMA-UBI), Universidade da Beira Interior, Rua Marquês d'Avila e Bolama, 6200-001 Covilhã, Portugal
2 Department of Physics, Qazvin Branch, Islamic Azad University, Qazvin 341851416, Iran
3 Departamento de Física, Universidade Federal de Pernambuco, Recife 52171-900, Brazil
* Correspondence: mrasouli@ubi.pt

Abstract: We present a brief description of noncompactified higher-dimensional theories from the perspective of general relativity. More concretely, the Space–Time–Matter theory, or Induced Matter theory, and the reduction procedure used to construct the modified Brans–Dicke theory and the modified Sáez–Ballester theory are briefly explained. Finally, we apply the latter to the Friedmann–Lemaître–Robertson–Walker (FLRW) cosmological models in arbitrary dimensions and analyze the corresponding solutions.

Keywords: higher dimensions; noncompactified Kaluza–Klein theories; induced–matter theory; space–time–matter theory; modified Brans–Dicke theory; modified Sáez–Ballester theory; noncompactified cosmology

1. Introduction

Since matter or the source of spacetime and fields are fundamental concepts in classical field theories, the Einstein tensor is expressed in terms of spacetime geometry and matter by the corresponding energy-momentum density tensor. The Einstein field equations connect these two fundamentals. As a result, in general relativity (GR), the distribution of matter determines how spacetime is shaped. On the other hand, one could interpret Einstein's field equations in a different way and assert that geometry creates matter. One of Einstein's objectives was to develop a gravitational theory in which the idea of matter is abandoned in favor of pure fields [1]. According to Einstein, unified field theory is a gravitational theory in which matter is absorbed into the field itself, leading to a set of homogeneous partial differential equations. Many extensions of Einstein's framework have been made to extract matter from pure geometry. One of Einstein's intriguing extensions is the suggestion that our four-dimensional spacetime (called a membrane or brane) is a submanifold embedded in a higher dimensional ambient space (bulk). This idea first appeared in papers by Kaluza and Klein, who proposed uniting gravity and electromagnetism. In Kaluza–Klein's (KK) theory, the extra dimension serves only a formal purpose, and the components of the ambient space metric tensor are independent of the coordinate associated with the extra dimension.

In the last two decades, there has been much appeal in the concept of extra dimensions where ordinary matter is constrained to a brane. Early examples of this methodology can be found in the works of Maia [2], Joseph [3], Akama [4], Rubakov–Shaposhnikov [5], and Visser [6]. Wesson's theory [7], which states that the geometry of the bulk space generates matter on the brane, is the basis for a revised KK approach to unified field theory. This theory differs from the traditional KK scenario due to the noncompact extra dimension and the absence of matter in the five-dimensional bulk space. This theory is called the Induced Matter Theory (IMT) because the effective four-dimensional matter results from the bulk's geometry. In other words, in IMT, the four-dimensional induced matter curves the four-dimensional hypersurface, while the five-dimensional bulk space is Ricci-flat.

Recently, instead of GR, by applying the scalar–tensor theories as underlying frameworks, another extended version of the noncomactified KK gravity has also been established [8,9], which will be the main content of the this review paper.

For other extensions of Wesson's theory that have assumed an arbitrary number of noncompact extra dimensions, see [10–12].

Furthermore, the authors of Ref. [13] have obtained interesting IMT cosmological solutions by assuming a conformally flat bulk space. The Weyl tensor of the bulk space vanishes in this case. As noted above, this restriction is in the spirit of IMT. The energy conditions associated with the model of Ref. [13] have been also investigated in detail in [14]. All matter fields in 4-dimensional spacetime are induced from the bulk, according to IMT, and the path of particles in the bulk space is null. Therefore, the bulk can be assumed to be empty. We can have some black holes in the bulk space if we only assume that the bulk is Ricci-flat. Consequently, since black holes have mass, we must have some kind of mass in the bulk, which contradicts the IMT axioms.

One of the consequences of IMT is that the mass of the particles varies from point-to-point in spacetime. Indeed, it is well known that in unified field theories, Mach's principle is satisfied [15]. Thus, the mass of particles may be affected by the distribution of matter fields in the Universe or the curvature of spacetime. As a result, it is not surprising that Wesson's IMT is a Machian theory [16] and the particle's mass is not constant. Wesson used dimensional analysis to introduce the relation $l = Gm/c^2$ (where G and c are the Newton gravitational constant and the speed of light, respectively) between the fifth coordinate, l, and the mass of the test particles, m, to demonstrate mass variation [17,18], and for this reason, he initially called his theory Space–Time–Matter (STM) theory [19]. The authors of Ref. [20] have shown that if we use the above equation to calculate the mass variation of the primordial nucleosynthetic particles and our time and compare it to the variation of mass obtained from nucleosynthesis bounds, the results do not agree. However, in Ref. [21], it was shown that if the induced mass is defined correctly, the variation of mass obtained from IMT agrees with the mass variation bound obtained from the Hot Big Bang.

Let us also mention one of the more intriguing futures of IMT based on quantum mechanics. Particles and waves, for example, have been shown to be merely different representations of the same underlying geometry and may be the same thing viewed in different ways [22]. Wesson [23] has also derived a form of the Heisenberg relation that applies to real and virtual particles using five-dimensional IMT. (For an extension of this idea to brane gravity, please see [24].) In this regard, the authors of [25] derived the induced Einstein equation on the perturbed brane by contracting the Gauss–Codazzi equations. They studied the Einstein equation on the perturbed brane of an FRW universe as a non-perturbed brane embedded in a five-dimensional flat bulk space. They demonstrated that the induced field equations correspond to the semiclassical Einstein equation. This means that the classical fluctuations of the perturbed brane can be interpreted as matter field quantum fluctuations.

Inspired by the IMT, the authors of [8,9] have shown that the field equations associated with two different types of scalar–tensor theories, namely the Sáez–Ballester (SB) and Brans–Dicke (BD) theories, in $(D+1)$-dimensions are equivalent to their D-dimensional counterparts with an effective matter field and a potential. These reduced gravitational models have been nominated for the modified Sáez–Ballester theory (MSBT) [9,26] and the modified Brans–Dicke theory (MBDT) [8], respectively. It is important to mention that in both frameworks, introducing a non-zero induced scalar potential whose shape is provided by the reduction process up to a constant of integration, is crucial to reconstruct the corresponding theory on the hypersurface.

Now let us focus on the BD theory and the MBDT. One of the attractive aspects of the BD theory is that the scalar field is not an ad hoc assumption but rather a fundamental component. In cosmological models that use BD theory as a background theory, the scalar field can act as quintessence or K-essence, resulting in an accelerated scale factor [27]. Such cosmological models, however, have flaws that contradict the principles of the original version of the BD theory, experimental data, and energy conditions. More concretely, in some of these models, the BD coupling parameter is assumed to be a variable constant [28]; in others, an ad hoc scalar potential is added [29]. Concerning related investigations within the standard BD theory as well as its modified version, see also [30–40]. Nevertheless, it should be noted that the MBDT is free of the shortcomings mentioned above. In particular, using the MBDT framework in cosmology does not necessitate using an ad hoc scalar potential or a variable BD coupling parameter to obtain an accelerated scale factor. Furthermore, unlike other modified BD cosmological models, the FLRW-MBDT cosmologies eliminate the inconsistency in the value of the BD coupling parameter associated with an accelerated expansion of a matter-dominated universe and a decelerated radiation-dominated epoch [8]. In this review paper, we will not present any cosmological application of the MBDT. We recommend readers look at the detailed study of such models in [41–46].

The main goal in constructing the Sáez–Ballester (SB) scalar–tensor theory [47] was to solve the problem of missing matter of the universe. The Einstein–Hilbert action was supplemented in this framework by a non-canonical kinetic term containing a coupling parameter. Both a scalar potential and a cosmological term are absent from the original SB theory. In one particular instance, it is possible to find appropriate transformations that allow the SB action to be reduced to the corresponding one with a canonical kinetic term [48]. In contrast to the latter, the SB theory includes the action of ordinary matter that is not coupled to the SB scalar field. Various cosmological models have been established in the context of the SB theory to study open problems in both the classical and quantum regimes [49–54]. Applying the similar reduction procedure used to construct the MBDT, without assuming the presence of the higher dimensional matter fields and imposing the cylindricity condition on the extra coordinate, it has been shown that the SB field equations associated to the bulk split into four sets of effective field equations on any hypersurface orthogonal to the extra dimension [9].

In the next section, we will review the IMT and present short discussion about this framework. In Section 3, we present a brief review of the MBDT and MSBT. In Section 4, as an application of the reduced SB theory, we will review the solutions of the FLRW-MSBT cosmology. Finally, in Section 5, in addition to a brief summary, we will include other useful discussions.

2. Five-Dimensional Ricci–Flat Space and the Effective Field Equations in Four Dimensions

In this section, we present an overview and important notes on the framework established in [55].

Assuming some postulates and applying an appropriate reduction procedure, the field equation of general relativity (GR) can be set up on a four-dimensional hypersurface. The five-dimensional manifold M_5

$$dS^2 = \mathcal{G}_{ab}(x^c)dx^a dx^b, \tag{1}$$

in which our universe is locally and isometrically embedded can be, at least locally, taken as

$$dS^2 = g_{\mu\nu}(x^\alpha, l)dx^\mu dx^\nu + \epsilon\psi^2(x^\alpha, l)dl^2. \tag{2}$$

In this section, the Latin and Greek indices run from zero to four, and to three, respectively; l is an extra noncompact coordinate, $\psi = \psi(x^\alpha, l)$ is a scalar field and $\epsilon = \pm 1$ (where $\epsilon^2 = 1$) is introduced such that we can take the extra dimension as either time-like or space-like.[1] Up to now, we have indeed expressed the '*Postulate I*' of the Wesson's scheme [56].

Remark. *Before reviewing the effective field equations, let us mention a few important points. In the IMT, similar to the Kaluza's approach, the same definitions for the five-dimensional Christoffel symbols and Ricci tensor have been used [7]. However, unlike the compactified KK theory, the authors* *of IMT did not want to geometrize the electromagnetic field. Moreover, the derivatives with respect to the extra coordinate were not assumed to be zero.*[2] *We must emphasize that the frameworks to be presented in this paper recover from a fully noncompactified approach without making any a priori assumptions about the nature of the extra-dimensional manifold.*

It is straightforward to show that the expressions for the $\alpha\beta-$, $\alpha 4-$ and $44-$parts of the Ricci tensor $\overset{(5)}{R}_{ab}$ are:

$$\overset{(5)}{R}_{\alpha\beta} = \overset{(4)}{R}_{\alpha\beta} - \frac{\mathcal{D}_\alpha \mathcal{D}_\beta \psi}{\psi} + \frac{\epsilon}{2\psi^2}\left(\frac{\overset{*}{\psi}\overset{*}{g}_{\alpha\beta}}{\psi} - \overset{**}{g}_{\alpha\beta} + g^{\lambda\mu}\overset{*}{g}_{\alpha\lambda}\overset{*}{g}_{\beta\mu} - \frac{1}{2}g^{\mu\nu}\overset{*}{g}_{\mu\nu}\overset{*}{g}_{\alpha\beta} \right), \tag{3}$$

$$\overset{(5)}{R}_{4\alpha} = \psi \mathcal{D}_\beta P^\beta{}_\alpha, \tag{4}$$

$$\overset{(5)}{R}_{44} = -\epsilon\psi\mathcal{D}^2\psi - \frac{1}{4}\overset{*}{g}{}^{\lambda\beta}\overset{*}{g}_{\lambda\beta} - \frac{1}{2}g^{\lambda\beta}\overset{**}{g}_{\lambda\beta} + \frac{\overset{*}{\psi}}{2\psi}g^{\lambda\beta}\overset{*}{g}_{\lambda\beta}, \tag{5}$$

where

$$P_{\alpha\beta} \equiv \frac{1}{2\psi}\left(\overset{*}{g}_{\alpha\beta} - g_{\alpha\beta}g^{\mu\nu}\overset{*}{g}_{\mu\nu}\right), \tag{6}$$

$\overset{*}{A} \equiv \frac{\partial A}{\partial l}$, $\mathcal{D}_\alpha$ is the covariant derivative on the hypersurface and $\mathcal{D}^2 \equiv \mathcal{D}^\alpha\mathcal{D}_\alpha$.

In the IMT [7,19,55], it was presumed that there is no higher-dimensional ordinary matter; therefore $\overset{(5)}{G}_{ab} = 0$ (where $\overset{(5)}{G}_{ab}$ are the components of the Einstein tensor in five dimensions), or equivalently, $\overset{(5)}{R}_{ab} = 0$, which is known as the *'Postulate II'* of the Wesson's scheme [56]. Therefore, by defining a hypersurface Σ_4, for which $l = l_0 =$ constant, and

$$g_{\mu\nu}(x^\alpha) = \overset{(4)}{\mathcal{G}}(x^\alpha, l_0), \tag{7}$$

Equations (3)–(5) reduce to

$$\overset{(4)}{R}_{\alpha\beta} = \frac{\mathcal{D}_\alpha \mathcal{D}_\beta \psi}{\psi} - \frac{\epsilon}{2\psi^2}\left[\frac{\overset{*}{\psi}\overset{*}{g}_{\alpha\beta}}{\psi} - \overset{**}{g}_{\alpha\beta} + g^{\lambda\mu}\overset{*}{g}_{\alpha\lambda}\overset{*}{g}_{\beta\mu} - \frac{1}{2}g^{\mu\nu}\overset{*}{g}_{\mu\nu}\overset{*}{g}_{\alpha\beta} \right], \tag{8}$$

$$\mathcal{D}_\beta P^\beta{}_\alpha = 0, \tag{9}$$

$$\epsilon\psi\mathcal{D}^2\psi = -\frac{1}{4}\overset{*}{g}{}^{\lambda\beta}\overset{*}{g}_{\lambda\beta} - \frac{1}{2}g^{\lambda\beta}\overset{**}{g}_{\lambda\beta} + \frac{\overset{*}{\psi}}{2\psi}g^{\lambda\beta}\overset{*}{g}_{\lambda\beta}. \tag{10}$$

Equations (8)–(10) *"form the basis of five-dimensional noncompactified KK theory"* [7]. The interpretation of these equations in four dimensions and their applications in cosmology and astrophysics have been extensively presented in the literature, see for instance, [7,19]. In what follows, let us briefly analyze one of the effective field equations.

Using Equations (8) and (10), and $(\delta^\mu_\nu)_{,4} = 0 = g^{\mu\beta}g^{\sigma\lambda}\overset{*}{g}_{\lambda\beta}\overset{*}{g}{}^{\mu\sigma} + \overset{*}{g}_{\mu\sigma}\overset{*}{g}_{\mu\sigma}$, we obtain an expression for the Ricci scalar $\overset{(4)}{R} = g^{\alpha\beta}\overset{(4)}{R}_{\alpha\beta}$:

$$\overset{(4)}{R} = \frac{\epsilon}{4\psi^2}\left[\overset{*}{g}{}^{\mu\nu}\overset{*}{g}_{\mu\nu} + \left(g^{\mu\nu}\overset{*}{g}_{\mu\nu}\right)^2\right]. \tag{11}$$

Defining an induced energy-momentum tensor in four dimensions as $T^{[\mathrm{IMT}]}_{\alpha\beta} \equiv \overset{(4)}{R}_{\alpha\beta} - 1/2\overset{(4)}{R}g_{\alpha\beta}$, from using Equations (8) and (11), one can easily obtain

$$T_{\alpha\beta}^{[\mathrm{IMT}]} \equiv \frac{\mathcal{D}_\alpha \mathcal{D}_\beta \psi}{\psi} - \frac{\epsilon}{2\psi^2}\left(\frac{\overset{*}{\psi}\overset{*}{g}_{\alpha\beta}}{\psi} - \overset{**}{g}_{\alpha\beta} + g^{\lambda\mu}\overset{*}{g}_{\alpha\lambda}\overset{*}{g}_{\beta\mu} - \frac{1}{2}g^{\mu\nu}\overset{*}{g}_{\mu\nu}\overset{*}{g}_{\alpha\beta}\right) - \frac{\epsilon g_{\alpha\beta}}{8\psi^2}\left[\overset{*}{g}^{\mu\nu}\overset{*}{g}_{\mu\nu} + \left(g^{\mu\nu}\overset{*}{g}_{\mu\nu}\right)^2\right]. \tag{12}$$

In summary, according to the *'Postulate III'* [56], we use the above expression as a four-dimensional energy momentum tensor associated with our universe; hence the Einstein field equations on a four-dimensional hypersurface, $G_{\alpha\beta}^{(4)} = T_{\alpha\beta}^{[\mathrm{IMT}]}$, are automatically contained in the corresponding five-dimensional vacuum equations $G_{ab}^{(5)} = 0$. In this regard, $T_{\alpha\beta}^{[\mathrm{IMT}]}$, which describes a matter as a manifestation of pure geometry in higher-dimensional spacetime, has been interpreted as the energy momentum tensor of an *induced-matter* in the KK theory.

3. Modified Scalar–Tensor Theories

In this section, we again take the line element (2) and the noncompact extra dimensions. Moreover, we will consider the Remark mentioned in the previous section as well as the following one

Remark. *Equations* (3)–(6) *are again valid for this framework as well as that which will be introduced in the next subsection. However, since in this and the subsequent subsection we relate the equations associated with* $(D+1)$*-dimensional spacetime to their corresponding D-dimensional counterparts, the indices* 4α *and* 44 *should be replaced by* $D\alpha$ *and* DD*, respectively. Moreover, the superscripts* (5) *should be replaced with* $(D+1)$*. Furthermore, due to the presence of the scalar field as well as the higher-dimensional matter in the bulk, equations* $G_{ab}^{(5)} = 0$ *or equivalently,* $R_{ab}^{(5)} = 0$ *are generally no longer valid. More precisely, as we will show below, not only are Equations* (8)–(12) *generalized, but we will also show that when the wave equation of each framework is reduced on the hypersurface, it involves induced scalar potential.*

In addition, we will consider the following generalizations. (i) Instead of deriving the modified field equations in four dimensions, we want to obtain them in arbitrary dimensions. Therefore, we assume that the Latin and Greek indices run from zero to D, and to $D-1$, respectively. Moreover, $\mathcal{G}$ and $R^{(D+1)}$, respectively, are the determinant and the Ricci scalar of the $(D+1)$-dimensional metric $\mathcal{G}_{ab}$; ∇_a stands for the covariant derivative in $(D+1)$-dimensional spacetime, and $\nabla^2 \equiv \nabla_a \nabla^a$. The Lagrangian $L_{\mathrm{matt}}^{(D+1)}$ describes ordinary matter in the $(D+1)$-dimensional spacetime. (ii) Rather than considering the GR as a background theory, let us consider two different types of scalar–tensor theories: in the first, the scalar field is minimally coupled to gravity, while in the second, the scalar field is non-minimally coupled. (iii) In order to apply a generalized reduction method, we consider, in addition to the presence of the scalar field in the action, a higher-dimensional ordinary matter[3] $L_{\mathrm{matt}}^{(D+1)}$.

3.1. Modified Sáez–Ballester Theory in Arbitrary Dimensions

Let us give a brief overview of the framework established in [9], see also [26,48]. We consider a generalized version[4] of the SB action proposed in [47] in $(D+1)$-dimensional spacetime as

$$\mathcal{S}_{\mathrm{SB}}^{(D+1)} = \int d^{D+1}x\sqrt{|\mathcal{G}|}\left[R^{(D+1)} - \mathcal{W}\phi^n \mathcal{G}^{ab}(\nabla_a\phi)(\nabla_b\phi) + \chi L_{\mathrm{matt}}^{(D+1)}\right], \tag{13}$$

where ϕ is the SB scalar field; $\mathcal{W}$, n are dimensionless parameters of the model; $\chi = 8\pi$, and we used the same units taken in [9]. We should emphasize that there is no scalar potential in action (13).

One can easily show that the equations of motion corresponding to (13) are:

$$G_{ab}^{(D+1)} = \mathcal{W}\phi^n\left[(\nabla_a\phi)(\nabla_b\phi) - \frac{1}{2}\mathcal{G}_{ab}(\nabla^c\phi)(\nabla_c\phi)\right] + \chi\, T_{ab}^{(D+1)} \tag{14}$$

and

$$2\phi^n\nabla^2\phi + n\phi^{n-1}(\nabla_a\phi)(\nabla^a\phi) = 0, \tag{15}$$

where $G_{ab}^{(D+1)}$ and $T_{ab}^{(D+1)}$, respectively, denote the Einstein tensor and the energy momentum tensor (of any ordinary matter field) in $(D+1)$-dimensions. We should emphasize that $T_{ab}^{(D+1)}$ does not depend on ϕ, and it is therefore identically conserved. Equation (14) leads to

$$R^{(D+1)} = \mathcal{W}\phi^n(\nabla_a\phi)(\nabla^a\phi) - \frac{2\chi}{D-1}T^{(D+1)}, \tag{16}$$

where $T^{(D+1)} = \mathcal{G}^{ab}T_{ab}^{(D+1)}$.

Before deriving the effective field equations, let us write some useful equations that will be used later:

$$\nabla_\mu\nabla_\nu\phi = \mathcal{D}_\mu\mathcal{D}_\nu\phi + \frac{\epsilon\overset{*}{\phi}\overset{*}{g}_{\mu\nu}}{2\psi^2}, \tag{17}$$

$$\nabla^2\phi = \mathcal{D}^2\phi + \frac{(\mathcal{D}_\alpha\psi)(\mathcal{D}^\alpha\phi)}{\psi} + \frac{\epsilon}{\psi^2}\left[\overset{**}{\phi} + \overset{*}{\phi}\left(\frac{g^{\mu\nu}\overset{*}{g}_{\mu\nu}}{2} - \frac{\overset{*}{\psi}}{\psi}\right)\right], \tag{18}$$

$$(\nabla^a\phi)(\nabla_a\phi) = (\mathcal{D}^\alpha\phi)(\mathcal{D}_\alpha\phi) + \epsilon\left(\frac{\overset{*}{\phi}}{\psi}\right)^2, \tag{19}$$

$$\nabla_D\nabla_D\phi = \epsilon\psi(\mathcal{D}_\alpha\psi)(\mathcal{D}^\alpha\phi) + \overset{**}{\phi} - \left(\frac{\overset{*}{\psi}}{\psi}\right)\overset{*}{\phi}. \tag{20}$$

Letting $a \to \mu$ and $b \to \nu$ in Equation (14) yields the D-dimensional counterpart of the corresponding $(D+1)$-dimensional quantity:

$$G_{\mu\nu}^{(D+1)} = \mathcal{W}\phi^n\left[(\mathcal{D}_\mu\phi)(\mathcal{D}_\nu\phi) - \frac{1}{2}g_{\mu\nu}(\mathcal{D}_\alpha\phi)(\mathcal{D}^\alpha\phi)\right] - \frac{\epsilon\mathcal{W}\phi^n}{2}\left(\frac{\overset{*}{\phi}}{\psi}\right)^2 g_{\mu\nu} + \chi T_{\mu\nu}^{(D+1)}, \tag{21}$$

where we have used (19).

Now we are going to derive the equations associated with the MSBT.[5]

1. Let us first obtain a dynamical equation for the scalar field ψ, i.e., an extended version of (10), which, in turn, will be applied to retrieve other modified equations. Letting $a \to D$ and $b \to D$ in Equation (14), we obtain

$$R_{DD}^{(D+1)} = \left(\frac{\epsilon\chi\psi^2}{1-D}\right)T^{(D+1)} + \chi T_{DD}^{(D+1)} + \mathcal{W}\phi^n\left(\overset{*}{\phi}\right)^2, \tag{22}$$

where we used Equation (16). Equating relations (5) (please see the Remarks) and (22), we retrieve

$$\frac{\mathcal{D}^2\psi}{\psi} = -\frac{\epsilon}{2\psi^2}\left[g^{\lambda\beta}\overset{**}{g}_{\lambda\beta} + \frac{1}{2}\overset{*}{g}^{\lambda\beta}\overset{*}{g}_{\lambda\beta} - \frac{g^{\lambda\beta}\overset{*}{g}_{\lambda\beta}\overset{*}{\psi}}{\psi}\right] - \epsilon\mathcal{W}\phi^n\left(\frac{\overset{*}{\phi}}{\psi}\right)^2 + \chi\left[\frac{T^{(D+1)}}{D-1} - \frac{\epsilon T_{DD}^{(D+1)}}{\psi^2}\right], \quad (23)$$

which is one of our effective equations. It can be seen that in a special case where $\phi =$ constant and the higher-dimensional ordinary matter is absent, Equation (23) reduces to its IMT counterpart.

2. Let us now construct the Einstein tensor on the hypersurface. Concretely, we want to retrieve the counterpart of (14) on a D-dimensional hypersurface. In this regard, by taking the SB theory into account, we first relate the Ricci scalars $R^{(D+1)}$ and $R^{(D)}$:

$$R^{(D+1)} = \mathcal{G}^{ab}R_{ab}^{(D+1)} = \mathcal{G}^{\alpha\beta}R_{\alpha\beta}^{(D+1)} + \mathcal{G}^{DD}R_{DD}^{(D+1)}. \quad (24)$$

Substituting $R_{\alpha\beta}^{(D+1)}$ and $R_{DD}^{(D+1)}$ from relations (3) and (5), respectively (where we respect the expressions presented in the Remarks) into Equation (24), and then using Equation (23), after some manipulations, we obtain

$$R^{(D+1)} = R^{(D)} - \frac{\epsilon}{4\psi^2}\left[\left(g^{\alpha\beta}\overset{*}{g}_{\alpha\beta}\right)^2 + \overset{*}{g}^{\alpha\beta}\overset{*}{g}_{\alpha\beta}\right] + \frac{2\epsilon\mathcal{W}\phi^n\overset{*}{\phi}^2}{\psi^2} + 2\chi\left(\frac{\epsilon T_{DD}^{(D+1)}}{\psi^2} - \frac{T^{(D+1)}}{D-1}\right). \quad (25)$$

Now, we proceed as follows. By substituting $R_{\mu\nu}^{(D+1)}$ and $R^{(D+1)}$ from relations (3) and (25) into

$$G_{\mu\nu}^{(D+1)} = R_{\mu\nu}^{(D+1)} - \frac{1}{2}\mathcal{G}^{\mu\nu}R^{(D+1)}, \quad (26)$$

and equating the result with (21), we can easily construct the Einstein tensor on a D-dimensional hypersurface:

$$\begin{aligned} G_{\mu\nu}^{(D)} &= \mathcal{W}\phi^n\left[(\mathcal{D}_\mu\phi)(\mathcal{D}_\nu\phi) - \frac{1}{2}g_{\mu\nu}(\mathcal{D}_\alpha\phi)(\mathcal{D}^\alpha\phi)\right] + \chi\left(F_{\mu\nu} + T_{\mu\nu}^{[\mathrm{MSBT}]}\right) - \frac{1}{2}g_{\mu\nu}V(\phi) \\ &\equiv \mathcal{W}\phi^n\left[(\mathcal{D}_\mu\phi)(\mathcal{D}_\nu\phi) - \frac{1}{2}g_{\mu\nu}(\mathcal{D}_\alpha\phi)(\mathcal{D}^\alpha\phi)\right] + \chi T_{\mu\nu}^{(D)[\mathrm{eff}]} - \frac{1}{2}g_{\mu\nu}V(\phi), \end{aligned} \quad (27)$$

where the induced scalar potential $V(\phi)$ should be obtained from the differential Equation (32).

Let us also introduce the effective matter in Equation (27):

(i) $F_{\mu\nu}$ is the effective matter induced from the $(D+1)$-dimensional ordinary energy momentum tensor:

$$F_{\mu\nu} \equiv T_{\mu\nu}^{(D+1)} + g_{\mu\nu}\left[\frac{\epsilon\, T_{DD}^{(D+1)}}{\psi^2} - \frac{T^{(D+1)}}{D-1}\right]. \quad (28)$$

Obviously, assuming a bulk without a higher-dimensional ordinary matter, i.e., $L_{\mathrm{matt}}^{(D+1)} = 0$, then $F_{\mu\nu}$ vanishes.

(ii) $T_{\mu\nu}^{[\mathrm{MSBT}]}$ is an induced energy momentum tensor associated with our herein MSBT framework, which, in turn, has three components:

$$\chi T_{\mu\nu}^{[\mathrm{MSBT}]} = T_{\mu\nu}^{[\mathrm{IMT}]} + T_{\mu\nu}^{[\phi]} + \frac{1}{2} g_{\mu\nu} V(\phi), \tag{29}$$

where

$$T_{\mu\nu}^{[\phi]} \equiv \left[\frac{1}{2}\epsilon \mathcal{W}\phi^n \left(\frac{\overset{*}{\phi}}{\psi}\right)^2\right] g_{\mu\nu}. \tag{30}$$

Here $T_{\mu\nu}^{[\mathrm{IMT}]}$ is exactly the same quantity introduced in the previous section, see Equation (12).

3. We now obtain the reduced wave equation on a D-dimensional hypersurface, i.e., the counterpart to Equation (15). Substituting $\nabla^2\phi$ and $(\nabla^c\phi)(\nabla_c\phi)$ from relations (18) and (19) into (15), we obtain

$$2\phi^n \mathcal{D}^2\phi + n\phi^{n-1}(\mathcal{D}_\alpha\phi)(\mathcal{D}^\alpha\phi) - \frac{1}{\mathcal{W}}\frac{dV(\phi)}{d\phi} = 0, \tag{31}$$

where

$$\frac{dV(\phi)}{d\phi} \equiv -\frac{2\mathcal{W}\phi^n}{\psi^2}\left\{\psi(\mathcal{D}_\alpha\psi)(\mathcal{D}^\alpha\phi) + \frac{n\epsilon}{2}\left(\frac{\overset{*}{\phi}{}^2}{\phi}\right) + \epsilon\left[\overset{**}{\phi} + \overset{*}{\phi}\left(\frac{1}{2}g^{\mu\nu}\overset{*}{g}_{\mu\nu} - \frac{\overset{*}{\psi}}{\psi}\right)\right]\right\}. \tag{32}$$

4. Finally, let us obtain the last equation, which is an extended version of (9). In this sense, substituting $a = D$ and $b = \mu$ into Equation (14), we obtain

$$\overset{(D+1)}{G}_{D\mu} = \overset{(D+1)}{R}_{D\mu} = \chi \overset{(D+1)}{T}_{D\mu} + \mathcal{W}\phi^n\overset{*}{\phi}(\mathcal{D}_\alpha\phi). \tag{33}$$

By equating (33) and (4) (which is obtained directly from metric (2); see also the second Remark), we obtain an extended dynamical equation for $P_{\alpha\beta}$ in MSBT:

$$\psi P^\beta{}_{\mu;\beta} = \chi \overset{(D+1)}{T}_{D\mu} + \mathcal{W}\phi^n\overset{*}{\phi}(\mathcal{D}_\mu\phi), \tag{34}$$

where $P_{\alpha\beta}$ is given by (6).

It is worth mentioning that Equations (27) and (31) are obtained from the action

$$\overset{(D)}{\mathcal{S}}_{\mathrm{SB}} = \int d^D x\sqrt{-g}\left[\overset{(D)}{R} - \mathcal{W}\phi^n g^{\alpha\beta}(\mathcal{D}_\alpha\phi)(\mathcal{D}_\beta\phi) - V(\phi) + \chi \overset{(D)}{L}_{\mathrm{matt}}\right], \tag{35}$$

which is a generalized action of the SB theory with a scalar potential and

$$\sqrt{-g}\, T_{\mu\nu}^{(D)[\mathrm{eff}]} \equiv 2\frac{\delta\mathcal{S}_{\mathrm{matt}}}{\delta g^{\mu\nu}}, \tag{36}$$

where $\mathcal{S}_{\mathrm{matt}} = \int d^D x\sqrt{-g}\, \overset{(D)}{L}_{\mathrm{matt}}$ is the action corresponding to the matter fields in D dimensions. We should emphasize that the induced energy momentum tensor is covariantly conserved, namely, $\mathcal{D}_\beta T_\alpha^{(D)[\mathrm{eff}]\beta} = 0$.

3.2. Modified Brans–Dicke Theory in Arbitrary Dimensions

In this subsection, we will present a brief overview of the modified Brans–Dicke theory (MBDT). The action in $(D+1)$ dimensions in the Jordan frame associated with the Brans–Dicke (BD) theory can be written as

$$\mathcal{S}_{\rm BD}^{(D+1)} = \int d^{D+1}x\sqrt{\left|\mathcal{G}\right|}\left[\varphi R^{(D+1)} - \frac{\omega}{\varphi}\mathcal{G}^{ab}\,(\nabla_a\varphi)(\nabla_b\varphi) + 16\pi\,L_{\rm matt}^{(D+1)}\right], \tag{37}$$

where φ and ω are the BD scalar field and an adjustable dimensionless parameter (called the BD coupling parameter), respectively[6].

The equations of motion corresponding to the action (37) can be written as

$$\begin{aligned} G_{ab}^{(D+1)} &= \frac{8\pi}{\varphi}T_{ab}^{(D+1)} + \\ &\quad \frac{\omega}{\varphi^2}\left[(\nabla_a\varphi)(\nabla_b\varphi) - \frac{1}{2}\mathcal{G}_{ab}(\nabla^c\varphi)(\nabla_c\varphi)\right] + \frac{1}{\varphi}\left(\nabla_a\nabla_b\varphi - \mathcal{G}_{ab}\nabla^2\varphi\right), \end{aligned} \tag{38}$$

and

$$\frac{2\omega}{\varphi}\nabla^2\varphi - \frac{\omega}{\varphi^2}\mathcal{G}^{ab}(\nabla_a\varphi)(\nabla_b\varphi) + R^{(D+1)} = 0. \tag{39}$$

From Equation (38), we easily obtain the Ricci scalar:

$$R^{(D+1)} = -\frac{16\pi\,T^{(D+1)}}{(D-1)\varphi} + \frac{\omega}{\varphi^2}(\nabla^c\varphi)(\nabla_c\varphi) + \frac{2D}{D-1}\frac{\nabla^2\varphi}{\varphi}, \tag{40}$$

where $T^{(D+1)} = \mathcal{G}^{ab}T_{ab}^{(D+1)}$.

Substituting (40) into (39) yields[7]

$$\nabla^2\phi = \frac{8\pi T^{(D+1)}}{(D-1)\omega + D}. \tag{41}$$

By applying a reduction procedure similar to that in the previous subsection, we can set up the effective equations associated with the D-dimensional hypersurface. For a detailed investigation of such a method in the context of the BD theory, see [8]. We therefore refrain from presenting the details of this approach and confine ourselves to a brief summary of the results. More precisely, considering the metric (2) and using an appropriate reduction procedure, it has been show that Equations (38) and (41) generate four sets of modified field equations on a D-dimensional hypersurface [8]. These field equations are:

1. An equation for the scalar field ψ is:

$$\begin{aligned} \frac{\mathcal{D}^2\psi}{\psi} &= -\frac{(\mathcal{D}_\alpha\psi)(\mathcal{D}^\alpha\varphi)}{\psi\varphi} - \frac{\epsilon}{2\psi^2}\left(g^{\lambda\beta}\overset{**}{g}_{\lambda\beta} + \frac{1}{2}\overset{*}{g}{}^{\lambda\beta}\overset{*}{g}_{\lambda\beta} - \frac{g^{\lambda\beta}\overset{*}{g}_{\lambda\beta}\overset{*}{\psi}}{\psi}\right) \\ &\quad - \frac{\epsilon}{\psi^2\varphi}\left[\overset{**}{\varphi} + \overset{*}{\varphi}\left(\frac{\omega\overset{*}{\varphi}}{\varphi} - \frac{\overset{*}{\psi}}{\psi}\right)\right] + \frac{8\pi}{\varphi}\left[\frac{(\omega+1)T^{(D+1)}}{(D-1)\omega + D} - \frac{\epsilon T_{DD}^{(D+1)}}{\psi^2}\right]. \end{aligned} \tag{42}$$

2. The other effective field equations are the counterpart Equations of (38) and (41):

$$\begin{aligned} G_{\mu\nu}^{(D)} &= \frac{8\pi T_{\mu\nu}^{(D)[\text{eff}]}}{\varphi} + \frac{\omega}{\varphi^2}\left[(\mathcal{D}_\mu\varphi)(\mathcal{D}_\nu\varphi) - \frac{1}{2}g_{\mu\nu}(\mathcal{D}_\alpha\varphi)(\mathcal{D}^\alpha\varphi)\right] \\ &\quad + \frac{1}{\varphi}\left(\mathcal{D}_\mu\mathcal{D}_\nu\varphi - g_{\mu\nu}\mathcal{D}^2\varphi\right) - g_{\mu\nu}\frac{V(\varphi)}{2\varphi}. \end{aligned} \tag{43}$$

In Equation (43), the induced scalar potential $V(\varphi)$ is obtained from a differential equation, see Equation (48); the effective energy-momentum tensor $T^{(D)[\mathrm{eff}]}_{\mu\nu}$ consists of two parts: $T^{(D)[\mathrm{eff}]}_{\mu\nu} \equiv E_{\mu\nu} + T^{[\mathrm{MBDT}]}_{\mu\nu}$ with

$$E_{\mu\nu} \equiv T^{(D+1)}_{\mu\nu} - g_{\mu\nu}\left[\frac{(\omega+1)T^{(D+1)}}{(D-1)\omega+D} - \frac{\epsilon\, T^{(D+1)}_{DD}}{\psi^2}\right], \tag{44}$$

$$\frac{8\pi}{\varphi}T^{[\mathrm{MBDT}]}_{\mu\nu} \equiv T^{[\mathrm{IMT}]}_{\mu\nu} + \frac{1}{\varphi}T^{[\varphi]}_{\mu\nu} + \frac{V(\varphi)}{2\varphi}g_{\mu\nu}. \tag{45}$$

We should note that in Equation (45) $T^{[\mathrm{IMT}]}_{\mu\nu}$ is exactly the same induced matter introduced in the IMT, but $T^{[\varphi]}_{\mu\nu}$ is given by

$$T^{[\varphi]}_{\mu\nu} \equiv \frac{\epsilon\overset{*}{\varphi}}{2\psi^2}\left[\overset{*}{g}_{\mu\nu} + g_{\mu\nu}\left(\frac{\omega\overset{*}{\varphi}}{\varphi} - g^{\alpha\beta}\overset{*}{g}_{\alpha\beta}\right)\right], \tag{46}$$

which depends on the first derivatives of the BD scalar field with respect to l. Concretely, it is another induced energy momentum tensor on the hypersurface arising due to the presence of φ.

3. The wave equation of the MBDT is:

$$\mathcal{D}^2\varphi = \frac{1}{(D-2)\omega+(D-1)}\left[8\pi T^{(D)[\mathrm{eff}]} + \left(\frac{D-2}{2}\right)\varphi\frac{dV(\varphi)}{d\varphi} - \frac{D}{2}V(\varphi)\right], \tag{47}$$

where $V(\varphi)$ is obtained from[8]

$$\begin{aligned}\varphi\frac{dV(\varphi)}{d\varphi} \equiv{}& -2(\omega+1)\left[\frac{(\mathcal{D}_\alpha\psi)(\mathcal{D}^\alpha\varphi)}{\psi} + \frac{\epsilon}{\psi^2}\left(\overset{**}{\varphi} - \frac{\overset{*}{\psi}\overset{*}{\varphi}}{\psi}\right)\right]\\ &- \frac{\epsilon\omega\overset{*}{\varphi}}{\psi^2}\left[\frac{\overset{*}{\varphi}}{\varphi} + g^{\mu\nu}\overset{*}{g}_{\mu\nu}\right] + \frac{\epsilon\varphi}{4\psi^2}\left[\overset{*}{g}{}^{\alpha\beta}\overset{*}{g}_{\alpha\beta} + (g^{\alpha\beta}\overset{*}{g}_{\alpha\beta})^2\right]\\ &+ 16\pi\left[\frac{(\omega+1)T^{(D+1)}}{(D-1)\omega+D} - \frac{\epsilon\, T^{(D+1)}_{DD}}{\psi^2}\right].\end{aligned} \tag{48}$$

4. A counterpart to the conservation Equation (9) introduced in the IMT is:

$$G^{(D+1)}_{\alpha D} = \psi\mathcal{D}_\beta P^\beta{}_\alpha = \frac{8\pi T^{(D+1)}_{\alpha D}}{\varphi} + \frac{\omega\overset{*}{\varphi}(\mathcal{D}_\alpha\varphi)}{\varphi^2} + \frac{\mathcal{D}_\alpha\overset{*}{\varphi}}{\varphi} - \frac{\overset{*}{g}_{\alpha\lambda}(\mathcal{D}^\lambda\varphi)}{2\varphi} - \frac{\overset{*}{\varphi}(\mathcal{D}_\alpha\psi)}{\varphi\psi}. \tag{49}$$

In summary, considering the metric (2) as the background geometry, by applying the reduction procedure introduced in [8], Equations (38) and (41) split into four sets of effective Equations (42), (43), (47) and (49) on a D-dimensional hypersurface. It should be noted that both the induced energy-momentum tensor and the induced scalar potential in the context of the MBDT are obtained from specific equations and therefore have specific types with respect to the phenomenological models. However, we should emphasize that these quantities can be considered as fundamental rather than some ad hoc phenomenological assumptions.

It is seen that Equations (43) and (47) are identical to those of the BD theory obtained from the action

$$\mathcal{S}_{\rm BD}^{(D)} = \int d^Dx\sqrt{-g}\left[\varphi R^{(D)} - \frac{\omega}{\varphi}g^{\alpha\beta}(\mathcal{D}_\alpha\varphi)(\mathcal{D}_\beta\varphi) - V(\varphi) + 16\pi L_{\rm matt}^{(D)}\right], \tag{50}$$

where specifically $\sqrt{-g}\left(E_{\mu\nu} + T_{\mu\nu}^{[\rm MBDT]}\right) \equiv 2\delta\left(\int d^Dx\sqrt{-g}\, L_{\rm matt}^{(D)}\right)/\delta g^{\alpha\beta}$ such that $T_{\mu\nu}^{(D)[\rm eff]} = E_{\mu\nu} + T_{\mu\nu}^{[\rm MBDT]}$ stands for the Lagrangian associated with the matter in D dimensions.

4. FLRW-MSBT Cosmology

Our main goal in this review paper has been to provide only a very brief overview of the noncompactified KK gravity frameworks and their applications in cosmology. Therefore, in this section, we confine ourselves to a single important cosmological application, namely the FLRW cosmology in one of our modified models. To study other applications, the reader is referred to the related work referenced in this paper, see, e.g., [13,41,43–45] and references therein.

As a cosmological application of the MSBT framework constructed in the previous section, let us present a review of the model investigated in [9]. The spatially flat FLRW universe in a $(D+1)$-dimensions was considered in Ref. [9]:

$$dS^2 = -dt^2 + a^2(t)\left[\sum_{i=1}^{D-1}\left(dx^i\right)^2\right] + \epsilon\psi^2(t)dl^2. \tag{51}$$

In Equation (51), t, x^i (where $i = 1, 2, \ldots, D-1$), and $a(t)$ stand for the cosmic time, the Cartesian coordinates, and the scale factor, respectively. Moreover, let us assume there is no higher-dimensional matter, and a, ϕ, and ψ to be depending on the cosmic time only.

Therefore, Equations (14) and (15) for the metric (51) yield

$$\frac{D-2}{2}\left(\frac{\dot a}{a}\right)^2 + \left(\frac{\dot a}{a}\right)\left(\frac{\dot\psi}{\psi}\right) = \frac{\mathcal{W}}{2(D-1)}\left(\phi^{\frac{n}{2}}\dot\phi\right)^2, \tag{52}$$

$$\frac{\ddot a}{a} + \frac{D-3}{2}\left(\frac{\dot a}{a}\right)^2 + \left(\frac{\dot a}{a}\right)\left(\frac{\dot\psi}{\psi}\right) + \frac{\ddot\psi}{\psi} = -\frac{\mathcal{W}}{2(D-2)}\left(\phi^{\frac{n}{2}}\dot\phi\right)^2, \tag{53}$$

$$\frac{\ddot a}{a} + \frac{D-2}{2}\left(\frac{\dot a}{a}\right)^2 = -\frac{\mathcal{W}}{2(D-1)}\left(\phi^{\frac{n}{2}}\dot\phi\right)^2, \tag{54}$$

$$\ddot\phi + \left[(D-1)\frac{\dot a}{a} + \frac{n}{2}\left(\frac{\dot\phi}{\phi}\right) + \frac{\dot\psi}{\psi}\right]\dot\phi = 0, \tag{55}$$

where $\dot A \equiv dA/dt$ for any arbitrary variable $A = A(t)$.

Only three of the Equations (52)–(55) are independent, from which we have to determine the three unknowns $a(t)$, $\phi(t)$, and $\psi(t)$.

Let us first focus on a simple case: if the SB scalar field ϕ takes only constant values, we can easily obtain a unique solution as

$$dS^2 = -dt^2 + \left(C_1 t^{\frac{2}{D}}\right)^2\left[\sum_{i=1}^{D-1}\left(dx^i\right)^2\right] + \epsilon\left(C_2 t^{\frac{2}{D}-1}\right)^2 dl^2, \tag{56}$$

where C_1 and C_2 are constants of integration. It is worth noting that (56) is the only solution in the context of general relativity when an empty universe is described by a $(D+1)$-dimensional spatially flat FLRW metric.

Whereas for the general case, employing Equations (52), (53) and (55), two constants of motion are retrieved:

$$a^{D-1}\phi^{\frac{n}{2}}\dot{\phi}\psi = c_1, \tag{57}$$

$$a^{D-1}\dot{\psi} = c_2, \tag{58}$$

where c_1 and c_2 are constants of integration. Then, from using Equations (57) and (58), we can determine ψ as a function of ϕ:

$$\psi(\phi) = \begin{cases} \psi_i \exp\left(\frac{2\beta}{n+2}\phi^{\frac{n+2}{2}}\right), & \text{for} \quad n \neq -2, \\ \psi_i \phi^{\beta}, & \text{for} \quad n = -2, \end{cases} \tag{59}$$

where $\beta \equiv \frac{c_2}{c_1}$, and ψ_i is an integration constant. We will assume $c_1 \neq 0$ and $a^{D-1}\psi \neq 0$. Moreover, to obtain a as a function of ϕ, we substitute ψ from Equation (59) into Equation (52):

$$a(\phi) = \begin{cases} a_i \exp\left[\frac{2\gamma}{n+2}\phi^{\frac{n+2}{2}}\right] & \text{for} \quad n \neq -2, \\ a_i \phi^{\gamma} & \text{for} \quad n = -2. \end{cases} \tag{60}$$

In Equation (60), a_i is a constant of integration and

$$\gamma \equiv \frac{1}{D-2}\left[-\beta \pm \sqrt{\beta^2 + \left(\frac{D-2}{D-1}\right)\mathcal{W}}\right], \tag{61}$$

which yields real values provided that

$$\mathcal{W} \geq -\left(\frac{D-1}{D-2}\right)\beta^2. \tag{62}$$

In order to obtain the unknowns in terms of the cosmic time, we substitute ψ and a, respectively, from Equations (59) and (60) into Equation (57). Such a procedure yields

$$\begin{cases} \dot{\phi}\phi^{\frac{n}{2}} \exp\left[\frac{2f}{n+2}\phi^{\frac{n+2}{2}}\right] = \frac{c_1 a_i^{1-D}}{\psi_i}, & \text{for} \quad n \neq -2, \\ \dot{\phi}\phi^{f} = \frac{c_1 a_i^{1-D}}{\psi_i}, & \text{for} \quad n = -2, \end{cases} \tag{63}$$

where

$$f \equiv (D-1)\gamma + \beta. \tag{64}$$

We should note that both γ and f depend on D. Depending on whether f is zero or not, we will have two different solutions for each of the above differential equations. In what follows, we will analyze them separately.

Case I: $f \equiv (D-1)\gamma + \beta = 0$

For this case, relation (64) yields

$$\gamma = -\frac{\beta}{D-1}, \qquad \mathcal{W} = -\left(\frac{D}{D-1}\right)\beta^2. \tag{65}$$

Moreover, from using Equation (63), we obtain

$$\phi(t) = \begin{cases} \left[\frac{(n+2)(1-D)h(t-t_i)}{2\beta}\right]^{\frac{2}{n+2}}, & \text{for} \quad n \neq -2, \\ \exp\left[\frac{(1-D)h(t-t_i)}{\beta}\right], & \text{for} \quad n = -2, \end{cases} \tag{66}$$

where t_i is an integration constant and

$$h \equiv \frac{c_1 \beta a_i^{1-D}}{(1-D)\psi_i}. \tag{67}$$

We should note that h does not depend on the cosmic time, but we emphasize that it is a function of D, i.e., $h = h(D)$. Moreover, substituting the scalar field from (66) into (59) and (60), we obtain the other unknowns in terms of the cosmic time:

$$a(t) = a_i \exp[h(t-t_i)], \qquad \forall n \tag{68}$$

$$\psi(t) = \psi_i \exp[(1-D)h(t-t_i)], \qquad \forall n. \tag{69}$$

For Case I, it is seen that among the unknowns, only $\phi(t)$ explicitly depends on n. Substituting $a(t)$ and $\psi(t)$ from Equations (68) and (69) into (51), we eventually obtain

$$dS^2 = -dt^2 + a_i^2 \exp[2(t-t_i)] \sum_{i=1}^{D-1} \left(dx^i\right)^2 + \epsilon\psi_i^2 \exp[2(1-D)h(t-t_i)]dl^2, \qquad \forall n. \tag{70}$$

Case II: $f \equiv (D-1)\gamma + \beta \neq 0$

For this case, from using Equation (61), $\mathcal{W}$ can be expressed as

$$\mathcal{W} = (D-1)\gamma[2\beta + (D-2)\gamma]. \tag{71}$$

In order to obtain the SB scalar field in terms of the cosmic time, we integrate both sides of Equation (63) over dt, which yields

$$\phi(t) = \begin{cases} \left\{\frac{n+2}{2f} \ln\left[\tilde{h}(t-t_i)\right]\right\}^{\frac{2}{n+2}}, & \text{for} \quad n \neq -2, \\ \left[\tilde{h}(t-t_i)\right]^{\frac{1}{f}}, & \text{for} \quad n = -2, \end{cases} \tag{72}$$

where

$$\tilde{h} \equiv \frac{c_1 f}{a_i^{D-1}\psi_i}. \tag{73}$$

Moreover, by substituting $\phi(t)$ from Equation (72) into Equations (59) and (60), we obtain

$$a(t) = a_i\left[\tilde{h}(t-t_i)\right]^r, \qquad \forall n \tag{74}$$

$$\psi(t) = \psi_i\left[\tilde{h}(t-t_i)\right]^m, \qquad \forall n, \tag{75}$$

where

$$r \equiv \frac{\gamma}{f}, \qquad m \equiv \frac{\beta}{f}, \qquad m + (D-1)r = 1. \tag{76}$$

Then, we can rewrite relation (71) as

$$\mathcal{W} = (D-1)f^2 r[2m + (D-2)r]. \tag{77}$$

In what follows, to construct the cosmological solutions on the D-dimensional hypersurface and to analyze the reduced cosmological dynamics, we will use the MSBT framework presented in the previous section.

Employing Equation (29) for metric (51), we can easily obtain the components of the induced energy momentum tensor:

$$\rho_{\text{SB}} \equiv -T^{0[\text{MSBT}]}_{\ 0} = \frac{1}{\chi}\left[\frac{\ddot{\psi}}{\psi} - \frac{V(\phi)}{2}\right], \tag{78}$$

$$p_{\text{SB}} \equiv T^{i[\text{MSBT}]}_{\ i} = -\frac{1}{\chi}\left[\frac{\dot{a}\dot{\psi}}{a\psi} - \frac{V(\phi)}{2}\right], \tag{79}$$

where $i = 1, 2, 3, \ldots, D-1$ (with no sum), and ρ_{SB} and p_{SB} are the induced energy density and pressure, respectively. Moreover, we use Equation (32) to obtain the induced scalar potential:

$$\left.\frac{dV}{d\phi}\right|_{\Sigma_o} = 2\mathcal{W}\phi^n\dot{\phi}\left(\frac{\dot{\psi}}{\psi}\right). \tag{80}$$

In order to eliminate $\dot{\phi}$ and $\dot{\psi}$ in favor of the other variables of the model, we use Equations (57) and (58):

$$\left.\frac{dV}{d\phi}\right|_{\Sigma_o} = 2c_1^2\beta\mathcal{W}a^{2(1-D)}\phi^{\frac{n}{2}}\psi^{-2}. \tag{81}$$

Finally, by substituting a and ψ from Equations (59) and (60) into Equation (81), we obtain

$$\left.\frac{dV}{d\phi}\right|_{\Sigma_o} = \begin{cases} V_0\phi^{-[1+2f]}, & \text{for} \quad n = -2, \\ V_0\phi^{\frac{n}{2}}\exp\left[-\frac{4f}{n+2}\phi^{\frac{n+2}{2}}\right], & \text{for} \quad n \neq -2, \end{cases} \tag{82}$$

where

$$V_0 \equiv 2c_1^2\beta\mathcal{W}a_i^{2(1-D)}\psi_i^{-2}. \tag{83}$$

One sees that the differential Equation (82) depends on the values that f takes on. More precisely, we must continue our discussions for two different cases, I and II, separately. Such a procedure will be carried out later. Here are some important comments.

Equations (27) and (31) corresponding to the D-dimensional spatially flat FLRW metric read

$$\frac{(D-1)(D-2)}{2}H^2 = \chi\rho_{\text{SB}} + \rho_\phi \equiv \rho_{\text{tot}}, \tag{84}$$

$$(D-2)\frac{\ddot{a}}{a} + \frac{(D-2)(D-3)}{2}H^2 = -(\chi p_{\text{SB}} + p_\phi) \equiv -p_{\text{tot}}, \tag{85}$$

$$2\phi^n\ddot{\phi} + 2(D-1)H\phi^n\dot{\phi} + n\phi^{n-1}\dot{\phi}^2 + \left.\frac{1}{\mathcal{W}}\frac{dV}{d\phi}\right|_{\Sigma_o} = 0, \tag{86}$$

where $H \equiv \dot{a}/a$ denotes the Hubble parameter, $\rho_{\rm SB}$ and $p_{\rm SB}$ and $\frac{dV}{d\phi}\Big|_{\Sigma_o}$ are given by Equations (78), (79) and (80), respectively. Moreover, ρ_ϕ and p_ϕ stand for the energy density and pressure associated with the SB scalar field ϕ:

$$\rho_\phi \equiv \frac{1}{2}\Big[\mathcal{W}\phi^n\dot{\phi}^2 + V(\phi)\Big], \tag{87}$$

$$p_\phi \equiv \frac{1}{2}\Big[\mathcal{W}\phi^n\dot{\phi}^2 - V(\phi)\Big]. \tag{88}$$

Employing Equations (84) and (85), one can easily show

$$\frac{\ddot{a}}{a} = -\frac{1}{(D-1)(D-2)}[(D-3)\rho_{\rm tot} + (D-1)p_{\rm tot}]. \tag{89}$$

For later usage, let us also introduce the following equation of state (EoS) and density parameters:

$$W_{\rm SB} \equiv \frac{p_{\rm SB}}{\rho_{\rm SB}}, \qquad W_\phi \equiv \frac{p_\phi}{\rho_\phi}, \qquad W_{\rm tot} \equiv \frac{p_{\rm tot}}{\rho_{\rm tot}} = \frac{\chi p_{\rm SB} + p_\phi}{\chi \rho_{\rm SB} + \rho_\phi}, \tag{90}$$

$$\Omega_{\rm SB} \equiv \frac{2\chi}{(D-1)(D-2)}\frac{\rho_{\rm SB}}{H^2}, \tag{91}$$

$$\Omega_\phi \equiv \frac{2}{(D-1)(D-2)}\frac{\rho_\phi}{H^2}. \tag{92}$$

Employing these parameters, it is seen that Equation (84) can be written as $\Omega_{\rm SB} + \Omega_\phi = 1$.

In the following, we will observe that each equation of (82), due to whether the quantity f vanishes or not, in turn generates two different functions of the scalar field, which are analyzed separately.

4.1. Case (I) D-Dimensional Solutions with $f = (D-1)\gamma + \beta = 0$

For this case, by solving Equation (82), we obtain

$$V(\phi) = \begin{cases} \frac{2V_0}{n+2}\phi^{\frac{n+2}{2}}, & \text{for} \quad n \neq -2, \\ V_0 \ln\left(\frac{\phi}{\phi_i}\right), & \text{for} \quad n = -2. \end{cases} \tag{93}$$

In Equation (93), ϕ_i is an integration constant and

$$V_0 = 2\beta D(1-D)h^2, \tag{94}$$

is obtained from using Equations (65), (67) and (83). Substituting $\phi(t)$ from Equation (66) into (93), we eventually obtain the corresponding induced scalar potential in terms of the cosmic time:

$$V(t) = V_0(1-D)h\beta^{-1}(t-t_i) = 2D(1-D)^2h^3(t-t_i), \qquad \forall n, \tag{95}$$

where h is given by Equation (67).

Now, having the scale factors $a(t)$ and $\psi(t)$ from (68) and (69) as well as the scalar potential in terms of the cosmic time, we substitute them into (78) and (79), which yields

$$\chi\rho_{\rm SB} \equiv -\chi T^{0[\rm MSBT]}_{\ 0} = (1-D)^2h^2[-Dh(t-t_i)+1], \tag{96}$$

$$\chi p_{\rm SB} \equiv \chi T^{i[\rm MSBT]}_{\ i} = (1-D)h^2[D(1-D)h(t-t_i)-1]. \tag{97}$$

Employing Equations (68), (96) and (97), it is easy to show that the quantity $\dot{\rho}_{\rm SB}+(D-1)H(\rho_{\rm SB}+p_{\rm SB})$ vanishes identically. More concretely, the conservation of the induced energy momentum tensor holds in MSBT, which is one of the distinctive features of the framework.

Moreover, substituting $\phi(t)$ and $V(t)$ from relations (66) and (95) into (87) and (88), we obtain

$$\rho_\phi = \frac{1}{2}D(1-D)h^2[1+2(1-D)h(t-t_i)], \quad \forall n, \tag{98}$$

$$p_\phi = \frac{1}{2}D(1-D)h^2[1-2(1-D)h(t-t_i)], \quad \forall n, \tag{99}$$

from which we obtain

$$W_\phi = \frac{1-2(D-1)h(t-t_i)}{1+2(D-1)h(t-t_i)}, \quad \forall n. \tag{100}$$

In summary, the SB cosmological model in D-dimensional hypersurface derived from the $(D+1)$-dimensional solutions associated with Case I are:

$$ds^2 = -dt^2 + a_i^2 \exp\left[2\sqrt{\frac{\Lambda}{D-1}}(t-t_i)\right]\left[\sum_{i=1}^{D-1}\left(dx^i\right)^2\right], \tag{101}$$

$$\rho_{\rm tot} = \frac{1}{2}(D-2)\Lambda, \quad p_{\rm tot} = -\frac{1}{2}(D-2)\Lambda, \quad W_{\rm tot} = -1, \quad \forall n, \tag{102}$$

where

$$\Lambda = \Lambda(D) \equiv \frac{1}{D-1}\left(\frac{c_1\beta}{a_i^{D-1}\psi_i}\right)^2 = \text{constant} > 0, \quad \text{for} \quad D>1. \tag{103}$$

We note that in deriving the total energy density, we have used Equations (84), (85) and (96)–(99).

Relation (103) implies that the quantity $\Lambda(D)>0$ can be interpreted as a cosmological constant, which, in turn, emerges from combining varying induced matter fields, see relations (96)–(99). Let us look at $\Lambda(D)$ from a different perspective. With Equation (69), the expression in parenthesis of Equation (103) can be replaced by $\dot{\psi}/\psi$, i.e., we obtain

$$\Lambda(D) = \frac{1}{D-1}\left(\frac{\dot{\psi}}{\psi}\right)^2 = \text{constant}, \quad \forall t, \tag{104}$$

which holds forever. Equation (104) implies that the value of the so-called cosmological constant depends not only on the number of spatial dimensions, but also on the squared expansion rate associated with the extra dimension. Finally, it should be mentioned that the solution belonging to Case I describes an exponentially expanding universe, which is analogous to the usual de Sitter solution.

It is worth making a few more comments about our solution here. Let us consider the canonical metric

$$dS^2 = \frac{l^2}{L^2}\tilde{g}_{\mu\nu}(x^\alpha, l)dx^\mu dx^\nu - dl^2, \tag{105}$$

where $L > 0$ is a constant, and $\tilde{g}_{\mu\nu} = \mathrm{diag}(1, -a^2(t), -a^2(t), \ldots, -a^2(t))$. We also assume a special case where $T^{(D+1)}_{\mu\nu} = 0$, and the SB scalar field depends only on t. Therefore, we obtain

$$\chi T^{(D)[\text{eff}]}_{\mu\nu} = \frac{1}{2L^2}\left(D^2 - 3D + 2 + l_0^2 V_0\right)\tilde{g}_{\mu\nu}, \tag{106}$$

where $V_0 =$ constant is obtained from (32), and $l_0 =$ constant is the value of l on the hypersurface. For this case, from Equation (27) we obtain

$$G^{(D)}_{\mu\nu} = \left(\frac{D^2 - 3D + 2}{2L^2}\right)\tilde{g}_{\mu\nu} + \mathcal{W}\phi^n\left[(\mathcal{D}_\mu\phi)(\mathcal{D}_\nu\phi) - \frac{l_0^2}{2L^2}(\mathcal{D}_\alpha\phi)(\mathcal{D}^\alpha\phi)\tilde{g}_{\mu\nu}\right]. \tag{107}$$

Applying the framework presented in the preceding section, the cosmological solutions corresponding to the metric (105) are

$$a(t) = a_i \exp\left[\frac{(D-1)(t-t_i)}{L}\right], \tag{108}$$

$$\phi(t) = \begin{cases} \left\{\frac{CL(n+2)}{2(1-D)}\exp\left[\frac{(1-D)(t-t_i)}{L}\right]\right\}^{\frac{2}{n+2}}, & \text{for} \quad n \neq -2, \\ \\ \left\{\exp\left[\exp\frac{(1-D)(t-t_i)}{L}\right]\right\}^{\frac{CL}{1-D}}, & \text{for} \quad n = -2, \end{cases} \tag{109}$$

where t_i and a_i are integration constants, and $C \equiv a^{D-1}\phi^{\frac{n}{2}}\dot{\phi}$ is a constant of motion. In a special case where $\phi =$ constant, from using Equation (107), we obtain $G_{\mu\nu} = \Lambda(D)\tilde{g}_{\mu\nu}$, where $\Lambda = \Lambda(D) \equiv (D^2 - 3D + 2)/(2L^2)$. To study the latter in four dimensions in more detail, see [59,60] and the references therein.

4.2. Case (II) D-Dimensional Solutions with $f \equiv \beta + (D-1)\gamma \neq 0$

The induced scalar potential assigned to this case is obtained from solving the differential Equation (82):

$$V(\phi) = \begin{cases} -\frac{V_0}{2f}\exp\left[\frac{-4f}{n+2}\phi^{\frac{n+2}{2}}\right], & \text{for} \quad n \neq -2, \\ \\ -\frac{V_0}{2f}\phi^{-2f}, & \text{for} \quad n = -2. \end{cases} \tag{110}$$

Equation (110) can be expressed in terms of t with a single relation for all values of n:

$$V(t) = -\frac{(D-1)mr[1+m-r]}{(t-t_i)^2}, \qquad \forall n, \tag{111}$$

where we have used (72).

Substituting $a(t)$, $\psi(t)$ and $V(t)$ from relations (74), (75), and (111) into (78) and (79), we obtain

$$\rho_{\text{SB}} = -\frac{D(D-1)mr^2}{2\chi(t-t_i)^2}, \qquad p_{\text{SB}} = -\frac{Dmr(1+m)}{2\chi(t-t_i)^2}, \qquad \forall n, \tag{112}$$

which yields a barotropic fluid:

$$p_{\text{SB}} = W_{\text{SB}}\rho_{\text{SB}}, \qquad W_{\text{SB}} = \frac{1+m}{(D-1)r}, \qquad \forall n. \tag{113}$$

Moreover, ρ_ϕ and p_ϕ are obtained from substituting $\phi(t)$ and $V(t)$ from (72) and (111) into (87) and (88):

$$\mathcal{W}\phi^n\dot{\phi}^2 = \frac{(D-1)r(1+m-r)}{2(t-t_i)^2}, \tag{114}$$

$$\rho_\phi = \frac{\mathcal{W}(1-m)}{2f^2(t-t_i)^2} = \frac{[(D-1)r]^2[2m+(D-2)r]}{2(t-t_i)^2}, \tag{115}$$

$$p_\phi = \frac{\mathcal{W}(1+m)}{2f^2(t-t_i)^2} = \frac{[(D-1)r][2m+(D-1)r][2m+(D-2)r]}{2(t-t_i)^2}, \tag{116}$$

where the first relation was written for later use. Therefore, we obtain

$$W_\phi = \frac{1+m}{1-m} = \frac{(D-1)r+2m}{(D-1)r}, \tag{117}$$

$$\Omega_\phi = \frac{(1-m)(1+m-r)}{(D-2)r} = \frac{(D-1)[(D-2)r+2m]}{(D-2)[(D-1)r+m]}. \tag{118}$$

Now, the total energy density and pressure are obtained easily by substituting $\rho_{\rm SB}$, $p_{\rm SB}$, ρ_ϕ, and p_ϕ from relations (112), (115), and (116) into corresponding definitions:

$$\rho_{\rm tot} = \frac{(D-1)(D-2)r^2}{2(t-t_i)^2}, \qquad p_{\rm tot} = \frac{(D-2)r(1+m)}{2(t-t_i)^2}, \qquad W_{\rm tot} = \frac{1+m}{(D-1)r} \qquad \forall n, \tag{119}$$

which implies that the total induced matter is also a barotropic obtained from adding two other barotropic matter fluids. Let us also write another useful equation:

$$\frac{\ddot{a}}{a} = -\frac{r[m+(D-2)r]}{(t-t_i)^2}, \qquad \forall n, \tag{120}$$

which has been obtained from substituting $\rho_{\rm tot}$ and $p_{\rm tot}$ from relations (119) into Equation (89).

In what follows, let us analyze the solutions of Case II.

- It is straightforward to show that the corresponding conservation equation is satisfied for our herein three matter fields. Namely,

$$\dot{\rho}_{\rm SB} + 3H(\rho_{\rm SB} + p_{\rm SB}) = 0, \quad \dot{\rho}_\phi + 3H(\rho_\phi + p_\phi) = 0, \quad \dot{\rho}_{\rm tot} + 3H(\rho_{\rm tot} + p_{\rm tot}) = 0. \tag{121}$$

- If we assume that the induced matter plays the role of an ordinary matter in the universe, it is better to check at least the satisfaction of the weak energy condition for it. Let us be more specific. Relations (112) imply that for satisfying $\rho_{SB} \geq 0$, m must take negative values. Moreover, in order to satisfy $\rho_{SB} + p_{SB} = -\frac{Dmr}{\chi}(t-t_i)^2 \geq 0$, r and m must take positive and negative values, respectively. Therefore, if $a(t)$ increases and $\psi(t)$ decreases with cosmic time, then the weak energy condition will be satisfied.
- It is seen that $W_{\rm tot} = W_{\rm SB} = W_\phi$, which can also be fulfilled under approximation conditions with $|p_\phi| \ll \chi|\rho_{\rm SB}|$ and/or $\chi|p_{\rm SB}| \ll |\rho_\phi|$.
- Let us now focus on Equation (120). Before continuing our discussions, we want to mention that we will focus on the solutions with $D > 3$. Moreover, we assume that the constant coefficients $a_i\tilde{h}^r$ and $\psi_i\tilde{h}^m$ that appeared in relations (74) and (75) always take positive values. Furthermore, we do not need to write $\forall n$ in front of the equations since they apply to all n.

  In what follows, we present two different cases.

  Case IIa: $\gamma < 0$, $\beta + (D-2)\gamma > 0$

  For this case, we obtain

$$-\frac{\beta}{D-2} < \gamma < 0, \qquad \beta > 0, \qquad \gamma < f,$$
$$2(D-1)\beta\gamma < \mathcal{W} < (D-1)\beta\gamma < 0. \tag{122}$$

Moreover, for this case, we find $-\beta/(D-2) < f < \beta$, namely, f takes positive as well as negative values. However, in order to have an expanding universe, relation (74) yields $r = \gamma/f > 0$. Therefore, from using (74) and (122), we find $f(D) < 0$. We eventually conclude that $f(D)$ must be constrained as

$$-\frac{\beta}{D-2} < f(D) < 0. \tag{123}$$

The above conditions lead to a cosmological model with $a > 0$ and $\ddot{a} > 0$. However, the extra dimension in such a model shrinks with cosmic time, which is favorable within the Kaluza–Klein frameworks [7].

Let us also determine the allowed ranges of the energy density, pressure, and density parameters. According to conditions (122) and (123) together with $f + \beta > 0$ (that is satisfied for $D > 3$), we find

$$\begin{array}{llll} \rho_{\text{SB}} > 0, & p_{\text{SB}} < 0, & W_{\text{SB}} < 0, & \Omega_{\text{SB}} > 0, \\ \rho_{\phi} < 0, & p_{\phi} > 0, & W_{\phi} < 0, & \Omega_{\phi} < 0, \\ \rho_{\text{tot}} > 0, & p_{\text{tot}} < 0, & W_{\text{tot}} < 0. & \end{array} \tag{124}$$

Moreover, in this case, both the potential $V(t)$ and the kinetic term $\mathcal{W}\phi^n\dot{\phi}^2$ take negative values forever.

Case IIb: $\gamma > 0, \beta + (D-2)\gamma < 0$

In this case, we obtain

$$0 < \gamma < -\tfrac{\beta}{D-2}, \qquad \beta < 0, \qquad \gamma > f,$$
$$2(D-1)\beta\gamma < \mathcal{W} < (D-1)\beta\gamma < 0. \tag{125}$$

Moreover, we find $\beta < f(D) < -\beta/(D-2)$, which implies that f takes both positive and negative values. Using the similar procedure mentioned above, we eventually obtain:

$$0 < f(D) < -\frac{\beta}{(D-2)}. \tag{126}$$

Admitting the above constraints together with assumption $D > 3$, we can easily show that the inequalities (124) hold among the corresponding physical quantities in this case.

In summary, FLRW-MSBT cosmological solutions associated with this case (Case II where $f \neq 0$), by removing the parameter m, can be written as

$$ds^2 = dS^2\Big|_{\Sigma_y} = -dt^2 + a_i^2\left[\tilde{h}(t-t_i)\right]^{2r}\left[\sum_{i=1}^{D-1}\left(dx^i\right)^2\right], \tag{127}$$

$$\rho_{\text{SB}} = -\frac{D(D-1)[1-(D-1)r]r^2}{2\chi(t-t_i)^2}, \qquad p_{\text{SB}} = W_{\text{SB}}\rho_{\text{SB}}, \tag{128}$$

$$V(t) = -\frac{(D-1)r(2-Dr)[1-(D-1)r]}{(t-t_i)^2}, \tag{129}$$

where scalar field $\phi(t)$ is given by (72). Moreover, employing relations (76), (77), and (113), W_{SB} and $\mathcal{W}$ are rewritten as

$$W_{SB} = \frac{2}{(D-1)r} - 1, \tag{130}$$

$$\mathcal{W} = \frac{(D-1)r\beta^2(2-Dr)}{[1-(D-1)r]^2}. \tag{131}$$

Concerning the power-law solution, we see that the scale factor accelerates whenever the deceleration parameter $q = -a\ddot{a}/(\dot{a})^2$ takes negative values, which for our model, we need to have $r > 1$. More concretely, all the three EoS parameters should be less than $(3-D)/(D-1)$. (For instance, for $D = 4$, we obtain $W_{SB}(= W_{tot} = W_\phi) < -1/3$.) Let us assume $D > 3$. We therefore find that both ρ_{SB} and ρ_{tot} take positive values, whilst the corresponding pressures take negative values. Consequently, both play the role of dark energy. (For investigations on dark energy models in the context of scalar–tensor theories, see, for instance, [61,62] and related papers.) On the other hand, we can easily show $\rho_{SB} + p_{SB} \geq 0$ an $\rho_{tot} + p_{tot} \geq 0$. The above expressions indicate that the weak energy condition is satisfied for both matters. However, regarding the matter associated with the scalar field, we obtain other properties. Let us be more precise. Assuming $r > 1$, from relation (131), we obtain $\mathcal{W} < 0$. Moreover, Equations (114)–(118) indicate $\mathcal{W}\phi^n\dot{\phi}^2 < 0$, $\rho_\phi < 0$, $p_\phi > 0$, and $\Omega_\phi < 0$, which can be considered as a dark energy [63]. It is worth noting that for this case (i.e., Case II where $f \neq 0$), m takes negative values, which implies that the extra dimension decreases with cosmic time.

Let us see under what conditions we obtain a decelerating scale factor. Assuming that the induced matter is analogous to an ordinary matter with an EoS parameter W constrained as $0 \leq W \leq 1$ (which in particular can be a matter-dominated, radiation-dominated, or stiff fluid with $W = 0, 1/(D-1), 1$ in a D-dimensional hypersurface). From relation (130), we therefore obtain $1/(D-1) \leq r \leq 2/(D-1)$, which is, for $D > 3$, associated with a decelerating universe. Moreover, we obtain $-1 \leq m \leq 0$, i.e., the extra dimension shrinks with cosmic time. It is easy to show that the weak energy condition is satisfied for both the induced and total matters. Whilst, according to (114)–(116), we find that the $\mathcal{W}$, ρ_ϕ and p_ϕ (the quantities associated with the SB scalar field) in the ranges $2/D < r \leq 2/(D-1)$ and $1/(D-1) \leq r < 2/D$ take negative and positive values, respectively, and they vanish when $r = 2/D$. In addition, note that inequality (62) is satisfied for the allowed ranges that yield the decelerating as well as accelerating scale factor, see [9].

5. Conclusions and Discussion

Since the theories discussed in this paper can be considered as alternative theories to GR, let us take a look at some other theories in this category. Concretely, there are the following methods to alter GR: by changing the matter source, the underlying geometry, the gravitational action functional, or all three, for example, by taking into account scalar field contributions or adding an exotic energy source component to the standard field equations. The latter approach recently attracted much interest and substituted any arbitrary function for the standard Einstein–Hilbert action. This function could be a Ricci scalar R ($f(R)$ gravity), a scalar torsion T ($f(T)$ gravity), a G ($f(G)$ gravity, and $f(R,G)$ gravity), or it could be the inclusion of another matter field Lagrangian along with some geometrical features. Because of the importance of $f(R)$ models in cosmology, the $f(R)$ theory of gravity is regarded as the best modification among the others. It has been hypothesized that cosmic acceleration can be achieved by replacing the Einstein–Hilbert action with a generic Ricci scalar function, $f(R)$. As an extension of $f(R)$ modified theories of gravity, the explicit coupling of any arbitrary function of the Ricci scalar with the matter Lagrangian density has been proposed [64]. Because of the interaction, the massive particles' motion is nongeodesic, and an extra force orthogonal to the four velocities results. Nojiri and Odintsov [65] analyzed many modified gravity theories that are viewed as gravitational alternatives to dark energy. Several authors [66–77] have examined $f(R)$, $f(G)$, $f(R,T)$, and $f(R,G)$ gravity in various contexts. Shamir [78] provided a theoretically viable $f(R)$ gravity model that illustrates the unification of early-time inflation and late-time acceleration. In addition, the extension of $f(R,T)$ gravity to KK spacetime has been investigated by several authors [79–83]. One may want to add Einstein–Dilaton–Gauss–Bonnet gravity [84–86].

Induced–Metter Theory is one of several ideas (for example, Weylian geometry, Finsler geometry, and braneworld gravity) for modifying the underlying geometry.

In 1921, Kaluza proposed a unification of electromagnetism and gravity in the context of a four-dimensional hypersurface wrapped in a five-dimensional bulk space. Klein imposed the cylindricity condition and completed the Kaluza theory by proposing a circular topology for the fifth dimension. The extra coordinate has no bearing on the elements of the five-dimensional metric tensor and serves only a formal purpose in the KK theory. To obtain much better performance of the unified theory of gravitation and electromagnetism, Einstein and Mayer thought of Kaluza's idea from the posture that the spacetime continuum could be a four-dimensional one but possessing vectors (and tensors) with a fifth component. In 1938, Einstein and Bergmann generalized the KK theory. During this work, the condition of cylindricity (resembling the existence of a five-dimensional Killing vector) is replaced by the idea that the space is periodically closed concerning the fifth coordinate.

In recent decades, the KK theory of extra dimensions, in which matter is confined within a lower-dimensional hypersurface, has attracted much attention. Wesson's induced matter theory, also known as the spacetime matter theory, is based on a revised KK approach. It is worthwhile to highlight that P. S. Wesson and J. Ponce de Leon's papers are among the original works [55,87,88]. Overduin and Wesson's work [7] and Wesson's monograph [19] are particularly significant. These studies demonstrated the possibility of dropping the cylindrical assumption, the revised KK theory's suitability for addressing cosmological issues, and the presence of an induced electromagnetic field on the brane. One of the most significant achievements of the induced matter theory is the elegant demonstration of the geometric origin of matter. More precisely, Paul Wesson and his colleagues considered our universe as a brane embedded in a five-dimensional bulk space and showed that the latter's geometry, which is warped but devoid of matter, induces the matter on the brane. The IMT includes papers on dark matter, dark energy [89,90], the induced unified theory of gravity and electromagnetism, successful cosmological models [91], cosmological models with a variable cosmological constant [92,93], and test particles in higher-dimensional models [94–98].

It is worth revieingw the significant results demonstrated in [99] regarding the equivalence between STM and Z_2-symmetric brane-world theories [100–102]. (i) In both frameworks, matter fields emerge only by assuming the metric as a function of the extra variables, so no matter arises for metrics that do not depend on l. Therefore, brane models incorporating the concept that matter can be viewed as an effect of geometry of bulk are the ultimate goal of STM theory. (ii) It was pointed out that the motion of the test particles has similar properties in both theories. (iii) From a theoretical perspective, both theories employ two opposing approaches to explain the same problem. More concretely, the goal of brane theory is to employ physical information from the brane to reconstruct the generating bulk, whereas in the STM theory, the physics on the hypersurface is constructed from bulk. (iv) From a practical point of view, employing a concrete example, it was demonstrated that solely the interpretation of the effective matter quantities is different in these frameworks [99]. It should also be noted that the main fundamental difference between these frameworks lies in their motivation to introduce an extra dimension.

If, instead of GR, the BD (SB) theory is considered as an underlying framework, then MBDT (MSBT) is established by the same reduction procedure used to construct the STM theory; for a detailed study see [8,9]. The two sets of the field equations associated with the BD (SB) theory in $(D+1)$-dimensional spacetime, by employing the reduction procedure, split into four sets on the any D-dimensional hypersurface orthogonal to the extra dimension. One pair of these sets, whose scalar potential and matter fields emerge from the geometry, are the modified version of their corresponding counterparts of the standard BD (SB) framework consisting a scalar potential. Another pair, which has no equivalent in the mentioned standard scalar–tensor theories, is associated with the scalar field of the extra dimension, ψ and the modified version of the conservation Equation introduced in the STM theory. Inspired by [99], the equivalence between the MBDT (MSBT) and the BD (SB) brane models can also be investigated.

Let us focus on some special cases. (i) As mentioned, the induced scalar potential is one of the fundamental and inseparable sector of the MBDT and MSBT, which vanishes only for a few special cases. For instance, in the MBDT, without loss of generality, it vanishes when $\omega = -1$, $T^{(D+1)}_{DD} = 0$ and l is a cyclic coordinate. Assuming the latter condition and $\psi =$ constant, we obtain a zero scalar potential in the MSBT. (ii) When $L^{(D+1)}_{\rm matt} = 0$ and the BD and SB scalar fields take constant values, then both the MSBT and MBDT reduce to the corresponding framework constructed in [103]. It is worth noting that the authors of [103] generalized the IMT to arbitrary dimensions. More concretely, they applied the same procedure of IMT to relate a $(D+1)$-dimensional vacuum space with a D-dimensional hypersurface with sources and eventually obtained the same effective field equations of IMT. Moreover, they employed this formalism to investigate the relation between lower-dimensional gravity frameworks and the four-dimensional vacuum GR. It has been claimed that the former might be more easily quantized than the usual GR [7,103]. (iii) In both the MBDT and MSBT, when the corresponding scalar fields take constant values and $T^{(D+1)}_{D\mu} = 0$, we obtain $P^{\beta}{}_{\mu;\beta} = 0$.

In this review paper, we detail only the FLRW-MSBT model among various cosmological models studied in the context of MBDT and MSBT. (For a very brief review of modified gravity and cosmology, see [104].) Regarding the FLRW-MBDT cosmology, let us mention that the accelerated epoch of the matter-dominated universe is not only consistent with the decelerated radiation-dominated epoch, but also with more recent cosmological data [8].

Finally, we should note that the MBDT and MSBT have been applied less as background framework in research. We believe that they have great potential to probe open problems, and we therefore will include them in our future research.

Author Contributions: Conceptualization, S.M.M.R., S.J. and P.M.; Formal analysis, S.M.M.R., S.J. and P.M.; Methodology, S.M.M.R., S.J. and P.M.; Writing–original draft, S.M.M.R., S.J. and P.M.; Writing–review and editing, S.M.M.R., S.J., and P.M. All authors have read and agreed to the published version of the manuscript.

Funding: This research received no external funding.

Institutional Review Board Statement: Not applicable.

Informed Consent Statement: Not applicable.

Data Availability Statement: Not applicable.

Acknowledgments: We appreciate the anonymous reviewers for their valuable comments. SMMR and PVM acknowledge the FCT grants UID-B-MAT/00212/2020 and UID-P-MAT/00212/2020 at CMA-UBI plus the COST Action CA18108 (Quantum gravity phenomenology in the multi–messenger approach).

Conflicts of Interest: The authors declare no conflict of interest.

Abbreviations

The following abbreviations have been used in this manuscript:

BD	Brans–Dicke
FLRW	Friedmann–Lemaître–Robertson–Walker
GR	General Relativity
IMT	Induced–Matter Theory
KK	Kaluza–Klein
MBDT	Modified Brans–Dicke Theory
MSBT	Modified Sáez–Ballester Theory
SB	Sáez–Ballester
STM	Space–Time–Matter

Notes

1 To keep in touch with the original works discussed in each (sub)section, let us apply the same units contained within. For example, in this section we use the same units as in [55].

2 It is worth noting that noncompact extra dimensions have also been adopted within compactified KK theory as an approach to incorporate chiral fermions into the theory and to organize a vanishing four-dimensional cosmological constant, see e.g., [57,58]. However, these frameworks have adapted the Klein's mechanism of harmonic expansion, i.e., a finite volume has been assumed for the compact manifold [7].

3 It is worth mentioning that in the IMT [55], an apparent vacuum bulk was taken into account. (In this paper, a 'vacuum' spacetime means that 'ordinary matter' does not exist.) As mentioned earlier, however, we will consider a non-vanishing energy momentum tensor, $L^{(D+1)}_{\rm matt} \neq 0$, to construct an extended version of the corresponding reduced framework. One is free to impose higher-dimensional ordinary matter fields that, in turn, invoke an intricate Kaluza–Klein framework. It should be emphasized that with such a procedure there is a risk that introduces numerous degrees of freedom, so that the corresponding model can no longer be tested. In this respect, we usually assume a vacuum bulk in cosmological applications.

4 In the original SB theory, scalar potential was not added to the action [47].

5 In [9], the general coordinate free framework has been used to construct the MSBT. However, in this paper, we employ a different particular approach.

6 In $(D+1)$–dimensions, assuming $\omega > -D/(D-1)$, it will be possible to go from the Jordan frame to the Einstein frame by conformal transformations.

7 We should note that Equations (38)–(40) and (38)–(41) are valid not only for the metric (2) but also for any other more generalized metrics.

8 We must emphasize that Equations (45), (47) and (48) (for $D \neq 4$) are the modified version of those introduced in [8].

References

1. Einstein, A.; Infeld, L. *The Evolution of Physics*; Simon and Schuster: New York, NY, USA, 1938; Volume 257.
2. Maia, M.D. Geometry of Kaluza-Klein theory. I. Basic setting. *Phys. Rev. D* **1985**, *31*, 262–267. [CrossRef]
3. Joseph, D.W. Coordinate Covariance and the Particle Spectrum. *Phys. Rev.* **1962**, *126*, 319–323. [CrossRef]
4. Akama, K. Pregeometry. In *Gauge Theory and Gravitation (G & G)*; Kikkawa, K., Nakanishi, N., Nariai, H., Eds.; Springer: Berlin/Heidelberg, Germany, 1983; Volume 176, p. 267. [CrossRef]
5. Rubakov, V.A.; Shaposhnikov, M.E. Do we live inside a domain wall? *Phys. Lett. B* **1983**, *125*, 136–138. [CrossRef]
6. Visser, M. An exotic class of Kaluza-Klein models. *Phys. Lett. B* **1985**, *159*, 22–25. [CrossRef]
7. Overduin, J.M.; Wesson, P.S. Kaluza-Klein gravity. *Phys. Rept.* **1997**, *283*, 303–380. [CrossRef]
8. Rasouli, S.M.M.; Farhoudi, M.; Vargas Moniz, P. Modified Brans–Dicke theory in arbitrary dimensions. *Class. Quant. Grav.* **2014**, *31*, 115002. [CrossRef]
9. Rasouli, S.M.M.; Pacheco, R.; Sakellariadou, M.; Moniz, P.V. Late time cosmic acceleration in modified Sáez–Ballester theory. *Phys. Dark Univ.* **2020**, *27*, 100446. [CrossRef]
10. Jalalzadeh, S.; Vakili, B.; Ahmadi, F.; Sepangi, H.R. Stabilization of test particles in induced matter Kaluza-Klein theory. *Class. Quant. Grav.* **2006**, *23*, 6015–6030. [CrossRef]
11. Rostami, T.; Jalalzadeh, S. Why the measured cosmological constant is small. *Phys. Dark Univ.* **2015**, *9–10*, 31–36. [CrossRef]
12. Jalalzadeh, S.; Rostami, T. Covariant extrinsic gravity and the geometric origin of dark energy. *Int. J. Mod. Phys. D* **2015**, *24*, 1550027. [CrossRef]
13. Doroud, N.; Rasouli, S.M.M.; Jalalzadeh, S. A class of cosmological solutions in induced matter theory with conformally flat bulk space. *Gen. Rel. Grav.* **2009**, *41*, 2637–2656. [CrossRef]
14. Rasouli, S.M.M.; Jalalzadeh, S. On the energy conditions in non-compact Kaluza-Klein gravity. *Ann. Phys.* **2010**, *19*, 276–280. [CrossRef]
15. Callaway, J. Mach's Principle and Unified Field Theory. *Phys. Rev.* **1954**, *96*, 778–780. [CrossRef]
16. Jalalzadeh, S. Non-integrability and Mach's principle in Induced Matter Theory. *Gen. Rel. Grav.* **2007**, *39*, 387. [CrossRef]

17. Wesson, P.S. A new approach to scale-invariant gravity/or: A variable-mass embedding for general relativity. *Astron. Astrophys.* **1983**, *119*, 145–152.
18. Wesson, P.S. A Physical Interpretation of Kaluza-Klein Cosmology. *Astrophys. J.* **1992**, *394*, 19. [CrossRef]
19. Wesson, P. *Space-Time-Matter: Modern Kaluza-Klein Theory*; World Scientific: Singapore, 1999.
20. Anchordoqui, L.A.; Torres, D.F.; Vucetich, H. Primordial nucleosynthesis as a test of variable rest masses five-dimensional cosmology. *Phys. Lett. A* **1996**, *222*, 43–46. [CrossRef]
21. Jalalzadeh, S.; Yazdani, A.M. Variation of mass in primordial nucleosynthesis as a test of Induced Matter Brane Gravity. *Phys. Lett. B* **2008**, *664*, 229–234. [CrossRef]
22. Wesson, P.S. Wave mechanics and general relativity: A Rapprochement. *Gen. Rel. Grav.* **2006**, *38*, 937–944. [CrossRef]
23. Wesson, P.S. Space-time uncertainty from higher dimensional determinism (Or: How Heisenberg was right in 4-D because Einstein was right in 5-D). *Gen. Rel. Grav.* **2004**, *36*, 451–457. [CrossRef]
24. Rasouli, S.M.M.; Bahrehbakhsh, A.F.; Jalalzadeh, S.; Farhoudi, M. Quantum mechanics and geodesic deviation in the brane world. *EPL* **2009**, *87*, 40006. [CrossRef]
25. Moyassari, P.; Jalalzadeh, S. Semiclassical corrections to the Einstein equation and induced matter theory. *Gen. Rel. Grav.* **2007**, *39*, 1467–1476. [CrossRef]
26. Rasouli, S.M.M.; Vargas Moniz, P. Modified Saez–Ballester scalar–tensor theory from 5D space-time. *Class. Quant. Grav.* **2018**, *35*, 025004. [CrossRef]
27. De Leon, J.P. Brans-Dicke Cosmology in 4D from scalar-vacuum in 5D. *J. Cosmol. Astropart. Phys.* **2010**, *3*, 30. [CrossRef]
28. Banerjee, N.; Pavon, D. Cosmic acceleration without quintessence. *Phys. Rev. D* **2001**, *63*, 043504. [CrossRef]
29. Sen, S.; Sen, A.A. Late time acceleration in Brans-Dicke cosmology. *Phys. Rev. D* **2001**, *63*, 124006. [CrossRef]
30. Rasouli, S.M.M.; Farhoudi, M.; Khosravi, N. Horizon Problem Remediation via Deformed Phase Space. *Gen. Rel. Grav.* **2011**, *43*, 2895–2910. [CrossRef]
31. Bahrehbakhsh, A.F.; Farhoudi, M.; Vakili, H. Dark Energy From Fifth Dimensional Brans-Dicke Theory. *Int. J. Mod. Phys. D* **2013**, *22*, 1350070. [CrossRef]
32. Rasouli, S.M.M.; Vargas Moniz, P. Noncommutative minisuperspace, gravity-driven acceleration, and kinetic inflation. *Phys. Rev. D* **2014**, *90*, 083533. [CrossRef]
33. Rasouli, S.M.M.; Ziaie, A.H.; Jalalzadeh, S.; Moniz, P.V. Non-singular Brans–Dicke collapse in deformed phase space. *Ann. Phys.* **2016**, *375*, 154–178. [CrossRef]
34. Rasouli, S.M.M.; Vargas Moniz, P. Gravity-Driven Acceleration and Kinetic Inflation in Noncommutative Brans-Dicke Setting. *Odessa Astron. Pub.* **2016**, *29*, 19. [CrossRef]
35. Kofinas, G.; Papantonopoulos, E.; Saridakis, E.N. Modified Brans–Dicke cosmology with matter-scalar field interaction. *Class. Quant. Grav.* **2016**, *33*, 155004. [CrossRef]
36. Rasouli, S.M.M.; Marto, J.; Vargas Moniz, P. Kinetic inflation in deformed phase space Brans–Dicke cosmology. *Phys. Dark Univ.* **2019**, *24*, 100269. [CrossRef]
37. Reyes, L.M.; Perez Bergliaffa, S.E. On the emergence of the Λ CDM model from self-interacting Brans–Dicke theory in $\mathbf{d} = \mathbf{5}$. *Eur. Phys. J. C* **2018**, *78*, 17. [CrossRef]
38. Akarsu, O.; Katırcı, N.; Özdemir, N.; Vázquez, J.A. Anisotropic massive Brans-Dicke gravity extension of the standard ΛCDM model. *Eur. Phys. J. C* **2020**, *80*, 32. [CrossRef]
39. Chakrabarti, S.; Dutta, K.; Said Levi, J. Screening mechanism and late-time cosmology: Role of a Chameleon-Brans-Dicke scalar field. *Mon. Not. R. Astron. Soc.* **2022**, *514*, 427–439. [CrossRef]
40. Ildes, M.; Arik, M. Analytic Solutions of Brans-Dicke Cosmology: Early Inflation and Late Time Accelerated Expansion. *arXiv* **2022**, arXiv:2204.10396.
41. Rasouli, S.M.M.; Farhoudi, M.; Sepangi, H.R. Anisotropic Cosmological Model in Modified Brans–Dicke Theory. *Class. Quant. Grav.* **2011**, *28*, 155004. [CrossRef]
42. Rasouli, S.M.M. Kasner Solution in Brans–Dicke Theory and Its Corresponding Reduced Cosmology. *Springer Proc. Math. Stat.* **2014**, *60*, 371–375. [CrossRef]
43. Rasouli, S.M.M.; Vargas Moniz, P. Exact Cosmological Solutions in Modified Brans-Dicke Theory. *Class. Quant. Grav.* **2016**, *33*, 035006. [CrossRef]
44. Rasouli, S.M.M.; Vargas Moniz, P. Extended anisotropic models in noncompact Kaluza-Klein theory. *Class. Quant. Grav.* **2019**, *36*, 075010. [CrossRef]
45. Rasouli, S.M.M.; Shojai, F. Geodesic deviation equation in Brans–Dicke theory in arbitrary dimensions. *Phys. Dark Univ.* **2021**, *32*, 100781. [CrossRef]
46. Amani, H.; Halpern, P. Energy conditions in a modified Brans-Dicke theory. *Gen. Rel. Grav.* **2022**, *54*, 64. [CrossRef]
47. Sáez, D.; Ballester, V. A simple coupling with cosmological implications. *Phys. Lett. A* **1986**, *113*, 467–470. [CrossRef]
48. Rasouli, S.M.M.; Sakellariadou, M.; Vargas Moniz, P. Geodesic deviation in Saez-Ballester theory. *arXiv* **2022**, arXiv:2203.00766.
49. Socorro, J.; Sabido, M.; Urena-Lopez, L.A. Classical and quantum Cosmology of the Saez-Ballester theory. *Fizika B* **2010**, *19*, 177–186.
50. Dixit, A.; Zia, R.; Pradhan, A. Anisotropic bulk viscous string cosmological models of the Universe under a time-dependent deceleration parameter. *Pramana* **2020**, *94*, 25. [CrossRef]

51. Rasouli, S.M.M. Noncommutativity, Sáez–Ballester Theory and Kinetic Inflation. *Universe* **2022**, *8*, 165. [CrossRef]
52. Mishra, R.K.; Dua, H. Bianchi type-I cosmological model in Sáez-Ballester theory with variable deceleration parameter. *Astrophys. Space Sci.* **2021**, *366*, 47. [CrossRef]
53. Daimary, J.; Roy Baruah, R. Five Dimensional Bianchi Type-I Anisotropic Cloud String Cosmological Model With Electromagnetic Field in Saez-Ballester Theory. *Front. Astron. Space Sci.* **2022**, *9*, 878653. [CrossRef]
54. Singh, P.S.; Singh, K.P. Vacuum Energy in Saez-Ballester Theory and Stabilization of Extra Dimensions. *Universe* **2022**, *8*, 60. [CrossRef]
55. Wesson, P.S.; Ponce de Leon, J. Kaluza–Klein equations, Einstein's equations, and an effective energy-momentum tensor. *J. Math. Phys.* **1992**, *33*, 3883–3887. [CrossRef]
56. Romero, C.; Tavakol, R.K.; Zalaletdinov, R. The embedding of general relativity in five-dimensions. *Gen. Rel. Grav.* **1996**, *28*, 365–376. [CrossRef]
57. Wetterich, C. Dimensional Reduction of Fermions in Generalized Gravity. *Nucl. Phys. B* **1984**, *242*, 473. [CrossRef]
58. Gell-Mann, M.; Zwiebach, B. Dimensional reduction of spacrtime induced by nonlinear scalar dynamics and noncompact extra dimensions. *Nucl. Phys. B* **1985**, *260*, 569–592. [CrossRef]
59. Mashhoon, B.; Wesson, P.; Liu, H.Y. Dynamics in Kaluza-Klein gravity and a fifth force. *Gen. Rel. Grav.* **1998**, *30*, 555–571. [CrossRef]
60. Wesson, P.S. Particle Masses and the Cosmological 'Constant' in Five Dimensions. *arXiv* **2011**, arXiv:1111.4698.
61. Konitopoulos, S.; Saridakis, E.N.; Stavrinos, P.C.; Triantafyllopoulos, A. Dark gravitational sectors on a generalized scalar-tensor vector bundle model and cosmological applications. *Phys. Rev. D* **2021**, *104*, 064018. [CrossRef]
62. Langlois, D. Dark energy and modified gravity in degenerate higher-order scalar–tensor (DHOST) theories: A review. *Int. J. Mod. Phys. D* **2019**, *28*, 1942006. [CrossRef]
63. Nemiroff, R.J.; Joshi, R.; Patla, B.R. An exposition on Friedmann Cosmology with Negative Energy Densities. *J. Cosmol. Astropart. Phys.* **2015**, 006. [CrossRef]
64. Bertolami, O.; Boehmer, C.G.; Harko, T.; Lobo, F.S.N. Extra force in f(R) modified theories of gravity. *Phys. Rev. D* **2007**, *75*, 104016. [CrossRef]
65. Nojiri, S.; Odintsov, S.D. Introduction to modified gravity and gravitational alternative for dark energy. *eConf* **2006**, *C0602061*, 6. [CrossRef]
66. Multamaki, T.; Vilja, I. Spherically symmetric solutions of modified field equations in f(R) theories of gravity. *Phys. Rev. D* **2006**, *74*, 064022. [CrossRef]
67. Multamaki, T.; Vilja, I. Static spherically symmetric perfect fluid solutions in f(R) theories of gravity. *Phys. Rev. D* **2007**, *76*, 064021. [CrossRef]
68. Nojiri, S.; Odintsov, S.D. Unified cosmic history in modified gravity: From F(R) theory to Lorentz non-invariant models. *Phys. Rept.* **2011**, *505*, 59–144. [CrossRef]
69. Ziaie, A.H.; Atazadeh, K.; Rasouli, S.M.M. Naked Singularity Formation In f(R) Gravity. *Gen. Rel. Grav.* **2011**, *43*, 2943–2963. [CrossRef]
70. Clifton, T.; Ferreira, P.G.; Padilla, A.; Skordis, C. Modified Gravity and Cosmology. *Phys. Rept.* **2012**, *513*, 1–189. [CrossRef]
71. Yousaf, Z.; Bhatti, M.Z.; Farhat, A. Various phases of irregular energy density in charged spheres. *Ann. Phys.* **2022**, *442*, 168935. [CrossRef]
72. Yousaf, Z.; Bhatti, M.Z.; Aman, H. Cosmic bounce with $\alpha(e^{-\beta G}-1)+2\lambda$ T model. *Phys. Scr.* **2022**, *97*, 055306. [CrossRef]
73. Astashenok, A.V.; Capozziello, S.; Odintsov, S.D.; Oikonomou, V.K. Maximum baryon masses for static neutron stars in f(R) gravity. *Europhys. Lett.* **2021**, *136*, 59001. [CrossRef]
74. Mishra, B.; Sahoo, P.K. Bianchi type VI_h perfect fluid cosmological model in f(R, T) theory. *Astrophys. Space Sci.* **2014**, *352*, 331–336. [CrossRef]
75. Sharif, M.; Zubair, M. Thermodynamics in f(R,T) Theory of Gravity. *J. Cosmol. Astropart. Phys.* **2012**, *3*, 28. [CrossRef]
76. Oikonomou, V.K. Universal inflationary attractors implications on static neutron stars. *Class. Quant. Grav.* **2021**, *38*, 175005. [CrossRef]
77. Odintsov, S.D.; Oikonomou, V.K. Neutron stars in scalar–tensor gravity with quartic order scalar potential. *Ann. Phys.* **2022**, *440*, 168839. [CrossRef]
78. Shamir, M.F. Some Bianchi Type Cosmological Models in f(R) Gravity. *Astrophys. Space Sci.* **2010**, *330*, 183–189. [CrossRef]
79. Moraes, P.H.R.S. Cosmological solutions from Induced Matter Model applied to 5D $f(R,T)$ gravity and the shrinking of the extra coordinate. *Eur. Phys. J. C* **2015**, *75*, 168. [CrossRef]
80. Moraes, P.H.R.S. Cosmology from induced matter model applied to 5D f(R, T) theory. *Astrophys. Space Sci.* **2014**, *352*, 273–279. [CrossRef]
81. Sahoo, P.K.; Mishra, B.; Tripathy, S.K. Kaluza-Klein cosmological model in $f(R,T)$ gravity with $\Lambda(T)$. *Indian J. Phys.* **2016**, *90*, 485–493. [CrossRef]
82. Bamba, K.; Nojiri, S.; Odintsov, S.D. Effective $F(T)$ gravity from the higher-dimensional Kaluza-Klein and Randall-Sundrum theories. *Phys. Lett. B* **2013**, *725*, 368–371. [CrossRef]
83. Fabris, J.C.; Popov, A.A.; Rubin, S.G. Multidimensional gravity with higher derivatives and inflation. *Phys. Lett. B* **2020**, *806*, 135458. [CrossRef]

84. Kanti, P.; Mavromatos, N.E.; Rizos, J.; Tamvakis, K.; Winstanley, E. Dilatonic black holes in higher curvature string gravity. *Phys. Rev. D* **1996**, *54*, 5049–5058. [CrossRef] [PubMed]
85. Antoniou, G.; Bakopoulos, A.; Kanti, P. Evasion of No-Hair Theorems and Novel Black-Hole Solutions in Gauss-Bonnet Theories. *Phys. Rev. Lett.* **2018**, *120*, 131102. [CrossRef] [PubMed]
86. Copeland, E.J.; Lidsey, J.E.; Wands, D. Cosmology of the type IIB superstring. *Phys. Rev. D* **1998**, *57*, 625–629. [CrossRef]
87. Wesson, P.S. The Effective properties of matter of Kaluza-Klein solitons. *Phys. Lett. B* **1992**, *276*, 299–302. [CrossRef]
88. Ponce de Leon, J.; Wesson, P.S. Exact solutions and the effective equation of state in Kaluza-Klein theory. *J. Math. Phys.* **1993**, *34*, 4080–4092. [CrossRef]
89. Liko, T.; Wesson, P.S. An Exact solution of the five-dimensional Einstein equations with four-dimensional de Sitter-like expansion. *J. Math. Phys.* **2005**, *46*, 062504. [CrossRef]
90. Mashhoon, B.; Wesson, P.S. Gauge dependent cosmological 'constant'. *Class. Quant. Grav.* **2004**, *21*, 3611. [CrossRef]
91. Ponce de Leon, J.; Wesson, P.S. A Class of Anisotropic Five-Dimensional Solutions for the Early Universe. *Europhys. Lett.* **2008**, *84*, 20007. [CrossRef]
92. Wesson, P.S.; Mashhoon, B.; Overduin, J.M. Cosmology with Decaying Dark Energy and Cosmological "constant". *Int. J. Mod. Phys. D* **2008**, *17*, 2527–2533. [CrossRef]
93. Overduin, J.M.; Wesson, P.S.; Mashhoon, B. Decaying Dark Energy in Higher-Dimensional Gravity. *Astron. Astrophys.* **2007**, *473*, 727. [CrossRef]
94. Jalalzadeh, S.; Vakili, B.; Sepangi, H.R. On extra forces from large extra dimensions. *Phys. Scr.* **2007**, *76*, 122–126. [CrossRef]
95. Dahia, F.; Monte, E.M.; Romero, C. Fifth force from fifth dimension: A Comparison between two different approaches. *Mod. Phys. Lett. A* **2003**, *18*, 1773–1782. [CrossRef]
96. Ponce de Leon, J. Does the force from an extra dimension contradict physics in 4-D? *Phys. Lett. B* **2001**, *523*, 311–316. [CrossRef]
97. Ponce de Leon, J. Extra force from an extra dimension: Comparison between brane theory, STM and other approaches. *Gen. Rel. Grav.* **2004**, *36*, 1333–1358. [CrossRef]
98. Ponce de Leon, J. Mass and charge in brane world and noncompact Kaluza-Klein theories in five-dimensions. *Gen. Rel. Grav.* **2003**, *35*, 1365–1384. [CrossRef]
99. Ponce De Leon, J. Equivalence between space-time matter and brane world theories. *Mod. Phys. Lett. A* **2001**, *16*, 2291–2304. [CrossRef]
100. Randall, L.; Sundrum, R. A Large mass hierarchy from a small extra dimension. *Phys. Rev. Lett.* **1999**, *83*, 3370–3373. [CrossRef]
101. Randall, L.; Sundrum, R. An Alternative to compactification. *Phys. Rev. Lett.* **1999**, *83*, 4690–4693. [CrossRef]
102. Maartens, R. Geometry and dynamics of the brane world. In Proceedings of the Spanish Relativity Meeting on Reference Frames and Gravitomagnetism (EREs2000), Valladolid, Spain, 6–9 September 2001. [CrossRef]
103. Rippl, S.; Romero, C.; Tavakol, R.K. D-dimensional gravity from (D+1)-dimensions. *Class. Quant. Grav.* **1995**, *12*, 2411–2422. [CrossRef]
104. Lazkoz, R. Cosmophysics of Modified Gravity. In *Modified Gravity and Cosmology: An Update by the CANTATA Network*; Saridakis, E.A., Ed.; Springer International Publishing: Cham, Switerland, 2021. [CrossRef]

universe

MDPI

Review

Progress in the Composite View of the Newton Gravitational Constant and Its Link to the Planck Scale

Espen Gaarder Haug

Christian Magnus Falsens vei 18, Norwegian University of Life Sciences, 1430 Ås, Norway; espen.haug@nmbu.no

Abstract: The Newtonian gravity constant G plays a central role in gravitational theory. Researchers have, since at least the 1980s, tried to see if the Newton gravitational constant can be expressed or replaced with more fundamental units, such as the Planck units. However, it was already pointed out in 1987 that this led to a circular problem; namely, that one must know G to find the Planck units, and that it is therefore of little or no use to express G through the Planck units. This is a view repeated in the literature in recent years, and is held by the physics' community. However, we will claim that the circular problem was solved a few years ago. In addition, when one expresses the mass from the Compton wavelength formula, this leads to the conclusion that the three universal constants of G, h, and c now can be replaced with only l_p and c to predict observable gravitational phenomena. While there have been several review papers on the Newton gravitational constant, for example, about how to measure it, we have not found a single review paper on the composite view of the gravitational constant. This paper will review the history of, as well as recent progress in, the composite view of the gravitational constant. This should hopefully be a useful supplement in the ongoing research for understanding and discussion of Newton's gravitational constant.

Keywords: Newton gravitational constant; Planck units; composite constant; gravity; mass; quantum gravity; cosmology

1. Short History on the Newton Gravitational Constant and the Planck Units

Newton's gravitational constant plays an important role in almost any gravity calculation. However, Newton actually never introduced or used a gravitational constant [1] in his gravitational force formula. His formula was stated by words in Principia [2] as $F = \frac{\bar{M}\bar{m}}{R^2}$. This is equivalent to today's gravitational force formula without the gravitational constant. Well, almost so, as we are, on purpose, using the notation $\bar{M}$ and $\bar{m}$ for the two masses, rather than M and m. This is because Newton had a quite different view of mass than the view held today. Even without a gravity constant, Newton was able to do many predictions, such as finding the relative mass between planets and the sun, see also Cohen [3]. He also found the relative density of Earth relative to the sun to be a value very close to what is known today. What he tried but failed to do was to find the density of Earth relative to a known substance, such as water, lead, or gold.

Modern physicists often consider Cavendish [4] in 1798 to be the first to measure the gravitational constant. However, Cavendish, in his paper, did not describe a gravitational constant or use one as also pointed out by Clotfelter [5]. What Cavendish did was to measure the density of Earth relative to the density of a known substance, which Newton had tried but failed to do. Cavendish succeeded by using a torsion balance, also known today as a Cavendish apparatus. The main point of using such an apparatus is to measure the gravity effects from a human-sized object; that is, due to the balls in the Cavendish apparatus, one can easily control what kind of substance the mass (the large spheres in the apparatus) are made of, for example, lead. Then by next comparing the gravitational effect measured in the apparatus with the gravitational effect of Earth, one can find the density of

Earth relative to the density of a known substance. Such an apparatus can also be used to extract the gravitational constant.

The so-called Newton gravitational constant was first introduced in 1873 by the two French physicists Cornu and Baille [6]. In their paper, they gave the formula as:

$$F = f\frac{Mm}{R^2} \tag{1}$$

where f is the gravitational constant. Big G as the notation of the gravity constant was likely first introduced by Boys [7] in 1894. It took many years before the notation G became standard in the international physics' community; for example, Max Planck [8] used f for the gravitational constant as late as 1928, and Einstein used notation k in 1916. Naturally, whether one uses f, k or G as a symbol for the gravitational constant is purely cosmetic. What is important to bear in mind is that the gravitational constant is relatively new (at least compared to Newton's Principia) and that it also came into existence at about the same time that the kilogram became the international standard mass. For a more in-depth historical perspective on the Newton gravitational constant, see [9].

In 1899, a few years after the invention of the Newtonian gravitational constant, Max Planck [10,11] introduced the Planck units. He assumed there were three important universal constants: G, c, and h, and then used dimensional analysis to derive a unique length $l_p = \sqrt{\frac{G\hbar}{c^3}}$, time $t_p = \sqrt{\frac{G\hbar}{c^5}}$, mass $m_p = \sqrt{\frac{\hbar c}{G}}$, and temperature $T_p = \sqrt{\frac{\hbar c^5}{Gk_b^2}}$. Today these are known as the Planck units. In 1916, Einstein [12] already suggest that the next step forward in gravity would be quantum gravity. Eddington [13] was, in 1918, suggesting that quantum gravity must be linked to the Planck scale or, in his own words:

> *But it is evident that this length (the Planck length) must be the key to some essential structure. It may not be an unattainable hope that someday a clearer knowledge of the process of gravitation may be reached?*

However, Eddington's idea was criticized by Bridgeman [14] in 1931. Bridgeman (who later received the Nobel prize in physics) thought the Planck units were more likely mathematical artifacts coming out of dimensional analysis rather than something fundamental and related to gravity. Today, most researchers working with quantum gravity theory seem to think the Planck units will play an important role in a final unified theory; see, for example, [15–17]. Others are more critical. Meschini [18] pointed out that the "*the significance of Planck's natural units in a future physical theory of spacetime is only a plausible, yet by no means certain*". The lack of certainty in the significance of the Planck units is because the Planck scale at that time could still only be found very indirectly by dimensional analysis. For example, Unzicker [19] still seems to hold on to the view of Bridgeman, that the Planck units are little more than mathematical artifacts from dimensional analysis and that they are of no use and can basically be seen as undetectable mathematical artifacts. The Planck units are almost rather like the ether; if there are no ways to detect the Planck units then why not simply abandon the idea that they will play a central role in physics?

These opposing views of the Planck units also play a historically important role when it comes to the gravity constant itself. Okun [20] in 1991 pointed out that "*The status of G and its derivatives, m_p, l_p , t_p, is at present different from that of c and $\hbar$, because the quantum theory of gravity is still under construction.*". So a better understanding of G can perhaps also bring us closer to understanding the Planck scale and even closer to a unified quantum gravity theory. Thus it is important to keep questioning the real meaning of G; something we will look at this paper, mainly by reviewing the existing literature on how G can potentially be linked to the Planck scale.

2. History of the Composite View of G and the Circular Problem

The gravity constant has SI units of $\mathrm{m}^3 \cdot \mathrm{kg}^{-1} \cdot \mathrm{s}^{-2}$. It would be strange if anything physical had units: meters cubed, divided by kilograms, and seconds squared. We can all imagine something that has length, for example, a cat, or something that has mass in terms of kilograms, for example, a cat, and we all have a sense of time being related to change, and it can also be measured with clocks. So, the output unit of Newton's gravitational constant is perhaps the first hint that it could be a universal composite constant that can actually be represented by some more fundamental constants that we can physically link more directly to something [21]. Still, as we will see, G is a composite constant has been discussed for more than 60 years without a resolution, until perhaps very recently. We will here review much of the important history and progress around how the gravitational constant can be expressed in the form of Planck units.

Thüring [22], in 1961, concluded that G had been introduced somewhat ad hoc and that it cannot be associated with a unique property of nature; see also Gillies [23]. Zee [24] in 1982, in a paper titled "*Calculation of Newton's Gravitational Constant in Infrared-Stable Yang-Mills Theories*" wrote:

> "*Is Newton's gravitational constant G a, fundamental parameter or is it calculable in terms of other fundamental parameters? In this paper I would like to argue the latter view and to present a calculation of G, unfortunately not in the real world, but in a toy world, just to demonstrate that G is indeed calculable.*"

Cahill [25,26] in 1984 was likely the first to suggest that instead of calculating the Planck mass from G, $\hbar$, and c, that perhaps G can be calculated from the Planck mass and suggested that G is given by:

$$G = \frac{\hbar c}{m_p^2} \tag{2}$$

This is nothing more than solving the Planck mass formula, $m_p = \sqrt{\frac{\hbar c}{G}}$, with respect to G. Chaill commented:

> "*The actual distribution of energy throughout space-time causes the tetrads to assume vacuum expected values of the order of the Planck mass, m_p. Thus the gravitational constant, $G = \frac{\hbar c}{m_p^2}$, may be viewed not as a fundamental constant, but as a mass scale that is dynamically determined by the large-scale structure of the Universe.*"

Cohen [27] suggested the same formula in 1987, and that he correctly pointed out also can be found from dimensional analysis or, in his own words,

> "*Dimensional analysis let us write $G = hc/m_{pl}^2$, where m_{pl} is the Planck mass 21.77×10^{-9} kg, but this is of no help of determining G since there are no independent determination of m_{pl}.*" (Page 74. Note that we will use notation m_p for the Planck mass, while several papers also use notation m_{pl})

This insight is of great importance and is what we will call the circular problem of the composite view of the Newton gravitational constant. Namely, this is the view that to express G from Planck units is of little or no use if one needs G to find the Planck units in the first place. Independently, a series [20,28–31] of researchers also later on suggested the same formula for G, likely without knowing about the paper of Cahill or Cohen, but none of these solved the circular problem. McCuloch [32] in 2016 again pointed out the circular problem with expressing the gravity constant from the Planck mass with the same formula as introduced by Cahill and Cohen or, in his own words,

> "*In the above gravitational derivation, the correct value for the gravitational constant G can only be obtained when it is assumed that the gravitational interaction occurs between whole multiples of the Planck mass, but this last part of the derivation involves some circular reasoning, since the Planck mass is defined using the value for G.*"

Again, this demonstrates that the circular problem of expressing G in the form of Planck units has existed for a very long time, and this is, in our view, directly linked to the problem that, in quantum gravity, there has been little or no progress in detecting the Planck scale, and therefore limited progress, as we soon will discuss.

Clark [33] in 2003 suggested the gravitational constant is given by:

$$G = \frac{a_g \hbar c}{u^2} \tag{3}$$

where $\frac{a_g}{u^2} = \frac{1}{m_p^2}$, so this is in many ways just an indirect way of writing the Cahill and Cohen formula, as we have $G = \frac{a_g \hbar c}{u^2} = \frac{\hbar c}{m_p}$. Independently, Zwiebach [34] and Nastasenko [35] both, in 2004, described the following formula to express G from the Planck units:

$$G = \frac{l_p^3}{t_p^2 m_p} \tag{4}$$

Zwiebach described this as a "*Planckian system of units*" but gives no indications that the Planck units can be found independently of G. Bruneton [36] suggested the same formula in 2013, and also his view that instead of G, $\hbar$, and c being the fundamental universal constants, they are just composites, and the Planck units are much more fundamental; see also [37,38]. This formula for G can, for example, be derived from dimensional analysis of G. The dimensions of G are $[G] = L^3 M^{-1} T^{-2}$; then simply replace L with l_p, and M with m_p, and T with t_p, and we get the formula $G = \frac{l_p^3}{t_p^2 m_p}$. However, if one needs to find G first to find the Planck units then one can naturally question the usefulness of this. The same formula was later used by, for example, Mercier [39] and Humpherys [40]. In 2013, Zivlak [41] suggested the formula of $G = \frac{c^2 l_p}{m_p}$, but without any suggestions for how to find l_p or m_p independent of G, thus leading one back to the circular problem, see also Eldred [42] 2019 that basically suggested the same formula.

In natural units, when first setting $c = h = 1$, we must have $G = 1/m_p^2$, as pointed out by Kiritsis [43] in 1997 as well as by Cerdeno and Munoz [44] in 1998 and later mentioned by, for example, [36,45–50]. We find others like Peebles [51] who in 1989 already pointed out that $m_p = G^{-1/2}$ when $\hbar = c = 1$, so one could claim he then also pointed out $G = 1/m_p^2$, as it is naturally trivial to turn the equation around. Still, writing $G = 1/m_p^2$ rather than $m_p = G^{-1/2}$ gives a hint or even a strong indication that perhaps we should think that the gravity constant is a function of the Planck units, and not only the Planck units can be a function of G as first suggested by Max Planck; this idea is what this paper focuses on. Further, in the natural units system, when $\hbar = c = 1$ we will then have $G = l_p^2$ as pointed out by Schwarzschild [52] in 2000 and also [48,50,53]. In addition, since $t_p = \frac{l_p}{c}$, we must naturally also have $G = t_p^2$ when $\hbar = c = 1$. When only $c = 1$ must we then have $\hbar = m_p l_p$ and we get $G = \frac{l_p}{m_p}$ as pointed out by Casadio [54] in 2009, and also discussed by [36,55–60].

We also have:

$$G = \frac{t_p^2 c^5}{\hbar} = \frac{c^5}{\hbar f_p^2} \tag{5}$$

where v_p is the Planck frequency $f_p = \frac{c}{l_p}$; this was likely first mentioned by Nastasenko [61] in 2013. Haug, in 2016, refs. [21,62,63] suggested that G is a universal composite gravitational constant of the form:

$$G = \frac{l_p^2 c^3}{\hbar} \tag{6}$$

This he arrived at from dimensional analysis by assuming the more fundamental constants are l_p, $\hbar$, and c and that the gravitational constant is simply a composite constant.

His argument is that the complex output units of G indicate it is a composite constant and, further, that the gravity constant coming before the Planck length does not mean the gravity constant is more fundamental than the Planck length. It is natural that we first understand the world more from the surface, before we understand the deeper aspects of it. Further, he shows how many of the Planck units can be simplified when one assumes G is such a composite. Still, none of the above-mentioned papers has solved the circular problem, so they are at best hypotheses that perhaps G can be expressed in the form of Planck units, but that there are unsolved problems to do so.

As we have seen, a series of ways to express the G in the form of Planck units have been expressed in the literature. Some authors have done this because they think G is a composite constant and that the Planck units are more real and fundamental, while others have mentioned G as a function of Planck units just so as to use in some calculations they have been done to achieve other results not directly related to the view that G is a composite constant.

Table 1 shows a series of ways to write G from Planck units and that we have found in the literature, and there are also many more additional ways. All these ways are valid mathematically, but again it is assumed one needs to know G to find the Planck units. A series of the formulas are marked as being presented first in this paper; we do not do this to indicate we have made any important new inventions simply by this, but simply to demonstrate that there are many ways to express G from Planck units. Basically, any Planck unit-related formula can be simply solved with respect to G. This is trivial mathematically; the big question is if it can lead to some significant new insight or not?

Table 1. The table shows various ways we can express the gravity constant from Planck units.

From	Gravity Constant Formula	Likely First Described[1]
Planck mass $m_p = \sqrt{\frac{\hbar c}{G}}$	$G = \frac{\hbar c}{m_p^2}$	Cahill [25] 1984 and Cohen[2] [27] 1987
Planck time $t_p = \sqrt{\frac{G\hbar}{c^5}}$	$G = \frac{t_p^2 c^5}{\hbar}$	Nastasenko [61] 2013
Planck length $l_p = \sqrt{\frac{G\hbar}{c^3}}$	$G = \frac{l_p^2 c^3}{\hbar}$	Haug [21] 2016
Planck energy $E_p = \sqrt{\frac{\hbar c^5}{G}}$	$G = \frac{\hbar c^5}{E_p^2}$	this paper Haug [64] 2020
Planck temperature $T_p = \sqrt{\frac{\hbar c^5}{G k_b^2}}$	$G = \frac{\hbar c^5}{T_p^2 k_b}$	this paper
Planck mass $a_g = \frac{m^2}{m_p^2}$	$G = \frac{a_g \hbar c}{u^2} = \frac{\hbar c}{m_p^2}$	Clark [33] 2003
Planck frequency $f_p = \sqrt{\frac{c^5}{G\hbar}}$	$G = \frac{c^5}{f_p^2 \hbar}$	Nastasenko [61] 2013
Planck acceleration $a_p = \sqrt{\frac{c^7}{G\hbar}}$	$G = \frac{c^7}{a_p^2 \hbar}$	this paper
Planck density $\rho_p = \frac{c^5}{\hbar G^2}$	$G = \sqrt{\frac{c^5}{\rho_p \hbar}}$	this paper
Planck momentum $p_p = \sqrt{\frac{\hbar c^3}{G}}$	$G = \frac{\hbar c^3}{p_p^2}$	this paper
Planck force $F_p = \frac{E_p}{l_p}$	$G = \frac{c^4}{F_p}$	this paper
Planck length, time and mass	$G = \frac{l_p^3}{m_p t_p^2}$	Zwiebach [34] 2004 and Nastasenko [35] 2004
Planck length and Planck time	$G = \frac{l_p c^2}{m_p}$	Zivlak [41] 2013
Planck mass and Planck time	$G = \frac{t_p c^3}{m_p}$	Eldred [42] 2019
Planck length, time and Planck energy	$G = \frac{l_p^3 c^2}{E_p t_p^2}$	this paper
Planck time and Planck length	$G = \frac{t_p l_p c^4}{\hbar}$	this paper
Planck frequency Planck mass	$G = \frac{c^3}{f_p m_p}$	this paper
Planck acceleration and mass	$G = \frac{c^4}{a_p m_p}$	this paper
Planck charge and Planck length	$G = \frac{l_p^2 c^2 10^7}{q_p^2}$	this paper
Planck charge and Planck mass	$G = \frac{10^7}{m_p^2 q_p^2}$	this paper
Planck charge and Planck time	$G = \frac{t_p^2 c^4 10^7}{q_p^2}$	this paper

Table 2 shows how to write G from Planck units when $\hbar = c = 1$ and when $c = 1$ and $\hbar = m_p l p$. So these formulas are simplified cases of the formulas in Table 1.

Table 2. The table shows various ways we can express the gravity constant from Planck units.

From	Gravity Constant Formula	Likely First Described
when $\hbar = c = 1$	$G = 1/m_p^2$	Kiritsis 1997 [43] and Cerdeno and Munoz 1998 [44]
when $\hbar = c = 1$	$G = l_p^2$	Schwarzschild 2000 [52]
when $\hbar = c = 1$	$G = t_p^2$	this paper
when $\hbar = c = 1$	$G = 1/a_p^2$	this paper
when $\hbar = c = 1$	$G = 1/E_p^2$	this paper
when $\hbar = c = 1$	$G = 1/p_p^2$	this paper
when $c = 1$	$G = l_p/m_p$	Casadio 2009 [54]
when $c = 1$	$G = t_p/m_p$	this paper
when $c = 1$	$G = l_p/E_p$	this paper
when $c = 1$	$G = t_p/E_p$	this paper
when $c = 1$	$G = l_p/a_p$	this paper
when $c = 1$	$G = t_p/a_p$	this paper

3. The Breakthrough in the Circular Problem

We have just looked at a long series of ways to express G in the form of Planck units. However, as long as one needs to know G to find the Planck units, this just leads to a circular problem as has been pointed out by a series of researchers, so at first glance this does not seem to help us understand G better. Still, we will claim that in recent years there has been a breakthrough in the circular problem. In 2017, Haug [65] showed a reliable way of find the Planck length independent of G, but still dependent on knowledge of $\hbar$ and c. This was done by using a Cavendish apparatus as described in the appendix of that paper. That one needs to use a Cavendish apparatus has nothing to do with one needing to know G. Haug derived the formula:

$$l_p = \sqrt{\frac{\hbar 2\pi^2 L r^2 \theta}{M T^2 c^3}} \tag{7}$$

where r is the distance between the centers of the large and small balls (when the balance is deflected), further, L is the distance between the small balls in the apparatus. M is the kilogram mass of the large ball in the apparatus that can be found, for example, with a standard letter weight as compared to the one-kilogram prototype mass. θ is the angle of deflection measured and T is the measured period of oscillation of the torsion balance. In other words, this way of finding the Planck length is only dependent on $\hbar$ and c, and not on prior knowledge of G. The formula above can be simplified further so we also get rid of the Planck constant, and then only depend on knowledge of c; this point we will soon return to.

In 2020, Haug [66,67] showed it is possible to find the Planck length and the Planck time without knowledge of both G and $\hbar$, but that to find the Planck mass (in kilograms) one needs to know $\hbar$ and c. Further, in 2021 Haug [66] showed an approach combined with a long list of gravity phenomena that can be used to find the Planck length independent of G and $\hbar$. In another paper [68], his main focus was on how to find the Planck time independent of G and $\hbar$. In 2022, Haug showed a way to find the Planck length and the Planck time without knowledge off G, $\hbar$, and c; see [69,70]. If one knows how to find the Planck length independent of G and $\hbar$, one naturally knows how to find the Planck time independent on G and $\hbar$ as the Planck time is simply the Planck length divided by the speed of light. However, the Planck time is also the Planck length divided by the speed of gravity, so if we can extract the speed of gravity from observable gravity phenomena only with no prior knowledge off c, then this must be the speed of gravity, and we have recently demonstrated that this is practically possible.

That the Planck units can be found without any knowledge of G means the gravity constant can indeed be expressed in the form of Planck units. This alone is a breakthrough, in our view. Still, what does it mean? This we will look more closely at in the next sections.

4. Putting the Pieces Together

We now know that the Planck units can be found without any knowledge of G. Table 3 shows a series of predictions from Newton and Einstein gravity simply re-written when we replace G with $G = \frac{\hbar c}{m_p^2}$. For example, the gravitational acceleration that can be predicted by $g = \frac{GM}{R^2}$ can now be re-written as:

$$g = \frac{GM}{R^2} = \frac{\hbar c}{m_p^2}\frac{M}{R^2} \tag{8}$$

This, in our view, gives little if any new insight or important results; one could even argue the formula is now even less intuitive than before. We can claim that this shows that gravity is related to the Planck mass and that it therefore gives some new insight, but it is not obvious why this should be the case. Still, G is replaced with an expression containing the Planck mass, and the Plank mass can be found independently of G, so this is a big step forward from the view held during the time of Max Planck and up until recently when researchers thought the Planck units could not be found without knowing G first. We could argue that this approach replaces three universal constants G, $\hbar$, and c with three new ones, namely m_p, $\hbar$, and c. Still, so far it seems that even after we have solved the circular problem in this composite view of G, this simply means we can replace G with another constant; namely m_p. This could be interesting on its own, as it indeed could indicate G is more of a human construct than something representing directly physical aspects of the depth of reality. The many formulas in Table 3, when re-written $G = \frac{\hbar c}{m_p^2}$, do not seem to make things more intuitive or, we could argue, it looks perhaps even less intuitive. It looks as though we still need three constants, but we have replaced G with m_p.

Table 3. The table shows the standard gravitational prediction formulas re-written when we assume $G = \frac{\hbar c}{m_p^2}$. We can see that the end results are likely even less intuitive than the existing results, and that we basically only have swapped one constant for a new one (G for m_p).

	Gravity with $G = \frac{\hbar c}{m_p^2}$:
Mass	M and m (kg)
Gravity force	$F = G\frac{Mm}{R^2} = \frac{\hbar c}{m_p^2}\frac{Mm}{R^2}$ $(\mathrm{kg \cdot m \cdot s^{-2}})$
Gravity acceleration	$g = \frac{GM}{R^2} = \frac{\hbar c M}{m_p^2 R^2}$
Orbital velocity	$v_o = \sqrt{\frac{GM}{R}} = \frac{1}{m_p}\sqrt{\frac{\hbar c M}{R}}$
Orbital time	$T = \frac{2\pi R}{\sqrt{\frac{GM}{R}}} = \frac{2\pi R m_p}{\sqrt{\frac{\hbar c M}{R}}}$
Periodicity pendulum[3] (clock)	$T = 2\pi\sqrt{\frac{L}{g}} = 2\pi R\sqrt{\frac{L}{GM}} = 2\pi R m_p\sqrt{\frac{L}{\hbar c M}}$
Frequency Newton spring	$f = \frac{1}{2\pi}\sqrt{\frac{k}{m}} = \frac{1}{2\pi R m_p}\sqrt{\frac{\hbar c M}{x}}$
Velocity ball Newton cradle[4]	$v_{out} = \sqrt{2\frac{GM}{R^2}H} = \frac{1}{Rm_p}\sqrt{2\hbar c M H}$
Predictions from GR:	
Advance of perihelion	$\sigma = \frac{6\pi GM}{a(1-e^2)c^2} = \frac{6\pi\hbar c M}{a(1-e^2)c^2 m_p^2}$
Gravitational redshift	$z = \frac{\sqrt{1-\frac{2GM}{R_1 c^2}}}{\sqrt{1-\frac{2GM}{R_2 c^2}}} - 1 = \frac{\sqrt{1-\frac{2\hbar M}{R_1 c m_p^2}}}{\sqrt{1-\frac{2\hbar M}{R_2 c m_p^2}}} - 1$
Time dilation	$T_R = T_f\sqrt{1 - \sqrt{\frac{2GM}{R}}^2 / c^2} =$
Deflection	$\delta = \frac{4GM}{c^2 R} = \frac{4\hbar M}{cRm_p^2}$
Microlensing	$\theta_E = \sqrt{\frac{4GM}{c^2}\frac{(d_S-d_L)}{d_S d_L}} = \sqrt{\frac{4\hbar M}{cm_p^2}\frac{(d_S-d_L)}{d_S d_L}}$

We could choose any of the other ways to express G from Planck units as shown in Table 1 or Table 2; for example, we could choose Haug's formula $G = \frac{l_p^2 c^3}{\hbar}$. This would, at first glance, seem to merely lead to G, $\hbar$, and c being replaced with l_p, $\hbar$, and c. In other words, after we know that the Planck units can be found without G, we can replace the three universal constants G, $\hbar$, and c with a chosen Planck unit plus c and $\hbar$. So then, one can question whether this is just a change of unit systems. This alone is interesting, but obviously not a big break through; we could even claim it is trivial information.

Another important step is needed before we can discover the great utility of the composite view of the gravitational constant. The mass in kilograms of any mass can be described as:

$$m = \frac{\hbar}{\bar{\lambda}} \frac{1}{c} \tag{9}$$

where $\bar{\lambda}$ is the reduced Compton wavelength. This expression for mass we simply get by solving the Compton [72] wavelength formula $\bar{\lambda} = \frac{\hbar}{mc}$ with respect to mass. One could claim that only elementary particles have Compton wavelength and that composite masses do not, or at least not such large objects as planets or suns. Only elementary particles likely have a "physical" Compton wavelength can be measured by Compton scattering, but larger masses consist of elementary particles and the aggregated Compton wavelength in the composite mass is given by [1,67]:

$$\bar{\lambda} = \frac{1}{\frac{1}{\bar{\lambda}_1} + \frac{1}{\bar{\lambda}_2} + \frac{1}{\bar{\lambda}_3} + \cdots \frac{1}{\bar{\lambda}_n}} \tag{10}$$

This aggregation is fully consistent with standard mass aggregation $m = m_1 + m_2 + m_3 + \cdots + m_n$, and can even be derived from it. It is also important to understand that one can find the Compton wavelength of any mass without knowing $\hbar$ and G. This mass addition formula is known to over-estimate the mass slightly, mostly due to nuclear binding energy, see [73,74]. However, one can easily adjust for this by treating the binding energy as mass equivalent and subtracting it.

Let us start with an electron. The Compton wavelength can be found by shooting a photon at an electron, and by measuring the photon wavelength before and after the impact with the electron, and also the angle between the incoming and outgoing photon, and thus we have:

$$\lambda_e = \frac{\lambda_2 - \lambda_1}{1 - \cos\theta} \tag{11}$$

The reduced Compton wavelength of the electron is simply thus divided by 2π as is well known. Next, we can find the reduced Compton wavelength of the proton by utilizing the fact that the cyclotron frequency ratio is proportional to the Compton wavelength ratio. This is because the charge on the electron and proton is the same, and the cyclotron frequency is given by:

$$f = \frac{qB}{2\pi m} \tag{12}$$

So, we must have

$$\frac{f_e}{f_P} = \frac{\frac{qB}{2\pi m_e}}{\frac{qB}{2\pi m_P}} = \frac{\bar{\lambda}_P}{\bar{\lambda}_e} \approx \frac{1}{1836.15} \tag{13}$$

So, if we know the electron Compton wavelength, we know the proton Compton wavelength as it is arrived at by taking the electron Compton wavelength and dividing it by 1836.15. Next, we can find the Compton wavelength of any larger mass by "simply" counting the number of atoms in the object of interest and then dividing the Compton wavelength of the proton by this count. To count atoms in a clump of matter is not easy, but fully possible. One way is to construct a precise silicon (^{28}Si) sphere. As one knows the crystal structure here very well and since it is very uniform, one can accurately calculate the number of atoms in such a sphere. This way of counting atoms has even been one of the recently suggested methods to re-define the kilogram; see [75–77]. There also exist other methods to count atoms [78,79], so this is fully possibly in practice, even if it takes some effort.

Based on that, we can write the formula of a mass as $M = \frac{\hbar}{\bar{\lambda}}\frac{1}{c}$ and we can replace the mass in Equation (7) with this mass and this gives us:

$$l_p = \sqrt{\frac{\hbar 2\pi^2 L r^2 \theta}{M T^2 c^3}}$$

$$l_p = \sqrt{\frac{\hbar 2\pi^2 L r^2 \theta}{\frac{\hbar}{\bar{\lambda}}\frac{1}{c} T^2 c^3}}$$

$$l_p = \sqrt{\frac{2\pi^2 L r^2 \theta \bar{\lambda}}{T^2 c^2}} \tag{14}$$

That is, the two Planck constants cancel each other out to find the Planck length. In other words, we do not need to know $\hbar$ or G to find the Planck length. All the other parameters in the formula we can easily find without knowledge of G or $\hbar$ using a Cavendish apparatus. The reason we use a Cavendish apparatus is because we can deal with sizes of matter where we can count the number of atoms, but similar methods for even much larger masses can be used [1].

Haug [1,67] has recently shown a practically feasible way to find the Compton wavelength independent of G and $\hbar$ for planets, stars, galaxies and even of the whole mass of the observable universe, see [80]. The main point here is that any mass in terms of kilograms can be expressed by the Formula (9). Next, let us multiply the composite $G = \frac{l_p^2 c^3}{\hbar}$ with the composite mass $M = \frac{\hbar}{\bar{\lambda}}\frac{1}{c}$ and we get:

$$GM = \frac{l_p^2 c^3}{\hbar} \times \frac{\hbar}{\bar{\lambda}}\frac{1}{c} = c^2 \frac{l_p^2}{\bar{\lambda}} \tag{15}$$

What is important to pay attention to here is that the two Planck constants actually cancel each other out, and we are left with two constants, c and l_p, and both these can be found without knowledge of G and $\hbar$. Table 4 shows a series of predicted gravitational phenomena that can actually be observed. As we see, in all the observable phenomena, we have GM and not GMm. The small mass m in the Newton gravitational force formula is only used in derivations of observable gravitational phenomena and then one of the two masses always cancels out. In real two mass gravity phenomena we have the gravity parameter $\mu = G(M_1 + M_2) = GM_1 + GM_1 = c^2 \frac{l_p^2}{\bar{\lambda}_1} + c^2 \frac{l_p^2}{\bar{\lambda}_2}$ so also in real two body gravitational phenomena, the Planck constant cancels out.

It is evident from Table 4 that a long series of observable gravity phenomena can be predicted by knowing only two constants, namely l_p and c and naturally a variable which is linked to the mass size, namely the reduced Compton wavelength of the gravitational object. As seen from the table, some observable gravity phenomena only needs one constant, namely the Planck length. Again, it has, in recent years, been demonstrated how to find the Planck length independently off G so this is a fully practical way to do gravity predictions, and is not just a hypothesis.

Table 4. The table shows that any observable gravity phenomena contains GM and not GMm and, further, that when assuming G is a composite, then we end up being able to predict all observable gravity phenomena only from l_p and c.

Mass	$M = \frac{\hbar}{\bar{\lambda}_M}\frac{1}{c}$ (kg)
Non observable (contains GMm)	
Gravitational constant	$G, \left(G = \frac{l_p^2 c^3}{\hbar}\right)$
Gravity force	$F = G\frac{Mm}{R^2}$ $(\mathrm{kg}\cdot\mathrm{m}\cdot\mathrm{s}^{-2})$
Observable predictions: (contains only GM)	
Gravity acceleration	$g = \frac{GM}{R^2} = \frac{c^2}{R^2}\frac{l_p^2}{\bar{\lambda}_M}$
Orbital velocity	$v_o = \sqrt{\frac{GM}{R}} = cl_p\sqrt{\frac{1}{R\bar{\lambda}_M}}$
Orbital time	$T = \frac{2\pi R}{\sqrt{\frac{GM}{R}}} = \frac{2\pi\sqrt{\bar{\lambda}_M R^3}}{cl_p}$
Periodicity pendulum[5] (clock)	$T = 2\pi\sqrt{\frac{L}{g}} = 2\pi R\sqrt{\frac{L}{GM}} = \frac{2\pi R}{cl_p}\sqrt{L\bar{\lambda}_M}$
Frequency Newton spring	$f = \frac{1}{2\pi}\sqrt{\frac{k}{m}} = \frac{1}{2\pi R}\sqrt{\frac{GM}{x}} = \frac{cl_p}{2\pi R}\sqrt{\frac{1}{\bar{\lambda}_M x}}$
Velocity ball Newton cradle[6]	$v_{out} = \sqrt{2\frac{GM}{R^2}H} = \frac{cl_p}{R}\sqrt{\frac{2H}{\bar{\lambda}_M}}$
Observable predictions (from GR): (contain only GM)	
Advance of perihelion	$\sigma = \frac{6\pi GM}{a(1-e^2)c^2} = \frac{6\pi}{a(1-e^2)}\frac{l_p^2}{\bar{\lambda}_M}$
Gravitational redshift	$z = \frac{\sqrt{1-\frac{2GM}{R_1c^2}}}{\sqrt{1-\frac{2GM}{R_2c^2}}} - 1 = \frac{\sqrt{1-\frac{2l_p^2}{R_1\bar{\lambda}_M}}}{\sqrt{1-\frac{2l_p^2}{R_2\bar{\lambda}_M}}} - 1$
Time dilation	$T_R = T_f\sqrt{1-\sqrt{\frac{2GM}{R}}^2/c^2} = T_f\sqrt{1-\frac{2l_p^2}{R\bar{\lambda}_M}}$
Deflection	$\delta = \frac{4GM}{c^2R} = \frac{4}{R}\frac{l_p^2}{\bar{\lambda}_M}$
Microlensing	$\theta_E = \sqrt{\frac{4GM}{c^2}\frac{(d_S-d_L)}{d_Sd_L}} = 2l_p\sqrt{\frac{d_S-d_L}{\bar{\lambda}_M(d_Sd_L)}}$

As all predictions of observable gravity phenomena contain GM, this leads to $GM = c^3\frac{l_p}{c}\frac{l_p}{\bar{\lambda}}$. Haug has, in a series of papers, suggested that c^3 can be used as a gravity constant and that real gravity mass should be re-defined as $\bar{m} = \frac{l_p}{c}\frac{l_p}{\bar{\lambda}}$; something we will soon get back to. This view is shown in Table 5, which gives all the exact same predictions as the standard gravity formulas, but without any G and also without any need for $\hbar$. One exception is for the gravity force itself, but the gravity force cannot be measured directly; we can only observe the consequences from it. Our new way of representing the gravity force formula gives the same predictions for observable phenomena, and is only linked to the Planck length and the speed of light. The speed of light is, in this context, the same as the speed of gravity ("gravitons"?). For example, Abbot et al. [81] in 2017 constrained *"the difference between the speed of gravity and the speed of light to be between* -3×10^{-15} *and* $+7 \times 10^{-16}$ *times the speed of light."*.

Table 5. The table shows that we can write the gravitational constant as c^3 when using, in our view, a more complete mass definition, $\bar{m} = \frac{l_p}{c}\frac{l_p}{\bar{\lambda}}$. That is, mass is related to time, or what Haug has called collision-time. Different mass sizes then only differ in different Compton wavelengths. Writing the gravitational force formula this way yields the same predictions as standard Newton gravity except we only rely on two constants, l_p and c, to describe mass and any observable gravity phenomena. In addition, in general relativity predictions, we can replace the mass with this mass definition if we replace G with c^3. The reason we can do this is that $c^3\bar{M} = GM$. This is clear when we understand that G is a composite constant and, in addition, understand that the kilogram mass can be written by simply solving the Compton wavelength formula with respect to m.

Mass	$M = \frac{\hbar}{\bar{\lambda}_M}\frac{1}{c}$ (kg)
Non observable :	
Gravitational constant	c^3
Gravity force	$F = c^3\frac{\bar{M}\bar{m}}{R^2}$ $(\text{kg}\cdot\text{m}\cdot\text{s}^{-2})$
Observable predictions:	
Gravity acceleration	$g = \frac{c^3\bar{M}}{R^2} = \frac{c^2}{R^2}\frac{l_p^2}{\bar{\lambda}_M}$
Orbital velocity	$v_o = \sqrt{\frac{c^3\bar{M}}{R}} = cl_p\sqrt{\frac{1}{R\bar{\lambda}_M}}$
Orbital time	$T = \frac{2\pi R}{\sqrt{\frac{c^3\bar{M}}{R}}} = \frac{2\pi\sqrt{\bar{\lambda}_M R^3}}{cl_p}$
Periodicity pendulum[7] (clock)	$T = 2\pi\sqrt{\frac{L}{g}} = 2\pi R\sqrt{\frac{L}{c^3\bar{M}}} = \frac{2\pi R}{cl_p}\sqrt{L\bar{\lambda}_M}$
Frequency Newton spring	$f = \frac{1}{2\pi}\sqrt{\frac{k}{M}} = \frac{1}{2\pi R}\sqrt{\frac{c^3\bar{M}}{x}} = \frac{cl_p}{2\pi R}\sqrt{\frac{1}{\bar{\lambda}_M x}}$
Velocity ball Newton cradle[8]	$v_{out} = \sqrt{2\frac{c^3\bar{M}}{R^2}H} = \frac{cl_p}{R}\sqrt{\frac{2H}{\bar{\lambda}_M}}$
Observable predictions (from GR):	
Advance of perihelion	$\sigma = \frac{6\pi c^3\bar{M}}{a(1-e^2)c^2} = \frac{6\pi}{a(1-e^2)}\frac{l_p^2}{\bar{\lambda}_M}$
Gravitational redshift	$z = \frac{\sqrt{1-\frac{2c^3\bar{M}}{R_1c^2}}}{\sqrt{1-\frac{2c^3\bar{M}}{R_2c^2}}} - 1 = \frac{\sqrt{1-\frac{2l_p^2}{R_1\bar{\lambda}_M}}}{\sqrt{1-\frac{2l_p^2}{R_2\bar{\lambda}_M}}} - 1$
Time dilation	$T_R = T_f\sqrt{1-\sqrt{\frac{2c^3\bar{M}}{R}}^2/c^2} = T_f\sqrt{1-\frac{2l_p^2}{R\bar{\lambda}_M}}$
Deflection	$\delta = \frac{4c^3\bar{M}}{c^2R} = \frac{4}{R}\frac{l_p^2}{\bar{\lambda}_M}$
Microlensing	$\theta_E = \sqrt{\frac{4c^3\bar{M}}{c^2}\frac{(d_S-d_L)}{d_Sd_L}} = 2l_p\sqrt{\frac{d_S-d_L}{\bar{\lambda}_M(d_Sd_L)}}$

5. Is the Inertial Mass Really Identical to the Gravitational Mass?

As shown for all the observational gravitational phenomena reported in Tables 4 and 5, we have GM and not GMm. Again, $GM = \frac{l_p^2c^3}{\hbar}\frac{\hbar}{\bar{\lambda}}\frac{1}{c} = c^3 t_p\frac{l_p}{\bar{\lambda}}$, where Haug has claimed in a series of papers [67,82] that c^3 can then be seen as a gravitational constant and $t_p\frac{l_p}{\bar{\lambda}}$ as a more complete mass definition. This mass definition he has coined collision-time, which again can be seen as the gravitational mass. Yes, this mass definition has dimensions of simply time. This mass is already indirectly embedded in standard gravity theory since one is multiplying the kilogram mass with the gravitational constant. But here the traditional view is that this is a gravity constant multiplied by a mass, and that they are two separate things. No one has yet figured out exactly, from a deeper perspective, why this must be done. Well, what is a gravitational mass and what is an inertial mass? A gravitational mass is linked to the mass caused by, and acted on, by a body by the force of gravity, so it has always been assumed that both the masses in Newton's formula represent gravitational masses. However, we will challenge that view here. This is because if the small mass m has insignificant impact on M then it cancels out in all derivations of direct observable gravitational phenomena, so we could even write:

$$ya = G\frac{My}{R^2} \quad (16)$$

and we would still get the correct predictions about measurable gravitational phenomena from this equation; that is, y on both sides of the equation could be replaced with basically anything. We could even define y as money. Money has naturally nothing to do with gravity, but since y is on both sides of the equation we can divide by y on both sides and we get $a = GM/R^2$, and we can measure both a and the gravitational acceleration. Our point is that even when putting in a completely wrong mass definition for m on both sides of $ma = G\frac{Mm}{R^2}$, these two m masses will cancel out in the derivation of anything observable. This is the case for derivation of any observable gravity phenomena, except for real two-body problems where one has $GM_1 + GM_2$, not GMm. The reminding kilogram mass M is always multiplied by the gravity constant. We will claim this is done (unknowingly) to correct an incomplete mass definition (the kilogram mass) into a more complete mass definition, so the real gravitational mass is $\frac{G}{c^3}M = \frac{l_p}{c}\frac{l_p}{\bar{\lambda}} = \bar{M}$. This was discussed in detail by Haug [67,82]. Newton naturally did not have this in mind when he developed his gravity theory and, as we have pointed out, he also never used a gravity constant. The gravity constant is a missing value constant simply found by calibration to observable gravitational phenomena when one has decided upon using the kilogram definition of mass. This is also at least part of the reason why the gravity constant came into existence about the same time as the kilogram mass became popular in Europe.

The mass linked to non-gravitational acceleration is often thought of as the inertial mass, and since it has been shown experimentally that the following relation seems to hold

$$m_i a = G\frac{Mm}{R^2} \tag{17}$$

we assume that the inertial mass m_i must be equal to the gravitational mass m since it seems to be an equivalence between standard acceleration (for example, in an elevator) and in a gravitational acceleration field, and we do not doubt this, we simply claim the mass m is not used directly for any predictions of any observable gravitational phenomena. That is, one is not measuring ma or $G\frac{Mm}{R^2}$, but is observing a and $g = a = G\frac{M}{R^2}$; in other words, after the two small masses have cancelled each other out in derivations for predictions of observable phenomena. In our view, there is only one type of mass and we have just defined it as $\bar{M} = \frac{G}{c^3}M$. Inputting this mass definition in all parts in the Newton formula would lead to:

$$\frac{G}{c^3}ma = c^3\frac{\frac{G}{c^3}M\frac{G}{c^3}m}{R^2} \tag{18}$$

and we would also now end up with $a = \frac{GM}{R^2}$, since $\frac{G}{c^3}m$ is on both sides and cancels out. And since $\bar{M} = \frac{G}{c^3}M$ and $\bar{m} = \frac{G}{c^3}m$ we can write this as:

$$\bar{m}a = c^3\frac{\bar{M}\bar{m}}{R^2} \tag{19}$$

Our point is that in the standard Newton gravity formula, in its modern form invented in 1873, one is likely unknowingly using two different masses: the standard kilogram mass multiplied by G; that is, GM which combined can be seen as the gravitational mass (collision-time mass) multiplied by c^3; that is, $GM = c^3\bar{M}$ and the other non-gravitational mass m is an incomplete kilogram mass that says nothing about gravity. When we want to just know, for example, the relation between mass and energy, then the standard kilogram mass will have enough information to do so. So, we can still use m, that is the kilogram mass definition without adjustments, in relations such as $E = mc^2$ or in $E = mc^2\gamma$, but the same mass definition cannot be used for calculating gravity effects from that mass without multiplying it with G or, better, by understanding that GM actually represents the real gravitational mass (divided by c^3). There is only one mass, but to describe gravity requires additional information that is lacking in the kilogram mass definition. The kilogram mass is incomplete, but good enough for calculations related to just energy and mass (the kilogram definition), but it is incomplete when also taking into account gravity. We can then either fix this mass ad hoc by using G or we can understand that G is a composite constant and, when combined with M, it gives a deeper insight into mass also related to gravity.

So the real gravitational mass even of the small mass is $\bar{m} = \frac{G}{c^3}m = t_p\frac{l_p}{\lambda} = \frac{l_p}{c}\frac{l_p}{\lambda}$, while the kilogram mass is given by $m = \frac{\hbar}{\lambda}\frac{1}{c}$. Since the Planck length, the speed of light, and the Planck constants are constants, and the only thing changing is the Compton wavelength (mass size) in both the gravitational mass (collision-time mass) and in the kilogram mass, they are proportional. So, the weak equivalence principle holds also under this view. This view does not change the output from predictions of observable phenomena, but it shows us how the Planck scale is already directly linked to gravity. Detection of gravity is, in our view, detection of the Planck scale. This view is new and controversial, but we think it should be taken seriously enough to also be carefully investigated by other researchers before being rejected prematurely.

At the same time, one is incorporating important aspects of mass linked to the Planck length. The same can easily be done with energy, as has been demonstrated in the papers referred to. We naturally do not ask other researchers to take this for granted, but simply suggest that, in our view, this looks like a promising path that we think requires further investigation.

We have, in this paper, claimed that it looks as though all observable gravity phenomena can be predicted by using the two constants: the speed of light and the Planck length, without dependence on G or $\hbar$. One could think such a view is wrong, as c and l_p only contain length and time, so how could this ever represent energy, which is normally presented as joule? This at first seems inconsistent. However, we have recently shown how energy can indeed be represented as a length (collision-length) and that mass can be represented as time (collision-time); see [82,83]. Energy is most often expressed as joule which again, in SI units, represents $\mathrm{kg} \cdot \mathrm{m}^2 \cdot \mathrm{s}^{-2}$. That is, energy contains kilogram. Kilogram is an arbitrary amount of matter that came into use in science around 1870, a few years before the introduction of the gravitational constant in 1873. Our recent research indicates this is an unnecessarily complex way to express energy and that it can be expressed as collision-length. One can easily move from the collision-length energy and back to the joule energy, simply by multiplying the collision-length energy with $\frac{c^4}{G} = \frac{\hbar}{l_p^2}c$ or, the other way around, multiplying the joule energy with $\frac{G}{c^4} = \frac{l_p^2}{\hbar c}$. This is discussed in great detail in [67,82]. The main point is that kilogram and joule both contain information about the kilogram, and the kilogram is an arbitrary, human-chosen clump of matter that gravity does not care about. Our new suggested way to look at mass and energy, that is already embedded in standard gravity theory through the use of G, strips the standard mass of unnecessary information about the kilogram.

When going from the collision-length energy, $\bar{E}$, to the joule energy, E, or from the joule energy to the collision-length energy, it seems (from the paragraph above) that we need to know the Planck constant. However, one can easely find the collision-length energy more directly; for example, from the gravitational acceleration field of the Earth that simply given by:

$$\bar{E} = \frac{gR^2}{c^2} = l_p\frac{l_p}{\bar{\lambda}} \tag{20}$$

So the only constant one needs to know to find the gravitational energy[9] is the speed of light (in $\frac{gR^2}{c^2}$). Neither the gravity constant nor the Planck constant are needed. To estimate the gravitational acceleration, one has $g = \frac{GM}{R^2}$, so one could mistakenly think one needs to know G to find $\bar{E}$. This is only if one wants to predict g from standard theory. One can also measure g without any knowledge of G; for example, by dropping a ball. The gravitational acceleration is then simply given by $g = \frac{2H}{T^2}$, where H is the height from which the ball is dropped, and T is the time it took for the ball to fall to the ground.

To find this energy for smaller masses, we can use a Cavendish apparatus; see [67,84] that provides a detailed discussion of the topic. Or alternatively, we can find the gravitational energy by first finding the Planck length and the reduced Compton wavelength of the mass in question without knowledge off G and $\hbar$, as we have demonstrated, in this and other papers, is fully possible.

The Planck constant is normally associated with the concept that energy comes in quanta, something that has been clearly demonstrated in experimental research. One might therefore think that a theory without the Planck constant must be incomplete with respect to describing energy. To our own surprise, even without $\hbar$ in the gravitational energy (collision-length), this is actually a quantum energy that comes in quanta. The collision-length energy comes in quanta of the Planck length. The quantity are given by $\frac{l_p}{\bar{\lambda}}$. This is the quanta of the gravitational energy in a Planck time observational window that we easily can find indirectly. For a Planck mass it is one, for a mass larger than a Planck mass it is a quantity larger than one. Here, the integer part will be the number of full Planck events, and the reminding fractional part is the probability for an additional one such event. For particles smaller than the Planck mass, this factor is always smaller than one and then represents the probability for the particle to be in a particular state that we, in previous papers. have called collision-state inside the Planck time window see [67,82]. It is not necessary to measure anything at the Planck time to measure this indirectly, as we have demonstrated in the papers just mentioned. The term $\frac{l_p}{\bar{\lambda}}$ embedded in the collision-length energy (quantum gravitational energy) can simply be seen as an aggregate of probabilities for Planck events, as is discussed in-depth in the papers just referred to.

6. The Gravity Constant Calculated from Cosmological Entities

Another line of thought in relation to the composite view of G has been that the Newtonian gravitational constant can perhaps be calculated from cosmological entities or constants. Already, in 1936, Milne [85] explicitly suggested that:

$$G = \frac{c^3 T_H}{M_u} \tag{21}$$

where T_H is one divided by the Hubble constant, and M_u is the mass of the universe or what Milne called the fictional mass of the universe, as he thought the universe was infinite, but that this was the mass inside the Hubble sphere. Unzicker [86] pointed out that such a relation indirectly and ad hoc was already hinted at by Einstein [87] in 1917 but at best obliquely so, and further that Schrödinger [88] in 1925, according to Unzicker, must have suspected this relation based on numerical calculations. Bleksley [89] suggested that the gravitational constant can be expressed as:

$$G = \frac{R_u c^2}{M_u} \tag{22}$$

where R_u is the radius of the observable universe and M_u is the mass of the observable universe. So this is basically the same as the Milne formula since: $R_u = \frac{c}{H_0} = cT_H$.

As the mass of the universe from the Friedmann equation published in 1922 (general relativity theory) requires knowledge of G, Bleksley could not use this mass to find G, but instead came up with his own way to calculate the universe mass, a way we think looks a bit like numerology or at least is very speculative. For example, he suggested that the number of protons in the universe must be $R_u^2/(4\rho^2)$ where ρ is the diameter of the proton. It is far from clear how he got to this formula, so we are questioning the validity of this approach. But $G = \frac{R_u c^2}{M_u}$ is still, at best, only a rough approximation for the Friedmann model. The correct formula in the Friedmann model is $M_c = \frac{c^3}{2GH_0}$, where M_c is the critical mass of the universe, so this leads to $G = \frac{R_u c^2}{2M_c}$ and not $G = \frac{R_u c^2}{M_c}$, as suggested by Milne and Bleksley. Mercier [39] in 2020 basically gave the same formula for G as Milne and Bleksley. He used a universe mass rooted in a paper by Carvalho [90]. Carvalho started with a relation between mass density and the Hubble constant that he claimed is given by Weinberg [91]

$$G\rho_0 = H_0^2 \tag{23}$$

And, from this, got

$$M_u \approx \frac{c^3}{GH} \tag{24}$$

Carvalho, in addition, derived a universe mass to be the same as given by this formula, but independently from G by some assumptions of π mesons. However, his derivation here seems quite speculative. Carvalho further claimed *"This is identical to expression derived in the context of Friedmann's cosmological model."*. This claim is not fully correct or, at least, not precise enough. Both the formula he presented, the one he claimed is from the Friedman model and the other universe mass he derived from π mesons, are both actually twice what one gets from the Friedmann model. The universe mass one gets from the Friedmann model is:

$$M_c = \frac{1}{2}\frac{c^3}{H_0 G} \tag{25}$$

as one also finds indirectly in the book of Weinberg as well as in a series of other independent sources (see, for example, [92,93]). Several authors (for example, Cook in 2011 [94] and Mercier [39]) have suggested that G can be expressed as:

$$G = \frac{c^3}{HM_u} = \frac{T_H c^2}{M_u} \tag{26}$$

This is identical to the Milne 1936 formula and is naturally equal to $G = \frac{R_H c^2}{M_u}$, as the Hubble radius is given by $R_H = \frac{c}{H_0}$, and $T_H = \frac{R_H}{c} = \frac{1}{H_0}$. Again, this formula is not fully consistent with the Friedmann model, as that would require $G = \frac{c^3}{2HM_u}$ and, since the Hubble time is given by $T_H = \frac{1}{H_0}$, this is naturally the same as $G = \frac{T_H c^2}{2M_u}$, but the Equation (26) is consistent when using the Haug [95] universe mass, which is predicted to be twice that of the Friedmann critical mass of the universe.

None of these authors have shown how to find the universe mass without already knowing G, except from what we would call relatively speculative approaches which we think lack a solid foundation, even if this naturally can be discussed further. Still, there is, as we will see, a way to find this critical mass of the universe without knowing G. First, if we solve the Friedmann critical mass equation (Equation (25)) with respect to H_0, then this gives the formula:

$$H_0 = \frac{1}{2}\frac{c^3}{M_c G} \tag{27}$$

And since any mass in kilogram can be written as $m = \frac{\hbar}{\bar{\lambda}}\frac{1}{c}$, and also because G can be written as $G = \frac{l_p^2 c^3}{\hbar}$, this means we have:

$$H_0 = \frac{1}{2}\frac{c^3}{\frac{\hbar}{\bar{\lambda}_c}\frac{1}{c}\frac{l_p^2 c^3}{\hbar}} = \frac{\bar{\lambda} c}{2 l_p^2} \tag{28}$$

where $\bar{\lambda}_c$ is the reduced Compton wavelength of the critical mass in the Friedmann universe (the critical universe). This also means we must have:

$$G = \frac{\bar{\lambda}_c c^4}{2H_0\hbar} \tag{29}$$

The Hubble constant can be found with no knowledge of G as also the Compton wavelength of the universe mass can be found without this knowledge of G as we [80] recently demonstrated. Further, as $H_0 = \frac{c\bar{\lambda}_c}{2l_p^2}$, this means the Equation (29) can be simplified further to:

$$G = \frac{\bar{\lambda}_c c^4}{2H_0\hbar} = \frac{l_p^2 c^3}{\hbar} \tag{30}$$

which is the same composite formula for G that one gets by solving Max Planck's Planck length formula; $l_p = \sqrt{\frac{G\hbar}{c^3}}$, with respect to G.

For G times the critical mass of the universe (the gravitational parameter of the universe), we must have:

$$\mu_c = GM_c = \frac{l_p^2 c^4}{\hbar}\frac{\hbar}{\bar{\lambda}_c}\frac{1}{c} = c^2\frac{l_p^2}{\bar{\lambda}_c} \tag{31}$$

where $\bar{\lambda}_c$ is the reduced Compton wavelength of the mass in the critical universe. To predict gravitational phenomena related to the mass of the critical universe, all we need is the Planck length and the speed of light; that is, two constants, and the reduced Compton wavelength of the critical mass of the universe. All these can be found with no knowledge off G or even $\hbar$; see [80]. Actually, the Hubble constant is given by:

$$H_0 = \frac{1}{\frac{l_p}{c}\frac{l_p}{\bar{\lambda}_u}} = \frac{1}{t_p\frac{l_p}{\bar{\lambda}_u}} \tag{32}$$

where λ_u is the reduced Compton wavelength of the mass in the Haug universe. Pay attention to how $t_p\frac{l_p}{\bar{\lambda}_c}$ is identical to what we call the collision-time mass. So, the Hubble constant, in this view, is nothing more than one divided by the collision-time mass of the observable universe [83].

If one knows the collision-time mass of the universe, then there is no need to multiply it with G to do gravitational predictions. This is why cosmological red-shift can be predicted simply by:

$$Z = \frac{Hd}{c} = \frac{d}{c\bar{M}_c} \tag{33}$$

where $\bar{M}_c = t_p\frac{l_p}{\bar{\lambda}}$ is the collision-time of the observable universe. If we use the critical mass of the universe in terms of kilogram, then we need to multiply it with G divided by c^3 to convert it into the real gravitational mass, so we have:

$$Z = \frac{d}{c\frac{G}{c^3 M_c}} = \frac{d}{c\bar{M}_c} = \frac{H_0 d}{c} \tag{34}$$

This means we can also predict cosmological phenomena from the Planck length and the speed of light. This strongly indicates there is a link between the largest and the smallest scales of the universe. This is not a very big surprise as the largest scales are built from the smallest. The rules of the smallest (quantum) somehow give us the rules for even the cosmic scales. Our view is that the Planck scale is actually indirectly detected in any (or at least most) gravitational observation, including also cosmological red-shift.

Table 6 shows some ways to express the gravity constant in the form of cosmological entities. All these are at the deepest level nothing else than $G = \frac{l_p^2 c^3}{\hbar}$. In addition, pay attention to how closely the formulas linked to the Hubble scale are linked to the formulas presented that are linked to the Schwarzschild radius and Haug radius. The reason for this is that the Hubble radius is identical to the Schwarzschild radius for the observable universe, and this is why in papers it has been considered whether the Hubble sphere is actually a gigantic black hole; see, for example, [96,97].

Table 6. The table shows various ways we can express the gravity constant from cosmological units, as well as from units related to black holes.

From	Gravity Formula	Comments
From universe mass and Hubble time	$G = \frac{c^3 T_H}{M_u}$	Milne 1936 [85]
From universe mass and Hubble radius	$G = \frac{R_u c^2}{M_u}$	Bleksley 1951 [89]
From universe mass and universe radius	$G = \frac{R_u c^2}{6M_u}$	Unzicker 2020 [19]
Hubble constant, Friedmann critical mass	$G = \frac{c^3}{2H_0 M_c}$	
Hubble radius, Friedmann critical mass	$G = \frac{R_H c^2}{2M_c}$	
Hubble constant, Friedmann critical mass	$G = \frac{T_H c^3}{2M_c}$	
Hubble time, Friedmann critical mass	$G = \frac{c^3}{2H_0 M_c}$	
Hubble radius, Hubble time, and Friedmann critical mass	$G = \frac{R_H^3}{2M_c T_H^2}$	
Hubble constant, Haug universe mass,	$G = \frac{c^3}{H_0 M_u}$	
Hubble radius Hubble time and Haug universe mass	$G = \frac{R_H c^2}{M_u}$	
Hubble radius, Haug universe mass,	$G = \frac{T_H c^3}{M_u}$	
Hubble time, Hubble time, and Haug universe mass	$G = \frac{T_H c^3}{M_u T_H^2}$	
Hubble constant, Friedmann critical mass,	$G = \frac{c^3}{2H_0 M_c}$	
Hubble time and Haug universe mass	$G = \frac{T_H c^3}{M_u}$	
Schwarzschild radius, mass,	$G = \frac{R_s c^2}{2M}$	$R_s = \frac{2GM}{c^2}$
Schwarzschild time, mass,	$G = \frac{T_s c^3}{2M}$	$T_s = \frac{R_s}{c}$
Haug escape velocity radius, mass,	$G = \frac{R_h c^2}{M}$	$R_h = \frac{GM}{c^2}$
Haug radius time, mass,	$G = \frac{T_h c^3}{M}$	$T_h = \frac{R_h}{c}$

7. The Composite View of G with Respect to Alternative Gravitational Theories

An interesting question is naturally how well or not the composite view of G also fits into alternative gravitational theories. There exist a long series of alternative or modified gravitational theories so it is outside the scope of this paper to discuss all of them or go in depth about them. Still, we can say something about a few of these theories. One set of modified gravitational theories are minimum acceleration theories. They try to explain galaxy rotation without dark matter by incorporating a minimum acceleration. The most well-known of these is modified Newton dynamics (MOND) introduced by Milgrom [98] in 1983. Here, we suspect it would just be to replace G with $G = \frac{l_p^2 c^3}{\hbar}$ and M with $M = \frac{\hbar}{\bar{\lambda}}\frac{1}{c}$ without any issues. Whether this can give deeper insight into the Milgrom model would need to be investigated. Another minimum acceleration modified gravitational theory is the so-called quantized inertia [30,32,99]. Here the composite view of G falls nicely together with this model, as it has even been suggested by McCulloch that G is a composite and that the Planck units are perhaps a deeper and more important level. However, back then he was then not able to find the Planck units without knowledge of G, as pointed out earlier in his paper. To be able to find the Planck units without first knowing G has been solved in recent years and strengthens the view of McCulloch that one can indeed write G as a composite constant from Planck units. This alone naturally does not give credit to his model. In most alternative gravity models, where G is considered constant, then the composite view of G should in general cause no problems, but likely only helps us to get deeper insight into these models. This naturally must be carefully investigated for each alternative gravitational model before any conclusions are made, and to do so is beyond this paper's scope. The recent collision space-time gravitational theory [67,83] is another alternative gravitational theory we already have mentioned and that fits very well with the composite view off G.

Several researchers have discussed and suggested gravity theories in which the gravity constant can vary. The suggestion that G is not a constant goes back at least to Milne [85] in 1936. In modern times, for example, Zee [100], Adler [101], and Smolin [102] discuss symmetry breaking in relation to gravity where it seems G needs to vary. Davis [103], Pollock [104], Raychaudhuri and Bagch [105], and Masood-ul-Alam [106] discusses gravity theories where G has to vary. If G is simply a composite constant that can be expressed on the form $G = \frac{l_p^2 c^3}{\hbar}$, then either the Planck length, or the speed of light, or the Planck constant has to vary for G to vary, or alternatively there are other, still- missing parts in gravity that our models do not take into account, and which these models have tried to express through varying G. Could c, or l_p, or $\hbar$ vary? The constancy and isotropic futures of the speed of light is one of the corner stones in special relativity theory. The round-trip speed of light has, through a long series of experiments, been tested to be constant and invariant (in vacuum). However, there are, to this day, discussions about how to measure the one-way speed of light and if it truly is invariant. It was Poincaré [107] that first pointed out that to measure the one-way speed of light, one has to have two clocks synchronized over distance, and to synchronize these clocks one has to know the one-way speed of light, so it led to a circular problem that seemed impossible to solve. Poincaré therefore suggested that, for continuity, one could also assume the one-way speed of light was constant for synchronization purposes.

Einstein went one step further and abandoned the ether, as detecting the ether is directly related to detecting anisotropy in the one-way speed of light, and since it could not be done, why not simply abandon something that adds complexity to the theory? However, if the one-way speed of light can somehow be measured, and whether it is invariant, is actually still an ongoing discussion, as is clear from recent publications; see, for example, Spavieri [108,109] and Kipreos and Balachandran [110,111].

There are also researchers discussing whether the Planck constant even could vary; see, for example, Masood-ul-Alam [106]. However, as we have shown, the Planck constant seems to cancel out between the embedded $\hbar$ in the gravity constant and in the mass, so this would likely not be helpful for gravitational theories needing variable gravitational constant.

The Planck length has been less investigated since until recently we have only been able to derive it from dimensional analysis. The possibility cannot be excluded that the Planck length can be variable; this could be due to relativistic effects such as length contraction, but several researchers think the Planck scale (Planck length and Planck time in particular) will break with Lorentz symmetry and that the Planck length is invariant. This despite Lorentz symmetry being a foundational principle in general relativity and the standard model; however, many expect new physics to be discovered at the Planck scale, which again is often linked to quantum gravity; see for example, Tasson [112], so there are questions clearly still open for debate. A possible explanation for why the Planck length can be invariant was suggested by Haug [69,82]. All we can say is that our composite view of the gravitational constant does not yet exclude the view that the gravitational constant could be variable and that further research is needed before any "final" conclusion is made.

8. Conclusions

The idea that the gravitational constant can be a composite constant, which is related to more fundamental Planck units, goes back to at least 1984. However, in 1987 it was already pointed out that expressing the gravitational constant through Planck units led to a circular problem; namely, that one had to know the gravity constant to find the Planck units. This view has been repeated by researchers as recently as 2016. However, in more recent years, a series of papers have shown how one can clearly find the Planck units without knowledge of G, and even without knowledge of G and $\hbar$, so the circular problem regarding G and the Planck units has been solved. An in-depth study showed that this leads to a reduction in universal constants from G, $\hbar$, and c to only c and l_p and, in addition, one needs other constants like the fine structure constants when describing electromagnetic phenomena, but the traditional three universal constants that Max Planck used can be reduced from three to two. To predict all observable gravity phenomena, one only needs knowledge of l_p and c, and both can be found without knowledge of G and $\hbar$. This paper has given an overview of much of what has been done in relation to the composite view of the gravitational constant, but we have also tried to tie it nicely together. The implications of this should be worth studying further as this seems to open doors of insight into relationships between macroscopic gravity phenomena and the Planck scale.

Funding: This research received no external funding.

Data Availability Statement: No data was used for this research.

Conflicts of Interest: The author declares no conflict of interest.

Notes

1 It is impossible for anyone today to know the full literature on physics, so there could be other authors already publishing these formulas; however, we have made a very serious attempt to search and find anyone who might have published these results first.

2 See also McCulloch [30,32].

3 The formula is a very good approximation when the angle of the pendulum is small, as it is in most pendulum clocks. It is not accurate for large angles, but is again exact for an angle of 360; that is to say, for full circle, see [71].

4 Where H is the height of the ball drop.

5 The formula is a very good approximation when the angle of the pendulum is small, as it is in most pendulum clocks. It is not accurate for large angles, but is again exact for an angle of 360; that is to say, for a full circle; see [71].

6 Where H is the height of the ball drop.

7 The formula is a very good approximation when the angle of the pendulum is small, as it is in most pendulum clocks. It is not accurate for large angles, but is again exact for an angle of 360; that is to say, for a full circle, see [71].

8 Where H is the height of the ball drop.

9 That not should be confused with gravitational potential energy.

References

1. Haug, E.G. Demonstration That Newtonian Gravity Moves at the Speed of Light and Not Instantaneously (Infinite Speed) as Thought! *J. Phys. Commun.* **2021**, *5*, 025005. [CrossRef]
2. Newton, I. *Philosophiae Naturalis Principia Mathematica*; Jussu Societatis Regiae ac Typis Josephi Streater: London, UK, 1686.
3. Cohen, I.B. Newton's Determination of the Masses and Densities of the Sun, Jupiter, Saturn, and the Earth. *Arch. Hist. Exact Sci.* **1998**, *53*, 83–95. [CrossRef]
4. Cavendish, H. Experiments to determine the Density of the Earth. *Philos. Trans. R. Soc. Lond. Part II* **1798**, *88*, 469.
5. Clotfelter, B.E. The Cavendish experiment as Cavendish knew it. *Am. J. Phys.* **1987**, *55*, 210. [CrossRef]
6. Cornu, A.; Baille, J.B. Détermination nouvelle de la constante de l'attraction et de la densité moyenne de la Terre. *Comptes Rendus Math. Acad. Sci. Paris* **1873**, *76*, 1.
7. Boys, C.V. On the Newtonian Constant of Gravitation. *Nature* **1894**, *5*, 330. [CrossRef]
8. Planck, M. *Einführung in die Allgemeine Mechanik*; Verlag von Hirtzel: Leipzig, Germany, 1928.
9. Haug, E.G. Newton Did Not Invent or Use the So-Called Newton's Gravitational Constant; *G*, It Has Mainly Caused Confusion. *J. Mod. Phys.* **2022**, *13*, 179. [CrossRef]
10. Planck, M. *Natuerliche Masseinheiten*; Der Königlich Preussischen Akademie Der Wissenschaften: Berlin, Germany, 1899; p. 479.
11. Planck, M. *Vorlesungen über Die Theorie der Wärmestrahlung*; Verlag Von Johann Ambrosius-Barth: Leipzig, Germany, 1923; p. 163.
12. Einstein, A. *Näherungsweise Integration der Feldgleichungen der Gravitation*; Sitzungsberichte der Königlich Preussischen Akademie der Wissenschaften: Berlin, Germany, 1916.
13. Eddington, A.S. *Report on the Relativity Theory of Gravitation*; The Physical Society of London, Fleetway Press: London, UK, 1918.
14. Bridgman, P.W. *Dimensional Analysis*; Yale University Press: New Haven, CT, USA, 1931.
15. Padmanabhan, T. Planck Length as the Lower Bound to all Physical Length Scales. *Gen. Relativ. Gravit.* **1985**, *17*, 215–221. [CrossRef]
16. Adler, R.J. Six Easy Roads to the Planck Scale. *Am. J. Phys.* **2010**, *78*, 925–932. [CrossRef]
17. Hossenfelder, S. Can we measure structures to a precision better than the Planck length? *Class. Quantum Grav.* **2012**, *29*, 115011. [CrossRef]
18. Meschini, D. Planck-Scale Physics: Facts and Beliefs. *Found. Sci.* **2007**, *12*, 277–294. [CrossRef]
19. Unzicker, A. *The Mathematical Reality: Why Space and Time Are an Illusion*; Independently Published: Chicago, IL, USA, 2020.
20. Okun, L.B. The fundamental constants of physics. *Sov. Phys. Usp.* **1991**, *34*, 818. [CrossRef]
21. Haug, E.G. The Gravitational Constant and the Planck Units. A Simplification of the Quantum Realm. *Phys. Essays* **2016**, *29*, 558–561. [CrossRef]
22. Thüring, B. The gravitational constant. *Ann. Acad. Sci. Fenn. A* **1961**, *61*, 269.

23. Gillies, G.T. The Newtonian gravitational constant: Recent measurements and related studies. *Rep. Prog. Phys.* **1997**, *60*, 151. [CrossRef]
24. Zee, A. Calculation of Newton's Gravitational Constant in Infrared-Stable Yang-Mills Theories. *Phys. Rev. Lett.* **1982**, *48*, 295. [CrossRef]
25. Cahill, K. The gravitational constant. *Lett. Nuovo Cim.* **1984**, *39*, 181–184. [CrossRef]
26. Cahill, K. Tetrads, Broken Symmetries, and the Gravitational Constant. *Z. Phys. C Part. Fields* **1984**, *23*, 353–356. [CrossRef]
27. Cohen, E.R. *Fundamental Physical Constants, in the Book Gravitational Measurements, Fundamental Metrology and Constants*; Sabbata, V., Melniko, V.N., Eds.; Springer: Dordrecht, The Netherland, 1987.
28. Sabbata, V.S.C. A minimal time and time-temperature uncertainty principle. *Found. Phys. Lett.* **1992**, *5*, 183–189. [CrossRef]
29. Tong, D. *String Theory*; University of Cambridge Part III Mathematical Tripos, Lecture Notes; University of Cambridge: Cambridge, UK, 2009.
30. McCulloch, M.E. Gravity from the uncertainty principle. *Astrophys. Space Sci.* **2014**, *349*, 957–959. [CrossRef]
31. Bari, P.D. *Cosmology and the Early Universe*; CRC Press: Boca Raton, FL, USA, 2018.
32. McCulloch, M.E. Quantised inertia from relativity and the uncertainty principle. *Europhys. Lett. EPL* **2016**, *115*, 69001. [CrossRef]
33. Clark, D. *Graviton Exchange and the Gravitational Constant in the Book the Gravitational Constant: Generalized Gravitational Theories and Experiments*; Sabbata, V., Melniko, V.N., Eds.; Kluwer Academic Publishers: Dordrecht, The Netherland, 2003.
34. Zwiebach, B. *A First Course in String Theory*; Cambridge University Press: Cambridge, UK, 2004.
35. Nastasenko, V.A. Otkrutie predelno vozmojnuh volnovuh parametrov: Theoria and tehnica peredachi, priema i pererabotki informacii. In Proceedings of the 10-ja Jubilejnaja Megdunarodnaja Conferencia, Kharkov, Ukraine, 30–31 January 2004; Volume 1, p. 30
36. Bruneton, J.P. Notes on several phenomenological laws of quantum gravity. *arXiv* **2013**, arXiv:1308.4044.
37. Casadio, R.; Micuc, O.; Scardigli, F. Quantum hoop conjecture Black hole formation by particle collisions. *Phys. Lett. B* **2014**, *732*, 105–109. [CrossRef]
38. Nastasenko, V. Selection and Justification of a New Initial Level of the Material World. *J. Appl. Math. Phys.* **2021**. *9*, 1089–1099. [CrossRef]
39. Mercier, C. Several Ways to Calculate the Universal Gravitational Constant *G* Theoretically and Cubic Splines to Verify Its Measured Value. *J. Mod. Phys.* **2020**, *11*, 103080. [CrossRef]
40. Humpherys, D. Natural Planck Units and the Structure of Matter and Radiation. *Quantum Specul.* **2021**, *3*, 1–20.
41. Zivlak, B. Universal Gravitational Constant Via Proton. *viXrA* **2013**. Available online: https://vixra.org/abs/1310.0018 (accessed on 15 September 2013).
42. Eldred, M. *Movement and Time in the Cyberworld: Questioning the Digital Cast of Being*; De Gruyter: Berlin, Germany, 2019.
43. Kiritsis, E. *Introduction to Super String Theory*; CERN-TH/97-218 Lectures presented at the Catholic University of Leuven and at the University of Padova during the Academic Year 1996–97; Cornell University Press: Ithaca, NY, USA, 1997.
44. Cerdeno, D.G.; Munoz, C. An Introduction to Supergravity. In Proceedings of the 6th Hellenic School and Workshop on Elementary Particle Physics (CORFU 1998), Corfu, Greece, 6–26 September 1998.
45. Langacker, P. Time Variation of Fundamental Constants as a Probe of New Physics. In *Advances in the Interplay Between Quantum and Gravity Physics*; Bergmann, P.G., de Sabbata, V., Eds.; Springer: Berlin, Germany, 2003.
46. Langacker, P. Time Variation of Fundamental Constants as a Probe of New Physics. *Int. J. Mod. Phys. A* **2004**, *19*, 155. [CrossRef]
47. Kazanas, D. Cosmological Inflation: A Personal Perspective. In *Chaos in Astronomy: Conference 2007*; Springer: Berlin/Heidelberg, Germany, 2007.
48. Ritter, R.C.; Gillies, G.T. Laboratory Searches for Effects Beyond Classical Gravity. In *Advances in Interplay between Quantum and Gravity Physics*; Bergmann, P.G., Sabbata, V., Eds.; Springer: Dordrecht, The Netherland, 2001.
49. Ellis, J. Beyond the Standard Model in the book: Astronomy, Cosmology and Fundamental Physics. In Proceedings of the ESO/CERN/ESA Symposium, Garching, Germany, 4–7 March 2002; ESO Astrophysics Symposia Shaver, P.A., Lella, L., Gimenez, A., Eds.; Springer: Dordrecht, The Netherland, 2002.
50. Hossenfelder, S. Minimal Length Scale Scenarios for Quantum Gravity. *Living Rev. Relativ.* **2013**, *16*, 2. [CrossRef]
51. Peebles, P.J.E. Inflation and the Baryon Insocurvator Model. In *Large Scale Structure and Motions in the Universe*; Mezzett, M., Giuricin, G., Mardirossian, F., Ramella, M., Eds.; Kluwer Academic Publishers: Dordrecht, The Netherland, 1989.
52. Schwarzschild, B. Theorists and Experimenters Seek to Learn Why Gravity Is So Weak. *Phys. Today* **2000**, *53*, 22–24. [CrossRef]
53. Padmanabhan, T. Atoms of Spacetime and the Nature of Gravity. *J. Phys. Conf. Ser.* **2016**, *701*, 012018. [CrossRef]
54. Casadio, R. Gravitational renormalization of quantum field theory: A "conservative" approach. *J. Phys. Conf. Ser.* **2009**, *174*, 012058. [CrossRef]
55. Casadio, R. Gravitational renormalization of quantum field theory. *J. Mod. Phys. A* **2012**, *27*, 1250186. [CrossRef]
56. Mück, W. On the Number of Soft Quanta in Classical Field Configurations. 2013. Available online: https://arxiv.org/pdf/1306.6245.pdf (accessed on 26 August 2013).
57. Casadio, R. Horizons and Non-Local Time Evolution of Quantum Mechanical Systems. 2014. Available online: https://arxiv.org/abs/1411.5848 (accessed on 29 November 2014).
58. Casadio, D. What is the Schwarzchild Radius of a quantum Mechanical Particle. In *1st Karl Schwarzschild Meeting on Gravitational Physics*; Nicolini, P., Kaminski, M., Mureika, J., Bleicher, M., Eds.; Springer International Publishing: Cham, Switzerland, 2016.

59. Roberto Casadio, R.; Giugno, A.; Giusti, A.; Micu, O. Horizon quantum mechanics of rotating black holes. *Eur. J. Phys. C* **2017**, *77*, 322. [CrossRef]
60. Ovalle, J.; Casadio, R. *Beyond Einstein Gravity: The Minimal Geometric Deformation Approach in the Brane-World*; Springer: Cham, Switzerland, 2020.
61. Nastasenko, V.A. *About of More Exact Definition of Value Gravitational Constant by Means of Calculation: Materials of the International Scientific and Technical Conference: High-Performance Computing Systems*; NTUU: Kiev, Ukraine, 2013; p. 266.
62. Haug, E.G. Planck Quantization of Newton and Einstein Gravitation. *Int. J. Astron. Astrophys.* **2016**, *6*, 206. [CrossRef]
63. Haug, E.G. Newton and Einstein's Gravity in a New Perspective for Planck Masses and Smaller Sized Objects. *Int. J. Astron. Astrophys.* **2018**, *8*, 6–23. [CrossRef]
64. Haug, E.G. The Gaussian Law of Gravitation under Collision Space-Time. Available online: https://hal.archives-ouvertes.fr/hal-03375883/document (accessed on 31 July 2022).
65. Haug, E.G. Can the Planck Length Be Found Independent of Big *G*? *Appl. Phys. Res.* **2017**, *9*, 58. [CrossRef]
66. Haug, E.G. Finding the Planck length multiplied by the speed of light without any knowledge of *G*, *c*, or *h*, using a Newton force spring. *J. Phys. Commun.* **2020**, *4*, 075001. [CrossRef]
67. Haug, E.G. Collision Space-Time: Unified Quantum Gravity. *Phys. Essays* **2020**, *33*, 46–78. [CrossRef]
68. Haug, E.G. Using a Grandfather Pendulum Clock to Measure the World's Shortest Time Interval, the Planck Time (With Zero Knowledge of G). *J. Appl. Math. Phys.* **2021**, *9*, 1076. [CrossRef]
69. Haug, E.G. Planck Speed: The Missing Speed of Physics? Absolute Still Without Breaking Lorentz Symmetry! *Eur. J. Appl. Phys.* **2022**, *4*, 11–18. [CrossRef]
70. Haug, E.G. Planck Length, Planck Time and Speed of Gravity When Taking into Account Relativistic Mass with No Knowledge of *G*, *h* or *c*. HAL Archives 2022. Available online: https://hal.archives-ouvertes.fr/hal-03669505/document (accessed on 16 May 2022).
71. Haug, E.G. The Huygens Formula is Exact for a Very Large Angle: The Moon-Earth System as a Gigantic Pendulum Clock. Working Paper Norwegian University of Life Sciences, HAL Archive. Available online: https://hal.archives-ouvertes.fr/hal-03334579/document (accessed on 4 September 2021).
72. Compton, A.H. A Quantum Theory of the Scattering of X-Rays by Light Elements. *Phys. Rev.* **1923**, *21*, 483. [CrossRef]
73. Walker, J.; Halliday, D.; Resnick, R. *Fundamentals of Physics*; Wiley: New York, NY, USA, 2007.
74. D'Auria, S. *Introduction to Nuclear and Particle Physics*; Springer: Cham, Switzerland, 2018.
75. Becker, P.; Bettin, H. The Avogadro constant: Determining the number of atoms in a single-crystal 28Si sphere. *Phil. Trans. R. Soc. A* **2011**, *369*, 3925–3935. [CrossRef] [PubMed]
76. Becker, P. The new kilogram definition based on counting the atoms in a ^{28}Si crystal. *Contemp. Phys.* **2012**, *53*, 461–479. [CrossRef]
77. Bartl, G.; Becker, P.; Beckhoff, B.; Bettin, H.; Beyer, E.; Borys, M.; Busch, I.; Cibik, L.; D'Agostino, G.; Darlatt, E.; et al. A new ^{28} Si Single Crystal: Counting the Atoms for the New Kilogram Definition. *Metrologica* **2017**, *54*, 693. [CrossRef]
78. Wang, Z.W.; Toikkanen, O.; Yin, F.; Li, Z.; Quinn, B.M.; Palmer, R.E. Counting the Atoms in Supported, Monolayer-Protected Gold Clusters. *J. Am. Chem. Soc.* **2010**, *132*, 2854–2855. [CrossRef]
79. Massam, E.; Mana, G. Counting Atoms. *Nat. Phys.* **2016**, *12*, 522. [CrossRef]
80. Haug, E.G. Extraction of the Planck Length From Cosmological redshift Without Knowledge off G or $\hbar$. *Int. J. Quantum Found. Suppl. Ser. Quantum Specul.* **2022**, *4*, 1.
81. Abbott, B.P.; Abbott, R.; Abbott, T.D.; Acernese, F.; Ackley, K.; Adams, C.; Adams, T.; Addesso, P.; Adhikari, R.X.; Adya, V.B.; et al. Gravitational Waves and Gamma-Rays from a Binary Neutron Star Merger: GW170817 and GRB 170817A. *Astrophys. J. Lett.* **2017**, *848*, L13. [CrossRef]
82. Haug, E.G. Quantum Gravity Hidden In Newton Gravity And How To Unify It With Quantum Mechanics. In *The Origin of Gravity from the First Principles*; Krasnoholovets, V., Ed.; NOVA Publishing: New York, NY, USA, 2021.
83. Haug, E.G. Unified Quantum Gravity Field Equation Describing the Universe from the Smallest to the Cosmological Scales. *Phys. Essays* **2022**, *35*, 61–71. [CrossRef]
84. Haug, E.G. Quantum Gravitational Energy Simplifies Gravitational Physics. HAL Archive, under Peer Review by Journal. 2022. Available online: https://hal.archives-ouvertes.fr/hal-03739599/ (accessed on 31 July 2022).
85. Milne, E.A. On the foundations of dynamics. *Proc. R. Soc. A* **1936**, *154*, 22–52.
86. Unzicker, A. *Einstein's Lost Key: How We Overlooked the Best Idea of the 20th Century*; CreateSpace Independent Publishing Platform: Scotts Valley, CA, USA, 2015.
87. Einstein, A. Cosmological Considerations in the General Theory of Relativity. *Sitzungsber. Preuss. Akad. Wiss Berl. Math. Phys.* **1917**, *1917*, 142–152.
88. Schrödinger, E. Die Erfüllbarkeit der Relativitätsforderung in der klassischen Mechanik. *Sitzungsber. Preuss. Akad. Wiss Berl. Math. Phys.* **1925**, *382*, 325–336. [CrossRef]
89. Bleksley, A. A new approach to cosmology: III. *S. Afr. J. Sci.* **1951**, *48*, 20.
90. Carvalho, J.C. Derivation of the mass of the observable universe. *Int. J. Theor. Phys. Vol.* **1995**, *34*, 2507–2509. [CrossRef]
91. Weinberg, S. *Gravitation and Cosmology*; Wiley: New York, NY, USA, 1972.
92. Hobson, M.P.; Efstathiou, G.; Lasenby, A.N. *General Relativity, An Introduction for Physicists*; Cambridge University Press: Cambridge, UK, 2014.

93. Guidry, M. *Modern General Relativity*; Cambridge University Press: Cambridge, UK, 2019.
94. Cook, N. *Quantum Gravity and the Standard Model*; Lulu Publisher: Morrisville, NC, USA, 2011.
95. Haug, E.G. A new full relativistic escape velocity and a new Hubble related equation for the universe. *Phys. Essays* **2021**, *34*, 502. [CrossRef]
96. Pathria, R.K. The Universe as a Black Hole. *Nature* **1972**, *240*, 298. [CrossRef]
97. Stuckey, W.M. The Observable Universe Inside a Black Hole. *Am. J. Phys.* **1994**, *62*, 788–795. [CrossRef]
98. Milgrom, M. A modification of the Newtonian dynamics as a possible alternative to the hidden mass hypothesis. *Astrophys. J.* **1983**, *270*, 365–370. [CrossRef]
99. McCulloch, M.E. Galaxy rotations from quantised inertia and visible matter only. *Astrophys. Space Sci.* **2017**, *362*, 149. [CrossRef]
100. Zee, B. Broken-Symmetric Theory of Gravity. *Phys. Rev. Lett.* **1979**, *42*, 417. [CrossRef]
101. Adler, S.L. Order-R Vacuum Action Functional in Scalar-Free Unified Theories with Spontaneous Scale Breaking. *Phys. Rev. Lett.* **1980**, *44*, 1567. [CrossRef]
102. Smolin, L. Towards a theory of spacetime structure at very short distances. *Nucl. Phys. B* **1979**, *160*, 253–268. [CrossRef]
103. Davies, P. Gauge theories, black hole evaporation and cosmic censorship. *Nucl. Phys. B* **1981**, *101*, 399–400. [CrossRef]
104. Pollock, M. On the proposed existence of an anti-gravity regime in the early universe. *Phys. Lett. B* **1982**, *108*, 386–388. [CrossRef]
105. Raychaudhuri, A.; Bagchi, B. Temperature dependent gravitational constant and black hole physics. *Phys. Lett. B* **1983**, *124*, 168–170. [CrossRef]
106. Masood-ul Alam, A.K.M. The variable Planck's constant due to imaginary gravitational temperature. *J. Physics Conf. Ser.* **2017**, *1051*, 012021. [CrossRef]
107. Poincaré, H. La mesure du temps. *Revue de Métaphysique et de Morale* **1898**, *6*, 1.
108. Spavieri, G.; Rodriguez, M.; Sanchez, A. Thought experiment discriminating special relativity from preferred frame theories. *J. Phys. Commun.* **2018**, *2*, 085009. [CrossRef]
109. Spavieri, G.; Rodriguez, M.; Sanchez, A. Testing absolute vs relative simultaneity with the spin-orbit interaction and the Sagnac effect. *Appl. Phys. Res.* **2019**, *11*, 59–69. [CrossRef]
110. Kipreos, E.T.; Balachandran, R.S. Optical data implies a null simultaneity test theory parameter in rotating frames. *Mod. Phys. Lett. A* **2021**, *36*, 2150131. [CrossRef]
111. Kipreos, E.T.; Balachandran, R.S. Assessment of the relativistic rotational transformations. *Mod. Phys. Lett. A* **2021**, *36*, 2150113. [CrossRef]
112. Tasson, J.D. What do we know about Lorentz invariance? *Rep. Prog. Phys.* **2014**, *77*, 062901. [CrossRef] [PubMed]

universe

MDPI

Article

Metric Gravity in the Hamiltonian Form—Canonical Transformations—Dirac's Modifications of the Hamilton Method and Integral Invariants of the Metric Gravity

Alexei M. Frolov

Department of Applied Mathematics, University of Western Ontario, London, ON N6H 5B7, Canada; alex1975frol@gmail.com

Abstract: Two different Hamiltonian formulations of the metric gravity are discussed and applied to describe a free gravitational field in the d dimensional Riemann space-time. Theory of canonical transformations, which relates equivalent Hamiltonian formulations of the metric gravity, is investigated in detail. In particular, we have formulated the conditions of canonicity for transformation between the two sets of dynamical variables used in our Hamiltonian formulations of the metric gravity. Such conditions include the ordinary condition of canonicity known in classical Hamilton mechanics, i.e., the exact coincidence of the Poisson (or Laplace) brackets which are determined for both the new and old dynamical Hamiltonian variables. However, in addition to this, any true canonical transformations defined in the metric gravity, which is a constrained dynamical system, must also guarantee the exact conservation of the total Hamiltonians H_t (in both formulations) and preservation of the algebra of first-class constraints. We show that Dirac's modifications of the classical Hamilton method contain a number of crucial advantages, which provide an obvious superiority of this method in order to develop various non-contradictory Hamiltonian theories of many physical fields, when a number of gauge conditions are also important. Theory of integral invariants and its applications to the Hamiltonian metric gravity are also discussed. For Hamiltonian dynamical systems with first-class constraints this theory leads to a number of peculiarities some of which have been investigated.

Keywords: gravity; Hamiltonian; momenta; constraints; Poisson brackets

1. Introduction

The main goal of this study is a deep analysis of canonical transformations of the Hamiltonian dynamical variables which are applied in the metric gravity. We investigate general properties of such transformations and formulate some criteria of canonicity. Another aim is to discuss modifications made by Dirac [1–3] in the classical Hamilton method. We want to show that Dirac's approach has many crucial advantages for the development and following improvement of various Hamiltonian formulations for the free gravitational field in metric gravity, where $2d$ additional gauge conditions exist. We also introduce the integral invariants of the metric gravity and discuss applications of these invariants to the current and new Hamiltonian formulations of the metric gravity.

In metric gravity the classical gravitational field is described as a symmetric tensor field which is defined in the d dimensional Riemann space-time. As is well known, the general theory of metric gravitational field(s) have been created more than 100 years ago by A. Einstein and it is based on his fundamental idea that the actual gravitational field has a tensor nature and it is defined in the four-dimensional (Riemann) space-time. Since then the gravitational field is designated as covariant components of the fundamental metric tensor $g_{\alpha\beta}$. In general, the metric gravity can be developed in the d dimensional space-time (or d-space), where $d \geq 3$, while the time is always one-dimensional. In other words, there is no need to restrict ourselves to the four-dimensional case only, where $d = 4$. In this study we also deal with the d-dimensional Riemann space-time. Everywhere below the notation $\overline{x}$, designates the d-dimensional vector which has the contravariant components x^{α}, while its covariant components are designated as x_{α}, where $\alpha = 0, 1, \ldots, d-1$. The temporal component (or time component) of the coordinate vector $\overline{x}$ is x^0, while its spatial components are x^k $(k = 1, 2, \ldots, d-1)$. The same rule is applied everywhere in this study: components of d-vectors are labelled by Greek indices, while spatial components of these vectors are denoted by Latin indices. In respect to this agreement all components of the covariant fundamental tensor (or metric tensor) are designated as $g_{\alpha\beta}$ (see above), while the notation $g^{\alpha\beta}$ stands for the components of contravariant fundamental (or metric) tensor [4,5]. The determinant of the $g_{\alpha\beta}$ is called the fundamental determinant and it is denoted by letter g. In the metric gravity the numerical value of g is always real and negative, but the $-g$ value is positive, which allows one to operate with the expression such as $\sqrt{-g}$ and consider functions of $\sqrt{-g}$. Any suffix with a comma before it denotes differentiation according to the general scheme $F_{,\mu} = \frac{dF}{dx^{\mu}}$. In particular, the temporal derivative are always designated as $F_{,0}\Big(= \frac{dF}{dx^0}\Big)$. For an arbitrary metric-dependent functional $F(g_{\alpha\beta})$ the notation $F_{,\gamma}$ means $F_{,\gamma} = \Big(\frac{\partial F}{\partial g_{\mu\nu}}\Big) g_{\mu\nu,\gamma}$, etc.

This paper has the following structure. The next two Sections play the role of 'introductory part' for our current analysis. In particular, in the next Section we introduce the regular $\Gamma - \Gamma$ Lagrangian of the metric gravity. Then, by using this $\Gamma - \Gamma$ Lagrangian we define the momenta $\pi^{\mu\nu}$ which are also the dynamical (Hamiltonian) variables. These momenta are considered as variables which are dynamically conjugate to the corresponding generalized coordinates $g_{\alpha\beta}$. In general, the momenta $\pi^{\mu\nu}$ are defined as the partial derivatives of the $\Gamma - \Gamma$ Lagrangian in respect to the 'velocities' $\frac{\partial g_{\mu\nu}}{\partial t} = g_{\mu\nu,0}$. The arising set of $2d$ dynamical variables $\{g_{\alpha\beta}, \pi^{\mu\nu}\}$ includes all variables which are needed to develop the Hamiltonian formulation of the metric gravity. In particular, both canonical H_C and total H_t Hamiltonians of the metric gravity are written as the quadratic functions of all space-like momenta π^{pq} and as a linear combination of temporal momenta $\pi^{0\mu}(= \pi^{\mu 0})$. This essentially means that the free gravitational field is a constrained dynamical (Hamiltonian) system [6,7] and all d temporal momenta $\pi^{0\mu}(= \pi^{\mu 0})$ of this field can rigorously be defined only as the primary constraints. The commutators (or Poisson brackets) of these primary constraints with the canonical and total Hamiltonians produce d secondary constraints. In general, the properly defined Poisson brackets always play a central role in Hamiltonian formulations of many physical theories. Briefly, we can say that the Poisson brackets is the most important working tool of any consistent Hamiltonian theory.

In Section 5 by using our Poisson brackets we derive the Hamiltonian equations of motion for all dynamical variables of the metric gravity. Analogous Hamilton equations for the primary and secondary first-class constraints are also derived and discussed. The Dirac closure of this Hamiltonian formulation is explicitly demonstrated. Then we consider another Hamiltonian formulation of the metric gravity developed by Dirac [1]. For this Hamiltonian formulation we also derive the corresponding Hamilton equations of motion and determine the Poisson brackets between all essential first-class constraints (see, also our Appendix A). Then we define the canonical transformations of the dynamical Hamiltonian variables which relate both these Hamiltonian formulations, i.e., [1,6]. The criteria of canonicity for arbitrary transformations of the dynamical variables are formulated in the Section 6. Then we discuss modifications made by Dirac in the classical Hamiltonian method [1–3]. We have shown explicitly that these Dirac's modifications allowed him to create a logically closed and transparent Hamiltonian approach which has many advantages to study actual motions of various physical fields when numerous gauge conditions must also be taken into account. Here we also formulate the new principle of the "complete reverse recovery" which must be applied to any Hamiltonian formulation of the metric gravity to check its validity and correct relations with the original Einstein's field equations. This simple and physically clear principle can be used to "separate the wheat from the chaff" (Matthew, 3:13) in Hamiltonian gravity. It allows us to operate only with the true Hamiltonian formulations of metric gravity and discard the fake ones.

In Section 8 we introduce integral invariants of the metric gravity. In some sense this is a central part of this study, since the method of integral invariants allows one to create a real foundation for the Hamiltonian formulations of the metric gravity. In particular, by applying the integral invariants of metric gravity one can easy perform all steps of the rigorous Hamiltonian procedure and formulate various criteria of canonicity which can be applied actual transformations of dynamical variables. Concluding remarks can be found in the last Section. This paper also includes three Appendixes. Appendix A is a 'pure technical' part, which contains derivations of some formulas. These formulas and expressions are of paramount importance for this study, but they could not be included in the main text, since this would damage the logic and harmony of our presentation. In Appendix B we discuss a number of tricky moments which traditionally complicate the correct definition of canonicity in classical and quantum mechanics. In Appendix C we explicitly derive some important formulas for the integral invariants. Along with the discussions of the latest achievements in Hamiltonian metric gravity, we also wanted to write a simple and transparent article which can be understood by any theoretical physicist who is familiar with the modern Hamiltonian methods developed for constrained dynamical systems.

2. Regular $\Gamma - \Gamma$ Lagrangian Density of the Metric Gravity

In this Section we introduce the regularized (or regular) Lagrangian density of metric gravity. As is well known (see, e.g., [8,9]) the original Lagrangian density of the metric gravity coincides with the integrand in the Einstein–Hilbert integral-action L_{EH} which equals to the product of scalar (or Gauss) curvature of the d-dimensional space $R = g^{\alpha\beta}R_{\alpha\beta}$ and factor $\sqrt{-g}$, which is Jacobian of the transformation from the flat space to the curved Riemann space (see, e.g., [8,9]). The invariant integral $\int R\sqrt{-g}d\Omega$ is called the gravitational action. The explicit form of Lagrangian density is $L_{EH} = \sqrt{-g}R = \sqrt{-g}g^{\alpha\beta}R_{\alpha\beta} = \sqrt{-g}g^{\alpha\beta}g^{\gamma\sigma}R_{\gamma\alpha\sigma\beta}$, where $R = g^{\alpha\beta}R_{\alpha\beta}$ is the scalar (or Gauss) curvature of d-dimensional space-time, while $R_{\alpha\beta}$ is the Ricci tensor

$$R_{\alpha\beta} = \frac{\partial\Gamma^{\gamma}_{\alpha\beta}}{\partial x^{\gamma}} - \frac{\partial\Gamma^{\gamma}_{\alpha\gamma}}{\partial x^{\beta}} + \Gamma^{\gamma}_{\alpha\beta}\Gamma^{\lambda}_{\gamma\lambda} - \Gamma^{\lambda}_{\alpha\gamma}\Gamma^{\gamma}_{\beta\lambda} \; , \; \text{or} \;\; R_{\alpha\beta} = g^{\mu\nu}R_{\mu\alpha\nu\beta} = g^{\nu\mu}R_{\nu\beta\mu\alpha} = R_{\beta\alpha} \tag{1}$$

In this equation and everywhere below in this study the notation $\Gamma^{\gamma}_{\alpha\beta} = \frac{1}{2}g^{\gamma\nu}\Big(\frac{\partial g_{\nu\alpha}}{\partial x^{\beta}} + \frac{\partial g_{\nu\beta}}{\partial x^{\alpha}} - \frac{\partial g_{\alpha\beta}}{\partial x^{\nu}}\Big)$ are the Cristoffel symbols of the second kind (see, e.g., [4]). The Ricci tensor $R_{\alpha\beta} = g^{\gamma\sigma}R_{\gamma\alpha\sigma\beta}$ is simply related to the Einstein tensor $G_{\alpha\beta} = g^{\gamma\sigma}R_{\alpha\gamma\sigma\beta}$, since $R_{\alpha\beta} = -G_{\alpha\beta}$ [4]. In this notation the governing equations of the free gravitational field (famous Einstein's equations) are written in one of the following forms: $R_{\alpha\beta} = 0 = G_{\alpha\beta}$. Any Hamiltonian formulation of the metric gravity must reproduce these original field equations exactly and unambiguously. This is the new fundamental principle of the "complete reverse recovery" and its applications to various new and old Hamiltonian formulations of the metric gravity allows one quickly "to separate the wheat from the chaff" (Mathew, 13, 24–30) (see below).

An alternative (but equivalent!) form of the same Lagrangian density is written as: $L_{EH} = \sqrt{-g}g^{\alpha\beta}g^{\gamma\sigma}R_{\gamma\alpha\sigma\beta}$, where the notation $R_{\alpha\beta\gamma\sigma}$ designates the Riemann curvature tensor (or Riemann-Cristoffel tensor) which is

$$R_{\alpha\beta\gamma\sigma} = \frac{1}{2}\Big[\frac{\partial^2 g_{\alpha\sigma}}{\partial x^{\beta}\partial x^{\gamma}} + \frac{\partial^2 g_{\beta\gamma}}{\partial x^{\alpha}\partial x^{\sigma}} - \frac{\partial^2 g_{\alpha\gamma}}{\partial x^{\beta}\partial x^{\sigma}} - \frac{\partial^2 g_{\beta\sigma}}{\partial x^{\alpha}\partial x^{\gamma}}\Big] + \Gamma_{\rho,\alpha\sigma}\Gamma^{\rho}_{\beta\gamma} - \Gamma_{\rho,\beta\sigma}\Gamma^{\rho}_{\alpha\gamma} \; , \tag{2}$$

where $\Gamma_{\lambda,\mu\nu} = \frac{1}{2}\Big(\frac{\partial g_{\nu\alpha}}{\partial x^{\beta}} + \frac{\partial g_{\nu\beta}}{\partial x^{\alpha}} - \frac{\partial g_{\alpha\beta}}{\partial x^{\nu}}\Big)$ are the Cristoffel symbols of the first kind. The Riemann–Cristoffel tensor defined in Equation (2) is a covariant tensor of the fourth rank. As follows from the last equation the Einstein–Hilbert Lagrangian density $L_{EH} =$

$\sqrt{-g}g^{\alpha\gamma}g^{\beta\sigma}R_{\alpha\beta\gamma\sigma}$ contains a number of second-order derivatives $\frac{\partial^2 g_{\alpha\beta}}{\partial x^\gamma \partial x^\lambda}$ and cannot be used directly in the principle of least action. However, as it follows from Equations (1) and (2) all these derivatives of the second order are included in the Lagrangian density L_{EH} only as a linear combination with the constant coefficients which do not contain any derivative of the metric tensor. Such a linearity of the invariant integral $\int R\sqrt{-g}d\Omega$ upon the second-order derivatives of the metric tensor can be used to transform this integral (by means of Gauss's theorem) to the integral which does not include any second-order derivative. After a few simple transformations the invariant integral S_g is reduced to the form

$$\int R\sqrt{-g}d\Omega = \int g^{\alpha\beta}\left(\Gamma^{\lambda}_{\alpha\gamma}\Gamma^{\gamma}_{\beta\lambda} - \Gamma^{\gamma}_{\alpha\beta}\Gamma^{\lambda}_{\gamma\lambda}\right)\sqrt{-g}d\Omega + \int \frac{\partial\left[\sqrt{-g}\left(g^{\alpha\beta}\Gamma^{\gamma}_{\alpha\beta} - g^{\alpha\gamma}\Gamma^{\beta}_{\alpha\beta}\right)\right]}{\partial x^\gamma}d\Omega\,, \tag{3}$$

where the integrand of the first integral in the right-hand side of this equation contains only products of different powers of components of the metric tensor and their first-order derivatives, while the second integral has the form of a divergence of the vector-like quantity $\sqrt{-g}\left(g^{\alpha\beta}\Gamma^{\gamma}_{\alpha\beta} - g^{\alpha\gamma}\Gamma^{\beta}_{\alpha\beta}\right)$. It is clear that the second integral can be transformed (with the help of Gauss's theorem) into an integral over a hyper-surface surrounding the d-dimensional volume over which the integration is carried out in other two integrals. When we vary the gravitational action S_g, the variation of this (second) term in the right-hand side of Equation (3) vanishes, since in respect to the principle of least action, the variation of the gravitational field at the limits of the region of integration must be equal zero.

Now, from Equation (3), one finds

$$\delta S_g = \delta\int L_{EH}d\Omega = \delta\int R\sqrt{-g}d\Omega = \delta\int L_{\Gamma-\Gamma}d\Omega\,, \text{ or } \frac{\delta L_{EH}}{\delta g_{\mu\nu}} = \frac{\delta\left(R\sqrt{-g}\right)}{\delta g_{\mu\nu}} = \frac{\delta L_{\Gamma-\Gamma}}{\delta g_{\mu\nu}} \tag{4}$$

where the notation δ means variation, while the notation $\frac{\delta F}{\delta g_{\mu\nu}}$ means the variational derivative (or Lagrange derivative) of the functional F. Furthermore, in this equation the symbol $L_{\Gamma-\Gamma} = \sqrt{-g}g^{\alpha\beta}\left(\Gamma^{\lambda}_{\alpha\gamma}\Gamma^{\gamma}_{\beta\lambda} - \Gamma^{\gamma}_{\alpha\beta}\Gamma^{\lambda}_{\gamma\lambda}\right)$ stands for the regularized (or regular) $\Gamma - \Gamma$ Lagrangian density of the metric gravity which plays a central role in numerous Hamiltonian approaches developed for the metric gravity. As follows from Equation (4) the variational derivative of the $L_{\Gamma-\Gamma}$ Lagrangian density is a true tensor, while the original $L_{\Gamma-\Gamma}$ Lagrangian density is not a true scalar. The equality, Equation (4), expresses the fact that we can replace the 'singular' Einstein–Hilbert Lagrangian density $L_{EH} = \sqrt{-g}R$ by the regular $\Gamma - \Gamma$ Lagrangian density $L_{\Gamma-\Gamma} = \sqrt{-g}g^{\alpha\beta}\left(\Gamma^{\lambda}_{\alpha\gamma}\Gamma^{\gamma}_{\beta\lambda} - \Gamma^{\gamma}_{\alpha\beta}\Gamma^{\lambda}_{\gamma\lambda}\right)$ which is variationally equivalent to the original Einstein–Hilbert Lagrangian density and contains no second-order derivative. This $\Gamma - \Gamma$ Lagrangian density is also written in the following form

$$\begin{aligned} L_{\Gamma-\Gamma} &= \frac{1}{4}\sqrt{-g}B^{\alpha\beta\gamma\mu\nu\rho}\left(\frac{\partial g_{\alpha\beta}}{\partial x^\gamma}\right)\left(\frac{\partial g_{\mu\nu}}{\partial x^\rho}\right) = \frac{1}{4}\sqrt{-g}B^{\alpha\beta\gamma\mu\nu\rho}g_{\alpha\beta,\gamma}g_{\mu\nu,\rho} \\ &= \frac{1}{4}\sqrt{-g}\left(g^{\alpha\beta}g^{\gamma\rho}g^{\mu\nu} - g^{\alpha\mu}g^{\beta\nu}g^{\gamma\rho} + 2g^{\alpha\rho}g^{\beta\nu}g^{\gamma\mu} - 2g^{\alpha\beta}g^{\gamma\mu}g^{\nu\rho}\right)g_{\alpha\beta,\gamma}g_{\mu\nu,\rho}\,, \end{aligned} \tag{5}$$

where $B^{\alpha\beta\gamma\mu\nu\rho}$ is a homogeneous cubic polynomial of the contravariant components of the metric tensor $g^{\alpha\beta}$. The explicit definition of the $B^{\alpha\beta\gamma\mu\nu\rho}$ quantities follows directly from Equation (5). Below, we shall deal with the $\Gamma - \Gamma$ Lagrangian density only. In order to simplify the following formulas we shall designate this Lagrangian density by the letter L, i.e., $L = L_{\Gamma-\Gamma}$. Now, by using the $\Gamma - \Gamma$ Lagrangian density we can derive the explicit expressions for all contravariant components of momenta $\pi^{\mu\nu}$ and obtain the closed expression for the Hamiltonian(s) of the metric gravity. These important steps are made in the next Section.

3. Momenta—Canonical and Total Hamiltonians of Metric Gravity

In the previous Section we have introduced the $\Gamma - \Gamma$ Lagrangian density, Equation (5), of the metric gravity. At the second step of any standard Hamiltonian procedure, by using the known Lagrangian density we have to define the corresponding momenta. Our current derivation of momenta in this study is based on the approaches developed in the two earlier papers [1,6] which still play a central role in all modern Hamiltonian formulations of the metric gravity. First, we need to re-write the formula, Equation (5), for the $\Gamma - \Gamma$ Lagrangian density to a slightly different form where all temporal derivatives of the covariant components of metric tensor, i.e., $g_{\alpha\beta,0}$, are explicitly separated from other similar derivatives (see, e.g., [6,10])

$$L = \frac{1}{4}\sqrt{-g}B^{\alpha\beta0\mu\nu0}g_{\alpha\beta,0}g_{\mu\nu,0} + \frac{1}{2}\sqrt{-g}B^{(\alpha\beta0|\mu\nu k)}g_{\alpha\beta,0}g_{\mu\nu,k} + \frac{1}{4}\sqrt{-g}B^{\alpha\beta k\mu\nu l}g_{\alpha\beta,k}g_{\mu\nu,l} \; , \quad (6)$$

where the notation $B^{(\alpha\beta\gamma|\mu\nu\rho)}$ means a 'symmetrical' $B^{\alpha\beta\gamma\mu\nu\rho}$ quantity which is symmetrized in respect to the permutation of two groups of indexes, i.e.,

$$\begin{aligned} B^{(\alpha\beta\gamma|\mu\nu\rho)} &= \frac{1}{2}\Big(B^{\alpha\beta\gamma\mu\nu\rho} + B^{\mu\nu\rho\alpha\beta\gamma}\Big) = g^{\alpha\beta}g^{\gamma\rho}g^{\mu\nu} - g^{\alpha\mu}g^{\beta\nu}g^{\gamma\rho} \\ &+ 2g^{\alpha\rho}g^{\beta\nu}g^{\gamma\mu} - g^{\alpha\beta}g^{\nu\rho}g^{\gamma\mu} - g^{\alpha\rho}g^{\beta\gamma}g^{\mu\nu} \; . \end{aligned} \quad (7)$$

The contravariant components of momentum $\pi^{\gamma\sigma}$ are defined as partial derivatives of the Lagrangian density, Equation (6), in respect to the corresponding velocities $g_{\gamma\sigma,0}$ (see, e.g., [3]). The expressions for the contravariant components of gravitational momenta (or momenta, for short) are

$$\pi^{\gamma\sigma} = \frac{\partial L}{\partial g_{\gamma\sigma,0}} = \frac{1}{2}\sqrt{-g}B^{((\gamma\sigma)0|\mu\nu0)}g_{\mu\nu,0} + \frac{1}{2}\sqrt{-g}B^{((\gamma\sigma)0|\mu\nu k)}g_{\mu\nu,k} \; , \quad (8)$$

where $B^{((\gamma\sigma)0|\mu\nu0)} = \frac{1}{2}\Big(B^{(\gamma\sigma0|\mu\nu0)} + B^{(\sigma\gamma0|\mu\nu0)}\Big)$ is the symmetrized linear combination of the two $B^{(\gamma\sigma0|\mu\nu0)}$ quantities. The first term in the right-hand side of this equation is written in the form

$$\frac{1}{2}\sqrt{-g}B^{((\gamma\sigma)0|\mu\nu0)}g_{\mu\nu,0} = \frac{1}{2}\sqrt{-g}g^{00}\Big(e^{\mu\nu}e^{\gamma\sigma} - e^{\mu\gamma}e^{\nu\sigma}\Big)g_{\mu\nu,0} = \frac{1}{2}\sqrt{-g}g^{00}E^{\mu\nu\gamma\sigma}g_{\mu\nu,0} \; , \quad (9)$$

where the notations $e^{\mu\nu}$ and $E^{\mu\nu\gamma\sigma}$ stands for the Dirac contravariant tensors of the second and fourth ranks, respectively. The explicit expressions for these tensors are

$$e^{\mu\nu} = g^{\mu\nu} - \frac{g^{0\mu}g^{0\nu}}{g^{00}} \; , \text{ and } \; E^{\mu\nu\gamma\rho} = e^{\mu\nu}e^{\gamma\rho} - e^{\mu\gamma}e^{\nu\rho} \; , \quad (10)$$

i.e., each component of these two tensors is a function of the contravariant components of the metric tensor only. For these tensors one finds the following symmetries in respect to permutations of their indexes: $e^{\mu\nu} = e^{\nu\mu}$ and $E^{\mu\nu\gamma\sigma} = E^{\gamma\sigma\mu\nu}$. Furthermore, as follows directly from the formulas, Equation (10), the tensor $e^{\mu\nu}$ equals zero, if either index μ, or index ν (or both) equals zero. Analogously, for the Dirac $E^{\mu\nu\gamma\sigma}$ tensor one finds that $E^{0\nu\gamma\sigma} = 0$, $E^{\mu0\gamma\sigma} = 0$, $E^{\mu\nu0\sigma} = 0$ and $E^{\mu\nu\gamma0} = 0$. Therefore, it is more productive to discuss the space-like quantities e^{mn} and E^{mnpq} only.

The space-like E^{pqkl} quantity is, in fact, the space-like Dirac tensor of the fourth rank. This space-like tensor E^{pqkl} do not have components which are equal zero identically. Furthermore, this space-like tensor E^{pqkl} is a positively defined, invertible tensor. Its inverse space-like tensor I_{mnpq} is also positively defined and invertible space-like tensor of the fourth rank which is written in the form [6]

$$I_{mnqp} = \frac{1}{d-2}g_{mn}g_{pq} - g_{mp}g_{nq} \; . \quad (11)$$

The relation between space-like tensors I_{mnpq} and E^{pqkl} is written in the form $I_{mnpq}E^{pqkl} = g^k_m g^l_n = \delta^k_m \delta^l_n$ (see, Appendix A), where $g^\alpha_\beta = \delta^\alpha_\beta$ is the substitution tensor [4] and the symbol δ^α_β denotes the Kroneker delta ($\delta^\alpha_\alpha = 1$ and $\delta^\alpha_\beta = 0$, if $\alpha \neq \beta$). From this definition one easily finds that $\delta^\alpha_\beta = \delta^\beta_\alpha$.

In general, for the $B^{((\gamma\sigma)0|\mu\nu 0)}$ coefficients in the formula, Equation (8) one finds from Equations (9)–(11) two possible situations. First, for $\gamma = p$ and $\sigma = q$ these coefficients are always different from zero. In this 'regular' case we obtain the following formula for space-like contravariant components of the momentum tensor

$$\begin{aligned} \pi^{pq} &= \frac{\partial L}{\partial g_{pq,0}} = \frac{1}{2}\sqrt{-g}B^{((pq)0|\mu\nu 0)}g_{\mu\nu,0} + \frac{1}{2}\sqrt{-g}B^{((pq)0|\mu\nu k)}g_{\mu\nu,k} \\ &= \frac{1}{2}\sqrt{-g}B^{((pq)0|mn0)}g_{mn,0} + \frac{1}{2}\sqrt{-g}B^{((pq)0|mnk)}g_{mn,k} \end{aligned} \quad (12)$$

for each pair of the spatial (pq)-indexes. In this case the $(pq;mn)$-matrix of the $\sqrt{-g}B^{((pq)0|mn0)} = g^{00}E^{pqmn}$ coefficients, which are located in front of the space-like $g_{mn,0}$ velocities in the right-hand side of this equation, is invertible (see above). Therefore, in this case the field-velocity $g_{mn,0}$ can be expressed as the linear combination of the space-like components π^{pq} of momentum tensor, Equation (11):

$$\begin{aligned} g_{mn,0} &= \frac{1}{g^{00}}\Big(\frac{2}{\sqrt{-g}}I_{mnpq}\pi^{pq} - I_{mnpq}B^{((pq)0|\mu\nu k)}g_{\mu\nu,k}\Big) = \frac{1}{g^{00}}I_{mnpq}\Big(\frac{2}{\sqrt{-g}}\pi^{pq} \\ &- B^{((pq)0|\mu\nu k)}g_{\mu\nu,k}\Big) \ , \end{aligned} \quad (13)$$

where the Dirac tensor I_{mnpq} is defined by Equation (11). As follows from Equations (12) and (13) for all space-like components of the metric tensor g_{pq} and corresponding momenta π^{mn} one essentially finds no principal difference with those systems in classical mechanics which have Lagrangians written as quadratic functions of the velocities. Indeed, in metric gravity all space-like components of momenta and velocities are always related to each other by a few simple, linear equations, which however, take a multi-dimensional, or matrix form. The method described above is the direct and transparent generalization of Legendre's dual transformation for the tensor fields.

In the second 'singular' case, when $\gamma = 0$ (or $\sigma = 0$) in Equation (8), the first term in the right-hand side of each of these equations equals zero. Therefore, these equations take the from of pure algebraic equations

$$\pi^{0\sigma} = \frac{\partial L}{\partial g_{0\sigma,0}} = \frac{1}{2}\sqrt{-g}B^{((0\sigma)0|\mu\nu k)}g_{\mu\nu,k} \ , \ \text{and} \ \ \pi^{\sigma 0} = \frac{\partial L}{\partial g_{\sigma 0,0}} = \frac{1}{2}\sqrt{-g}B^{((\sigma 0)0|\mu\nu k)}g_{\mu\nu,k} \quad (14)$$

for $\sigma = 0, 1, \ldots, d-1$. From this equations one finds finds $\pi^{0\sigma} = \pi^{\sigma 0}$. Note also that these equations contain no velocities at all, i.e., we cannot express the $g_{0\sigma,0}$ velocities in terms of the momenta $\pi^{0\sigma}$ and vice versa. Each of the equations, Equation (14), directly determines the momentum $\pi^{0\sigma}$ as a cubic polynomial of the contravariant components of the metric tensor $g^{\alpha\beta}$ which is multiplied by an additional factor $\sqrt{-g}g_{\mu\nu,k}$. In other words, the following d functions

$$\phi^{0\sigma} = \pi^{0\sigma} - \frac{1}{2}\sqrt{-g}B^{((0\sigma)0|\mu\nu k)}g_{\mu\nu,k} = \phi^{\sigma 0} \ , \quad (15)$$

where $\sigma = 0, 1, \ldots, d-1$, must be equal zero during actual physical motions (or time-evolution) of the free gravitational field. In other words, during any actual motion (or time-evolution) of the free gravitational field the d additional conditions (or constraints) $\phi^{0\sigma} = 0$ must be obeyed for the Hamiltonian dynamical variables, since otherwise such a motion is not possible. We have to emphasize that the equations $\phi^{0\sigma} = 0$ are correct only on the true Hamiltonian trajectories (or curves) $(x^0, g_{\alpha\beta}(x^0), \pi^{\mu\nu}(x^0))$ of the free gravitational field. Outside these trajectories, only some, or even none of these equations are satisfied. In [1] Dirac proposed to write these additional conditions in the symbolic form $\phi^{0\sigma} \approx 0$ (for $\sigma = 0, 1, \ldots, d-1$) with a different sign $\approx$ from the usual (=). These 'weak' equations are the primary constraints of the given Hamiltonian formulation (see, e.g., [1,3]). In other words, during time-evolution of the free gravitational field we always have d primary constraints for the $d(d+1)$ Hamiltonian variables $\{g_{\alpha\beta}, \pi^{\mu\nu}\}$ and this number d never changes if one applies canonical transformations of the Hamiltonian dynamical variables (see below). This is a partial case of the law of inertia for the first-class constraints which is discussed below.

Now, by applying the Legendre transformation to the known $\Gamma - \Gamma$ Lagrangian density L, of the metric gravity, Equation (6), and excluding all space-like velocities $g_{mn,0}$, we can derive the explicit formulas for the total H_t and canonical H_C Hamiltonians of the metric GR. Formally, these quantities are the Hamiltonian densities, but in this study we try to avoid any mention of Hamiltonian densities, since constant play with words 'Hamiltonians' and 'Hamiltonian densities' substantially complicates explanations and often leads to various confusions. In particular, the total Hamiltonian H_t of the gravitational field in metric gravity derived from the $\Gamma - \Gamma$ Lagrangian density L, Equation (5), is written in the form

$$\begin{aligned} H_t &= g_{\alpha\beta,0}\pi^{\alpha\beta} - L = g_{pq,0}\pi^{pq} + g_{0\sigma,0}\phi^{0\sigma} - L = g_{pq,0}\pi^{pq} - L + g_{0\sigma,0}\phi^{0\sigma} \\ &= H_C + g_{00,0}\phi^{00} + g_{0k,0}\phi^{0k} + g_{k0,0}\phi^{k0} = H_C + g_{00,0}\phi^{00} + 2g_{0k,0}\phi^{0k}, \end{aligned} \tag{16}$$

where $\phi^{0\sigma} = \pi^{0\sigma} - \frac{1}{2}\sqrt{-g}B^{((0\sigma)0|\mu\nu k)}g_{\mu\nu,k}$ are the primary constraints, while $g_{0\sigma,0}$ are the σ velocities (or temporal velocities) and H_C is the canonical Hamiltonian of the metric gravity

$$\begin{aligned} H_C &= \frac{1}{\sqrt{-g}g^{00}} I_{mnpq}\pi^{mn}\pi^{pq} - \frac{1}{g^{00}} I_{mnpq}\pi^{mn}B^{(pq0|\mu\nu k)}g_{\mu\nu,k} \\ &+ \frac{1}{4}\sqrt{-g}\Big[\frac{1}{g^{00}} I_{mnpq}B^{((mn)0|\mu\nu k)}B^{(pq0|\alpha\beta l)} - B^{\mu\nu k\alpha\beta l}\Big]g_{\mu\nu,k}g_{\alpha\beta,l}\ , \end{aligned} \tag{17}$$

which does not contain any primary constraint $\phi^{0\sigma}$. The total Hamiltonian $H_t = H_C + g_{0\sigma,0}\phi^{0\sigma}$ is a scalar function, which is defined in the $d(d+1)$ even-dimensional phase space $\left\{g_{\alpha\beta}, \pi^{\mu\nu}\right\}$, where all components of the metric $g_{\alpha\beta}$ and momentum $\pi^{\mu\nu}$ tensors have been chosen as the basic Hamiltonian variables. The corresponding $d(d+1)$ dimensional space of Hamiltonian variables must be endowed with a symplectic (or anti-symmetric) bilinear form (Poisson brackets), which turns this space into a symplectic, even-dimensional phase space. The definition of Poisson brackets between all basic dynamical variables, i.e., between coordinates $g_{\alpha\beta}$ and momenta $\pi^{\mu\nu}$, is discussed in the next Section. At the same time the spatial (covariant) components of the metric tensor g_{mn} and spatial (contravariant) components of momenta π^{pq} form another $d(d-1)$ dimensional space, which is also transformed (by the same Poisson brackets) into a symplectic, even-dimensional phase space of the $d(d-1)$ space-like Hamiltonian variables $\left\{g_{mn}, \pi^{pq}\right\}$.

4. Poisson Brackets

In general, the Poisson brackets (or PB, for short) are the fundamental and crucially important tools of any correct Hamiltonian theory. The correct definition of these Poisson brackets is the central part of numerous Hamiltonian formulations developed for different physical systems of particles, fields and their combinations. As is well known (see, e.g., [11–14]) the Poisson bracket is an antisymmetric, bi-linear form defined in the $2M$-dimensional phase space which is, in fact, a cotangent space to an M-dimensional manifold located in the position space. More accurate definition of the Poisson brackets can be found, e.g., in [14]. For arbitrary vectors X, Y, Z from this phase space we can define the bi-linear form which is designated below as $[X, Y]$ and it obeys the four following rules (or axioms)

$$[X, Y] = -[Y, X] \quad \text{(antisymmetry)} ,$$
$$[a_1 X_1 + a_2 X_2, Y] = a_1 [X_1, Y] + a_2 [X_2, Y] \quad \text{(linearity in either member)} ,$$
$$[XY, Z] = [X, Z]Y + X[Y, Z] \quad \text{(the product law)} ,$$
$$[X, [Y, Z]] + [Y, [Z, X]] + [Z, [X, Y]] = 0 \quad \text{(Jacobi identity)} ,$$

where each of the X, Y and Z vectors belongs to the $2M$ dimensional phase space. In metric gravity the $d(d+1)$-dimensional phase space includes the $\frac{d(d+1)}{2}$ generalized coordinates $g_{\alpha\beta}$ and $\frac{d(d+1)}{2}$ momenta $\pi^{\mu\nu}$, or contravariant components of the momentum tensor π. An additional (or fifth) fundamental rule for the Poisson brackets, which is often called the 'time-evolution' of the Poisson bracket, is written in the form

$$\frac{\partial}{\partial t}[X, Y] = [\frac{\partial X}{\partial t}, Y] + [X, \frac{\partial Y}{\partial t}] \ , \ \text{and} \ \ \frac{\partial}{\partial b}[X, Y] = [\frac{\partial X}{\partial b}, Y] + [X, \frac{\partial Y}{\partial b}]$$

where t is the temporal variable (or time, for short), while b is an arbitrary numerical parameter.

In the Hamiltonian version of metric gravity the basic dynamical variables are the generalized 'coordinates' $g_{\alpha\beta}$ and momenta $\pi^{\mu\nu}$ defined above. In respect to this the Poisson brackets between the two functions of these dynamical variables are defined as follows:

$$[f_1, f_2] = \frac{\partial f_1}{\partial g_{\alpha\beta}} \frac{\partial f_2}{\partial \pi^{\alpha\beta}} - \frac{\partial f_2}{\partial g_{\alpha\beta}} \frac{\partial f_1}{\partial \pi^{\alpha\beta}} = \frac{\partial f_1}{\partial g_{\alpha\beta}} \frac{\partial f_2}{\partial \pi^{\alpha\beta}} - \frac{\partial f_1}{\partial \pi^{\alpha\beta}} \frac{\partial f_2}{\partial g_{\alpha\beta}} \ . \quad (18)$$

The Poisson brackets between the generalized coordinates and momenta have the fundamental value for the purposes of this study. They are:

$$[g_{\alpha\beta}, \pi^{\mu\nu}] = \frac{\partial g_{\alpha\beta}}{\partial g_{\gamma\sigma}} \frac{\partial \pi^{\mu\nu}}{\partial \pi^{\gamma\sigma}} - \frac{\partial g_{\alpha\beta}}{\partial \pi^{\gamma\sigma}} \frac{\partial \pi^{\mu\nu}}{\partial g_{\gamma\sigma}} = \Delta^{\mu\nu}_{\alpha\beta} = \frac{1}{2}\left(g^{\mu}_{\alpha} g^{\nu}_{\beta} + g^{\nu}_{\alpha} g^{\mu}_{\beta}\right) = \frac{1}{2}\left(\delta^{\mu}_{\alpha} \delta^{\nu}_{\beta} + \delta^{\nu}_{\alpha} \delta^{\mu}_{\beta}\right) \ , \quad (19)$$

where $g^{\mu}_{\alpha} = \delta^{\mu}_{\alpha} (= \delta^{\alpha}_{\mu})$ is the substitution tensor [4] and symbol δ^{μ}_{β} is the Kronecker delta, while the notation $\Delta^{\mu\nu}_{\alpha\beta}$ stands for the gravitational (or tensor) delta-symbol. The three following properties of this delta-symbol are obvious and very useful in calculations of many Poisson brackets: (1) 'horizontal' index symmetry $\Delta^{\mu\nu}_{\alpha\beta} = \Delta^{\nu\mu}_{\alpha\beta} = \Delta^{\mu\nu}_{\beta\alpha} = \Delta^{\nu\mu}_{\beta\alpha}$, (2) 'vertical' index symmetry $\Delta^{\mu\nu}_{\alpha\beta} = \Delta^{\alpha\nu}_{\mu\beta} = \Delta^{\mu\alpha}_{\nu\beta} = \ldots = \Delta^{\alpha\beta}_{\mu\nu}$, and (3) the product property: $\Delta^{\nu\mu}_{\rho\sigma} \Delta^{\rho\sigma}_{\alpha\beta} = \Delta^{\mu\nu}_{\alpha\beta}$. By using these properties we can write that $[g_{\alpha\beta}, \pi^{\mu\nu}] = \Delta^{\mu\nu}_{\alpha\beta} = \Delta^{\alpha\beta}_{\mu\nu} = [g_{\mu\nu}, \pi^{\alpha\beta}]$. Note again that the total number of dynamical Hamiltonian variables in metric gravity is always even and equals $d(d+1)$. The Poisson bracket, Equation (19), explains why in some papers the gravitational momenta $\pi^{\mu\nu}$ are called and considered as conjugate dynamical variables for the corresponding covariant components of metric tensor $g_{\mu\nu}$ (our coordinates).

Other fundamental Poisson brackets between basic dynamical variables of the metric gravity equal zero identically, i.e., $[g_{\alpha\beta}, g_{\mu\nu}] = 0$ and $[\pi^{\alpha\beta}, \pi^{\mu\nu}] = 0$. In general, our dynamical variables depend upon one temporal and $(d-1)$ spatial coordinates $x^0, x^1, \ldots, x^{d-1} = (x_0, \overline{x})$. In this case we have to apply the following definition of the Poisson brackets, e.g.,

$$[g_{\alpha\beta}(\bar{x}, t), \pi^{\mu\nu}(\bar{x}', t)] = \Delta^{\mu\nu}_{\alpha\beta} \delta^{d-1}(\bar{x} - \bar{x}') \ , \quad (20)$$

where $\delta^{d-1}(\bar{y})$ is the usual delta-function in the position $(d-1)$-space. Such a generalization of the Poisson brackets is straightforward and simple, but in this study we do not want to complicate our system of notations. In respect to this, below we shall always deal with the Poisson brackets of two quantities taken at the same spatial point.

The explicit form of the fundamental Poisson brackets, Equation (19), allows one to derive the following formulas for slightly different Poisson brackets

$$[g^{\alpha\beta}, \pi^{\mu\nu}] = -\frac{1}{2}\Big(g^{\alpha\mu}g^{\beta\nu} + g^{\alpha\nu}g^{\beta\mu}\Big) = -g^{\alpha\gamma}\Delta^{\mu\nu}_{\gamma\sigma}g^{\beta\sigma} = -[\pi^{\mu\nu}, g^{\alpha\beta}] \text{ and } [g^{\alpha\beta}, g_{\mu\nu}] = 0. \quad (21)$$

which contain the contravariant components of the metric tensor $g^{\alpha\beta}$. Now, by using the Poisson brackets, Equations (19) and (21), defined above we can determine the Poisson brackets of more complicated quantities and functions. As the first example we calculate the following Poisson bracket

$$[g_{\alpha\beta}g^{\lambda\sigma}, \pi^{\mu\nu}] = [g_{\alpha\beta}, \pi^{\mu\nu}]g^{\lambda\sigma} + g_{\alpha\beta}[g^{\lambda\sigma}, \pi^{\mu\nu}] = \Delta^{\mu\nu}_{\alpha\beta}g^{\lambda\sigma} - \frac{1}{2}g_{\alpha\beta}\Big(g^{\lambda\mu}g^{\sigma\nu} + g^{\lambda\nu}g^{\sigma\mu}\Big) . \quad (22)$$

Let us assume that in this formula $\lambda = \beta$. In this case $g_{\alpha\beta}g^{\beta\sigma} = g^{\sigma}_{\alpha} = \delta^{\sigma}_{\alpha}$ and it is clear that $[g^{\sigma}_{\alpha}, \pi^{\mu\nu}] = 0$. On the other hand, if $\lambda = \beta$, then for the right-hand side of Equation (22) one finds:

$$\Delta^{\mu\nu}_{\alpha\beta}g^{\beta\sigma} - \frac{1}{2}g_{\alpha\beta}\Big(g^{\beta\mu}g^{\sigma\nu} + g^{\beta\nu}g^{\sigma\mu}\Big) = \frac{1}{2}\Big(\delta^{\mu}_{\alpha}g^{\nu\sigma} + \delta^{\nu}_{\alpha}g^{\mu\sigma}\Big) - \frac{1}{2}\Big(\delta^{\mu}_{\alpha}g^{\nu\sigma} + \delta^{\nu}_{\alpha}g^{\mu\sigma}\Big) = 0 , \quad (23)$$

which means that for $\lambda = \beta$ the equation, Equation (22), is written in the form $0 = 0$ and we have no contradiction here. Now, consider the following Poisson bracket $[g_{\alpha\beta}g^{\alpha\beta}, \pi^{\mu\nu}]$. As is well known from tensor calculus (see, e.g., [4]) $g_{\alpha\beta}g^{\alpha\beta} = d$, where d is the dimension of tensor space. Therefore, this Poisson bracket is reduced to the equation

$$0 = [g_{\alpha\beta}, \pi^{\mu\nu}]g^{\alpha\beta} + g_{\alpha\beta}[g^{\alpha\beta}, \pi^{\mu\nu}] \text{ or } \Delta^{\mu\nu}_{\alpha\beta}g^{\alpha\beta} - \frac{1}{2}g_{\alpha\beta}\Big(g^{\alpha\mu}g^{\beta\nu} + g^{\alpha\nu}g^{\beta\nu}\Big) = 0 ,$$

which is easily transformed to an obvious identity $g^{\mu\nu} - g^{\nu\mu} = 0$. Analogously, it is easy to find a number of remarkable relations between the Poisson brackets $[g_{\alpha\beta}, \pi^{\mu\nu}]$ and $[g^{\alpha\beta}, \pi^{\mu\nu}]$, temporal derivatives of the covariant and contravariant components of the metric tensor and Poisson brackets of these components with the canonical Hamiltonian H_C which directly follows from Equation (22) (see, also our 'technical' Appendix A):

$$[g^{\sigma\gamma}, \pi^{\mu\nu}] = -g^{\alpha\sigma}[g_{\alpha\beta}, \pi^{\mu\nu}]g^{\beta\gamma} = -g^{\alpha\sigma}\Delta^{\mu\nu}_{\alpha\beta}g^{\beta\gamma} = -\frac{1}{2}\Big(g^{\sigma\mu}g^{\gamma\nu} + g^{\sigma\nu}g^{\gamma\mu}\Big) , \quad (24)$$

$$\frac{dg^{\sigma\gamma}}{dt} = -g^{\alpha\sigma}\Big(\frac{dg_{\alpha\beta}}{dt}\Big)g^{\beta\gamma} \;, \quad [g^{\sigma\gamma}, H_C] = -g^{\alpha\sigma}[g_{\alpha\beta}, H_C]g^{\beta\gamma} . \quad (25)$$

The second example is of great interest for analytical calculations of the Poisson brackets between components of momenta and some functions of the coordinates only. Let $F(g_{\mu\nu}, g^{\lambda\sigma})$ be an arbitrary analytical function (or functional) of the co- and contravariant components of the metric tensor. The Poisson bracket of this $F(g_{\mu\nu}, g^{\lambda\sigma})$ function and components of momentum tensor $\pi^{\alpha\beta}$ takes the form

$$\begin{aligned}
[\pi^{\alpha\beta}, F(g_{\mu\nu}, g^{\lambda\sigma})] &= -\frac{\partial F}{\partial g_{\mu\nu}}[g_{\mu\nu}, \pi^{\alpha\beta}] - \frac{\partial F}{\partial g^{\sigma\lambda}}[g^{\sigma\lambda}, \pi^{\alpha\beta}] \\
&= \frac{1}{2}\Big(\frac{\partial F}{\partial g^{\sigma\lambda}}\Big)\Big(g^{\sigma\alpha}g^{\lambda\beta} + g^{\sigma\beta}g^{\lambda\alpha}\Big) - \Big(\frac{\partial F}{\partial g_{\mu\nu}}\Big)\Delta^{\alpha\beta}_{\mu\nu} . \quad (26)
\end{aligned}$$

In particular, for the $F(g_{\mu\nu})$ function (or functional) of the covariant components of the metric tensor only this Poisson bracket can be written in the form

$$[\pi^{\alpha\beta}, F(g_{\mu\nu})] = -\frac{\partial F}{\partial g_{\mu\nu}}[g_{\mu\nu}, \pi^{\alpha\beta}] = -\Big(\frac{\partial F}{\partial g_{\mu\nu}}\Big)\Delta^{\alpha\beta}_{\mu\nu} . \quad (27)$$

In the case when $F = \Phi(g_{\mu\nu})g_{\rho\sigma,k}$ we obtain

$$[\pi^{\alpha\beta}, \Phi(g_{\mu\nu})g_{\rho\sigma,k}] = -\Big(\frac{\partial\Phi}{\partial g_{\mu\nu}}\Big)\Delta^{\alpha\beta}_{\mu\nu}g_{\rho\sigma,k} - \Delta^{\alpha\beta}_{\rho\sigma}\Big(\Phi\Big)_{,k} = -\Big(\frac{\partial\Phi}{\partial g_{\alpha\beta}}\Big)g_{\rho\sigma,k} - \Big(\Phi\Big)_{,k}\Delta^{\alpha\beta}_{\rho\sigma} \quad (28)$$

Here we have applied the integration by parts (see, also discussion at the end of this Section). The third example includes Poisson brackets between space-like components of the momenta π^{mn} and components of the Dirac space-like tensor e_{pq} and/or e^{pq}. They are

$$[\pi^{mn}, e_{pq}] = -\Delta^{mn}_{pq} \quad , \quad \text{and} \quad [\pi^{mn}, e^{pq}] = \frac{1}{2}\Big(g^{pm}g^{qn} + g^{pn}g^{qm}\Big) \quad . \quad (29)$$

As follows from the first PB in Equation (29) and two other groups of Poisson brackets: $[\pi^{mn}, \pi^{pq}] = 0$, $[e_{mn}, e_{pq}] = 0$ these $d(d-1)$ variables e_{mn} and π^{pq} are the canonical Hamiltonian variables in the $d(d-1)$-dimensional subspace of space-like dynamical variables of metric gravity. These PB play an important role in this study (see below).

From the formula, Equation (27), one also finds

$$[\pi^{\alpha\beta}, F(g_{\mu\nu})] = -\Big(\frac{\partial F}{\partial g_{\mu\nu}}\Big)\Delta^{\alpha\beta}_{\mu\nu} = -\Big(\frac{\partial F}{\partial g_{\alpha\beta}}\Big)\Delta^{\mu\nu}_{\alpha\beta} = [\pi^{\mu\nu}, F(g_{\alpha\beta})] \quad , \quad (30)$$

or simply $[\pi^{\alpha\beta}, F(g_{\mu\nu})] = [\pi^{\mu\nu}, F(g_{\alpha\beta})]$. The principal moment here is the presence of tensor Δ-symbol in these Poisson brackets. This equality simplifies calculations of a large number of Poisson brackets which are needed to show canonicity of different sets of Hamiltonian dynamical variables. Moreover, by using the formula, Equation (27), we can determine another group of important Poisson brackets between momenta and analytical functions of the fundamental determinant g and its square root $\sqrt{g}$ (or $\sqrt{-g}$ as it is designated in the metric gravity). The general expression for the Poisson bracket between such a $F(g)$ function and $\pi^{\alpha\beta}$ is derived as follows

$$[F(g), \pi^{\alpha\beta}] = \Big(\frac{\partial F}{\partial g}\Big)\Big(\frac{\partial g}{\partial g_{\mu\nu}}\Big)[g_{\mu\nu}, \pi^{\alpha\beta}] = \Big(\frac{\partial F}{\partial g}\Big)gg^{\mu\nu}\Delta^{\alpha\beta}_{\mu\nu} = \Big(\frac{\partial F}{\partial g}\Big)gg^{\alpha\beta} \quad . \quad (31)$$

or $[\pi^{\alpha\beta}, F(g)] = -\Big(\frac{\partial F}{\partial g}\Big)gg^{\alpha\beta}$. Now, if $F(g) = g^x$, then one finds $[\pi^{\alpha\beta}, F(g)] = -xF(g)g^{\alpha\beta}$. In particular, if $F(g) = \sqrt{-g}$ and $F(g) = \frac{1}{\sqrt{-g}}$ we obtain from the last equation

$$[\sqrt{-g}, \pi^{\alpha\beta}] = -\frac{1}{2\sqrt{-g}}gg^{\alpha\beta} = \frac{1}{2}\sqrt{-g}g^{\alpha\beta} \quad \text{and} \quad [\frac{1}{\sqrt{-g}}, \pi^{\alpha\beta}] = -\frac{1}{2\sqrt{-g}}g^{\alpha\beta} \quad , \quad (32)$$

respectively. Another Poisson bracket is often needed in operations with both primary and secondary constraints:

$$[\pi^{0\gamma}, \frac{g^{\sigma\lambda}}{g^{00}}] = \frac{1}{2g^{00}}\Big(g^{0\sigma}g^{\gamma\lambda} + g^{\gamma\sigma}g^{0\lambda} - 2g^{\sigma\lambda}g^{0\gamma}\Big) \quad . \quad (33)$$

If $\lambda = 0$ here, then one finds

$$[\pi^{0\gamma}, \frac{g^{\sigma 0}}{g^{00}}] = \frac{1}{2g^{00}}\Big(g^{\gamma\sigma}g^{00} - g^{0\sigma}g^{0\gamma}\Big) \quad . \quad (34)$$

Now, if we assume that $\sigma = 0$ here, then this Poisson bracket equals zero identically (as expected).

To conclude our discussion of the Poisson brackets let us make the two following remarks. First, as it was shown in [6,15] in metric gravity the Poisson bracket(s) between two primary constraints $\phi^{0\sigma}$ and $\phi^{0\gamma}$, Equation (15), are always equal zero, i.e., $[\phi^{0\sigma}, \phi^{0\gamma}] = 0$. In fact, the explicit derivation of this formula is very good and relatively simple exercise in calculations of Poisson brackets (see, also discussion below). Thus, in the metric gravity all primary constraints commute with each other. This drastically simplifies many important steps of our procedure which is described below. Second, we have to explain calculations of the Poisson brackets between momenta and expressions which include some spatial derivatives of the metric tensor such as $\Phi(g_{\mu\nu})g_{\gamma\lambda,k}$, $\Phi(g_{\mu\nu})g_{\gamma\lambda,k}g_{\rho\sigma,m}$, $g_{\gamma\lambda,k}g_{\rho\sigma,m}$, etc. These and other similar expressions arise very often in actual Hamiltonian formulations, and they can be found, e.g., in operations with both canonical and total Hamiltonians, primary and secondary constraints and other expressions. Analytical calculations of such Poisson brackets by using 'integration by parts' (see, the text around Equation (28)). To explain a few hidden details of such calculations let us consider the following Poisson brackets

$$[\pi^{\alpha\beta}, \Phi(g_{\mu\nu})g_{\gamma\lambda,p}g_{\rho\sigma,q}] = -\frac{\partial\Phi(g_{\mu\nu})}{\partial g_{\mu\nu}}\Delta^{\alpha\beta}_{\mu\nu}g_{\gamma\lambda,p}g_{\rho\sigma,q} + \Big[\Phi(g_{\mu\nu})g_{\rho\sigma,q}\Big]_{,p}\Delta^{\alpha\beta}_{\gamma\lambda} + \Big[\Phi(g_{\mu\nu})g_{\gamma\lambda,p}\Big]_{,q}\Delta^{\alpha\beta}_{\rho\sigma}$$
$$= -\frac{\partial\Phi}{\partial g_{\mu\nu}}g_{\gamma\lambda,p}g_{\rho\sigma,q}\Delta^{\alpha\beta}_{\mu\nu} + \Phi_{,p}g_{\rho\sigma,q}\Delta^{\alpha\beta}_{\gamma\lambda} + \Phi g_{\rho\sigma,qp}\Delta^{\alpha\beta}_{\gamma\lambda} + \Phi_{,q}g_{\gamma\lambda,p}\Delta^{\alpha\beta}_{\rho\sigma} + \Phi g_{\gamma\lambda,pq}\Delta^{\alpha\beta}_{\rho\sigma}\ , \quad (35)$$

where $\Phi(x)$ is a scalar function of tensor argument(s). This formula can be simplified even further, but our goal here is to illustrate analytical computations of the Poisson brackets of momenta and some special functions and expressions which contain spatial derivatives of the metric tensor. In particular, the formula, Equation (35), explains the presence of second-order spatial derivatives of the metric tensor in some formulas below.

All Poisson brackets mentioned above are crucially important for the goals of this study, since they define the unique symplectic structure which is closely related to our $d(d+1)$-dimensional (tensor) phase space $\{g_{\alpha\beta}, \pi^{\mu\nu}\}$, which is closely related to the original d-dimensional Riemann space in the metric gravity. In other words, such a simplectic structure is uniformly determined by the Poisson brackets between the covariant components of the fundamental metric tensor $g_{\alpha\beta}$ and contravariant components $\pi^{\mu\nu}$ of the momentum tensor. Finally, we have to note that there is an alternative approach to develop Hamiltonian formulations of the metric gravity which is based on the use of covariant components of momenta $\pi_{\mu\nu}$. In some sense this new approach is simpler than the method discussed above, but its applications lead to re-consideration of fundamental principles of the classical Hamiltonian procedure, operations in the dual phase space and analysis of combinations of both straight and dual phase spaces for the tensor fields. This alternative approach and arising dual phase space are briefly considered below.

Covariant Components of Momenta—On the Dual Phase Space

In actual physical theories of tensor fields an arbitrary tensor can be represented either by its covariant, or contravariant components. For an arbitrary Riemann space relations between co- and contravariant components of the same vector, or tensor-like quantity are determined by the co- and contravariant components of the fundamental metric tensor $g_{\alpha\beta}$ and $g^{\alpha\beta}$. Therefore, one can always represent the same tensor equations in both covariant and contravariant forms. In general, many problems from tensor calculus can be simplified (even substantially), if they are re-written in the contravariant components of the same tensors and vice versa (some examples are considered in [4,5]). The metric gravity can be one of such problems, since both canonical and total Hamiltonians, Equations (17) and (16), contain multiple products of many contravariant components of the fundamental tensor $g^{\alpha\beta}$. Therefore, if we can properly define the covariant components of momenta $\pi_{\lambda\sigma}$, then our original problem can drastically be simplified.

Let us define the covariant components of momenta by the relation $\pi_{\lambda\sigma} = g_{\lambda\mu}\pi^{\mu\nu}g_{\nu\sigma}$. Note here that in any Hamiltonian formulation of the metric gravity, the role of fundamental tensor $g_{\alpha\beta}$ is always twofold. First, it is traditionally used to raise and lower indices in vector and tensor expressions. On the other hand, in all Hamiltonian formulations of the metric gravity the components of the fundamental tensor $g_{\alpha\beta}$ are traditionally chosen as the generalized coordinates, i.e., dynamical variables which are dynamically conjugate to the corresponding momenta $\pi^{\mu\nu}$. Such a twofold role of the fundamental tensor $g_{\alpha\beta}$ (and $g^{\alpha\beta}$) in Hamiltonian metric gravity leads to an additional problem, since the momenta $\pi^{\mu\nu}$ do not commute with the coordinates $g_{\alpha\beta}$. In turn, this means that the following definitions of covariant components of momenta: $\pi_{\lambda\sigma} = g_{\lambda\mu}g_{\nu\sigma}\pi^{\mu\nu}, \pi_{\lambda\sigma} = g_{\lambda\mu}\pi^{\mu\nu}g_{\nu\sigma}$ and $\pi_{\lambda\sigma} = \pi^{\mu\nu}g_{\lambda\mu}g_{\nu\sigma}$ are not equivalent to each other. Indeed, it is easy to show that $\pi_{\lambda\sigma} = g_{\lambda\mu}g_{\nu\sigma}\pi^{\mu\nu} \neq g_{\lambda\mu}\pi^{\mu\nu}g_{\nu\sigma}$, since $[g_{\nu\sigma}, \pi^{\mu\nu}] = \Delta^{\mu\nu}_{\nu\sigma} = \frac{1}{2}\Big(\delta^{\mu}_{\nu}\delta^{\nu}_{\sigma} + \delta^{\mu}_{\sigma}\Big) \neq 0$ in the general case. To avoid repetitive discussions of similar problems in this study we shall always define the covariant components of momenta by the relation: $\pi_{\lambda\sigma} = g_{\lambda\mu}\pi^{\mu\nu}g_{\nu\sigma}$ mentioned above.

By using this definition of covariant momenta we can determine the following Poisson brackets

$$[g_{\alpha\beta}, \pi_{\mu\nu}] = \frac{1}{2}\Big(g_{\alpha\mu}g_{\beta\nu} + g_{\alpha\nu}g_{\beta\mu}\Big) \text{ and } [g^{\alpha\beta}, \pi_{\mu\nu}] = -\frac{1}{2}\Big(g^{\alpha}_{\mu}g^{\beta}_{\nu} + g^{\alpha}_{\nu}g^{\beta}_{\mu}\Big) = -\Delta^{\alpha\beta}_{\mu\nu} \tag{36}$$

and also $[g^{\alpha\beta}, g^{\mu\nu}] = 0$ and $[g_{\alpha\beta}, g^{\mu\nu}] = 0$. The formulas Equations (31), (32) and others can be re-derived for the covariant components of momentum $\pi_{\alpha\beta}$:

$$[F(g), \pi_{\alpha\beta}] = -\Big(\frac{\partial F}{\partial g}\Big)g g_{\alpha\beta}\;,\; [\sqrt{-g}, \pi_{\alpha\beta}] = -\frac{1}{2}\sqrt{-g}g_{\alpha\beta}\;,\; [\frac{1}{\sqrt{-g}}, \pi_{\alpha\beta}] = -\frac{1}{2\sqrt{-g}}g_{\alpha\beta}\;. \tag{37}$$

These Poisson brackets are also important to perform analytical calculations in the Hamiltonian formulation of the metric gravity. As follows from these formulas the dynamical Hamiltonian variables $\{g^{\alpha\beta}, \pi_{\mu\nu}\}$ form another set of Hamiltonian dynamical variables which is often called the dual set of Hamiltonian (dynamical) variables. In general, this dual set of Hamiltonian variables $\{g^{\alpha\beta}, \pi_{\mu\nu}\}$ can also be used to develop another Hamiltonian formulation of the metric gravity which is simpler than the approach described above. Thus, for the tensor field in metric gravity we always have two sets of canonical Hamiltonian variables: (a) straight (or natural) set $\{g_{\alpha\beta}, \pi^{\mu\nu}\}$, and (b) dual set of dynamical variables $\{g^{\alpha\beta}, \pi_{\mu\nu}\}$ [10]. Further analysis [10] shows that the two similar sets of dynamical Hamiltonian variables (straight and dual sets) always arise and exist in any Hamiltonian formulation of the tensor field theory and they are related to each other by a special canonical transformation. Moreover, there is a beautiful formula [10] for the Poisson brackets which unites both straight and dual sets of dynamical variables defined for the same Hamiltonian system

$$[g_{\alpha\beta}, \pi^{\mu\nu}] = \Delta^{\mu\nu}_{\alpha\beta} = [\pi_{\alpha\beta}, g^{\mu\nu}] \quad . \tag{38}$$

Another interesting Poisson bracket in the metric gravity is

$$[\pi_{\alpha\beta}, \pi^{\mu\nu}] = \frac{1}{2}\Big(\delta^{\mu}_{\alpha}\pi^{\nu}_{\beta} + \delta^{\nu}_{\alpha}\pi^{\mu}_{\beta} + \delta^{\mu}_{\beta}\pi^{\nu}_{\alpha} + \delta^{\nu}_{\beta}\pi^{\mu}_{\alpha}\Big) = -[\pi^{\mu\nu}, \pi_{\alpha\beta}] \quad , \tag{39}$$

where $\pi^{\rho}_{\kappa} = g^{\rho\lambda}\pi_{\lambda\kappa} = g_{\kappa\lambda}\pi^{\lambda\rho}$. The last equality means that the co- and contravariant components of the momentum tensor do not commute with each other. On the other hand, if they commuted, then the direct and dual sets of simplectic dynamical variables in metric gravity would be equivalent to each other and there would be no real need to keep these two sets of dynamical variables (straight and dual). Indeed, in this case one can always express one set of dynamic variables in terms of another set and vice versa. Such cases include all Hamiltonian theories developed for the truly scalar fields and those fields which are represented by affine vectors and tensors. However, this is not true for the metric gravity and for other theories developed for actual tensor fields in multi-dimensional Riemann spaces of non-zero curvature. In general, to develop the truly correct and covariant Hamiltonian formulation for many dynamical system of tensor fields it is much better to deal with the mixed set of $2d(d+1)$ Hamiltonian dynamical variables. This big set is a unification of the two different $d(d+1)$-dimensional sets of Hamiltonian dynamical variables: (a) the straight set $\{g_{\alpha\beta}, \pi^{\mu\nu}\}$, and (b) the corresponding dual set $\{g^{\alpha\beta}, \pi_{\mu\nu}\}$. Applications of the two sets of dynamical variables makes our Hamiltonian formulation complete, truly covariant and physically transparent. In addition to this, an instant use of the direct and dual sets of Hamiltonian dynamical variables allows one to write canonical transformations of the Hamiltonian dynamical variables in the most general and powerful form.

5. Hamilton Equations of Motion

The main goal of any Hamiltonian formulation of some physical theory is to derive the correct Hamilton equations of motion by following the well established and physically transparent Hamilton procedure which has its internal logic based on Stokes's theorem in multi-dimensions. In general, the Hamilton method always provides a remarkable simplicity and universality in applications to actual dynamical systems and fields. Each of the Hamilton equations describes the complete time-evolution of one of the Hamiltonian dynamical variables. These correct Hamilton equations (or canonical equations) for the metric gravity are written in the following form

$$\frac{dg_{\alpha\beta}}{dx^0} = [g_{\alpha\beta}, H_t] \quad \text{and} \quad \frac{d\pi^{\mu\nu}}{dx^0} = [\pi^{\mu\nu}, H_t] \ , \tag{40}$$

where x_0 is the temporal variables and expressions in the right-hand sides of both equations are the Poisson brackets. In other words, the first-order time derivative of each of the Hamiltonian variables is proportional to the corresponding Poisson bracket of this variable with the total Hamiltonian H_t, Equation (16). The explicit form of these Hamiltonian equations and their solutions are discussed in [10]. In particular, for space-like components of the metric tensor g_{ij} one finds the following system of Hamilton equations [10]:

$$\begin{aligned} \frac{dg_{ij}}{dx^0} &= [g_{ij}, H_t] = [g_{ij}, H_C] = \frac{2}{\sqrt{-g}g^{00}} I_{(ij)pq}\pi^{pq} - \frac{1}{g^{00}} I_{(ij)pq} B^{(pq0|\mu\nu k)} g_{\mu\nu,k} \\ &= \frac{2}{\sqrt{-g}g^{00}} I_{(ij)pq}\Big[\pi^{pq} - \frac{1}{2}\sqrt{-g}B^{(pq0|\mu\nu k)} g_{\mu\nu,k}\Big] , \end{aligned} \tag{41}$$

where the notation $I_{(ij)pq}$ designates the (ij)-symmetrized value of the I_{ijpq} tensor defined in Equation (11), i.e.,

$$I_{(ij)pq} = \frac{1}{2}\Big(I_{ijpq} + I_{jipq}\Big) = \frac{1}{d-2} g_{ij}g_{pq} - \frac{1}{2}(g_{ip}g_{jq} + g_{iq}g_{jp}) \ . \tag{42}$$

Now, let us consider the Poisson brackets for the covariant components $g_{0\sigma}$ of the fundamental tensor. It is clear that the Poisson bracket of any $g_{0\sigma}$ component with the canonical Hamiltonian H_C, Equation (17), equals zero identically. Therefore, the Hamilton equations of motion for the covariant $g_{0\sigma}(= g_{\sigma 0})$ components of the metric tensor take the form

$$\frac{dg_{0\sigma}}{dx_0} = [g_{0\sigma}, H_t] = [g_{0\sigma}, H_t - H_C] = g_{0\sigma,0} \tag{43}$$

and analogous equations for the $g_{\gamma 0}$ components. These formulas are, in fact, the definitions of the σ-velocities, where $\sigma = 0, 1, \ldots, d-1$, which essentially coincide with the coefficients in front of the primary constraints in the total Hamiltonian, Equation (16). As follows from Equation (25) there is no need to derive and solve the equations of motion for the covariant components of the metric tensor $g^{\alpha\beta}$. Indeed, if we know the time evolution of all covariant components $g_{\mu\nu}$, then from Equation (25) one easily finds the exact description of time evolution for each $g^{\alpha\beta}$ component.

In general, the Hamilton equations of motion for the contravariant components of momenta $\pi^{\alpha\beta}$ are significantly more complicated (see, e.g., [10]) than analogous equations for the $g_{\alpha\beta}$ components (our coordinates). However, all these complications are purely technical and they are mainly related to a large number of Poisson brackets which must be determined in order to describe the complete time-evolution of all momentum variables $\pi^{\mu\nu}$. To understand the scale of this problem let us present here the Hamiltonian equations of motion for the contravariant space-like components of the momentum tensor π^{ab}:

$$\frac{d\pi^{ab}}{dx_0} = [\pi^{ab}, H_t] = [\pi^{ab}, H_C] = -\frac{1}{g^{00}}\Big[\frac{I_{mnpq}}{\sqrt{-g}}, \pi^{ab}\Big]\pi^{mn}\pi^{pq}$$
$$+ \frac{1}{g^{00}}\Big[I_{mnpq}, \pi^{ab}\Big]\pi^{mn}B^{(pq0|\mu\nu k)}g_{\mu\nu,k} + \frac{1}{g^{00}}I_{mnpq}\pi^{mn}\Big[B^{(pq0|\mu\nu k)}, \pi^{ab}\Big]g_{\mu\nu,k} + \dots . \quad (44)$$

This formula indicates clearly that in the Hamilton equations in metric gravity which describe time-evolution of momenta are significantly more complicated than analogous equations for time-evolution of the generalized coordinates $g_{\alpha\beta}$. In general, the Poisson bracket $[\pi^{ab}, H_t]$ is determined term-by-term.

As follows from the Hamilton equations presented above the Hamilton method itself has a number of problems when it is applied to the metric gravity, or other dynamical systems with first-class constraints. First, we note that to write Hamilton equations of actual motion we need only the canonical Hamiltonian H_C (not the total Hamiltonian H_t). Indeed, these equations are:

$$\frac{dg_{mn}}{dx_0} = [g_{mn}, H_C] \quad \text{and} \quad \frac{d\pi^{pq}}{dx_0} = [\pi^{pq}, H_C] \quad (45)$$

and there are $d(d-1)$ of these Hamilton equations of actual motion. Second, our Hamilton equations mentioned above, Equation (45), do not contain temporal momenta $\pi^{0\mu}$ and/or $\pi^{\nu 0}$ at all. This means that in these frames we cannot describe time-evolution of the temporal components of metric tensor $g_{0\mu}$ and $g_{\nu 0}$ (our coordinates). Moreover, it is not entirely clear where we can obtain these equations, since all these temporal momenta are included in our Hamiltonian formulation of the metric gravity only as primary constraints. Formally, to solve this problem introduce in the new Hamilton equations either the total Hamiltonian H_t, or the difference $H_t - H_C$, which is a linear combination of the primary constraints $\phi^{0\gamma}$, Equation (15). However, the coefficients in this linear combination are the σ-velocities, which are, in fact, arbitrary parameters of the method, rather than its dynamical variables. These arguments lead to an unambiguous conclusion that the Hamiltonian method itself must substantially be modified, if we want to apply it successfully to constrained dynamical systems, including the metric gravity. Such a modification was carried out by Dirac in his papers [1–3] and will be analyzed in detail in Section 7, but here we just want to mention its main steps.

First of all, Dirac accepted all Hamilton equations from Equations (45) as the equations which correctly describe the actual motions in our dynamical system. Thus, these $d(d-1)$ Hamilton equations have been incorporated in the new Dirac's modification of the classical Hamilton method. At the second step, Dirac rejected the d equations, Equation (43), that are formally correct but practically useless. Instead, these equations have been replaced by an equal number of equations which describe time-evolution of the primary constraints $\phi^{0\sigma}$ and define the new secondary constraints $\chi^{0\sigma}$, i.e., $\frac{d\phi^{0\sigma}}{dx_0} = [\phi^{0\sigma}, H_t] = [\phi^{0\sigma}, H_C] = \chi^{0\sigma}$. Here we apply the fact that all primary constraints commute with each other, i.e., $[\phi^{0\sigma}, \phi^{0\gamma}] = 0$ (see above). At the next (third) step Dirac explicitly derived the Hamilton equations which describe time-evolution of all d secondary constraints $\chi^{0\sigma}$: $\frac{d\chi^{0\sigma}}{dx_0} = [\chi^{0\sigma}, H_t] = [\chi^{0\sigma}, H_C] = D^{\sigma}_{c} = a^{\alpha}_{\mu}(g)\chi^{0\mu} + b^{\alpha}_{\mu}(g)\Big(f^k(g)\chi^{0\mu}\Big)_{,k}$, where the function (or functional) D^{σ}_{c} is the σ-component of Dirac closure which is a quasi-linear combination of the same secondary constraints and total spatial derivatives of expressions which contain the same secondary constraints. All other temporal derivatives of the Dirac closure will produce only similar quasi-linear combinations of secondary constraints and a few total spatial derivatives of them. The process of time-evolution is formally closed, since we will never see anything new in this chain. Briefly, this means that in Dirac's modification of the classical Hamiltonian method all d primary and d secondary constraints are considered as the new Hamiltonian dynamical variables. Note that the equations which describe the time-evolution of these new dynamical variables are written in a manifestly Hamilton form.

Let us show how this procedure works in the case of metric gravity. The Poisson brackets between the primary constraints $\phi^{0\sigma}$ and canonical Hamiltonian are [6]:

$$\begin{aligned}
[\quad & \phi^{0\sigma}, H_C] = -\frac{g^{0\sigma}}{2\sqrt{-g}g^{00}} I_{mnpq}\pi^{mn}\pi^{pq} + \frac{g^{0\sigma}}{2g^{00}} I_{mnpq}\pi^{mn} U^{(pq0|\mu\nu k)} g_{\mu\nu,k} + \Big[\pi^{\sigma k}_{,k} + \Big(\pi^{pk}e^{q\sigma} \\
- \quad & \frac{1}{2}\pi^{pq}e^{k\sigma}\Big)g_{pq,k}\Big] - \frac{\sqrt{-g}}{8}\Big(\frac{g^{0\sigma}}{g^{00}} I_{mnpq} B^{((mn)0|\mu\nu k)} B^{(pq0|\alpha\beta t)} - g^{0\sigma} B^{\mu\nu k\alpha\beta t}\Big) g_{\mu\nu,k} g_{\alpha\beta,t} \\
+ \quad & \frac{\sqrt{-g}}{4g^{00}} I_{mnpq} B^{((mn)0|\mu\nu k)} g_{\mu\nu,k} g_{\alpha\beta,t} \Big[g^{\sigma t}\Big(g^{00}g^{p\alpha}g^{q\beta} + g^{pq}g^{0\alpha}g^{0\beta} - 2g^{\alpha q}g^{0p}g^{0\beta}\Big) \\
- \quad & g^{\sigma p}\Big(2g^{00}g^{q\alpha}g^{t\beta} - g^{00}g^{\alpha\beta}g^{qt} + g^{\alpha\beta}g^{0q}g^{0t} - 2g^{q\alpha}g^{0\beta}g^{0t} - 2g^{t\alpha}g^{0\beta}g^{0q} + 2g^{qt}g^{0\alpha}g^{0\beta}\Big) \\
+ \quad & g^{0\sigma}(2g^{\beta t}g^{\alpha p}g^{0q} - 2g^{p\alpha}g^{q\beta}g^{0t} - 2g^{pq}g^{t\beta}g^{0\alpha} + 2g^{pt}g^{q\beta}g^{0\alpha} + g^{pq}g^{\alpha\beta}g^{0t} - g^{tp}g^{\alpha\beta}g^{0q})\Big] \\
- \quad & \frac{\sqrt{-g}}{4} g_{\mu\nu,k} g_{\alpha\beta,t}\Big[g^{\sigma t}(g^{\alpha\mu}g^{\beta\nu}g^{0k} + g^{\mu\nu}g^{\alpha t}g^{0\beta} - 2g^{\mu\alpha}g^{k\nu}g^{0\beta}) \\
+ \quad & g^{0\sigma}(2g^{\alpha t}g^{\beta\mu}g^{\nu k} - 3g^{t\mu}g^{\nu k}g^{\alpha\beta} - 2g^{\mu\alpha}g^{\nu\beta}g^{kt} + g^{\mu\nu}g^{kt}g^{\alpha\beta} + 2g^{\mu t}g^{\nu\beta}g^{k\alpha}) \\
+ \quad & g^{\sigma\mu}\Big((g^{\alpha\beta}g^{\nu t} - 2g^{\nu\alpha}g^{t\beta})g^{0k} + 2(g^{\beta\nu}g^{kt} - g^{\beta k}g^{t\nu})g^{0\alpha} + (2g^{k\beta}g^{\alpha t} - g^{\alpha\beta}g^{kt})g^{0\nu}\Big)\Big] \\
- \quad & \frac{\sqrt{-g}g^{00}}{2} E^{pqt\sigma}\Big(\frac{1}{g^{00}} I_{mnpq} B^{((mn)0|\mu\nu k)} g_{\mu\nu,k}\Big)_{,t} - \frac{\sqrt{-g}}{2} B^{((\sigma 0)k|\alpha\beta t)} g_{\alpha\beta,kt}\,, \qquad (46)
\end{aligned}$$

where $U^{(pq0|\mu\nu k)}$ is the symmetrized form of the following expression

$$U^{\alpha\beta 0\mu\nu k} = B^{(\alpha\beta 0|\mu\nu k)} - g^{0k}E^{\alpha\beta\mu\nu} + 2g^{0\mu}E^{\alpha\beta k\nu} \qquad (47)$$

and $\sigma = 0, 1, \ldots, d-1$. Thus, the corresponding Poisson brackets of the primary constraints with the canonical Hamiltonians H_C are the new functions of generalized coordinates $g_{\alpha\beta}$ (or $g^{\sigma\gamma}$) and momenta $\pi^{\mu\nu}$. In respect to the original terminology introduced by Dirac (see, e.g., [3]) these $\chi^{0\sigma}$ functions are the secondary constraints of this Hamiltonian formulation. Briefly, the definition of secondary constraints is written in the form: $\chi^{0\sigma} = [\phi^{0\sigma}, H_C]$, where $\sigma = 0, 1, \ldots, d-1$. This means that in metric gravity we always have d secondary constraints $\chi^{0\sigma}$ ($= \chi^{\sigma 0}$). On actual Hamiltonian trajectories (and only on these trajectories) of the free gravitational field these secondary constraints must be equal zero, i.e., we can write the following weak equations $\chi^{0\sigma} \approx 0$ for $\sigma = 0, 1, \ldots, d-1$. Note also that the Poisson brackets between the primary and secondary constraints are $[\phi^{0\gamma}, \chi^{0\sigma}] = \frac{1}{2}g^{\gamma\sigma}\chi^{00}$ [10]. It can also be shown that all primary $\phi^{0\lambda}$ and secondary constraints $\chi^{0\sigma}$ which arise during this Hamiltonian formulation of the metric gravity are the first-class constraints [3].

At the next step of the original Dirac procedure [1–3] we have to determine the temporal derivatives of all secondary constraints, i.e., $\frac{d\chi^{0\sigma}}{dx_0} = [\chi^{0\sigma}, H_t] = [\chi^{0\sigma}, H_C] + [\chi^{0\sigma}, g_{00,0}\phi^{00} + 2g_{0k,0}\phi^{0k}]$. The first Poisson brackets is

$$\begin{aligned}
\frac{d\chi^{0\sigma}}{dx_0} \quad &= \quad [\chi^{0\sigma}, H_C] = -\frac{2}{\sqrt{-g}} I_{mnpq}\pi^{mn}\Big(\frac{g^{\sigma q}}{g^{00}}\Big)\chi^{0p} + \frac{1}{2}g^{\sigma k}g_{00,k}\chi^{00} + \delta^{\sigma}_{0}\chi^{0k}_{,k} \\
&+ \quad \Big[-2\frac{1}{\sqrt{-g}} I_{mnpk}\pi^{mn}\frac{g^{\sigma p}}{g^{00}} + I_{mkpq}\Big(\frac{g^{\sigma m}}{g^{00}}\Big) U^{(pq)0\mu\nu l} g_{\mu\nu,l}\Big]\chi^{0k} \\
&- \quad \Big[g^{0\sigma}g_{00,k} + 2g^{n\sigma}g_{0n,k} + \frac{g^{n\sigma}g^{0m}}{g^{00}}(g_{mn,k} + g_{km,n} - g_{kn,m})\Big]\chi^{0k} = D^{\sigma}_{c}\,, \qquad (48)
\end{aligned}$$

where D_c^σ is the σ-component of the Dirac closure, $U^{(pq)0\mu\nu k}$ is the quantity $U^{pq0\mu\nu k}$ from Equation (47) which is symmetrized upon all $p \leftrightarrow q$ permutations. The second Poisson bracket in the original expression for $\frac{d\chi^{0\sigma}}{dx_0}$ takes the form

$$[\chi^{0\sigma}, g_{00,0}\phi^{00} + 2g_{0k,0}\phi^{0k}] = -\frac{1}{2}g^{0\sigma}g_{00,0}\chi^{00} - g^{0k}g_{0k,0}\chi^{00} = -\frac{1}{2}\Big(g^{0\sigma}g_{00,0} + 2g^{0k}g_{0k,0}\Big)\chi^{00} \tag{49}$$

and it is proportional to the secondary constraint χ^{00}. Thus, the Poisson brackets $[\chi^{0\sigma}, H_t]$ and $[\chi^{0\sigma}, H_C]$ are written as a linear combinations with field-dependent coefficients (we call them quasi-linear combinations) of the secondary constraints $\chi^{0\gamma}$ only. The $[\chi^{0\sigma}, H_C]$ Poisson bracket is called the σ-component of Dirac closure D_c^σ, or the Dirac σ-closure for the Hamiltonian formulation of metric gravity. In some old papers the Dirac closure has been defined as the $[\chi^{0\sigma}, H_t]$ Poisson bracket. The difference between these two definitions is proportional to the secondary constraint χ^{00} (see, Equation (49)), and we do not have any principal contradiction between these two definitions of Dirac's closure. Furthermore, note that this expression for Dirac closure, Equation (48), written in terms of secondary constraints only, is one of three possible results in the original Dirac procedure [2,3]. Briefly, this means that our Hamiltonian formulation of metric gravity does not lead either to any constraints of higher order, e.g., tertiary constraints, or to any inconsistency which can be fatal for the whole theory based on the $\Gamma - \Gamma$ Lagrangian, Equation (5) [3]. Finally, we need to say that in metric gravity the Dirac closure is a d-vector-like quantity in contrast with the Maxwell d-dimensional electrodynamics of the free EM-field, where the Dirac closure is a scalar which equals zero for the free EM-field [3,16].

Thus, in the metric gravity each primary constraint generates one secondary constraint and the Dirac's chain of first-class constraints ends at the secondary constraints. Finally, we have d primary and d secondary first-class constraints, i.e., the total number of the first-class constraints in metric gravity equals $2d$. In this sense there is an obvious similarity between the Hamiltonian approach for the Maxwell theory of multi-dimensional electromagnetic field (see, e.g., [3,16]) and Hamiltonian formulation of the metric gravity. Furthermore, all Hamiltonian formulations of different physical fields, which contain equal numbers of the primary and secondary first-class constraints, are quite similar to each other. The source of such a similarity can be traced back to the fact that the original Lagrangian density ($L_{\Gamma-\Gamma}$, Equation (5), in our case) is written as a quadratic-linear combination of velocities (or field-velocities).

In conclusion we want to note that there is a direct relation which allows one to express the canonical Hamiltonian H_C in terms of the secondary constraints $\chi^{0\sigma}$ and total spatial derivatives

$$H_C = -2g^{0\sigma}\chi^{0\sigma} + \Big(2g_{0m}\pi^{mk}\Big)_{,k} + \Big[\sqrt{-g}\Big(g_{0\gamma}B^{((0\gamma)k|\alpha\beta m)} - g_{0n}B^{((nk)0|\alpha\beta m)}\Big)g_{\alpha\beta,m}\Big]_{,k} \,. \tag{50}$$

This formula relates the canonical Hamiltonian H_C which depends upon the space-like momenta π^{mn} (they belong to the pure dynamical $d(d-1)$-dimensional space) and secondary first-class constraints $\chi^{0\sigma}$ which belong to the pure constraint, or non-dynamical $2d$ dimensional subspace (see below). From this point of view the equation, Equation (50), is the 'additionality' relation between the dynamical and constraint parts of the total Hamiltonian of the metric gravity.

6. Canonical Transformations

One of the main advantages of the Hamiltonian formulation(s) of any physical theory is a possibility to apply various canonical transformations of the Hamiltonian dynamical variables. In general, such canonical transformations can be used to simplify either the canonical Hamiltonian H_C, or to reduce this Hamiltonian to some special forms, e.g., to its natural form [10]. In the Hamiltonian formulations of metric gravity the canonical transformations of Hamiltonian dynamical variables are often used to simplify the explicit form of secondary constraints. Indeed, the secondary constraints derived above in the form of Equation (46) are very complex. Applications and even simple operations with secondary constraints written in such a form are very difficult. For instance, calculations of the Poisson brackets between primary and secondary constraints, of between any pair of secondary constraints produce formulas which are extremely cumbersome. For the first time, this has been noticed by Dirac in his fundamental paper [1]. To resolve these issues he used some canonical transformation of the original (Hamiltonian) dynamical variables which were originally introduced in [15]. At that time nobody performed similar transformations in metric gravity. This explains why Dirac in [1] started his transformations from the original $\Gamma - \Gamma$ Lagrangian density, Equation (5), which is also an ultimate source of the Hamiltonian theory. As is well known in classical mechanics we can always add the total temporal derivative to our original Lagrangian density without any change in the Lagrange equations of motion. The same rule is true in the general relativity and metric gravity, where we have also to take care about general covariance of all our formulas and expressions.

Briefly, the relation between the Dirac Lagrangian density introduced in [1] and our $L_{\Gamma-\Gamma}$ Lagrangian density, Equation (5), is written in the form $L_D = L_{\Gamma-\Gamma} - L^{\star}$ [17], where the additional Lagrangian density $L^{\star}$ takes the manifestly covariant form

$$L^{\star} = \Big[\big(\sqrt{-g}g^{00}\big)_{,\alpha}\frac{g^{0\alpha}}{g^{00}}\Big]_{,0} - \Big[\big(\sqrt{-g}g^{00}\big)_{,0}\frac{g^{0\alpha}}{g^{00}}\Big]_{,\alpha} = \Big[\big(\sqrt{-g}g^{00}\big)_{,k}\frac{g^{0k}}{g^{00}}\Big]_{,0} - \Big[\big(\sqrt{-g}g^{00}\big)_{,0}\frac{g^{0k}}{g^{00}}\Big]_{,k} . \qquad (51)$$

This equation is reduced to the form

$$\begin{aligned} L^{\star} = \frac{1}{2}\sqrt{-g}A^{\alpha\beta0\mu\nu k}g_{\alpha\beta,0}g_{\mu\nu,k} &= \frac{1}{2}\sqrt{-g}\Big(e^{\alpha\beta}e^{k\mu}g^{0\nu} - e^{\mu\nu}e^{k\alpha}g^{0\beta} + e^{k\alpha}\frac{g^{0\mu}g^{0\nu}g^{0\beta}}{g^{00}} \\ &- e^{k\mu}\frac{g^{0\alpha}g^{0\nu}g^{0\beta}}{g^{00}}\Big)g_{\alpha\beta,0}g_{\mu\nu,k} \cdot \end{aligned} \qquad (52)$$

The $A^{\alpha\beta0\mu\nu k}$ coefficients defined in this equation have a few following symmetries. First, these coefficients are symmetric upon the $\alpha\beta \leftrightarrow \beta\alpha$ and $\mu\nu \leftrightarrow \nu\mu$ permutations. Second, the important property of the $A^{\alpha\beta0\mu\nu k}$ coefficients is their anti-symmetry with respect to interchange of the two pairs of Greek indices, i.e., $A^{\alpha\beta0\mu\nu k} = -A^{\mu\nu0\alpha\beta k}$. Third, these coefficients are linearly related with the coefficients $B^{(\alpha\beta0|\mu\nu k)}$ and Dirac tensor $E^{\alpha\beta\gamma\sigma}$ (both defined in Section 3). The explicit form of this relation is

$$A^{\alpha\beta0\mu\nu k} = B^{(\alpha\beta0|\mu\nu k)} - g^{0k}E^{\alpha\beta\mu\nu} + 2g^{0\mu}E^{\alpha\beta k\nu} . \qquad (53)$$

Finally, the relation between the Dirac's Lagrangian L_D and our original $L_{\Gamma-\Gamma}$ Lagrangian of the metric gravity (see above) is written in the form [17]

$$L_D = L_{\Gamma-\Gamma} - L^{\star} = L_{\Gamma-\Gamma} - \frac{1}{2}\sqrt{-g}A^{\alpha\beta0\mu\nu k}g_{\alpha\beta,0}g_{\mu\nu,k} \text{ , where } L^{\star} = \frac{1}{2}\sqrt{-g}A^{\alpha\beta0\mu\nu k}g_{\alpha\beta,0}g_{\mu\nu,k} \cdot \qquad (54)$$

From this equation one easily finds the following expression for the momenta $p^{\gamma\sigma}$ in the Dirac Hamiltonian formulation of the metric gravity

$$\frac{\partial L_D}{\partial g_{\gamma\sigma,0}} = \frac{\partial L_{\Gamma-\Gamma}}{\partial g_{\gamma\sigma,0}} - \frac{\partial L^{\star}}{\partial g_{\gamma\sigma,0}} \text{ , or } p^{\gamma\sigma} = \pi^{\gamma\sigma} - \frac{1}{2}\sqrt{-g}A^{(\gamma\sigma)0\mu\nu k}g_{\mu\nu,k} \text{ ,} \qquad (55)$$

where $p^{\gamma\sigma}$ are the new momenta (or Dirac's momenta), while $\pi^{\gamma\sigma}$ are the old momenta defined above in Section 3. The last equation in Equation (55) is, in fact, the explicit definition of the Dirac's momenta which is conveniently to write in the two following forms:

$$p^{pq} = \pi^{pq} - \frac{1}{2}\sqrt{-g}A^{(pq)0\mu\nu k}g_{\mu\nu,k} \text{ and } p^{0\sigma} = \pi^{0\sigma} - \frac{1}{2}\sqrt{-g}A^{(0\sigma)0\mu\nu k}g_{\mu\nu,k} \text{ ,} \qquad (56)$$

where $A^{(0\sigma)0\mu\nu k}g_{\mu\nu,k} = \frac{1}{2}\Big(B^{(\alpha\beta0|\mu\nu k)} + B^{(\beta\alpha0|\mu\nu k)}\Big)$ and $p^{0\sigma} = p^{\sigma0}$. Thus, we have the two sets of Hamiltonian dynamical variables for the two different Hamiltonian formulations of the metric gravity: $\{g_{\alpha\beta}, \pi^{\mu\nu}\}$ (the old set) and $\{g_{\alpha\beta}, p^{\mu\nu}\}$ (the new set). Since these two sets of dynamical variables are related to each other by a canonical transformation, then the three following conditions for the Poisson brackets must be obeyed: $[g_{\alpha\beta}, g_{\mu\nu}] = 0$, $[g_{\alpha\beta}, p^{\mu\nu}] = \Delta^{\mu\nu}_{\alpha\beta}$ and $[p^{\alpha\beta}, p^{\mu\nu}] = 0$, where all new variables are written in terms of the old variables. For old variables we already know that the following equations are true: $[g_{\alpha\beta}, g_{\mu\nu}] = 0$, $[g_{\alpha\beta}, \pi^{\mu\nu}] = \Delta^{\mu\nu}_{\alpha\beta}$ and $[\pi^{\alpha\beta}, \pi^{\mu\nu}] = 0$. In reality, applications of these canonicity conditions needs some additional explanations, since for all Hamiltonian systems such conditions are always derived and formulated in a different form which is based on the 'alternative' Laplace (not Poisson!) brackets. Here we have to make one step aside and discuss the general canonicity conditions for an arbitrary transformation of the Hamiltonian dynamical variables.

6.1. General Conditions of Canonicity for Transformations of the Dynamical Variables

Let us assume that some Hamiltonian system is described by the $2n$ independent dynamical variables $\{q_k, p_k\}$, where $k = 1, \ldots, n$. In general, it is possible to replace these 'old' dynamical variables by the new dynamical variables $\{\tilde{q}_i, \tilde{p}_i\}$, where $i = 1, \ldots, n$:

$$\tilde{q}_i = \phi_i(t, q_k, p_k) \quad \tilde{p}_i = \psi_i(t, q_k, p_k) \ , \tag{57}$$

but after such a transformation of variables we want to be sure that the new Hamiltonian system will be 'dynamically equivalent' to our original Hamiltonian system. Transformations of the dynamical variables each of which transforms one Hamiltonian system into another Hamiltonian system, which is completely and unambiguously 'dynamically equivalent' to the original system, are defined as the canonical transformations. In general, all canonical transformations of any Hamiltonian system form the closed algebraic structure, or group, for short (see, e.g., [10,11]). It was shown (by Jacobi) that for any time-dependent canonical transformation of the dynamical variables, Equation (57), the following canonicity condition (below, the main canonicity condition) must be obeyed

$$\sum_{k=1}^{n} \tilde{p}_k d\tilde{q}_k - \tilde{H}\delta t = c\Big(\sum_{k=1}^{n} p_k dq_k - H\Big)\delta t - \delta F(t, q_k, p_k) \ , \tag{58}$$

where $c(\neq 0)$ is some real number which does not depend upon the time t. The function $F(t, q_k, p_k)$ is the Jacobi generating function, i.e., the function which generates this canonical transformation. Vice versa, one can easily show that, if Equation (58) holds for some transformation of the dynamical variables, then this transformation is canonical. For better understanding of equations from this subsection we use the explicit sign of summation.

Moreover, since the valence $c(\neq 0)$ does not depend upon the time, then by establishing the criteria of canonicity, we can always restrict ourselves (for more details, see, e.g., [11]) to the time-independent canonical transformations only, i.e.,

$$\tilde{q}_i = \phi_i(q_k, p_k) \quad \tilde{p}_i = \psi_i(q_k, p_k) \ . \tag{59}$$

For a canonical time-independent transformation the main condition, Equation (58), is written in the form

$$\sum_{k=1}^{n} \tilde{p}_k d\tilde{q}_k = c \sum_{k=1}^{n} p_k dq_k - \delta K(q_k, p_k) \ , \tag{60}$$

where $\tilde{q}_k, \tilde{p}_k$ ($k = 1, \ldots, n$) are the new generalized coordinates and momenta, while q_i, p_i ($i = 1, \ldots, n$) are the old coordinates and momenta (old dynamical variables). Moreover, in this equation $K(q_k, p_k) = F(\bar{t}, q_k, p_k)$, i.e., it is a short Jacobi generating function of the coordinates and momenta only, which coincides with the Jacobi generating function $F(\bar{t}, q_k, p_k)$ taken at some fixed time $t = \bar{t}$. The variation of $K(q_k, p_k)$ is written in the form

$$\delta K = -\sum_{i=1}^{n}\Big(\Phi_i \delta q_i + \Psi_i \delta p_i\Big) \; . \tag{61}$$

On the other hand, by using the formula $\delta\tilde{q}_k = \frac{\partial \tilde{q}_k}{\partial q_i}\delta q_i + \frac{\partial \tilde{q}_k}{\partial p_i}\delta p_i$ in Equation (60) one finds the following expression for the $\delta K(q_k, p_k)$ variation:

$$\delta K = -\sum_{i=1}^{n}\Big[\sum_{k=1}^{n}\Big(\tilde{p}_k \frac{\partial \tilde{q}_k}{\partial q_i}\Big) - c p_i\Big]\delta q_i - \sum_{i=1}^{n}\Big[\sum_{k=1}^{n}\Big(p_k \frac{\partial \tilde{q}_k}{\partial p_i}\Big)\Big]\delta p_i \; . \tag{62}$$

By comparing Equations (61) and (62) one finds

$$\Phi_i = \sum_{k=1}^{n} \tilde{p}_k \frac{\partial \tilde{q}_k}{\partial q_i} - c p_i \quad \text{and} \quad \Psi_i = \sum_{k=1}^{n} p_k \frac{\partial \tilde{q}_k}{\partial p_i} \; . \tag{63}$$

For canonical transformation(s) the expression in the left side of Equation (61) must be a total differential. From here one finds three following conditions:

$$\frac{\partial \Phi_i}{\partial q_j} = \frac{\partial \Phi_j}{\partial q_i} \;\; , \;\; \frac{\partial \Psi_i}{\partial p_j} = \frac{\partial \Psi_j}{\partial p_i} \;\; , \;\; \frac{\partial \Phi_i}{\partial p_j} = \frac{\partial \Psi_j}{\partial q_i} \;\; , \;\; . \tag{64}$$

By substituting the functions Φ_i and Ψ_i in these equations by their expressions from Equation (63) one finds after a few additional and simple transformations:

$$\sum_{k=1}^{n}\Big(\frac{\partial \tilde{q}_k}{\partial q_i}\frac{\partial \tilde{p}_k}{\partial q_j} - \frac{\partial \tilde{q}_k}{\partial q_i}\frac{\partial \tilde{q}_k}{\partial q_j}\Big) = 0 \;\; , \text{ or } \;\; \{q_i, q_j\} = 0 \;\; , \tag{65}$$

$$\sum_{k=1}^{n}\Big(\frac{\partial \tilde{q}_k}{\partial p_i}\frac{\partial \tilde{p}_k}{\partial p_j} - \frac{\partial \tilde{q}_k}{\partial p_i}\frac{\partial \tilde{q}_k}{\partial p_j}\Big) = 0 \;\; , \text{ or } \;\; \{p_i, p_j\} = 0 \;\; , \tag{66}$$

$$\sum_{k=1}^{n}\Big(\frac{\partial \tilde{q}_k}{\partial q_i}\frac{\partial \tilde{p}_k}{\partial p_j} - \frac{\partial \tilde{q}_k}{\partial p_i}\frac{\partial \tilde{q}_k}{\partial q_j}\Big) = c\delta_{ij} \;\; , \text{ or } \;\; \{q_i, p_j\} = c\delta_{ij} \; , \tag{67}$$

where δ_{ij} is the Kronecker symbol and c is some numerical constant. The constructions (or sums) which appear in these three equations are the Laplace brackets which are well known in classical mechanics (see, e.g., [11,12]). The standard notation for the Laplace brackets (see, e.g., [11,12]) is $\{,\}$. Each of these sums includes $2n$ functions ($\tilde{q}_k$ and $\tilde{p}_k$) and two variables only, e.g., either q_i, q_j, or p_i, p_j, or q_i, p_j. As follows from Equations (65)–(67) some transformation of the dynamical variables will be canonical if (and only if) the three groups of following conditions are obeyed: $\{q_i, q_j\} = 0, \{p_i, p_j\} = 0$ and $\{q_i, p_j\} = c\delta_{ij}$, where $c \neq 0$ and $(i, j) = 1, \ldots, n$, for all $2n$ new dynamical variables $\tilde{q}_k, \tilde{p}_k$ $(k = 1, \ldots, n)$.

In reality, the original Laplace brackets are not convenient in applications. However, as follows from the Appendix B these brackets can be replaced by the Poisson brackets, each of which is the adjoint to the corresponding Laplace bracket. In terms of the Poisson brackets the same criteria of canonicity are written in a different form (for more details, see our Appendix B): $[\tilde{q}_i, \tilde{q}_j] = 0 \;\; , \;\; [\tilde{p}_i, \tilde{p}_j] = 0 \;\; , \;\; [\tilde{q}_i, \tilde{p}_j] = c\delta_{ij}$, where $(i, j) = 1, \ldots, n$ and c is the valence of this canonical transformation. These numerical values of the Poisson brackets taken for $c = 1$ are used below as the criteria of canonicity for the transformation of dynamical variables. To simplify the text below we shall call these brackets by the canonical, univalent set of the Poisson brackets, or CUSPB, for short.

6.2. Applications to the Metric Gravity

Let us apply the formulas derived above to the metric gravity by considering a transformation from the old set of dynamical variables $\{g_{\alpha\beta}, \pi^{\mu\nu}\}$ to the new set of such variables

$\{g_{\alpha\beta}, p^{\mu\nu}\}$. First, we note that the generalized coordinates $g_{\alpha\beta}$ are identical in both sets. For the brackets defined in the previous subsection this means that $\{g_{\alpha\beta}, g_{\gamma\rho}\} = 0$ and $[g_{\alpha\beta}, g_{\gamma\rho}] = 0$. Furthermore, for the univalent ($c = 1$) transformations of dynamical variables we have for the new momenta $p^{\mu\nu} = \pi^{\mu\nu} + f^{\mu\nu}(g_{\gamma\rho})$, where $f^{\mu\nu}(g_{\gamma\rho})$ is a tensor function of generalized coordinates only. From here one finds that $[g_{\alpha\beta}, p^{\mu\nu}] = [g_{\alpha\beta}, \pi^{\mu\nu}] = \Delta^{\mu\nu}_{\alpha\beta}$. In other words, the first and last Poisson brackets from CUSPB are obeyed automatically for our transformation of the Hamiltonian dynamical variables. The only non-trivial bracket in CUSPB (see, Equation (A17) in the Appendix B) is the second Poisson bracket between two new momenta which takes the following form in our tensor notations:

$$[p^{\alpha\beta}, p^{\mu\nu}] = 0 \ , \ \text{or} \ [\pi^{\alpha\beta} - \frac{1}{2}\sqrt{-g}A^{(\alpha\beta)0\sigma\rho k}g_{\sigma\rho,k}, \pi^{\mu\nu} - \frac{1}{2}\sqrt{-g}A^{(\mu\nu)0\lambda\kappa k}g_{\lambda\kappa,k}] = 0 \ , \quad (68)$$

which is instantly reduced to the equation

$$[\pi^{\alpha\beta}, \sqrt{-g}A^{(\mu\nu)0\sigma\rho m}g_{\sigma\rho,m}] = [\pi^{\mu\nu}, \sqrt{-g}A^{(\alpha\beta)0\sigma\rho m}g_{\sigma\rho,m}] \ . \quad (69)$$

The transformation of dynamical variables will be canonical, if (and only if) this equation is obeyed. To prove the validity of this equation one has to perform direct calculations of the Poisson brackets in both sides of Equation (69). In reality, both sides of Equation (69) are compared with each other and identical terms (in both sides) are cancelled. Finally, this equation is reduced to the form of identity such as $0 = 0$.

In those cases when either $\alpha = 0$, or $\beta = 0$ (or both) one obtains from Equations (53) and (54) the following equation:

$$p^{0\gamma} = \pi^{0\gamma} - \frac{1}{2}\sqrt{-g}B^{((0\gamma)0|\mu\nu k)}g_{\mu\nu,k}, \quad (70)$$

which defines the momenta with one (or two) temporal component(s), or primary constraints $p^{0\gamma} \approx 0$ in the Dirac's Hamiltonian formulation of the metric gravity. For these momenta the canonicity conditions, Equation (68), must also be obeyed. After a few simple transformations the essential canonicity conditions for the $p^{0\gamma}$ and $p^{0\sigma}$ momenta take one of the following forms

$$[p^{0\gamma}, p^{0\sigma}] = 0 \ , \ \text{or} \ [\pi^{0\gamma}, \sqrt{-g}B^{((0\sigma)0|\mu\nu k)}g_{\mu\nu,k}] = [\pi^{0\sigma}, \sqrt{-g}B^{((0\gamma)0|\mu\nu k)}g_{\mu\nu,k}] \ , \quad (71)$$

which simply means that all primary constraints in the Dirac's Hamiltonian formulations commute with each other. The same statement is true for the original Hamiltonian formulation of metric GR [6] discussed above. This fact has been checked in [15]. On the other hand, we have to note that the fact that all primary constraints in the metric gravity commute with each other follows directly from the canonicity of the complete Dirac's set of Hamiltonian dynamical variables.

At this point it is very convenient to introduce the universal notation $\phi^{\mu\nu}$ for the momenta, or for contravariant components of the momenta. In Dirac's Hamiltonian formulation these momenta are $\phi^{\mu\nu} = p^{\mu\nu}$, while in the Hamiltonian formulation from [6] these momenta are $\phi^{\mu\nu} = \pi^{\mu\nu} - \frac{1}{2}\sqrt{-g}A^{(\mu\nu)0\alpha\beta k}g_{\alpha\beta,k}$. In this notation the canonical Hamiltonians H_C in both formulations of metric gravity are represented in the same 'universal' form [17]

$$\begin{aligned} H_C = & \frac{1}{\sqrt{-g}g^{00}}I_{mnpq}\phi^{mn}\phi^{pq} - \frac{1}{g^{00}}\phi^{mn}\left(g^{0l}g_{mn,l} - 2g^{0\alpha}g_{\alpha n,m}\right) \\ + & \frac{1}{4}\sqrt{-g}\Big[\frac{1}{g^{00}}\left(g^{0k}E^{(mn)\mu\nu} - 2g^{0\mu}E^{(mn)k\nu}\right)\left(g^{0l}g^{\alpha}_{m}g^{\beta}_{n} - 2g^{0\alpha}g^{l}_{m}g^{\beta}_{n}\right) - B^{\mu\nu k\alpha\beta l}\Big]g_{\mu\nu,k}g_{\alpha\beta,l} \ , \end{aligned} \quad (72)$$

where $g^{\alpha}_{\beta} = \delta^{\alpha}_{\beta}$ is the substitution tensor defined above. In both formulations the primary constraints commute with each other, i.e., $[\phi^{0\gamma}, \phi^{0\sigma}] = 0$. The knowledge of the canonical Hamiltonian H_C and all primary constraints allows one to restore the total Hamiltonian H_t:

$$H_t = H_C + g_{00,0}\phi^{00} + 2g_{0k,0}\phi^{0k} = H_C + g_{00,0}p^{00} + 2g_{0k,0}p^{0k} \ . \tag{73}$$

In particular, in the Dirac's Hamiltonian formulation we obtain $H_t = H_C + g_{00,0}p^{00} + 2g_{0k,0}p^{0k}$. It appears that the total Hamiltonian H_t does not change during canonical transformations of the dynamical variables, i.e., $H_t^{K\&K}(g_{\alpha\beta}, \pi^{\mu\nu}) = H_t^{Dir}(g_{\alpha\beta}, p^{\mu\nu})$ [17]. In other words, the total Hamiltonian is an obvious and unique invariant of this theory. In respect to this, the Hamilton equations do not change its form during canonical transformations and we can write, e.g.,

$$g_{\alpha\beta,0} = [g_{\alpha\beta}, H_t^{K\&K}] \ , \ \pi^{\mu\nu}_{,0} = [\pi^{\mu\nu}, H_t^{K\&K}] \ \Leftrightarrow \ g_{\alpha\beta,0} = [g_{\alpha\beta}, H_t^{Dir}] \ , \ p^{\mu\nu}_{,0} = [p^{\mu\nu}, H_t^{Dir}] \tag{74}$$

i.e., these two sets of Hamilton equations are equivalent to each other. In other words, the Hamilton equations conserve their form during canonical transformation of the dynamical variables. In fact, this was the first definition (or criterion) of canonicity for the transformations of dynamical variables which has been formulated by Sir William R. Hamilton himself in 1834 and 1835. We have shown that his criterion works for the Hamiltonian approach to the metric gravity. However, the metric gravity is a dynamical system with constraints. It is clear that the Hamilton criterion, as well as other criteria of canonicity known for the transformations of dynamical variables in classical mechanics, must be supplemented by some statement(s) about the algebra of constraints (see below). Therefore, we need to derive the explicit expressions for the secondary constraints, their Poisson brackets with the canonical and/or total Hamiltonians, primary constraints, etc.

All secondary constraints in the Dirac's Hamiltonian formulation are derived from the equations $\chi^{0\sigma} = [\phi^{0\sigma}, H_t] = [\phi^{0\sigma}, H_C]$. The explicit expressions are

$$\chi^{0\sigma} = -\frac{g^{0\sigma}}{2\sqrt{-g}g^{00}} I_{mnpq}\phi^{mn}\phi^{pq} + g^{\sigma}_{m}\Big[(\phi^{mk})_{,k} + \Big(\phi^{pk}e^{qm} - \frac{1}{2}\phi^{pq}e^{km}\Big)g_{pq,k}\Big]$$
$$+\frac{1}{2}\sqrt{-g}g^{0\sigma}\Big[-g_{mn,kl}E^{mnkl} + \frac{1}{4}g_{mn,k}g_{pq,l}\Big(-E^{mnpq}e^{kl} + 2E^{klpn}e^{mq} + E^{pqnl}e^{mk}\Big)\Big] \ . \tag{75}$$

This formula is very compact and contains only two lines (compare with the formula, Equation (46)). It indicates clearly that Dirac's idea to apply canonical transformations of the Hamiltonian dynamical variables in order to simplify secondary first-class constraints works perfectly. Now, by using the explicit form of the primary $\phi^{0\gamma} = p^{0\gamma}$ and secondary $\chi^{0\sigma}$ constraints in Dirac's formulation one finds

$$[\phi^{0\gamma}, \chi^{0\sigma}]_{Dirac} = [p^{0\gamma}, \chi^{0\sigma}]_{Dirac} = \frac{1}{2}g^{\gamma\sigma}\chi^{00}_{Dirac} \ , \tag{76}$$

i.e., the formula which exactly coincides (by its form) with the formula $[\phi^{0\gamma}, \chi^{0\sigma}]_{K\&K} = \frac{1}{2}g^{\gamma\sigma}\big(\chi^{00}\big)_{K\&K}$ mentioned above. It is very interesting, since the explicit forms of all primary and secondary constraints are substantially different in these two formulations. This and other similar facts directly follow from the canonicity of our transformation of the Hamiltonian dynamical variables. The time-evolution of the secondary constraints leads to the following formula

$$\frac{d\chi^{0\sigma}}{dx_0} = \chi^{0\sigma}_{,0} = [\chi^{0\sigma}, H_C] = -\Big[\frac{2}{\sqrt{-g}g^{00}} I_{pqmk}g^{\sigma m}\phi^{pq} + g^{0\sigma}g_{00,k} + 2g^{\sigma p}g_{0p,k}$$
$$+ \ \Big(\frac{g^{\sigma p}g^{0q}}{g^{00}}\Big) \ \Big(g_{pq,k} + g_{qk,p} - g_{pk,q}\Big)\Big]\chi^{0k} - g^{\sigma}_{0}(\chi_{0k})_{,k} + \frac{1}{2}g^{\sigma k}g_{00,k}\chi^{00} = D^{\sigma}_{c} \ , \tag{77}$$

where D_c^σ is the σ-component of the Dirac closure derived in the Dirac's Hamiltonian formulation of the metric gravity. All components of the Dirac closure are quasi-linear combination of secondary constraints and some total spatial derivatives of these secondary constraints. Again, this formula is very compact and transparent. Finally, we want to present the formula, which allows one to express the canonical Hamiltonian H_C in terms of secondary constraints and some (total) spatial derivatives, Equation (50), can also be derived in the Dirac Hamiltonian formulation. The formula takes the form, which is slightly different from Equation (50) above:

$$H_C = -2g_{0\lambda}\chi^{0\lambda} + \left(2g_{0m}\phi^{mk}\right)_{,k} - \Big[\sqrt{-g}E^{mnpq}g_{mn,q} - \sqrt{-g}g_{\mu\nu,k}\left(\frac{g^{0\mu}}{g^{00}}\right)\left(g^{\nu p}g^{0k} - g^{\nu k}g^{0p}\right)\Big]_{,p} \tag{78}$$

This formula also represents the canonical Hamiltonian H_C (in the Dirac formulation) written as a quasi-linear combination of the secondary constraints $\chi^{0\sigma}$ and a few total spatial derivatives of some expressions which include the same secondary constraints.

Now, we can complete our discussion of canonical transformations in the metric gravity. There are three general rules which regulate changes in the primary and secondary constraints during such transformations. The first rule is simple and it is called the law of inertia for the first-class constraints. Indeed, by performing a number of canonical transformations between different sets of dynamical variables we have found that the total number of the primary $\phi^{0\sigma}$ constraints N_p never changes during such transformations. The same statement is true for the total number of secondary $\chi^{0\sigma}$ constraints N_s and for the sum $N_p + N_s$. We have to emphasize here that all primary and secondary constraints which arise in the Hamiltonian formulations of metric gravity are first-class. The second rule of 'form-invariance' is even simpler: the internal structure of all first-class constraints must be conserved during canonical transformations of the Hamiltonian dynamical variables. The preservation of form-invariance for all first-class constraints is crucial to prove that any two Hamiltonian formulations developed for the same constrained dynamical system are equivalent to each other. The third rule essentially follows from the second rule: all Poisson brackets between the first-class constraints and canonical/total Hamiltonians, other constraints, etc, must also be form-invariant during canonical transformations. For simple Poisson brackets, e.g., for the $[\phi^{0\sigma}, \chi^{0\gamma}]$ brackets, this rule leads to the exact coincidence of corresponding expressions. The three rules mentioned here essentially mean preservation of the algebra of first-class constraints. Thus, the canonical transformations in the metric gravity must guarantee a complete preservation of the form-invariance for the total Hamiltonian H_t and for the algebra of first-class constraints.

The formulas derived in this Section allow one to apply the Dirac's Hamiltonian formulation of metric gravity to analyze and solve various gravitational problems. In some cases, however, one needs to know analytical expressions for other Poisson brackets, e.g., the Poisson bracket between two secondary constraints $[\chi^{0\gamma}, \chi^{0\sigma}]$ is of great interest, but it has never been obtained in previous papers. This Poisson bracket is determined in our 'technical' Appendix A. In general, calculations of this and other Poisson brackets can be performed with the use of our formulas and method described in Section 4.

7. Dirac's Modifications of the Classical Hamilton Method

In this Section we want to reconsider modifications which were made by Dirac in the classical Hamilton method [1–3]. This will eventually lead us to the new universal criterion of canonicity for Hamiltonian formulations of metric gravity. First, we note again that in any Hamiltonian formulation of metric gravity we always have $\frac{d(d+1)}{2}$ generalized coordinates $g_{\alpha\beta}$ and $\frac{d(d+1)}{2}$ momenta $p^{\mu\nu}$. These coordinates and momenta are the Hamiltonian dynamical variables of our problem (metric gravity). The total number of these variables equals $d(d+1)$ which is an even number for any d-dimensional Riemann space-time. Note also that our original Lagrangian $L_{\Gamma-\Gamma}$, Equation (6), is a quadratic function upon velocities of the space-like components of the metric tensor g_{mn}, i.e., upon $g_{mn,0}$. On the other hand, the same $L_{\Gamma-\Gamma}$ Lagrangian is a linear function of the d remaining $g_{0\gamma,0}$ (= $g_{\gamma 0,0}$) velocities which are also called the temporal velocities.

By using the standard Legendre transformation (see Section 3) one can pass from the Lagrangian $L_{\Gamma-\Gamma}$ to the Hamiltonian H_C which is quadratic function of space-like momenta π^{mn}. This Hamiltonian is called the canonical Hamiltonian, and it is an explicit function of the space-like dynamical $\{g_{mn}, p^{pq}\}$ variables (there are $d(d-1)$ of such variables) and d 'temporal' coordinates $g_{0\sigma}$ only. The Hamiltonian H_C is of great interest for the whole metric gravity, since it describes the actual motions of a free gravitational field. However, we have to note that this Hamiltonian H_C does not depend upon any of the temporal momenta, i.e., it does not include any of the $p^{0\sigma}$ (or $p^{\sigma 0}$) momenta. This means that all Poisson brackets such as $[g_{0\sigma}, H_C] = 0$ and $[g_{\sigma 0}, H_C] = 0$ equal zero identically, and canonical Hamiltonian H_C does not describe time-evolution of the temporal coordinates $g_{0\sigma}$ and/or $g_{\sigma 0}$ coordinates, in principle. Briefly, we can say that in the canonical Hamiltonian H_C these temporal coordinates $g_{0\mu}$ (and $g_{\mu 0}$) are rather parameters than actual dynamical variables.

For normal applications of the Hamilton method we must have a Hamiltonian that contains all momenta, including temporal ones. Such a complete, or total Hamiltonian H_t will describe time-evolution of all $d(d+1)$ dynamical variables of the problem $\{g_{\alpha\beta}, p^{\mu\nu}\}$ and all functions of these variables, including the canonical Hamiltonian H_C, new coordinates and momenta, which can be introduced by some canonical transformations, etc. Formally, this total Hamiltonian can be derived from our quadratic-linear Lagrangian $L_{\Gamma-\Gamma}$ by using the Legendre transform which is described in detail in Section 3. However, for our quadratic-linear Lagrangian $L_{\Gamma-\Gamma}$, Equation (6), the Legendre transform works with some singularities. The two main singularities must be mentioned here, since they play crucial roles in Dirac's modification of the classical Hamilton method. First, as follows from the definition of momenta and from the general technique of Legendre transformations, we cannot obtain, in principle, the explicit expressions of the velocities $v_\gamma (= g_{0\gamma,0})$ written in terms of momenta $p^{0\gamma}$ and vice versa. Instead, we obtain the following algebraic equations: $p^{0\gamma} \approx f(g_{\alpha\beta}, g^{0\gamma})$, or $\phi^{0\gamma} = p^{0\gamma} - f(g_{\alpha\beta}, g^{0\gamma}) \approx 0$ which are called the primary constraints (see, e.g., [3] and references therein). Second, in respect to the procedure of Legendre transformation, this moment must be multiplied by the corresponding velocity $g_{0\gamma,0} (= v_\gamma)$, which is not a dynamical variable of our Hamiltonian method. This velocity is rather a parameter (arbitrary parameter) of the updated Legendre procedure.

Thus, we have derived the total Hamiltonian H_t, Equation (16), which is written as the sum of the canonical Hamiltonian H_C and primary constraints $\phi^{0\alpha}$. The coefficients in front of the primary constraints equal to the corresponding velocities v_α, i.e., $H_t = H_C + g_{00,0}\phi^{00} + 2g_{0k,0}\phi^{0k} = H_C + v_\alpha \phi^{0\alpha}$. Now, the time-evolution of any dynamical variable, or any function/functional, or quantity, which depends upon the complete set of dynamical variables $\{g_{\alpha\beta}, p^{\mu\nu}\}$, are determined by the Poisson bracket of this variable (or function) with the total Hamiltonian H_t, e.g., $g_{\alpha\beta,0} = [g_{\alpha\beta}, H_t]$. Note that this new (total) Hamiltonian of the metric gravity acts in the $d(d+1)$ dimensional space of the dynamical variables, in contrast with the canonical Hamiltonian H_C which formally operates in the $d(d-1)$ dimensional space of the space-like dynamical variables. In general, introduction of the new Hamiltonian H_t always brings some new motions that did not exist in the original Hamiltonian system with the canonical Hamiltonian H_C.

Immediately, the two following questions arise: (1) what is the sense of these 'additional' motions, and (2) how can they affect the actual motions of our field, which are determined by the canonical Hamiltonian H_C. To understand this and answer the questions raised, let us consider the time-evolution of the canonical Hamiltonian H_C. First of all, we can write the following general formula which describes time-evolution of the canonical Hamiltonian

$$H_C(t+\Delta) = H_C(t) + \frac{\Delta}{1!}\Big(\frac{dH_C}{dt}\Big) + \frac{\Delta^2}{2!}\Big(\frac{d^2H_C}{dt^2}\Big) + \frac{\Delta^3}{3!}\Big(\frac{d^3H_C}{dt^3}\Big) + \dots , \tag{79}$$

where Δ is a small time interval. In this equation the first-order time derivative of H_C is written in the form:

$$\frac{dH_C}{dt} = [H_C, H_t] = [H_C, H_C + v_\alpha \phi^{0\alpha}] = v_\alpha [H_C, \phi^{0\alpha}] = -v_\alpha \chi^{0\alpha} , \tag{80}$$

where $\phi^{0\alpha}$ and $\chi^{0\alpha}$ are the primary and secondary first-class constraints, respectively. The explicit formulas for the $\phi^{0\alpha}$ and $\chi^{0\alpha}$ constraints are presented above (see, Equations (70) and (75)). Moreover, in this equation and everywhere below the notation $v_\alpha (= g_{0\alpha,0})$ is an arbitrary, in principle, velocity of the temporal (0α)-component of the metric tensor. In Dirac's theory this and other similar velocities, e.g., $v_\beta (= g_{0\beta,0}), v_\gamma (= g_{0\gamma,0})$, etc, are considered as arbitrary parameters of the method.

The second time-derivative of the canonical Hamiltonian H_C is

$$\frac{d^2 H_C}{dt^2} = \Big[\frac{dH_C}{dt}, H_t\Big] = [-v_\alpha \chi^{0\alpha}, H_C + v_\beta \phi^{0\beta}] = -v_\alpha D^\alpha_c - \frac{1}{2}\, v_\alpha v_\beta\, g^{\alpha\beta} \chi^{00} \tag{81}$$

where D^α_c is the α component of the Dirac closure (see, Equation (77)), while χ^{00} is the secondary constraint defined above (see, Equation (75)). The third time-derivative of the canonical Hamiltonian takes the form

$$\begin{aligned} \frac{d^3 H_C}{dt^3} = \Big[\frac{d^2 H_C}{dt^2}, H_t\Big] = -v_\alpha [D^\alpha_c, H_C] \; &+ \; \frac{1}{2} v_\alpha v_\beta [g^{\alpha\beta} \chi^{00}, H_C] + \frac{1}{2} v_\alpha v_\beta v_\gamma [g^{\alpha\beta} \chi^{00}, \phi^{0\gamma}] \\ &- \; v_\alpha v_\gamma [D^\alpha_c, \phi^{0\gamma}] \; . \end{aligned} \tag{82}$$

In principle, such a chain of time derivatives $\frac{d^n H_C}{dt^n}$ is infinite (in contrast with the n-dimensional Maxwell electrodynamics [16]), but we have to note that all values in the right-hand sides of these equations are always represented as finite, linear (or quasi-linear) combinations of the secondary, first-class constraints only. Furthermore, the coefficients in front of each term in these expressions depends upon the $v_\alpha, v_\beta, v_\gamma$ and other similar velocities, which "are completely arbitrary and at our disposal" [3]. In other words, these velocities are arbitrary parameters in the Dirac's modification of the classical Hamilton method. It is clear that similar transformations which depend upon arbitrary parameters cannot affect the actual (Hamiltonian) motion of the original dynamical system, e.g., a free gravitational field, in our case. Instead, they produce some changes in the Hamiltonian dynamical variables, which do not correspond to a change of physical state. Generators of such 'fictional' transformations are the secondary first-class constraints (as Dirac predicted in [3]). This follows directly from Equations (80)–(82) and other similar equations for higher-order derivatives in that chain. In field theory similar transformations of the dynamical variables are well known, and in earlier papers they were called gauge transformations, or simply gauges. Dirac could obtain and write (see, e.g., [3]) all essential equations for the actual motion and for the corresponding gauge generators in the united form of Hamilton equations. This explains the overall significance of Dirac's modification of the classical Hamilton method.

In Dirac method the complete system of Hamiltonian equations for the metric gravity is written in the form

$$\frac{dg_{pq}}{dt} = g_{pq,0} = [g_{pq}, H_C] \quad \text{and} \quad \frac{dp^{mn}}{dt} = \Big(p^{mm}\Big)_{,0} = [p^{pq}, H_C] \quad . \tag{83}$$

These $d(d-1)$ Hamilton equations describe the actual motion of a free gravitational field. Solutions of these equations cannot become v-dependent at any moment of time-evolution (see discussion above). In addition to these equations we also have d Hamilton equations which describe time-evolution of the primary constraints:

$$\frac{d\phi^{0\alpha}}{dt} = [\phi^{0\alpha}, H_C] = \chi^{0\alpha} , \tag{84}$$

where $\chi^{0\alpha}$ are the secondary constraints. The following group of d Hamilton equations describe time-evolution of the secondary constraints:

$$\frac{d\chi^{0\alpha}}{dt} = [\chi^{0\alpha}, H_C] = D_c^\alpha \quad , \tag{85}$$

where $D_c^\alpha = a_\mu^\alpha(g)\chi^{0\mu} + b_\mu^\alpha(g)\Big(f^k(g)\chi^{0\mu}\Big)_{,k}$ is the α-component of the Dirac closure, which is a quasi-linear combination of the secondary constraints $\chi^{0\sigma}$ and some total spatial derivatives of expressions which also include the same secondary constraints. All classical theory of the free gravitational field (in metric gravity) is summed up in these Dirac's equations, Equations (83)–(85), written here in a manifestly Hamiltonian form. Note also that there is a simple procedure which allows one to simplify (drastically) the explicit form of Dirac closure in the metric gravity. Indeed, in metric gravity on the shell of primary constraints we can always determine d field-dependent coefficients $C_\beta^\alpha(g)$ for which the following equations are satisfied:

$$[\chi^{0\alpha} + C_\beta^\alpha \phi^{0\beta}, H_C] = \Lambda_\alpha \chi^{0\alpha} \quad , \tag{86}$$

where $\alpha = 0, 1, \ldots, d-1, \beta = 0, 1, \ldots, d-1$ and $\Lambda_\alpha(g)$ are some algebraic, field-dependent expressions, which are often called 'eigenvalues' (or factor-eigenvalues) of the Dirac closure. Derivation of equations for the unknown $C_\beta^\alpha(g)$ coefficients in Equation (86) is straightforward. In this procedure the Dirac closure becomes 'diagonal' and each component of Dirac closure, e.g., D_c^α always contains only one secondary constraint, e.g., $\chi^{0\alpha}$ in Equation (86). All secondary constraints in this procedure are uniquely determined as factor-eigenvectors which are defined on the shell of primary constraints. In this version of Dirac's approach we do not need to say many words to describe the internal structure of Dirac closure.

It is also important to remember that in metric gravity we always have $[\phi^{0\alpha}, \phi^{0\beta}] = 0$, which means the pair-wise commutativity of the primary constraints. These generalized Hamilton equations, Equations (83)–(85), form a complete and unambiguous set of equations, which govern the behaviour of a free gravitational field in the $d(d+1)$-dimensional space of dynamical variables, or in the original d-dimensional Riemann space. The Hamiltonian equations from the first group, Equation (83), are the canonical Hamilton equations of actual motion for true dynamical variables. Analogous equations from the second group, Equations (84) and (85), are the Hamilton equations for gauge generators. These equations determine the actual gauge generators for the given dynamical system, i.e., for the free gravitational field in our case. All equations from the second group describe certain changes in the dynamical variables, i.e., coordinates and momenta, which do not affect the real physical state.

Thus, our original $d(d+1)$-dimensional space of dynamical variables in the metric gravity splits into the $d(d-1)$-dimensional space of dynamical variables, which describe actual motions, and 2 d-dimensional space of variables, which are transformed in some way with time, but this does not make any changes in the real physical state. Formally, we can write this in the form: $S[d(d+1)] = S[d(d-1)] \oplus S[2\ d]$, where all spaces are even-dimensional. If an additional temporal variable t is introduced in our analysis, then all these three spaces become odd-dimensional and Hamilton method works perfectly in each of these spaces. Note that the Hamilton equations in the form of Equations (83)–(85) are more useful and informative for the field people, than the equivalent original system of the $d(d+1)$ Hamilton equations:

$$\frac{dg_{\alpha\beta}}{dt} = g_{\alpha\beta,0} = [g_{\alpha\beta}, H_t] \quad \text{and} \quad \frac{dp^{\mu\nu}}{dt} = \Big(p^{\mu\nu}\Big)_{,0} = [p^{\mu\nu}, H_t] \quad . \tag{87}$$

The replacement of this system of Hamilton equations by much more useful system of slightly different Hamilton equations, Equations (83)–(85), is the main advantage of the Dirac's modifications made in the classical Hamilton method. Another advantage follows from the fact that the governing equations for all gauge generators are also written in the form of Hamilton equations. The third advantage is obvious: from now on all calculations in the metric gravity are reduced to analytical calculations of the Poisson brackets only.

Finally, we can formulate the complete and pure formal criterion of canonicity for some transformation between any two equivalent Hamiltonian formulations of the metric gravity. Based on arguments and equations presented in this and previous Sections, the universal criterion of canonicity for the metric gravity can be formulated in the following form. Some transformation of the dynamical Hamilton variables in metric gravity is canonical if (and only if) it transforms our original system of Hamilton equations, Equations (83)–(85), into a new system of similar Hamilton equations:

$$\frac{d\tilde{g}_{pq}}{dt} = \tilde{g}_{pq,0} = [\tilde{g}_{pq}, \tilde{H}_C] \quad \text{and} \quad \frac{d\tilde{p}^{mn}}{dt} = \left(\tilde{p}^{mm}\right)_{,0} = [\tilde{p}^{pq}, \tilde{H}_C] \ , \tag{88}$$

$$\frac{d\tilde{\phi}^{0\alpha}}{dt} = [\tilde{\phi}^{0\alpha}, \tilde{H}_C] = \tilde{\chi}^{0\alpha} \quad \text{and} \quad \frac{d\tilde{\chi}^{0\alpha}}{dt} = [\tilde{\chi}^{0\alpha}, \tilde{H}_C] = \tilde{D}^{\alpha}_{c} \ , \tag{89}$$

where the sign ˜ means the new variable and/or function, while all new functions $\tilde{H}_C, \tilde{\chi}^{0\alpha}$ and $\tilde{D}^{\alpha}_{c} = \tilde{a}^{\alpha}_{\mu}(\tilde{g})\tilde{\chi}^{0\mu} + \tilde{b}^{\alpha}_{\mu}(\tilde{g})\left(\tilde{f}^{k}(g)\tilde{\chi}^{0\mu}\right)_{,k}$, which appear in these equations, must have the same structure as the old functions $H_C, \chi^{0\alpha}$ and D^{α}_{c} in Equations (83)–(85). This new system of equations represents the form-invariance of the Hamilton equations derived by Dirac for the metric gravity, which is a constrained dynamical system with the first-class constraints only. Furthermore, for the true canonical transformation in the metric gravity the following equations for the Poisson brackets must be obeyed:

$$[\tilde{g}_{\alpha\beta}, \tilde{g}_{\mu\nu}] = 0 \ , \ [\tilde{g}_{\alpha\beta}, \tilde{p}^{mn}] = \Delta^{mn}_{\alpha\beta} \ , \ [\tilde{p}^{mn}, \tilde{p}^{pq}] = 0 \ , \ [\tilde{\phi}^{0\gamma}, \tilde{\phi}^{0\sigma}] = 0$$
$$[\tilde{g}_{\mu\nu}, \tilde{\phi}^{0\sigma}] = \Delta^{0\sigma}_{\mu\nu} \ , \ [\tilde{p}^{mn}, \tilde{\phi}^{0\sigma}] = 0 \ . \tag{90}$$

This criterion of canonicity can be generalized to other Hamiltonian dynamical systems with first-class constraints.

Note also that in this Section we discuss only one version of the complete Dirac's approach [1], which has been developed to deal with a free gravitational field in the metric gravity. Generalization of this procedure to other fields with non-trivial gauge invariance is also possible. For instance, the same approach works perfectly for a free electromagnetic field even in multi-dimensions (see, e.g., [3,7,16]). Our preliminary results indicate clearly that the quantum version of this approach is applicable (with some changes) to the modern united electroweak theory.

On Complete Reverse Recovery of the Original Field Equations

In the previous Sections, we have carefully derived the Hamiltonian equations for a free gravitational field and all primary and secondary first-class constraints. The main purpose of our analysis was to obtain the correct equations of motion (or time-evolution) of a free gravitational field and obtain all important gauge conditions. Here the following question immediately arises: what are these correct Hamiltonian equations of motion? Where and how was the criterion of correctness established? The answer is clear, and we have to recognize as correct only such Hamiltonian equations and first-class constraints which uniformly lead us back to the original (or maternal) equations of motion already known for our field. For a free gravitational field the maternal field equations are the Einstein's equations $G_{\alpha\beta} = 0$ (or $R_{\alpha\beta} = 0$) mentioned in Section 2. For a free electromagnetic field (or EM-field, for short) the maternal equations are the Maxwell equations in vacuum. Therefore, any correct Hamiltonian approach for EM-field must be able to produce the governing Maxwell equations at any spatial point $\mathbf{x}$ and at any moment of time t. To explain how this works, let us consider the Hamiltonian formulation for the Maxwell electromagnetic field in the $(n+1)$-dimensional (flat) space-time.

We can start directly form the explicit form of the corresponding EM-Hamiltonian (all missing details, definitions and notations can be found in [3,16]). Moreover, form the definition of momenta $B^{\mu} = F^{\mu 0}$ to this point we already have one primary constraint $\phi = B^0 \approx 0$, since both $F^{\mu\nu}$ and $F_{\mu\nu}$ tensors are always antisymmetric. The fundamental Poisson brackets are: $[A_\mu(\mathbf{x}), B^\nu(\mathbf{x}')] = g_\mu^\nu \delta^n(\mathbf{x}-\mathbf{x}')$, $[A_\mu(\mathbf{x}), A_\nu(\mathbf{x}')] = 0$ and $[B^\mu(\mathbf{x}), B^\nu(\mathbf{x}')] = 0$. The Hamiltonian of a free electromagnetic field in the $(n+1)$-dimensional space-time is written in the form [3,16]:

$$H = \int \Big(\frac{1}{4}F^{pq}F_{pq} - \frac{1}{2}F^{p0}F_{p0} + F^{q0}A_{0,q}\Big) d^n\mathbf{x} = \int \Big(\frac{1}{4}F^{pq}F_{pq} + \frac{1}{2}B^p B^p - A_0 B^p_{,p}\Big) d^n\mathbf{x} \;, \quad (91)$$

where all notations are exactly the same as in [3,10]. The corresponding Hamiltonian density takes the form

$$\mathcal{H} = \frac{1}{4}F^{pq}F_{pq} + \frac{1}{2}B^p B^p - A_0 B^p_{,p} \; . \quad (92)$$

Integration by parts of the first term in the Hamiltonian, Equation (91), leads to the following expression for the Hamiltonian density Equation (92):

$$\mathcal{H} = -\frac{1}{2}\Big(F^{pq}\Big)_{,q} A_p + \frac{1}{2}B^p B^p - A_0 B^p_{,p} = \frac{1}{2}\Big(\frac{\partial^2 A_q}{\partial x_p \partial x_q} - \frac{\partial^2 A_p}{\partial x_q \partial x_q}\Big) A_p + \frac{1}{2}B^p B^p - A_0 B^p_{,p} \;, \quad (93)$$

where $p = 1, 2, \ldots, n$ and $q = 1, 2, \ldots, n$, i.e., all these indexes are space-like. First of all, by determining the Poisson bracket $[B^0, \mathcal{H}] = B^p_{,p}$ one finds the secondary constraint $\chi = B^p_{,p} \approx 0$. In standard notation this means that $\frac{\partial}{\partial t}\Big(\frac{\partial A_p}{\partial x_p}\Big) \approx 0$, or $\frac{\partial A_p}{\partial x_p} = C$, where C is a numerical constant which does not depend upon $\mathbf{x}$ and/or t. This secondary constraints χ commute with the Hamiltonian density $\mathcal{H}$ and Dirac closure equals zero identically.

By using the Hamiltonian density $\mathcal{H}$, Equation (93), we obtain the following system of canonical Hamilton equations

$$\frac{dA_p}{dt} = [A_p, \mathcal{H}] = \frac{\partial \mathcal{H}}{\partial B^p} = \frac{1}{2}\,(2B^p) = B^p \quad (94)$$

and

$$\frac{dB^p}{dt} = [B^p, \mathcal{H}] = -\frac{\partial \mathcal{H}}{\partial A_p} = -\frac{1}{2}\Big[2\Big(\frac{\partial^2 A_q}{\partial x_q \partial x_p} - \frac{\partial^2 A_p}{\partial x_q \partial x_q}\Big)\Big] = \frac{\partial^2 A_p}{\partial x_q \partial x_q} - \frac{\partial^2 A_q}{\partial x_q \partial x_p} \; . \quad (95)$$

Combination of these two equations one finds

$$\frac{d^2 A_p}{dt^2} = \frac{\partial^2 A_p}{\partial x_q \partial x_q} - \frac{\partial^2 A_q}{\partial x_q \partial x_p} \; . \quad (96)$$

Now, taking into account the condition which follows from secondary constraint: $\frac{\partial A_q}{\partial x_q} = C$, we can reduce this equation to the form of n-dimensional wave equation:

$$\frac{\partial^2 A_p}{\partial t^2} - \frac{\partial^2 A_p}{\partial x_q \partial x_q} = 0 \;\;, \text{ or } \;\; \frac{\partial^2 \mathbf{A}}{\partial t^2} - \Delta \mathbf{A} = 0 \;, \quad (97)$$

where $\mathbf{A} = (A_1, A_2, \ldots, A_n)$ is the n-dimensional vector potential of the EM-field. This is the Maxwell equations for a free electromagnetic field in the $(n+1)$-dimensional space-time. The n-dimensional Laplace operator Δ in this equation is $\Delta = \frac{\partial^2}{\partial x_q \partial x_q}$. Thus, we have recovered all Maxwell equations of the free radiation field. Note here that if someone does not recognize any constraints at all, or these constraints were determined with mistakes, then such a Hamiltonian formulation of the electromagnetic theory does not allow one to recover the corresponding Maxwell equations. Formally, in this case any relation with the original Maxwell theory of radiation will be lost. In reality, this means that our Hamiltonian, Equation (91), does not describe the Maxwell EM-field in vacuum, or this Hamiltonian formulation is not valid for the Maxwell electromagnetic field and cannot be used in applications to this field. This is the principle of complete reverse recovery of the original field equations in application to the Maxwell theory of EM-field.

Now, consider the Hamiltonian formulation of the metric gravity which has been developed by Dirac in [1]. In contrast with a free Maxwell EM-field, for a free gravitational field, everything becomes significantly more complicated, but our principle of the complete reverse recovery works in this case too. Recently, we have shown that the equations for the second-order temporal derivatives of the space-like co-variant components g_{mn} of the metric tensor, which follow from the Hamilton equations obtained in the Dirac Hamiltonian formulation of the metric gravity, essentially coincide with the corresponding Einstein's equations for the same components. However, at that time this article was already completed and it was not possible to add a few new chapters into it. In addition to this, it takes a long time to transform a set of difficult formulas into a logically perfect text. Therefore, our results in this direction will be published some time later and elsewhere. Here we just want to present a few important details of the procedure used.

Note that there are three peculiarities in the Einstein's equations for a free gravitational field ($R_{\alpha\beta} = 0$ plus d additional conditions $R^{\gamma}_{\alpha,\gamma} = \frac{1}{2}\frac{\partial R}{\partial x^{\alpha}}$), which are crucially important for our present purposes. First, these Einstein's equations are written as a system of differential equations which contains the first- the second-order temporal derivatives of the metric tensor $g_{\alpha\beta}$. Second, all second-order temporal derivatives from the $g_{0\beta}$ and $g_{\beta 0}$ components of metric tensor cancel out from these Einstein's equations. Third, the temporal second-order derivatives of the spatial components of metric tensor are explicitly included in the Einstein's equations. In fact, each second-order derivative $\frac{d^2 g_{mn}}{dt^2} = \frac{d^2 g_{mn}}{dx_0^2}$, arises in Einstein's equations only from the R_{0m0n} components of the Riemann curvature tensor. From here it is easy to find that all these second-order temporal derivatives $\frac{d^2 g_{mn}}{dt^2}$ are included in Einstein's equations only as separated terms, and each of these terms has the same numerical coefficient $-\frac{1}{2}$ in front. As follows from here one can reduce the Einstein's equations for the g_{mn} components to the form $\frac{d^2 g_{mn}}{dt^2} = Q(g_{pq}, \frac{dg_{pq}}{dt}, g_{0\alpha}, \frac{dg_{0\alpha}}{dt})$. These equations must coincide (or be equivalent) with the analogous equations for the $\frac{d^2 g_{mn}}{dt^2}$ derivatives which follow from the Hamilton equations in the Dirac formulation of the Hamiltonian metric gravity. In fact, we have to calculate the Poisson bracket $\frac{d^2 g_{mn}}{dt^2} = [[g_{mn}, H_t], H_t] = [[g_{mn}, H_C], H_t]$ and exclude all momenta by using the Hamilton equations, primary and secondary first-class constraints derived in the Dirac Hamiltonian formulation (see above). Such calculations are quite complex, extremely time-consuming and very sensitive, since any mistake made either in the Hamiltonian, or in one of the first-class constraints substantially complicates further calculations. After such a mistake one can lose any relation with the original (or maternal) theory and cannot move forward, until this mistake is found and corrected. Nevertheless, after many weeks of calculations, we are happy to report that Hamiltonian formulation of the metric gravity, which has been developed by Dirac in [1], successfully passed this our test of recovery (at least partially). Now, by using this Hamiltonian formulation we are able to recover the original field equations for all covariant space-like components g_{mn} of the metric tensor. An alternative Hamiltonian formulation of metric gravity which obviously fails this test is mentioned in Section 9.

To conclude this discussion we have to note that any correct Hamiltonian formulation of an arbitrary, in principle, field theory must reproduce (exactly and unambiguously) the original governing equations of this field. The correct Hamilton equations of motions and explicit form of all first-class constraints are crucially important to reach this goal. If this is not the case, then such a Hamiltonian theory is wrong and has nothing to do with the maternal field theory.

8. Invariant Integrals of the Metric Gravity

This Section is a central part of our study, since here we define a number of integral invariants of the metric gravity, i.e., we reach an absolute top of the classical Hamilton mechanics. Obviously, in one Section we cannot even outline the main problems which exist in the theory of integral invariants and its applications to the Hamiltonian metric gravity. Therefore, the following presentation of this theory will be very brief. Moreover, here we restrict ourselves only to a description of the extension of integral invariants for dynamical (Hamiltonian) systems with constraints. More details of the theory of integral invariants and its applications to the Hamiltonian formulations of metric gravity will be presented in our next article [18].

First of all we need to define the one-dimensional, or line integrals in the metric gravity. In general, the one-dimensional integral in multi-dimensional Riemann spaces is defined as follows

$$\begin{aligned}\int \pi^{\alpha\beta}\, dg_{\alpha\beta} &= \int \pi^{\alpha\beta} \Big(\frac{\partial g_{\alpha\beta}}{\partial x^{\beta}}\Big) dx^{\beta} = \int \pi^{\alpha\beta} \Big(\Gamma_{\alpha,\beta\gamma} + \Gamma_{\beta,\alpha\gamma}\Big) dx^{\gamma} \\ &= \int \pi^{\alpha\beta} \Big(g_{\alpha\lambda}\Gamma^{\lambda}_{\beta\gamma} + g_{\beta\lambda}\Gamma^{\lambda}_{\alpha\gamma}\Big) dx^{\gamma} \;, \end{aligned} \tag{98}$$

where $\Gamma_{\alpha,\beta\gamma}$ are the Cristoffel symbols of the first kind, while $\Gamma^{\alpha}_{\beta\gamma}$ are the Cristoffel symbols of the second kind (see, e.g., [4]). The integrand in this integral is not a tensor. This means that the line (or one-dimensional) integral substantially depends on the curve along which it is calculated and also on the initial and final points chosen on this curve. If the start and end points coincide with each other, then such an integral is called a closed loop integral, or an integral taken along a closed loop. Below, similar closed loop integrals are designated by the sign $\oint$. In general, the complete theory of line (or one-dimensional) integrals in multi-dimensional Riemann spaces is very complex. However, for our current analysis of the integral invariants in the Hamiltonian formulations of metric gravity we do need to use the formula, Equation (98). In fact, all line integrals can be considered in the $\frac{d(d+1)}{2}$-dimensional pseudo-Euclidean (orthogonal) space, which is formally identical (or isomorphic) to our original d-dimensional Riemann space-time (for more details, see, e.g., [4,5]).

This is very good news, since all integrals and integral forms defined in multi-dimensional pseudo-Euclidean spaces can be handled in a familiar way (see, e.g., [19,20]). In particular, by applying the usual definition of the closed loop integrals in pseudo-Euclidean spaces we can consider the two following integrals in the $\frac{d(d-1)}{2}$-dimensional space:

$$I = \oint \Big[\pi^{mn}\, dg_{mn} - H_C dt\Big] \quad \text{and} \quad I_P = \oint \pi^{mn}\, dg_{mn} \;, \tag{99}$$

where π^{mn} are the space-like components of momenta, while H_C is the canonical Hamiltonian which has been defined in Section 3. For now we restrict ourselves to the consideration of the space-like components of momenta π^{mn} and coordinates g_{mn}. In other words, below we shall deal with the $\frac{d(d-1)}{2}$ dimensional Euclidean position space (and $d(d-1)$-dimensional phase space) instead of the original $(d-1)$-dimensional sub-space in our original d-dimensional Riemann space-time. The coordinates in this position space coincide with the covariant components of the fundamental space-like tensor g_{mn}. The first integral in Equation (99) is called the Poincare–Cartan integral invariant, while the second integral is the Poincare integral invariant [21,22] which is often called the main integral invariant of mechanics (or Hamiltonian mechanics). This Poincare integral invariant has a fundamental value for the Hamiltonian formulation of metric gravity as well as for the general theory of canonical transformations in metric gravity and for analysis and solution of other gravitational problems. Indeed, it is relatively easy to prove the following statement. If for some system of first-order differential equations written for the space-like components of the metric tensor g_{mn} and momenta π_{mn}:

$$\frac{dg_{mn}}{dt} = Q_{mn}(t, g_{ab}, \pi^{pq}) \quad , \quad \frac{d\pi^{mn}}{dt} = P^{mn}(t, g_{ab}, \pi^{pq}) \tag{100}$$

the Poincare integral I_P, Equation (99), is invariant, then this system of equations, Equation (100), is Hamiltonian in the moment of time t which is located between $t - \delta$ and $t + \delta$, where δ is a very small positive number. This term 'Hamiltonian' means here that the functions $Q_{mn}(t, g_{ab}, \pi^{pq})$ and $P^{mn}(t, g_{ab}, \pi^{pq})$ from the right-hand side of Equations (100) are represented as the partial derivatives (or Poisson brackets) of some scalar function H, i.e.,

$$Q_{mn}(t, g_{ab}, \pi^{pq}) = \frac{\partial H}{\partial \pi^{mn}} = [g_{mn}, H] \quad , \quad P^{mn}(t, g_{ab}, \pi^{pq}) = -\frac{\partial H}{\partial g_{mn}} = [\pi^{mn}, H] \quad , \tag{101}$$

where the notation $[a, b]$ stands for the Poisson bracket defined by Equation (18). A uniform reconstruction of the explicit form of this function H (or Hamiltonian) is not an easy task, but if we know that the Poincare–Cartan integral is also an integral invariant, then the unknown Hamiltonian exactly coincides [21] with the canonical Hamiltonian H_C mentioned in the first integral from Equation (99).

Let us discuss the following fundamental question. We shall assume that the integral I, Equation (99), is an integral invariant for our dynamical system and H_C is our Hamiltonian which describes the actual motion of this system, i.e., the equations of time-evolution take the form of Hamilton equations, Equation (83), for our $d(d-1)$ dynamical variables $\{g_{mn}, \pi^{pq}\}$, where $[g_{mn}, g_{pq}] = 0, [\pi^{mn}, \pi^{pq}] = 0$ and $[g_{mn}, \pi^{pq}] = \Delta^{pq}_{mn}$. Now, we want to extend our phase space by adding a set of $2d$ new dynamical variables $\{g_{0\gamma}, \pi^{0\gamma}\}$. Here we assume the usual permutation symmetry for all 'additional' coordinates and momenta: $g_{0\gamma} = g_{\gamma 0}$ and $\pi^{0\gamma} = \pi^{\gamma 0}$. The total dimension of this new phase space will be $d(d+1)$, which corresponds to the d-dimensional Riemann space-time, and it is the main working space for the general relativity and metric gravity. We require that in this new extended phase space the integral defined by the last expression in Equation (102) must also be an integral invariant with the new Hamiltonian H_t. Furthermore, the new 'extended' Hamiltonian H_t must be closely related with the canonical Hamiltonian H_C from Equation (99).

Based on the formulas derived in Section 3 we can transform the Poincare–Cartan integral invariant from Equation (99) into the following form

$$\begin{aligned} I = \oint \Big[\pi^{mn}\, dg_{mn} \;-\; H_C dt\Big] &= \oint \Big\{ \Big[\pi^{mn}\, dg_{mn} + \pi^{0\gamma}\Big(\frac{dg_{0\gamma}}{dt}\Big)dt\Big] - \Big[H_C + \pi^{0\gamma}\Big(\frac{dg_{0\gamma}}{dt}\Big)\Big]dt \Big\} \\ &= \oint \Big[\pi^{\alpha\beta}\, dg_{\alpha\beta} - \Big(H_C + v_\gamma \pi^{0\gamma}\Big)dt\Big] = \oint \Big[\pi^{\alpha\beta}\, dg_{\alpha\beta} - H_t dt\Big] \quad , \end{aligned} \tag{102}$$

where $H_t = H_C + v_\gamma \pi^{0\gamma} = H_C + g_{0\gamma,0}\pi^{0\gamma}$ and $g_{0\gamma,0} = v_\gamma$ ($\gamma = 0, 1, \ldots, d-1$) are the corresponding velocities, while $\pi^{0\gamma}$ are the temporal momenta which must be equal zero along each Hamilton trajectory of the actual motion. In other words, we have d primary constraints $\pi^{0\gamma} \approx 0$ in the metric gravity. Otherwise, i.e., if $\pi^{0\gamma} \neq 0$ for some γ, then Equation (102) does not hold. Another crucial fact which has been used to transform, Equation (102), follows from the formulas for canonical Hamiltonian H_C, Equations (17) and (72), which do not contain any of the temporal momenta $\pi^{0\gamma}$, but it may include some of the $g_{0\mu}$ and/or $g_{\mu 0}$ coordinates. If these conditions are obeyed, then from Equation (102) we can derive the following equality:

$$\oint \Big[\pi^{mn}\, dg_{mn} - H_C dt\Big] = I = \oint \Big[\pi^{\alpha\beta}\, dg_{\alpha\beta} - H_t dt\Big] \quad , \tag{103}$$

which means that both of these integrals, i.e., integrals on the left and right sides of this equation, are the true integral invariants and their numerical values equal to each other. In other words, these two integral invariants coincide with each other, i.e., they are not independent, and for constrained dynamical systems we always have to deal with this complication.

The formula, Equation (103), has a number of consequences for Hamiltonian formulations of the metric gravity, but here we consider just one of them. First, as follows from the left-hand side of Equation (103) the set of dynamical variables $\{g_{mn}, \pi^{pq}\}$ will be canonical, if the Poisson brackets between these dynamical variables coincide with CUSPB, i.e., $[g_{mn}, g_{pq}] = 0, [g_{mn}, \pi^{pq}] = \Delta^{pq}_{mn}, [\pi^{mn}, \pi^{pq}] = 0$ for all possible (mn)- and (pq)-pairs. In other words, it is a necessary and sufficient condition of canonicity in this case. However, if we apply the same arguments to the integral in right-hand side of Equation (103) we can only say that the exact coincidence of the Poisson brackets between dynamical variables $[g_{\alpha\beta}, g_{\mu\nu}] = 0, [g_{\alpha\beta}, \pi^{\mu\nu}] = \Delta^{\mu\nu}_{\alpha\beta}, [\pi^{\alpha\beta}, \pi^{\mu\nu}] = 0$ with the standard CUSPB values is only necessary (but not sufficient!) condition of canonicity. In order to obtain the sufficient conditions of canonicity we must also guarantee that all temporal momenta $\pi^{0\gamma}$ and/or $\pi^{\gamma 0}$ do not change with time t along the true Hamilton trajectories. This means that all time-derivatives of the temporal momenta must be equal zero at all times, i.e., we have a number of additional equations such as $\pi^{0\gamma} \approx 0, [\pi^{0\gamma}, H_t] \approx 0, [[\pi^{0\gamma}, H_t], H_t] \approx 0$, etc. To obey all these equations for the primary, secondary and other constraints we have to follow Dirac's modifications made in the classical Hamilton method for constrained dynamical systems (see above). Otherwise, if some of these conditions do not hold, then the numerical value of the integral in the right-hand side of Equation (103) will be different from I, i.e., this integral is not invariant in this case and we have an obvious contradiction here. This explains why the criteria of canonicity derived for constrained dynamical systems always include two parts: (a) coincidence of the Poisson brackets between dynamical variables with the standard CUSPB values, and (b) conservation of the algebra of first-class constraints. This presumes the form-invariance of all first-class constraints and Poisson brackets of these constraints with each other, and with the canonical/total Hamiltonians and other essential functions of dynamical variables.

9. Conclusions

Thus, we have investigated the two different Hamiltonian formulations [1,6] of the metric gravity in d-dimensional Riemann space, where $d \geq 3$. These two Hamiltonian formulations are related to each other by a canonical transformation of dynamical variables in the $d(d+1)$-dimensional phase space and each of them allows one to restore the complete d-dimensional diffeomorphism as the correct (and well-known) gauge invariance of the free gravitational field in the metric gravity. By using the known canonical transformation between these two Hamiltonian formulations of the metric gravity we have investigated the basic properties of other similar canonical transformations and derived some useful criteria of canonicity for an arbitrary transformation of the Hamiltonian dynamical variables in the $d(d+1)$-dimensional phase space. The results of our study are important in numerous applications, since in metric gravity canonical transformations of Hamiltonian dynamical variables are often used to simplify either the canonical H_C and/or total Hamiltonian(s) H_t, or secondary constraints, or to reduce the canonical Hamiltonian H_C to some special form, e.g., to its normal form, which is well known in classical mechanics.

In general, all criteria of canonicity for transformations of dynamical variables in the metric gravity require the exact coincidence of the Poisson (or Laplace) brackets for the new and old dynamical (Hamiltonian) variables. Briefly, if the Poisson brackets of the new dynamical variables (expressed in the old dynamical variables) do not coincide with their canonical values, then such a transformation is not canonical. This is the universal criterion of canonicity which is known from classical mechanics of Hamiltonian dynamical systems. However, in all Hamiltonian formulations of metric gravity we always deal with the constrained dynamical systems. Therefore, all criteria of canonicity, which are valid for such systems, must contain the second part which deals with the algebra of first-class constraints, form-invariance of the canonical/total Hamiltonian(s) and/or form-invariance of the Hamilton equations. For instance, the true canonical transformation in the metric gravity must keep form-invariance of the Hamilton equations derived in the Dirac's modification of the classical Hamilton method. It can be illustrated by the transformation of Equations (83)–(85) into Equations (88) and (89) during our canonical transformation of the Hamiltonian dynamical variables. This is the first criterion of canonicity in the metric which is relatively simple and ready to be applied to actual problems. The second similar criterion [17] requires the exact coincidence of the total Hamiltonian H_t and preservation of the algebra of constraints for both (old and new) Hamiltonian formulations of the metric gravity.

We have also reconsidered modifications made by Dirac [1–3] in the classical Hamilton approach. It is shown that these modifications are crucial to improve overall efficiency of the new Hamiltonian method for dynamical systems with constraints, including various physical fields with additional gauge conditions, or gauges, for short. The main advantage of the new Dirac's approach is a possibility to write all governing equations in the united, manifestly Hamilton form (see Equations (83)–(85) and Equations (88) and (89) above). The original Dirac's idea that all motions in Hamiltonian dynamical systems with first-class constraints can always be separated into actual motions and special motions along constraints (or gauge-consistent motions) was extremely productive. Now, by using this Dirac's modification of the classical Hamilton method we can describe time-evolution of a large number of actual and model fields. Furthermore, we can make a conjecture that the free fields which represents all currently known fundamental interactions can unambiguously be described by the this version of Hamiltonian method, which was originally developed and later modified by Dirac.

In this study we have also considered the method of integral invariants and applied it to investigate canonical transformations between different Hamiltonian formulations of the metric gravity. This method was originally proposed and developed by Poincare and Cartan [21]. Since then it was transformed into a very powerful approach, which currently is an absolute tool in the Hamilton classical mechanics. In reality, the invariance of the Poincare–Cartan integral can be chosen as a foundation of the whole Hamiltonian mechanics. Indeed, if this integral is invariant for some dynamical system, then such a system is Hamiltonian and its time-evolution is described by a system of Hamilton equations. For Hamiltonian dynamical systems with constraints the general theory of integral invariants must be modified, but its overall power still remains outstanding.

Unfortunately, the limited space of this article did not allow us to discuss other important directions of the Hamiltonian formulations of metric gravity. In particular, we could not consider the explicit derivation of the gauge generators which are defined by chain of the first-class constraints [6,23] (see also [10]). Moreover, in this study, we did not even mention various non-canonical quasi-Hamiltonian formulations of the metric gravity. However, we can make a reference to an excellent review article [24] which contains a detailed analysis of this problem and a large number of references to papers published up to the beginning of 2011. Here we want to note that any of these non-canonical Hamiltonian formulations uses a set of dynamical ADM-variables, which were introduced in [25]. This fact has been noticed and criticized by Bergmann, Dirac and many others. Our calculations of the corresponding Poisson brackets can be found in [10,17]. However, since early 1960's there were no explanations of this remarkable fact and its consequences neither from ADM people, nor from their followers (see, e.g., [26–28]). Then, in 1985 it was suddenly detected that ADM formulation of the Hamiltonian metric gravity cannot restore, in principle, the total four-dimensional diffeomorphism [29] which is the correct and well-known gauge symmetry of a free gravitational field in four-dimensional space-time. Recently, we have found another crucial problem for ADM gravity and similar non-canonical 'Hamiltonian' formulations. Indeed, in Dirac's Hamiltonian formulation, we could restore the original Einstein's equations for a free gravitational field. Analogous ADM

Hamiltonian formulation uses, in part, the same dynamical variables (12 of 20 variables), but there are some fundamental mistakes in all secondary constraints. Therefore, the extra terms which present in the restored field equations for ADM formulation do not cancel each other (as they do in Dirac's formulation), but remain and even multiply. Finally, in Dirac's Hamiltonian formulation we obtain the maternal Einstein equations with no additional terms, while for ADM Hamiltonian formulation we have similar equations with many extra terms in them. As follows from this fact the ADM Hamiltonian formulation either describes some different (i.e., non-Einstein's) field, or it is an absolutely wrong theoretical construction which does not represent and real and/or model field (if the arising system of extended Einstein-like equations is not closed). In the future, under better circumstances, we plan to discuss these (and other) issues which currently exist in the Hamiltonian formulations of the metric gravity.

Funding: This research received no external funding.

Acknowledgments: I am grateful to my friends N. Kiriushcheva, S.V. Kuzmin and D.G.C. (Gerry) McKeon (all from the University of Western Ontario, London, Ontario, Canada) for helpful discussions and inspiration.

Conflicts of Interest: The author declares no conflict of interest.

Appendix A. Some Useful Relations in the Hamiltonian Version of Metric Gravity

In this 'pure technical' Appendix we derive a few equations and relations which are crucially important for our Hamiltonian formulation of the metric gravity. First, let us define the space-like tensor I_{mnpq} by the equation $I_{mnpq} = a g_{mn} g_{pq} - g_{mp} g_{nq}$, Here we want to show that $I_{mnpq} E^{pqkl} = \delta^k_m \delta^l_n$ in the case when $a = \frac{1}{d-2}$. First of all, we note that $g_{\alpha\beta} g^{\alpha\beta} = d$, where d is the total dimension of our Riemann space. From here one finds

$$g_{\alpha\beta} e^{\alpha\beta} = g_{\alpha\beta} g^{\alpha\beta} - g_{\alpha\beta} \frac{g^{0\alpha} g^{0\beta}}{g^{00}} = d - g^0_\alpha \frac{g^{0\alpha}}{g^{00}} = d - 1 \ . \tag{A1}$$

On the other hand, since $e^{0\beta} = 0, e^{\alpha 0} = 0$ and $e^{00} = 0$, we can write the last equality in a different form

$$d - 1 = g_{\alpha\beta} e^{\alpha\beta} = g_{pq} e^{pq} + g_{p0} e^{p0} + g_{0q} e^{0q} + g_{00} e^{00} = g_{pq} e^{pq} \ . \tag{A2}$$

In other words, we obtain $g_{pq} e^{pq} = d - 1 = g_{\alpha\beta} e^{\alpha\beta}$. Furthermore, one can derive a similar rule to lower the index in the $e^{\alpha\beta}$ tensor: $g_{\alpha\beta} e^{\beta\gamma} = e^\gamma_\alpha = g_{pq} e^{qm} = e^m_p = g^m_p = \delta^m_p$.

Now, we can prove the statement formulated above. The formula for the $I_{mnpq} E^{pqkl}$ product takes the form

$$\begin{aligned} I_{mnpq} E^{pqkl} &= (a g_{mn} g_{pq} - g_{mp} g_{nq}) \left(e^{pq} e^{kl} - e^{pk} e^{ql} \right) = a(d-1) g_{mn} e^{kl} - g_{mp} \delta^p_n e^{kl} \\ &- a g_{mn} e^{pk} \delta^l_p + \delta^k_m \delta^l_n = [a(d-1) - 1 - a] g_{mn} e^{kl} + \delta^k_m \delta^l_n \ . \end{aligned} \tag{A3}$$

From here one finds that if $a(d-2) = 1$ (or $a = \frac{1}{d-2}$), then the first term in the last equation equals zero identically and $I_{mnpq} E^{pqkl} = \delta^k_m \delta^l_n$.

Another important relation which we want to prove here is the connection between the $[g_{\alpha\beta}, \pi^{\mu\nu}]$ and $[g^{\alpha\beta}, \pi^{\mu\nu}]$ Poisson brackets. Formally, the covariant components of the metric tensor $g_{\alpha\beta}$ are the only generalized coordinates in the metric gravity. However, there is an obvious relation between covariant and contravariant components of the metric tensor: $g_{\alpha\beta} g^{\beta\gamma} = g^\gamma_\alpha = \delta^\gamma_\alpha$. From here one finds the following relation between the two Poisson brackets

$$[\pi^{\mu\nu}, g_{\alpha\beta}] g^{\beta\gamma} + g_{\alpha\beta} [\pi^{\mu\nu}, g^{\beta\gamma}] = 0 \ \text{ or } \ [g_{\alpha\beta}, \pi^{\mu\nu}] g^{\beta\gamma} = - g_{\alpha\beta} [g^{\beta\gamma}, \pi^{\mu\nu}] \ . \tag{A4}$$

By multiplying both sides of this equation by the tensor $g^{\alpha\sigma}$ we obtain the following relation

$$g^{\alpha\sigma}[g_{\alpha\beta}, \pi^{\mu\nu}]g^{\beta\gamma} = -g_{\alpha\beta}g^{\alpha\sigma}[g^{\beta\gamma}, \pi^{\mu\nu}] = -g^{\sigma}_{\beta}[g^{\beta\gamma}, \pi^{\mu\nu}] = -[g^{\gamma\sigma}, \pi^{\mu\nu}] \; , \tag{A5}$$

or

$$[g^{\gamma\sigma}, \pi^{\mu\nu}] = -g^{\alpha\sigma}\, \Delta^{\mu\nu}_{\alpha\beta}\, g^{\beta\gamma} = -\frac{1}{2}\Big(g^{\mu\sigma}g^{\nu\gamma} + g^{\mu\gamma}g^{\nu\sigma}\Big) = -[\pi^{\mu\nu}, g^{\gamma\sigma}] \; , \tag{A6}$$

i.e., the result which exactly coincides with the formulas, Equations (21) and (24), from the main text. The derivation of two other formulas from Equations (25) is absolutely analogous.

The last formula, which we want to derive in this Appendix, is the Poisson brackets between two secondary, fist-class constraints of the metric gravity, i.e., $[\chi^{0\sigma}, \chi^{0\gamma}]$. This formula is needed to complete our Hamiltonian formulation of the metric gravity. Furthermore, it is of great interest in a number of gravitational problems. This formula has never been produced in earlier studies, since its direct derivation is not an easy task. Below, we apply a different approach [10] which is based on the Jacobi identity:

$$\begin{aligned}[][\chi^{0\sigma}, \chi^{0\gamma}] &= [\chi^{0\sigma}, [\phi^{0\gamma}, H_C]] = -[\phi^{0\gamma}, [H_C, \chi^{0\sigma}]] - [H_C, [\chi^{0\sigma}, \phi^{0\gamma}]] \\ &= [\phi^{0\gamma}, [\chi^{0\sigma}, H_C]] + [H_C, [\phi^{0\gamma}, \chi^{0\sigma}]] = [\phi^{0\gamma}, D^{\sigma}_{C}] - \frac{1}{2}[g^{\gamma\sigma}, H_C] \; ,\end{aligned} \tag{A7}$$

where D^{σ}_{C} is the σ-component of the Dirac's closure, Equation (77), while H_C is the canonical Hamiltonian of the metric gravity, Equation (72). Here we apply the expressions derived in the Dirac's Hamiltonian formulation of metric gravity.

Analytical calculations of both terms in the last equation from Equation (A7) are relatively easy, since only a few Poisson brackets really contribute. Here we just present the final result for the second term in Equation (A7):

$$\begin{aligned}-\frac{1}{2}[g^{\gamma\sigma}, H_C] &= \frac{1}{4\sqrt{-g}g^{00}} I_{mnpq}\Big[\Big(g^{\gamma m}g^{\sigma n} + g^{\gamma n}g^{\sigma m}\Big)\phi^{pq} + \phi^{mn}\Big(g^{\gamma p}g^{\sigma q} + g^{\gamma q}g^{\sigma p}\Big)\Big] \\ &- \frac{1}{4g^{00}}\Big(g^{\gamma m}g^{\sigma n} + g^{\gamma n}g^{\sigma m}\Big)\Big(g^{0k}g_{mn,k} - 2g^{0\alpha}g_{\alpha n,m}\Big) \; .\end{aligned} \tag{A8}$$

This expression can be simplified, e.g., by using the identities such as $g_{mn}g^{m\gamma} = g_{\alpha n}g^{\alpha\gamma} - g_{0n}g^{0\gamma} = g^{\gamma}_{n} - g_{0n}g^{0\gamma}$, but here we do not want to make similar simplifications.

In order to determine the first Poisson bracket in Equation (A7) we note that the σ-component of the Dirac closure can be written in the compact form of quasi-linear combination of secondary first-class constraints $D^{\sigma}_{C} = V^{\sigma}_{\lambda}\chi^{0\lambda} = V^{\sigma}_{k}\chi^{0k} + V^{\sigma}_{0}\chi^{00}$, where $V^{\sigma}_{\lambda} = V^{\sigma}_{\lambda}(g_{pq}, \phi^{mn}, g^{0\mu})$ is the structure functions (or functional) [17]. Therefore, for the Poisson bracket in Equation (A8) we can write $[\phi^{0\gamma}, D^{\sigma}_{C}] = [\phi^{0\gamma}, V^{\sigma}_{\lambda}]\ \chi^{0\lambda} + V^{\sigma}_{\lambda}\ [\phi^{0\gamma}, \chi^{0\lambda}]$, where $\lambda = (0, k)$. In these notations the explicit formula for the second term is written in the form:

$$\begin{aligned}V^{\sigma}_{\lambda}[\phi^{0\gamma}, \chi^{0\lambda}] &= -\Big[\frac{g^{\sigma m}}{\sqrt{-g}g^{00}} I_{pqmk}\phi^{pq} + \frac{1}{2}g^{\sigma 0}g_{00,k} + g^{\sigma p}g_{0p,k} + \frac{g^{\sigma p}g^{0q}}{2g^{00}}\Big(g_{pq,k} + g_{qk,p} \\ &- g_{pk,q}\Big)\Big]g^{\gamma k}\chi^{00} - \frac{1}{2}g^{\sigma}_{0}(g^{\gamma k}\chi^{00})_{,k} + \frac{1}{2}g^{0\gamma}g^{\sigma k}g_{00,k}\chi^{00} \; ,\end{aligned} \tag{A9}$$

where $g^{\alpha}_{\beta} = \delta^{\alpha}_{\beta}$ is the substitution tensor. An analogous formula for the first term is derived either by applying the following expression $[\phi^{0\gamma}, V^{\sigma}_{\lambda}] = -\frac{\partial V^{\sigma}_{\lambda}}{\partial g_{0\gamma}}$, or directly. The final expression takes the form

$$\begin{aligned} [\phi^{0\gamma}, V^{\sigma}_{\lambda}]\chi^{0\lambda} &= \Big[\frac{1}{\sqrt{-g}g^{00}}\Big(g^{0\sigma}g^{\gamma m} + g^{\gamma\sigma}g^{0m} - 3g^{0\gamma}g^{\sigma m}\Big)I_{pqmk}\phi^{pq}\Big]\chi^{0k} \\ &- \delta^{\gamma}_{0}\Big(g^{0\sigma}\chi^{0k}\Big)_{,k} - \delta^{\gamma}_{p}\Big(g^{\sigma p}\chi^{0k}\Big)_{,k} + \frac{1}{2}\delta^{\gamma}_{0}\Big(g^{\sigma k}\chi^{0k}\Big)_{,k} \end{aligned} \tag{A10}$$

The complete formula for the $[\chi^{0\sigma}, \chi^{0\gamma}]$ Poisson bracket is the algebraic sum of three expressions from Equations (A8)–(A10).

Appendix B. On Canonicity of the Hamiltonian Dynamical Variables

Let us briefly discuss the criteria of canonicity for the different sets of dynamical variables which describe the same Hamiltonian system. In almost all applications of similar criteria in classical mechanics it is important to know that the new set of dynamical variables will be canonical, if the old set of dynamical variables was canonical. For simplicity, below the new dynamical variables $\tilde{q}_i$ and $\tilde{p}_j$ are designated by the upper signs $\tilde{}$, while old dynamical variables are denoted as q_i and p_j. Now, we can define the following $2n \times 2n$ Jacobi matrix $\hat{M}$ of some transformation of the dynamical variables:

$$\hat{M} = \begin{pmatrix} \frac{\partial \tilde{q}_1}{\partial q_1} & \ldots & \frac{\partial \tilde{q}_1}{\partial q_n} & \frac{\partial \tilde{q}_1}{\partial p_1} & \ldots & \frac{\partial \tilde{q}_1}{\partial p_n} \\ \ldots & \ldots & \ldots & \ldots & \ldots & \ldots \\ \frac{\partial \tilde{q}_n}{\partial q_n} & \ldots & \frac{\partial \tilde{q}_n}{\partial q_n} & \frac{\partial \tilde{q}_n}{\partial p_1} & \ldots & \frac{\partial \tilde{q}_n}{\partial p_n} \\ \frac{\partial \tilde{p}_1}{\partial q_1} & \ldots & \frac{\partial \tilde{p}_1}{\partial q_n} & \frac{\partial \tilde{p}_1}{\partial p_1} & \ldots & \frac{\partial \tilde{p}_1}{\partial p_n} \\ \ldots & \ldots & \ldots & \ldots & \ldots & \ldots \\ \frac{\partial \tilde{p}_n}{\partial q_n} & \ldots & \frac{\partial \tilde{p}_n}{\partial q_n} & \frac{\partial \tilde{p}_n}{\partial p_1} & \ldots & \frac{\partial \tilde{p}_n}{\partial p_n} \end{pmatrix} = \begin{pmatrix} \frac{\partial \tilde{\mathbf{q}}}{\partial \mathbf{q}} & \frac{\partial \tilde{\mathbf{q}}}{\partial \mathbf{p}} \\ \frac{\partial \tilde{\mathbf{p}}}{\partial \mathbf{q}} & \frac{\partial \tilde{\mathbf{p}}}{\partial \mathbf{p}} \end{pmatrix}. \tag{A11}$$

where $\frac{\partial \tilde{\mathbf{q}}}{\partial \mathbf{q}}, \frac{\partial \tilde{\mathbf{q}}}{\partial \mathbf{p}}, \frac{\partial \tilde{\mathbf{p}}}{\partial \mathbf{q}}$ and $\frac{\partial \tilde{\mathbf{p}}}{\partial \mathbf{p}}$ are the $n \times n$ matrices. Another matrix which we need here is the unit simplectic $2n \times 2n$ matrix $\hat{J}$:

$$\hat{J} = \begin{pmatrix} 0 & \ldots & 0 & -1 & \ldots & 0 \\ \ldots & \ldots & \ldots & \ldots & \ldots & \ldots \\ 0 & \ldots & 0 & 0 & \ldots & -1 \\ 1 & \ldots & 0 & 0 & \ldots & 0 \\ \ldots & \ldots & \ldots & \ldots & \ldots & \ldots \\ 0 & \ldots & 1 & 0 & \ldots & 0 \end{pmatrix} = \begin{pmatrix} 0 & -E \\ E & 0 \end{pmatrix}. \tag{A12}$$

where E is the $n \times n$ unit matrix. It easy to show that the matrix $\hat{J}$ is invertable and it obeys the following equation: $\hat{J}^2 = -\hat{J}\hat{J}^{-1}$, or $\hat{J}^{-1} = -\hat{J}$. This fundamental property of the unit simplectic matrix $\hat{J}$ substantially determines many known properties of the canonical transformations and predicts a number of necessary steps in the Hamilton method. As follows from Equation (A12) the matrix $\hat{J}$ is not self-adjoint, but its product with the imaginary unit $\imath$, i.e., the matrix $\imath\hat{J}$ is a truly self-adjoint matrix. This fact is crucial for correct definition of the Poisson brackets in quantum mechanics (see below).

It can be shown that for the matrix $\hat{M}$, which represents some canonical transformation of the Hamilton dynamical variables, the following condition is always obeyed: $\hat{M}'\hat{J}\hat{M} = c\hat{J}$, where $c(\neq 0)$ is a real and/or complex number which is called the valence of this transformation. Additionally, in this equation $\hat{J}$ is the unit simplectic matrix and $\hat{M}'$ is the matrix adjoint to the original matrix $\hat{M}$. Vice versa, if some matrix of differential transformation of variables $\hat{M}$ obeys the equation $\hat{M}'\hat{J}\hat{M} = c\hat{J}$, then such a differential transformation of dynamical variables, which is represented by the matrix $\hat{M}$, is canonical with the valence c. Thus, we have formulated the criterion of canonicity in terms of the

Jacobi matrix which is easily determined for an arbitrary transformation of the Hamiltonian dynamical variables. Let us discuss this matrix criterion with some additional details. First of all, by using the properties of the $\hat{J}$ matrix mentioned above it is possible to show (in two steps) that the two following equations (or conditions) follow from each other

$$\hat{M}'\hat{J}\hat{M} = c\hat{J} \iff \hat{M}\hat{J}\hat{M}' = c\hat{J} \ . \tag{A13}$$

which means that these two equations are equivalent to each other. Indeed, at the first step we multiply both sides of the equation $\hat{M}'\hat{J}\hat{M} = c\hat{J}$ by the $[\hat{M}']^{-1}$ matrix from the left and by the $[\hat{M}]^{-1}$ matrix from the right. These operations lead to the new equation $\frac{1}{c}\hat{J} = [\hat{M}']^{-1}\hat{J}[\hat{M}]^{-1}$. At the second step we just need to reverse both sides of this equation. By taking into account that $\hat{J}^{-1} = -\hat{J}, [\hat{M}^{-1}]^{-1} = \hat{M}$ and $\{[\hat{M}']^{-1}\}^{-1} = \hat{M}'$, we find the equality: $c\hat{J} = \hat{M}\hat{J}\hat{M}'$, which exactly coincides with the second equation in Equations (A13). On the other hand it is easy to check that, if we start from the second equation in Equations (A13), then the two analogous and simple steps allow one to derive the first equation in Equations (A13).

The matrix $\hat{M}$ (and/or $\hat{M}'$) which obeys any of these equations, Equations (A13), is called the true simplectic matrix with numerical valence c. In general, all even-dimensional (non-singular) simplectic matrices form the closed simplectic group $Sp(k,R)$ (or $Sp(k,C)$), where $k = 2n$. Finally, we arrive to the following theorem: some non-singular transformation of Hamilton dynamical variables $\tilde{q}_i = \tilde{q}_i(t, q_i, p_i), \tilde{p}_i = \tilde{p}_i(t, q_i, p_i)$, where $i = 1, \ldots, n$, will be canonical, if (and only if) its Jacobi matrix $\hat{M}$, Equation (A11), is the true symplectic matrix with the valence c. In this case the condition, Equation (A13), must be obeyed identically for all old dynamical variables and time t.

Further investigation shows that the matrix elements of the $\hat{M}'\hat{J}\hat{M}$ matrix coincide with the corresponding Lagrange brackets written in old coordinates and momenta, while the matrix elements of the $\hat{M}\hat{J}\hat{M}'$ matrix coincide with the corresponding Poisson brackets which are also written in old coordinates and momenta. Thus, we have an obvious duality between the Lagrange and Poisson brackets, which can be illustrated by a simple rule which is applied to form the (ij)-matrix elements of the adjoint matrix $\hat{M}'$ (from $\hat{M}$) and vice versa. This rule is simple: in each partial derivative, which is included in the Jacobi matrix of the canonical transformation (or its adjoint), the letters and indices at the top and bottom are swapped, while the symbol $\tilde{}$ always stays at the top. Let us consider the following example. As mentioned above the matrix equation $\hat{M}'\hat{J}\hat{M} = c\hat{J}$ for the canonical transformation is equivalent to the following system of equations:

$$\{q_i, q_j\} = 0 \ , \ \{p_i, p_j\} = 0 \ , \ \{q_i, p_j\} = c\delta_{ij} \ , \tag{A14}$$

where the notation $\{a, b\}$ means the Laplace bracket defined in the main text. In other words, the matrix elements of the $\hat{M}'\hat{J}\hat{M}$ matrix always coincide with the corresponding Laplace brackets, while numerical values of these brackets are determined from the matrix equation $\hat{M}'\hat{J}\hat{M} = c\hat{J}$. Now, by taking adjoint of Equation (A14) one finds the following matrix equation: $\hat{M}\hat{J}\hat{M}' = c\hat{J}$, which leads to the three group of equalities for the adjoint Laplace brackets:

$$\{q_i, q_j\}^\star = 0 \ , \ \{p_i, p_j\}^\star = 0 \ , \ \{q_i, p_j\}^\star = c\delta_{ij} \ , \tag{A15}$$

where the sign '$\star$' means that inside of Lagrange brackets we have to apply our 'swap of variables' described above. By using the explicit formulas for the Lagrange brackets and our recipes to construct the adjoint matrix we can write

$$\{q_i, q_j\}^\star = \Big[\sum_{k=1}^{n}\Big(\frac{\partial \tilde{q}_k}{\partial q_i}\frac{\partial \tilde{p}_k}{\partial q_j} - \frac{\partial \tilde{q}_k}{\partial q_j}\frac{\partial \tilde{p}_k}{\partial q_i}\Big)\Big]^\star = \sum_{k=1}^{n}\Big(\frac{\partial \tilde{q}_i}{\partial q_k}\frac{\partial \tilde{q}_j}{\partial p_k} - \frac{\partial \tilde{q}_j}{\partial q_k}\frac{\partial \tilde{q}_i}{\partial p_k}\Big) = [\tilde{q}_i, \tilde{q}_j] \tag{A16}$$

where the notation $[a, b]$ stands for the 'regular' Poisson bracket. Analogous expressions can be derived for other fundamental Lagrange brackets $\{q_i, p_j\}^\star = [\tilde{q}_i, \tilde{p}_j]$ and $\{p_i, p_j\}^\star = [\tilde{p}_i, \tilde{p}_j]$, where all Poisson brackets are calculated in the old dynamical variables. These formulas indicate clear that the adjoint of the Laplace bracket equals (remarkably and unambiguously) to the corresponding Poisson bracket. Briefly, this means that we have reduced calculations of the adjoints of the Laplace brackets to computations of the corresponding Poisson brackets. Note also that the right-hand sides of Equations (A15) do not change during the $\star$ procedure and we obtain the following numerical values for the fundamental Poisson brackets

$$[\tilde{q}_i, \tilde{q}_j] = 0 \; , \; [\tilde{p}_i, \tilde{p}_j] = 0 \; , \; [\tilde{q}_i, \tilde{p}_j] = c\delta_{ij} \; , \tag{A17}$$

which coincide with the expected numerical values. Thus, we have shown that the Poisson brackets coincide with the adjoints of the corresponding Laplace brackets (and vice versa). In other words, these two systems of brackets are closely related to each other and each of these brackets can equally be used to check and prove the canonicity of some new set of Hamilton dynamical variables. Based on these facts it is relatively easy (for simplicity, we choose $c = 1$ here) to prove the following theorem [12]: if $u_1, u_2, \ldots, u_{2n}$ are the $2n$ independent functions of the variables $q_1, \ldots, q_n, p_1, \ldots, p_n$, then the two equations

$$\sum_{k=1}^{2n} \{u_k, u_i\}[u_k, u_j] = \sum_{k=1}^{2n} \{u_i, u_k\}[u_j, u_k] = \delta_{ij} \; , \tag{A18}$$

are always obeyed for these functions. This equation(s) explicitly shows a very close relation between the Laplace and Poisson brackets and they essentially follow from the definitions of these two brackets.

The last remark, which we want to make here, describes the main difference which arises in definitions of the canonical transformations in the classical and quantum mechanics. The Jacobi matrix $\hat{M}$, which describes the canonical transformation of the Hamiltonian dynamical variables, can also be defined (with a few additional tricks) in quantum mechanics. However, the governing equation Equation (A11) for the Jacobi $\hat{M}$ matrix in quantum mechanics includes the self-adjoint unit matrix $\imath\hat{J}$, which is not the unit simplectic $\hat{J}$ matrix as it was in classical mechanics. This means that in quantum mechanics both newly-defined Poisson brackets and their numerical values will always include (explicitly) the imaginary unit $\imath$. In the fundamental Poisson bracket we can introduce the 'new' momenta in the coordinate representation by including $\imath$ in its definition. On the other hand, the numerical value of the corresponding Poisson bracket must also include the imaginary unit $\imath$. Let us consider the following example. Suppose we have a point non-relativistic particle with the mass m which is located at the point with the Cartesian coordinates (x, y, z) and has the velocity (v_x, v_y, v_z). One can introduce the momenta for this particle $p_i = mv_i$, where $i = (x, y, z)$. In classical mechanics we have three fundamental Poisson brackets: $[x_i, p_j] = \delta_{ij}, [x_i, x_j] = 0$ and $[p_i, p_j] = 0$. The same momenta and numerical values of all non-zero Poisson brackets in quantum mechanics are defined as follows: $p_k = -\imath\hbar\frac{\partial}{\partial x_k}$ and $[p_i, x_j] = -\imath\hbar\delta_{ij}$, respectively [30]. Here $(i, j) = (x, y, z)$ and $\hbar = \frac{h}{2\pi} \approx 1.054571817 \cdot 10^{-34}$ $J \cdot sec$ is the reduced Planck constant which is also called the Dirac constant. An additional trick in this case is the explicit form of the coordinate operator x_k in momentum representation: $x_k = -\imath\hbar\frac{\partial}{\partial p_k}$, or $\frac{\partial}{\partial p_k} = \frac{\imath}{\hbar}x_k$, where $k = x, y, z$.

Note that both quantum momenta and numerical values of the Poisson brackets contain the imaginary unit $\imath$. Actual presence of the self-adjoint matrix $\imath\hat{J}$ in the governing equation $\hat{M}'(\imath\hat{J})\hat{M} = \imath\hat{J}$ for the quantum Jacobi matrix $\hat{M}$ allows one to define the canonical transformations in quantum mechanics as the unitary transformations. In other words, canonical transformations of the Hamiltonian dynamical variables in quantum mechanics are represented by the unitary matrices $\hat{M}$ and $\hat{M}'$ only. Thus, we have found the following relation between the classical and quantum Poisson brackets $[f, g]_{class} = \frac{\imath}{\hbar}[\hat{f}, \hat{g}]_{quant}$. This relation can be considered as a canonical transformation between the classical and quantum mechanics with the imaginary valence $c = \frac{\imath}{\hbar}$. The imaginary valence of this canonical transformation means that all dynamical variables in quantum mechanics must be represented by self-adjoint operators and canonical transformations between different sets of dynamical variables in quantum mechanics are always performed by the unitary matrices, which differ from the simplectic matrices used for the same purposes in classical mechanics.

Appendix C. Variational Derivation of Integral Invariants for Hamiltonian Systems

As is well known, the Hamilton method has a number of significant advantages over other methods which are used to solve the same problems in classical mechanics, e.g., over the Lagrange method. One of these advantages is the method of integral invariants which is a 'hidden' part of any true Hamilton approach. The method of integral invariants allows one to analyze and solve many problems in mechanics. Here we describe a variational derivation of integral invariants for arbitrary, in principle, Hamiltonian dynamical systems. First, let us consider variations of the two following actions (or action integrals) W_L and W_H:

$$W_L = \int_{t_0}^{t_1} L\Big(t, q_i(t,\alpha), \dot{q}_i(t,\alpha)\Big)dt \;\;\text{and}\;\; W_H = \int_{t_0(\alpha)}^{t_1(\alpha)} L\Big(t, q_i(t,\alpha), \dot{q}_i(t,\alpha)\Big)dt \; , \qquad \text{(A19)}$$

where the notation α stands for the parameter. All coordinates $q_i(t,\alpha)$ and velocities $\dot{q}_i(t,\alpha)$ in these two actions depend on this parameter. In the second action (W_H) the lower and upper limits in the time-integral also depend upon this parameter α. During variations of these two integrals we can always interchange the sings of variations δ and time derivative $\frac{d}{dt}$, since we can write

$$\delta\dot{q}_i = \delta\Big(\frac{d}{dt}q_i(t,\alpha)\Big) = \frac{\partial}{\partial\alpha}\Big[\frac{d}{dt}q_i(t,\alpha)\Big]\delta\alpha = \frac{d}{dt}\Big[\frac{\partial}{\partial\alpha}q_i(t,\alpha)\delta\alpha\Big] = \frac{d}{dt}\delta q_i(t,\alpha) \qquad \text{(A20)}$$

By using this equation we can derive the following formulas for variations of these two actions W_L and W_H:

$$\delta W_L = \int_{t_0}^{t_1} \sum_{i=1}^{n} \Big[\frac{\partial L}{\partial q_i} - \frac{d}{dt}\Big(\frac{\partial L}{\partial \dot{q}_i}\Big)\Big]\delta q_i dt \qquad \text{(A21)}$$

and

$$\delta W_H = \Big\{\sum_{i=1}^{n}\Big(\frac{\partial L}{\partial \dot{q}_i}\Big)\delta q_i - \Big[\sum_{i=1}^{n}\Big(\frac{\partial L}{\partial \dot{q}_i}\Big)\dot{q}_i - L\Big]\delta t\Big\}\Bigg|_{t_0(\alpha)}^{t_1(\alpha)} + \int_{t_0(\alpha)}^{t_1(\alpha)} \sum_{i=1}^{n}\Big[\frac{\partial L}{\partial q_i} - \frac{d}{dt}\Big(\frac{\partial L}{\partial \dot{q}_i}\Big)\Big]\delta q_i dt \, . \qquad \text{(A22)}$$

Now, by introducing the following notations: $\frac{\partial L}{\partial \dot{q}_i} = p_i$ and $\sum_{i=1}^{n}\Big(\frac{\partial L}{\partial \dot{q}_i}\Big)\dot{q}_i - L = H$ we can re-write the last expression in the form:

$$\delta W_H = \Big[\sum_{i=1}^{n} p_i\delta q_i - H\delta t\Big]\Bigg|_{t_{0(\alpha)}}^{t_1(\alpha)} + \int_{t_0(\alpha)}^{t_1(\alpha)} \sum_{i=1}^{n}\Big[\frac{\partial L}{\partial q_i} - \frac{d}{dt}\Big(\frac{\partial L}{\partial \dot{q}_i}\Big)\Big]\delta q_i dt \; , \qquad \text{(A23)}$$

where the function H is Hamiltonian of the system, while p_i ($i = 1, \ldots, n$) are the momenta.

Let us define the optimal (or shortest) path between two spatial points M_0 and M_1. The optimal path satisfies the Lagrange equation $\frac{\partial L}{\partial q_i} - \frac{d}{dt}\Big(\frac{\partial L}{\partial \dot{q}_i}\Big) = 0$ and variation of the Lagrange action W_L is always equal zero for the shortest path. It is clear that in this case the optimal path does not depend upon the parameter α, i.e., for this path we can write the equation $q_i = q_i(t)$, where $i = 1, \ldots, n$ and all $q_i(t)$ functions do not depend on the parameter α. If we consider the variation of the Hamilton action W_H, then the corresponding optimal path is also determined by the Lagrange equation. However, since in this case $t_0 = t_0(\alpha)$ and $t_1 = t_1(\alpha)$, then we obtain the whole α-dependent family of optimal paths $q_i = q_i(t,\alpha)$, where $i = 1,\dots,n$. On each of these optimal α-dependent paths, the variation of W_H is written in the form

$$\delta W_H(\alpha) = \delta \int_{t_0(\alpha)}^{t_1(\alpha)} L\Big(t, q_i(t,\alpha), \dot{q}_i(t,\alpha)\Big) dt = \Big(\frac{\partial W_H(\alpha)}{\partial \alpha}\Big)\delta\alpha = \Big[\sum_{i=1}^{n} p_i \delta q_i - H\delta t\Big]\Bigg|_{t_{0(\alpha)}}^{t_1(\alpha)} . \tag{A24}$$

A different derivation of this formula can be found, e.g., [31]. For actual (or optimal) motion of the Hamiltonian system the variation of W_H must be equal zero, i.e., $\delta W_H(\alpha) = 0$.

Instead of the extended $(n+1)$-dimensional coordinate space $\{q_i(\alpha), t(\alpha)\}$, let us introduce the extended $(2n+1)$-dimensional phase space of the generalized coordinates q_i, momenta p_i and time t which we shall designate as the $\{q_i(\alpha), p_i(\alpha), t(\alpha)\}$ space. In this phase space, we choose an arbitrary closed curve C_0, which is described by the equations: $q_i(0) = q_i(t_0,\alpha), p_i(0) = p_i(t_0,\alpha), t = t_0(\alpha)$, where $i = 1,\dots,n$ and $0 \le \alpha \le \ell$. Note that for $\alpha = 0$ and for $\alpha = \ell$ we have the same point on the curve. This means that $q_i(0) = q_i(\ell)$ and $p_i(0) = p_i(\ell)$. At the next step, from each point of this closed curve C_0, as from the initial point, we draw the corresponding optimal path. Such an optimal path is uniquely determined (after we set the initial point on the curve C_0) from the system of Hamiltonian canonical equations. By choosing different initial points on the closed curve C_0 we obtain a tube of optimal (or shortest) paths each of which is also α-dependent. The explicit equations for this tube of optimal paths are: $q_i = q_i(t,\alpha), p_i = p_i(t,\alpha)$, where $i = 1,\dots,n, 0 \le \alpha \le \ell$ and also $q_i(t,0) = q_i(t,\ell)$ and $p_i(t,0) = p_i(t,\ell)$ for $i = 1,\dots,n$. Formally, each optimal path on the surface of this tube can be considered as some generatrix of the tube.

On this tube, we select the second closed curve C_1, which also covers the tube and has only one common point with each generatrix. The equations of this closed curve C_1 are written in the form: $q_i(1) = q_i(t_1,\alpha), p_i(1) = p_i(t_1,\alpha), t = t_1(\alpha)$, where $i = 1,\dots,n$ and $0 \le \alpha \le \ell$. Now, by integrating the equation $\delta W_H(\alpha) = 0$ over the parameter α in the range from $\alpha = 0$ to $\alpha = \ell$, one finds

$$\begin{aligned} 0 &= W_H(\ell) - W_H(0) = \int_0^{\ell} \Big[\sum_{i=1}^{n} p_i\delta q_i - H\delta t\Big]\Bigg|_{t_0}^{t_1} = \int_0^{\ell} \Big[\sum_{i=1}^{n} p_i^1 \delta q_i^1 - H_1 \delta t_1\Big] \\ &- \int_0^{\ell} \Big[\sum_{i=1}^{n} p_i^0 \delta q_i^0 - H_0\delta t_0\Big] = \oint_{C_1} \Big[\sum_{i=1}^{n} p_i\delta q_i - H\delta t\Big] - \oint_{C_0} \Big[\sum_{i=1}^{n} p_i\delta q_i - H\delta t\Big] . \end{aligned} \tag{A25}$$

or

$$\oint_{C_1} \Big[\sum_{i=1}^{n} p_i\delta q_i - H\delta t\Big] = \oint_{C_0} \Big[\sum_{i=1}^{n} p_i\delta q_i - H\delta t\Big] . \tag{A26}$$

This means that the closed loop integral $I = \oint \Big[\sum_{i=1}^{n} p_i\delta q_i - H\delta t\Big]$ taken along an arbitrary closed contour does not change its value during an arbitrary displacement along the tube of straight paths. In other words, this integral I is invariant, which is called the integral invariant of Poincare–Cartan (see, Equation (99), in the main text) for the Hamiltonian dynamical system with the Hamiltonian H.

In pure mathematical language we have created and investigated the rotor tube for the differential form $\omega^1 = \mathbf{p}\; d\mathbf{q} - Hdt$, where $\mathbf{p} = (p_1,\dots,p_n)$ and $\mathbf{q} = (q_1,\dots,q_n)$. The rotor lines, each of which is the generatrix of this tube, are uniformly determined by the canonical Hamiltonian equations. In other words, the true Hamiltonian trajectories (or integral curves of the canonical Hamilton equations) are the generatrixes of the rotor tube for the differential form $\omega^1 = \mathbf{p}\; d\mathbf{q} - Hdt$. Note that about integral invariants for Hamiltonian systems it is possible to talk endlessly, but for now it is better to stop here.

The last thing which we want to present in this Appendix is the explicit formula for the differential of the ω^1 form, which is the 2-form $d\omega^1$ [20]:

$$d\omega^1 = \sum_{i=1}^{n} \Big(dp_i \wedge dq_i - \frac{\partial H}{\partial p_i} dp_i \wedge dt - \frac{\partial H}{\partial q_i} dq_i \wedge dt \Big) = 0 \ , \tag{A27}$$

where $\wedge$ is the standard notation of exterior product [19,20]. The matrix $\hat{A}$ of this 2-form $d\omega^1$ in the $(2n+1)$-dimensional phase space is:

$$\hat{A} = \begin{pmatrix} 0 & -E & \frac{\partial H}{\partial \mathbf{p}} \\ E & 0 & \frac{\partial H}{\partial \mathbf{q}} \\ -\frac{\partial H}{\partial \mathbf{p}} & -\frac{\partial H}{\partial \mathbf{q}} & 0 \end{pmatrix} , \text{ where } \begin{pmatrix} 0 & -E \\ E & 0 \end{pmatrix} = \hat{J} \tag{A28}$$

and $\hat{J}$ is the unit simplectic $2n \times 2n$ matrix which has been defined and discussed in the Appendix B. The rank of this matrix $\hat{A}$ equals $2n$ (the rank of $\hat{J}$ matrix), i.e., it has one zero-eigenvalue. The corresponding eigenvector is $(-\frac{\partial H}{\partial \mathbf{q}}, \frac{\partial H}{\partial \mathbf{p}}, 1)$ and it determines the direction of rotor lines for the differential form $\omega^1 = \mathbf{p}\, d\mathbf{q} - Hdt$. More about the uniqueness of integral invariants, their classification and integral invariants of higher orders can be found, e.g., in [32].

References

1. Dirac, P.A.M. The theoty of gravitation in Hamiltonian form. *Proc. Roy. Soc.* **1958**, *246*, 333.
2. Dirac, P.A.M. Generalized Hamiltonian Dynamics. *Can. J. Math.* **1950**, *2*, 129. [CrossRef]
3. Dirac, P.A.M. *Lectures on Quantum Mechanics*; Befler Graduate School of Sciences, Yeshiva University: New York, NY, USA, 1964.
4. Kochin, N.E. *Vector Calculus and the Principles of Tensor Calculus*, 9th ed.; USSR Acad. of Sciences Publishing: Moscow, Russia, 1965; Chpt. IV.
5. Dashevskii, P.K. *Riemannian Geometry and Tensor Analysis*, 3rd ed.; Nauka: Moscow, Russia, 1967.
6. Kiriushcheva, N.; Kuzmin, S.V.; Rankor, C.; Valluri, S.R. Diffeomorphism invariance in the Hamilton formulation of General Relativity. *Phys. Lett. A* **2008**, *327*, 5101. [CrossRef]
7. Gitman, D.M.; Tyutin, I.V. *Canonical Quantization of Fields with Constraints*; Springer: Berlin/Heidelberg, Germany, 1990.
8. Carmeli, M. *Classical Fields: General Relativity and Gauge Theory*; World Scientific Publ. Co.: Singapore, 2002.
9. Landau, L.D.; Lifshitz, E.M. *The Classical Theory of Fields*, 4th ed.; Pergamon Press: Oxford, UK, 1975; Chpt. 11.
10. Frolov, A.M. General Principles of Hamiltonian Formulations of the Metric Gravity. *Phys. Atomic Nuclei* **2021**, *84*, 750. [CrossRef]
11. Gantmakher, F.R. *Lectures on Analytical Mechanics*; Chpts. III and IV (The Best Course on Hamilton Methods in Classical Mechanics); Nauka: Moscow, Russia, 1966.
12. Goldstein, H. *Classical Mechanics*; Addison-Wesley Press, Inc.: Cambridge, MA, USA, 1950; Chpts. 7–9.
13. Lanczos, C. *The Variational Principles of Mechanics*; University of Toronto Press: Toronto, ON, Canada, 1970; Chpts. V–VIII.
14. Arnold, V.I. *Mathematical Methods of Classical Mechanics*, 2nd ed.; Nauka: Moscow, Russia, 1979; Part III.
15. Pirani, F.A.E.; Schild, A.; Skinner, R. Quantization of Einstein's Gravitational Field Equations. II. *Phys. Rev.* **1952**, *87*, 452. [CrossRef]
16. Frolov, A.M. On Maxwell electrodynamics in multi-dimensional spaces. *Universe* **2022**, *8*, 20. [CrossRef]
17. Frolov, A.M.; Kiriushcheva, N.; Kuzmin, S.V. On Canonical Transformations between Equivalent Hamiltonian Formulations of General Relativity. *Gravit. Cosmol.* **2011**, *17*, 314. [CrossRef]
18. Frolov, A.M.; Department of Applied Mathematics, University of Western Ontario, London, ON N6H 5B7, Canada. Metric gravity in the Hamiltonian Form. Integral Invariants. 2023; in preparation.
19. Cartan, A. *Differential Calculus and Differential Forms*; In Russian These Two Well Known Books of H. Caratan Were Published in One Volume; Mir Publ.: Moscow, Russia, 1971.
20. Flanders, H. *Differential Forms with Applications to the Physical Sciences*; Dover Publications, Inc.: New York, NY, USA, 1989.
21. Cartan, E. *Lessons on Integral Invariants*; Delphenich, D.H., Translator; Hermann Editors for the Sciences and Arts: Paris, France, 1922.
22. Poincaré, H. *New Methods of Celestial Mechanics*; Springer: Berlin/Heidelberg, Germany, 1977.
23. Castellani, L. Symmetries in constrained Hamiltonian systems. *Ann. Phys.* **1982**, *143*, 357. [CrossRef]
24. Kiriushcheva, N.; Kuzmin, S.V. The Hamiltonian formulation of general relativity: Myths and reality. *Cent. Eur. J. Phys.* **2011**, *9*, 576. [CrossRef]
25. Arnowitt, R.; Deser, S.; Misner, C. Canonical Variables for General Relativity. *Phys. Rev.* **1960**, *117*, 1595. [CrossRef]
26. Regge, T.; Teitelboim, C. Role of surface integrals in the Hamiltonian formulation of general relativity. *Ann. Phys.* **1974**, *88*, 286. [CrossRef]

27. Lee, J.; Wald, R.M. Local symmetries and constraints. *J. Math. Phys.* **1990**, *31*, 725. [CrossRef]
28. Wald, R.M.; Zoupas, A. General definition of "conserved quantitie" in general relativity and other theories of gravity. *Phys. Rev. D* **2000**, *61*, 084027. [CrossRef]
29. Isham, C.J.; Kuchar, K.V. Representations of spacetime diffeomorphisms. I. Canonical parametrized field theories. *Ann. Phys.* **1985**, *164*, 473. [CrossRef]
30. Dirac, P.A.M. *The Principles of Quantum Mechanics*, 4th ed.; Oxford at Clarendon Press: Oxford, UK, 1958.
31. Gelfand, I.M.; Fomin, S.V. *Calculus of Variations;* Dover Publ., Inc.: Mineola, NY, USA, 1990; Chpt. 3.
32. Lee, H.-C. Invariants of Hamilton system and applications to the theory of canonical transformations. *Proc. Roy. Soc. Edinbourgh Ser. A* **1947**, *LXII*, 237.

universe

MDPI

Article

Sign Switching Dark Energy from a Running Barrow Entropy

Sofia Di Gennaro [1] and Yen Chin Ong [1,2,*]

[1] Center for Gravitation and Cosmology, College of Physical Science and Technology, Yangzhou University, 180 Siwangting Road, Yangzhou 225002, China
[2] Shanghai Frontier Science Center for Gravitational Wave Detection, School of Aeronautics and Astronautics, Shanghai Jiao Tong University, Shanghai 200240, China
* Correspondence: ycong@yzu.edu.cn

Abstract: Barrow proposed that the area law of the entropy associated with a horizon might receive a "fractal correction" due to quantum gravitational effects—in place of $S \propto A$, we have instead $S \propto A^{1+\delta/2}$, where $0 \leqslant \delta \leqslant 1$ measures the deviation from the standard area law ($\delta = 0$). Based on black hole thermodynamics, we argue that the Barrow entropy should run (i.e., energy scale dependent), which is reasonable given that quantum gravitational corrections are expected to be important only in the high-energy regime. When applied to the Friedmann equation, we demonstrate the possibility that such a running Barrow entropy index could give rise to a dynamical effective dark energy, which is asymptotically positive and vanishing, but negative at the Big Bang. Such a sign switching dark energy could help to alleviate the Hubble tension. Other cosmological implications are discussed.

Keywords: Barrow entropy; running Barrow index; modified Friedmann equation; varying gravitational constant; cosmological constant

1. Introduction: Barrow Entropy and Quantum Gravity

One of the most striking properties of gravity is that black holes—and by extension, apparent horizons—have entropy that scales with the surface area [1,2]. Specifically, the Bekenstein–Hawking entropy is

$$S = \frac{A}{4}\frac{k_B c^3}{G\hbar}, \tag{1}$$

where k_B, c, and G are the Boltzmann constant, the speed of light, and Newton's gravitational constant, respectively.

In different approaches to quantum gravity, this expression receives different forms of correction (see Section 2). Recently, Barrow [3] proposed that, due to quantum gravity corrections, black holes might exist with extremely wrinkled surfaces, such that the event horizon is like, e.g., a Koch fractal surface. The entropy expression becomes

$$S = \left(\frac{A}{A_0}\right)^{1+\frac{\delta}{2}}, \tag{2}$$

where $0 \leqslant \delta \leqslant 1$ is the parameter that governs how "fractalized" the surface has become, and we shall refer to it as the "Barrow entropy index" (BEI). It is not quite clear whether we should have $S = \frac{k_B}{4}\left(\frac{A}{\ell_p^2}\right)^{1+\delta/2}$ or $S = k_B\left(\frac{A}{4\ell_p^2}\right)^{1+\delta/2}$. Barrow worked with $S \propto A/4 \approx A$ from the beginning (before the correction), and thus, this issue was not discussed. However, the constant factor $1/4$, whether it is raised to the power of $1+\delta/2$ or not, does not affect the results much, except possibly when dealing with microscopic black holes, since the area would be very large anyway, so we use the form of Equation (2), in line with most literature, for easy comparison. However, for definiteness, we set $A_0 = 4\ell_p^2$, where ℓ_p is the Planck length $\ell_p = \sqrt{G\hbar/c^3}$. Hereinafter, we employ the Planck units, so that $\hbar = G = c = k_B = 1$.

Cosmological and black hole shadow constraints (assuming a fixed δ) have put an upper bound on δ, which is typically $\delta \lesssim O(10^{-3})$ or $O(10^{-4})$ [4–9].

In this work, we argue that δ has to be a function of the energy scale and study the implications of Barrow entropy for cosmology (by the expansion of the universe, δ is thus a function of cosmic time). Such a possibility was also raised in [7,10,11]. We expect that, as quantum gravitational effects become more pronounced near the Big Bang, $\delta \to 1$. As the universe expands and cools, $\delta \to 0$ monotonically. Incidentally, in the context of an asymptotically flat Schwarzschild black hole, this would mean that, as the black hole evaporates and its temperature increases, δ would approach unity (the effect of a running BEI on Hawking evaporation will be studied elsewhere).

The running Barrow entropy index gives rise to an additional term in the modified Friedmann equation, which we propose to be identified with an effective dynamical dark energy, which is asymptotically zero (hence, small at late time). Surprisingly, the effective dark energy has a negative sign right about the Big Bang, i.e., the Universe started out as an anti-de Sitter-like spacetime. Interestingly, this may help to relax the Hubble tension—the mismatch between the locally measured expansion rate of the universe and the inferred rate from early universe via the cosmic microwave background (CMB) [12,13].

In addition, following a recent work of Sheykhi [14], which considered the fixed δ scenario, we shall identify the correction on the right-hand side (r.h.s.) of the Friedmann equation as an effective modification on the gravitational constant, instead of the matter field. This has the advantage that the resulting fixed δ cosmology satisfies the generalized second law (GSL), whereas if the matter sector is modified instead of the geometry, then the GSL fails [15]. The viability of the GSL in the context of the interacting Barrow holographic dark energy model was also investigated in [16].

2. Barrow Entropy Index Is Energy Scale Dependent

In a recent work [17], Chen et al. studied black holes in asymptotically safe gravity, in which the gravitational constant also runs [18] (see [19,20] for cosmological studies). They showed that the area law receives a well-known logarithmic correction in the quantum gravity literature:

$$S = \frac{A}{4} + C\ln(A); \tag{3}$$

see [21] for a review. This at least suggests that the reverse implication—a modification of the Bekenstein–Hawking entropy leading to a varying effective gravitational constant (as in the BEI case)—is perhaps not so surprising.

Here, we remark that, although the Barrow entropy is supposed to quantify a quantum gravitational correction, it is *not* the—arguably more understood—quantum gravity effect that gives rise to Equation (3). (In fact, the effects of having both corrections together were considered in [22].) The sign of the constant C in front of the logarithmic term in Equation (3) is usually negative [21], but positive in some works, such as [17,23]. In [24], Gour argued that a positive constant might be required if the area fluctuation is taken into account, whereas for a fixed area, the corrections due to the number of microstates (that describe the black hole) would give a negative constant. We have not much to add to this discussion. Instead, we would like to point out that the Barrow entropy, Equation (2), if expanded out as a series assuming small δ (when the quantum gravity correction is small), yields, up to the first order in δ,

$$S = \frac{A}{4} + \frac{A}{8}\ln\left(\frac{A}{4}\right)\delta. \tag{4}$$

Note the presence of the area as a coefficient of the logarithmic correction. Thus, there is a risk that the "correction" term can be of the same order as the leading term. In fact, for any fixed $\delta \ll 1$, the expression will still deviate significantly from the Bekenstein–Hawking area law for a sufficiently *large* black hole, precisely when we expect quantum gravity corrections to be *small*. This provides a strong argument for a running BEI—δ should at least scale inversely proportional to $A\ln(A)$ to keep the subleading term small at $O(1)$. Such a running BEI would guarantee that quantum gravity correction is small when the black hole is large (i.e., when the energy scale is small, since the Hawking temperature is inversely proportional to the mass). Another option is to consider a δ that runs like $1/A$, then the Barrow entropy would just be the same as the standard logarithmic correction (at least up to the first order in δ), but in this work, we take the former view, with the hope of uncovering new physics.

3. Dark Energy From a Running Barrow Entropy Index

Sheykhi derived the correction to the Friedmann equation due to the Barrow entropy (with a fixed BEI) in [14]:

$$\left(H^2+\frac{k}{a^2}\right)^{1-\frac{\delta}{2}}=\frac{8\pi}{3}\rho\left[\frac{2-\delta}{2+\delta}\frac{A_0^{1+\frac{\delta}{2}}}{4(4\pi)^{\frac{\delta}{2}}}\right]=:\frac{8\pi\rho}{3}G_{\text{eff}}. \tag{5}$$

In other words, the matter field feels an effective gravitational constant[1] G_{eff} instead of G. This is one of the main results of [14].

This was obtained by applying the first law of thermodynamics to the energy flux through the apparent horizon of the Friedmann–Lemaître–Robertson–Walker (FLRW) universe, instead of the black hole horizon [23]. The expression for the horizon is

$$r_{\text{AH}}=\left(H^2+\frac{k}{a^2}\right)^{-\frac{1}{2}}. \tag{6}$$

In the following, we assumed a spatially flat universe ($k=0$).

Allowing the Barrow entropy index to run, the change in entropy would be:

$$\begin{aligned} dS &= d\left(\frac{4\pi r^2}{A_0}\right)^{1+\frac{\delta}{2}} \\ &= \left(\frac{4\pi r^2}{A_0}\right)^{1+\frac{\delta}{2}}\left[\frac{(2+\delta)\dot{r}}{r}+\ln\left(\frac{4\pi r^2}{A_0}\right)\frac{\dot{\delta}}{2}\right]dt. \end{aligned} \tag{7}$$

Then, using the first law of thermodynamics and the continuity equation, we can show, following [14], that

$$\begin{aligned} dE &= -\frac{1}{2\pi r}\left(\frac{4\pi r^2}{A_0}\right)^{1+\frac{\delta}{2}}\left[\frac{(2+\delta)\dot{r}}{r}+\ln\left(\frac{4\pi r^2}{A_0}\right)\frac{\dot{\delta}}{2}\right]dt \\ &= -4\pi r^3 H(\rho+p)dt, \end{aligned} \tag{8}$$

and, consequently,

$$\frac{(4\pi)^{\frac{\delta}{2}}}{2\pi A_0^{1+\frac{\delta}{2}}}\left[(2+\delta)r^{\delta-3}+\ln\left(\frac{4\pi r^2}{A_0}\right)\frac{r^{\delta-2}}{2}\frac{d\delta}{dr}\right]dr=-\frac{\dot{\rho}}{3}dt. \tag{9}$$

We note that $r_{\text{AH}}=1/H$. In a sensible cosmology, H decreases with time, so r_{AH} is an increasing function of time. In other words, a larger r corresponds to a later time. From the black hole thermodynamics argument as per Section 2, we want δ to at least decrease as $\delta\sim\text{const.}/r^2\ln(r^2)$ at large r. At early time, it should approach unity. For definiteness, we can take a function that is always smaller than the asymptotic behavior $\text{const.}/r^2\ln(r^2)$, so a natural choice is $\delta(r)=e^{-r}$.

The differential form of the modified Friedmann equation is equivalent to

$$(2-\delta)r^{\delta-3}dr + \frac{2-\delta}{2+\delta}\ln\left(\frac{4\pi r^2}{A_0}\right)\frac{r^{\delta-2}}{2}\frac{d\delta}{dr}dr$$
$$= -\frac{8\pi}{3}\underbrace{\left[\frac{2-\delta}{2+\delta}\frac{A_0^{1+\frac{\delta}{2}}}{4(4\pi)^{\frac{\delta}{2}}}\right]}_{=:G_{\text{eff}}}d\rho, \tag{10}$$

where G_{eff} is the effective gravitational constant, the same as defined in the fixed BEI case [14], as in Equation (5). We immediately notice that G_{eff} is monotonically decreasing as $\delta \to 0$, with $G_{\text{eff}}(\delta = 1) = 1/(3\sqrt{\pi}) \approx 0.189$ near the Big Bang, which is about 20%, the percent value of unity (in Planck units). We defer the possible cosmological implications of this varying G_{eff} to the Discussion Section.

For now, we note that, since δ is a function of H, the Friedmann equation in the case of a running BEI cannot be easily solved by direct integration. We can, however, still analyze the late time and early time behavior separately. At late time, we chose some r^* sufficiently large such that $1/r^2\ln(r^2) \ll 1$ is as small as we wish. We have $\delta \sim 0$ (more precisely, we may say that δ is small enough that it is slowly varying compared to r, so we can take some small value of fixed δ when integrating with respect to r). Then, with $A_0 = 4$ in Planck units, the first term yields (with $R > r^*$)

$$2\int_{\tilde{r}}^{R} r^{-3}dr = 2\left[-\frac{1}{2r^2}\right]_{r^*}^{R} = \frac{1}{r^{*2}} - \frac{1}{R^2}. \tag{11}$$

The r.h.s. of Equation (10) integrates to give the matter density

$$-\frac{8\pi}{3}[\rho]_{\rho^*}^{\rho(R)}. \tag{12}$$

Thus, in the absence of the second term in Equation (10), this will just give the usual Friedmann equation (as we took $R \to \infty$; note $\rho(R) \to 0$ in this limit, because matter density decreases as the universe expands):

$$H^2 = \frac{8\pi\rho}{3} \tag{13}$$

evaluated at the time that corresponds to r^*.

Therefore, we can interpret the second term in Equation (10) as the differential form of the effective dark energy (so that Equation (10) yields the form[2] $H^2 - \Lambda/3 = 8\pi\rho/3$):

$$\frac{d\Lambda_{\text{eff}}}{3} = \frac{2-\delta}{2+\delta}\ln\left(\frac{4\pi r^2}{A_0}\right)\frac{r^{\delta-2}}{2}\frac{d\delta}{dr}dr. \tag{14}$$

If we use the ansatz $1/r^2\ln(r^2)$ for δ, then

$$\frac{d\Lambda_{\text{eff}}}{3} = -\frac{2-\delta}{2+\delta}\ln\left(\frac{4\pi r^2}{A_0}\right) \tag{15}$$
$$\times\frac{r^{\delta-2}}{2}\left[\frac{2}{r^3\ln(r^2)} + \frac{2}{r^3(\ln(r^2))^2}\right]dr.$$

At late time, $\delta \sim 0$, we have an upper bound:

$$-\frac{\Lambda_{\text{eff}}}{3}\bigg|_{\Lambda(r^*)}^{\Lambda(R)} \sim \frac{1}{2}\int_{r^*}^{R} \frac{\ln(\pi r^2)}{r^2}\left[\frac{2}{r^3\ln(r^2)} + \frac{2}{r^3(\ln(r^2))^2}\right]dr \tag{16}$$

$$< \frac{1}{2}\int_{r^*}^{R} \frac{\ln(\pi r^2)}{r^2}\frac{4}{r^3\ln(r^2)}dr \tag{17}$$

$$= 2\int_{r^*}^{R} \frac{1}{r^5}\frac{1}{\ln(r^2)}[\ln\pi + \ln(r^2)]dr \tag{18}$$

$$< 2\int_{r^*}^{R}\left(\frac{1}{r^5}+\frac{1}{r^5}\right)dr = 4\int_{r^*}^{R}\frac{1}{r^5}dr \tag{19}$$

$$= \left[-\frac{1}{r^4}\right]_{r^*}^{R} = \frac{1}{r^{*4}} - \frac{1}{R^4}. \tag{20}$$

That is,

$$\frac{\Lambda_{\text{eff}}(r^*)}{3} < \frac{1}{r^{*4}} \tag{21}$$

as $R \to \infty$. Since r^* was chosen to be very large, Λ_{eff} must be very small, and furthermore, it becomes smaller at later time. In fact, the integral Equation (16) yields

$$\Lambda_{\text{eff}}(r^*) \sim \frac{3}{4}\left[\frac{1}{r^{*4}}\frac{\ln(\pi r^{*2})}{\ln(r^{*2})} - 2(1-\ln(\pi))\text{Ei}(-4\ln(r^*))\right],$$

where Ei is the exponential integral function. One can check numerically that it is positive and decreases towards 0 very rapidly. Such an asymptotically vanishing effective cosmological constant scenario was previously discussed in [26–31]. Next, we show that, although the universe is late time de Sitter (dS)-like, it is anti-de Sitter (AdS)-like in the very early universe. Such a behavior had been observed, for example, in non-commutative quantum field theory [32].

In the vicinity of the Big Bang, the effective dark energy term depends on the profile of δ (how δ approaches $1/r^2\ln(r^2)$). For example, if $\delta = 1$ for some time after the Big Bang, then $d\delta/dr = 0$, and there is no effective dark energy. However, if we chose as we did previously $\delta = e^{-r}$, then the effective dark energy at the Big Bang is (with $\delta \sim 1$ and $0 < \varepsilon \ll 1$)

$$\frac{\Lambda_{\text{eff}}}{3}\bigg|_{\Lambda_{\text{eff}}(\varepsilon)}^{\Lambda_{\text{eff}}(2\varepsilon)} \sim -\frac{1}{3}\int_{\varepsilon}^{2\varepsilon} \ln(\pi r^2)\frac{e^{-r}}{2r}dr. \tag{22}$$

The integration interval is kept small near the Big Bang to ensure that $\delta = e^{-r}$ is nearly constant, so that we can approximate as follows:

$$\Lambda_{\text{eff}}\big|_{\Lambda_{\text{eff}}(\varepsilon)}^{\Lambda_{\text{eff}}(2\varepsilon)} \sim -\int_{\varepsilon}^{2\varepsilon} \ln(\pi r^2)\frac{e^{-r}}{2r}dr \sim -\varepsilon\ln(\pi\varepsilon^2)\frac{e^{-\varepsilon}}{2\varepsilon}. \tag{23}$$

That is,

$$\frac{\Lambda_{\text{eff}}(2\varepsilon) - \Lambda_{\text{eff}}(\varepsilon)}{\varepsilon} \sim -\ln(\pi\varepsilon^2)\frac{e^{-\varepsilon}}{2\varepsilon}. \tag{24}$$

In the limit $\varepsilon \to 0$, this yields $d\Lambda_{\text{eff}}/d\varepsilon \to \infty$. It follows that[3] $\Lambda_{\text{eff}} \to -\infty$ at the Big Bang with a log divergence.[4] It is worth commenting that [5] gives a strong constraint for the smallness of δ (assuming δ to be fixed) during the Big Bang nucleosynthesis (BBN) epoch. Thus, we know that δ cannot be slowly varying in the early universe if δ starts out near unity at the Big Bang. In fact, this is partly the reason why we chose $\delta = e^{-r}$ in our toy model, as it decays rather quickly at the beginning. Here, however, we used the mathematical fact that any function is nearly constant if the interval of its domain is sufficiently small.

An infinitely large negative cosmological "constant" is not good (the universe might immediately re-collapse), though we should also note that this is a logarithmic divergence

$\sim O(\ln \varepsilon)$, which is still "smaller" in magnitude than the $1/\varepsilon$-divergence in the modified Hubble term[5] (from the first term in Equation (10)), which is proportional to

$$\int_{\varepsilon}^{2\varepsilon} r^{-2}dr \sim \varepsilon\frac{1}{\varepsilon^2} = \frac{1}{\varepsilon}. \tag{25}$$

Of course, we could in principle choose other functions to obtain a finite nonzero value. Regardless, the integral:

$$\Lambda_{\text{eff}}(\varepsilon) \sim -\int_{\varepsilon}^{2\varepsilon} \ln\left(\pi r^2\right)\frac{1}{2r}\frac{d\delta}{dr}dr. \tag{26}$$

is always negative because $d\delta/dr < 0$ *and* $\ln(\pi r^2) < 0$ for small r, as long as δ is a strictly decreasing function. Therefore, Λ_{eff} started out negative and crossed zero at some point in time (at $r_H = \sqrt{1/\pi} \approx 0.564$), and at late time, it is small and positive. By continuity, this suggests that Λ_{eff} reached a maximum value at some point in the past, before decreasing towards its current small value 10^{-122}. Another possibility is that δ contains a factor that cancels the $\ln(\pi r^2)$ term; then, in principle, the transition for AdS to dS can happen at an even later time, depending on the zero of the function. Similar scenarios have been proposed in the literature [33–40], with [35,36], that the transition from the AdS phase to the dS phase occurred around the recombination epoch, while [37] argued that the transition occurred as late as at redshift $z \approx 2.32$, which triggered the late time acceleration. It is also worth mentioning that [35,41] proposed that the AdS–dS transition could help to alleviate the Hubble tension. In fact, [41] argued that the S_8 tension [42,43] (equivalently, the σ_8 tension) is also alleviated, among other improvements; see also [44]. In [45], the authors concluded that solving both the H_0 and σ_8 tensions would require that $G_{\text{eff}} < G$ at some redshift z, if the equation of state satisfies $w(z) \leqslant -1$ (otherwise, $w(z)$ must cross the phantom divide). Thus, it is worth mentioning that a running BEI also causes a varying gravitational constant, and furthermore, from Equation (5), our G_{eff}, in the Planck units, always satisfies

$$\frac{1}{3\sqrt{\pi}} < G_{\text{eff}} = \frac{2-\delta}{2+\delta}\frac{1}{\pi^{\frac{\delta}{2}}} < 1 = G. \tag{27}$$

for all values of admissible $\delta \in [0,1]$. However, there are also arguments that state that this is unlikely to alleviate the Hubble tension by itself [46]; see also the subtleties discussed in the next section.

4. Discussion

In this work, we argued that if the Barrow entropy does encode some form of quantum gravitational correction, then in order to be consistent with the black hole thermodynamics of large black holes (for which quantum gravity correction should be small), the BEI must vary according to the energy scale. In terms of the horizon area, it should decay at least as fast as $1/(A\ln A)$.

When applied to the cosmological context, we showed that the modified Friedmann equation has many surprising properties: the universe started out with an effective negative cosmological constant, but at late time, it became positive and small, decaying towards zero asymptotically. As already explored in the literature, such a scenario could help to alleviate the Hubble tension and the S_8 tension. See, however, in ref. [47,48].

Despite the fact that the BEI is, thus far, only a phenomenological parametrization of quantum gravity effects, it is interesting to see that it could potentially explain some cosmological mysteries. In fact, although the explicit form of δ can only be determined once we have a better understanding of how to derive the Barrow entropy from a theory of quantum gravity, the features we obtained are quite generic as long as δ is assumed to be monotonically decreasing with the energy scale. The shortcoming is that we still do not know the exact behavior of δ, especially at early time. Another caveat is that in deriving Equation (8), we used the first law as $dE = -TdS = \dot{\rho}Vdt$, in which the volume change was neglected, which amounts to the assumption that $\dot{r} \ll Hr$. This holds at late time with exponential late time expansion, but may not hold near the Big Bang; this will depend on the explicit solution of the scale factor, which in turn depends on the profile of δ.

Lastly, we comment on the varying effective gravitational constant. Whenever there is a varying G, one might worry that the continuity equation (the one employed in Equation (8) is the standard one $\dot{\rho} = -3H(\rho + p)$ that follows from $\nabla_\mu T^{\mu\nu} = 0$) may be modified since $\nabla_\mu(GT^{\mu\nu})$ may no longer imply $\nabla_\mu(T^{\mu\nu}) = 0$. However, the "bare" gravitational constant G that appears at the level of the action is not the same as the effective G_{eff} that appears in the modified Friedmann Equation (10). Furthermore, the Barrow entropy itself contains $A_0 = 4G$, which is the "bare" G. The fact that G_{eff} is smaller in the early universe could suggest that we may be able to ameliorate the arrow of time problem [49–58] following the same line of thought as Greene et al. [59], as well as Sloan and Ellis [60]. (Note that to ameliorate the H_0 tension, one usually requires that $G_{\text{eff}} < G$ at late time. However, in our case, this condition always holds, even in the early universe.) However, this would require a very careful examination of the structure formation process by studying the perturbation equation. Typically, the effective gravitational constant that governs structure formation may not be the same as the G_{eff} that appears in the background Friedmann equation, which may also differ from the "bare" G. A concrete example was provided in [61]. Structure formation in the case of a fixed BEI was recently studied in [10]. In fact, a varying gravitational constant would cause many issues, and early time cosmology would need to be re-examined in close detail [62] (for example, the CMB angular power spectrum would be modified [46,62]).

In conclusion, the effects of a running BEI requires a deeper investigation. There is also a need to better understand how Barrow entropy can possibly arise from a theory of quantum gravity. Does the fractalized geometry only hold for horizons or any spacetime hypersurface in general? What about the spacetime manifold itself? Barrow entropy is not the first proposal that involves fractal geometry in gravity. It has previously been proposed that the spacetime dimension becomes fractalized and decreases towards the Planck scale [63–68], which is a different modification (a two-sphere in such a fractalized geometry has a *lower* dimension [63], not higher, as per Barrow's proposal), but perhaps, one could obtain some ideas about how to derive Barrow entropy from these different approaches. Connections with fractional quantum mechanics [69] and other notions of entropies in relation to holographic dark energy [11,70–72] should also be further investigated.

Author Contributions: Conceptualization, Y.C.O.; validation, Y.C.O.; formal analysis, S.D.G. and Y.C.O.; investigation, S.D.G.; writing–original draft preparation, Y.C.O.; writing—review and editing, S.D.G.; supervision, Y.C.O.; funding acquisition, Y.C.O. All authors have read and agreed to the published version of the manuscript.

Funding: This research was funded by the National Natural Science Foundation of China grant number 11922508.

Acknowledgments: Y.C.O. thanks Brett McInnes for useful discussions.

Conflicts of Interest: The authors declare no conflict of interest.

Notes

1 We remark that [25], on the contrary, incorporated the extra term in a new energy density ρ_{DE}, which is attributed to dark energy, instead of modifying the gravitational constant. We can check that the two are equivalent by taking Equation (2.12) of [25] and substituting their expressions for Λ and β. The difference between the two expressions is given by the fact that [25] considered only the case $k = 0$ and kept the integration constant C, which in [14] was set to 0 or considered as part of the total energy density ρ. In our work, we use the form of [14]. As we shall see below, when allowing the BEI to run, we would have an extra term that can be interpreted as a dynamical dark energy, different from the identification in [25] for the fixed index case.

2 $\frac{1}{3}\int_{\Lambda_{\text{eff}}(r^*)}^{\Lambda_{\text{eff}}(R)} d\Lambda_{\text{eff}} = \frac{1}{3}[\Lambda_{\text{eff}}(R) - \Lambda_{\text{eff}}(r^*)]$. This means that the observed value of the dark energy would be $\Lambda_{\text{eff}}(r^*) - \Lambda_{\text{eff}}(\infty)$ in the limit $R \to \infty$. Without loss of generality, we can take $\Lambda_{\text{eff}}(\infty) \equiv 0$, so that $\Lambda_{\text{eff}}(r^*)$ is the observed value of the effective cosmological constant.

3 As per Footnote 4, the observed value of the effective cosmological constant over the interval $(\varepsilon, 2\varepsilon)$ is $\Lambda_{\text{eff}}(\varepsilon) - \Lambda_{\text{eff}}(2\varepsilon) < 0$. Since they are both comparable in magnitude (negative and with the same divergence properties), we can just refer to $\Lambda(\varepsilon)$ as the effective cosmological constant at the Big Bang.

4 Indeed, a function $f(\varepsilon) = \ln(\pi\varepsilon^2)e^{-\varepsilon} < 0$ for small ε, satisfies $f(2\varepsilon) - f(\varepsilon) \to \infty$ and $f'(\varepsilon) \to \infty$. Another example to illustrate the same phenomenon, albeit with a different divergence, is to consider $f(x) = -1/x$. This function satisfies $f(2\varepsilon) - f(\varepsilon) \to \infty$ and $f'(\varepsilon) = 1/\varepsilon^2 \to \infty$. Colloquially, f climbs out from $-\infty$ at the origin with an infinite positive slope.

5 Since $H \to \infty$, there is still a Big Bang singularity in this model.

References

1. Wang, B.; Gong, Y.; Abdalla, E. Thermodynamics of an Accelerated Expanding Universe. *Phys. Rev. D* **2006**, *74*, 083520. [CrossRef]
2. Gong, Y.; Wang, A. Friedmann Equations and Thermodynamics of Apparent Horizons. *Phys. Rev. Lett.* **2007**, *99*, 211301. [CrossRef] [PubMed]
3. Barrow, J.D. The Area of a Rough Black Hole. *Phys. Lett. B* **2020**, *808*, 135643. [CrossRef]
4. Anagnostopoulos, F.K.; Basilakos, S.; Saridakis, E.N. Observational Constraints on Barrow Holographic Dark Energy. *Eur. Phys. J. C* **2020**, *80*, 826. [CrossRef]
5. Barrow, J.D.; Basilakos, S.; Saridakis, E.N. Big Bang Nucleosynthesis Constraints on Barrow Entropy. *Phys. Lett. B* **2021**, *815*, 136134. [CrossRef]
6. Leon, G.; Magaña, J.; Hernàndez-Almada, A.;García-Aspeitia, M.A.; Verdugo, T.; Motta, V. Barrow Entropy Cosmology: An Observational Approach With a Hint of Stability Analysis. *J. Cosmol. Astropart. Phys.* **2021**, *12*, 032. [CrossRef]
7. Luciano, G.G.; Saridakis, E.N. Baryon Asymmetry From Barrow Entropy: Theoretical Predictions and Observational Constraints. *arXiv* **2022**, arXiv:2203.12010.
8. Jusufi, K.; Azreg-Aïnou, M.; Jamil, M.; Saridakis, E.N. Constraints on Barrow Entropy From M87* And S2 Star Observations. *Universe* **2022**, *8*, 102. [CrossRef]
9. Vagnozzi, S.; Roy, R.; Tsai, Y.; Visinelli, L. Horizon-Scale Tests of Gravity Theories and Fundamental Physics From the Event Horizon Telescope Image of Sagittarius A*. *arXiv* **2022**, arXiv:2205.07787.
10. Farsi, B.; Sheykhi, A. Growth of Perturbations in Tsallis and Barrow Cosmology. *arXiv* **2022**, arXiv:2205.04138.
11. Nojiri, S.; Odintsov, S.D.; Paul, T. Barrow Entropic Dark Energy: A Member of Generalized Holographic Dark Energy Family. *Phys. Lett. B* **2022**, *825*, 136844. [CrossRef]
12. Verde, L.; Treu, T.; Riess, A.G. Tensions Between the Early and Late Universe. *Nat. Astron.* **2019**, *3*, 891. [CrossRef]
13. di Valentino, E.; Mena, O.; Pan, S.; Visinelli, L.; Yang, W.; Melchiorri, A.; Mota, D.F.; Riess, A.G.; Silk, J. In the Realm of the Hubble Tension—A Review of Solutions. *Class. Quantum Grav.* **2021**, *38*, 153001. [CrossRef]
14. Sheykhi, A. Barrow Entropy Corrections to Friedmann Equations. *Phys. Rev. D* **2021**, *103*, 123503. [CrossRef]
15. Saridakis, E.N.; Basilakos, S. The Generalized Second Law of Thermodynamics With Barrow Entropy. *Eur. Phys. J. C* **2021**, *81*, 7. [CrossRef]
16. al Mamon, A.; Paliathanasis, A.; Saha, S. Dynamics of an Interacting Barrow Holographic Dark Energy Model and its Thermodynamic Implications. *Eur. Phys. J. Plus* **2021**, *136*, 134. [CrossRef]
17. Chen, C.; Chen, Y.; Ishibashi, A.; Ohta, N.; Yamaguchi, D. Running Newton Coupling, Scale Identification and Black Hole Thermodynamics. *arXiv* **2022**, arXiv:2204.09892.
18. Bonanno, A.; Reuter, M. Renormalization Group Improved Black Hole Spacetimes. *Phys. Rev. D* **2000**, *62*, 043008. [CrossRef]
19. Cai, Y.-F.; Easson, D.A. Asymptotically Safe Gravity as a Scalar-Tensor Theory and Its Cosmological Implications. *Phys. Rev. D* **2011**, *84*, 103502. [CrossRef]
20. Gubitosi, G.; Ooijer, R.; Ripken, C. Frank Saueressig, Consistent Early and Late Time Cosmology From the RG Flow of Gravity. *J. Cosmol. Astropart. Phys.* **2018**, *12*, 004. [CrossRef]
21. Page, D.N. Hawking Radiation and Black Hole Thermodynamics. *New J. Phys.* **2005**, *7*, 203. [CrossRef]
22. Abreu, E.M.C.; Neto, J.A. Thermal Features of Barrow Corrected-Entropy Black Hole Formulation. *Eur. Phys. J. C* **2020**, *80*, 776. [CrossRef]
23. Cai, R.-G.; Cao, L.-M.; Hu, Y.-P. Corrected Entropy-Area Relation and Modified Friedmann Equations. *J. High Energy Phys.* **2008**, *8*, 090. [CrossRef]
24. Gour, G. Algebraic Approach to Quantum Black Holes: Logarithmic Corrections to Black Hole Entropy. *Phys. Rev. D* **2002**, *66*, 104022. [CrossRef]
25. Saridakis, E.N. Modified Cosmology Through Spacetime Thermodynamics and Barrow Horizon Entropy. *J. Cosmol. Astropart. Phys.* **2020**, *7*, 031. [CrossRef]
26. Wetterich, C. An Asymptotically Vanishing Time-Dependent Cosmological 'Constant'. *Astron. Astrophys.* **1995**, *301*, 321.
27. Wetterich, C. The Cosmological Constant and Higher Dimensional Dilatation Symmetry. *Phys. Rev. D* **2010**, *81* 103507. [CrossRef]
28. Wetterich, C. Graviton Fluctuations Erase the Cosmological Constant. *Phys. Lett. B* **2017**, *773*, 6. [CrossRef]
29. Overduin, J.M.; Wesson, P.S.; Mashhoon, B. Decaying Dark Energy in Higher-Dimensional Gravity. *Astron. Astrophys.* **2007**, *473*, 727. [CrossRef]
30. Mashhoon, B.; Wesson, P.S. Gauge-Dependent Cosmological 'Constant'. *Class. Quant. Grav.* **2004**, *21*, 3611. [CrossRef]

31. Babourova, O.; Frolov, B. The Solution of the Cosmological Constant Problem: The Cosmological Constant Exponential Decrease in the Super-Early Universe. *Universe* **2020**, *6*, 230. [CrossRef]
32. Available online: https://link.springer.com/article/10.1140/epjc/s10052-020-7918-6 (accessed on 13 October 2022).
33. Prokopec, T. Negative Energy Cosmology and the Cosmological Constant. *arXiv* **2011**, arXiv:1105.0078.
34. Biswas, T.; Koivisto, T.; Mazumdar, A. Could Our Universe Have Begun With Negative Lambda? *arXiv* **2011**, arXiv:1105.2636.
35. Ye, G.; Piao, Y.-S. Is the Hubble Tension a Hint of AdS Phase Around Recombination? *Phys. Rev. D* **2020**, *101*, 083507. [CrossRef]
36. Ye, G.; Piao, Y.-S. T_0 Censorship of Early Dark Energy and AdS Vacua. *Phys. Rev. D* **2020**, *102*, 083523. [CrossRef]
37. Akarsu, O.; Barrow, J.D.; Escamilla, L.A.; Vazquez, J.A. Graduated Dark Energy: Observational Hints of a Spontaneous Sign Switch in the Cosmological Constant. *Phys. Rev. D* **2020**, *101*, 063528. [CrossRef]
38. Gupt, B.; Singh, P. Non-singular AdS-DS Transitions in a Landscape Scenario. *Phys. Rev. D* **2014**, *89*, 063520. [CrossRef]
39. Dutta, K.; Ruchika; Roy, A.; Sen, A.A.; Sheikh-Jabbari, M.M. Beyond ΛCDM with Low and High Redshift Data: Implications for Dark Energy. *Gen. Relativ. Gravit.* **2020**, *52*, 15. [CrossRef]
40. Sen, A.A.; Adil, S.A.; Sen, S. Do Cosmological Observations Allow a Negative Λ? *arXiv* **2021**, arXiv:2112.10641.
41. Akarsu, O.; Kumar, S.; Ozulker, E.; Vazquez, J.A. Relaxing Cosmological Tensions With a Sign Switching Cosmological Constant. *Phys. Rev. D* **2021**, *104*, 123512. [CrossRef]
42. Battye, R.A.; Charnock, T.; Moss, A. Tension Between the Power Spectrum of Density Perturbations Measured on Large and Small Scales. *Phys. Rev. D* **2015**, *91*, 103508. [CrossRef]
43. Benisty, D. Quantifying the S_8 Tension With the Redshift Space Distortion Data Set. *Phys. Dark Univ.* **2021**, *1*, 100766. [CrossRef]
44. Benevento, G.; Hu, W.; Raveri, M. Can Late Dark Energy Transitions Raise the Hubble Constant. Phys. Rev. D **2020**, *101*, 103517. [CrossRef]
45. Heisenberg, L.; Villarrubia-Rojo, H.; Zosso, J. Simultaneously Solving the H_0 and σ_8 Tensions With Late Dark Energy. *arXiv* **2022**, arXiv:2201.11623.
46. Sakr, Z.; Sapone, D. Can Varying the Gravitational Constant Alleviate the Tensions? *J. Cosmol. Astropart. Phys.* **2022**, *3*, 034. [CrossRef]
47. Visinelli, L.; Vagnozzi, S.; Danielsson, U. Revisiting a Negative Cosmological Constant From Low-Redshift Data. *Symmetry* **2019**, *11*, 1035. [CrossRef]
48. Available online: https://journals.aps.org/prd/abstract/10.1103/PhysRevD.106.L041301 (accessed on 13 October 2022).
49. Feynman, R.; Sands, M.; Leighton, R.B. *The Feynmann Lectures on Physics*; Addison Wesley Publishing Co.: Reading, MA, USA, 1963; pp. 46-1–46-9.
50. Price, H. Cosmology, Time's Arrow, and That Old Double Standard. In *Time's Arrows Today*; Savitt, S., Ed.; Cambridge University Press: Cambridge, UK, 1994.
51. Price, H. The Thermodynamic Arrow: Puzzles and Pseudo-puzzles. *arXiv* **2004**, arXiv:physics/0402040.
52. Vaas, R. Time After Time—Big Bang Cosmology and the Arrows of Time. *Fundam. Theor. Phys.* **2012**, *172*, 5.
53. Price, H. Time's Arrow and Eddington's Challenge. *Prog. Math. Phys.* **2013**, *63*, 187.
54. Penrose, R. Singularities and Time-Asymmetry. In *General Relativity, An Einstein Centenary Survey*; Hawking, S.T., Israel, W., Eds.; Cambridge University Press: Cambridge, UK, 1979.
55. McInnes, B. Arrow of Time in String Theory. *Nucl. Phys. B* **2007**, *782*, 1. [CrossRef]
56. Penrose, R. Before the Big Bang: An Outrageous New Perspective and Its Implications for Particle Physics. Contribution to: 10th European Particle Accelerator Conference (EPAC 06). *Conf. Proc.* C **2006**, *060626*, 2759.
57. McInnes, B. The Arrow Of Time In The Landscape. *arXiv* **2007**, arXiv:0711.1656.
58. Carroll, S.M. In What Sense Is the Early Universe Fine-Tuned? *arXiv* **2007**, arXiv:1406.3057.
59. Greene, B.; Hinterbichler, K.; Judes, S.; Parikh, M.K. Smooth Initial Conditions From Weak Gravity. *Phys. Lett. B* **2011**, *97*, 178. [CrossRef]
60. Sloan, D.; Ellis, G. Solving the Cosmological Entropy Issue with a Higgs Dilaton. *Phys. Rev. D* **2019**, *99*, 063518. [CrossRef]
61. Maeda, K.-i.; Panpanich, S. Cuscuta-Galileon Cosmology: Dynamics, Gravitational Constants and the Hubble Constant. *Phys. Rev. D* **2022**, *105*, 104022. [CrossRef]
62. Uzan, J.-P. Varying Constants, Gravitation and Cosmology. *Living Rev. Rel.* **2011**, *14*, 2. [CrossRef] [PubMed]
63. Benedetti, D. Fractal Properties of Quantum Spacetime. *Phys. Rev. Lett.* **2009**, *102*, 111303. [CrossRef]
64. Carlip, S. The Small Scale Structure of Spacetime. *arXiv* **2010**, arXiv:1009.1136.
65. Alencar, G.; Bezerra, V.B.; Cunha, M.S.; Muniz, C.R. On Effective Spacetime Dimension in the Hořava-Lifshitz Gravity. *Phys. Lett. B* **2015**, *747*, 536. [CrossRef]
66. Calcagni, G. ABC of Multi-Fractal Spacetimes and Fractional Sea Turtles. *Eur. Phys. J. C* **2016**, *76*, 181. [CrossRef]
67. Lok Hu, B.L. Fractal Spacetimes in Stochastic Gravity?—Views from Anomalous Diffusion and the Correlation Hierarchy. *J. Phys. Conf. Ser.* **2017**, *880*, 012004.
68. Carlip, S. Dimension and Dimensional Reduction in Quantum Gravity. *Universe* **2019**, *5*, 83. [CrossRef]
69. Jalalzadeh, S.; da Silva, F.R.; Moniz, P.V. Prospecting Black Hole Thermodynamics With Fractional Quantum Mechanics. *Eur. Phys. J. C* **2021**, *81*, 632. [CrossRef]
70. Nojiri, S.; Odintsov, S.D.; Paul, T. Different Faces of Generalized Holographic Dark Energy. *arXiv* **2021**, arXiv:2105.08438.

71. Nojiri, S.; Odintsov, S.D.; Paul, T. Early and Late Universe Holographic Cosmology From a New Generalized Entropy. *arXiv* **2022**, arXiv:2205.08876.
72. Nojiri, S.; Odintsov, S.D.; Faraoni, V. How Fundamental Is Entropy? From Non-extensive Statistics and Black Hole Physics to the Holographic Dark Universe. *arXiv* **2022**, arXiv:2201.02424.

universe

MDPI

Article

Quintom Fields from Chiral K-Essence Cosmology

José Socorro [1,*,†], Sinuhé Pérez-Payán [2,†], Rafael Hernández-Jiménez [3,†], Abraham Espinoza-García [2,†] and Luis Rey Díaz-Barrón [2,†]

1 Departamento de Física, DCeI, Universidad de Guanajuato-Campus León, León C.P. 37150, Guanajuato, Mexico
2 Unidad Profesional Interdisciplinaria de Ingeniería, Campus Guanajuato del Instituto Politécnico Nacional, Av. Mineral de Valenciana No. 200, Col. Fraccionamiento Industrial Puerto Interior, Silao de la Victoria C.P. 36275, Guanajuato, Mexico
3 Departamento de Física, Centro Universitario de Ciencias Exactas e Ingenierías, Universidad de Guadalajara, Av. Revolución 1500, Colonia Olímpica, Guadalajara C.P. 44430, Jalisco, Mexico
* Correspondence: socorro@fisica.ugto.mx
† These authors contributed equally to this work.

Abstract: In this paper, we present an analysis of a chiral cosmological scenario from the perspective of K-essence formalism. In this setup, several scalar fields interact within the kinetic and potential sectors. However, we only consider a flat Friedmann–Robertson–Lamaître–Walker universe coupled minimally to two quintom fields: one quintessence and one phantom. We examine a classical cosmological framework, where analytical solutions are obtained. Indeed, we present an explanation of the "big-bang" singularity by means of a "big-bounce". Moreover, having a barotropic fluid description and for a particular set of parameters, the phantom line is in fact crossed. Additionally, for the quantum counterpart, the Wheeler–DeWitt equation is analytically solved for various instances, where the factor-ordering problem has been taken into account (measured by the factor Q). Hence, this approach allows us to compute the probability density of the previous two classical subcases. It turns out that its behavior is in effect damped as the scale factor and the scalar fields evolve. It also tends towards the phantom sector when the factor ordering constant $Q \ll 0$.

Keywords: chiral cosmology; quintom fields; K-essence; exact solutions; bounce cosmology

1. Introduction

Over the past decades, various cosmological surveys have suggested that two stages of accelerated expansion have occurred during the evolution of the universe [1–6]. The first of these epochs, the so-called inflation [5,6], would have happened in a very early stage of the expansion of the cosmos, whilst the second one would be taking place at late times. Additionally, the consensus is that this accelerated expansion is caused by dark energy (DE) [7–9]. To account for these phenomena, several cosmological frameworks incorporate scalar fields into their prescriptions and, in fact, they play a preponderant role. Moreover, one of the most studied scenarios in the literature is the quintessence model, which is a fluctuating, homogeneous scalar field that rolls down its scalar potential [10–17]. Different avenues have been explored, broadening the spectrum of scalar field models. For instance, the relevant proposals are the phantom [18–20], quintom [21–26], and Chiral fields [27–34], and there are many more [35–41].

However, despite many efforts [7,42–47], the nature of dark energy has not yet been deciphered, except for its negative pressure. Accordingly, the main characteristic of DE is given by its equation of state (EoS), defined by the ratio of the pressure-to -energy density, that is, $\omega_{DE} \equiv P_{DE}/\rho_{DE}$. This definition allows us to classify the cosmological models mentioned above, according to the behavior of the EoS, namely, quintessence $w_Q \geq -1$ [11,48]; phantom $w_P \leq -1$ [49,50]; and quintom [51], where the latter is able to evolve across the cosmological constant boundary. In [21], the authors have shown that a single scalar field model does not reproduce the quintom scenario, thus opening a window to new paradigms where additional degrees of freedom can be considered (for non conventional approaches into this matter, we refer the reader to [52,53]).

Our aim is to study a quintom cosmological model. The most basic construction of a quintom model can be achieved by considering a pair of scalar fields, namely, a canonical one and a phantom one, endowed with their respective scalar potentials; within this line of research, different schemes have been considered [21–26]. These multi-scalar components bring us additional degrees of freedom; thus, various physical phenomena can be addressed such as primordial, hybrid [54–58], or assisted inflation [59,60], as well as perturbations analysis [61,62].

In this work, we present an analysis of a chiral cosmological scenario from the perspective of K-essence formalism (following the scheme presented in [63]). In this prescription, scalar fields interact within the kinetic and potential sectors. We consider a Friedmann–Robertson–Lamaître–Walker (FRLW) universe coupled minimally to two quintom fields: a quintessence and a phantom. We examine a classical cosmological framework, where exact solutions are obtained. In fact, some of them may indicate that the cosmological singularity is resolved via a "big-bounce". Moreover, we show that the phantom line is crossed. Lastly, for the quantum counterpart, the Wheeler–DeWitt (WDW) equation is obtained, where the factor-ordering problem takes into account the introduction of the parameter Q, and analytical solutions are presented employing the same relevant cases that appear in the classical scheme. We show that the probability density is in fact damped as the scale factor and the scalar fields evolve.

The paper is laid out as follows. Section 2 is devoted to the analysis of the classical multi-scalar field cosmological model, and analytical solutions are obtained considering different cases. In Section 3, the quantum counterpart is addressed; in this formalism, different cases are analyzed and their corresponding solutions are presented. Section 4 is devoted to the final remarks.

2. Classical Approach

We start by considering the action of the chiral cosmological model from the K-essence perspective, which reads

$$S = \int \sqrt{-g}\Big[R - M^{ab}(\phi_c)\mathcal{G}(\xi_{ab}) + C(\phi_c)\Big]d^4x, \tag{1}$$

where R is the Ricci scalar; $M^{ab}(\phi_c)$ is a matrix related to the kinetic energy mixed terms; $C(\phi_c)$ is the scalar potential, which depends on k scalar fields ($c = 1, 2, \cdots, k$); and $\mathcal{G}[\xi_{ab}(\phi_c)]$ is a functional in terms of the chiral kinetic energy $\xi_{ab}(\phi_c, g^{\mu\nu}) = -\frac{1}{2}g^{\mu\nu}\nabla_\mu\phi_a\nabla_\nu\phi_b$. Note that we are working with the reduced Planck units since $8\pi G = 1$, so this eliminates the $8\pi G$ term from the expression (1). An action similar to (1) also appears in modified theories of gravity [64], and more recently in [65]. Making the variation of the action (1) with respect to the fields $(g^{\mu\nu}, \phi_c)$, we obtain

$$\begin{aligned}
\delta S &= \int \delta[\sqrt{-g}R]d^4x + \int \delta\sqrt{-g}\Big[-M^{ab}(\phi_c)\mathcal{G}(\xi_{ab}) + C(\phi_c)\Big]d^4x \\
&\quad + \int \sqrt{-g}\Big[-\delta M^{ab}(\phi_c)\mathcal{G}(\xi_{ab}) - M^{ab}(\phi_c)\delta\mathcal{G}(\xi_{ab}) + \delta C(\phi_c)\Big]d^4x, \\
&= \int \sqrt{-g}\,G_{\lambda\theta}\delta g^{\lambda\theta}d^4x - \int \sqrt{-g}\frac{1}{2}\Big[-M^{ab}(\phi)\mathcal{G}(\xi_{ab}) + C(\phi_c)\Big]g_{\lambda\theta}\delta g^{\lambda\theta}d^4x + \\
&\quad + \int \sqrt{-g}\left[-\frac{\partial M^{ab}(\phi_c)}{\partial\phi_c}\delta\phi_c\mathcal{G}(\xi_{ab}) - M^{ab}(\phi_c)\frac{\partial\mathcal{G}(\xi_{ab})}{\partial\xi_{ab}}\delta\xi_{ab} + \frac{\partial C(\phi_c)}{\partial\phi}\delta\phi_c\right]d^4x,
\end{aligned} \tag{2}$$

where $G_{\lambda\theta} = R_{\lambda\theta} - Rg_{\lambda\theta}/2$, $R_{\lambda\theta}$, and $g_{\lambda\theta}$ are the Einstein, Ricci, and metric tensors, respectively. The variation of the functional $\mathcal{G}(\xi_{ab})$ is

$$\begin{aligned}\delta\xi_{ab}(\phi_c) &= -\frac{1}{2}\nabla_\mu\phi_a\nabla_\nu\phi_b\delta g^{\mu\nu} - \frac{1}{2}g^{\mu\nu}\nabla_\mu\delta\phi_a\nabla_\nu\phi_b - \frac{1}{2}g^{\mu\nu}\nabla_\mu\phi_a\nabla_\nu\delta\phi_b, \\ &= -\frac{1}{2}\nabla_\mu\phi_a\nabla_\nu\phi_b\delta g^{\mu\nu} + \frac{1}{2}\nabla_\mu[g^{\mu\nu}\nabla_\nu\phi_b]\delta\phi_a + \frac{1}{2}\nabla_\nu[g^{\mu\nu}\nabla_\mu\phi_a]\delta\phi_b \\ &+ \nabla_\mu\left[\frac{1}{2}\delta\phi_a g^{\mu\nu}\nabla_\nu\phi_b\right] + \nabla_\nu\left[\frac{1}{2}\delta\phi_b g^{\mu\nu}\nabla_\nu\phi_a\right].\end{aligned}$$

Thus, finally we have

$$\begin{aligned}\delta S = \int \sqrt{-g}\Bigg\{ G_{\mu\nu} + \frac{1}{2}M^{ab}(\phi_c)\left[\nabla_\mu\phi_a\nabla_\nu\phi_b\frac{\partial\mathcal{G}(\xi_{ab})}{\partial\xi_{ab}} + g_{\mu\nu}\mathcal{G}(\xi_{ab})\right] \\ -\frac{1}{2}g_{\mu\nu}C(\phi_c)\Bigg\}\delta g^{\mu\nu}d^4x,\end{aligned}$$

and since δS vanishes ($\delta S = 0$) for arbitrary variations $\delta g^{\mu\nu}$, we are led to the field equations

$$G_{\mu\nu} = -\frac{1}{2}M^{ab}(\phi_c)\left[\nabla_\mu\phi_a\nabla_\nu\phi_b\frac{\partial\mathcal{G}(\xi_{ab})}{\partial\xi_{ab}} + g_{\mu\nu}\mathcal{G}(\xi_{ab})\right] + \frac{1}{2}g_{\mu\nu}C(\phi_c). \tag{3}$$

Then, the energy-momentum tensor in this setup becomes

$$T_{\mu\nu}(\phi_c) = +\frac{1}{2}M^{ab}(\phi_c)\left[\nabla_\mu\phi_a\nabla_\nu\phi_b\frac{\partial\mathcal{G}(\xi_{ab})}{\partial\xi_{ab}} + g_{\mu\nu}\mathcal{G}(\xi_{ab})\right] - \frac{1}{2}g_{\mu\nu}C(\phi_c). \tag{4}$$

Moreover, we consider the energy-momentum tensor of a barotropic perfect fluid $T_{\alpha\beta}(\phi_c) = (\rho + P)u_\alpha(\phi_c)u_\beta(\phi_c) + P g_{\alpha\beta}$ (where the four-velocity is given by $u_\alpha u_\beta = \nabla_\alpha\phi_a\nabla_\beta\phi_b/2\xi_{ab}$). Hence, the pressure P and the energy density ρ of the scalar fields take the following form:

$$P(\phi_c) = \frac{1}{2}M^{ab}(\phi_c)\mathcal{G} - \frac{1}{2}C(\phi_c), \qquad \rho(\phi_c) = \frac{1}{2}M^{ab}\left[2\xi_{ab}\frac{\partial\mathcal{G}}{\partial\xi_{ab}} - \mathcal{G}\right] + \frac{1}{2}C(\phi_c). \tag{5}$$

Additionally, the barotropic parameter $\omega_{\xi_{ab}}$ becomes

$$\omega_{\xi_{ab}} = \frac{P(\phi_c)}{\rho(\phi_c)} = \frac{M^{ab}(\phi_c)\mathcal{G} - C(\phi_c)}{M^{ab}\left[2\xi_{ab}\frac{\partial\mathcal{G}}{\partial\xi_{ab}} - \mathcal{G}\right] + C(\phi_c)}. \tag{6}$$

On the other hand, taking the variation of the action (1) with respect to the scalar field ϕ_c, we obtain

$$\begin{aligned}\delta S = \int \sqrt{g}\Bigg\{ -\frac{\partial M^{ab}(\phi_c)}{\partial\phi_c}\mathcal{G}(\xi_{ab}) - \frac{1}{2}M^{cb}(\phi_c)\frac{\partial\mathcal{G}(\xi_{ab})}{\partial\xi_{ab}}\nabla_\nu\nabla^\nu\phi_b \\ -\frac{1}{2}M^{ac}(\phi_c)\frac{\partial\mathcal{G}(\xi_{ab})}{\partial\xi_{ab}}\nabla_\nu\nabla^\nu\phi_a + \frac{\partial C(\phi_c)}{\partial\phi_c}\Bigg\}\delta\phi_c d^4x,\end{aligned}$$

where a Klein-Gordon-like equation can be written as follows:

$$\frac{\partial M^{ab}(\phi_c)}{\partial\phi_c}\mathcal{G}(\xi_{ab}) - M^{ac}(\phi_c)\frac{\partial\mathcal{G}(\xi_{ab})}{\partial\xi_{ab}}\nabla_\nu\nabla^\nu\phi_a - \frac{\partial C(\phi_c)}{\partial\phi_c} = 0. \tag{7}$$

Note that Equations (2)–(7) represent the general framework; however, we will particularize to the case $\mathcal{G} = \xi$, therefore obtaining the standard chiral Einstein field equations

$$G_{\mu\nu} = -\frac{1}{2}M^{ab}(\phi_c)\left[\nabla_\mu\phi_a\nabla_\nu\phi_b - \frac{1}{2}g_{\mu\nu}g^{\alpha\beta}\nabla_\alpha\phi_a\nabla_\beta\phi_b\right] + \frac{1}{2}g_{\mu\nu}C(\phi_c). \tag{8}$$

Now, if we consider that M^{ab} is a constant matrix, we obtain

$$M^{ac}\nabla_\nu\nabla^\nu\phi_a - \frac{\partial C(\phi_c)}{\partial\phi_c} = 0. \tag{9}$$

All of the aforementioned results can be employed to consider a two-field cosmological model: a **quin**tessence and a phan**tom** field, with their corresponding scalar potentials. Setting $M^{ab}(\phi_c) = m^{ab}$ as a constant matrix, in (1), we obtain

$$\mathcal{L} = \sqrt{-g}\left(R - \frac{1}{2}g^{\mu\nu}m^{ab}\nabla_\mu\phi_a\nabla_\nu\phi_b + V(\phi_1,\phi_2)\right), \tag{10}$$

where $C(\phi_c) = V(\phi_1,\phi_2)$ is the combined scalar field potential; ϕ_1 and ϕ_2 are the quintessence and phantom fields, respectively; and m^{ab} is a 2×2 constant matrix of the form $m^{ab} = \begin{pmatrix} 1 & m^{12} \\ m^{12} & -1 \end{pmatrix}$.

Thus, the Einstein–Klein–Gordon field Equations (8) and (9) are

$$G_{\alpha\beta} = -\frac{1}{2}m^{ab}\left(\nabla_\alpha\phi_a\nabla_\beta\phi_b - \frac{1}{2}g_{\alpha\beta}g^{\mu\nu}\nabla_\mu\phi_a\nabla_\nu\phi_b\right) + \frac{1}{2}g_{\alpha\beta}V(\phi_1,\phi_2), \tag{11}$$

$$m^{cb}\nabla_\nu\nabla^\nu\phi_b - \frac{\partial C(\phi_c)}{\partial\phi_c} = 0, \tag{12}$$

where $a,b,c = 1,2$. From (11), the energy-momentum tensor of the scalar fields (ϕ_1,ϕ_2) is given by

$$8\pi G T_{\alpha\beta}(\phi_1,\phi_2) = -\frac{1}{2}m^{ab}\left(\nabla_\alpha\phi_a\nabla_\beta\phi_b - \frac{1}{2}g_{\alpha\beta}g^{\mu\nu}\nabla_\mu\phi_a\nabla_\nu\phi_b\right) + \frac{1}{2}g_{\alpha\beta}V(\phi_1,\phi_2), \tag{13}$$

then, using (6), the barotropic index ω_{ϕ_a,ϕ_b} is given by

$$\omega_{\phi_a\phi_b} = \frac{-\frac{1}{2}m^{ab}\nabla^\mu\phi_a\nabla_\mu\phi_b - C(\phi_c)}{-\frac{1}{2}m^{ab}\nabla^\mu\phi_a\nabla_\mu\phi_b + C(\phi_c)}. \tag{14}$$

In our analysis, the background spacetime to be considered is a spatially flat FRLW with line element

$$ds^2 = -N(t)^2dt^2 + e^{2\Omega(t)}\left[dr^2 + r^2(d\theta^2 + sin^2\theta d\phi^2)\right], \tag{15}$$

where N represents the lapse function, $A(t) = e^{\Omega(t)}$ is the scale factor in the Misner parametrization, and Ω is a scalar function whose interval is $(-\infty,\infty)$. Choosing $C(\phi_c) = V_1(\phi_1) + V_2(\phi_2) = V_{01}e^{-\lambda_1\phi_1} + V_{02}e^{-\lambda_2\phi_2}$, the mixed Einstein field equations are

$$3\frac{\dot{\Omega}^2}{N^2}-\frac{1}{2}\left[\frac{1}{2}\frac{\dot{\phi}_1^2}{N^2}+V_1(\phi_1)\right]-\frac{1}{2}\left[-\frac{1}{2}\frac{\dot{\phi}_2^2}{N^2}+V_2(\phi_2)\right]-\frac{m^{12}}{2}\frac{\dot{\phi}_1}{N}\frac{\dot{\phi}_2}{N}=0, \tag{16}$$

$$2\frac{\ddot{\Omega}}{N^2}+3\frac{\dot{\Omega}^2}{N^2}-2\frac{\dot{\Omega}}{N}\frac{\dot{N}}{N^2}+\frac{1}{2}\left[\frac{1}{2}\frac{\dot{\phi}_1^2}{N^2}-V_1(\phi_1)\right]+\frac{1}{2}\left[-\frac{1}{2}\frac{\dot{\phi}_2^2}{N^2}-V_2(\phi_2)\right]+\frac{m^{12}}{2}\frac{\dot{\phi}_1}{N}\frac{\dot{\phi}_2}{N}=0, \tag{17}$$

$$\dot{\phi}_1\left(-3\frac{\dot{\Omega}}{N}\frac{\dot{\phi}_1}{N}+\frac{\dot{N}}{N^2}\frac{\phi_1}{N}-\frac{\ddot{\phi}_1}{N^2}\right)+m^{12}\dot{\phi}_1\left(-3\frac{\dot{\Omega}}{N}\frac{\dot{\phi}_2}{N}+\frac{\dot{N}}{N^2}\frac{\phi_2}{N}-\frac{\ddot{\phi}_2}{N^2}\right)-\dot{V}_1(\phi_1)=0, \tag{18}$$

$$m^{12}\dot{\phi}_2\left(-3\frac{\dot{\Omega}}{N}\frac{\dot{\phi}_1}{N}+\frac{\dot{N}}{N^2}\frac{\phi_1}{N}-\frac{\ddot{\phi}_1}{N^2}\right)+\dot{\phi}_2\left(3\frac{\dot{\Omega}}{N}\frac{\dot{\phi}_2}{N}-\frac{\dot{N}}{N^2}\frac{\phi_2}{N}+\frac{\ddot{\phi}_2}{N^2}\right)-\dot{V}_2(\phi_2)=0, \tag{19}$$

where " $\cdot$ " represents a time derivative. By plugging the line element (15) into the energy-momentum tensor of the scalar fields (13), the energy density, and the pressure, the following form is taken:

$$8\pi G\rho_{\phi_1\phi_2}=\frac{1}{2}\left[\frac{1}{2}\dot{\phi}_1^2+N^2V_1(\phi_1)\right]+\frac{1}{2}\left[-\frac{1}{2}\dot{\phi}_2^2+N^2V_2(\phi_2)\right]+\frac{m^{12}}{2}\dot{\phi}_1\dot{\phi}_2, \tag{20}$$

$$8\pi GP_{\phi_1\phi_2}=\frac{1}{2}\left[\frac{1}{2}\frac{\dot{\phi}_1^2}{N^2}-V_1(\phi_1)\right]+\frac{1}{2}\left[-\frac{1}{2}\frac{\dot{\phi}_2^2}{N^2}-V_2(\phi_2)\right]+\frac{m^{12}}{2}\frac{\dot{\phi}_1}{N}\frac{\dot{\phi}_2}{N}, \tag{21}$$

having these two quantities at hand, the barotropic parameter will be written as

$$\omega_{\phi_1\phi_2}=\frac{P_{\phi_1\phi_2}}{\rho_{\phi_1\phi_2}}=\frac{\left[\frac{1}{2}\frac{\dot{\phi}_1^2}{N^2}-V_1(\phi_1)\right]+\left[-\frac{1}{2}\frac{\dot{\phi}_2^2}{N^2}-V_2(\phi_2)\right]+m^{12}\frac{\dot{\phi}_1}{N}\frac{\dot{\phi}_2}{N}}{\left[\frac{1}{2}\dot{\phi}_1^2+N^2V_1(\phi_1)\right]+\left[-\frac{1}{2}\dot{\phi}_2^2+N^2V_2(\phi_2)\right]+m^{12}\dot{\phi}_1\dot{\phi}_2}. \tag{22}$$

Now we are in position to construct the corresponding Lagrangian and Hamiltonian densities for this cosmological model. Using Hamilton's approach, classical solutions to EKG (16)–(19) can be found; additionally, the quantum counterpart can be established and solved. Taking these ideas into consideration, putting back the metric (15) into (10), the Lagrangian density reads

$$\mathcal{L}=e^{3\Omega}\left(\frac{6\dot{\Omega}^2}{N}-\frac{\dot{\phi_1}^2}{2N}+\frac{\dot{\phi_2}^2}{2N}-\frac{m^{12}\dot{\phi}_1\dot{\phi}_2}{N}+NV_{01}e^{-\lambda_1\phi_1}+NV_{02}e^{-\lambda_2\phi_2}\right). \tag{23}$$

The resulting momenta are given by

$$\begin{aligned}
\Pi_\Omega &= 12\frac{e^{3\Omega}}{N}\dot{\Omega}, & \dot{\Omega} &= \frac{Ne^{-3\Omega}}{12}\Pi_\Omega,\\
\Pi_{\phi_1} &= -\frac{e^{3\Omega}}{N}\left(\dot{\phi}_1+m^{12}\dot{\phi}_2\right), & \dot{\phi}_1 &= -\frac{Ne^{-3\Omega}}{\triangle}\left(\Pi_{\phi_1}+m^{12}\Pi_{\phi_2}\right),\\
\Pi_{\phi_2} &= -\frac{e^{3\Omega}}{N}\left(m^{12}\dot{\phi}_1-\dot{\phi}_2\right), & \dot{\phi}_2 &= \frac{Ne^{-3\Omega}}{\triangle}\left(m^{12}\Pi_{\phi_1}-\Pi_{\phi_2}\right),
\end{aligned} \tag{24}$$

where $\triangle = 1+\left(m^{12}\right)^2$. In order to obtain a Hamiltonian density, we write (23) in a canonical form, i.e., $\mathcal{L}_{can}=\Pi_q\dot{q}-N\mathcal{H}$; then, we perform the variation with respect to the lapse function N, $\delta\mathcal{L}_{can}/\delta N=0$, yielding the Hamiltonian constraint $\mathcal{H}=0$, that is,

$$\mathcal{H}=\frac{e^{-3\Omega}}{24}\left[\Pi_\Omega^2-\frac{12}{\triangle}\Pi_{\phi_1}^2+\frac{12}{\triangle}\Pi_{\phi_2}^2-24\frac{m^{12}}{\triangle}\Pi_{\phi_1}\Pi_{\phi_2}-24V_1e^{-\lambda_1\phi_1+6\Omega}-24V_2e^{-\lambda_2\phi_2+6\Omega}\right]. \tag{25}$$

The fact that $\mathcal{H} = 0$ guarantees us that its solutions are unique and well defined. Putting forward the following canonical transformation on the variables $(\Omega, \phi_1, \phi_2) \leftrightarrow (\xi_1, \xi_2, \xi_3)$ and fixing the gauge $N = 24e^{3\Omega}$, we obtain

$$\begin{aligned} \xi_1 &= 6\Omega - \lambda_1\phi_1, \\ \xi_2 &= 6\Omega - \lambda_2\phi_2, \\ \xi_3 &= 6\Omega + \lambda_1\phi_1 + \lambda_2\phi_2, \end{aligned} \qquad \longleftrightarrow \qquad \begin{aligned} \Omega &= \frac{\xi_1 + \xi_2 + \xi_3}{18}, \\ \phi_1 &= \frac{-2\xi_1 + \xi_2 + \xi_3}{3\lambda_1}, \\ \phi_2 &= \frac{\xi_1 - 2\xi_2 + \xi_3}{3\lambda_2}, \end{aligned} \tag{26}$$

leading us to obtain a new set of conjugate momenta (P_1, P_2, P_3)

$$\begin{aligned} \Pi_\Omega &= 6P_1 + 6P_2 + 6P_3, \\ \Pi_{\phi_1} &= \lambda_1(-P_1 + P_3), \\ \Pi_{\phi_2} &= \lambda_2(-P_2 + P_3), \end{aligned} \tag{27}$$

therefore, the Hamiltonian density can be written as

$$\begin{aligned} \mathcal{H} = &\, 12(3 - \Lambda_1)P_1^2 + 12(3 + \Lambda_2)P_2^2 + 12(3 - 2\Lambda_{12} + \Lambda_2 - \Lambda_1)P_3^2 \\ &+ 24[(3 + \Lambda_1 + \Lambda_{12})P_1 + (3 + \Lambda_{12} - \Lambda_2)P_2]P_3 \\ &+ 24(3 - \Lambda_{12})P_1P_2 - 24\left(V_1e^{\xi_1} + V_2e^{\xi_2}\right), \end{aligned} \tag{28}$$

where $\Lambda_1 = \lambda_1^2/\triangle$, $\Lambda_2 = \lambda_2^2/\triangle$, and $\Lambda_{12} = m^{12}\,\lambda_1\lambda_2/\triangle$. In the end, even if the Hamiltonian density (28) exhibits an intricate form, this configuration will indeed allow us to compute various relevant scenarios. Thus, the Hamilton equations become

$$\begin{aligned} \dot{\xi}_1 &= 24(3 - \Lambda_1)P_1 + 24(3 - \Lambda_{12})P_2 + 24(3 + \Lambda_1 + \Lambda_{12})P_3, \\ \dot{\xi}_2 &= 24(3 + \Lambda_2)P_2 + 24(3 - \Lambda_{12})P_1 + 24(3 - \Lambda_2 + \Lambda_{12})P_3, \\ \dot{\xi}_3 &= 24(3 + \Lambda_1 + \Lambda_{12})P_1 + 24(3 - \Lambda_2 + \Lambda_{12})P_2 + 24(3 + \Lambda_2 - \Lambda_1 - 2\Lambda_{12})P_3, \\ \dot{P}_1 &= 24V_1e^{\xi_1}, \\ \dot{P}_2 &= 24V_2e^{\xi_2}, \\ \dot{P}_3 &= 0. \end{aligned} \tag{29}$$

Right away, we can see that $P_3 = p_3 = constant$. Moreover, the end game of this analysis is to find solutions to the variables (Ω, ϕ_1, ϕ_2). Hence, we simplify our expression. First, we drop the mixed momenta P_1 and P_2 from $\dot{\xi}_1$ and $\dot{\xi}_2$ (Equation (29)) by setting their coefficients to zero: $3 - \Lambda_{12} = 0$. Therefore, we can obtain one relation among the parameters $(m^{12}, \lambda_1\lambda_2)$, where the matrix element m^{12} satisfies the constraint

$$m^{12} = \frac{\lambda_1\lambda_2}{6}\left[1 \pm \sqrt{1 - \left(\frac{6}{\lambda_1\lambda_2}\right)^2}\right]. \tag{30}$$

Additionally, we set the second term inside the square root of (30) to be a real number and consider $\lambda_1 > 0$, $\lambda_2 > 0$, thus yielding the relation $\lambda_1\lambda_2 \geq 6$, ensuring that m^{12} is always positive.

2.1. Classical Exact Solutions

In this section, we will calculate the exact solutions of (Ω, ϕ_1, ϕ_2), where different cases will appear due to the parameters (λ_1, λ_2). Recall the master Hamiltonian density

$$\begin{aligned}\mathcal{H} &= 12\eta_1 P_1^2 + 12\eta_2 P_2^2 + 12(-9+\eta_1+\eta_2)p_3^2 + 24[(9-\eta_1)P_1 + (9-\eta_2)P_2]p_3 \\ &- 24\left(V_1 e^{\xi_1} + V_2 e^{\xi_2}\right), \end{aligned} \tag{31}$$

with $\eta_1 = 3 - \Lambda_1$ and $\eta_2 = 3 + \Lambda_2$. Then, Hamilton equations for these new coordinates ξ_i are

$$\begin{aligned} \dot{\xi}_1 &= 24\eta_1 P_1 + 24(9-\eta_1)p_3, \\ \dot{\xi}_2 &= 24\eta_2 P_2 + 24(9-\eta_2)p_3, \\ \dot{\xi}_3 &= 24(9-\eta_1)P_1 + 24(9-\eta_2)P_2 + 24(-9+\eta_1+\eta_2)p_3, \end{aligned} \tag{32}$$

and equations for $\dot{P}_1$ and $\dot{P}_2$ are still given by (29). In the following sections, we will obtain analytical solutions for differents values of λ_1 and λ_2.

2.1.1. Case: $\lambda_1 = \lambda_2 = \sqrt{6}$

For these particular values, we have $\Lambda_1 = \Lambda_2 = 3$ with $\eta_1 = 0$ and $\eta_2 = 6$; then, the Hamilton equations are reduced to

$$\begin{aligned} \dot{\xi}_1 &= 216p_3, \\ \dot{\xi}_2 &= 144P_2 + 72p_3, \\ \dot{\xi}_3 &- 216P_1 + 72P_2 - 72p_3. \end{aligned} \tag{33}$$

From the last set of equations, we can see that the solution for $\dot{\xi}_1$ will be given by

$$\xi_1 = a_1 + 216p_3 t, \tag{34}$$

where a_1 is an integration constant. Then, taking the time derivative of $\dot{\xi}_2$ results in $\ddot{\xi}_2 = 3456V_2\, e^{\xi_2}$, whose solution is

$$\xi_2 = Ln\left(\frac{\alpha_2^2}{1728\, V_2}\right) + Ln\left[Csch^2(\alpha_2\, t - \beta_2)\right]. \tag{35}$$

Now we know the functional form of ξ_2, we can compute the remaining momenta, yielding

$$\begin{aligned} P_1(t) &= p_1 + \frac{V_1}{9p_3}\, e^{a_1 + 216p_3\, t}, \\ P_2(t) &= p_2 - \frac{\alpha_2}{72}\, Coth(\alpha_2\, t - \beta_2). \end{aligned} \tag{36}$$

Plugging back P_1 and P_2, given by (36), into the Hamiltonian constraint $\mathcal{H} = 0$, we found that $p_2 = -p_3/2$ and $3888p_3^2 - 15{,}552p_1p_3 - \alpha_2^2 = 0$; solving for p_3 gives $p_3 = 2p_1 \pm \frac{\sqrt{3}}{108}\sqrt{\alpha_2^2 + 15{,}552\, p_1^2}$.

With these results, the variable ξ_3 becomes

$$\xi_3 = a_3 + (216p_1 - 108p_3)t + \frac{V_1}{9p_3^2}e^{a_1 + 216p_3 t} + Ln[Csch(\alpha_2\, t - \beta_2)]. \tag{37}$$

where a_3 is an integration constant. Having found ξ_1, ξ_2, and ξ_3 and then applying the inverse transformation (26), we can present the solutions in the original variables

$$\Omega(t) = \frac{a_1 + a_3}{18} + Ln\left[\frac{\alpha_2}{24\sqrt{3}\,V_2}\right]^{\frac{1}{9}} + Ln\left[Csch^{\frac{1}{6}}(\alpha_2\, t - \beta_2)\right] + (12p_1 + 6p_3)t + \frac{V_1}{162p_3^2}e^{a_1+216p_3t}, \tag{38}$$

$$\phi_1(t) = \frac{-2a_1 + a_3}{3\lambda_1} + Ln\left(\frac{\alpha_2^2}{1728\,V_2}\right)^{\frac{1}{3\lambda_1}} + \frac{1}{\lambda_1}\left[(72p_1 - 180p_3)t + \frac{V_1}{27p_3^2}e^{a_1+216p_3t} + Ln[Csch(\alpha_2\, t - \beta_2)]\right], \tag{39}$$

$$\phi_2(t) = \frac{a_1 + a_3}{3\lambda_2} + Ln\left(\frac{\alpha_2^2}{1728\,V_2}\right)^{-\frac{2}{3\lambda_2}} + \frac{1}{\lambda_2}\left[(72p_1 + 36p_3)t + \frac{V_1}{27p_3^2}e^{a_1+216p_3t} + Ln[Sinh(\alpha_2\, t - \beta_2)]\right], \tag{40}$$

Recalling that the scale factor is given by $A(t) = e^{\Omega(t)}$, we have

$$A(t) = e^{\frac{a_1+a_3}{18}}\left[\frac{\alpha_2}{24\sqrt{3V_2}}\right]^{\frac{1}{9}} Csch^{\frac{1}{6}}(\alpha_2\, t - \beta_2)\; Exp\left[\frac{V_1}{162p_3^2}e^{a_1+216p_3t}\right]e^{(12p_1+6p_3)t}. \tag{41}$$

In Figure 1, we present the behaviour of the scale factor $A = A(t)$, the Hubble parameter $H = H(t)$, and the barotropic parameter $\omega_{\phi_1\phi_2} = \omega_{\phi_1\phi_2}(t)$. From the upper left graph, we can see that A grows very rapidly as time goes by; it can also be seen that this solution avoids the singularity by means of a bounce, where H does cross the horizontal axis. In the panel at the bottom, the barotropic parameter $\omega_{\phi_1\phi_2}$ is presented, and it can be seen that the EoS parameter crosses the "-1" boundary, which is in fact a characteristic of the quintom models.

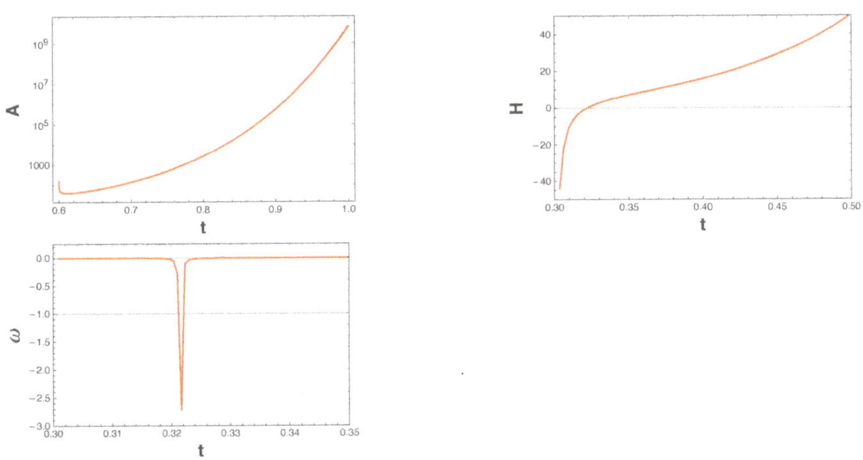

Figure 1. This figure shows the time $(0.3 \leq t \leq 1.0)$ evolution of the scale factor $A(t)$, the Hubble parameter $H(t)$, and the barotropic parameter $\omega_{\phi_1\phi_2}(t)$. We use arbitrary units, namely, $V_1 = 6.0, V_2 = 0.1, \alpha_2 = 3.0, a_1 = -6.548, a_3 = -2.0$, and $p_1 = 0.001$. Recall that $\lambda_1 = \lambda_2 = \sqrt{6}$; the remaining constants can be obtained from the aforementioned values. Note that time is measured in reduced Planck units since $8\pi G = 1$.

2.1.2. Case: $\lambda_1\lambda_2 = 6$

Now, we have $m^{12} = 1$, and for $\lambda_2 = \sqrt{6}$ we obtain the previous case; therefore, we devote this section to carrying out an analysis of values $\lambda_2 \neq \sqrt{6}$ and explore whether the phantom or quintessence scheme prevails under the domain of the scalar potential. On the one hand, when $\lambda_2 \ll \lambda_1$ the phantom sector dominates. On the other hand, when $\lambda_2 \gg \lambda_1$, the quintessence counterpart becomes the relevant scenario. Then, we consider the Hamilton Equations (32) and take the time derivative of $\dot{\xi}_1$, which reads

$$\ddot{\xi}_1 = 576\eta_1 V_1 e^{\xi_1}, \tag{42}$$

where we also resort to the equation for $\dot{P}_1$. Solutions of (42) strongly depend on λ_1, which has the form

$$e^{\xi_1} = \frac{r_1^2}{288|\eta_1| V_1} \begin{cases} Sech^2(r_1 t - q_1) & \lambda_1 > \sqrt{6} \text{ corresponding at } \eta_1 < 0 \\ Csch^2(r_1 t - q_1) & \lambda_1 < \sqrt{6} \text{ corresponding at } \eta_1 > 0 \end{cases} \tag{43}$$

From (32), we can see that both $\dot{\xi}_2$ and $\dot{\xi}_1$ have the same functional structure when $\eta_1 > 0$, and since $\eta_2 > 0$ for all values of λ_2, the solution of ξ_2 is

$$e^{\xi_2} = \frac{r_2^2}{288\eta_2 V_2} Csch^2(r_2 t - q_2), \tag{44}$$

where in (43) and (44), r_i and q_i (with $i = 1, 2$) are integration constants. In the next segments, we will examine the two cases: $\lambda_1 > \sqrt{6}$ and $\lambda_1 < \sqrt{6}$.

2.1.3. Phantom Domination: $\lambda_1 > \sqrt{6}$ and $\lambda_2 < \sqrt{6}$

Considering this setup, we start by reinserting the solutions for $\lambda_1 > \sqrt{6}$ ($\eta_1 < 0$) and for $\lambda_2 < \sqrt{6}$ into the Hamilton equations for the momenta, obtaining

$$P_1 = p_1 + \frac{r_1}{12|\eta_1|} Tanh(r_1 t - q_1), \tag{45}$$

$$P_2 = p_2 - \frac{r_2}{12\eta_2} Coth(r_2 t - q_2), \tag{46}$$

where p_1 and p_2 are integration constants. Now, with the aid of Equations (45) and (46), the Hamiltonian is identically zero when

$$p_1 = \frac{|\eta_1| + 9}{|\eta_1|} p_3, \qquad p_2 = \frac{\eta_2 - 9}{\eta_2} p_3, \qquad p_3 = +\frac{1}{36}\sqrt{\frac{\eta_2 r_1^2 - |\eta_1| r_2^2}{3[|\eta_1|\eta_2 - 3|\eta_1| + 3\eta_2]}}. \tag{47}$$

Consequently, the solutions of ξ_i become

$$\xi_1 = Ln\left(\frac{r_1^2}{288|\eta_1| V_1}\right) + Ln\left[Sech^2(r_1 t - q_1)\right], \tag{48}$$

$$\xi_2 = Ln\left(\frac{r_2^2}{288\eta_2 V_2}\right) + Ln\left[Csch^2(r_2 t - q_2)\right], \tag{49}$$

$$\xi_3 = a_3 + 648\frac{|\eta_1|\,\eta_2 - 3|\eta_1| + 3\eta_2}{|\eta_1|\,\eta_2} p_3 t + \frac{9 + |\eta_1|}{|\eta_1|} Ln\left[Cosh^2(r_1 t - q_1)\right]$$
$$+ \frac{\eta_2 - 9}{\eta_2} Ln\left[Sinh^2(r_2 t - q_2)\right], \tag{50}$$

where a_3 is an integration constant. To arrive at the solutions in terms of the original variables (Ω, ϕ_1, ϕ_2), we apply the inverse canonical transformation (26), obtaining the following:

$$\begin{aligned}\Omega &= \Omega_0 + Ln\left[Cosh^{\beta_1}(r_1\, t - q_1)Csch^{\beta_2}(r_2\, t - q_2)\right] \\ &\quad + 36\frac{|\eta_1|\,\eta_2 - 3|\eta_1| + 3\eta_2}{|\eta_1|\,\eta_2} p_3 t\,, \end{aligned} \tag{51}$$

$$\begin{aligned}\phi_1 &= \phi_{10} + Ln\left[Cosh^{\frac{2(|\eta_1|+3)}{\lambda_1|\eta_1|}}(r_1\, t - q_1)Csch^{\frac{6}{\lambda_1\eta_2}}(r_2\, t - q_2)\right] \\ &\quad + 216\frac{|\eta_1|\,\eta_2 - 3|\eta_1| + 3\eta_2}{\lambda_1\,|\eta_1|\,\eta_2} p_3 t\,, \end{aligned} \tag{52}$$

$$\begin{aligned}\phi_2 &= \phi_{20} + Ln\left[Cosh^{\frac{6}{\lambda_2|\eta_1|}}(r_1\, t - q_1)Sinh^{\frac{2(\eta_2-3)}{\lambda_2\eta_2}}(r_2\, t - q_2)\right] \\ &\quad + 216\frac{|\eta_1|\,\eta_2 - 3|\eta_1| + 3\eta_2}{\lambda_2\,|\eta_1|\,\eta_2} p_3 t\,, \end{aligned} \tag{53}$$

where $\beta_1 = 1/|\eta_1|$, $\beta_2 = 1/\eta_2$, and the constants Ω_0, ϕ_{10} and ϕ_{20} are given by

$$\begin{aligned}\Omega_0 &= Ln\left[\frac{r_1\, r_2}{288\sqrt{|\eta_1|\eta_2 V_1\, V_2}}\right]^{\frac{1}{9}} + \frac{a_3}{18}, \\ \phi_{10} &= Ln\left[\frac{12\sqrt{2}r_2\,|\eta_1|V_1}{r_1^2\sqrt{\eta_2 V_2}}\right]^{\frac{2}{3\lambda_1}} + \frac{a_3}{3\lambda_1}, \\ \phi_{20} &= Ln\left[\frac{12\sqrt{2}r_1\eta_2 V_2}{r_2^2\sqrt{|\eta_1|V_1}}\right]^{\frac{2}{3\lambda_2}} + \frac{a_3}{3\lambda_2}. \end{aligned} \tag{54}$$

For this case, the scale factor becomes

$$\begin{aligned}A(t) = &\left[\frac{r_1 r_2}{288\sqrt{|\eta_1|\eta_2 V_1 V_2}}\right]^{\frac{1}{9}} e^{\frac{a_3}{18}} Cosh^{\beta_1}(r_1\, t - q_1)Csch^{\beta_2}(r_2\, t - q_2) \\ &\times Exp\left[36\frac{|\eta_1|\,\eta_2 - 3|\eta_1| + 3\eta_2}{|\eta_1|\,\eta_2} p_3 t\right]. \end{aligned} \tag{55}$$

In Figure 2, we can appreciate the evolution of the scale factor, the Hubble parameter, and the barotropic parameter, with respect to time. First, we can once again observe a bouncing A, which consolidates our previous outcome. In fact, this behavior was claimed recently in [66], using a dynamical system approach. Additionally, in the upper right plot, H crosses the horizontal axis (at the bounce of A). Then, in the panel at the bottom, once again $\omega_{\phi_1\phi_2}$ traverses the phantom divide line "−1", an upshot consistent with the quintom description .

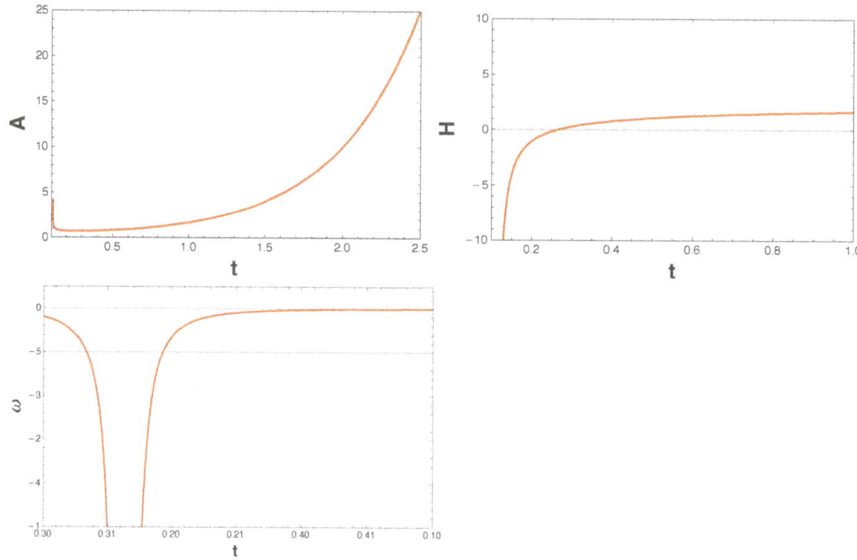

Figure 2. Phantom domination. This figure shows the time ($0.1 \leq t \leq 2.5$) evolution of the scale factor $A(t)$, the Hubble parameter $H(t)$, and the barotropic parameter $\omega_{\phi_1\phi_2}(t)$. We use arbitrary units, namely, $V_1 = V_2 = 1.0$, $r_1 = 1.5$, $r_2 = 0.9$, $q_1 = q_2 = 0.1$, $a_3 = -6.0$, $\lambda_2 = \sqrt{2}$, and $\lambda_1 = 6/\lambda_2$. The remaining constants can be obtained from the aforementioned values. Note that time is measured in reduced Planck units since $8\pi G = 1$.

2.1.4. Quintessence Domination: $\lambda_1 < \sqrt{6}$ and $\lambda_2 > \sqrt{6}$

We reinsert the solutions of $\lambda_1 < \sqrt{6}$ ($\eta_1 > 0$) and $\lambda_2 > \sqrt{6}$ into the Hamilton equations for the momenta, leading to

$$P_1 = p_1 - \frac{r_1}{12\eta_1} Coth(r_1 t - q_1), \tag{56}$$

$$P_2 = p_2 - \frac{r_2}{12\eta_2} Coth(r_2 t - q_2), \tag{57}$$

where p_1 and p_2 are integration constants. We use (56) and (57) to obtain a null Hamiltonian when

$$p_1 = \frac{\eta_1 - 9}{\eta_1} p_3, \qquad p_2 = \frac{\eta_2 - 9}{\eta_2} p_3, \qquad p_3 = \pm\frac{1}{36}\sqrt{\frac{\eta_2 r_1^2 + \eta_1 r_2^2}{3[3\eta_1 + 3\eta_2 - \eta_1\eta_2]}}. \tag{58}$$

As a consequence, the solutions of ξ_i take the following form:

$$\xi_1 = Ln\left(\frac{r_1^2}{288\eta_1 V_1}\right) + Ln\left[Csch^2(r_1 t - q_1)\right], \tag{59}$$

$$\xi_2 = Ln\left(\frac{r_2^2}{288\eta_2 V_2}\right) + Ln\left[Csch^2(r_2 t - q_2)\right], \tag{60}$$

$$\xi_3 = a_3 - 648\frac{-\eta_1\eta_2 + 3\eta_1 + 3\eta_2}{\eta_1\eta_2} p_3 t + \frac{\eta_1 - 9}{\eta_1} Ln\left[Sinh^2(r_1 t - q_1)\right] + \frac{\eta_2 - 9}{\eta_2} Ln\left[Sinh^2(r_2 t - q_2)\right], \tag{61}$$

with an integration constant a_3. Then, we apply the inverse transformation (26) to arrive at the solutions in terms of the original variables, which read

$$\begin{aligned}
\Omega &= Ln\left[\frac{r_1 r_2}{288\sqrt{\eta_1\eta_2 V_1 V_2}}\right]^{\frac{1}{9}} + \frac{a_3}{18} + Ln\left[Csch^{\beta_1}(r_1 t - q_1)Csch^{\beta_2}(r_2 t - q_2)\right] \\
&\quad - 36\frac{3\eta_1 + 3\eta_2 - \eta_1\eta_2}{\eta_1\eta_2}p_3 t, \\
\phi_1 &= \phi_{10} + Ln\left[Cosh^{\frac{2(|\eta_1|+3)}{\lambda_1|\eta_1|}}(r_1 t - q_1)Csch^{\frac{6}{\lambda_1\eta_2}}(r_2 t - q_2)\right] + 216\frac{|\eta_1|\eta_2 - 3|\eta_1| + 3\eta_2}{\lambda_1|\eta_1|\eta_2}p_3 t, \\
\phi_2 &= \phi_{20} + Ln\left[Cosh^{\frac{6}{\lambda_2|\eta_1|}}(r_1 t - q_1)Sinh^{\frac{2(\eta_2-3)}{\lambda_2\eta_2}}(r_2 t - q_2)\right] \\
&\quad + 216\frac{|\eta_1|\eta_2 - 3|\eta_1| + 3\eta_2}{\lambda_2|\eta_1|\eta_2}p_3 t,
\end{aligned}$$

where $\beta_1 = 1/\eta_1, \beta_2 = 1/\eta_2$, and the constants Ω_0, ϕ_{10}, and ϕ_{20} are those in (54). With these solutions, we can write the scale factor in the following form:

$$\begin{aligned}
A(t) &= \left[\frac{r_1 r_2}{288\sqrt{\eta_1\eta_2 V_1 V_2}}\right]^{\frac{1}{9}} e^{\frac{a_3}{18}} Csch^{\beta_1}(r_1 t - q_1)Csch^{\beta_2}(r_2 t - q_2) \\
&\quad \times Exp\left[-36\frac{3\eta_1 + 3\eta_2 - \eta_1\eta_2}{\eta_1\eta_2}p_3 t\right].
\end{aligned} \tag{62}$$

Immediately, one can observe that to obtain an increasing scale factor with respect to time, the constant p_3 must be negative. However, none of the parameters considered in this scenario lead to $p_3 < 0$; therefore, this solution is not physically relevant.

3. Quantum Formalism

To present the quantum mechanical version of the classical model, in (25) we promote the classical momenta to operators making the replacement $\Pi_{q^\mu} = -i\hbar\partial_{q^\mu}$, obtaining the following Hamiltonian density:

$$\mathcal{H} = \Pi_\Omega^2 + Qi\hbar\Pi_\Omega - 12\Lambda_2\Pi_{\phi_1}^2 + 12\Lambda_1\Pi_{\phi_2}^2 - 24\Lambda_0\Pi_{\phi_1}\Pi_{\phi_2} - 24V_1 e^{-\lambda_1\phi_1+6\Omega} - 24V_2 e^{-\lambda_2\phi_2+6\Omega}. \tag{63}$$

To obtain Equation (63), we have substituted $e^{-3\Omega}\Pi_\Omega^2 \to e^{-3\Omega}\left[\Pi_\Omega^2 + Qi\hbar\Pi_\Omega\right]$ since one has to take into account the factor-ordering problem between the $e^{-3\Omega}$ and its momentum Π_Ω; hence, Q is a number that measures such ambiguity. In order to have a more manageable functional form of (63), we take the constraint of the matrix element m^{12} (Equation (30)); then, we apply the canonical transformation on variables $(\Omega, \phi_1, \phi_2) \leftrightarrow (\xi_1, \xi_2, \xi_3)$ (Equations (26) and (27)), as well as the gauge $N = 24e^{3\Omega}$. Therefore, we obtain

$$\begin{aligned}
\mathcal{H} &= 12\eta_1 P_1^2 + 12\eta_2 P_2^2 + 12(-9 + \eta_1 + \eta_2)P_3^2 \\
&\quad + 24P_3[(9 - \eta_1)P_1 + (9 - \eta_2)P_2] + 6Qi\hbar(P_1 + P_2 + P_3) \quad - 24\left(V_1 e^{\xi_1} + V_2 e^{\xi_2}\right),
\end{aligned} \tag{64}$$

with $\eta_1 = 3 - \Lambda_1$ and $\eta_2 = 3 + \Lambda_2$. Recall that the Hamiltonian density is identically zero $\mathcal{H} = 0$; hence, the quantum counterpart of (64) is obtained by applying the same prescription used to obtain (63). Having this at hand, we can write down the Wheeler–DeWitt (WDW) equation, which reads

$$
\begin{aligned}
\hat{\mathcal{H}}\Psi(\xi_i) &= -12\hbar^2\eta_1\frac{\partial^2\Psi}{\partial\xi_1^2} - 12\hbar^2\eta_2\frac{\partial^2\Psi}{\partial\xi_2^2} - 12\hbar^2(-9+\eta_1+\eta_2)\frac{\partial^2\Psi}{\partial\xi_3^2} \\
&+ 6Q\hbar^2\left(\frac{\partial\Psi}{\partial\xi_1}+\frac{\partial\Psi}{\partial\xi_2}+\frac{\partial\Psi}{\partial\xi_3}\right) - \hbar^2 24\left[(9-\eta_1)\frac{\partial^2\Psi}{\partial\xi_1\partial\xi_3}+(9-\eta_2)\frac{\partial^2\Psi}{\partial\xi_2\partial\xi_3}\right] \\
&- 24\left(V_1 e^{\xi_1}+V_2 e^{\xi_2}\right)\Psi = 0 .
\end{aligned} \tag{65}
$$

In order to solve the WDW equation, we propose the following solution for the wave function $\Psi(\xi_1,\xi_2,\xi_3) = e^{p_3\xi_3}\mathcal{G}(\xi_1,\xi_2)$ with $p_3 = constant$. Additionally, we take as an ansatz $\mathcal{G}(\xi_1,\xi_2) = G_1(\xi_1)G_2(\xi_2)$; upon substitution in (65), we obtain the following.

$$
\begin{aligned}
&-12\eta_1 G_2\frac{\partial^2 G_1}{\partial\xi_1^2} + 6(Q-4p_3(9-\eta_1))G_2\frac{\partial G_1}{\partial\xi_1} \\
&+3\left[p_3(Q-2p_3(9-\eta_1+\eta_2)) - 8\frac{V_1}{\hbar^2}e^{\xi_1}\right]G_1G_2 + \\
&-12\eta_2 G_1\frac{\partial^2 G_2}{\partial\xi_2^2} + 6(Q-4p_3(9-\eta_2))G_1\frac{\partial G_2}{\partial\xi_2} \\
&+3\left[p_3(Q-2p_3(9-\eta_1+\eta_2)) - 8\frac{V_2}{\hbar^2}e^{\xi_2}\right]G_1G_2 = 0 ,
\end{aligned} \tag{66}
$$

finally, we factorize G_1G_2. Thus, two ordinary differential equations for the functions G_1 and G_2 emerge

$$
\begin{aligned}
&-12\frac{\eta_1}{G_1}\frac{\partial^2 G_1}{\partial\xi_1^2} + 6(Q-4p_3(9-\eta_1))\frac{1}{G_1}\frac{\partial G_1}{\partial\xi_1} \\
&\qquad +3\left[p_3(Q-2p_3(9-\eta_1+\eta_2)) - 8\frac{V_1}{\hbar^2}e^{\xi_1}\right] - \nu^2 = 0,
\end{aligned} \tag{67}
$$

$$
\begin{aligned}
&-12\frac{\eta_2}{G_2}\frac{\partial^2 G_2}{\partial\xi_2^2} + 6[Q-4p_3(9-\eta_2)]\frac{1}{G_2}\frac{\partial G_2}{\partial\xi_2} \\
&\qquad +3\left[p_3(Q-2p_3(9-\eta_1+\eta_2)) - 8\frac{V_2}{\hbar^2}e^{\xi_2}\right] + \nu^2 = 0,
\end{aligned} \tag{68}
$$

where ν^2 is an arbitrary constant. These last two equations can be written as $y'' + ay' + (be^{\kappa x} + c)y = 0$, and their solutions are of the form [67]

$$
Y(x) = Exp\left(-\frac{ax}{2}\right)Z_\rho\left(\frac{2\sqrt{b}}{\kappa}e^{\frac{\kappa x}{2}}\right), \tag{69}
$$

here, Z_ρ are the generic Bessel functions with the order $\rho = \sqrt{a^2-4c}/\kappa$. If $\sqrt{b}$ is real, Z_ρ becomes the ordinary Bessel function; otherwise, the solutions will be given in terms of the modified Bessel functions. In the next sections, we will show quantum solutions separated into two classes, according to η_1 and $\lambda_1\lambda_2 = 6$.

3.1. Quantum Solution for $\eta_1 > 0$ and $\lambda_1 < \sqrt{6}$

First, we identify the following expressions for Equation (67):

$$\begin{aligned} \kappa &= 1, \qquad a = -\frac{Q - 4p_3(9-\eta_1)}{2\eta_1}, \\ b &= \frac{2V_2}{\eta_1\hbar^2}, \qquad c = -\frac{p_3[Q - 2p_3(9-\eta_1+\eta_2)]}{4\eta_1} + \frac{\nu^2}{12\eta_1}, \end{aligned} \tag{70}$$

and for (68)

$$\begin{aligned} \kappa &= 1, \qquad a = -\frac{Q - 4p_3(9-\eta_2)}{2\eta_2}, \\ b &= \frac{2V_1}{\eta_2\hbar^2}, \qquad c = -\frac{p_3[Q - 2p_3(9-\eta_1+\eta_2)]}{4\eta_2} - \frac{\nu^2}{12\eta_2}. \end{aligned} \tag{71}$$

Note that in both cases, $\sqrt{b}$ is real; then, the solutions are written in terms of the ordinary Bessel functions $Z_{\rho_i} = J_{\rho_i}$. Thus, the wave function becomes the following:

$$\mathcal{B}_{\rho_1\rho_2} = \mathcal{B}_0\, J_{\rho_1}\left[\frac{2}{\hbar}\sqrt{\frac{2V_1}{\eta_1}}e^{\frac{\xi_1}{2}}\right] J_{\rho_2}\left[\frac{2}{\hbar}\sqrt{\frac{2V_2}{\eta_2}}e^{\frac{\xi_2}{2}}\right] e^{\theta},$$

where

$$\theta = \frac{Q - 4p_3(9-\eta_1)}{4\eta_1}\xi_1 + \frac{Q - 4p_3(9-\eta_2)}{4\eta_2}\xi_2,$$

and $\mathcal{B}_0$ is an integration constant. Additionally, the order of the two Bessel functions are

$$\rho_1 = \sqrt{\left(-\frac{Q - 4p_3(9-\eta_1)}{2\eta_1}\right)^2 + \frac{p_3(Q - 2p_3(9-\eta_1+\eta_2))}{\eta_1} - \frac{\nu^2}{3\eta_1}}, \tag{72}$$

$$\rho_2 = \sqrt{\left(-\frac{Q - 4p_3(9-\eta_2)}{2\eta_2}\right)^2 + \frac{p_3(Q - 2p_3(9-\eta_1+\eta_2))}{\eta_2} + \frac{\nu^2}{3\eta_2}}. \tag{73}$$

Hence, the wave function Ψ in the original variables becomes

$$\Psi_{\rho_1\rho_2} = \Psi_0\, A^{6\alpha} Exp[\alpha_1\lambda_1\phi_1 + \alpha_2\lambda_2\phi_2] J_{\rho_1}\left[\frac{2}{\hbar}\sqrt{\frac{2V_1}{\eta_1}}A^3 e^{\frac{-\lambda_1\phi_1}{2}}\right] J_{\rho_2}\left[\frac{2}{\hbar}\sqrt{\frac{2V_2}{\eta_2}}A^3 e^{-\frac{\lambda_2\phi_2}{2}}\right], \tag{74}$$

where Ψ_0 is a normalization constant, and

$$\begin{aligned} \alpha &= \frac{Q(\eta_2+\eta_1)}{4\eta_1\eta_2} - \frac{p_3(9-\eta_1)}{\eta_1} - \frac{p_3(9-\eta_2)}{\eta_2} + p_3, \\ \alpha_1 &= -\frac{Q - 4p_3(9-\eta_1)}{4\eta_1} + p_3, \quad \alpha_2 = -\frac{Q - 4p_3(9-\eta_2)}{4\eta_2} + p_3. \end{aligned} \tag{75}$$

By analyzing solution (74), we could not find any set of parameter values for which the probability density function (defined by the wave function (74)) is bounded. This unwanted behavior prevents us from directly implementing the standard interpretation of quantum mechanics in order to draw meaningful physical conclusions. This setback is tempered by the fact that the corresponding classical solution (given essentially by (62)) is not of physical relevance, and so no further analysis will be performed regarding this case.

3.2. *Quantum Solution When $\eta_1 < 0$ and $\lambda_1 > \sqrt{6}$*

We set up the corresponding parameters for Equation (67)

$$\begin{aligned} \kappa &= 1 \quad a = \frac{Q - 4p_3(9 + |\eta_1|)}{2|\eta_1|}, \\ b &= -\frac{2V_2}{|\eta_1|\hbar^2} \quad c = \frac{p_3[Q - 2p_3(9 + |\eta_1| + \eta_2)]}{4|\eta_1|} - \frac{\nu^2}{12|\eta_1|}, \end{aligned} \tag{76}$$

and for (68)

$$\begin{aligned} \kappa &= 1 \quad a = -\frac{Q - 4p_3(9 - \eta_2)}{2\eta_2}, \\ b &= \frac{2V_1}{\eta_2\hbar^2}, \quad c = -\frac{p_3[Q - 2p_3(9 + |\eta_1| + \eta_2)]}{4\eta_2} - \frac{\nu^2}{12\eta_2}, \end{aligned} \tag{77}$$

note that we have inverted the sign of the previous formulas. Hereby, we introduce $|\eta_1|$. Then, the first case (76) yields an imaginary $\sqrt{b}$; therefore, its solution must be in terms of the modified Bessel function $Z_{\rho_1} = K_{\rho_1}$ (contrary to the second case, where the proper function is $Z_{\rho_2} = J_{\rho_2}$). Hence, we have

$$\mathcal{B}_{\rho_1\rho_2} = \mathcal{B}_0\, K_{\rho_1}\left[\frac{2}{\hbar}\sqrt{\frac{2V_1}{|\eta_1|}}e^{\frac{\xi_1}{2}}\right] J_{\rho_2}\left[\frac{2}{\hbar}\sqrt{\frac{2V_2}{\eta_2}}e^{\frac{\xi_2}{2}}\right] e^{\theta_2}, \tag{78}$$

here

$$\theta_2 = -\frac{Q - 4p_3(9 + |\eta_1|)}{4|\eta_1|}\xi_1 + \frac{Q - 4p_3(9 - \eta_2)}{4\eta_2}\xi_2, \tag{79}$$

and the order of both Bessel functions are

$$\rho_1 = \sqrt{\left(\frac{Q - 4p_3(9 + |\eta_1|}{2|\eta_1|}\right)^2 - \frac{p_3(Q - 2p_3(9 + |\eta_1| + \eta_2))}{|\eta_1|} + \frac{\nu^2}{3|\eta_1|}}, \tag{80}$$

$$\rho_2 = \sqrt{\left(-\frac{Q - 4p_3(9 - \eta_2}{2\eta_2}\right)^2 + \frac{p_3(Q - 2p_3(9 + |\eta_1| + \eta_2))}{\eta_2} + \frac{\nu^2}{3\eta_2}}. \tag{81}$$

Finally, the wave function in the original variables is given by

$$\Psi_{\rho_1\rho_2} = \Psi_0 A^{6\beta} Exp[\alpha_1\lambda_1\phi_1 + \alpha_2\lambda_2\phi_2] K_{\rho_1}\left[\frac{2}{\hbar}\sqrt{\frac{2V_1}{|\eta_1|}}A^3 e^{\frac{-\lambda_1\phi_1}{2}}\right] J_{\rho_2}\left[\frac{2}{\hbar}\sqrt{\frac{2V_2}{\eta_2}}A^3 e^{-\frac{\lambda_2\phi_2}{2}}\right], \tag{82}$$

where

$$\beta = -\frac{Q(\eta_2 - |\eta_1|)}{4|\eta_1|\eta_2} + \frac{p_3(9 + |\eta_1|)}{|\eta_1|} - \frac{p_3(9 - \eta_2)}{\eta_2} + p_3, \tag{83}$$

$$\alpha_1 = \frac{Q - 4p_3(9 + |\eta_1|)}{4|\eta_1|} + p_3, \quad \alpha_2 = -\frac{Q - 4p_3(9 - \eta_2)}{2\eta_2} + p_3, \tag{84}$$

and a normalization constant Ψ_0. The behaviour of the probability density can be seen in Figure 3. Observe that in all panels, the probability density dies away as the scale factor and scalar field evolve, an expected outcome already reported in [68–70]. On the other hand, we vary the factor ordering constant Q, in order to show how $|\Psi|^2$ behaves. We can see that whilst $Q \ll 0$, the probability density tends to the phantom sector. In [68], the authors showed that the parameter Q acts a retarder of the wave function and compresses the length on the axis where the field evolves; however, they analysed the case of two quintessence fields.

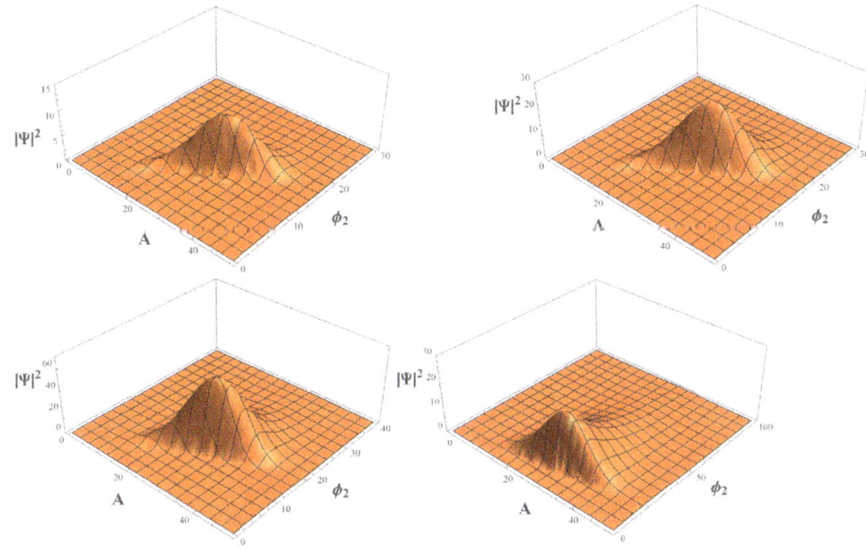

Figure 3. Phantom scenario. These figures show the probability density of the wave function (82) for the values of $Q = 2$ and $Q = 0$ (top panels from left to right, respectivley), and $Q = -2$ and $Q = -4$ (bottom panels from left to right, respectively). We use arbitrary units, namely, $\nu = 10$, $\lambda_1 = 10.5$, $\lambda_2 = 6/\lambda_1$, $V_1 = 0.1$, $V_2 = 10^{-5}$, $r_1 = 49$, $r_2 = 1.5$, $a_3 = -2.3$, and $p_3 = 0.326878$, and the bounce in the quintessence field $\phi_1 = 1.455$. The remaining constants can be obtained from the aforementioned values. Additionally, for $Q = 2, 0$ we take $\Psi_0 = 10^{-3}, 10^{-2}$ respectively; then, for $Q = -2, -4$ we chose $\Psi_0 = 10^{-1}, 1/\sqrt{10}$ respectively. Note that the probability density tends toward the phantom sector when the factor ordering constant $Q \ll 0$.

3.3. Quantum Solution When $\lambda_1 = \lambda_2 = \sqrt{6}$, Therefore $\eta_1 = 0$ and $\eta_2 = 6$

In this final case, we take $\lambda_1 = \lambda_2 = \sqrt{6}$; therefore, $\eta_1 = 0$ and $\eta_2 = 6$. Hence, the Equations (67) and (68) can be reduced to

$$6(Q - 36p_3)\frac{1}{G_1}\frac{\partial G_1}{\partial \xi_1} + 3\left[p_3(Q - 30p_3) - 8\frac{V_1}{\hbar^2}e^{\xi_1}\right] - \nu^2 = 0, \tag{85}$$

$$-\frac{72}{G_2}\frac{\partial^2 G_2}{\partial \xi_2^2} + 6[Q - 12p_3]\frac{1}{G_2}\frac{\partial G_2}{\partial \xi_2} + 3\left[p_3(Q - 30p_3) - 8\frac{V_2}{\hbar^2}e^{\xi_2}\right] + \nu^2 = 0. \tag{86}$$

The solution of (85) is given by

$$G_1 = G_0 Exp\left[\frac{\frac{\nu^2}{3} - p_3(Q - 30p_3)}{2(Q - 36p_3)}\xi_1 + \frac{4V_1}{\hbar^2(Q - 36p_3)}e^{\xi_1}\right], \tag{87}$$

where G_0 in an integration constant. Then, for G_2 we have the following ordinary Bessel function:

$$G_2 = Exp\left[\frac{Q - 12p_3}{24}\xi_2\right] J_{\rho_2}\left[\frac{\sqrt{V_2}}{\hbar}e^{\frac{\xi_2}{2}}\right], \tag{88}$$

here, the order is

$$\rho_2 = \sqrt{\left(\frac{Q - 12p_3}{12}\right)^2 + \frac{1}{6}\left[\frac{\nu^2}{3} + p_3(Q - 30p_3)\right]}. \tag{89}$$

Remarkably for this case, we can obtain a parameter space of Q, ν, and p_3 where the order can be real or imaginary. Hence, we have

$$\mathcal{B} = \mathcal{B}_0\, J_{\rho_2}\left[\frac{\sqrt{V_2}}{\hbar}e^{\frac{\xi_2}{2}}\right] e^{\theta_3}, \tag{90}$$

with

$$\theta_3 = \frac{\frac{\nu^2}{3} - p_3(Q-30p_3)}{2(Q-36p_3)}\xi_1 + \frac{Q-12p_3}{24}\xi_2 + \frac{4V_1}{\hbar^2(Q-36p_3)}e^{\xi_1},$$

Finally, in the original variables the wave function is

$$\Psi = \psi_0 A^{6\eta} J_{\rho_2}\left[\frac{\sqrt{V_2}}{\hbar}A^3 e^{\frac{-\lambda_2\phi_2}{2}}\right] e^{\theta_1}, \tag{91}$$

where

$$\theta_1 = \frac{4V_1}{\hbar^2(Q-36p_3)}A^6 e^{-\lambda_1\phi_1} + \alpha_1\lambda_1\phi_1 + \alpha_2\lambda_2\phi_2, \tag{92}$$

and

$$\eta = \frac{\frac{\nu^2}{3} - p_3(Q-30p_3)}{2(Q-36p_3)} + \frac{Q-12p_3}{24} + p_3$$

$$\alpha_1 = -\frac{\frac{\nu^2}{3} - p_3(Q-30p_3)}{2(Q-36p_3)} + p_3, \quad \alpha_2 = -\frac{Q-12p_3}{24} + p_3, \tag{93}$$

and a normalization constant ψ_0. For completeness of the above classical solutions, we include this case; however, once more the probability density function is not bounded since $|\Psi|^2$ does not fade as the scale factor and scalar field evolve. We recall that the standard interpretation of quantum mechanics becomes troublesome to realize due to this nuisance behavior. Therefore, the wave function (91) is not physically relevant.

4. Final Remarks

In this work, we have studied a chiral cosmological model from the point of view of a K-essence formalism. The background geometry was a flat FLRW universe minimally coupled to quintom fields: one quintessence and one phantom. In this approach, the scalar fields interact within the kinetic and potential sectors.

In the classical framework, we established the Hamiltonian density (31), which in turn allows one to find exact solutions for different sets of values of the free parameters. We highlight two cases: the first when $\lambda_1 = \lambda_2 = \sqrt{6}$, and the second where phantom domination is the relevant factor, namely, $\lambda_1 > \sqrt{6}$ and $\lambda_2 < \sqrt{6}$. In the two scenarios, the scale factor grows very rapidly and the big-bang singularity is avoided via a bounce. We call it the "big bounce". In fact, this claim is also supported by the behavior of both the scale factor and the Hubble parameter. Finally, we show that the barotropic parameter is capable of transiting from a quintessence phase to a phantom one, i.e., it crosses the phantom divide line. In Figures 1 and 2, we show the behavior of these quantities as a function of time.

On the other hand, using the canonical quantization procedure, we were able to establish the quantum counterpart of the classical model and compute the Wheeler–DeWitt equation. Once again, we solve it for various scenarios given by different sets of values of the free parameters. In particular, we found exact solutions for three distinct cases: $\eta_1 > 0$ and $\lambda_1 < \sqrt{6}$, $\eta_1 < 0$ and $\lambda_1 > \sqrt{6}$, and $\lambda_1 = \lambda_2 = \sqrt{6}$; therefore, $\eta_1 = 0$ and $\eta_2 = 6$. Figure 3 shows the behavior of the probability density as a function of the scale factor and scalar field, for the phantom case, i.e., $\eta_1 < 0$ and $\lambda_1 > \sqrt{6}$. The probability density exhibits a damped behavior as the scale factor and scalar fields evolve. An expected result has already been reported in [68–70]. Lastly, we note that by varying Q, specifically when $Q \ll 0$, the probability density evolves towards the phantom sector. This outcome contrasts with that reported in [68], where the authors showed that the parameter Q delays the evolution of the wave function and compresses the length on the axis where the field evolves; however, they analyzed the case of two quintessence fields.

Author Contributions: : Conceptualization, J.S., S.P.-P., R.H.-J., A.E.-G., and L.R.D.-B.; Methodology, J.S., S.P.-P., R.H.-J., A.E.-G., and L.R.D.-B.; Writing—Original Draft, J.S., S.P.-P., R.H.-J., A. E.-G., and L.R.D.-B.; Writing—Review and Editing, J.S., S.P.-P., R.H.-J., A.E.-G., and L.R.D.-B.; Visualization, J.S., S.P.-P. All authors have read and agreed to the published version of the manuscript.

Funding: This work was partially supported by PROMEP grants UGTO-CA-3. J.S. and L. R. D. B. were partially supported SNI-CONACyT. R.H.J is supported by CONACyT Estancias posdoctorales por México, Modalidad 1: Estancia Posdoctoral Académica.

Data Availability Statement: Not applicable.

Acknowledgments: This work is part of the collaboration within the Instituto Avanzado de Cosmología and Red PROMEP: Gravitation and Mathematical Physics, under project *Quantum aspects of gravity in cosmological models, phenomenology, and geometry of space-time*. Many calculations where done by Symbolic Program REDUCE 3.8.

Conflicts of Interest: The authors declare no conflict of interest.

References

1. Perlmutter, S.; Aldering, G.; Goldhaber, G.; Knop, R.A.; Nugent, P.; Castro, P.G.; Couch, W.J. Measurements of Ω and Λ from 42 high redshift supernovae. *Astrophys. J.* **1999**, *517*, 565–586. [CrossRef]
2. Riess, A.G.; Filippenko, A.V.; Challis, P.; Clocchiatti, A.; Diercks, A.; Garnavich, P.M.; Tonry, J. Observational evidence from supernovae for an accelerating universe and a cosmological constant. *Astron. J.* **1998**, *116*, 1009–1038. [CrossRef]
3. Garnavich, P.M.; Kirshner, R.P.; Challis, P.; Tonry, J.; Gilliland, R.L.; Smith, R.C.; Wells, L. Constraints on cosmological models from Hubble Space Telescope observations of high z supernovae. *Astrophys. J. Lett.* **1998**, *493*, L53–L57. [CrossRef]
4. Komatsu, E.; Dunkley, J.; Nolta, M.R.; Bennett, C.L.; Gold, B.; Hinshaw, G.; Wright, E.L. Five-Year Wilkinson Microwave Anisotropy Probe (WMAP) Observations: Cosmological Interpretation. *Astrophys. J. Suppl.* **2009**, *180*, 330–376. [CrossRef]
5. Guth, A.H. The Inflationary Universe: A Possible Solution to the Horizon and Flatness Problems. *Phys. Rev. D* **1981**, *23*, 347–356. [CrossRef]
6. Linde, A.D. A New Inflationary Universe Scenario: A Possible Solution of the Horizon, Flatness, Homogeneity, Isotropy and Primordial Monopole Problems. *Phys. Lett. B* **1982**, *108*, 389–393. [CrossRef]
7. Copeland, E.J.; Sami, M.; Tsujikawa, S. Dynamics of dark energy. *Int. J. Mod. Phys. D* **2006**, *15*, 1753–1936. [CrossRef]
8. Clifton, T.; Ferreira, P.G.; Padilla, A.; Skordis, C. Modified Gravity and Cosmology. *Phys. Rept.* **2012**, *513*, 1–189. [CrossRef]
9. Nojiri, S.; Odintsov, S.D.; Oikonomou, V.K. Modified Gravity Theories on a Nutshell: Inflation, Bounce and Late-time Evolution. *Phys. Rept.* **2017**, *692*, 1–104. [CrossRef]
10. Urena-Lopez, L.A.; Matos, T. A New cosmological tracker solution for quintessence. *Phys. Rev. D* **2000**, *62*, 081302. [CrossRef]
11. Ratra, B.; Peebles, P.J.E. Cosmological Consequences of a Rolling Homogeneous Scalar Field. *Phys. Rev. D* **1988**, *37*, 3406. [CrossRef]
12. Harko, T.; Lobo, F.S.N.; Mak, M.K. Arbitrary scalar field and quintessence cosmological models. *Eur. Phys. J. C* **2014**, *74*, 2784. [CrossRef]
13. Rubano, C.; Barrow, J.D. Scaling solutions and reconstruction of scalar field potentials. *Phys. Rev. D* **2001**, *64*, 127301. [CrossRef]
14. Sahni, V.; Starobinsky, A. The Case for a positive cosmological Lambda term. *Int. J. Mod. Phys. D* **2000**, *9*, 373–444. [CrossRef]
15. Sahni, V.; Wang, L.M. A New cosmological model of quintessence and dark matter. *Phys. Rev. D* **2000**, *62*, 103517. [CrossRef]
16. Paliathanasis, A.; Tsamparlis, M.; Basilakos, S.; Barrow, J.D. Dynamical analysis in scalar field cosmology. *Phys. Rev. D* **2015**, *91*, 123535. [CrossRef]
17. Dimakis, N.; Karagiorgos, A.; Zampeli, A.; Paliathanasis, A.; Christodoulakis, T.; Terzis, P.A. General Analytic Solutions of Scalar Field Cosmology with Arbitrary Potential. *Phys. Rev. D* **2016**, *93*, 123518. [CrossRef]
18. Fang, W.; Lu, H.Q.; Huang, Z.G.; Zhang, K.G. The evolution of the universe with the B-I type phantom scalar field. *Int. J. Mod. Phys. D* **2006**, *15*, 199–214. [CrossRef]
19. Cataldo, M.; Arevalo, F.; Mella, P. Canonical and phantom scalar fields as an interaction of two perfect fluids. *Astrophys. Space Sci.* **2013**, *344*, 495–503. [CrossRef]
20. Nojiri, S.; Odintsov, S.D.; Oikonomou, V.K.; Saridakis, E.N. Singular cosmological evolution using canonical and ghost scalar fields. *JCAP* **2015**, *9*, 044. [CrossRef]
21. Cai, Y.F.; Saridakis, E.N.; Setare, M.R.; Xia, J.Q. Quintom Cosmology: Theoretical implications and observations. *Phys. Rept.* **2010**, *493*, 1–60. [CrossRef]
22. Setare, M.R.; Saridakis, E.N. Quintom Cosmology with General Potentials. *Int. J. Mod. Phys. D* **2009**, *18*, 549–557. [CrossRef]

23. Lazkoz, R.; Leon, G.; Quiros, I. Quintom cosmologies with arbitrary potentials. *Phys. Lett. B* **2007**, *649*, 103–110. [CrossRef]
24. Leon, G.; Paliathanasis, A.; Morales-Martínez, J.L. The past and future dynamics of quintom dark energy models. *Eur. Phys. J. C* **2018**, *78*, 753. [CrossRef]
25. Dimakis, N.; Paliathanasis, A. Crossing the phantom divide line as an effect of quantum transitions. *Class. Quant. Grav.* **2021**, *38*, 075016. [CrossRef]
26. Elizalde, E.; Nojiri, S.; Odintsov, S.D.; Saez-Gomez, D.; Faraoni, V. Reconstructing the universe history, from inflation to acceleration, with phantom and canonical scalar fields. *Phys. Rev. D* **2008**, *77*, 106005. [CrossRef]
27. Chervon, S.V. On the chiral model of cosmological inflation. *Russ. Phys. J.* **1995**, *38*, 539–543. [CrossRef]
28. Chervon, S.V. Chiral Cosmological Models: Dark Sector Fields Description. *Quant. Matt.* **2013**, *2*, 71–82. [CrossRef]
29. Christodoulidis, P.; Roest, D.; Sfakianakis, E.I. Scaling attractors in multi-field inflation. *JCAP* **2019**, *12*, 059. [CrossRef]
30. Beesham, A.; Chervon, S.V.; Maharaj, S.D.; Kubasov, A.S. An Emergent Universe with Dark Sector Fields in a Chiral Cosmological Model. *Quant. Matt.* **2013**, *2*, 388–395. [CrossRef]
31. Chervon, S.V.; Abbyazov, R.R.; Kryukov, S.V. Dynamics of Chiral Cosmological Fields in the Phantom-Canonical Model. *Russ. Phys. J.* **2015**, *58*, 597–605. [CrossRef]
32. Fomin, I.V. The chiral cosmological models with two components. *J. Phys. Conf. Ser.* **2017**, *918*, 012009. [CrossRef]
33. Fomin, I.V. Two-Field Cosmological Models with a Second Accelerated Expansion of the Universe. *Moscow Univ. Phys. Bull.* **2018**, *73*, 696–701. [CrossRef]
34. Paliathanasis, A.; Leon, G.; Pan, S. Exact Solutions in Chiral Cosmology. *Gen. Rel. Grav.* **2019**, *51*, 106. [CrossRef]
35. Scherrer, R.J. Purely kinetic k-essence as unified dark matter. *Phys. Rev. Lett.* **2004**, *93*, 011301. [CrossRef]
36. Bandyopadhyay, A.; Gangopadhyay, D.; Moulik, A. The *k*-essence scalar field in the context of Supernova Ia Observations. *Eur. Phys. J. C* **2012**, *72*, 1943. [CrossRef]
37. Armendariz-Picon, C.; Damour, T.; Mukhanov, V.F. K-inflation. *Phys. Lett. B* **1999**, *458*, 209–218. [CrossRef]
38. Damour, T.; Esposito-Farese, G. Tensor multiscalar theories of gravitation. *Class. Quant. Grav.* **1992**, *9*, 2093–2176. [CrossRef]
39. Horndeski, G.W. Second-order scalar-tensor field equations in a four-dimensional space. *Int. J. Theor. Phys.* **1974**, *10*, 363–384. [CrossRef]
40. Deffayet, C.; Esposito-Farese, G.; Vikman, A. Covariant Galileon. *Phys. Rev. D* **2009**, *79*, 084003. [CrossRef]
41. Coley, A.A.; van den Hoogen, R.J. The Dynamics of multiscalar field cosmological models and assisted inflation. *Phys. Rev. D* **2000**, *62*, 023517. [CrossRef]
42. Peebles, P.J.E.; Ratra, B. The Cosmological Constant and Dark Energy. *Rev. Mod. Phys.* **2003**, *75*, 559–606. [CrossRef]
43. Padmanabhan, T. Cosmological constant: The Weight of the vacuum. *Phys. Rept.* **2003**, 380, 235–320. [CrossRef]
44. Albrecht, A.; Bernstein, G.; Cahn, R.; Freedman, W.L.; Hewitt, J.; Hu, W.; Huth, J.; Kamionkowski, M.; Kolb, E.W.; Knox, L. Report of the Dark Energy Task Force. *arXiv* **2003**, arXiv:astro-ph/0609591.
45. Linder, E.V. Mapping the Cosmological Expansion. *Rept. Prog. Phys.* **2008**, *71*, 056901. [CrossRef]
46. Frieman, J.; Turner, M.; Huterer, D. Dark Energy and the Accelerating Universe. *Ann. Rev. Astron. Astrophys.* **2008**, *46*, 385–432. [CrossRef]
47. Caldwell, R.R.; Kamionkowski, M. The Physics of Cosmic Acceleration. *Ann. Rev. Nucl. Part. Sci.* **2009**, *59*, 397–429. [CrossRef]
48. Wetterich, C. Cosmology and the Fate of Dilatation Symmetry. *Nucl. Phys. B* **1988**, *302*, 668–696. [CrossRef]
49. Caldwell, R.R. A Phantom menace? *Phys. Lett. B* **2002**, *545*, 23–29. [CrossRef]
50. Caldwell, R.R.; Kamionkowski, M.; Weinberg, N.N. Phantom energy and cosmic doomsday. *Phys. Rev. Lett.* **2003**, *91*, 071301. [CrossRef]
51. Feng, B.; Wang, X.L.; Zhang, X.M. Dark energy constraints from the cosmic age and supernova. *Phys. Lett. B* **2005**, *607*, 35–41. [CrossRef]
52. Vikman, A. Can dark energy evolve to the phantom? *Phys. Rev. D* **2005**, *71*, 023515. [CrossRef]
53. Deffayet, C.; Pujolas, O.; Sawicki, I.; Vikman, A. Imperfect Dark Energy from Kinetic Gravity Braiding. *JCAP* **2010**, *10*, 26. [CrossRef]
54. Chimento, L.P.; Forte, M.I.; Lazkoz, R.; Richarte, M.G. Internal space structure generalization of the quintom cosmological scenario. *Phys. Rev. D* **2009**, *79*, 043502. [CrossRef]
55. Lindle, A.D. Hybrid inflation. *Phys. Rev. D* **1994**, *49*, 784.
56. Copeland, E.J.; Liddle, A.R.; Lyth, D.H.; Stewart, E.D.; Wands, D. False vacuum inflation with Einstein gravity. *Phys. Rev. D* **1994**, *49*, 6410–6433. [CrossRef]
57. Kim, S.A.; Liddle, A.R. Nflation: Multi-field inflationary dynamics and perturbations. *Phys. Rev. D* **2006**, *74*, 023513. [CrossRef]
58. Socorro, J.; Núñez, O.E. Scalar potentials with Multi-scalar fields from quantum cosmology and supersymmetric quantum mechanics. *Eur. Phys. J. Plus* **2017**, *132*, 168. [CrossRef]
59. Liddle, A.R.; Mazumdar, A.; Schunck, F.E. Assisted inflation. *Phys. Rev. D* **1998**, 58, 061301. [CrossRef]
60. Copeland, E.J.; Mazumdar, A.; Nunes, N.J. Generalized assisted inflation. *Phys. Rev. D* **1999**, *60*, 083506. [CrossRef]
61. Yokoyama, S.; Suyama, T.; Tanaka, T. Primordial Non-Gaussianity in Multi-Scalar Inflation. *Phys. Rev. D* **2008**, *77*, 083511. [CrossRef]
62. Chiba, T.; Yamaguchi, M. Extended Slow-Roll Conditions and Primordial Fluctuations: Multiple Scalar Fields and Generalized Gravity. *JCAP* **2009**, *901*, 19. [CrossRef]

63. Socorro, J.; Pimentel, L.O.; Espinoza-García, A. Classical Bianchi type I cosmology in K-essence theory. *Adv. High Energy Phys.* **2014**, *2014*, 805164. [CrossRef]
64. Chervon, S.V.; Fomin, I.V.; Pozdeeva, E.O.; Sami, M.; Vernov, S.Y. Superpotential method for chiral cosmological models connected with modified gravity. *Phys. Rev. D* **2019**, *100*, 063522. [CrossRef]
65. Fomin, I.V.; Chervon, S.V. New method of exponential potentials reconstruction based on given scale factor in phantonical two-field models. *arXiv* **2021**, arXiv:2112.09359.
66. Tot, J.; Yildirim, B.; Coley, A.; Leon, G. The dynamics of scalar-field quintom cosmological models. *arXiv* **2022**, arXiv:2204.06538.
67. Zaitsev, V.F.; Polyanin, A.D. *Handbook of Exact Solutions for Ordinary Differential Equations*, 2nd ed.; Chapman & Hall/CRC: London, UK, 2003.
68. Socorro, J.; Pérez-Payán, S.; Hernández-Jiménez, R.; Espinoza-García, A.; Díaz-Barrón, L.R. Classical and quantum exact solutions for a FRW in chiral like cosmology. *Class. Quant. Grav.* **2021**, *38*, 135027. [CrossRef]
69. Socorro, J.; Núñez, O.E.; Hernández-Jiménez, R. Classical and quantum exact solutions for the anisotropic Bianchi type I in multi-scalar field cosmology with an exponential potential driven inflation. *Phys. Lett. B* **2020**, *809*, 135667. [CrossRef]
70. Socorro, J.; Núñez, O.E.; Hernández-Jiménez, R. Classical and Quantum Exact Solutions for a FRW Multiscalar Field Cosmology with an Exponential Potential Driven Inflation. *Adv. Math. Phys.* **2018**, *2018*, 3468381. [CrossRef]

universe

MDPI

Article

Galactic Wormhole under Lovelock Gravity

Koushik Chakraborty [1], Farook Rahaman [2,*], Saibal Ray [3], Banashree Sen [4] and Debabrata Deb [5]

1 Department of Physics, Government College of Education, Burdwan 713102, West Bengal, India
2 Department of Mathematics, Jadavpur University, Kolkata 700032, West Bengal, India
3 Centre for Cosmology, Astrophysics and Space Science (CCASS), GLA University, Mathura 281406, Uttar Pradesh, India
4 Department of Applied Mathematics, Maulana Abul Kalam Azad University of Technology, Haringhata 741249, West Bengal, India
5 Department of Physics, The Institute of Mathematical Sciences, Chennai 600113, Tamil Nadu, India
* Correspondence: rahaman@associates.iucaa.in

Abstract: We explore wormhole geometry in spiral galaxies under the third order Lovelock gravity. Using the cubic spline interpolation technique, we find the rotational velocity of test particles in the halo region of our spiral galaxy from observed values of radial distances and rotational velocities. Taking this value of the rotational velocity, we are able to show that it is possible to present a mathematical model regarding viable existence of wormholes in the galactic halo region of the Milky Way under the Lovelock gravity. A very important result that we obtain from the present investigation is that galactic wormhole in the halo region can exist with normal matter as well as exotic matter.

Keywords: galatic halo; wormhole; Lovelock gravity

PACS: 04.40.Nr; 04.20.Jb; 04.20.Dw

1. Introduction

The seed of a hypothetical geometric object connecting two asymptotically flat spacetime regions was featured in the works of Weyl [1], and Einstein and Rosen [2]. Much later, it was termed as Wormhole (WH) [3] and the existence of such Einstein–Rosen bridge in the form of traversable wormhole was first ever firmly proposed by Morris and Thorne [4]. They assumed a stress–energy tensor for these wormholes that violates the standard energy conditions. Since matter that violates null energy condition is exotic matter, traversable wormhole solutions need to assume the existence of exotic matter. An important condition for wormholes is that they have no event horizon. There exists extensive literature exploring aspects of traversable wormholes (See [5,6] for review).

Studying WH in modified and higher dimensional gravity theories is also an active field of research since long time. Shang and Xu [7] confirmed the existence of WH solution in general in Lovelock gravity. Bhawal and Kar [8] found the static WH solution in second order Lovelock gravity, i.e., Einstein–Gauss–Bonnet gravity. Bandypadhyay and Chakraborty [9] constructed spherically symmetric thin shell WH in Einstein–Yang–Mills–Gauss–Bonnet gravity. They showed that for certain choices of parameters involved ordinary matter is sufficient for the formation of thin shell WH. Lobo and Oliviera [10] explored possible WH solutions in the context of $f(R)$ modified theories of gravity. Considering violation of Weak Energy Condition (WEC) they obtained the possible range of values for the coupling parameter ω. Dehghani and Dayyani [11] obtained n-diemnsional Lorentzian WH solutions of third order Lovelock gravity. They found that the WH throat radius has a lower limit that depends on the Lovelock coefficients, the dimensionality of the spacetime and the shape function. They also derived the region in which the throat could be constituted from the normal matter.

Böhmer et al. [12] explored the possibility of static spherically symmetric traversable WH solutions in modified teleparallel gravity.

On the other hand, Mehdizadeh and Riazi [13] proposed dynamic WH solutions in the framework of Lovelock gravity with compact extra dimensions. Harko et al. [14] investigated the WH solutions in which the matter constituting the wormhole throat satisfied all the energy conditions in the context of modified gravity theories. In particular they explored $f(R)$ gravity, the curvature–matter coupling and the $f(R, \mathcal{L}_m)$ generalization. In their article Mehdizadeh, Zangeneh and Lobo [15] obtained thin shell WH solutions in third order Lovelock gravity using cut and paste technique. They explored different situations depending on choices of second and third order Lovelock gravity coefficients. Zangeneh et al. [16] explored the possibilities for traversable WHs in third order Lovelock gravity with a cosmological constant term in an n-dimensional spacetime $\mathcal{M}^4 \times \mathcal{K}^{n-4}$ where $\mathcal{K}^{n-4}$ is a constant curvature space. Mehedizadeh and Lobo [17] studied wormhole geometries in third order Lovelock gravity with specifically chosen redshift function and equation of state. There are plethora of articles exploring various aspects of WH geometry in different higher dimensional and modified gravity theories [18–28].

In the recent past, Rahaman et al. [29] considering the Navarro–Frenk–White (NFW) [30,31] density profile confirmed the possible existence of WH spacetime in the outer regions of galactic halo from general relativistic framework. Similar observation was also reported by Kuhfittig [32]. In another seminal paper, Rahaman et al. [33] used the Universal Rotation Curve (URC) [34] dark matter model in the galactic halo region to obtain analogous results for the central parts of the halo. The authors generalized the result to predict possible existence of wormholes in most of the spiral galaxies. In yet other papers, Rahaman et al. [35,36] taking NFW density profile as well as the URC and wormhole such as line element as input calculated the tangential velocities v^ϕ of the test particles in the galactic halo. They reported a satisfactory matching of the theoretical and observational plot in the range $9\ kpc \leq r \leq 100\ kpc$.

In the present paper, we extend the works of Rahaman et al. [29,33,35,36] to construct galactic wormhole model in the context of Lovelock gravity. Under this motivation our plan is as follows: In the Section 2 we discuss the basic principles of Lovelock gravity theories. In Section 3, formulation of the problem is provided, while the results and discussions are presented in Section 4. Finally, in Section 5, we conclude with specific comments on the results of the study and future prospects.

2. Brief Outline of Lovelock Gravity Theory

The action in the framework of third-order Lovelock gravity, is given by

$$I = \int d^n x \sqrt{-g}(\mathcal{L}_1 + \alpha'_2 \mathcal{L}_2 + \alpha'_3 \mathcal{L}_3) \tag{1}$$

where we assumed $8\pi G_n = 1$, G_n being the n-dimensional gravitational constant. Here α'_2 and α'_3 are the second (Gauss–Bonnet) and third order Lovelock coefficients, g is the determinant of the metric, $\mathcal{L}_1 = R$ is the Einstein–Hilbert Lagrangian, the term $\mathcal{L}_2$ is the Gauss–Bonnet Lagrangian given by

$$\mathcal{L}_2 = R_{ijkl}R^{ijkl} - 4R_{ij}R^{ij} + R^2, \tag{2}$$

and the third order Lovelock Lagrangian $\mathcal{L}_3$ is defined as

$$\begin{aligned} \mathcal{L}_3 \;=\;& 2R^{ijkl}R_{klmn}R^{mn}{}_{ij} + 8R^{ij}{}_{km}R^{kl}{}_{jn}R^{mn}{}_{il} \\ & +24R^{ijkl}R_{kljm}R^{m}{}_{i} + 3RR^{ijkl}R_{klij} + 24R^{ijkl}R_{ki}R_{lj} \\ & +16R^{ij}R_{jk}R^{k}{}_{i} - 12RR^{ij}R_{ij} + R^3. \end{aligned} \tag{3}$$

In Lovelock theory, for an n-dimensional space, only terms with order less than $\left[\frac{(n+1)}{2}\right]$ contribute to the field equations. Here, we used the notation that $\left[\frac{n}{2}\right]$ will give biggest integer less than $\frac{n}{2}$. Since we are considering third order Lovelock gravity, its effects will be apparent for $n \geq 7$.

Thus, varying the action (1) with respect to the metric we obtain the field equations up to third order as follows:

$$G^{E}_{ij} + \alpha'_2 G^{(2)}_{ij} + \alpha'_3 G^{(3)}_{ij} = T_{ij}, \tag{4}$$

where T_{ij} is the energy–momentum tensor, G^{E}_{ij} is the Einstein tensor whereas $G^{(2)}_{ij}$ and $G^{(3)}_{ij}$ are given by

$$\begin{aligned}
G^{(2)}_{ij} &= 2(-R_{ikln}R^{lnk}{}_{j} - 2R_{imjk}R^{mk} - 2R_{ik}R^{k}{}_{j} + RR_{ij}) \\
&\quad -\frac{1}{2}\mathcal{L}_2 g_{ij}, \\
G^{(3)}_{ij} &= -3(4R^{nmkl}R_{klpm}R^{p}{}_{jni} - 8R^{nm}{}_{pk}R^{kn}{}_{ni}R^{p}{}_{jml} \\
&\quad +2R_{j}^{\ nkl}R_{klpm}R^{pm}{}_{ni} - R^{nmkl}R_{klnm}R_{ji} \\
&\quad +8R^{n}{}_{jkm}R^{kl}{}_{ni}R^{m}{}_{l} + 8R^{k}{}_{jnl}R^{nm}{}_{ki}R^{l}{}_{m} \\
&\quad +4R_{j}^{\ nkl}R_{klim}R^{m}{}_{n} - 4R_{j}^{\ nkl}R_{klnm}R^{m}{}_{i} \\
&\quad +4R^{nmkl}R_{klni}R_{jm} + 2RR_{j}^{\ lnm}R_{nmli} \\
&\quad +8R^{n}{}_{jim}R^{m}{}_{k}R^{k}{}_{n} - 8R^{k}{}_{jnm}R^{n}{}_{k}R^{m}_{i} \\
&\quad -8R^{nm}_{ki}R^{k}{}_{n}R_{jm} - 4RR^{n}{}_{jim}R^{m}{}_{n} + 4R^{nm}R_{mn}R_{ji} \\
&\quad -8R^{n}{}_{j}R_{nm}R^{m}{}_{i} + 4RR_{jm}R^{m}{}_{i} \\
&\quad -R^2R_{ji}) - \frac{1}{2}\mathcal{L}_3 g_{ij}.
\end{aligned}$$

3. Basic Equations and Their Solutions

The n-dimensional traversable wormhole metric is given by

$$ds^2 = -e^{2\phi(r)}dt^2 + \left(1 - \frac{b(r)}{r}\right)^{-1} dr^2 + r^2 d\Omega^2_{n-2}, \tag{5}$$

using units in which $c = G = 1$. Here $\phi(r)$ is the red-shift function which must be everywhere finite to prevent the event horizon, $b(r)$ is the shape function and $d\Omega^2_{n-2}$ is the metric on the surface of a $(n-2)$-sphere. The shape function of the wormhole essentially satisfies the condition $b(r_{th}) = r_{th}$ at $r = r_{th}$ where r_{th} is the throat of the wormhole. This condition is commonly known as the flare-out condition which gives at the throat $b'(r_{th}) < 1$ while $b(r) < r$ near the throat.

The energy–momentum tensor is given by

$$T^{\mu}_{\nu} = diag[-\rho(r), p_r(r), p_t(r), p_t(r), \ldots]. \tag{6}$$

where $\rho(r)$ is the energy density, $p_r(r)$ is the radial pressure and $p_t(r)$ gives the transverse pressure.

The Einstein equations for the above mentioned metric are as follows [17]

$$\rho(r) = \frac{(n-2)}{2r^2}\left[-\mathcal{B}_2(r)\frac{(b - rb')}{r}\right]$$

$$+\frac{(n-2)b}{2r^3}\left[(n-3)+(n-5)\frac{\alpha_2 b}{r^3}+(n-7)\frac{\alpha_3 b^2}{r^6}\right], \tag{7}$$

$$p_r(r)=\frac{(n-2)}{r}\left[\mathcal{B}_1(r)\mathcal{B}_2(r)\phi'\right]$$
$$-\frac{(n-2)b}{2r^3}\left[(n-3)+(n-5)\frac{\alpha_2 b}{r^3}+(n-7)\frac{\alpha_3 b^2}{r^6}\right], \tag{8}$$

$$p_t(r)=\mathcal{B}_1(r)\mathcal{B}_2(r)\left[\phi''+\phi'^2+\frac{(b-rb')\phi'}{2r(r-b)}\right]$$
$$-\frac{2\phi'}{r^4}\mathcal{B}_1(r)(b-b'r)\left(\alpha_2+3\alpha_3\frac{b}{r^3}\right)$$
$$+\mathcal{B}_1(r)\mathcal{B}_3(r)\left[(n-3)+(n-5)\frac{2\alpha_2 b}{r^3}+(n-7)\frac{3\alpha_3 b^2}{r^6}\right]$$
$$-\frac{b}{2r^3}(n-3)(n-4)+(n-5)(n-6)\frac{\alpha_2 b^2}{2r^6}$$
$$+(n-7)(n-8)\frac{\alpha_3 b^3}{2r^9}, \tag{9}$$

where $\alpha_2=(n-3)(n-4)\alpha_2'$, $\alpha_3=(n-3)(n-4)(n-5)(n-6)\alpha_3'$, $\mathcal{B}_1(r)=\left(1-\frac{b}{r}\right)$, $\mathcal{B}_2(r)=\left(1+\frac{2\alpha_2 b}{r^3}+\frac{3\alpha_3 b^2}{r^6}\right)$ and $\mathcal{B}_3(r)=\left(\frac{\phi'}{r}+\frac{(b-rb')}{2r^2(r-b)}\right)$. In the above equations, the prime denotes the derivative with respect to r.

From the stable circular geodesic motion in the equatorial plane, the tangential component of the rotational velocity of the neutral hydrogen clouds in the galactic halo region can be found out from the gravitational potential corresponding to the flat rotation curves in the halo region as [29,33,35,36]

$$(v^t)^2=r\phi'(r). \tag{10}$$

Logically, it is plausible that if an observer is sitting in the plane, θ_3 = constant, θ_4 = constant, ..., θ_n = constant, then the observer recognizes all characteristics as a (3 + 1) dimensional picture. In this sense, one can use all the relevant data which are approximately the same for both space-times [37].

Moreover, it is generally believed that interactions other than gravity are confined to $3+1$ brane. However, gravity can propagate in all extra dimensions ($n \geq 4$) [38,39]. These "gravity-only" dimensions (GODs) may be compactified [40,41]. In the present study of wormholes, we used observational data of tangential velocities of neutral hydrogen cloud. These velocities are determined by the effective gravitational potential in the galactic halo. It must be influenced by the existence of higher dimensions, $n \geq 4$. Therefore, there must not be any problem in using the aforementioned data in the present study. Similar observational data was being used in many other important studies of higher dimensions [42,43].

We would also like to add here that in the literature [44–47] some works are available where the authors try to find out a higher dimensional connection with microphysics and/or macroastrophysics. Especially, Barnafoldi et al. [46] displayed the detection of electro-magnetic and particle radiation from the direction of $Cygnus X3$, which raise the question of the existence of special long-lived, neutral particles. After investigation of the origin of these particles they concluded that the source object may contain compactified extra dimension and these particles are messengers of this state.

Taking into consideration of all the aforementioned points, we use the data in Table 1 in the context of third order Lovelock gravity. From the best fit with the observational data, we obtain the following expression for v^t as

$$v^t = \sum_{i=0}^{i=5} c_i r^i, \tag{11}$$

where $c_i, i = 0 \ldots 5$ are constants having values $c_5 = 0.0000011$, $c_4 = -0.0003$, $c_3 = 0.031$, $c_2 = -1.4$, $c_1 = 28$ and $c_0 = 64$.

Table 1. The radial distance from galactic centre r in kpc and velocity v^t in km/s of objects in the galactic halo region with total virial mass $3 \times 10^{12} M_\odot$ [35,36,48,49].

R (kpc)	v^t (km/s)	*R* (kpc)	v^t (km/s)	*R* (kpc)	v^t (km/s)
0.1	10.053	34.1	234.623	68.1	234.293
1.1	74.467	35.1	234.225	69.1	234.317
2.1	118.223	36.1	233.891	70.1	234.334
3.1	151.113	37.1	233.390	71.1	234.345
4.1	176.445	38.1	233.213	72.1	234.349
5.1	196.099	39.1	233.079	73.1	234.347
6.1	211.331	40.1	232.982	74.1	234.339
7.1	223.057	41.1	232.918	75.1	234.324
8.1	231.975	42.1	232.883	76.1	234.303
9.1	238.634	43.1	232.873	77.1	234.303
10.1	243.475	44.1	232.884	78.1	234.276
11.1	246.857	45.1	232.913	79.1	234.243
12.1	249.072	46.1	232.957	80.1	234.205
13.1	250.362	47.1	233.013	81.1	234.160
14.1	250.927	48.1	233.078	82.1	234.109
15.1	250.930	49.1	233.151	83.1	234.054
16.1	250.509	50.1	233.230	84.1	233.993
17.1	249.774	51.1	233.313	85.1	233.927
18.1	248.817	52.1	233.397	86.1	233.856
19.1	247.712	53.1	233.483	87.1	233.779
20.1	246.519	54.1	233.568	88.1	233.698
21.1	245.285	55.1	233.652	89.1	233.612
22.1	244.048	56.1	233.733	90.1	233.522
23.1	242.836	57.1	233.811	91.1	233.427
24.1	241.671	58.1	233.733	92.1	233.328
25.1	240.568	59.1	233.811	93.1	233.225
26.1	239.539	60.1	233.886	94.1	233.118
27.1	238.589	61.1	233.956	95.1	233.007
28.1	237.724	62.1	234.021	96.1	232.892
29.1	236.943	63.1	234.081	97.1	232.774
30.1	236.245	64.1	234.136	98.1	232.652
31.1	235.628	65.1	234.184	99.1	232.527
32.1	235.089	66.1	234.227	0	0
33.1	234.623	67.1	234.263	0	0

4. Results and Discussion

Here, we shall calculate the expression of redshift function $\phi(r)$ by using the expression for v^t given in (11). With the help of Equations (10) and (11), we obtain

$$\phi'(r) = c_5^2 r^9 + 2c_4c_5 r^8 + \left(c_4^2 + 2c_3c_5\right) r^7 + 2(c_3c_4 + c_2c_5) r^6$$

$$+\left(c_3^2 + 2c_2c_4 + 2c_1c_5\right) r^5 + 2(c_2c_3 + c_1c_4 + c_0c_5) r^4 +$$

$$\left(c_2^2 + 2c_1c_3 + 2c_0c_4\right) r^3 + 2(c_1c_2 + c_0c_3) r^2$$

$$+\left(c_1^2+2c_0c_2\right)r+\frac{c_0^2}{r}+2c_0c_1, \tag{12}$$

$$\begin{aligned}\phi(r) &= \frac{1}{10}c_5^2r^{10}+\frac{2}{9}c_4c_5r^9+\frac{1}{8}\left(c_4^2+2c_3c_5\right)r^8\\ &+\frac{2}{7}(c_3c_4+c_2c_5)r^7+\frac{1}{6}\left(c_3^2+2c_2c_4+2c_1c_5\right)r^6\\ &+\frac{2}{5}(c_2c_3+c_1c_4+c_0c_5)r^5+\frac{1}{4}\left(c_2^2+2c_1c_3+2c_0c_4\right)r^4\\ &+\frac{2}{3}(c_1c_2+c_0c_3)r^3+\frac{1}{2}\left(c_1^2+2c_0c_2\right)r^2\\ &+2c_0c_1r+c_0^2\log(r),\end{aligned} \tag{13}$$

$$\begin{aligned}\phi''(r) &= 9c_5^2r^8+16c_4c_5r^7+7\left(c_4^2+2c_3c_5\right)r^6\\ &+12(c_3c_4+c_2c_5)r^5+5\left(c_3^2+2c_2c_4+2c_1c_5\right)r^4\\ &+8(c_2c_3+c_1c_4+c_0c_5)r^3+3\left(c_2^2+2c_1c_3+2c_0c_4\right)r^2\\ &-\frac{c_0^2}{r^2}+4(c_1c_2+c_0c_3)r+c_1^2+2c_0c_2.\end{aligned} \tag{14}$$

Now, we can substitute Equations (12)–(14) in Equations (7)–(9). We obtain very nonlinear equations of the function $b(r)$. In the next two sections, therefore, we take some specific forms of the shape function and study the nature of the spacetime under consideration. The 4-dimensional present spacetime structure is believed to be the self-compactified form of manifold with multidimensional spacetime. It is observed that cosmic string as well as superstring theories and hence M-theory reproduce higher dimensional general relativity at low energy so that scientists have been argued that theories of unification tend to require extra spatial dimensions to be consistent with the physically viable models [50–55]. Several authors have shown that some features of higher dimensional black holes differ significantly from four dimensional black holes which is due to the fact that higher dimensional analysis provides a much wider avenue to black hole solutions in comparison to 4-dimensional counterparts [56–59]. In the present investigation of wormholes, we considered third order Lovelock gravity. Therefore, we took $n \geq 7$. Cases for $n = 8, 9, 10, 11$ are studied. In these studies, all extra dimensions are taken to be compactified.

4.1. Case I: Physical Features for the Shape Function $b(r) = r\left(\frac{r_0}{r}\right)^k$

For general physical reasons, the shape function $b(r)$ must obey the following conditions to ensure a traversability of wormholes [26,60]:

(i) Throat condition: at the throat (i.e., $r = r_0$), the shape function should satisfy the condition $b(r_0) = r_0$, and for $r > r_0$ one should obtain $1 - \frac{b(r)}{r} > 0$.
(ii) Flaring out condition: this condition dictates that $b'(r_0) < 1$.
(iii) Asymptotically flatness condition: for this condition we should have $\frac{b(r)}{r} \to o$ as $r \to \infty$.

Under the above mentioned three conditions, we consider the shape function as follows:

$$b(r) = r\left(\frac{r_0}{r}\right)^k, \tag{15}$$

where r_0 is throat radius and k is arbitrary parameter. From the flare out condition of the wormhole, it may be found out that $k > 0$. We assume $k = 2$ in the present case as a representative value. With this choice of the shape function, let us look into the physical features of various quantities of interest:

(1) Now, the expression of the redshift function calculated from the consideration of the galactic flat rotation curves (See Figure 1) as given in Equation (12) remains finite everywhere. This clearly implies that there would be no event horizon which is an important criterion for the existence of wormholes.
(2) Plot on the left panel of Figure 2 show that the present shape function satisfies the condition $\frac{b(r)}{r} < 1$ for $r > r_{th}$. This indicates that the flare out condition of the wormhole $b'(r_{th}) < 1$ is satisfied.
(3) In terms of principal pressures the Null Energy Condition (NEC) is given by following inequations:

$$\rho + p_r \geq 0, \tag{16}$$
$$\rho + p_t \geq 0, \tag{17}$$

whereas Weak Energy Condition (WEC) is given by following inequations:

$$\rho \geq 0, \tag{18}$$
$$\rho + p_r \geq 0, \tag{19}$$
$$\rho + p_t \geq 0. \tag{20}$$

Thorne [4] observed that the traversable wormholes violate the Null Energy Condition (NEC) near the throat. In Figure 3, $\rho + p_t$ is plotted for four possible choices of α_2 and α_3, i.e., ($\alpha_2 > 0$, $\alpha_3 > 0$), ($\alpha_2 > 0$, $\alpha_3 < 0$), ($\alpha_2 < 0$, $\alpha_3 > 0$) and ($\alpha_2 < 0$, $\alpha_3 < 0$). The plots clearly show that $\rho + p_t > 0$ for $0 \leq r \leq 50$ kpc and n does not influence the nature of the plots. Thus, the NEC is satisfied in the galactic halo region. In Figure 4 similar plots are noted for $\rho + p_r$. No violation of the NEC is noted here also. Moreover, since $\rho > 0$ in the galactic halo region, the matter threading the wormhole satisfies the WEC (Figure 5). Studies investigating traversable wormhole in the galactic halo region under general theory of relativity, reported violation of the NEC near the throat indicating the existence of exotic matter [29,32,33,35,36]. In the context of Lovelock gravity; however, there are exceptions. Dehghani and Dayyani [11] reported wormhole solutions in third order Lovelock gravity satisfying the WEC. They further showed that positivity of the density and $p + \rho$ depends on the Lovelock coefficients. Mehdizadeh and Riazi [13] studied wormhole solution under Lovelock gravity supported by normal matter. There are numerous other studies concluding the existence of normal matter near the throat of the wormhole in second or third order Lovelock gravity [17,61].
(4) Figure 5 shows the variation of density function with distance in the galactic halo region. The plots show very highly dense region near the centre of the galaxy and its value very quickly decreases to a small value with increasing distance. The variation is nearly inverse square in the outer region of the galactic halo. However, the density function remains positive in the galactic halo region. These features of the density profile are in accordance with the predictions by many earlier studies of the galactic halos [62–65].

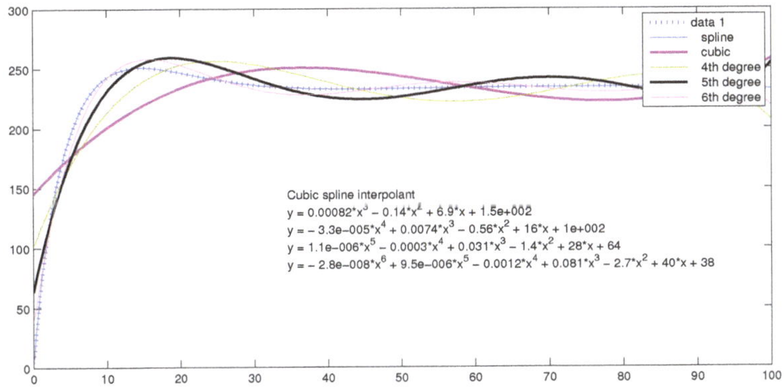

Figure 1. Polynomials fitted against the observed data set of rotational curve in galactic halo region (Table 1). The horizontal axis represents distance in kpc and the vertical axis represents the rotational velocity in galactic halo in km/s.

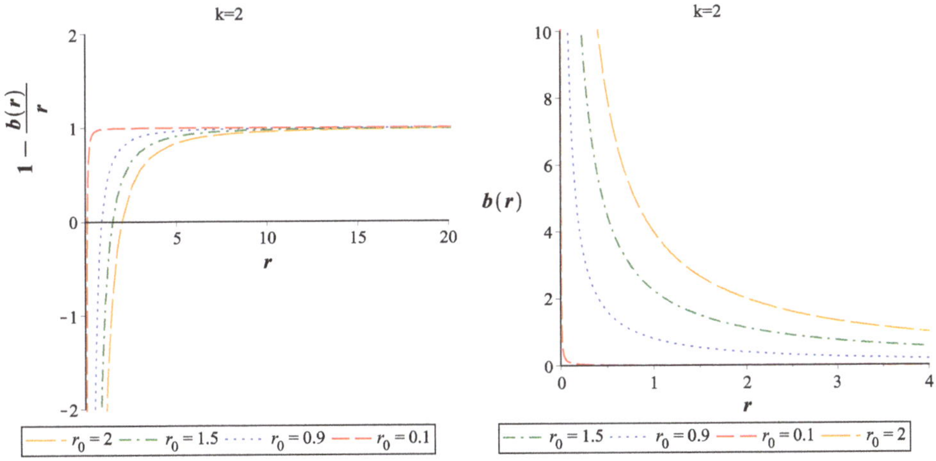

Figure 2. Case I: Plot to study $(1 - \frac{b(r)}{r})$ and $b(r)$ for different choices of r_0.

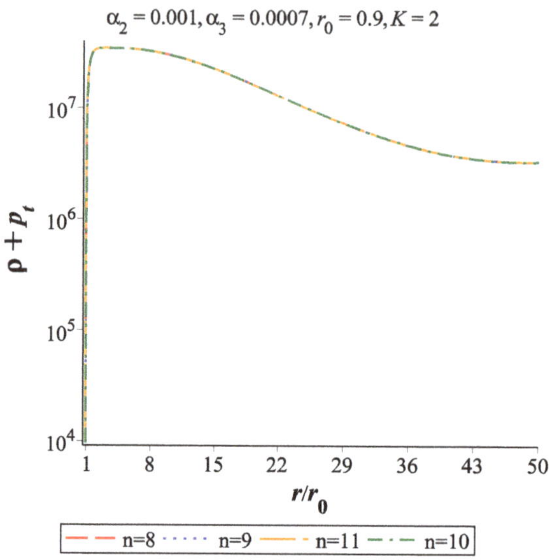

Figure 3. Case I: Plot to study $\rho + p_t$ for $n = 8$, $n = 9$, $n = 10$, $n = 11$. It is verified that plots are the same for all sets of values of the constants α, r_0 and K.

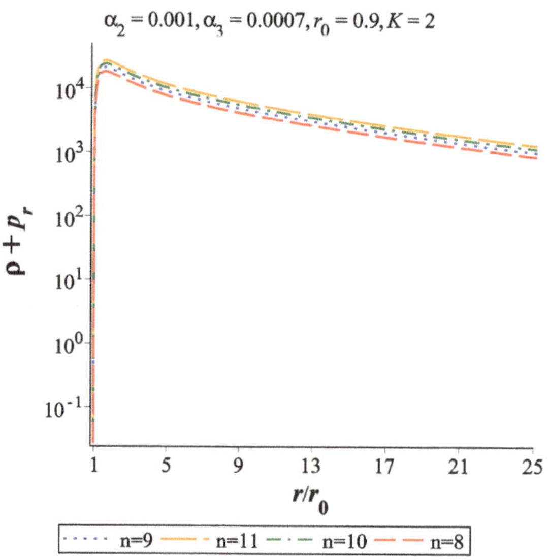

Figure 4. Case I: Plot to study $\rho + p_r$ for $n = 8$, $n = 9$, $n = 10$, $n = 11$. It is verified that plots are the same for all sets of values of the constants α, r_0 and K.

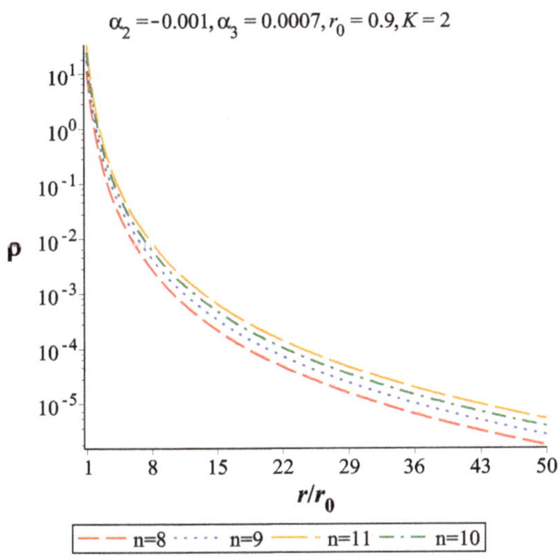

Figure 5. Case I: Plot to study ρ for $n = 8$, $n = 9$, $n = 10$, $n = 11$. It is verified that plots are the same for all sets of values of the constants α, r_0 and K.

4.2. Case II: Physical Features for the Shape Function $b(r) = r_0\left(\frac{r}{r_0}\right)^k$

Keeping the previously mentioned conditions in mnd, let us consider another form of the shape function which is as follows:

$$b(r) = r_0\left(\frac{r}{r_0}\right)^k, \tag{21}$$

where r_0 and k are arbitrary parameters as mentioned in the earlier case. Here, the flare out condition of the wormhole implies that $k < 1$.

The present form of the shape function is an increasing function of the radial coordinate r (See Figure 6). In this sense, it is opposite in nature to the shape function in Case I (See Figure 2). Now the physical features that can be noted from Figures 6–9 are as follows:

(1) The throat of the wormhole may be taken to be at $r = r_0$. The plot of $(1 - \frac{b(r)}{r})$ against the radial coordinate r (Figure 6) clearly shows that $\frac{b(r)}{r} < 1$ for $r > r_0$. Moreover, $b(r_0) = r_0$ where in the plot under consideration $r_0 = 1.72$. These results essentially point out that the flare out condition is satisfied near the throat. It has already been pointed out that there is no horizon in the spacetime.

(2) The density function (vide Figure 7) again shows a inverse square fall with radial coordinate in the galactic halo region and $\rho > 0$.

(3) Contrary to the previous case, in the present Case II, the Figures 8 and 9 show that $\rho + p_r \leq 0$ and $\rho + p_t \leq 0$ near the throat indicating the existence of exotic matter that violates the NEC. The plots $\rho + p_t$ remains the same for $n = 8, 9, 10, 11$. Plots are unaffected by different choices of α_2 and α_3. The plots of $\rho + p_r$ remains similar in nature for $n = 8, 9, 10, 11$ irrespective of the choices of α_2 and α_3. Thus, we note here a very interesting result that a galactic wormhole in the halo region can exist with normal matter as well as exotic matter under the framework of Lovelock gravity.

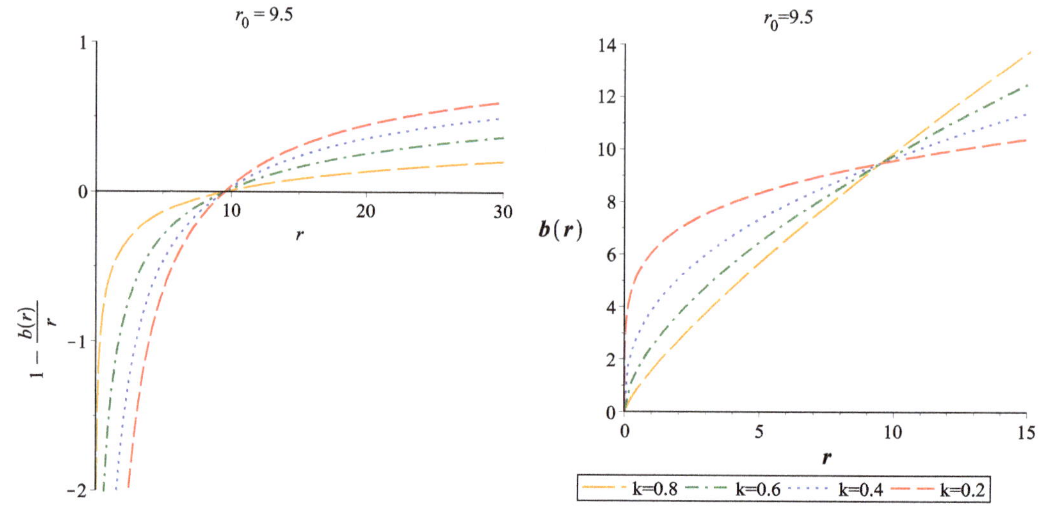

Figure 6. Case II: Plot to study $(1 - \frac{b(r)}{r})$ and $b(r)$ for $k = 0.2, 0.4, 0.6, 0.8$.

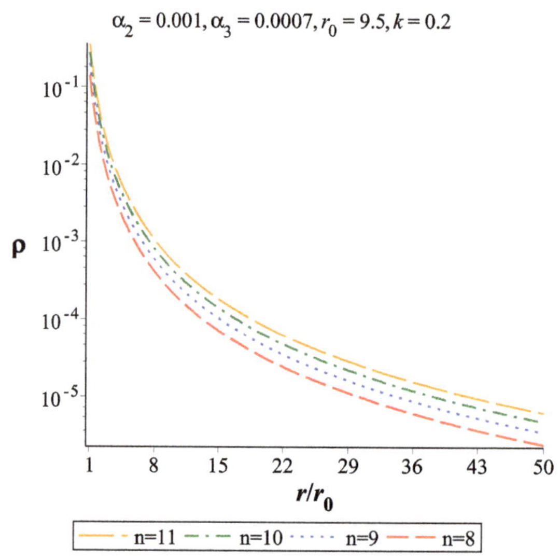

Figure 7. Case II: Plot to study ρ for $n = 8, n = 9, n = 10, n = 11$. It is verified that plots are the same for all sets of values of the constants α, r_0 and K.

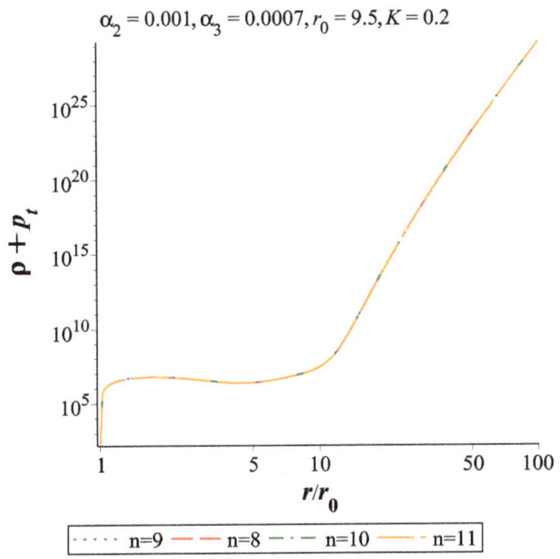

Figure 8. Case II: Plot to study $\rho + p_t$ for $n = 8, n = 9, n = 10, n = 11$. It is verified that plots are the same for all sets of values of the constants α, r_0 and K.

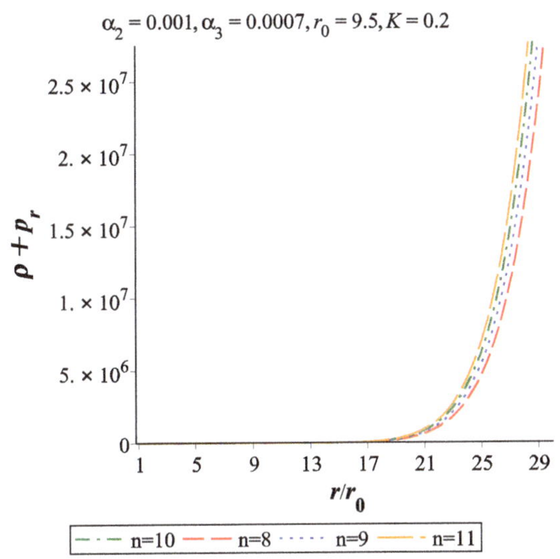

Figure 9. Case II: Plot to study $\rho + p_r$ for $n = 8, n = 9, n = 10, n = 11$. It is verified that plots are the same for all sets of values of the constants α, r_0 and K.

5. Concluding Remarks

In this paper, our motivation was to extend the works of Rahaman et al. [29,33,35,36] to construct WH model in the galactic halo region on the platform of Lovelock gravity. To do, so we specifically employed the third order Lovelock gravity. By using the cubic spline interpolation technique (see the Appendix A), the rotational velocity of test particles has been found out in the halo region of our galaxy from the observed values of the radial distances and rotational velocities. Considering this value of the rotational velocity, we present possible existence of WH in galactic halo region under the Lovelock gravity.

In favour of this successful execution we are now putting forward some of the salient features based on the results obtained from the proposed model as follows:

(i) The redshift function calculated from the consideration of the flat rotation curve of the galaxy (Figure 1) remains finite everywhere which clearly implies that there would be no event horizon and thus provide an important criterion for the existence of WH.

(ii) It is interesting to note that plots on the left panel of Figure 2 (Case I) exhibit presence of the shape function which satisfies the condition $\frac{b(r)}{r} < 1$ for the constraint $r > r_{th}$ and thus indicates that the flare out condition of the WH $b'(r_{th}) < 1$ is fulfilled. For the other form of the shape function (Case II), we observe an increasing function of the radial coordinate r (Figure 6) and hence the feature is opposite in nature to the shape function of Case I (i.e., Figure 2). If we consider the throat of the wormhole to be at $r = r_0$ then the plots of $(b(r) - r)$ against the radial coordinate r demonstrate that $\frac{b(r)}{r} < 1$ for $r > r_0$ which is in confirmation of fulfilling of the flare out condition near the throat.

(iii) Figures 5 and 7 show very highly dense region near the centre of the galaxy and its value very quickly decreases to a small value with increasing distance which is in accordance with the earlier predictions [62–65].

(iv) Figures 3 and 4 show that the NEC is satisfied in the galactic halo region which is in contradiction to the result under general relativity [29,32,33,35,36]. This may be considered as exception for Lovelock gravity [11,13,17,61]. On the other hand, Figures 8 and 9 show that the NEC is violated in the galactic halo region which is in accordance with the result under general relativity [29,32,33,35,36]. Thus, we note a very important result in the present investigation that galactic wormhole in the halo region can exist with normal matter as well as exotic matter.

However, the present results intrinsically raise the following awkward issues: (i) is exotic matter really indispensable for constructing traversable WH, and (ii) if not, as suggestions evolve from the modified theories of gravities that galactic dynamics of massive test particles can be explained without introducing any exotic dark energy [66–72], then before coming to a conclusive decision, further studies are needed to perform. These issues may be addressed in a future project under Lovelock gravity as well as other modified gravities in a series of exclusive investigations. This endeavor may shed light on the debate as to whether exotic matter (as GR advocates) is indispensable for constructing a traversable wormhole or alternative gravity theories (especially Lovelock gravity under the present treatment) are suffice to predict WH.

Another point to comment on here is related to evidence of WH in reality, which at present seems belongs to futuristic observational projects only. However, recent detection of black holes [73–76] obviously provides glimpse of hope and aspiration regarding observational signatures for WH also. As a special note, we would like to present here the work of Piotrovich et al. [77] where they have proposed a test for WH, considering an AGN at the centre of a gaclactic core as one mouth of a WH, to perform a comparative study between the γ ray spectrum originating from the jets of a supermassive black hole and that from a WH acting as mouth of an AGN. Therefore, under this proposal, they conclude that an observation of such radiation would serve as evidence of the existence of wormholes. A very recent work on the possibility of generalized wormhole formation in the galactic halo due to dark matter using the observational data within the matter coupling gravity formalism also can serve a purpose in this detection issue [78].

Author Contributions: Conceptualization, K.C.; methodology, F.R.; formal analysis, S.R.; writing—original draft preparation, B.S.; writing—review and editing, D.D. All authors have read and agreed to the published version of the manuscript.

Funding: This research received no external funding.

Data Availability Statement: Not applicable.

Acknowledgments: K.C., F.R. and S.R. are thankful to the authority of Inter-University Centre for Astronomy and Astrophysics, Pune, India for providing them Visiting Associateship under which a part of this work was carried out.

Conflicts of Interest: The authors declare no conflict of interest.

Appendix A. Cubic Spline Interpolation Method

Cubic Spline Interpolation Method is used to interpose a specific type of piecewise polynomial called spline. Spline interpolation follows the procedure of fitting a curve inside the trend of observational data minimizing the error within a given set of data point. This technique interpolate with minimized error when lower degree polynomial is fitted and the same can be made huge when we use higher degree polynomial. The cubic spline method is constructed with piecewise 3rd order polynomial which have to pass the observational data set called control points avoiding the Runge's phenomenon where oscillation occurs between points. There are advantages of cubic spline interpolation here. Firstly, it improves fitting of a curve by increasing the degree of fitted polynomial. Moreover, with respect to a linear interpolation cubic spline interpolation produces a rather sharp result if the data sets are not well spaced. In the present article, the cubic spline method is constructed based on the 5th degree polynomial, which pass through the observed data set. We used the MATLAB directly to implement the cubic spline interpolation based on the observation. In Figure 1, blue dotted line is for given set of data points for rotational velocity curve (v^t) and different splines are arranged to fit against the observed data in the mentioned figure. The fifth degree polynomial is found to be the best fit.

References

1. Weyl, H. Feld und Materie. *Ann. Phys.* **1921**, *65*, 541–563. [CrossRef]
2. Einstein, A.; Rosen, N. Can Quantum-Mechanical Description of Physical Reality Be Considered Complete? *Phys. Rev.* **1935**, *48*, 777–780. [CrossRef]
3. Misner, C.W.; Wheeler, J.A. Classical physics as geometry. *Ann. Phys.* **1957**, *2*, 525–603. [CrossRef]
4. Morris, M.S.; Thorne, K.S. Wormholes in Spacetime and Their Use for Interstellar Travel: A Tool for Teaching General Relativity. *Am. J. Phys.* **1988**, *56*, 395–412. [CrossRef]
5. Visser, M. *Lorentzian Wormholes: From Einstein to Hawking*; American Institute of Physics: New York, NY, USA, 1995.
6. Lobo, F.S.N. General class of wormhole geometries in conformal Weyl gravity. *Class. Quantum Gravit.* **2008**, *25*, 175006. [CrossRef]
7. Shang, Y.-W.; Xu, J.-J. Quantum Entanglement of Two Atoms Inside an Optical Cavity. *Chin. Phys. Lett.* **1999**, *16*, 985. [CrossRef]
8. Bhawal, B.; Kar, S. Lorentzian wormholes in Einstein-Gauss-Bonnet theory. *Phys. Rev. D* **1992**, *46*, 2464–2468. [CrossRef]
9. Bandyopadhyay, T.; Chakraborty, S. Cosmic No-Hair Conjecture in brane scenarios: Constraint on bulk matter. *Class. Quantum Gravit.* **2009**, *26*, 2461–2467. [CrossRef]
10. Lobo, F.S.N.; Oliveira, M.A. Wormhole geometries in $f(R)$ modified theories of gravity. *Phys. Rev. D* **2009**, *80*, 104012. [CrossRef]
11. Dehghani, M.H.; Dayyani, Z. Lorentzian wormholes in Lovelock gravity. *Phys. Rev. D* **2009**, *79*, 064010. [CrossRef]
12. Böhmer, C.G.; Harko, T.; Lobo, F.S.N. Wormhole geometries in modified teleparallel gravity and the energy conditions. *Phys. Rev. D* **2012**, *85*, 044033. [CrossRef]
13. Mehdizadeh, M.R.; Riazi, N. Cosmological wormholes in Lovelock gravity. *Phys. Rev. D* **2012**, *85*, 124022. [CrossRef]
14. Harko, T.; Lobo, F.S.N.; Mak, M.K.; Sushkov, S.V. Modified-gravity wormholes without exotic matter. *Phys. Rev. D* **2013**, *87*, 067504. [CrossRef]
15. Mehdizadeh, M.R.; Zangeneh, M.K.; Lobo, F.S.N. Higher-Dimensional Thin-Shell Wormholes in Third-Order Lovelock Gravity. *Phys. Rev. D* **2015**, *92*, 044022. [CrossRef]
16. Zangeneh, M.K.; Lobo, F.S.N.; Dehghani, M.H. Traversable wormholes satisfying the weak energy condition in third-order Lovelock gravity. *Phys. Rev. D* **2015**, *92*, 124049. [CrossRef]
17. Mehdizadeh, M.R.; Lobo, F.S.N. Novel third-order Lovelock wormhole solutions. *Phys. Rev. D* **2016**, *93*, 124014. [CrossRef]
18. Sharif, M.; Ikram, A. Wormholes supported by $f(G)$ gravity. *Int. J. Mod. Phys. D* **2015**, *24*, 1550003. [CrossRef]
19. Taser, D.; Doğru, M.U. Conformal and traversable wormholes with monopole and perfect fluid in f(R)-gravity. *Int. J. Mod. Phys. D* **2016**, *25*, 1650017. [CrossRef]
20. Sharif, M.; Nazir, K. Study of galactic halo $F(T, TG)$ wormhole solutions. *Int. J. Mod. Phys. D* **2018**, *27*, 1750170. [CrossRef]
21. Bhattacharya, S.; Halde, S.; Chakraborty, S. A non-canonical scalar field cosmological model: Stability and bifurcation analysis. *Mod. Phys. Lett. A* **2019**, *34*, 1950200. [CrossRef]
22. Mishra, A.K.; Sharma, U.K.; Dubey, V.C.; Pradhan, A. Traversable wormholes in $f(R, T)$ gravity. *Astrophys. Space Sci.* **202**, *365*, 1. [CrossRef]

23. Banerjee, A.; Pradhan, A.; Tangphati, T.; Rahaman, F. Wormhole geometries in f(Q) gravity and the energy conditions. *Eur. Phys. J. C* **2021**, *81*, 1031. [CrossRef]
24. Chawla, C.; Dixit, A.; Pradhan, A. Modeling of traversable wormholes in exponential $f(R,T)$ gravity. *Can. J. Phys.* **2021**, *99*, 634. [CrossRef]
25. Godani, N.; Singh, D.V.; Samanta, G.C. Stability of thin-shell wormhole in 4D Einstein-Gauss-Bonnet gravity. *Phys. Dark Univ.* **2022**, *35*, 100952. [CrossRef]
26. Mustafa, G.; Maurya, S.K.; Ray, S. Wormhole solutions in the galactic halo under symmetric teleparallel gravity. *Fortschritte Phys.—Prog. Phys.* **2022**, *under review*.
27. Mustafa, G.; Javed, F.; Maurya, S.K.; Ray, S. Thinshell around Wormhole Geometry. *Astrophys. J.* **2022**, *under review*.
28. Mustafa, G.; Maurya, S.K.; Ray, S. Relativistic wormhole surrounded by dark matter halos in symmetric teleparallel gravity. *Fortschritte Phys.—Prog. Phys.* **2022**, *under revision*.
29. Rahaman, F.; Kuhfittig, P.K.F.; Ray, S.; Islam, N. Possible existence of wormholes in the galactic halo region. *Eur. Phys. J. C* **2014**, *74*, 2750. [CrossRef]
30. Navarro, J.F.; Frenk, C.S.; White, S.D.M. The Structure of Cold Dark Matter Halos. *Astrophys. J.* **1996**, *462*, 563. [CrossRef]
31. Navarro, J.F.; Frenk, C.S.; White, S.D.M. A Universal Density Profile from Hierarchical Clustering. *Astrophys. J.* **1997**, *490*, 493. [CrossRef]
32. Kuhfittig, P.K.F. Gravitational lensing of wormholes in the galactic halo region. *Eur. Phys. J. C* **2014**, *74*, 2818. [CrossRef]
33. Rahaman, F.; Salucci, P.; Kuhfittig, P.K.F.; Ra, S.; Rahaman, M. Possible existence of wormholes in the central regions of halos. *Ann. Phys.* **2014**, *350*, 561. [CrossRef]
34. Castignani, G.; Frusciante, N.; Vernieri, D.; Salucci, P. The density profiles of dark matter halos in spiral galaxies. *Nat. Sci.* **2012**, *4*, 265. [CrossRef]
35. Rahaman, F.; Shit, G.C.; Sen, B.; Ray, S. Could wormholes form in dark matter galactic halos? *Astrophys. Space Sci.* **2016**, *361*, 37. [CrossRef]
36. Rahaman, F.; Sen, B.; Chakraborty, K.; Shit, G.C. Study of galactic rotation curves in wormhole spacetime. *Astrophys. Space Sci.* **2016**, *361*, 90. [CrossRef]
37. Bhar, P.; Manna, T.; Rahaman, F.; Ra, S.; Khadekar, G.S. Charged perfect fluid sphere in higher-dimensional spacetime. *Ind. J. Phys.* **2020**, *94*, 1679. [CrossRef]
38. Arkani-Hamed, N.; Dimopoulos, S.; Dvali, G. The hierarchy problem and new dimensions at a millimeter.*Phys. Lett. B* **1998**, *429*, 263. [CrossRef]
39. Antoniadis, I.; Arkani-Hamed, N.; Dimopoulos, S. New dimensions at a millimeter to a fermi and superstrings at a TeV. *Phys. Lett. B* **1998**, *436*, 257. [CrossRef]
40. Hanhart, C.; Phillips, D.R.; Reddy, S.; Savage, M.J. Extra dimensions, SN1987a, and nucleon–nucleon scattering data. *Nuc. Phys. B* **2001**, *595*, 335. [CrossRef]
41. DeBenedictis, A.; Das, A. Higher dimensional wormhole geometries with compact dimensions. *Nucl. Phys. B* **2003**, *653*, 279. [CrossRef]
42. Hannestad, S.; Raffelt, G.G. New Supernova Limit on Large Extra Dimensions: Bounds on Kaluza-Klein Graviton Production. *Phys. Rev. Lett.* **2001**, *87*, 051301. [CrossRef] [PubMed]
43. Hannestad, S.; Raffelt, G.G. Stringent Neutron-Star Limits on Large Extra Dimensions. *Phys. Rev. Lett.* **2002**, *88*, 071301. [CrossRef] [PubMed]
44. Liddle, A.R.; Moorhouse, R.G.; Henriques, A.B. Neutron stars and extra dimensions. *Class. Quantum Gravit.* **1990**, *7*, 1009. [CrossRef]
45. Barnaföldi, G.G.; Lèvai, P.; Lukács, B. *New Worlds in Astroparticle Physics*; World Scientific: Singapore, 2003; pp. 133–142. [CrossRef]
46. Barnaföldi, G.G.; Lèvai, P.; Lukács, B. Searching Extra Dimensions in Compact Stars. *Astron. Nachr.* **2007**, *328*, 809–812. [CrossRef]
47. Böhmer, C.G.; Lèvai, P.; Lukács, B. Compact stars in Kaluza–Klein World. *J. Phys. Conf. Ser.* **2010**, *218*, 012010. [CrossRef]
48. Salucci, P.; Lapi, A.; Tonini, C.; Gentile, G.; Yegorova, I.; Klein, U. The universal rotation curve of spiral galaxies—II. The dark matter distribution out to the virial radius. *MNRAS* **2007**, *378*, 41. [CrossRef]
49. Persic, M.; Salucci, P.; Stel, F. The universal rotation curve of spiral galaxies—I. The dark matter connection. *MNRAS* **1996**, *281*, 27. [CrossRef]
50. Schwarz, J.H. *Superstings*; World Scientific: Singapore, 1985. [CrossRef]
51. Weinberg, S. *Strings and Superstrings*; World Scientific: Singapore, 1986.
52. Duff, M.J.; Liu, J.T.; Minasian, R. Eleven-dimensional origin of string/string duality: A one-loop test. *Nucl. Phys. B* **1995**, *452*, 261. [CrossRef]
53. Polchinski, J. *String Theory*; Cambridge University Press: Cambridge, UK, 1998. [CrossRef]
54. Hellerman, S.; Swanson, I. Dimension-changing exact solutions of string theory. *J. High Energy Phys.* **2007**, *0709*, 096. [CrossRef]
55. Aharony, O.; Silverstein, E. Supercritical stability, transitions, and (pseudo)tachyons. *Phys. Rev. D* **2007**, *75*, 046003. [CrossRef]
56. Emparan, R.; Reall, H.S. Black holes in higher dimensions. *Liv. Rev. Relativ.* **2008**, *11*, 6. [CrossRef] [PubMed]
57. Cavaglia, M. Black hole and brane production in Tev gravity: A review. *Int. J. Mod. Phys. A* **2003**, *18*, 1843. [CrossRef]
58. Kanti, P. Black holes in theories with large extra dimensions: A review. *Int. J. Mod. Phys. A* **2004**, *19*, 4899. [CrossRef]

59. Aharony, O.; Gubser, S.S.; Maldacena, J.; Ooguri, H.; Oz, Y. Large N field theories, string theory and gravity. *Phys. Rept.* **2000**, *323* 183–386. [CrossRef]
60. Mishra, B.; Agrawal, A.S.; Tripathy, S.K.; Ray, S. Wormhole solutions in $f(R)$ gravity. *Int. J. Mod. Phys. D* **2021**, *30*, 2150061. [CrossRef]
61. Dotti, G.; Oliva, J.; Troncoso, R. Exact solutions for the Einstein-Gauss-Bonnet theory in five dimensions: Black holes, wormholes, and spacetime horns. *Phys. Rev. D* **2007**, *76*, 064038. [CrossRef]
62. Roberts, M.S.; Rots, A.H. Comparison of Rotation Curves of Different Galaxy Types. *Astron. Astrophys.* **1973**, *26*, 483.
63. Ostriker, P.; Peebles, P.J.E.; Yahill, A. The Size and Mass of Galaxies, and the Mass of the Universe. *ApJ* **1974**, *193*, L1. [CrossRef]
64. Rubin, V.C.; Thonnard, N.; Ford, W.K., Jr. Extended rotation curves of high-luminosity spiral galaxies. IV. Systematic dynamical properties, Sa -> Sc. *ApJ* **1978**, *225*, L107. [CrossRef]
65. Rubin, V.C.; Ford, W.K., Jr. Rotation of the Andromeda Nebula from a Spectroscopic Survey of Emission Regions. *ApJ* **1970**, *159*, 379. [CrossRef]
66. Carroll, S.M.; Duvvuri, V.; Trodden, M.; Turner, M.S. Is cosmic speed-up due to new gravitational physics? *Phys. Rev. D* **2004**, *70*, 043528. [CrossRef]
67. Nojiri, S.; Odintsov, S.D. Unifying inflation with LambdaCDM epoch in modified $f(R)$ gravity consistent with Solar System tests. *Phys. Lett. B* **2007**, *657*, 238. [CrossRef]
68. Nojiri, S.; Odintsov, S.D. Modified $f(R)$ gravity unifying R^m inflation with the ΛCDM epoch. *Phys. Rev. D* **2008**, *77*, 026007. [CrossRef]
69. Nojiri, S.; Odintsov, S.D. Unified cosmic history in modified gravity: From $F(R)$ theory to Lorentz non-invariant models. *Phys. Supple.* **2011**, *190*, 155. [CrossRef]
70. Cognola, G.; Elizalde, E.; Nojiri, S.; Odintsov, S.D.; Sebastiani, L.; Zerbini, S. Class of viable modified $f(R)$ gravities describing inflation and the onset of accelerated expansion. *Phys. Rev. D* **2008**, *77*, 046009. [CrossRef]
71. Elizalde, E.; Nojiri, S.; Odintsov, S.D.; Sebastiani, L.; Zerbini, S. Nonsingular exponential gravity: A simple theory for early- and late-time accelerated expansion. *Phys. Rev. D* **2011**, *83*, 086006. [CrossRef]
72. Deb, D.; Rahaman, F.; Ray, S.; Guha, B.K. Strange stars in f(R,T)gravity. *J. Cosmol. Astropart. Phys.* **2018**, *03*, 044. [CrossRef]
73. Oldham, L.J.; Auger, M.W. Galaxy structure from multiple tracers – II. M87 from parsec to megaparsec scales. *Mont. Not. R. Astron. Soc.* **2016**, *457*, 421. [CrossRef]
74. Abbott, B.P. (LIGO Scientific Collaboration and Virgo Collaboration). Observation of Gravitational Waves from a Binary Black Hole Merger. *Phys. Rev. Lett.* **2016**, *116*, 061102. [CrossRef]
75. Bouman, K.L.; Johnson, M.D.; Zoran, D.; Fish, V.L.; Doeleman, S.S.; Freeman, W.T. Computational Imaging for VLBI Image Reconstruction. In Proceedings of the 2016 IEEE Conference on Computer Vision and Pattern Recognition (CVPR), Las Vegas, NV, USA, 27–30 June 2016; pp. 913–922. [CrossRef]
76. Akiyama, K.; Alberdi, A.; Alef, W.; Asada, K.; Azulay, R.; Baczko, A.-K.; Ball, D.; Baloković, M.; Barrett, J.; Bintley, D. et al. First M87 Event Horizon Telescope Results. I. The Shadow of the Supermassive Black Hole. *Astrophys. J.* **2019**, *875*, L1. [CrossRef]
77. Piotrovich, M.Y.; Krasnikov, S.V.; Buliga, S.D.; Natsvlishvili, T.M. Search for wormhole candidates in active galactic nuclei: Radiation from colliding accreting flows. *MNRAS* **2020**, *498*, 3684–3686. Available online: https://academic.oup.com/mnras/article/498/3/3684/5896471 (accessed on 27 September 2022). [CrossRef]
78. Mustafa, G.; Maurya, S.K.; Ray, S. On the possibility of generalized wormhole formation in the galactic halo due to dark matter using the observational data within the matter coupling gravity formalism. *ApJ* **2022**, *accepted*.

universe

MDPI

Article

Weak Field Limit for Embedding Gravity

Stanislav Kuptsov [1], Mikhail Ioffe [2], Sergey Manida [2] and Sergey Paston [2,*]

1 Saint Petersburg Department of Steklov Mathematical Institute of RAS, 191023 Saint Petersburg, Russia
2 Department of High Energy and Elementary Particle Physics, Saint Petersburg State University, 199034 Saint Petersburg, Russia
* Correspondence: pastonsergey@gmail.com

Abstract: We study a perturbation theory for embedding gravity equations in a background for which corrections to the embedding function are linear with respect to corrections to the flat metric. The remaining arbitrariness after solving the linearized field equations is fixed by an assumption that the solution is static in the second order. A nonlinear differential equation is obtained, which allows for finding the gravitational potential for a spherically symmetric case if a background embedding is given. An explicit form of a spherically symmetric background parameterized by one function of radius is proposed. It is shown that this function can be chosen in such a way that the gravitational potential is in a good agreement with the observed distribution of dark matter in a galactic halo.

Keywords: modified gravity; embedding gravity; isometric embeddings; embedding theory; Regge–Teitelboim gravity; weak field limit

1. Introduction

Embedding gravity (also called embedding theory) [1] is a modified theory of gravity based on the idea of our spacetime as a 4D surface in a 10D flat space. From a geometric point of view, this approach renders the theory of gravity similar to string theory. In this case, the space–time metric is considered to be induced and has the form

$$g_{\mu\nu} = (\partial_\mu y^a)(\partial_\nu y^b)\eta_{ab}, \tag{1}$$

where η_{ab} is a flat metric of the ambient space, and $y^a(x^\mu)$ is an embedding function defining the shape of the surface. Here and further $\mu, \nu, \ldots = 0, \ldots, 3$ and $a, b, \ldots = 0, \ldots, 9$. After the pioneering work of [1], the ideas of the embedding approach were repeatedly used in the works of various authors to describe gravity, including the problem of its quantization; see, for example, [2–11].

The embedding gravity equations of motion, called Regge–Teitelboim equations, are more general in comparison with Einstein's equations $G^{\mu\nu} = \varkappa T^{\mu\nu}$ and have the following form:

$$(G^{\mu\nu} - \varkappa T^{\mu\nu})b^a_{\mu\nu} = 0, \tag{2}$$

where $b^a_{\mu\nu}$ is a second fundamental form of a 4D surface defined by an embedding function $y^a(x^\mu)$. All "Einsteinian" (i.e., satisfying Einstein's equations) solutions satisfy the Regge–Teitelboim equations, but they do not exhaust the full set of solutions. There may be additional solutions for which $G^{\mu\nu} \neq \varkappa T^{\mu\nu}$; nevertheless, (2) is satisfied. Initially, this was seen as a disadvantage of the approach, since embedding gravity was understood as some reformulation of GR, potentially more convenient for quantization due to the presence of a flat ambient space. Therefore, it was proposed to additionally introduce Einstein constraints into the theory, rendering it truly equivalent to GR [1,12]. However, the presence of extra solutions in the theory can be considered to be an advantage.

If we assume that just the Regge–Teitelboim equations describe the gravitational interaction, from the point of view of describing observations in terms of the usual GR, extra solutions manifest themselves as an additional contribution to the right side of Einstein's equations. It can be interpreted as the presence of some additional fictitious *embedding matter* that has nothing to do with ordinary matter. This contribution depends solely on the gravitational variables (which, in embedding theory, are $y^a(x^\mu)$), and its appearance is connected only with an attempt to reformulate a new theory in the old language. Since the direct detection of dark matter does not currently yield results [13,14], it is of interest to try to treat the *embedding matter* of the embedding theory as *dark matter* (and possibly as dark energy). Then, the effects associated with dark matter and energy are purely gravitational. In this way, observational problems that have no explanation within the framework of GR can be solved. The Regge–Teitelboim Equation (2) can be rewritten as a pair of equations

$$G^{\mu\nu} = \varkappa(T^{\mu\nu} + \tau^{\mu\nu}), \tag{3}$$

$$\tau^{\mu\nu} b^a_{\mu\nu} = 0, \tag{4}$$

i.e., in the form of a set of Einstein's equations with the contribution of an energy–momentum tensor $\tau^{\mu\nu}$ of fictitious embedding matter and its equations of motion. This was first indicated in [15]. There are also modified theories of gravity which are alternative to the embedding gravity, equations of motion of which can be written in the form of Einstein's equations supplemented by other equations. The most famous is mimetic gravity [16,17].

In order to obtain the properties of the embedding matter generated by embedding gravity (and thereby to understand whether they are similar to the properties of dark matter), it is necessary to investigate solutions of Equation (2). In general, due to their nonlinearity, this is too difficult a mathematical problem. So, we limit ourselves to the most physically interesting case of weak gravity, when metric $g_{\mu\nu}$ is close to flat metric $\eta_{\mu\nu}$. In this case, the problem arises in choosing a background value $\bar{y}^a(x^\mu)$ of the embedding function that would correspond to the flat metric. The simplest choice in the form of $\bar{y}^a(x^\mu)$, which defines a 4D plane in the ambient space, is unsuitable since, in such a background, Regge–Teitelboim Equations (2) are not linearized [2]. To linearize the equations, it is necessary to choose "unfolded" embedding [18] as the background, which means that the second fundamental form of the surface is nondegenerate in some sense.

In this paper, we use the unfolded embedding of the Minkowski metric, which is the product of a timelike line $y^I = const$ (we use indices $I, K, \ldots = 1, \ldots, 9; i, k, \ldots = 1, 2, 3$) on 9D unfolded embedding $\bar{y}^I(x^i)$ of the euclidean 3D metric, i.e.,

$$\bar{y}^a = \begin{pmatrix} x^0 \\ \bar{y}^I(x^i) \end{pmatrix}, \qquad \partial_i \bar{y}^I \partial_k \bar{y}^I = \delta_{ik}. \tag{5}$$

With such a choice of background embedding, a nonrelativistic motion of embedding matter is possible [19].

The purpose of this work is to study Regge-Teitelboim Equation (2), linearized in the background of (5). We look for a solution that corresponds to a galaxy that rotates so slowly that the effect of the rotation can be neglected, i.e., the distribution of ordinary matter is static and spherically symmetric on average. At the same time, we assume that metric (1) is also static (and spherically symmetric), which corresponds to the time independence of value $\tau^{\mu\nu}$ describing embedding matter. The resulting solution determines the dependence of the gravitational potential on the distance to the center of the galaxy, and it can be compared with observations of the rotation curves of galaxies.

In Sections 2 and 3, we obtain linearized equations and find their solution. In Section 4, the influence of the assumption of the exact static nature of the solution on its behavior in a linear approximation is investigated. In Section 5, we study how the problem is simplified in the case of spherical symmetry. We propose an explicit form of a spherically symmetric background embedding in Section 6. In Section 7, we study the possibility of choosing this embedding in such a way that the corresponding gravitational potential is in agreement with the observed rotation curves of galaxies.

2. Linearization of Regge–Teitelboim Equations

Let us recall some formulas of embedding theory. All convolutions by Latin indices are carried out using the flat metric of ambient space η_{ab}. The induced metric is expressed in terms of the embedding function by Formula (1). We use space–time signature $(-,+,+,+)$. The signature is changed by changing the sign of η_{ab}, and the induced metric changes the sign as a consequence. The second fundamental form of a 4D surface is expressed in terms of the covariant derivative D_μ of the embedding function, consistent with the metric or through projector $\Pi_{\perp}{}^{a}_{b}$ on a subspace transverse to the surface (see, for example, [12]):

$$b^a_{\mu\nu} = D_\mu \partial_\nu y^a = \Pi_{\perp}{}^{a}_{b} \partial_\mu \partial_\nu y^b. \tag{6}$$

We mark with a line the values corresponding to background embedding function $\bar{y}^a$ (5); for example, $\bar{b}^a_{\mu\nu}$ is the second fundamental form of the background surface. We raise and lower the 4D indices of values with a line using the background metric. Since it is flat, the background connection is zero, and the covariant derivative in (6) is reduced to the usual one. The second derivatives of $\bar{y}^0$ and the derivatives by x^0 of $\bar{y}^I$ are zero; hence, the nonzero components of $\bar{b}^a_{\mu\nu}$ are:

$$\bar{b}^I_{ij} = \partial_i \partial_j \bar{y}^I. \tag{7}$$

Through index I, this value is transverse to the 3D surface in 9D space described by embedding function $\bar{y}^I(x^i)$. Thus, index I at each point effectively runs through 6 values, exactly like multi-index $\{ij\}$. Therefore, $\bar{b}^I_{ij}$ can be represented as a 6×6 matrix. Since this matrix is nonsingular for "unfolded" embeddings (see details in [18]), you can introduce symmetric by l, m and transverse by I value $\bar{\alpha}^{lm}_I$, which is inverse to $\bar{b}^I_{ij}$ in the matrix sense:

$$\bar{\alpha}^{lm}_I \bar{b}^I_{ij} = \frac{1}{2}\left(\delta^l_i \delta^m_j + \delta^l_j \delta^m_i\right). \tag{8}$$

We look for solutions of the equations in the form of Systems (3) and (4) as a perturbation theory series in degrees of the gravitational constant $\varkappa$ in the background (5):

$$y^a = \bar{y}^a + \varkappa \overset{(1)}{y}{}^a + \varkappa^2 \overset{(2)}{y}{}^a + \dots. \tag{9}$$

We write similar expansions in a series of $\varkappa$ for any other quantities, including $\tau^{\mu\nu}$ and the energy–momentum tensor of ordinary matter $T^{\mu\nu}$. Substituting expansions into Systems (3) and (4), in the first nonvanishing order, we obtain

$$\overset{(1)}{G}{}^{\mu\nu} = \bar{T}^{\mu\nu} + \bar{\tau}^{\mu\nu}, \tag{10}$$

$$\bar{\tau}^{ik} \bar{b}^I_{ik} = 0. \tag{11}$$

Due to the nonsingularity of the (7) mentioned above, the last equation corresponds to Condition

$$\bar{\tau}^{ik} = 0. \tag{12}$$

Of the 10 linearized Einstein Equations (10), only six (for $\mu = i, \nu = k$) are equations for the first correction $\overset{(1)}{y}{}^a$ to the embedding function and hence for the first correction $\overset{(1)}{g}{}_{\mu\nu}$ to the metric, as follows from (1). The remaining four equations define the components of $\bar{\tau}^{0\mu}$; therefore, they do not restrict the embedding function.

Only the transverse contributions to the background surface in $\overset{(1)}{y}{}^a$ change the shape of the surface, so that only six components of $\overset{(1)}{y}{}^a$ make physical sense. The others four components correspond to the longitudinal deformations of the surface, which are equivalent to diffeomorphisms. Therefore, in order to uniquely find all 10 components of $\overset{(1)}{y}{}^a$, it is necessary to impose some coordinate conditions.

3. Solving of the Linearized Equations

Assume that at least in the zero order the distribution of ordinary matter is static, i.e., $\partial_0 \bar{T}^{\mu\nu} = 0$. We look for solutions of linearized Equation (10) that correspond to the static addition to the metric, i.e., $\partial_0 \overset{(1)}{g}_{\mu\nu} = 0$. This corresponds to the fact that, in the main order, the distribution of embedding matter is also static, i.e., $\partial_0 \bar{\tau}^{\mu\nu} = 0$.

Instead of solving only 6 out of 10 Equations (10) (which can be performed and gives the same result), it is convenient to solve all 10 equations, treating the quantities of $\bar{\tau}^{0\mu}$ as independent variables describing embedding matter. Then, the linearized Regge–Teitelboim equations are simply reduced to linearized Einstein Equation (10). We assume that ordinary matter is dustlike (as can be performed for the matter in galaxies) and is at rest in zero order; then,

$$\bar{T}_{\mu\nu} = \bar{\rho}(x^i)\delta^0_\mu \delta^0_\nu. \tag{13}$$

Equation (10) can be considered to be an equation not directly for $\overset{(1)}{y}{}^a$, but as an equation for the correction to metric $\overset{(1)}{g}_{\mu\nu}$, from which $\overset{(1)}{y}{}^a$ is then found by solving embedding Equation (1).

The equations for $\overset{(1)}{g}_{\mu\nu}$ take a simple form:

$$\Delta \overset{(1)}{g}_{00} = -(\bar{\rho}(x^i) + \bar{\tau}_{00}), \qquad \Delta \overset{(1)}{g}_{jk} = -(\bar{\rho}(x^i) + \bar{\tau}_{00})\delta_{jk}, \tag{14}$$

$$\Delta \overset{(1)}{g}_{0k} = -2\bar{\tau}_{0k} \tag{15}$$

(where Δ is the Laplacian operator) if we choose harmonic coordinates and assume that $\overset{(1)}{g}_{\mu\nu}$ does not depend on time. In addition, we assume that $\overset{(1)}{g}_{\mu\nu}$ is decreasing at spatial infinity. The conditions for the harmonic coordinates are reduced to only one additional equation:

$$\partial_k \overset{(1)}{g}_{0k} = 0. \tag{16}$$

Introducing gravitational potential $\varphi(x^i)$ (we did not include the gravitational constant $\varkappa$ in it for convenience) as a solution of Poisson equation

$$\Delta\varphi(x^i) = \frac{1}{2}(\bar{\rho}(x^i) + \bar{\tau}_{00}) \tag{17}$$

and taking into account that the solutions are decreasing at spatial infinity, we obtain

$$\overset{(1)}{g}_{00} = -2\varphi(x^i), \qquad \overset{(1)}{g}_{jk} = -2\varphi(x^i)\delta_{jk}. \tag{18}$$

Now, we find correction $\overset{(1)}{y}{}^a$ to the embedding function that corresponds to an arbitrary correction $\overset{(1)}{g}_{\mu\nu}$ to the metric. From Formula (1), in the first order by $\varkappa$, we have:

$$\overset{(1)}{g}_{\mu\nu} = (\partial_\mu \bar{y}^a)(\partial_\nu \overset{(1)}{y}_a) + (\partial_\nu \bar{y}^a)(\partial_\mu \overset{(1)}{y}_a). \tag{19}$$

Let us introduce a parameterization for the correction to the embedding function as the sum of a tangent and orthogonal to the background components:

$$\overset{(1)}{y}{}^a = \overset{(1)}{y}{}^\gamma_\parallel \partial_\gamma \bar{y}^a + \overset{(1)}{y}{}^a_\perp. \tag{20}$$

The structure of Background Embedding (5) leads to relation $\overset{(1)}{y}{}^0 = \overset{(1)}{y}{}^0_\parallel$. Substituting Representation (20) in (19) yields a system of differential equations for $\overset{(1)}{y}{}^\gamma_\parallel$ and $\overset{(1)}{y}{}^a_\perp$:

$$\eta_{\gamma\nu}\partial_\mu \overset{(1)}{y}{}^\gamma_\parallel + \eta_{\gamma\mu}\partial_\nu \overset{(1)}{y}{}^\gamma_\parallel - 2\overset{(1)}{y}{}^I_\perp \bar{b}^I_{\mu\nu} = \overset{(1)}{g}_{\mu\nu}. \tag{21}$$

Writing out its components, we obtain a system of equations:

$$\begin{aligned} -2\partial_0 \overset{(1)}{y}{}^0 &= \overset{(1)}{g}_{00}, \\ -\partial_k \overset{(1)}{y}{}^0 + \partial_0 \overset{(1)}{y}{}^k_\parallel &= \overset{(1)}{g}_{0k}, \\ \partial_i \overset{(1)}{y}{}^j_\parallel + \partial_j \overset{(1)}{y}{}^i_\parallel - 2\overset{(1)}{y}{}^I_\perp \bar{b}^I_{ij} &= \overset{(1)}{g}_{ij}. \end{aligned} \tag{22}$$

This can be solved by integrating the equations over time and taking into account the static $\overset{(1)}{g}_{\mu\nu}$. Denoting the integration constants as $\tilde{y}^a(x^i)$, we write the answer as follows:

$$\begin{aligned} \overset{(1)}{y}{}^0 &= -\frac{1}{2}\overset{(1)}{g}_{00} x^0 + \tilde{y}^0(x^i) = \varphi(x^i)x^0 + \tilde{y}^0(x^i), \\ \overset{(1)}{y}{}^k_\parallel &= \int \left(\overset{(1)}{g}_{0k} + \partial_k \overset{(1)}{y}{}^0 \right) dx^0 + \tilde{y}^k(x^i), \\ \overset{(1)}{y}{}^I_\perp &= \frac{1}{2}\bar{\varkappa}^{kj}_I \left(-\overset{(1)}{g}_{kj} + 2\partial_k \overset{(1)}{y}{}^j_\parallel \right) = \bar{\varkappa}^{kj}_I \left(\varphi(x^i)\delta_{kj} + \partial_k \overset{(1)}{y}{}^j_\parallel \right), \end{aligned} \tag{23}$$

where (18) was also used.

Substituting these expressions into (20), we lastly obtain

$$\begin{aligned} \overset{(1)}{y}{}^a = {} & \delta^a_0 \left(\varphi(x^i)x^0 + \tilde{y}^0(x^i) \right) + \\ & + \delta^a_I \left(\left(\int \left(\overset{(1)}{g}_{0k} + \partial_k \overset{(1)}{y}{}^0 \right) dx^0 + \tilde{y}^k(x^i) \right) \partial_k \bar{y}^I + \bar{\varkappa}^{kj}_I \left(\varphi(x^i)\delta_{kj} + \partial_k \overset{(1)}{y}{}^j_\parallel \right) \right). \end{aligned} \tag{24}$$

This expression gives the solution of the linearized Regge–Teitelboim equations. It is parameterized by the gravitational potential $\varphi(x^i)$ that, via Equation (17), is associated with the zero-order $\bar{\tau}_{00}$ of the density of embedding matter, and via the components of metric $\overset{(1)}{g}_{0k}$ that, via Equation (15), are related to the zero-order $\bar{\tau}_{0k}$ of the flux density of this matter. Due to (15) and (16), the components of $\bar{\tau}_{0k}$ obey condition $\partial_k \bar{\tau}_{0k} = 0$ (it has the meaning of the embedding matter conservation condition in the case under consideration), so only two of these three components can be considered as independent.

It is interesting to compare this result with the one obtained in [19] in the framework of the nonrelativistic approximation. For this purpose, let us write down the time derivative of Solution (24):

$$\begin{aligned} \partial_0 \overset{(1)}{y}{}^0 &= \varphi(x^i), \\ \partial_0 \overset{(1)}{y}{}^I &= \left(\overset{(1)}{g}_{0k} + \partial_k \overset{(1)}{y}{}^0 \right) \partial_k \bar{y}^I + \bar{\varkappa}^{ij}_I \partial_i \left(\overset{(1)}{g}_{0j} + \partial_j \overset{(1)}{y}{}^0 \right). \end{aligned} \tag{25}$$

The second equation reproduces Equation (62) from [19], taking into account the assumption $\overset{(1)}{g}_{0j} = 0$ there. However, the first equation differs from Equation (61) of [19] through the absence of a quadratic contribution. Of course, this is not surprising, since the solution of (24) is obtained in the linear approximation and hence with the rejection of the quadratic contributions. However, when analyzing the nonrelativistic limit in [19], it was taken into account that for the large intervals of time x^0 (when time x^0 is replaced by nonrelativistic time $t = x^0/c$, where c is the speed of light) quadratic corrections can give a comparable contribution to the linear ones.

4. Accounting for Equations in the Next Order

The solution obtained above in the first order of perturbation theory depends on the choice of the distribution of embedding matter in zero order characterized by $\bar{\tau}_{0\mu}$. Let us find restrictions on this distribution, assuming that both $\tau^{\mu\nu}$ characterizing embedding matter and the energy–momentum tensor of ordinary matter $T^{\mu\nu}$ are static not only in the zero approximation, but also exactly. In particular, it follows that the metric is also exactly static. When considering a galaxy, this means that the configuration of both ordinary and embedding matter at some point passed into a stationary state. We investigate just this particular case.

Let us write down Equation (4) in the next order in $\varkappa$, compared to already studied Equation (11), and use it:

$$\overset{(1)}{\tau}{}^{jk}\bar{b}^a_{jk} + \bar{\tau}^{00}\overset{(1)}{b}{}^a_{00} + 2\bar{\tau}^{0k}\overset{(1)}{b}{}^a_{0k} = 0. \tag{26}$$

Due to (6) and the properties of Background Embedding (5), we have $\overset{(1)}{b}{}^a_{0\mu} = \bar{\Pi}_{\perp}{}^a_b\partial_0\partial_\mu\overset{(1)}{y}{}^b$. This allows for us to express $\overset{(1)}{\tau}{}^{jk}$ from (26) in the form:

$$\overset{(1)}{\tau}{}^{jk} = -\left(\bar{\tau}^{00}\partial_0\partial_0\overset{(1)}{y}{}^I + 2\bar{\tau}^{i0}\partial_i\partial_0\overset{(1)}{y}{}^I\right)\bar{\alpha}^{jk}_I. \tag{27}$$

Therefore, the assumed time independence of $\tau^{\mu\nu}$ leads to additional restrictions on the quantities in the right-hand side. The time derivative of Equation (27) with (25) and condition $\partial_0\overset{(1)}{\tau}{}^{jk} = 0$ provide the equation

$$\bar{\tau}^{i0}\left(\left(\delta^j_i\delta^k_l + \delta^j_l\delta^k_i\right)\partial_l\varphi + 2\,\partial_i\left(\bar{\alpha}^{lm}_I\partial_l\partial_m\varphi\right)\bar{\alpha}^{jk}_I\right) = 0. \tag{28}$$

Then, we can also write down Equation (3) in the next order, and it provides the next correction $\overset{(2)}{g}_{\mu\nu}$ to the metric. However, it is possible to write down a consequence of this equation in which the determined correction does not enter. This is the corresponding order of equation $D_\mu\tau^{\mu\nu} = 0$, reflecting the conservation (with covariant corrections) of embedding matter. In the lowest order, this equation has the form $\partial_\mu\bar{\tau}^{\mu\nu} = 0$ and is satisfied given (12), (15) and (16). In the next order, it looks like

$$\partial_\mu\overset{(1)}{\tau}{}^{\mu\nu} + \overset{(1)}{\Gamma}{}^\mu_{\mu\gamma}\bar{\tau}^{\gamma\nu} + \overset{(1)}{\Gamma}{}^\nu_{\mu\gamma}\bar{\tau}^{\mu\gamma} = 0. \tag{29}$$

Taking $\nu = k$ in this equation, and using Equation (27) and the assumed time independence of $\tau^{\mu\nu}$ in all orders, from condition $\partial_0\overset{(1)}{\tau}{}^{0k} = 0$ we obtain an equation:

$$\begin{aligned}\partial_j&\left(\left(\bar{\tau}^{00}\partial_0\partial_0\overset{(1)}{y}{}^I + 2\bar{\tau}^{i0}\partial_i\partial_0\overset{(1)}{y}{}^I\right)\bar{\alpha}^{jk}_I\right) - \\ &-\bar{\tau}^{0k}\eta^{\mu\alpha}(\partial_\alpha\bar{y}_a)\partial_\mu\partial_0\overset{(1)}{y}{}^a - \bar{\tau}^{00}(\partial_k\bar{y}_a)\partial_0\partial_0\overset{(1)}{y}{}^a - 2\bar{\tau}^{j0}(\partial_k\bar{y}_a)\partial_j\partial_0\overset{(1)}{y}{}^a = 0,\end{aligned} \tag{30}$$

where (12) and an expression for connectivity are used (see, for example, [12])

$$\Gamma^\nu_{\mu\gamma} = g^{\nu\alpha}(\partial_\alpha y_a)\partial_\mu\partial_\gamma y^a, \tag{31}$$

from which it can be found that

$$\overset{(1)}{\Gamma}{}^{\nu}_{\mu 0} = \eta^{\nu\alpha}(\partial_\alpha \bar{y}_a)\partial_\mu \partial_0 \overset{(1)}{y}{}^a. \tag{32}$$

Using Formulas (25) in (30), we obtain an equation that, together with (28), restricts the choice of the background embedding $\bar{y}^a$ and $\bar{\tau}^{0\mu}$, which parameterize the solution of the linear approximation found in the previous section. From the remaining Equation (29) at $\nu = 0$, variable $\overset{(1)}{\tau}{}^{k0}$ is not excluded, so this equation does not provide a new restrictions on the above quantities.

The background embedding of the form (5) is parameterized by three functions, since nine components of $\bar{y}^I(x^i)$ are imposed by six conditions that the metric of the surface described by this function is flat Euclidean. At first sight, the resulting number of equations for the background and $\bar{\tau}^{0\mu}$ seems too large—six Equation (28) and three Equation (30) for six unknowns (as noted after (24)), only two of the three components $\bar{\tau}^{0k}$ are independent). However, it is easy to see that a solution nevertheless exists, since the six Equations (28) can be satisfied by choosing three quantities $\bar{\tau}^{0k}$ as

$$\bar{\tau}^{0k} = 0. \tag{33}$$

We study the solutions corresponding just to such a choice. Physically, it means that, in the zero approximation, embedding matter is at rest. Using Background (5) leads to the fact that time independence $\bar{\tau}^{\mu\nu}$ actually leads to (33) (otherwise, there are fewer unknowns than equations), i.e., to Embedding Matter at rest in the zero approximation. In the analysis of the nonrelativistic limit in [19], such a background was found when discussing the nonrelativistic character of motion of embedding matter. Taking into account (15) and assuming decreasing corrections to the metric at 3D infinity, it follows from (33) that $\overset{(1)}{g}{}_{0k} = 0$.

Given that $\bar{\tau}^{0k} = 0$, the remaining Equations (30) are simplified. Using (25), we can now rewrite them as

$$\partial_j\Big((2\Delta\varphi - \bar{\rho})\bar{\alpha}^{jk}_I \bar{\alpha}^{lm}_I \partial_l \partial_m \varphi\Big) - (2\Delta\varphi - \bar{\rho})\partial_k \varphi = 0, \tag{34}$$

where $\bar{\tau}^{00}$ was expressed from (17).

These are three equations for four quantities, namely, three that parameterize the background embedding function $\bar{y}^I(x^i)$ and one describing the density of embedding matter $\bar{\tau}^{00}$. Thus, linearized Regge–Teitelboim equations, supplemented by the condition of exact statics of solutions, leave an arbitrariness in the choice of one function of coordinates x^i. The physical meaning of this arbitrariness lies in the presence of arbitrariness in the general case when specifying the distribution of embedding matter at the initial moment of time.

5. The Case of Spherical Symmetry

Let us consider the situation when ordinary matter is distributed in a spherically symmetric way. Then, the solution can also be sought in the class of spherically symmetric functions. In this case, both gravitational potential φ and the distribution density of ordinary matter $\bar{\rho}$ depend only on the radial coordinate $r = \sqrt{x^i x^i}$.

The question of how to formulate a condition of any kind of symmetry for an embedding function is rather nontrivial; see, for example, [20]. Since the background embedding function $\bar{y}^I(x^i)$ enters Equation (34) only as a quantity

$$A^{lm,jk} = \bar{\alpha}^{lm}_I \bar{\alpha}^{jk}_I, \tag{35}$$

it is possible to directly impose symmetry constraints on this. Taking into account all index permutation symmetries, in the case of spherical symmetry, quantity A can be written in the form

$$A^{lm,jk} = f_1(r)\delta^{lm}\delta^{jk} + \frac{1}{2}f_2(r)\left(\delta^{lj}\delta^{mk} + \delta^{lk}\delta^{mj}\right) + \frac{1}{2r^2}f_3(r)\left(\delta^{lm}x^jx^k + \delta^{jk}x^lx^m\right) + \\ + \frac{1}{4r^2}f_4(r)\left(\delta^{lj}x^mx^k + \delta^{lk}x^mx^j + \delta^{jm}x^kx^l + \delta^{mk}x^jx^l\right) + \frac{1}{r^4}f_5(r)x^lx^mx^jx^k. \quad (36)$$

Here, functions $f_1 \dots f_5$ are characteristics of the particular background embedding used. If we substitute this representation into (34), a common multiplier x^k appears, and the three equations turn into one scalar equation. Introducing notations for different combinations of functions $f_1 \dots f_5$, the resulting equation can be written in the following form:

$$\left(\left(\varphi'' + \frac{2}{r}\varphi' - \frac{1}{2}\bar{\rho}\right)\left(\varphi'' F_1 + \frac{1}{r}\varphi' F_2\right)\right)' + \frac{2}{r}\left(\varphi'' + \frac{2}{r}\varphi' - \frac{1}{2}\bar{\rho}\right)\left(\varphi'' F_3 + \frac{1}{r}\varphi' F_4\right) = 0, \quad (37)$$

where

$$F_1 = f_1 + f_2 + f_3 + f_4 + f_5, \quad (38)$$
$$F_2 = 2f_1 + f_3, \quad (39)$$
$$F_3 = f_2 + \frac{1}{2}f_3 + f_4 + f_5, \quad (40)$$
$$F_4 = -f_2 + f_3 - \frac{r^2}{2}. \quad (41)$$

Equation (37) allows for us to determine gravitational potential $\varphi(r)$ if the distribution of ordinary matter $\bar{\rho}(r)$ and the background embedding function defining $f_1 \dots f_5$ are known. Since (37) is a nonlinear differential equation, in the general case, its solution can only be sought numerically. First of all, it is necessary to find all the functions $f_1 \dots f_5$ included in the equation using the given embedding function, which requires first finding the value of $\bar{\alpha}_I^{lm}$.

Finding it directly with Formula (8) is a nontrivial task for specific embeddings. Let us find a simpler way that allows for us to find functions $f_1 \dots f_5$ included in the expansion of (36) without explicit knowledge of $\bar{\alpha}_I^{lm}$. When writing this expansion, we took into account three types of symmetry by permutations of indices following from the definition of (35): inside the first pair, inside the second, and when replacing the first pair with the second. As a consequence, we have five independent coefficients $f_1 \dots f_5$. However, in fact, there are fewer independent variables parameterizing A. To show this, we define quantity B:

$$B_{jk,is} = \bar{b}^I_{jk}\bar{b}^I_{is} = (\partial_j\partial_k\bar{y}^I)\partial_i\partial_s\bar{y}^I. \quad (42)$$

It is easy to see that this quantity is symmetric through the permutation of any two indices due to the orthogonality of the first and second derivatives of $\bar{y}^I$:

$$B_{jk,is} = (\partial_j\partial_k\bar{y}^I)\partial_i\partial_s\bar{y}^I = -(\partial_i\partial_j\partial_k\bar{y}^I)\partial_s\bar{y}^I = (\partial_i\partial_k\bar{y}^I)\partial_j\partial_s\bar{y}^I = B_{ik,js}. \quad (43)$$

Hence, in the spherically symmetric case, it can be parameterized not by five but only by three radial functions:

$$B_{jk,is} = \frac{1}{3}g_1(r)\left(\delta^{jk}\delta^{is} + \delta^{js}\delta^{ik} + \delta^{ji}\delta^{sk}\right) + \frac{1}{6r^2}g_2(r)\left(\delta^{jk}x^ix^s + \delta^{ji}x^kx^s + \\ + \delta^{js}x^ix^k + \delta^{ik}x^jx^s + \delta^{is}x^jx^k + \delta^{sk}x^jx^i\right) + \frac{1}{r^4}g_3(r)\,x^jx^kx^ix^s. \quad (44)$$

There is a connection between A and B that follows from (8):

$$A^{lm,jk}B_{jk,is} = \frac{1}{2}\left(\delta_i^l\delta_s^m + \delta_s^l\delta_i^m\right). \tag{45}$$

It can be used to express five functions $f_1 \dots f_5$ through three functions $g_1 \dots g_3$:

$$\begin{aligned}
f_1 &= \frac{-24g_1^2 + (-24g_2 - 36g_3)g_1 + 3g_2^2}{4(20g_1^2 + (20g_2 + 24g_3)g_1 - g_2^2)g_1},\\
f_2 &= \frac{3}{2g_1},\\
f_3 &= \frac{(12g_2 + 36g_3)g_1 - 3g_2^2}{2(20g_1^2 + (20g_2 + 24g_3)g_1 - g_2^2)g_1},\\
f_4 &= \frac{-3g_2}{g_1(2g_1 + g_2)},\\
f_5 &= \frac{-360g_1^2g_3 + (138g_2^2 + 108g_2g_3)g_1 - 3g_2^3}{4(20g_1^2 + (20g_2 + 24g_3)g_1 - g_2^2)g_1(2g_1 + g_2)}.
\end{aligned} \tag{46}$$

As a result, to find the functions $f_1 \dots f_5$ corresponding to a specific background embedding, we first need to calculate the second fundamental form $\bar{b}^I_{ij} = \partial_i\partial_j\bar{y}^I$. Then, we have to determine quantity $B_{jk,is}$ according to (42). Comparing the result with the (44) expansion, we can find functions $g_1 \dots g_3$, and then find functions $f_1 \dots f_5$ by (46).

6. Explicit Unfolded Spherically Symmetric Embedding

Let us present an embedding $\bar{y}^I(x^i)$ of the flat Euclidean 3D metric into 9D Euclidean space that would be spherically symmetric in the sense discussed in [20] and would also be unfolded in terms of [18]. The latter means that, for such a surface, the second fundamental form (7) is nonsingular as a 6×6 matrix; see the text after (7).

We consider basis λ^A_{ij} in the space of symmetric traceless matrices, numbered by index A, which runs through values $4, \dots, 8$. We subject this basis to the orthonormality condition in the space of such matrices:

$$\lambda^A_{ij}\lambda^A_{lm} = \frac{1}{2}\left(\delta_{il}\delta_{jm} + \delta_{im}\delta_{jl}\right) - \frac{1}{3}\delta_{ij}\delta_{lm}, \qquad \lambda^A_{ij}\lambda^B_{ij} = \delta^{AB}. \tag{47}$$

We take the embedding function $\bar{y}^I(x^i)$ with components

$$\begin{aligned}
\bar{y}^k(x^i) &= f(r)x^k,\\
\bar{y}^A(x^i) &= g(r)\lambda^A_{lm}x^l x^m,\\
\bar{y}^9(x^i) &= h(r),
\end{aligned} \tag{48}$$

it is parameterized by three radial functions. One can check that this embedding function describes a surface with spherical symmetry. This means that there is such a representation (generally speaking, reducible) of the group of 3D rotations under the action of which in the space $\bar{y}^I$ the surface transforms to itself (see [20] for details). The three independent functions f, g, h are related to its irreducible parts, and the reducible representation as a whole can be described as "3+5+1".

Let us select, among all the embeddings of type (48), those that correspond to a flat metric. Calculating for (48) the components of the induced metric, we obtain the equations

$$\begin{aligned}
&f^2 + 2r^2g^2 = 1,\\
&(rf)'^2 + \frac{2}{3}\left(r^2g\right)'^2 + h'^2 = 1.
\end{aligned} \tag{49}$$

Expressing from these equations functions g and h through f we obtain the desired spherically symmetric embedding of the flat Euclidean 3D metric into 9D Euclidean space. It is parameterized by choosing one function of radius $f(r)$. The fact that such an embedding is unfolded by a non-special choice of this function can be checked directly by computing the second fundamental form $\partial_i\partial_j\bar{y}^I$ (7) and making sure that it is nonsingular.

Calculating this quantity and Expression (42) constructed from it lead to the following expressions for the coefficients $g_1 \dots g_3$ of its expansion (44):

$$g_1 = 6g^2, \tag{50}$$

$$g_2 = 6\left(f'^2 + 2r^2g'^2 + 4rgg'\right), \tag{51}$$

$$g_3 = r^2\left(\frac{h'}{r}\right)'^2 + r^2f''^2 + 2rf'f'' - 3f'^2 + \frac{2}{3}r^4g''^2 + 4r^3g'g'' + 4r^2gg'' - 2r^2g'^2 - 4rgg', \tag{52}$$

where the functions g and h must be expressed through a single arbitrary function $f(r)$ according to (49).

7. Inverse Problem

Equation (37) obtained above in Section 5 allows for searching for the gravitational potential $\varphi(r)$ if the distribution of ordinary matter $\bar{\rho}(r)$ and the background embedding function $\bar{y}^I(x^i)$ are known. It is easy to see that, among the solutions of this equation, there is an Einsteinian solution for which

$$\varphi'' + \frac{2}{r}\varphi' - \frac{1}{2}\bar{\rho} = 0. \tag{53}$$

However, there are other solutions too. In the galactic halo region where the contribution of $\bar{\rho}$ can be neglected, we have a Newtonian behavior of the gravitational potential $\varphi \sim 1/r$ for the Einsteinian solution. Such a behavior does not agree well with the data of the rotation curves of galaxies. To find other solutions in this area, it is sufficient to know background embedding $\bar{y}^I(x^i)$.

The background embedding constructed in the previous section has ambiguity in the choice of one function $f(r)$. There is no highlighted natural way to choose this function. As a consequence, there is also no possibility to uniquely find the gravitational potential $\varphi(r)$ as a result of solving Equation (37), since changing $f(r)$ also changes $\varphi(r)$. Therefore, we focus on investigating the inverse problem by posing the question: can we find such a background embedding $\bar{y}^I(x^i)$ (i.e., such a function $f(r)$) so that the solution of Equation (37) corresponds to the gravitational potential in good agreement with galactic halo observations?

As a good fitting of observational data, we take the Burkert profile [21] of the dark matter density; see, for example, review [22]:

$$\rho_B(r) = \frac{\rho_0 r_0^3}{(r+r_0)(r^2+r_0^2)}, \tag{54}$$

where r_0 is the core radius, ρ_0 is the density at the center of the galaxy. In order to find a behavior of the gravitational potential that agrees well with the observations, one needs to solve the equation

$$\varphi'' + \frac{2}{r}\varphi' = \frac{1}{2}(\bar{\rho} + \rho_B). \tag{55}$$

The solution of this equation in the galactic halo has the form:

$$\varphi(r) = \frac{M}{r} + \frac{\rho_0}{2}\left(\left(1+\frac{r_0}{r}\right)\left(\arctan\frac{r}{r_0} - \ln\left(1+\frac{r}{r_0}\right)\right) + \frac{1}{2}\left(1-\frac{r_0}{r}\right)\ln\left(1+\frac{r^2}{r_0^2}\right)\right), \tag{56}$$

where M is the mass of ordinary matter in the galaxy. To avoid introducing this quantity as an additional parameter, we further assume that the proportional contribution to M to the gravitational potential in the halo region can be neglected compared with the contribution of dark matter. The trivial substitution of Function (56) with $M = 0$ into Equation (37) leads to a very complicated (so we do not show it here) third-order nonlinear differential equation for the unknown function $f(r)$ parameterizing the background embedding. This equation fails to be solved analytically. We numerically solved the Cauchy problem for Equation obtained, choosing different values of the initial data at the value of the radius $r = r_0/2$, which is usually about the visual size of the galaxy. We solve Equation (37) with a zero contribution of ordinary matter, so it is applicable only in the region of the galactic halo.

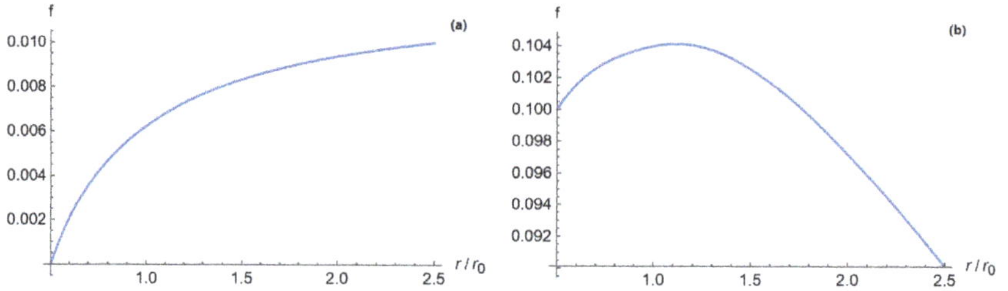

Figure 1. Numerically constructed solutions with different initial data: (**a**) $f = 0$, $f' = 0.025$, $f'' = -0.1$; (**b**) $f = 0.1$, $f' = 0.02$, $f'' = -0.1$.

A typical view of the obtained solutions is shown in Figure 1. The existence of such solutions shows the possibility to choose parameterizing embedding (48) function $f(r)$ in a special way. For this choice, equations of motion in the framework of linearization of the embedding theory in this background lead to the same gravitational potential as the dark matter with Burkert profile (54) in the framework of ordinary GR. With a significant change in the chosen values of the initial data, the algorithm used for the numerical solution of the nonlinear differential equation no longer made it possible to extend the solution into the region $r \gtrsim 0.95 r_0$. This can be explained by the fact that in the differential equation being solved, the coefficient at the major derivative (which itself is a compound quantity that also depends on the minor derivatives) may become too small, which prevents the correct behavior of the algorithm.

8. Conclusions

We considered the problem of constructing a perturbation theory for Regge-Teitelboim Equation (2) for some background. The background was chosen to be (5), which corresponded to a 4D surface that is a direct product of the timelike straight line on the 9D unfolded embedding of the euclidean 3D metric. This formulation of the problem is a special case of the weak gravitational field approximation in the embedding theory, in which the corrections to the embedding function are linear to the corrections to the metric; see (23). The analysis is simplified if the Regge–Teitelboim equations are rewritten as a set of Einstein Equation (3) with the contribution of fictitious embedding matter and Equation (4) describing the behavior of this matter.

First, the Einstein equations were linearized in the usual way and solved using harmonic coordinates. The result depended on the energy-momentum tensor of embedding matter specified arbitrarily in the lower order. Further, we restricted ourselves to the class of solutions for which both ordinary matter and embedding matter are in a stationary state, which corresponded to the description of an already fully formed galaxy. Taking this condition into account in the second order of perturbation theory leads to additional restrictions on the solutions found in the linear one. As a result, Equation (37) arises in the case of spherical symmetry, which allows for finding the gravitational potential $\varphi(r)$ if the background embedding is given. The nonlinear form of this equation is explained by the fact that it arises as a result of accounting for the static character of the metric in the second (following the linear) order.

It is possible to find a class of spherically symmetric surfaces (48) and (49) that have all necessary properties for their use as a background embedding. These embeddings are parameterized by a single function $f(r)$. This function could be chosen, so that the solution of Equation (37) led to a gravitational potential that corresponded well (if we neglected deviations from spherical symmetry for real galaxies) to the observed distribution of dark matter in the galactic halo.

In our analysis we used the following assumptions. We assumed that ordinary matter has, on average, static spherically symmetric distribution. As a consequence, the energy–momentum tensor of ordinary matter $T^{\mu\nu}$ has exactly spherically symmetric and static character (we use this fact in the first two orders of perturbation theory: for $\bar{T}^{\mu\nu}$ and $\overset{(1)}{T}{}^{\mu\nu}$). We further assumed that the metric had an exactly spherically symmetric and static character, and we used this for corrections to the metric $\overset{(1)}{g}{}_{\mu\nu}$ and $\overset{(2)}{g}{}_{\mu\nu}$. Given the restrictions on $T^{\mu\nu}$, imposing such an assumptions on the metric is the same as imposing them on the energy-momentum tensor of embedding matter in the first two orders: for $\bar{\tau}^{\mu\nu}$ and $\overset{(1)}{\tau}{}^{\mu\nu}$. As a background embedding function $\bar{y}^a(x^\mu)$ that corresponded to the flat metric $\eta_{\mu\nu}$, we took the Product (5) of a timelike line on 9D unfolded and the spherically-symmetric embedding (48) of the euclidean 3D metric. The assumptions were internally consistent, but for real galaxies, they are satisfied only approximately. The strongest ones are the deviations from the spherical symmetry.

We discussed the behavior of the gravitational potential $\varphi(r)$ only in the galactic halo. In order to study the behavior of $\varphi(r)$ inside the galaxy where the influence of ordinary matter cannot be neglected, one must additionally find an explicit form of its distribution, which is beyond the scope of this paper. At the same time, deviations of embedding gravity predictions from GR arise at scales where energy–momentum tensor $\tau^{\mu\nu}$ of embedding matter becomes significant compared to the energy–momentum tensor $T^{\mu\nu}$ of ordinary matter. We assumed that, at the scale of the galaxy, these two quantities were comparable, but at smaller distances, the influence of the embedding matter could be neglected, and the laws of GR would start to be satisfied with good accuracy.

Author Contributions: Conceptualization, S.K. and S.P.; investigation, S.K., M.I., S.M. and S.P.; writing, S.K., M.I., S.M. and S.P. All authors have read and agreed to the published version of the manuscript.

Funding: The work of S.S. Kuptsov was supported by Leonhard Euler International Mathematical Institute grant no. 075-15-2019-1620. The work of S.A. Paston was supported by RFBR grant no. 20-01-00081.

Data Availability Statement: Not applicable.

Conflicts of Interest: The authors declare no conflict of interest.

References

1. Regge, T.; Teitelboim, C. General relativity à la string: A progress report. In Proceedings of the First Marcel Grossmann Meeting, Trieste, Italy, 7–12 July 1975; pp. 77–88.
2. Deser, S.; Pirani, F.A.E.; Robinson, D.C. New embedding model of general relativity. *Phys. Rev. D* **1976**, *14*, 3301–3303. [CrossRef]
3. Tapia, V. Gravitation a la string. *Class. Quant. Grav.* **1989**, *6*, L49. [CrossRef]
4. Maia, M.D. On the integrability conditions for extended objects. *Class. Quant. Grav.* **1989**, *6*, 173–183. [CrossRef]
5. Franke, V.A.; Tapia, V. The ADM Lagrangian in extrinsic gravity. *Nuovo Cimento B* **1992**, *107*, 611. [CrossRef]
6. Estabrook, F.B.; Robinson, R.S.; Wahlquist, H.R. Constraint-free theories of gravitation. *Class. Quant. Grav.* **1999**, *16*, 911–918. [CrossRef]
7. Karasik, D.; Davidson, A. Geodetic Brane Gravity. *Phys. Rev. D* **2003**, *67*, 064012. [CrossRef]
8. Cordero, R.; Molgado, A.; Rojas, E. Ostrogradski approach for the Regge-Teitelboim type cosmology. *Phys. Rev. D* **2009**, *79*, 024024. [CrossRef]
9. Paston, S.A. Gravity as a field theory in flat space-time. *Theor. Math. Phys.* **2011**, *169*, 1611–1619. [CrossRef]

10. Faddeev, L.D. New dynamical variables in Einstein's theory of gravity. *Theor. Math. Phys.* **2011**, *166*, 279–290. [CrossRef]
11. Paston, S.A.; Sheykin, A.A. From the Embedding Theory to General Relativity in a result of inflation. *Int. J. Mod. Phys. D* **2012**, *21*, 1250043. [CrossRef]
12. Paston, S.A.; Franke, V.A. Canonical formulation of the embedded theory of gravity equivalent to Einstein's general relativity. *Theor. Math. Phys.* **2007**, *153*, 1582–1596. [CrossRef]
13. Undagoitia, T.M.; Rauch, L. Dark matter direct-detection experiments. *J. Phys. G Nucl. Part. Phys.* **2015**, *43*, 013001. [CrossRef]
14. Gaskins, J.M. A review of indirect searches for particle dark matter. *Contemp. Phys.* **2016**, *57*, 496–525. [CrossRef]
15. Pavsic, M. On The Quantization Of Gravity By Embedding Space-Time In A Higher Dimensional Space. *Class. Quant. Grav.* **1985**, *2*, 869. [CrossRef]
16. Chamseddine, A.H.; Mukhanov, V. Mimetic dark matter. *J. High Energy Phys.* **2013**, *2013*, 135. [CrossRef]
17. Golovnev, A. On the recently proposed mimetic Dark Matter. *Phys. Lett. B* **2014**, *728*, 39–40. [CrossRef]
18. Paston, S.A.; Zaitseva, T.I. Nontrivial isometric embeddings for flat spaces. *Universe* **2021**, *7*, 477. [CrossRef]
19. Paston, S.A. Non-relativistic limit of embedding gravity as General Relativity with dark matter. *Universe* **2020**, *6*, 163. [CrossRef]
20. Paston, S.A.; Sheykin, A.A. Embeddings for Schwarzschild metric: Classification and new results. *Class. Quant. Grav.* **2012**, *29*, 095022. [CrossRef]
21. Burkert, A. The Structure of Dark Matter Halos in Dwarf Galaxies. *Astrophys. J.* **1995**, *447*, L25. [CrossRef]
22. Di Paolo, C.; Salucci, P. Fundamental properties of the dark and the luminous matter from Low Surface Brightness discs. *arXiv* **2020**, arXiv:2005.03520.

universe MDPI

Article

Cooling Process of White Dwarf Stars in Palatini $f(R)$ Gravity

Surajit Kalita [1,*], Lupamudra Sarmah [2,3] and Aneta Wojnar [4]

1 High Energy Physics, Cosmology & Astrophysics Theory (HEPCAT) Group, Department of Mathematics & Applied Mathematics, University of Cape Town, Cape Town 7700, South Africa
2 Indian Institute of Astrophysics, Bengaluru 560034, India
3 Department of Physics, Indian Institute of Technology (BHU), Varanasi 221005, India
4 Laboratory of Theoretical Physics, Institute of Physics, University of Tartu, W. Ostwaldi 1, 50411 Tartu, Estonia
* Correspondence: surajit.kalita@uct.ac.za

Abstract: A simple cooling model of white dwarf stars is re-analyzed in Palatini $f(R)$ gravity. Modified gravity affects the white dwarf structures and consequently their ages. We find that the resulting super-Chandrasekhar white dwarfs need more time to cool down than sub-Chandrasekhar ones, or when compared to the Newtonian models.

Keywords: cooling process; white dwarf; modified gravity; Chandrasekhar limit

1. Introduction

Gravity, one of the most fundamental interactions that are readily experienced in everyday life, remains enigmatic. To date, general relativity (GR) continues to be a successful and well-accepted theory for explaining gravitational phenomena. Fulfilling the shortcomings of Newtonian gravity, GR can quite accurately explain a plethora of phenomena, starting from the perihelion precession of Mercury and gravitational redshift to the prediction of gravitational waves [1]. Even though Einstein's theory of gravity is well-tested in the weak-field regime with remarkable accuracy [2], it might break down in the very strong gravity regime. This, along with recent cosmological evidences [3,4], suggests that GR may not be adequate to explain the universe and demands appropriate modifications to this theory. Several cosmological observations indicate that ordinary matter comprises only a tiny percentage of the total energy component of the universe and is inadequate to explain the two phases of cosmic acceleration [3–7]. In order to make GR compatible with these cosmological data and other observations related to the rotation curve of the galaxies [8], dark matter and dark energy were postulated to exist [9,10]. The popular cosmological model, viz. ΛCDM model [11], can fit the new observations with remarkable accuracy. However, this model is burdened with well-known cosmological problems and is unable to explain the origin of inflation or dark matter [12]. Although the nature and origin of dark matter still remain a mystery, a scalar field with a slowly varying potential can be considered a candidate for inflation as well as dark energy [13,14]. On the other hand, a pure geometric contribution also seems able to play the same role as it does the scalar field [15–28].

The last approach means that one modifies Einstein's theory of gravity [29–32] in order to solve the aforementioned problems. Though there are several ways of extending and modifying GR, the simplest one is the $f(R)$ gravity [33,34], in which the Lagrangian density is considered to be an arbitrary function of the scalar curvature R. In order to obtain the modified field equations, one can apply either of the two variational principles to the action: the standard metric variation or the Palatini variation. The former approach involves varying the action with respect to the metric only, leading to the fourth-order field equations for the metric[1]. On the other hand, the latter is the variation of the action with respect to the metric as well as the connection, waiving thus the assumption that those two objects are related to each other, as it happens in GR or in the metric formulation. It gives rise to the second-order field equations [18,35]. It is worth mentioning that both approaches are equivalent only in the context of GR when $f(R)$ is a linear function of R. Otherwise, for a general $f(R)$ term in the action, they might lead to entirely different theories and spacetime structures. This is so because the connection in the Palatini approach depends on the particular $f(R)$ form[2] Moreover, being more general in nature and comparatively easier to work with, the Palatini $f(R)$ gravity is, therefore, increasingly used to study a wide variety of phenomena in recent times. Apart from the cosmological models based on the Palatini $f(R)$ gravity to study cosmic acceleration [19,36–39], several astrophysical systems, including white dwarfs (WDs), neutron stars [40,41], stellar [42–44] and substellar objects [45–50] have been extensively studied using this theory (for review, see [51,52]).

WDs represent the final evolutionary stage of progenitor stars with masses less than $(10 \pm 2)M_\odot$ [53,54]. Although the less massive WDs with mass being less than about $0.4M_\odot$, seem to be predominantly composed of helium, the other WDs, in general, might contain heavier elements such as carbon and oxygen [54]. Moreover, in all such stars, the core is surrounded by a thin layer of helium, which in turn is surrounded by an even thinner layer of hydrogen. In order to support this structure of WDs, the inward gravitational force is balanced by the electron degeneracy pressure acting outward due to the Pauli exclusion principle. Since these stars cannot derive energy from thermonuclear reactions, their evolution can be described simply as a cooling process. The degenerate core of the WDs acts as a reservoir of energy, whereas the outer non-degenerate layers are responsible for energy outflows. The simplest models assume that WDs possess temperatures substantially lower than the Fermi temperature. As a result, zero temperature calculations seem to suffice to obtain their structures, however temperature does play an important role to obtain their ages. By studying cooling processes, one can calculate the lifetime of these stars, which can provide important information about the age of different galactic components, stellar formation rate and past galactic history.

In recent times, with the observations of various over- and under-luminous-type Ia supernovae, WDs are extensively being studied in the framework of $f(R)$ gravity [55–60] and other theories of gravity or fundamental physics [61–69,69–79]. Other than a violation of the Chandrasekhar limit, we also expect a modification in the physical properties of the WDs, including the cooling rate, and hence, their ages. Therefore, this work aims at studying a part of the cooling process in WDs using Palatini $f(R)$ gravity. This will allow estimating the modifications in their lifetime, brought on by the particular theory of gravity.

The following is a breakdown of how this article is structured. In Section 2, we discuss the basic formalism of Palatini $f(R)$ gravity and briefly review how it modifies the hydrostatic balance equations. We further employ these equations in Section 3 to derive the modified temperature gradient equation in Palatini $f(R)$ gravity and thereby obtain the modified cooling age of WDs. In Section 4, we numerically calculate the masses and radii of the modified gravity-inspired WDs and their ages. Finally, in Section 5, we put forward our concluding remarks on this work.

2. Basic Formalism of Palatini $f(R)$ Gravity and Hydrostatic Balance Equations

In this paper, the metric signature convention is $(-,+,+,+)$. For a spacetime metric $g_{\mu\nu}$, the action for $f(R)$ gravity has the following form [80]

$$S[g,\Gamma,\Psi] = \frac{1}{2\mathrm{k}} \int \sqrt{-g} f(R(g,\Gamma))\, \mathrm{d}^4x + S_{\mathrm{m}}[g,\Psi], \tag{1}$$

where $\mathrm{k} = -8\pi G/c^4$, $g = \det(g_{\mu\nu})$, G is Newton's gravitational constant, c is the speed of light and S_{m} is the matter action depending on the spacetime metric and matter fields Ψ only. On the other hand, the Ricci–Palatini curvature scalar is constructed with two independent objects: the connection Γ and the metric $g_{\mu\nu}$, such that $R = R_{\mu\nu}(\Gamma)g^{\mu\nu}$.

Varying this action with respect to $g_{\mu\nu}$ results in the modified field equation, given by [80]

$$f'(R)R_{\mu\nu} - \frac{1}{2} f(R) g_{\mu\nu} = \mathrm{k} T_{\mu\nu}, \tag{2}$$

where $f'(R) = \mathrm{d}f(R)/\mathrm{d}R$ and $\mathcal{T}_{\mu\nu}$ is the energy-momentum tensor defined as usually $\mathcal{T}_{\mu\nu} = -(2/\sqrt{-g})\delta S_m/\delta g^{\mu\nu}$. Let us notice that for the spherical-symmetric WD in the statistical equilibrium, the microscopic description of Fermion gas comes down to the perfect-fluid description with the Chandrasekhar equation of state [81,82]. Moreover, the Ricci–Palatini tensor $R_{\mu\nu}$ must be symmetric since its antisymmetric part introduces instabilities [27,83,84].

In contrast, varying the action with respect to Γ and performing some small algebraic transformation, gives the following equation:

$$\nabla_\lambda(\sqrt{-g}f'(R)g^{\mu\nu}) = 0, \tag{3}$$

where ∇_λ is the covariant derivative ruled by Γ. Now, defining a new metric tensor $\bar{g}_{\mu\nu} = f'(R)g_{\mu\nu}$, the above equation can be recast as

$$\nabla_\lambda(\sqrt{-\bar{g}}\bar{g}^{\mu\nu}) = 0, \tag{4}$$

providing that Γ is the Levi–Civita connection with respect to $\bar{g}_{\mu\nu}$. Let us notice that it is conformally related to the spacetime metric g, allowing us to rewrite the field equations in a much simpler form in order to perform rather tedious calculations [85–87]. Moreover, the trace of Equation (2) with respect to $g_{\mu\nu}$ is given by

$$f'(R)R - 2f(R) = \kappa\mathcal{T}, \tag{5}$$

where $\mathcal{T} = g^{\mu\nu}\mathcal{T}_{\mu\nu}$. It is clear now that the curvature is not dynamical, and it can be expressed by the trace of the energy-momentum tensor for a given functional $f(R)$. In other words, all the modifications introduced by Palatini $f(R)$ gravity are functions of the matter fields.

In this paper, we work with the Starobinsky model of $f(R)$ gravity, i.e., $f(R) = R + \alpha R^2$ with α being the model parameter. Hence, the trace equation reduces to the relation between the scalar curvature and the trace of the energy momentum tensor, which resembles the one known from GR, given by

$$R = -\kappa\mathcal{T}. \tag{6}$$

In our previous study [59], we analyzed hydrostatic equilibrium equations in both Jordan and Einstein frames separately. We use the Einstein frame's equations to perform rather tedious calculations in the case of the Jordan ones, and then we transform them back to the Jordan frame, in which we are equipped with the equation of state. The frames are related by the conformal transformation $\bar{g}_{\mu\nu} = f'(R)g_{\mu\nu}$. Thereby we obtained the structure of WDs and later analyzed their stabilities with respect to the radial perturbations. We considered the weak-gravity regime for those equations[3] in a spherically symmetric case, that is, the pressure $\tilde{P}$ is insignificant with respect to the density $\tilde{\rho}$: $\tilde{P}(\tilde{r}) \ll \tilde{\rho}(\tilde{r})c^2$ with $4\pi\tilde{r}^3\tilde{P}/c^2 \ll m(\tilde{r})$, as well as for the geometric contribution, that is, $2Gm(\tilde{r})/\tilde{r}c^2 \ll 1$, where m is the mass function. The 'tilde' denotes variables in the Einstein frame which are related to the physical ones by the tranformations $\tilde{r}^2 = \Phi r^2$, $\tilde{\rho} = \rho/\Phi^2$ and $\tilde{P} = P/\Phi^2$, where $\Phi := f'(R)$. Hence, for the considered Starobinsky model, we have

$$\Phi = 1 + 2\alpha R = 1 + 2\alpha\kappa\rho c^2. \tag{7}$$

Thus, when we take into account the above simplifications, the hydrostatic balance equations in the Einstein frame, are given by

$$\frac{\mathrm{d}\tilde{P}}{\mathrm{d}\tilde{r}} = -\frac{Gm\tilde{\rho}}{\Phi\tilde{r}^2}, \tag{8}$$

$$\frac{\mathrm{d}m}{\mathrm{d}\tilde{r}} = 4\pi\tilde{r}^2\tilde{\rho}. \tag{9}$$

Now, performing the conformal transformation, one can obtain the hydrostatic balance equations in the Jordan frame, given by

$$\frac{dP}{dr} = -\frac{Gm\rho}{\Phi^{\frac{3}{2}}r^2}\left(1+\frac{1}{2}r\frac{\Phi'}{\Phi}\right)+2P\frac{\Phi'}{\Phi}, \tag{10}$$

$$\frac{dm}{dr} = \frac{4\pi r^2\rho}{\Phi^{\frac{1}{2}}}\left(1+\frac{1}{2}r\frac{\Phi'}{\Phi}\right), \tag{11}$$

where we define $\Phi' = d\Phi/dr = 2\alpha\kappa c^2 d\rho/dr$.

In order to solve the above system of differential equations, one needs to choose an equation of state (EoS). It relates the microscopic variables defining a physical system, as well as carries information about additional forces between particles, dependence on the temperature or phase transitions points, to mention just some of them. We consider a quite simple EoS, that is, the barotropic EoS of the form $P = P(\rho)$, neglecting its dependence on the temperature and other thermodynamic variables. This is so because, in a WD, temperature is much lower than the Fermi temperature due to its high density. Since in this paper, we focus on the non-rotating and non-magnetized WD stars, which can be well modeled as spherical-symmetric balls consisting of degenerate electrons, the EoS describing the microscopic properties of such a system is given by the Chandrasekhar one, written in the parametric form as [89]

$$P = \frac{\pi m_e^4 c^5}{3h^3}\left[x_F\left(2x_F^2-3\right)\sqrt{x_F^2+1}+3\sinh^{-1}x_F\right],$$
$$\rho = \frac{8\pi\mu_e m_H (m_e c)^3}{3h^3}x_F^3, \tag{12}$$

where $x_F = p_F/m_e c$ with p_F being the Fermi momentum, m_e the mass of an electron, h the Planck's constant, μ_e the mean molecular weight per electron and m_H the mass of a hydrogen atom. Solving Equations (10) and (11) simultaneously along with this EoS, one can obtain the mass–radius relation of WDs in the considered model of gravity.

3. Temperature Gradient Equation and Cooling Timescale of White Dwarfs in $f(R)$ Gravity

Let us consider a very simple model of the cooling process. In what follows, we assume that the WD star radiates its energy away and does not have any other energy sources. Therefore, its cooling process depends only on the atmosphere properties with the energy transports through the star and the model of gravity. In the case of GR, a similar analysis is given in [53,90].

Moreover, we assume the WD behaves as a perfect black body, such that the equation for radiative energy transport in the Einstein frame is given by

$$\frac{dT}{d\tilde{r}} = -\frac{3L\kappa\tilde{\rho}}{4\pi\tilde{r}^2 4acT^3}, \tag{13}$$

where $T(\tilde{r})$ is the temperature of the WD at a radius $\tilde{r}$, L is the luminosity at that radius, κ is the opacity and $a = 8\pi^5 k_B^4/15c^3h^3 \approx 7.6\times10^{-15}\,\mathrm{erg\,cm^{-3}\,K^{-4}}$ is the radiation constant with k_B being the Boltzmann constant. Moreover, for the perfect black body, the spatial variation of L is given by

$$\frac{dL}{d\tilde{r}} = 4\pi\tilde{r}^2\tilde{\rho}(\tilde{r})\epsilon(\tilde{r}), \tag{14}$$

where ϵ is the power produced per unit mass of stellar material. Now, combining Equations (8) and (13), we obtain

$$\frac{\partial T}{\partial\tilde{P}} = \frac{3L\kappa\Phi}{16\pi acGmT^3}. \tag{15}$$

Thus, defining $\nabla = \partial \ln T / \partial \ln \tilde{P}$, we obtain

$$\tilde{\nabla} = \frac{3L\kappa\tilde{P}\Phi}{16\pi acGmT^4}. \tag{16}$$

In Jordan frame, using Equation (13), the temperature gradient equation is given by

$$\frac{\mathrm{d}T}{\mathrm{d}r} = -\frac{3L\kappa\rho}{4\pi r^2 4acT^3\Phi^{5/2}}\left(1 + \frac{1}{2}r\frac{\Phi'}{\Phi}\right). \tag{17}$$

Therefore, combining Equations (10) and (17), we obtain

$$\frac{\partial T}{\partial P} = \frac{3L\kappa}{16\pi acGmT^3\Phi} - \frac{3L\kappa\rho\left(1 + \frac{1}{2}r\frac{\Phi'}{\Phi}\right)}{32\pi acr^2T^3P\Phi^{3/2}\Phi'}, \tag{18}$$

and thereby[4]

$$\nabla = \frac{\partial \ln T}{\partial \ln P} = \frac{3L\kappa P}{16\pi acGmT^4\Phi} - \frac{3L\kappa\rho\left(1 + \frac{1}{2}r\frac{\Phi'}{\Phi}\right)}{32\pi acr^2T^4\Phi^{3/2}\Phi'}. \tag{19}$$

This factor ∇ determines the dynamical stability against convective processes in the modified gravity-inspired stars [43]. For a fluid parcel inside a star, if the $\nabla < \nabla_{\mathrm{ad}}$, where ∇_{ad} the adiabatic gradient, it is convectively stable. This is famously known as the Schwarzschild stability condition. For a degenerate gas following $P \propto \rho^{\Gamma}$, $\nabla_{\mathrm{ad}} = 1 - 1/\Gamma$. Near the surface of a WD, generally the EoS is non-relativistic with $\Gamma = 5/3$ (can be obtained from Equation (12) when the non-relativistic limit is taken), and hence in this case, ∇_{ad} turns out to be $2/5$.

We assume Kramer's opacity throughout the paper, i.e., $\kappa = \kappa_0\rho T^{-3.5}$ with $\kappa_0 = 4.34 \times 10^{24} Z(1+X)\,\mathrm{cm^2\,g^{-1}}$, where X is the mass-fraction of hydrogen and Z is the mass-fraction of metals (elements other than hydrogen and helium). Therefore, from Equation (18), we have

$$\frac{\mathrm{d}P}{\mathrm{d}T} = \frac{16\pi acGmT^{6.5}\Phi}{3L\kappa_0\rho} - \frac{32\pi acr^2T^{6.5}P\Phi^{3/2}\Phi'}{3L\kappa_0\rho^2\left(1 + \frac{1}{2}r\frac{\Phi'}{\Phi}\right)}. \tag{20}$$

Near the surface, it is reasonable to assume the ideal gas EoS

$$P = \frac{\rho k_{\mathrm{B}} T}{\mu m_{\mathrm{u}}}, \tag{21}$$

where m_{u} is the atomic mass unit and μ is the mean molecular weight. Moreover, near the surface, m can be replaced by the total mass of the WD, that is, $m(r \approx \mathcal{R}) = M$. Therefore, combining Equations (20) and (21), we obtain

$$P\frac{\mathrm{d}P}{\mathrm{d}T} = \frac{16\pi acGMk_{\mathrm{B}}T^{7.5}\Phi}{3L\kappa_0\mu m_{\mathrm{u}}} - \frac{32\pi ac\mathcal{R}^2k_{\mathrm{B}}^2T^{8.5}\Phi^{3/2}\Phi'}{3L\kappa_0\mu^2m_{\mathrm{u}}^2\left(1 + \frac{1}{2}\mathcal{R}\frac{\Phi'}{\Phi}\right)}. \tag{22}$$

Integrating this equation with the boundary condition $P = 0$ at $T = 0$, we obtain

$$P = \left(\frac{2}{8.5}\frac{4ac\Phi}{3}\frac{4\pi GM}{\kappa_0 L}\right)^{1/2}\left(\frac{k_{\mathrm{B}}}{\mu m_{\mathrm{u}}}\right)^{1/2}T^{4.25} - \left(\frac{4}{9.5}\frac{4ac\Phi^{3/2}\Phi'}{3\left(1 + \frac{1}{2}\mathcal{R}\frac{\Phi'}{\Phi}\right)}\frac{4\pi\mathcal{R}^2}{\kappa_0 L}\right)^{1/2}\left(\frac{k_{\mathrm{B}}}{\mu m_{\mathrm{u}}}\right)T^{4.75}. \tag{23}$$

Equating this pressure with the ideal gas pressure in Equation (21), we obtain

$$\rho = \left(\frac{2}{8.5}\frac{4ac\Phi}{3}\frac{4\pi GM}{\kappa_0 L}\frac{\mu m_{\rm u}}{k_{\rm B}}\right)^{1/2} T^{3.25} - \left(\frac{4}{9.5}\frac{4ac\Phi^{3/2}\Phi'}{3\left(1+\frac{1}{2}\mathcal{R}\frac{\Phi'}{\Phi}\right)}\frac{4\pi\mathcal{R}^2}{\kappa_0 L}\right)^{1/2} T^{3.75}. \tag{24}$$

An order of magnitude calculation for a WD shows that the first term on the right hand side is at least 3–4 orders of magnitude larger than the second one. Thus, for further calculations, we can safely drop the second term. Near the surface, the non-relativistic EoS is followed, and hence in Equation (12), assuming $x_{\rm F} \ll 1$, we obtain $P \approx 1.0 \times 10^{13}(\rho/\mu_{\rm e})^{5/3}$. Substituting P in the ideal gas EoS (21), we obtain the density ρ_* and temperature T_* at the interface radius inside which temperature nearly remains constant. Thus, equating degenerate pressure and ideal gas pressure, we obtain $\rho_* = 2.4 \times 10^{-8}\mu_{\rm e}T_*^{3/2}\,{\rm g\,cm^{-3}}$. Substituting it in Equation (24), and after some rearrangements, luminosity is given by

$$L \approx \left(5.8 \times 10^5\,{\rm erg\,s^{-1}}\right)\frac{\mu}{\mu_{\rm e}^2}\frac{\Phi}{Z(1+X)}\frac{M}{M_\odot}T_*^{3.5}. \tag{25}$$

Assuming a carbon–oxygen WD with the surface containing 90% of helium and 10% other heavy elements, we have $\mu_{\rm e} = 2$, $X = 0$, $Y(\text{helium mass-fraction}) = 0.9$, $Z = 0.1$ and $\mu = 1.4$. Thus, the above luminosity expression reduces to

$$L = \left(2 \times 10^6\,{\rm erg\,s^{-1}}\right)\Phi\frac{M}{M_\odot}T_*^{3.5}. \tag{26}$$

Because luminosity is defined as the energy E radiated per unit time t, it is given by

$$L = -\frac{{\rm d}E}{{\rm d}t}. \tag{27}$$

Defining a constant C such that $CM_\odot = 2 \times 10^6\,{\rm erg\,s^{-1}}$ and substituting L from Equation (26), we obtain

$$-\frac{\rm d}{{\rm d}t}\left(\frac{3k_{\rm B}TM}{2Am_{\rm u}}\right) = CM\Phi T_*^{7/2}. \tag{28}$$

Integrating this equation from time t_0 to the present time t such that T_0 was the initial temperature and T_* is the present temperature of the WD, we obtain

$$\frac{3k_{\rm B}}{5Am_{\rm u}}\left(T_*^{-5/2} - T_0^{-5/2}\right) = C\Phi(t - t_0). \tag{29}$$

Assuming $T_0 \gg T_*$ and defining cooling time as $\tau = t - t_0$ and denoting L_* is the present surface luminosity of the WD, the use of the Equation (26) allows us to approximate the cooling timescale as a function of mass and luminosity for a WD as follows:

$$\tau = \frac{3}{5}\frac{k_{\rm B}T_*M}{Am_{\rm u}L_*} = \frac{3}{5}\frac{k_{\rm B}}{Am_{\rm u}C\Phi T_*^{5/2}} = \frac{3}{5}\frac{k_{\rm B}}{Am_{\rm u}C^{2/7}\Phi^{2/7}}\left(\frac{M}{L_*}\right)^{5/7}. \tag{30}$$

In the next section, we explore the numerical values of this timescale for different WDs. However, it is easy to notice from the above equation that the Palatini gravity effect on the cooling time can be negligible in such a simplified model. It is so because the only direct modification is given by $\Phi^{2/7} \approx 1$, while the indirect one is hidden in the value of the mass M. Notice that WDs with limiting masses larger (or smaller) than the Chandrasekhar limit exist only for $\alpha \neq 0$. See the discussion in the following section as well as Figure 1.

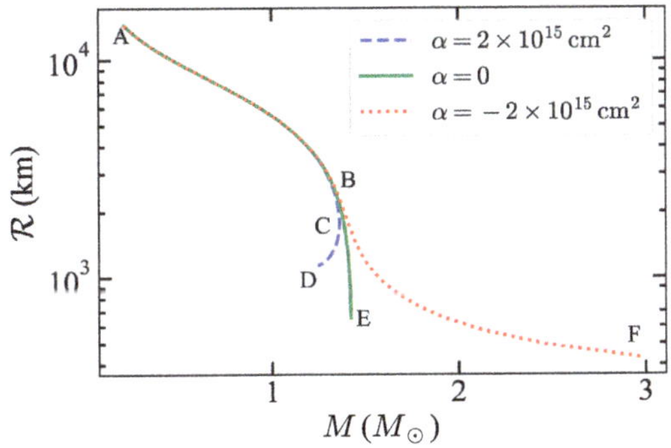

Figure 1. Mass–radius relation of $f(R)$ gravity inspired WDs. The GR case is represented by $\alpha = 0$.

4. Mass–Radius Relation and Cooling Age of White Dwarfs

Let us firstly solve the hydrostatic balance equations to obtain the mass–radius relation of the WDs. We perform these calculations numerically and present them as the mass–radius curves, which will be helpful for analyzing our cooling model. In our previous paper [59], we already obtained these curves for both the Einstein and Jordan frames, and showed that in a WD system, the results do not differ significantly. It is so because in the case of non-relativistic equations, the additional contribution enters via functions of Φ, which does not differ a lot from 1 for the low-density regimes; that is, $\Phi \approx 1$, in the considered Palatini model. However, at high-density regimes, it has some important consequences.

We solve Equations (10) and (11) simultaneously along with the Chandrasekhar EoS with the boundary conditions being $m(r=0)=0$, $\rho(r=0)=\rho_c$, $m(r=\mathcal{R})=M$ and $\rho(r=\mathcal{R})=0$, where M is the mass of a WD with radius $\mathcal{R}$ and central density ρ_c. Figure 1 depicts the mass–radius curves of WDs for different values of α. Note that $\alpha=0$ is the Chandrasekhar result, with mass-limit being approximately $1.44\,M_\odot$. Modified gravity is prominent only in the high-density regime. Hence, the AB branch overlaps with the Chandrasekhar original mass–radius curve for $\alpha = \pm 2\times 10^{15}\,\text{cm}^2$. Beyond point B, the curves differ from the original curve. For $\alpha > 0$, the curve reaches a maximum mass at point C and then turns back. It was already shown that the CD branch is unstable under radial perturbation, and hence point C corresponds to the limiting mass, which turns out to be sub-Chandrasekhar. On the other hand, for $\alpha < 0$, the curve turns and reaches the point F. This BF branch is more stable than the BE branch under radial perturbation, and hence points lying in this branch represent the super-Chandrasekhar WDs. Thus, positive and negative values of the parameter α can respectively explain the existence of sub- and super-Chandrasekhar limiting mass WDs. Let us also note that more significant deviations are expected in the more realistic description, with improved interior's and atmosphere's modeling.

Let us now explore the cooling timescale for sub- and super-Chandrasekhar WDs, and compare them with the conventional ones for the same ρ_c. We know $\Phi = 1 - 16\pi G\rho/c^2$, and hence near the surface, where $\rho \approx 0$, we have $\Phi \approx 1$. Thus, from Equation (30), we have $\tau \propto T_*^{-5/2}$, which means if the surface temperatures of the WDs are almost the same, then their cooling timescale remains indistinguishable irrespective of whether it is a sub- or super- or conventional WD, provided their chemical compositions are identical. Figure 2 shows the variation of τ as a function of M for the luminosity $L_* = 10^{-3}L_\odot$. Since $\tau \propto M^{5/7}$, it is evident that as mass increases, the star's age also does. Thus, it is expected that the cooling timescale is longer for the super-Chandrasekhar WDs and shorter for the sub-Chandrasekhar WDs than conventional ones. However, because $L \propto M$ and τ depend only on the surface temperature in this simple model, we find that the ages are indistinguishable for the same masses for all the three mass–radius tracks. Moreover, Figure 3 shows the cooling timescale as a function of the luminosity for WDs with different masses. We can see that the higher the mass of the WDs, the longer the cooling timescale of the WD. Because

$\tau \propto L^{-5/7}$, this timescale decreases with the increase in the luminosity. This is so because, with the higher luminosity, a WD can quickly release its internal energy and becomes cold. In summary, cooling is faster for heavier WDs with a larger surface luminosity.

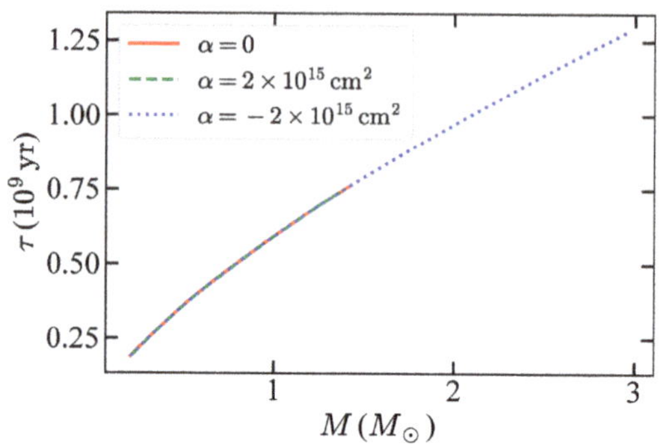

Figure 2. Cooling timescale as a function of the mass of the WDs for $L_* = 10^{-3} L_\odot$. Notice that modified gravity allows the white dwarf stars to have higher masses than the Chandrasekhar limit.

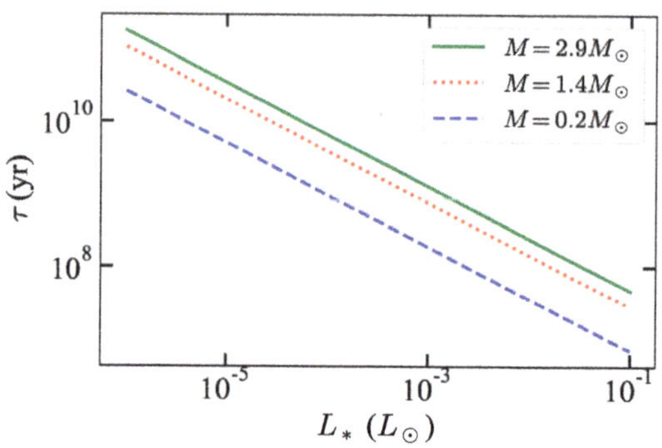

Figure 3. Cooling timescale as a function of the luminosity for WDs with different masses. Note that 2.9 $M_\odot$ WD is possible only when $\alpha < 0$. It is worth mentioning that the α value is not required explicitly to obtain this plot, because from Figure 2, it is evident that for a specific luminosity, cooling timescale depends only on the mass, and not the radius. Moreover, the 0.2 $M_\odot$ WD can be obtained regardless of α being positive, negative or zero, and hence it is not necessary to precisely specify a value of α in order to determine its cooling timescale.

5. Conclusions

The present study has been designed to determine the effect of modifications to the Newtonian hydrostatic equilibrium equation, introduced by the Palatini $f(R)$ gravity, on the cooling model of white dwarf stars. Since it is the first study related to these topics, it was limited to the set of assumptions and simplifications. Firstly, we have assumed that such an object can be modeled as a spherical-symmetric ball of degenerate electrons. It allows us to describe its interior with an analytical equation of state. The second approximation is related to the opacity, which we have assumed to be Kramer's one. Because it also has an analytical form, we could carry our calculations without the need of simulations, or without more advanced numerical methods. Subsequently using the temperature gradient expression, we were able to integrate the hydrostatic balance equation. Applying the ideal gas equation of state, which describes well the behavior of the particles on the surface in our modeling, we have derived the luminosity as a function of mass, temperature and chemical composition in the presence of modified gravity, which is given by Equation (25). As already mentioned, all those considerations are made in the non-relativistic limit of the theory. Moreover, in such a simple model, it is assumed that there is no energy generation in the stars' interior (such as, the latent heat, which is a product of the crystallization process, or further gravitational contraction). Therefore, the white dwarf radiates the whole stored energy away, which means that it is cooling down with time. This fact allows us to write the expression for the cooling timescale as a function of the star's mass and its luminosity as provided by Equation (30).

Recalling our previous results on the stability with similar assumptions on the white dwarf model allowed us to immediately notice that we do not expect any significant differences in our cooling for the low-density regime [59]. We have found there that $\alpha > 0$ gives the sub-Chandrasekhar limiting mass white dwarfs and $\alpha < 0$ gives the super-Chandrasekhar ones. It seems that among the various theories and models, modified gravity provides a mechanism, which is able to justify both mass ranges self-consistently, which can further explain the origins of peculiar under- and and over-luminous type Ia supernovae. Let us now discuss how the previous work helps us interpret the current findings.

In this paper, we have found that the cooling timescale is longer for the super-Chandrasekhar white dwarfs than the Newtonian or sub-Chandrasekhar ones. Note that independently of the mass of the dead star, varying the parameter in the given range does not provide any significant differences in the cooling timescale. However, the very massive ones can exist only in modified gravity[5]. It can be surprising because in the less compact objects, such as the main-sequence stars, brown dwarfs and even planets [43,44,47], the effects of Palatini gravity manifest in the stellar evolution, even in the cooling processes [45,46]. This result has an explanation: the physics we used in our modeling is too poor with respect to the models of non-relativistic stars. In the mentioned works, the authors employed much more realistic equations of state, which included information not only on the electron degeneracy, but also how it evolves with time. This is also a theory-dependent process [45,98]. These models also took into account the corrections related to the finite temperature of the gas, mixtures of highly degenerate fluid with the ideal one as well as ionized elements, and phase transitions in some cases [99]. Including the missing physics in order to describe a realistic white dwarf, we also expect to see differences with respect to the Newtonian or GR description. As the most important improvement we would like to analyze is taking into account crystallization processes and effects of gravity on matter properties [82]. This is, however, out of the scope of this paper. We will present results along these lines in the near future.

Author Contributions: A.W. conceptualized the project; S.K. and L.S. carried out the formal analysis and original writing; A.W. supervised the project and edited the draft. All authors have read and agreed to the published version of the manuscript.

Funding: This research received no external funding.

Institutional Review Board Statement: Not applicable.

Informed Consent Statement: Not applicable.

Data Availability Statement: The data presented in this study are available on a reasonable request to the corresponding author.

Acknowledgments: S.K. would like to acknowledge support from the South African Research Chairs Initiative of the Department of Science and Technology and the National Research Foundation. A.W. was supported by the EU through the European Regional Development Fund CoE program TK133 "The Dark Side of the Universe".

Conflicts of Interest: The authors declare no conflict of interest.

Notes

1 Although, one can rewrite the equations as second-order differential equations for the metric, and the additional one for the curvature scalar, which arises to a dynamical field in this theory.

2 This fact arises as a conclusion from the field equations and it will be evident in the next section, whereas in the metric formalism, the connection is assumed to be the Levi–Civita one of the spacetime metric.

3 To see the relativistic hydrostatic equilibrium equation for Palatini $f(R)$ gravity, see [88].

4 Let us notice that this form differs slightly from the one obtained in [43]. This is so because of different assumptions on the matter description and its behavior under the conformal transformation.

5 Of course, there are other processes, such as magnetic field [91,92], noncommutative geometry [93,94], ungravity effect [95], consequence of total lepton number violation [96], generalized Heisenberg uncertainty principle [97] and many more which can also explain massive white dwarfs but they are failing to explain the sub-Chandrasekhar mass-limit.

References

1. Carroll, S.M. *Spacetime and Geometry: An Introduction to General Relativity*; Cambridge University Press: Cambridge, UK, 2004. [CrossRef]
2. Poisson, E.; Will, C.M. *Gravity: Newtonian, Post-Newtonian, Relativistic*; Cambridge University Press: Cambridge, UK, 2014.
3. Perlmutter, S.; Aldering, G.; Goldhaber, G.; Knop, R.A.; Nugent, P.; Castro, P.G.; Deustua, S.; Fabbro, S.; Goobar, A.; Groom, D.E.; et al. Measurements of Ω and Λ from 42 High-Redshift Supernovae. *Astron. J.* **1999**, *517*, 565–586. [CrossRef]
4. Riess, A.G.; Filippenko, A.V.; Challis, P.; Clocchiatti, A.; Diercks, A.; Garnavich, P.M.; Gilliland, R.L.; Hogan, C.J.; Jha, S.; Kirshner, R.P.; et al. Observational Evidence from Supernovae for an Accelerating Universe and a Cosmological Constant. *Astron. J.* **1998**, *116*, 1009–1038. [CrossRef]
5. Guth, A.H. Inflationary universe: A possible solution to the horizon and flatness problems. *Phys. Rev. D* **1981**, *23*, 347–356. [CrossRef]
6. Coles, P. Book Review: Cosmological inflation and large-scale structure/Cambridge University Press, 2000. *Astron. Geophys.* **2003**, *44*, 32.
7. Huterer, D.; Turner, M.S. Prospects for probing the dark energy via supernova distance measurements. *Phys. Rev. D* **1999**, *60*, 081301. [CrossRef]
8. Battaner, E.; Florido, E. The Rotation Curve of Spiral Galaxies and its Cosmological Implications. *Fund. Cosmic. Phys.* **2000**, *21*, 1–154.
9. Gaskins, J.M. A review of indirect searches for particle dark matter. *Contemp. Phys.* **2016**, *57*, 496–525. . 00107514.2016.1175160. [CrossRef]
10. Taoso, M.; Bertone, G.; Masiero, A. Dark matter candidates: A ten-point test. *J. Cosmol. Astropart. Phys.* **2008**, *2008*, 22. [CrossRef]
11. Bullock, J.S.; Boylan-Kolchin, M. Small-Scale Challenges to the ΛCDM Paradigm. *arXiv* **2017**, arXiv:1707.04256.
12. Del Popolo, A.; Le Delliou, M. Small Scale Problems of the ΛCDM Model: A Short Review. *Galaxies* **2017**, *5*, 17. [CrossRef]
13. Guzmán, F.S.; Matos, T.; Villegas, H.B. Scalar fields as dark matter in spiral galaxies: Comparison with experiments. *Astron. Nachrichten* **1999**, *320*, 97. [CrossRef]
14. Gutiérrez-Luna, E.; Carvente, B.; Jaramillo, V.; Barranco, J.; Escamilla-Rivera, C.; Espinoza, C.; Mondragón, M.; Núñez, D. Scalar field dark matter with two components: Combined approach from particle physics and cosmology. *Phys. Rev. D* **2022**, *105*, 083533. [CrossRef]
15. Borowiec, A.; Kamionka, M.; Kurek, A.; Szydłowski, M. Cosmic acceleration from modified gravity with Palatini formalism. *J. Cosmol. Astropart. Phys.* **2012**, *2012*, 27. [CrossRef]
16. Koivisto, T.; Kurki-Suonio, H. Cosmological perturbations in the Palatini formulation of modified gravity. *Class. Quantum Gravity* **2006**, *23*, 2355. [CrossRef]
17. Flanagan, E.E. Palatini form of 1/R gravity. *Phys. Rev. Lett.* **2004**, *92*, 071101. [CrossRef] [PubMed]
18. Fay, S.; Tavakol, R.; Tsujikawa, S. $f(R)$ gravity theories in Palatini formalism: Cosmological dynamics and observational constraints. *Phys. Rev. D* **2007**, *75*, 063509. [CrossRef]
19. Sotiriou, T.P. Unification of inflation and cosmic acceleration in the Palatini formalism. *Phys. Rev. D* **2006**, *73*, 063515. [CrossRef]
20. Dioguardi, C.; Racioppi, A.; Tomberg, E. Slow-roll inflation in Palatini $F(R)$ gravity. *J. High Energy Phys.* **2022**, *6*, 106. [CrossRef]
21. Gialamas, I.D.; Karam, A.; Racioppi, A. Dynamically induced Planck scale and inflation in the Palatini formulation. *J. Cosmol. Astropart. Phys.* **2020**, *11*, 14. [CrossRef]
22. Dimopoulos, K.; Karam, A.; Sánchez López, S.; Tomberg, E. Palatini R^2 Quintessential Inflation. *J. Cosmol. Astropart. Phys.* **2022**, *2022*, 76. [CrossRef]
23. Szydłowski, M.; Stachowski, A.; Borowiec, A.; Wojnar, A. Do sewn up singularities falsify the Palatini cosmology? *Eur. Phys. J. C* **2016**, *76*, 567. [CrossRef]
24. Szydłowski, M.; Stachowski, A.; Borowiec, A. Emergence of running dark energy from polynomial $f(R)$ theory in Palatini formalism. *Eur. Phys. J. C* **2017**, *77*, 603. [CrossRef]
25. Stachowski, A.; Szydłowski, M.; Borowiec, A. Starobinsky cosmological model in Palatini formalism. *Eur. Phys. J. C* **2017**, *77*, 406. [CrossRef]
26. Allemandi, G.; Borowiec, A.; Francaviglia, M.; Odintsov, S.D. Dark energy dominance and cosmic acceleration in first order formalism. *Phys. Rev. D* **2005**, *72*, 063505. [CrossRef]

27. Allemandi, G.; Borowiec, A.; Francaviglia, M. Accelerated cosmological models in Ricci squared gravity. *Phys. Rev. D* **2004**, *70*, 103503. [CrossRef]
28. Allemandi, G.; Borowiec, A.; Francaviglia, M. Accelerated cosmological models in first order nonlinear gravity. *Phys. Rev. D* **2004**, *70*, 043524. [CrossRef]
29. Capozziello, S.; de Laurentis, M. Extended Theories of Gravity. *Phys. Rep.* **2011**, *509*, 167–321. [CrossRef]
30. Saridakis, E.N.; Lazkoz, R.; Salzano, V.; Moniz, P.V.; Capozziello, S.; Beltrán Jiménez, J.; De Laurentis, M.; Olmo, G.J. *Modified Gravity and Cosmology; An Update by the CANTATA Network*; Springer: Cham, Switzerland, 2021. [CrossRef]
31. Heisenberg, L. Scalar-vector-tensor gravity theories. *J. Cosmol. Astropart. Phys.* **2018**, *2018*, 54. [CrossRef]
32. Ferraro, R.; Fiorini, F. Modified teleparallel gravity: Inflation without an inflaton. *Phys. Rev. D* **2007**, *75*, 084031. [CrossRef]
33. Buchdahl, H.A. Non-linear Lagrangians and cosmological theory. *Mon. Not. R. Astron. Soc.* **1970**, *150*, 1–8. [CrossRef]
34. Sotiriou, T.P.; Faraoni, V. $f(R)$ theories of gravity. *Rev. Mod. Phys.* **2010**, *82*, 451–497. [CrossRef]
35. Sotiriou, T.P. Constraining $f(R)$ gravity in the Palatini formalism. *Class. Quantum Gravity* **2006**, *23*, 1253–1267. [CrossRef]
36. Nojiri, S.; Odintsov, S.D. Modified Gravity with ln R Terms and Cosmic Acceleration. *Gen. Relativ. Gravit.* **2004**, *36*, 1765–1780. [CrossRef]
37. Amarzguioui, M.; Elgarøy, Ø.; Mota, D.F.; Multamäki, T. Cosmological constraints on $f(R)$ gravity theories within the Palatini approach. *Astron. Astrophys.* **2006**, *454*, 707–714. [CrossRef]
38. Borowiec, A.; Stachowski, A.; Szydłowski, M.; Wojnar, A. Inflationary cosmology with Chaplygin gas in Palatini formalism. *J. Cosmol. Astropart. Phys.* **2016**, *2016*, 40. [CrossRef]
39. Järv, L.; Karam, A.; Kozak, A.; Lykkas, A.; Racioppi, A.; Saal, M. Equivalence of inflationary models between the metric and Palatini formulation of scalar-tensor theories. *Phys. Rev. D* **2020**, *102*, 044029. [CrossRef]
40. Teppa Pannia, F.A.; García, F.; Perez Bergliaffa, S.E.; Orellana, M.; Romero, G.E. Structure of compact stars in R-squared Palatini gravity. *Gen. Relativ. Gravit.* **2017**, *49*, 25. [CrossRef]
41. Herzog, G.; Sanchis-Alepuz, H. Neutron stars in Palatini $R + \alpha R^2$ and $R + \alpha R^2 + \beta Q$ theories. *Eur. Phys. J. C* **2021**, *81*, 888. [CrossRef]
42. Olmo, G.J.; Rubiera-Garcia, D.; Wojnar, A. Minimum main sequence mass in quadratic Palatini $f(R)$ gravity. *Phys. Rev. D* **2019**, *100*, 044020. [CrossRef]
43. Wojnar, A. Early evolutionary tracks of low-mass stellar objects in modified gravity. *Phys. Rev. D* **2020**, *102*, 124045. [CrossRef]
44. Wojnar, A. Lithium abundance is a gravitational model dependent quantity. *Phys. Rev. D* **2021**, *103*, 044037. [CrossRef]
45. Benito, M.; Wojnar, A. Cooling process of brown dwarfs in Palatini $f(\mathcal{R})$ gravity. *Phys. Rev. D* **2021**, *103*, 064032. [CrossRef]
46. Wojnar, A. Jupiter and jovian exoplanets in Palatini $f(\overline{R})$ gravity. *Phys. Rev. D* **2021**, *104*, 104058. [CrossRef]
47. Wojnar, A. Giant planet formation in Palatini gravity. *Phys. Rev. D* **2022**, *105*, 124053. [CrossRef]
48. Kozak, A.; Wojnar, A. Metric-affine gravity effects on terrestrial exoplanet profiles. *Phys. Rev. D* **2021**, *104*, 084097. [CrossRef]
49. Kozak, A.; Wojnar, A. Non-homogeneous exoplanets in metric-affine gravity. *Int. J. Geom. Meth. Mod. Phys.* **2022**, *19*, 2250157. [CrossRef]
50. Kozak, A.; Wojnar, A. Interiors of Terrestrial Planets in Metric-Affine Gravity. *Universe* **2021**, *8*, 3. [CrossRef]
51. Olmo, G.J.; Rubiera-Garcia, D.; Wojnar, A. Stellar structure models in modified theories of gravity: Lessons and challenges. *Phys. Rept.* **2020**, *876*, 1–75. [CrossRef]
52. Wojnar, A. Stellar and substellar objects in modified gravity. *arXiv* **2022**, arXiv: 2205.08160.
53. Shapiro, S.L.; Teukolsky, S.A. *Black Holes, White Dwarfs and Neutron Stars: The Physics of Compact Objects*; Wiley: Hoboken, NJ, USA, 2008.
54. Lauffer, G.R.; Romero, A.D.; Kepler, S.O. New full evolutionary sequences of H- and He-atmosphere massive white dwarf stars using MESA. *Mon. Not. R. Astron. Soc.* **2018**, *480*, 1547–1562. [CrossRef]
55. Kalita, S.; Mukhopadhyay, B. Modified Einstein's gravity to probe the sub- and super-Chandrasekhar limiting mass white dwarfs: A new perspective to unify under- and over-luminous type Ia supernovae. *J. Cosmol. Astropart. Phys.* **2018**, *2018*, 7. [CrossRef]
56. Kalita, S.; Mukhopadhyay, B. Gravitational Wave in $f(R)$ Gravity: Possible Signature of Sub- and Super-Chandrasekhar Limiting-mass White Dwarfs. *Astron. J.* **2021**, *909*, 65. [CrossRef]
57. Wojnar, A. White dwarf stars in modified gravity. *Int. J. Geom. Methods Mod. Phys.* **2021**, *18*, 2140006. [CrossRef]
58. Kalita, S.; Sarmah, L. Weak-field limit of $f(R)$ gravity to unify peculiar white dwarfs. *Phys. Lett. B* **2022**, *827*, 136942. [CrossRef]
59. Sarmah, L.; Kalita, S.; Wojnar, A. Stability criterion for white dwarfs in Palatini $f(R)$ gravity. *Phys. Rev. D* **2022**, *105*, 024028. [CrossRef]
60. Das, U.; Mukhopadhyay, B. Modified Einstein's gravity as a possible missing link between sub- and super-Chandrasekhar type Ia supernovae. *J. Cosmol. Astropart. Phys.* **2015**, *2015*, 45. [CrossRef]
61. Das, U.; Mukhopadhyay, B. Imprint of modified Einstein's gravity on white dwarfs: Unifying Type Ia supernovae. *Int. J. Mod. Phys. D* **2015**, *24*, 1544026. [CrossRef]
62. García-Berro, E.; Hernanz, M.; Isern, J.; Mochkovitch, R. The rate of change of the gravitational constant and the cooling of white dwarfs. *Mon. Not. R. Astron. Soc.* **1995**, *277*, 801–810. [CrossRef]
63. Althaus, L.G.; Córsico, A.H.; Torres, S.; Lorén-Aguilar, P.; Isern, J.; García-Berro, E. The evolution of white dwarfs with a varying gravitational constant. *Astron. Astrophys.* **2011**, *527*, A72. [CrossRef]

64. Córsico, A.H.; Althaus, L.G.; García-Berro, E.; Romero, A.D. An independent constraint on the secular rate of variation of the gravitational constant from pulsating white dwarfs. *J. Cosmol. Astropart. Phys.* **2013**, *2013*, 32. [CrossRef]
65. Benvenuto, O.G.; Garcia-Berro, E.; Isern, J. Asteroseismological bound on $\dot{G}/G$ from pulsating white dwarfs. *Phys. Rev. D* **2004**, *69*, 082002. [CrossRef]
66. Saltas, I.D.; Sawicki, I.; Lopes, I. White dwarfs and revelations. *J. Cosmol. Astropart. Phys.* **2018**, *2018*, 28. [CrossRef]
67. Liu, H.L.; Lü, G.L. Properties of white dwarfs in Einstein-Λ gravity. *J. Cosmol. Astropart. Phys.* **2019**, *2019*, 40. [CrossRef]
68. Carvalho, G.A.; Lobato, R.V.; Moraes, P.H.R.S.; Arbañil, J.D.V.; Otoniel, E.; Marinho, R.M.; Malheiro, M. Stellar equilibrium configurations of white dwarfs in the $f(R,T)$ gravity. *Eur. Phys. J. C* **2017**, *77*, 871. [CrossRef]
69. Eslam Panah, B.; Liu, H.L. White dwarfs in de Rham-Gabadadze-Tolley like massive gravity. *Phys. Rev. D* **2019**, *99*, 104074. [CrossRef]
70. Biesiada, M.; Malec, B. A new white dwarf constraint on the rate of change of the gravitational constant. *Mon. Not. R. Astron. Soc.* **2004**, *350*, 644–648. [CrossRef]
71. Benvenuto, O.; Althaus, L.; Torres, D.F. Evolution of white dwarfs as a probe of theories of gravitation: The case of Brans—Dicke. *Mon. Not. R. Astron. Soc.* **1999**, *305*, 905–919. [CrossRef]
72. Bienaymé, O.; Turon, C.; Isern, J.; GarcíaBerro, E.; Salaris, M. White dwarfs as tools of fundamental physics: The gravitational constant case. *Eur. Astron. Soc. Publ. Ser.* **2002**, *2*, 123–128.
73. Babichev, E.; Koyama, K.; Langlois, D.; Saito, R.; Sakstein, J. Relativistic stars in beyond Horndeski theories. *Class. Quantum Gravity* **2016**, *33*, 235014. [CrossRef]
74. Crisostomi, M.; Lewandowski, M.; Vernizzi, F. Vainshtein regime in scalar-tensor gravity: Constraints on degenerate higher-order scalar-tensor theories. *Phys. Rev. D* **2019**, *100*, 024025. [CrossRef]
75. Wibisono, C.; Sulaksono, A. Information-entropic method for studying the stability bound of nonrelativistic polytropic stars within modified gravity theories. *Int. J. Mod. Phys. D* **2018**, *27*, 1850051. [CrossRef]
76. Biesiada, M.; Malec, B. White dwarf cooling and large extra dimensions. *Phys. Rev. D* **2002**, *65*, 043008. [CrossRef]
77. Panotopoulos, G.; Lopes, I. White dwarf cooling via gravity portals. *Phys. Rev. D* **2020**, *101*, 023017. [CrossRef]
78. Isern, J.; García-Berro, E. White dwarf stars as particle physics laboratories. *Nucl. Phys. B Proc. Suppl.* **2003**, *114*, 107–110. [CrossRef]
79. Banerjee, S.; Shankar, S.; Singh, T.P. Constraints on modified gravity models from white dwarfs. *J. Cosmol. Astropart. Phys.* **2017**, *2017*, 4. [CrossRef]
80. De Felice, A.; Tsujikawa, S. $f(R)$ Theories. *Living Rev. Relativ.* **2010**, *13*, 3.
81. Chavanis, P.H. Statistical mechanics of self-gravitating systems in general relativity: I. The quantum Fermi gas. *Eur. Phys. J. Plus* **2020**, *135*, 1–79. [CrossRef]
82. Wojnar, A. Fermi gas and modified gravity. *arXiv* **2022**, arXiv:2208.04023.
83. Borowiec, A.; Ferraris, M.; Francaviglia, M.; Volovich, I. Universality of Einstein equations for the Ricci squared Lagrangians. *Class. Quant. Grav.* **1998**, *15*, 43–55. [CrossRef]
84. Beltrán Jiménez, J.; Delhom, A. Ghosts in metric-affine higher order curvature gravity. *Eur. Phys. J. C* **2019**, *79*, 656. [CrossRef]
85. Afonso, V.I.; Olmo, G.J.; Rubiera-Garcia, D. Mapping Ricci-based theories of gravity into general relativity. *Phys. Rev. D* **2018**, *97*, 021503. [CrossRef]
86. Afonso, V.I.; Olmo, G.J.; Orazi, E.; Rubiera-Garcia, D. Mapping nonlinear gravity into General Relativity with nonlinear electrodynamics. *Eur. Phys. J. C* **2018**, *78*, 1–11. [CrossRef] [PubMed]
87. Afonso, V.I.; Olmo, G.J.; Orazi, E.; Rubiera-Garcia, D. Correspondence between modified gravity and general relativity with scalar fields. *Phys. Rev. D* **2019**, *99*, 044040. [CrossRef]
88. Wojnar, A. On stability of a neutron star system in Palatini gravity. *Eur. Phys. J. C* **2018**, *78*, 421. [CrossRef]
89. Chandrasekhar, S. The highly collapsed configurations of a stellar mass (Second paper). *Mon. Not. R. Astron. Soc.* **1935**, *95*, 207–225.
90. Mestel, L. On the theory of white dwarf stars. I. The energy sources of white dwarfs. *Mon. Not. R. Astron. Soc.* **1952**, *112*, 583. [CrossRef]
91. Das, U.; Mukhopadhyay, B. New Mass Limit for White Dwarfs: Super-Chandrasekhar Type Ia Supernova as a New Standard Candle. *Phys. Rev. Lett.* **2013**, *110*, 071102. [CrossRef]
92. Kalita, S.; Mukhopadhyay, B. Continuous gravitational wave from magnetized white dwarfs and neutron stars: Possible missions for LISA, DECIGO, BBO, ET detectors. *Mon. Not. R. Astron. Soc.* **2019**, *490*, 2692–2705. [CrossRef]
93. Kalita, S.; Mukhopadhyay, B.; Govindarajan, T.R. Significantly super-Chandrasekhar mass-limit of white dwarfs in noncommutative geometry. *Int. J. Mod. Phys. D* **2021**, *30*, 2150034. [CrossRef]
94. Kalita, S.; Govindarajan, T.R.; Mukhopadhyay, B. Super-Chandrasekhar limiting mass white dwarfs as emergent phenomena of noncommutative squashed fuzzy spheres. *Int. J. Mod. Phys. D* **2021**, *30*, 2150101. [CrossRef]
95. Bertolami, O.; Mariji, H. White dwarfs in an ungravity-inspired model. *Phys. Rev. D* **2016**, *93*, 104046. [CrossRef]
96. Belyaev, V.B.; Ricci, P.; Šimkovic, F.; Adam, J.; Tater, M.; Truhlík, E. Consequence of total lepton number violation in strongly magnetized iron white dwarfs. *Nucl. Phys. A* **2015**, *937*, 17–43. [CrossRef]
97. Ong, Y.C. Generalized uncertainty principle, black holes, and white dwarfs: A tale of two infinities. *J. Cosmol. Astropart. Phys.* **2018**, *9*, 15. [CrossRef]

98. Kozak, A.; Soieva, K.; Wojnar, A. Cooling process of substellar objects in scalar-tensor gravity. *arXiv* **2022**, arXiv:2205.12812.
99. Auddy, S.; Basu, S.; Valluri, S.R. Analytic models of brown dwarfs and the substellar mass limit. *Adv. Astron.* **2016**, *2016*, 5743272. [CrossRef]

universe MDPI

Article

Bulk Viscous Fluid in Symmetric Teleparallel Cosmology: Theory versus Experiment

Raja Solanki [1], Simran Arora [1], Pradyumn Kumar Sahoo [1,*] and Pedro H. R. S. Moraes [2]

1 Department of Mathematics, Birla Institute of Technology and Science-Pilani, Hyderabad Campus, Hyderabad 500078, India
2 Centro de Ciências Naturais e Humans (CCNH), Universidade Federal do ABC (UFABC), Avenida dos Estados 5001, Santo André 09210-580, SP, Brazil
* Correspondence: pksahoo@hyderabad.bits-pilani.ac.in

Abstract: The standard formulation of General Relativity Theory, in the absence of a cosmological constant, is unable to explain the responsible mechanism for the observed late-time cosmic acceleration. On the other hand, by inserting the cosmological constant in Einstein's field equations, it is possible to describe the cosmic acceleration, but the cosmological constant suffers from an unprecedented fine-tuning problem. This motivates one to modify Einstein's spacetime geometry of General Relativity. The $f(Q)$ modified theory of gravity is an alternative theory to General Relativity, where the non-metricity scalar Q is the responsible candidate for gravitational interactions. In the present work, we consider a Friedmann–Lemâitre–Robertson–Walker cosmological model dominated by bulk viscous cosmic fluid in $f(Q)$ gravity with the functional form $f(Q) = \alpha Q^n$, where α and n are free parameters of the model. We constrain our model with the Pantheon supernovae dataset of 1048 data points, the Hubble dataset of 31 data points, and the baryon acoustic oscillations dataset consisting of 6 data points. We find that our $f(Q)$ cosmological model efficiently describes the observational data. We present the evolution of our deceleration parameter with redshift, and it properly predicts a transition from decelerated to accelerated phases of the universe's expansion. Furthermore, we present the evolution of density, bulk viscous pressure, and the effective equation of state parameter with redshift. Those show that bulk viscosity in a cosmic fluid is a valid candidate to acquire the negative pressure to drive the cosmic expansion efficiently. We also examine the behavior of different energy conditions to test the viability of our cosmological $f(Q)$ model. Furthermore, the statefinder diagnostics are also investigated in order to distinguish among different dark energy models.

Keywords: cosmological constant; $f(Q)$ gravity; energy conditions; statefinder diagnostic

1. Introduction

The acceleration of the universe's expansion is one of the most active discoveries of modern cosmology. Observational studies include Type Ia supernovae [1,2], large-scale structure [3,4], baryon acoustic oscillations [5,6], and cosmic microwave background radiation [7,8]. The reason behind the late-time acceleration is a mystery. Several models hypothesize the existence of a component called *dark energy* , which makes up around 70% of the entire universe and could possess the feature of speeding up the universe's expansion.

The cosmological constant Λ in the General Relativity (GR) field equations plays the role of dark energy, i.e., a fluid with constant energy density and high negative pressure. There are some issues with the cosmological constant model, such as the cosmic coincidence problem [9], which is the fact that the density of non-relativistic matter and dark energy is the same order today. A more delicate issue surrounding the cosmological constant is the so-called *cosmological constant problem*, which is the high discrepancy between the astronomically observed value of Λ [1,2] and the particle physics theoretically predicted value of the quantum vacuum energy [10].

With the main purpose of solving the above cosmological issues, dynamical (time-varying) dark energy models such as the Chaplygin gas model [11,12], k-essence [13,14], quintessence [15,16], and decaying vacuum models [17–20] have been proposed for some time in the literature.

Modified theories of gravity have also been intensively investigated to understand the origin of the cosmic acceleration, as well as to address the cosmological constant model problems. It is possible to predict late-time cosmic acceleration by modifying GR action. Some possibilities can be seen within the $f(R)$ [21–23], $f(G)$ [24,25], $f(R,\mathcal{T})$ [26], and $f(T)$ [27–30] theories of gravitation, with R, G, $\mathcal{T}$, and T being, respectively, the Ricci, Gauss–Bonnet, energy–momentum, and torsion scalars.

In the present article, we will work with the recently introduced $f(Q)$ theory of gravity [31], for which Q is the non-metricity scalar, to be presented below. The $f(R)$ gravity is an extension of GR that describes the spacetime geometry by the non-vanishing curvature in the absence of torsion and non-metricity, whereas the $f(\mathcal{T})$ gravity generalizes the teleparallel equivalent of GR (TEGR), in which the gravitational interactions are attributed to the non-zero torsion with vanishing curvature and non-metricity. Finally, the $f(Q)$ theory of gravity, which will be described in Section 2, is an extension of the symmetric teleparallel equivalent of GR (STEGR), where the complete charge of gravitational interactions attributed to the non-metricity term with zero curvature and torsion. The non-metricity formulation was discussed earlier by Hehl and Ne'eman (see References [32–35]). The symmetric teleparallel gravity is proven in the so-called coincidence gauge by imposing that the connection is symmetric [36]. The Weyl geometry is also observed to be a particular example of the Weyl–Cartan geometry, in which torsion disappears. The non-metricity is interpreted as a massless spin three-field in the case of symmetric connections [37,38]. Furthermore, it is noted in the literature that, due to the appearance of non-metricity, the light cone structure is not preserved during parallel transport [39]. Further, fermions are an issue in TEGR because they couple to the axial contorsion of the Weitzenbock connection. This difficulty is eliminated in STEGR since Dirac fermions only couple to the completely antisymmetric component of the affine connection and are unaffected by any disformation piece.

Although recently proposed, the $f(Q)$ gravity theory already presents some interesting and valuable applications in the literature [40–49]. The first cosmological solutions in the $f(Q)$ gravity appear in References [50,51], while the $f(Q)$ cosmography and energy conditions can, respectively, be seen in [52,53].

Here, we are going to consider the $f(Q)$ cosmology in the presence of a viscous fluid. When a cosmic fluid expands too fast, the recovering of thermodynamic equilibrium generates an effective pressure. The high viscosity in a cosmic fluid is the manifestation of such an effective pressure [54,55].

Basically, there are two viscosity coefficients, namely shear viscosity and bulk viscosity. Shear viscosity is related to velocity gradients in the fluid, and by considering the universe as described by the homogeneous and isotropic Friedmann–Lemaître–Robertson–Walker (FLRW) background, it can be omitted. Anyhow, by dropping the assumption of the homogeneity and isotropy of the universe, several cosmological models with shear viscosity fluid have been constructed, as one can check, for instance, in [56–59]. On the other hand, bulk viscosity, which we are going to consider here, introduces damping associated with volumetric straining. To become familiar with bulk viscous fluid cosmological models, one can check References [60–67]. Moreover, some interesting applications of bulk viscous cosmology in black holes are presented in [68,69].

Researchers examine dark energy reconstruction with numerous observations as the data increase. The majority of studies have been concentrated on observable evidence from Type Ia supernovae, cosmic microwave background, and baryon acoustic oscillations (BAOs), which are known to be helpful in constraining cosmological models. The Hubble parameter dataset shows the intricate structure of the expansion of the universe. The ages of the most massive and slowly evolving galaxies offer direct measurements of the Hubble parameter $H(z)$ at various redshifts z, resulting in the development of a new form of standard cosmological probe [70].

In our present work, we include 31 measurements of Hubble expansion spanned using the differential age method [71] and BAO data consisting of six points [72]. Scolnic et al. published a large Type Ia supernovae sample named Pantheon, with 1048 points and covering the redshift range $0.01 < z < 2.3$ [73]. Our analysis uses the $H(z)$, BAO, and Pantheon samples to constrain the cosmological model.

This work aims to describe the recently observed late-time acceleration with the help of the bulk viscosity of the cosmic fluid (without including any dark energy component) in the framework of the $f(Q)$ theory of gravity. The manuscript is organized as follows: in Section 2, we discuss the $f(Q)$ gravity formalism. In Section 3, we describe the FLRW universe dominated by bulk viscous non-relativistic matter and derive the expression for the Hubble parameter and the deceleration parameter. Further, in Section 4, we analyze the observational data to find the best-fit ranges for the parameters using the Hubble dataset containing 31 points, the BAO sample, and the Pantheon dataset of 1048 samples. Moreover, we analyze the behavior of different cosmological parameters such as Hubble, density, effective pressure, deceleration parameter, and effective equation of state (EoS) parameter. In Section 5, we investigate the consistency of our bulk viscous fluid model by analyzing the different energy conditions. In Section 6, we analyze the behavior of statefinder parameters on the values constrained by the observational data to differentiate between dark energy models. Lastly, we discuss our results in Section 7.

2. Fundamental Formulations in $f(Q)$ Gravity

In $f(Q)$ gravity theory, the spacetime is established with the help of the non-metricity and symmetric teleparallelism condition, i.e., $\nabla_\alpha g_{\mu\nu} \neq 0$ and $R^{\rho}_{\sigma\mu\nu} = 0$ in an torsionless environment. Hence, the associated affine connection is given by

$$\Upsilon^{\alpha}{}_{\mu\nu} = \Gamma^{\alpha}{}_{\mu\nu} + L^{\alpha}{}_{\mu\nu}, \tag{1}$$

with

$$\Gamma^{\alpha}{}_{\mu\nu} \equiv \frac{1}{2} g^{\alpha\lambda}(g_{\mu\lambda,\nu} + g_{\lambda\nu,\mu} - g_{\mu\nu,\lambda}), \tag{2}$$

$$L^{\alpha}{}_{\mu\nu} \equiv \frac{1}{2}(Q^{\alpha}{}_{\mu\nu} - Q_{\mu}{}^{\alpha}{}_{\nu} - Q_{\nu}{}^{\alpha}{}_{\mu}), \tag{3}$$

being the Christoffel symbols and the distortion tensor, respectively, where

$$Q_{\alpha\mu\nu} \equiv \nabla_\alpha g_{\mu\nu}, \tag{4}$$

is the non-metricity tensor.

The non-metricity tensor given by Equation (4) has the following two traces,

$$Q_\alpha = Q_\alpha{}^{\mu}{}_{\mu}, \tag{5}$$

$$\tilde{Q}_\alpha = Q^{\mu}{}_{\alpha\mu}. \tag{6}$$

In addition, the superpotential tensor or non-metricity conjugate is given by

$$\begin{aligned} 4P^{\lambda}{}_{\mu\nu} &= -Q^{\lambda}{}_{\mu\nu} + Q_{\mu}{}^{\lambda}{}_{\nu} + Q_{\nu}{}^{\lambda}{}_{\mu} + (Q^{\lambda} - \tilde{Q}^{\lambda})g_{\mu\nu} \\ &- \frac{1}{2}(\delta^{\lambda}{}_{\mu}Q_{\nu} + \delta^{\lambda}{}_{\nu}Q_{\mu}). \end{aligned} \tag{7}$$

Then, the trace of the non-metricity tensor can be acquired as [74]

$$Q = -Q_{\lambda\mu\nu}P^{\lambda\mu\nu}. \tag{8}$$

The definition of the energy–momentum tensor for matter is

$$\mathcal{T}_{\mu\nu} = \frac{-2}{\sqrt{-g}} \frac{\delta(\sqrt{-g}L_m)}{\delta g^{\mu\nu}}. \tag{9}$$

Furthermore, one can obtain the following relation between the curvature tensor $R^{\rho}_{\sigma\mu\nu}$ and $\mathring{R}^{\rho}_{\sigma\mu\nu}$ corresponding to the connection Υ and Γ as

$$R^{\rho}_{\sigma\mu\nu} = \mathring{R}^{\rho}_{\sigma\mu\nu} + \mathring{\nabla}_{\mu} L^{\rho}_{\nu\sigma} - \mathring{\nabla}_{\nu} L^{\rho}_{\mu\sigma} + L^{\rho}_{\mu\lambda} L^{\lambda}_{\nu\sigma} - L^{\rho}_{\nu\lambda} L^{\lambda}_{\mu\sigma} \tag{10}$$

and so,

$$R_{\sigma\nu} = \mathring{R}_{\sigma\nu} + \frac{1}{2}\mathring{\nabla}_{\nu} Q_{\sigma} + \mathring{\nabla}_{\rho} L^{\rho}_{\nu\sigma} - \frac{1}{2} Q_{\lambda} L^{\lambda}_{\nu\sigma} - L^{\rho}_{\nu\lambda} L^{\lambda}_{\rho\sigma} \tag{11}$$

$$R = \mathring{R} + \mathring{\nabla}_{\lambda} Q^{\lambda} - \mathring{\nabla}_{\lambda} \tilde{Q}^{\lambda} - \frac{1}{4} Q_{\lambda} Q^{\lambda} + \frac{1}{2} Q_{\lambda} \tilde{Q}^{\lambda} - L_{\rho\nu\lambda} L^{\lambda\rho\nu} \tag{12}$$

Now, the connection (1) can be parameterized as [31]

$$\Upsilon^{\alpha}{}_{\mu\beta} = \frac{\partial x^{\alpha}}{\partial \xi^{\rho}} \partial_{\mu} \partial_{\beta} \xi^{\rho}. \tag{13}$$

Here, $\xi^{\alpha} = \xi^{\alpha}(x^{\mu})$ is an invertible relation. Hence, it is always possible to find a coordinate system so that the connection $\Upsilon^{\alpha}_{\mu\nu}$ vanishes. This situation is called the coincident gauge, and the covariant derivative ∇_{α} reduces to the partial one ∂_{α}. However, in any other coordinate system in which this affine connection does not vanish, the metric evolution will be affected and result in a completely different theory [75,76]. Thus, in the coincident gauge coordinate, we have

$$Q_{\alpha\mu\nu} = \partial_{\alpha} g_{\mu\nu} \tag{14}$$

while, in an arbitrary coordinate system,

$$Q_{\alpha\mu\nu} = \partial_{\alpha} g_{\mu\nu} - 2\Upsilon^{\lambda}_{\alpha(\mu} g_{\nu)\lambda}. \tag{15}$$

Locally, GR does not distinguish between gravitational and inertial effects; however, by invoking frame fields, it is possible to covariantly define gravitational energy in the teleparallel approach [77]. The canonical frame is identified in the absence of both curvature and torsion and the canonical coordinates in the absence of inertial effects. New physics can emerge from such a formalism. Symmetric teleparallel gravity is broadly derived in three gravity theories based on the coordinate transformations [78]. In this case, we use the spatially flat case, i.e., the $f(Q)$ gravity, whose field equations are much easier to understand.

The $f(Q)$ gravity is described by the action [31]:

$$S = \int \frac{1}{2} f(Q) \sqrt{-g} d^4x + \int L_m \sqrt{-g} d^4x, \tag{16}$$

in which $f(Q)$ is an arbitrary function of the non-metricity, $g = \det g_{\mu\nu}$, and L_m is the Lagrangian density of matter. One can check in Reference [31] that the functional form $f(Q) = -Q$ corresponds to the symmetric teleparallel equivalent to General Relativity (STEGR) limit.

The gravitational field equations obtained by varying the action (16) with respect to the metric are

$$\frac{2}{\sqrt{-g}} \nabla_{\lambda}(\sqrt{-g} f_Q P^{\lambda}{}_{\mu\nu}) + \frac{1}{2} g_{\mu\nu} f + f_Q (P_{\mu\lambda\beta} Q_{\nu}{}^{\lambda\beta} - 2 Q_{\lambda\beta\mu} P^{\lambda\beta}{}_{\nu}) = -\mathcal{T}_{\mu\nu}, \tag{17}$$

in which we define $f_Q = df/dQ$.

3. The Cosmological Model

We consider the flat FLRW metric for our analysis [79], such that:

$$ds^2 = -dt^2 + a^2(t)(dx^2 + dy^2 + dz^2). \tag{18}$$

In Equation (18) above, $a(t)$ is the cosmic scale factor, as usual.

From now on, we will fix the coincident gauge so that connection becomes trivial and the metric is only a fundamental variable.

We are going to assume a bulk viscous fluid, and below, we present some considerations favoring such an assumption.

Firstly, considering bulk viscosity in a fluid can be seen as an attempt to refine its description, minimizing its ideal properties. This can be checked, for instance, in the stellar astrophysics realistic models in References [80–82].

Under the conditions of spatial homogeneity and isotropy (which refers to the cosmological principle, as one can check, for instance, in Reference [83]), the bulk viscous pressure is the unique admissible dissipative phenomenon. In a gas dynamical model, the existence of an effective bulk pressure can be traced back to a non-standard self-interacting force on the particles of the gas [63]. The bulk viscosity contributes negatively to the total pressure, as one can check, for instance, in References [84–86].

Due to spatial isotropy, the bulk viscous pressure is the same in all spatial directions and, hence, proportional to the volume expansion $\theta = 3H$, with $H = \dot{a}/a$ being the Hubble parameter and the dot representing the time derivative.

The effective pressure of the cosmic fluid becomes [87–89]

$$\bar{p} = p - \zeta\theta = p - 3\zeta H, \tag{19}$$

in which p is the usual pressure and $\zeta > 0$ is the bulk viscosity coefficient, which we will assume as a free parameter of the model.

The corresponding energy–momentum tensor is given by

$$\mathcal{T}_{\mu\nu} = (\rho + \bar{p})u_\mu u_\nu + \bar{p}g_{\mu\nu}, \tag{20}$$

in which ρ is the matter–energy density and the four-velocity u^μ is such that its components are $u^\mu = (1,0,0,0)$.

The relation between normal pressure and matter–energy density follows [90] $p = (\gamma - 1)\rho$, with γ being a constant lying in the range $0 \leq \gamma \leq 2$. Then, the effective equation of state for the bulk viscous fluid is given by the following:

$$\bar{p} = (\gamma - 1)\rho - 3\zeta H. \tag{21}$$

The Friedmann-like equations for our $f(Q)$ gravitational model are obtained from the substitution of Equations (18)–(21) into Equation (11) and read as follows (check, for instance, References [74,91]):

$$3H^2 = \frac{1}{2f_Q}\left(-\rho + \frac{f}{2}\right), \tag{22}$$

$$\dot{H} + \left(3H + \frac{\dot{f}_Q}{f_Q}\right)H = \frac{1}{2f_Q}\left(\bar{p} + \frac{f}{2}\right). \tag{23}$$

In particular, for $f(Q) = -Q$, we retrieve the usual Friedmann equations [91], as expected, since, as we have mentioned above, this particular choice for the functional form of the function $f(Q)$ is the STEGR limit of the theory.

For our investigation of the bulk viscosity fluid cosmological model, we consider the following $f(Q)$ functional form:

$$f(Q) = \alpha Q^n, \tag{24}$$

with $\alpha \neq 0$ and constant n. This particular functional form for $f(Q)$ was motivated by a polynomial form applied, for instance, in Reference [53]. Note that the parameter n considered in our model is not the spectral index; rather, it is a free model parameter.

For the above choice of the $f(Q)$ function (Equation (24)), we rewrite Equations (22)–(23) as follows

$$\rho = \alpha 6^n \left(\frac{1}{2} - n\right) H^{2n}, \tag{25}$$

$$\dot{H} + \frac{3}{2n} H^2 = \frac{6^{1-n}\bar{p}}{2\alpha n(2n-1)} H^{2(1-n)}, \tag{26}$$

in which, for the former, we isolated ρ.

From Equations (26) and (21), we have the following:

$$\dot{H} + \frac{3\gamma}{2n} H^2 = -\frac{6^{2-n}\zeta}{4\alpha n(2n-1)} H^{3-2n}. \tag{27}$$

Now, we replace the term d/dt by $d/dlna$ via the expression $d/dt = Hd/dlna$, such that Equation (27) becomes

$$\frac{dH}{dlna} + \frac{3\gamma}{2n} H = -\frac{6^{2-n}\zeta}{4\alpha n(2n-1)} H^{2(1-n)}. \tag{28}$$

The integration of Equation (28) yields the following solution:

$$H(a) = \left\{ (H_0 a^{-\frac{3\gamma}{2n}})^{2n-1} + \frac{6^{1-n}\zeta}{\gamma\alpha(2n-1)^2} [a^{-\frac{3\gamma(2n-1)}{2n}} - 1] \right\}^{\frac{1}{2n-1}}, \tag{29}$$

with H_0 being a constant of integration to be found below.

We obtain the Hubble parameter in terms of redshift by using relation [83] $a(t) = 1/(1+z)$ in Equation (30). By making $z = 0$ in (30), we find that $H(0) = H_0$. The deceleration parameter is defined as $q = -\ddot{a}a/\dot{a}^2 = -\ddot{a}/(H^2 a)$. Henceforth, from Equation (29), we have

$$H(z) = \left\{ [H_0(1+z)^{\frac{3\gamma}{2n}}]^{2n-1} + \frac{6^{1-n}\zeta}{\gamma\alpha(2n-1)^2} [(1+z)^{\frac{3\gamma(2n-1)}{2n}} - 1] \right\}^{\frac{1}{2n-1}}, \tag{30}$$

$$q(z) = \frac{3}{2n} \left\{ \frac{\zeta}{\alpha 6^{n-1}(2n-1)\left\{ [H_0(1+z)^{\frac{3\gamma}{2n}}]^{2n-1} + \frac{6^{1-n}\zeta}{\gamma\alpha(2n-1)^2}[(1+z)^{\frac{3\gamma(2n-1)}{2n}} - 1] \right\}} + \gamma \right\} - 1. \tag{31}$$

4. Observational Constraints

To examine the observational features of our cosmological model, we used the cosmic Hubble and supernovae observations. We used the 31 points of the Hubble datasets, the 6 points of the BAO datasets, and the 1048 points from the Pantheon supernovae samples. We applied the Bayesian analysis and likelihood function along with the Markov chain Monte Carlo (MCMC) method in the `emcee` python library [92].

4.1. Hubble Datasets

The Hubble parameter can be expressed as $H(z) = -dz/[dt(1+z)]$. As dz is derived from a spectroscopic survey, the model-independent value of the Hubble parameter may be calculated by measuring the quantity dt.

We incorporated the set of 31 data points that are measured from the differential age approach [93] to avoid extra correlation with the BAO data. The mean values of the model parameters ζ, α, γ and n are calculated using the chi-squared function as follows:

$$\chi^2_H(\zeta, \alpha, \gamma, n) = \sum_{k=1}^{31} \frac{[H_{th}(\zeta, \alpha, \gamma, n, z_k) - H_{obs}(z_k)]^2}{\sigma^2_{H(z_k)}}. \tag{32}$$

Here, H_{th} is the Hubble parameter value predicted by the model, H_{obs} represents its observed value, and the standard error in the observed value of H is $\sigma_{H(z_k)}$.

Figure 1 shows the error bar plot of the considered model and the ΛCDM or standard cosmological model, with cosmological constant density parameter $\Omega_{\Lambda_0} = 0.7$, matter density parameter $\Omega_{m_0} = 0.3$, and $H_0 = 69$ km/s/Mpc.

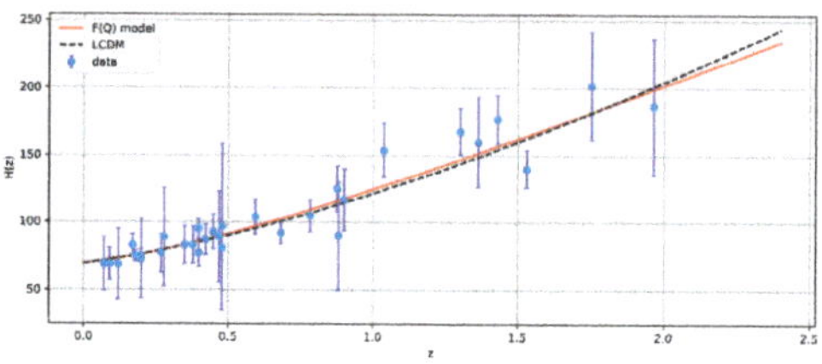

Figure 1. The error bar plot of H versus z for the considered $f(Q)$ model. The solid red line is the curve for the $f(Q)$ model, whereas the black dotted line represents the ΛCDM model. The blue dots depict the 31 points of the Hubble data.

4.2. BAO Datasets

The BAO distance dataset, which includes the 6dFGS, SDSS, and WiggleZ surveys, comprises BAO measurements at six different redshifts in Table 1. The characteristic scale of BAO is ruled by the sound horizon r_s at the epoch of photon decoupling z_*, which is given by the following relation:

$$r_s(z_*) = \frac{c}{\sqrt{3}} \int_0^{\frac{1}{1+z_*}} \frac{da}{a^2 H(a) \sqrt{1 + (3\Omega_{b0}/4\Omega_{\gamma 0})a}}. \tag{33}$$

Here, Ω_{b0} and $\Omega_{\gamma 0}$ correspond to the present densities of baryons and photons, respectively.

The following relations are used in the BAO measurements:

$$\triangle\theta = \frac{r_s}{d_A(z)}, \tag{34}$$

$$d_A(z) = \int_0^z \frac{dz'}{H(z')}, \tag{35}$$

$$\triangle z = H(z) r_s, \tag{36}$$

where $\triangle\theta$ represents the measured angular separation, d_A is the angular diameter distance, and $\triangle z$ represents the measured redshift separation of the BAO feature in the two-point correlation function of the galaxy distribution on the sky along the line of sight.

In this work, the BAO datasets of six points for $d_A(z_*)/D_V(z_{BAO})$ is taken from References [5,6,72,94–96], where the redshift at the epoch of photon decoupling is taken as $z_* \approx 1091$ and $d_A(z)$ is the co-moving angular diameter distance together with the dilation scale $D_V(z) = \left[d_A(z)^2 z / H(z)\right]^{1/3}$. The chi-squared function for the BAO dataset is taken to be [96]

$$\chi^2_{BAO} = X^T C^{-1} X, \tag{37}$$

$$X = \begin{pmatrix} \frac{d_A(z_\star)}{D_V(0.106)} - 30.95 \\ \frac{d_A(z_\star)}{D_V(0.2)} - 17.55 \\ \frac{d_A(z_\star)}{D_V(0.35)} - 10.11 \\ \frac{d_A(z_\star)}{D_V(0.44)} - 8.44 \\ \frac{d_A(z_\star)}{D_V(0.6)} - 6.69 \\ \frac{d_A(z_\star)}{D_V(0.73)} - 5.45 \end{pmatrix}, \tag{38}$$

Table 1. Values of $d_A(z_*)/D_V(z_{BAO})$ for distinct values of z_{BAO}.

z_{BAO}	0.106	0.2	0.35	0.44	0.6	0.73
$\frac{d_A(z_*)}{D_V(z_{BAO})}$	30.95 ± 1.46	17.55 ± 0.60	10.11 ± 0.37	8.44 ± 0.67	6.69 ± 0.33	5.45 ± 0.31

The inverse covariance matrix C^{-1} is defined in [96]

$$C^{-1} = \begin{pmatrix} 0.48435 & -0.101383 & -0.164945 & -0.0305703 & -0.097874 & -0.106738 \\ -0.101383 & 3.2882 & -2.45497 & -0.0787898 & -0.252254 & -0.2751 \\ -0.164945 & -2.454987 & 9.55916 & -0.128187 & -0.410404 & -0.447574 \\ -0.0305703 & -0.0787898 & -0.128187 & 2.78728 & -2.75632 & 1.16437 \\ -0.097874 & -0.252254 & -0.410404 & -2.75632 & 14.9245 & -7.32441 \\ -0.106738 & -0.2751 & -0.447574 & 1.16437 & -7.32441 & 14.5022 \end{pmatrix}. \tag{39}$$

4.3. Pantheon Datasets

Scolnic et al. [73] put together the Pantheon samples consisting of 1048 Type Ia supernovae in the redshift range $0.01 < z < 2.3$. The PanSTARSS1 Medium Deep Survey, SDSS, SNLS, and numerous low-z and HST samples contribute to it. The empirical relation used to calculate the distance modulus of SNeIa from the observation of light curves is given by $\mu = m_B^* + \alpha X_1 - \beta C - M_B + \Delta_M + \Delta_B$, where X_1 and C denote the stretch and color correction parameters, respectively [73,97], m_B^* represents the observed apparent magnitude, and M_B is the absolute magnitude in the B-band for SNeIa. The parameters α and β are the two nuisance parameters describing the luminosity stretch and luminosity color relations, respectively. Further, the distance correction factor is Δ_M, and Δ_B is a distance correction based on the predicted biases from simulations.

The nuisance parameters in the Tripp formula [98] were reconstructed using a novel technique called BEAMS with bias corrections [99,100], and the observed distance modulus was reduced to the difference between the corrected apparent magnitude m_B and the absolute magnitude M_B, which is $\mu = m_B - M_B$. We shall avoid marginalizing over the nuisance parameters α and β, but marginalize over the Pantheon data for M_B. Hence, we ignored the values of α and β for the present investigation of the model.

The luminosity distance read as

$$D_L(z) = (1+z)\int_0^z \frac{cdz'}{H(z')}, \tag{40}$$

with c being the speed of light.

The χ^2 function for Type Ia supernovae is obtained by correlating the theoretical distance modulus:

$$\mu(z) = 5log_{10}D_L(z) + \mu_0, \tag{41}$$

$$\mu_0 = 5log(1/H_0Mpc) + 25, \tag{42}$$

such that

$$\chi^2_{SN}(\zeta, \alpha, \gamma, n) = \sum_{k=1}^{1048} \frac{[\mu_{obs}(z_k) - \mu_{th}(\zeta, \alpha, \gamma, n, z_k)]^2}{\sigma^2(z_k)}, \tag{43}$$

where μ_{th} is the theoretical value of the distance modulus and μ_{obs} is the observed value, whereas $\sigma^2(z_k)$ is the standard error in the observed value.

Figure 2 shows the error bar plot of $\mu(z)$ for the considered model and the ΛCDM or standard cosmological model, with cosmological constant density parameter $\Omega_{\Lambda_0} = 0.7$, matter density parameter $\Omega_{m_0} = 0.3$, and $H_0 = 69$ km/s/Mpc.

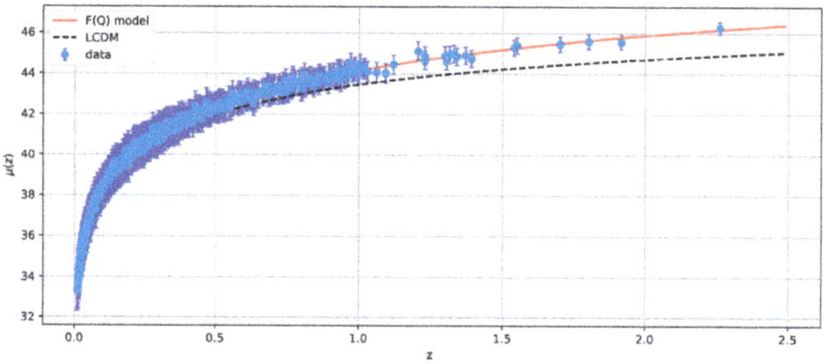

Figure 2. The error bar plot of $\mu(z)$ versus z for the considered $f(Q)$ model. The solid red line is the curve for the $f(Q)$ model, whereas the black dotted line represents the ΛCDM model. The blue dots depict the 1048 points of the Pantheon data.

The $1-\sigma$ and $2-\sigma$ likelihood contours for the model parameters using the Hubble, BAO, and Pantheon datasets is presented in Figure 3. The obtained best-fit values are presented in Table 2.

Table 2. The table shows the constraints on the model parameters corresponding to the Hubble, BAO, and Pantheon datasets.

Data	ζ	α	γ	n
Hubble	0.69 ± 0.11	$-0.0145^{+0.0077}_{-0.0039}$	$1.39^{+0.13}_{-0.11}$	0.982 ± 0.024
BAO	0.69 ± 0.13	$-0.0141^{+0.0055}_{-0.0035}$	$1.157^{+0.064}_{-0.071}$	$0.997^{+0.010}_{-0.014}$
Pantheon	0.69 ± 0.13	$-0.0118^{+0.0062}_{-0.0028}$	1.36 ± 0.14	0.994 ± 0.025

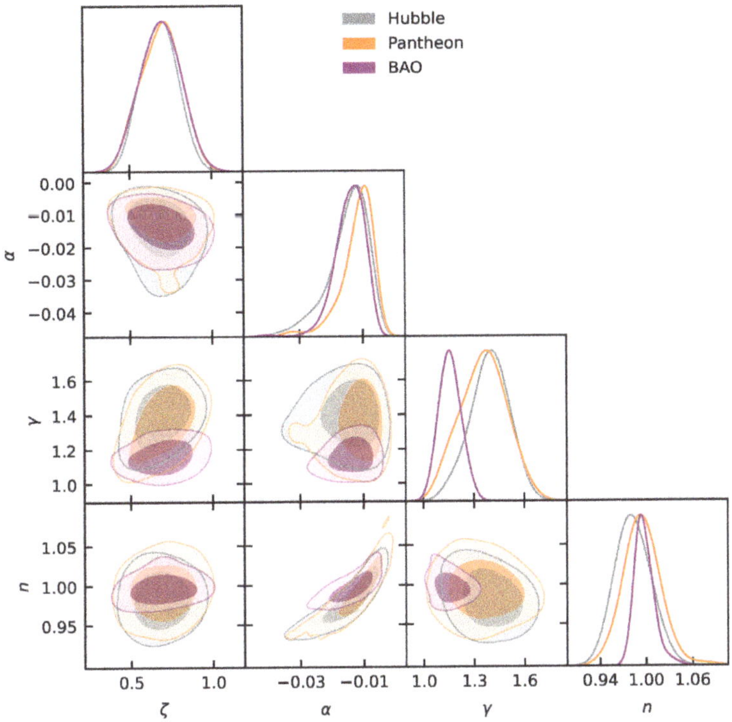

Figure 3. The $1-\sigma$ and $2-\sigma$ likelihood contours for the model parameters using the Hubble, BAO, and the Pantheon datasets.

4.4. Cosmological Parameters

The evolution of the Hubble parameter, deceleration parameter, energy density, pressure with the bulk viscosity, and the effective EoS parameter for the redshift range $-1 < z < 8$ are presented below, in order to test the late-time cosmic expansion history and the future of the expanding universe [101]. In order to do so, we used the set of values constrained by the Hubble, BAO, and Pantheon datasets for the model parameters.

From Figure 4, it is clear that the deceleration parameter shows the transition from a decelerated ($q > 0$) to an accelerated ($q < 0$) phase of the universe's expansion for the constrained values of the model parameters. The transition redshift is $z_t \approx 0.236$, $z_t \approx 0.691$, and $z_t \approx 0.428$ corresponding to the Hubble, BAO, and Pantheon datasets, respectively. The present value of the deceleration parameter is, respectively, $q_0 = -0.211$, $q_0 = -0.384$, and $q_0 = -0.346$.

From Figures 5 and 6, it is clear that the Hubble and density parameters show the positive behavior for all the constrained values of the model parameters, which is expected.

Figure 7 indicates that the bulk viscous cosmic fluid exhibits, for lower redshifts, the negative pressure that makes the bulk viscosity a viable candidate to drive the cosmic acceleration. Furthermore, the effective EoS parameter presented in Figure 8 indicates that the cosmic viscous fluid behaves like quintessence dark energy. The present values of the EoS parameter corresponding to the Hubble, BAO, and Pantheon samples are $\omega_0 = -0.48$, $\omega_0 = -0.59$, and $\omega_0 = -0.56$.

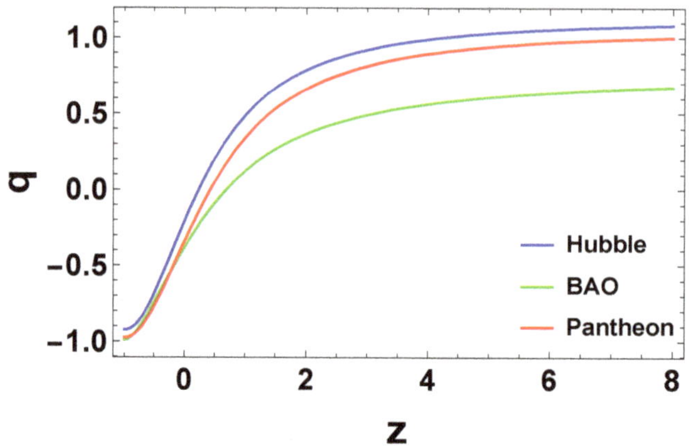

Figure 4. Profile of the deceleration parameter for the given model corresponding to the values of the parameters constrained by the Hubble, BAO, and Pantheon data point sets.

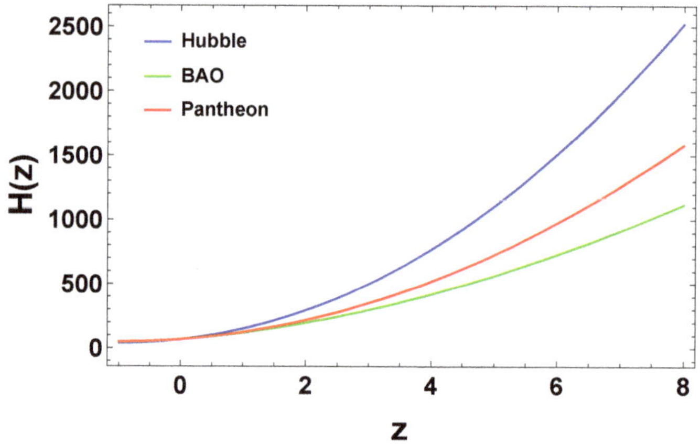

Figure 5. Profile of the Hubble parameter for the given model corresponding to the values of the parameters constrained by the Hubble, BAO, and Pantheon data point sets.

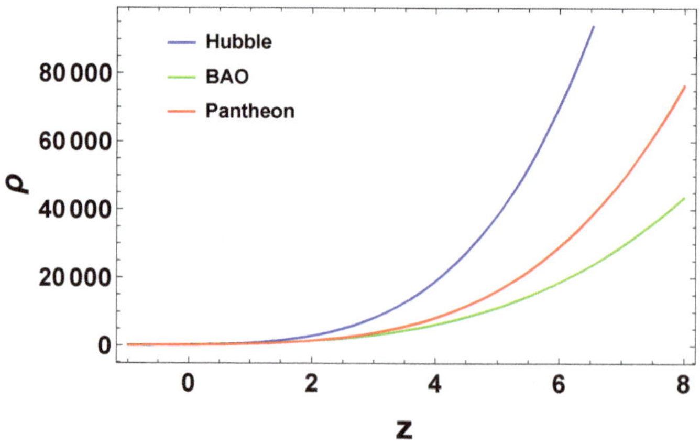

Figure 6. Profile of the density parameter for the given model corresponding to the values of the parameters constrained by the Hubble, BAO, and Pantheon data point sets.

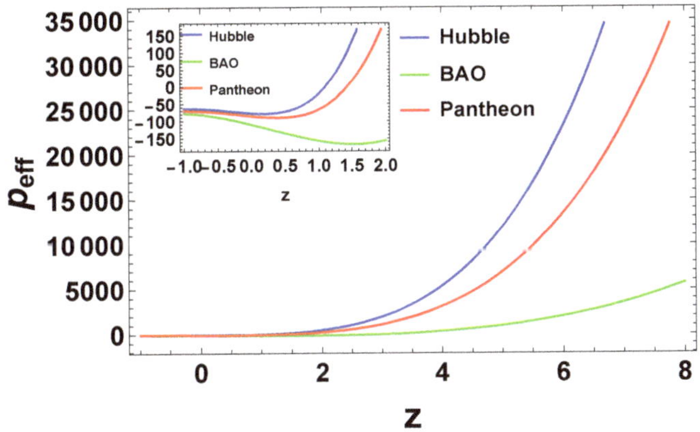

Figure 7. Profile of the pressure for the given model corresponding to the values of the parameters constrained by the Hubble, BAO, and Pantheon data point sets.

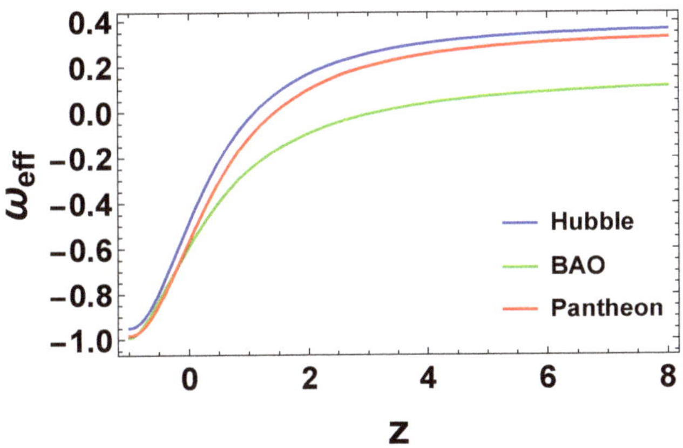

Figure 8. Profile of the EoS parameter for the given model corresponding to the values of the parameters constrained by the Hubble, BAO, and Pantheon data point sets.

5. Energy Conditions

In the present section, we are going to construct the energy conditions for the solutions of the present model. The energy conditions are the relations applied to the matter energy–momentum tensor with the purpose of satisfying positive energy. The energy conditions are derived from the Raychaudhuri equation and are written as [102]:

- **Null energy condition (NEC):** $\rho_{eff} + p_{eff} \geq 0$;
- **Weak energy condition (WEC):** $\rho_{eff} \geq 0$ and $\rho_{eff} + p_{eff} \geq 0$;
- **Dominant energy condition (DEC):** $\rho_{eff} \pm p_{eff} \geq 0$;
- **Strong energy condition (SEC):** $\rho_{eff} + 3p_{eff} \geq 0$.

with ρ_{eff} being the effective energy density.

In Figures 9 and 10, it is evident that the NEC and DEC exhibit the positive behavior for all the constrained values of the model parameters. As the WEC is the combination of energy density and the NEC, we conclude that the NEC, DEC, and WEC are all satisfied in the entire domain of redshift. Figure 11 indicates that the SEC exhibits, for lower redshifts, the negative behavior that is related to cosmic acceleration [53,103]. This is also reflected in the deceleration parameter behavior in Figure 4.

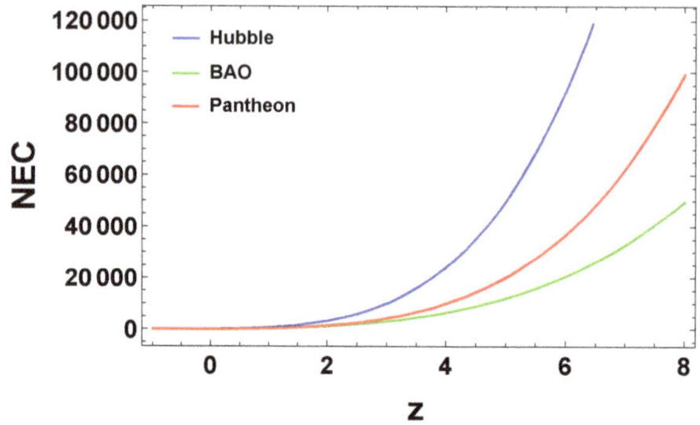

Figure 9. Profile of the null energy condition for the given model corresponding to the values of the parameters constrained by the Hubble, BAO, and Pantheon data point sets.

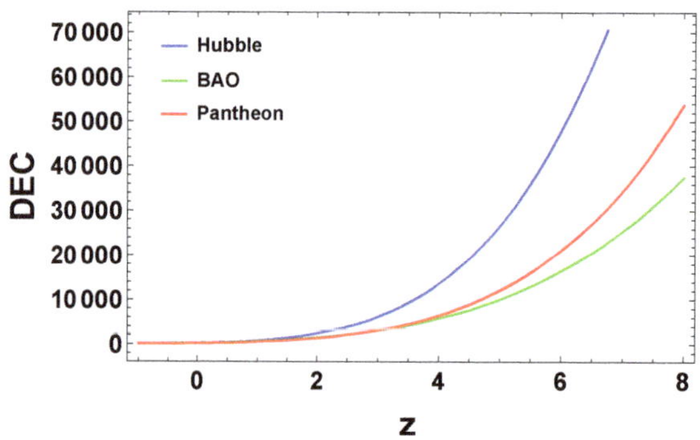

Figure 10. Profile of the dominant energy condition for the given model corresponding to the values of the parameters constrained by the Hubble, BAO, and Pantheon data point sets.

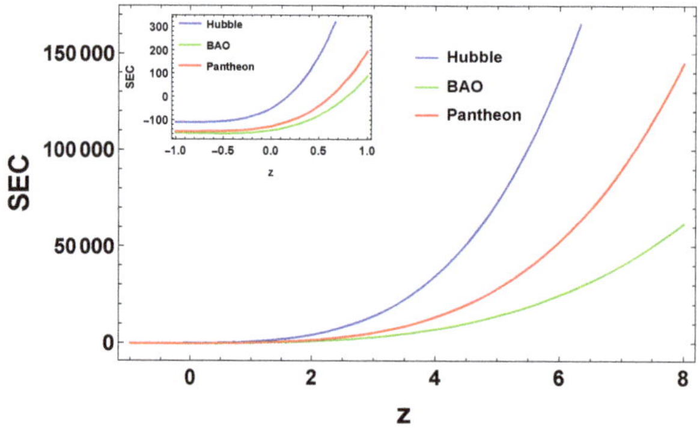

Figure 11. Profile of the strong energy condition for the given model corresponding to the values of the parameters constrained by the Hubble, BAO, and Pantheon data point sets.

6. Statefinder Analysis

The cosmological constant Λ suffers from two major drawbacks, namely the aforementioned cosmological constant and cosmic coincidence problems. To surmount these problems, dynamic models of dark energy have been introduced in the literature, as we have also mentioned in the Introduction. To discriminate between these time-varying dark energy models, an appropriate tool is required. In this direction, V. Sahni et al. introduced a new pair of geometrical parameters known as *statefinder parameters* (r, s) [104]. The statefinder parameters are defined as

$$r = \frac{\dddot{a}}{aH^3}, \tag{44}$$

$$s = \frac{(r-1)}{3(q-\frac{1}{2})}. \tag{45}$$

The parameter r can be rewritten as $r = 2q^2 + q - \frac{\dot{q}}{H}$.

For different values of the statefinder pair (r, s), it represents the following dark energy models:
$r = 1, s = 0$ represents the ΛCDM model;
$r > 1, s < 0$ represents the Chaplygin gas model;
$r < 1, s > 0$ represents the quintessence model.

In Figures 12 and 13, we plot the $s - r$ and $q - r$ diagrams for our cosmological model by taking the values of the parameters constrained by the Hubble, BAO, and Pantheon datasets.

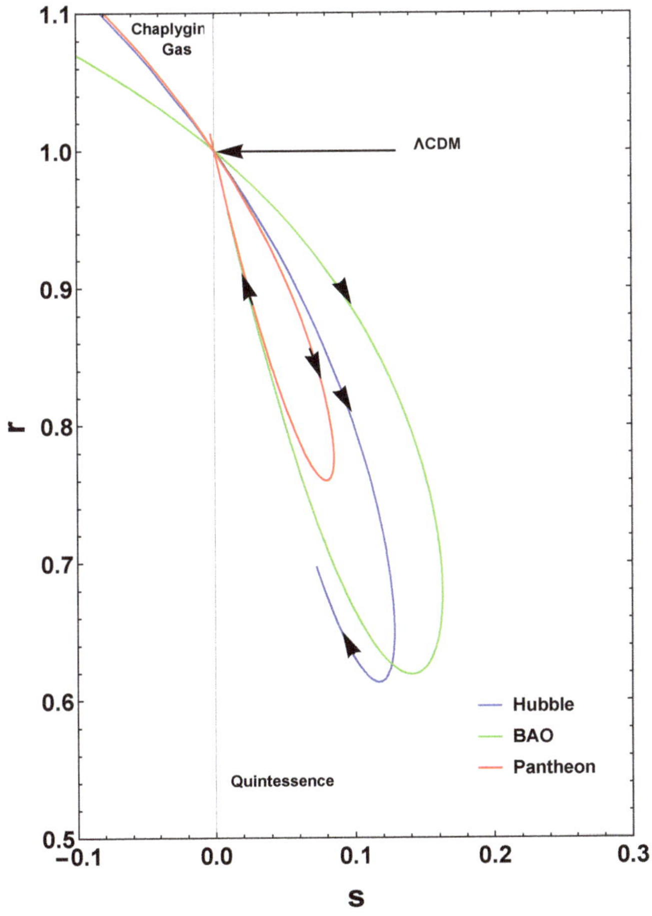

Figure 12. Plot of the trajectories in the $r - s$ plane for the given cosmological model corresponding to the parameter values constrained by the Hubble, BAO, and Pantheon datasets.

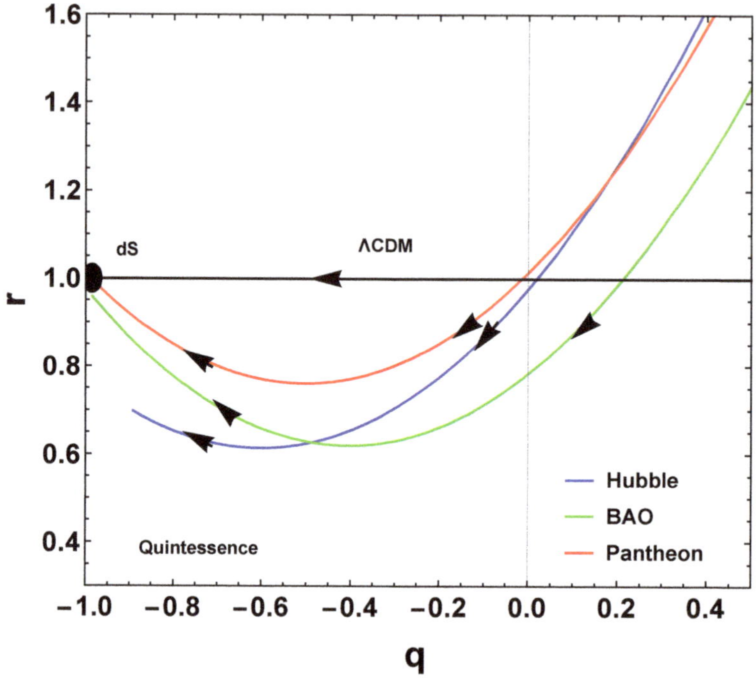

Figure 13. Plot of the trajectories in the $q - r$ plane for the given cosmological model corresponding to the parameter values constrained by the Hubble, BAO, and the Pantheon datasets.

Figures 12 and 13 show that our bulk viscous model lies in the quintessence region. Furthermore, the evolutionary trajectories of our model depart from the ΛCDM point. The present values of the statefinder parameters corresponding to the values of the model parameters constrained by the Hubble, BAO, and Pantheon samples are $r_0 = 0.828$ and $s_0 = 0.08$, $r_0 = 0.592$ and $s_0 = 0.15$, and $r_0 = 0.773$ and $s_0 = 0.089$, respectively.

7. Discussions and Conclusions

In the present section, we will discuss the results obtained in Sections 4–6 for the bulk viscous symmetric teleparallel cosmological model here developed and presented.

Cosmology has been on the agenda mainly for two reasons: dark energy and dark matter. While dark energy was deeply discussed throughout the paper, dark matter is predicted within the ΛCDM model as a sort of matter that does not interact electromagnetically, so that it cannot be seen, but its gravitational effects otherwise can well be detected. Still, we have not yet detected or even associated dark matter with a particle of the standard model or beyond [105–107]. Modified (or alternative) theories of gravity have also been used to describe dark matter's effects [108,109]. In these cases, dark matter is simply an effect of the modification of gravity.

Returning to the dark energy question, it is highly counter-intuitive that the expansion of the universe is actually accelerating. Although the vacuum quantum energy can well explain this dynamical effect via the cosmological constant in Einstein's field equations of GR, the aforementioned important and persistent problems related to Λ supply the search for alternative explanations.

In the present article, as an attempt to describe dark energy, we assumed the symmetric teleparallel gravity as the underlying gravity theory.

The $f(Q)$ gravity was recently proposed by Jiménez et al. in [31] as a ramification of the geometric trinity, which says that the spacetime manifold can be described by curvature, torsion, or non-metricity. In particular the symmetric teleparallel gravity describes gravitational interactions via the non-metricity scalar, with null curvature and torsion.

Our $f(Q)$ cosmological model was based on a spatially homogeneous and isotropic flat metric and an energy–momentum tensor describing a bulk viscous fluid. We let the $f(Q)$ function be a power of n as $f(Q) \sim Q^n$, with n a free parameter.

In Section 4, we started testing our cosmological solutions. We started confronting the Hubble parameter (30) with 31 data points that are measured from the differential age approach.We then plotted Figures 1 and 2, in which $H(z)$ and $\mu(z)$ for our $f(Q)$ model are confronted with cosmological data and compared with the ΛCDM prediction. We can see the $f(Q)$ model describes observations with good agreement, and it is clear that it provides a better fit when compared to the ΛCDM model. Further, in Figure 3, we obtained the best-fit values for the model free parameters presented in Table 2.

From Figures 5 and 6, we found that the Hubble and density parameters show the expected positive behavior for all the constrained values of the model parameters. Figure 7 indicates that the bulk viscous cosmic fluid exhibits the negative pressure that makes the bulk viscosity a viable candidate to drive the cosmic acceleration. This is also reflected in the deceleration parameter behavior in Figure 4, which shows a transition from decelerated to accelerated phases of the universe's expansion. Furthermore, the effective EoS parameter presented in Figure 8 indicates that the cosmic viscous fluid behaves like quintessence dark energy.

In Section 5, we investigated the consistency of our model by analyzing the different energy conditions. We found that the NEC, DEC, and WEC are all satisfied in the entire domain of redshift (presented in Figures 9 and 10), while the SEC, presented in Figure 11, is violated for lower redshifts, which implies the cosmic acceleration, and satisfied for higher redshifts, which implies a decelerated phase of the universe.

In Section 6, Figures 12 and 13 show that the evolutionary trajectories of our model depart from the ΛCDM fixed point $r = 1, s = 0$. In the present epoch, they lie in the quintessence region $r < 1, s > 0$. The present model is, therefore, a good alternative to explain the universe's dynamics, particularly with no necessity of invoking the cosmological constant.

Author Contributions: Conceptualization, R.S.; Data curation, R.S. and S.A.; Formal analysis, P.H.R.S.M. and P.K.S.; Investigation, R.S. and S.A.; Methodology, R.S. and S.A.; Project administration, P.K.S.; Software, R.S. and S.A.; Supervision, P.H.R.S.M. and P.K.S.; Validation, P.H.R.S.M. and P.K.S.; Visualization, R.S. and S.A.; Writing—original draft, R.S. and S.A.; Writing—review & editing, P.H.R.S.M. and P.K.S. All authors have read and agreed to the published version of the manuscript.

Funding: This research received no external funding.

Institutional Review Board Statement: Not applicable.

Informed Consent Statement: Not applicable.

Data Availability Statement: There are no new data associated with this article.

Acknowledgments: RS acknowledges University Grants Commission (UGC), New Delhi, India for awarding Junior Research Fellowship (UGC-Ref. No.: 191620096030). SA acknowledges CSIR, New Delhi, India for SRF. PKS acknowledges CSIR, New Delhi, India for financial support to carry out the Research project [No.03(1454)/19/EMR-II Dt.02/08/2019] and IUCAA, Pune, India for providing support through visiting associateship program. We are very much grateful to the honorable referees and to the editor for the illuminating suggestions that have significantly improved our work in terms of research quality, and presentation.

Conflicts of Interest: The authors declare no conflict of interest.

References

1. Riess, A.G.; Filippenko, A.V.; Challis, P.; Clocchiatti, A.; Diercks, A.; Garnavich, P.M.; Gilliland, R.L.; Hogan, C.J.; Jha, S.; Kirshner, R.P.; et al. Observational Evidence from Supernovae for an Accelerating Universe and a Cosmological Constant. *Astron. J.* **1998**, *116*, 1009. [CrossRef]
2. Perlmutter, S.; Aldering, G.; Goldhaber, G.; Knop, R.A.; Nugent, P.; Castro, P.G.; Deustua, S.; Fabbro, S.; Goobar, A.; Groom, D.E.; et al. Measurements of Ω and Λ from 42 High-Redshift Supernovae. *Astrophys. J.* **1999**, *517*, 565. [CrossRef]
3. Koivisto, T.; Mota, D.F. Dark energy anisotropic stress and large scale structure formation. *Phys. Rev. D* **2006**, *73*, 083502. [CrossRef]
4. Daniel, S.F. Large scale structure as a probe of gravitational slip. *Phys. Rev. D* **2008**, *77*, 103513. [CrossRef]
5. Eisenstein, D.J.; Zehavi, I.; Hogg, D.W.; Scoccimarro, R.; Blanton, M.R.; Nichol, R.C.; Scranton, R.; Seo, H.-J.; Tegmark, M.; Zheng, Z.; et al. Detection of the Baryon Acoustic Peak in the Large-Scale Correlation Function of SDSS Luminous Red Galaxies. *Astrophys. J.* **2005**, *633*, 560. [CrossRef]
6. Percival, W.J.; Reid, B.A.; Eisenstein, D.J.; Bahcall, N.A.; Budavari, T.; Frieman, J.A.; Fukugita, M.; Gunn, J.E.; Ivezic, Z.; Knapp, G.R.; et al. Baryon acoustic oscillations in the Sloan Digital Sky Survey Data Release 7 galaxy sample. *Mon. Not. R. Astron. Soc.* **2010**, *401*, 2148–2168. [CrossRef]
7. Caldwell, R.R.; Doran, M. Cosmic microwave background and supernova constraints on quintessence: Concordance regions and target models. *Phys. Rev. D* **2004**, *69*, 103517. [CrossRef]
8. Huang, Z.Y.; Wang, B.; Abdalla, E.; Suet, R.K. Holographic explanation of wide-angle power correlation suppression in the cosmic microwave background radiation. *J. Cosm. Astrop. Phys.* **2006**, *0605*, 013. [CrossRef]
9. Dalal, N.; Abazajian, K.; Jenkins, E.; Manoharet, A.V. Testing the Cosmic Coincidence Problem and the Nature of Dark Energy. *Phys. Rev. Lett.* **2001**, *87*, 141302. [CrossRef]
10. Weinberg, S. The cosmological constant problem. *Rev. Mod. Phys.* **1989**, *61*, 1. [CrossRef]
11. Bento, M.C.; Bertolami, O.; Sen, A.A. Generalized Chaplygin gas, accelerated expansion, and dark-energy-matter unification. *Phys. Rev. D* **2002**, *66*, 043507. [CrossRef]
12. Kamenshchik, A.Y.; Moschella, U.; Pasquier, V. An alternative to quintessence. *Phys. Lett. B* **2001**, *511*, 265. [CrossRef]
13. Chiba, T.; Okabe, T.; Yamaguchi, M. Kinetically driven quintessence. *Phys. Rev. D* **2000**, *62*, 023511. [CrossRef]
14. Armendariz-Picon, C.; Mukhanov, V.; Steinhardt, P.J. Dynamical Solution to the Problem of a Small Cosmological Constant and Late-Time Cosmic Acceleration. *Phys. Rev. Lett.* **2000**, *85*, 4438. [CrossRef]
15. Carroll, S.M. Quintessence and the Rest of the World: Suppressing Long-Range Interactions. *Phys. Rev. Lett.* **1998**, *81*, 3067. [CrossRef]
16. Fujii, Y. Origin of the gravitational constant and particle masses in a scale-invariant scalar-tensor theory. *Phys. Rev. D* **1982**, *26*, 2580. [CrossRef]
17. Xu, L.; Wang, Y.; Tong, M.; Noh, H. CMB temperature and matter power spectrum in a decay vacuum dark energy model. *Phys. Rev. D* **2011**, *84*, 123004. [CrossRef]
18. Tong, M.; Noh, H. Observational constraints on decaying vacuum dark energy model. *Eur. Phys. J. C* **2011**, *71*, 1586. [CrossRef]
19. Freese, K.; Adams, F.C.; Frieman, J.A.; Mottola, E. Cosmology with decaying vacuum energy. *Nucl. Phys. B* **1987**, *287*, 797. [CrossRef]
20. Abdel-Rahman, A.-M.M. Nonsingular decaying-vacuum cosmology and baryonic matter. *Phys. Rev. D* **1992**, *45*, 3497. [CrossRef]
21. Appleby, S.; Battye, R. Do consistent $F(R)$ models mimic General Relativity plus Λ. *Phys. Lett. B* **2007**, *654*, 7. [CrossRef]
22. Amendola, L.; Gannouji, R.; Polarski, D.; Tsujikawa, S. Conditions for the cosmological viability of $f(R)$ dark energy models. *Phys. Rev. D* **2007**, *75*, 083504. [CrossRef]
23. Saffari, R.; Rahvar, S. $f(R)$ gravity: From the Pioneer anomaly to cosmic acceleration. *Phys. Rev. D* **2008**, *77*, 104028. [CrossRef]
24. Cognola, G.; Elizalde, E.; Nojiri, S.; Odintsov, S.D.; Zerbini, S. Dark energy in modified Gauss-Bonnet gravity: Late-time acceleration and the hierarchy problem. *Phys. Rev. D* **2006**, *73*, 084007. [CrossRef]
25. Li, B.; Barrow, J.D.; Mota, D.F. Cosmology of modified Gauss-Bonnet gravity. *Phys. Rev. D* **2007**, *76*, 044027. [CrossRef]
26. Moraes, P.H.R.S.; Santos, J.R.L. A complete cosmological scenario from $f(R, T^{\phi})$ gravity theory. *Eur. Phys. J. C* **2016**, *76*, 60.
27. Ren, X.; Wong, T.H.T.; Cai, Y.F.; Saridakis, E.N. Data-driven Reconstruction of the Late-time Cosmic Acceleration with f(T) Gravity. *Phys. Dark Univ.* **2021**, *32*, 100812. [CrossRef]
28. Nashed, G.L.; Capozziello, S. Charged Anti-de Sitter BTZ black holes in Maxwell-f(T) gravity. *Gen. Rel. Grav.* **2015**, *47*, 75. [CrossRef]
29. Setare, M.R.; Mohammadipour, N. Can f(T) gravity theories mimic ΛCDM cosmic history. *J. Cosm. Astrop. Phys.* **2013**, *01*, 015. [CrossRef]
30. Bahamonde, S.; Dialektopoulos, K.; Escamilla-Rivera, C.; Farrugia, G.; Gakis, V.; Hendry, M.; Hohmann, M.; Said, J.; Mifsud, J.; Valentino, E.D. Teleparallel Gravity: From Theory to Cosmology. *arXiv* **2021**, arXiv:2106.13793.
31. Jiménez, J.B.; Heisenberg, L.; Koivisto, T. Coincident general relativity. *Phys. Rev. D* **2018**, *98*, 044048. [CrossRef]
32. Hehl, F.W.; McCrea, J.D.; Mielke, E.W.; Neéman, Y. Metric-affine gauge theory of gravity: Field equations, Noether identities, world spinors, and breaking of dilation invariance. *Phys. Rep.* **1995**, *258*, 1–171. [CrossRef]
33. Hehl, F.W.; Neéman, Y.; Nitsch, J.; Heyde, P.V.D. Short-range confining component in a quadratic poincare gauge theory of gravitation. *Phys. Lett. B* **1978**, *78*, 102–106. [CrossRef]

34. Hehl, F.W.; Obukhov, Y.N. How does the electromagnetic field couple to gravity, in particular to metric, nonmetricity, torsion, and curvature ? *arXiv* **2000**, arXiv:0001010.
35. Ne'eman, Y.; Hehl, F.W. Matter Particled—Patterns, Structure and Dynamics. *Class. Quant. Grav.* **1997**, *14*, 251–260
36. Nester, J.M.; Yo, H.-J. Symmetric teleparallel general relativity. *Chin. J. Phys.* **1999**, *37*, 113.
37. Boulanger, N.; Kirsch, I. Higgs mechanism for gravity. II. Higher spin connections *Phys. Rev. D* **2006**, *73*, 124023.
38. Baekler, P.; Boulanger, N.; Hehl, F.W. Linear connections with a propagating spin-3 field in gravity. *Phys. Rev. D* **2006**, *74*, 125009. [CrossRef]
39. Weyl, H. Gravitation and electricity. *Sitzungsber. Preuss. Akad. Wiss.* **1918**, *465*, 1. [CrossRef]
40. Dialektopoulos, K.F.; Koivisto, T.S.; Capozziello, S. Noether symmetries in symmetric teleparallel cosmology. *Eur. Phys. J. C* **2019**, *79*, 606. [CrossRef]
41. Frusciante, N. Signatures of $f(Q)$ gravity in cosmology. *Phys. Rev. D* **2021**, *103*, 044021. [CrossRef]
42. Barros, B.J.; Barreiro, T.; Koivisto, T.; Nunes, N.J. Testing $F(Q)$ gravity with redshift space distortions. *Phys. Dark Univ.* **2020**, *30*, 100616. [CrossRef]
43. Ferreira, J.; Barreiro, T.; Mimoso, J.; Nunes, N.J. Forecasting $F(Q)$ cosmology with ΛCDM background using standard sirens. *Phys. Rev. D* **2022**, *105*, 123531. [CrossRef]
44. Albuquerque, I.S.; Frusciante, N. A designer approach to f(Q) gravity and cosmological implications. *Phys. Dark Univ.* **2022**, *35*, 100980. [CrossRef]
45. Capozziello, S.; D'Agostino, R. Model-independent reconstruction of f(Q) non-metric gravity. *Phys. Lett. B* **2022**, *832*, 137229. [CrossRef]
46. Atayde, L.; Frusciante, N. Can $f(Q)$ gravity challenge ΛCDM *Phys. Rev. D* **2022**, *104*, 064052. [CrossRef]
47. Ayuso, I.; Lazkoz, R.; Salzano, V. Observational constraints on cosmological solutions of $f(Q)$ theories. *Phys. Rev. D* **2021**, *103*, 063505. [CrossRef]
48. Khyllep, W. Cosmology in f(Q) gravity: A unified dynamical system analysis at background and perturbation levels. *arXiv* **2022**, arXiv:2207.02610.
49. Sahlu, S.; Tsegaye, E. Linear Cosmological perturbations in $F(Q)$ Gravity. *arXiv* **2022**, arXiv:2206.02517.
50. Jiménez, J.B.; Heisenberg, L.; Koivisto, T.; Pekar, S. Cosmology in $f(Q)$ geometry. *Phys. Rev. D* **2020**, *101*, 103507. [CrossRef]
51. Khyllep, W.; Paliathanasis, A.; Dutta, J. Cosmological solutions and growth index of matter perturbations in $f(Q)$ gravity. *Phys. Rev. D* **2021**, *103*, 103521. [CrossRef]
52. Mandal, S.; Wang, D.; Sahoo, P.K. Cosmography in $f(Q)$ gravity. *Phys. Rev. D* **2020**, *102*, 124029. [CrossRef]
53. Mandal, S.; Sahoo, P.K.; Santos, J.R.L. Energy conditions in $f(Q)$ gravity. *Phys. Rev. D* **2020**, *102*, 024057. [CrossRef]
54. Wilson, J.R.; Mathews, G.J.; Fuller, G.M. Bulk viscosity, decaying dark matter, and the cosmic acceleration. *Phys. Rev. D* **2007**, *75*, 043521. [CrossRef]
55. Okumura, H.; Yonezawa, F. New expression of the bulk viscosity. *Phys. A* **2003**, *321*, 207. [CrossRef]
56. Bali, R.; Jain, D.R. Some expanding and shearing viscous fluid cosmological models in general relativity. *Astrophys. Spa. Sci.* **1988**, *141*, 207. [CrossRef]
57. Bali, R.; Jain, D.R. A gravitationally non-degenerate cosmological model with expanding and shearing viscous fluid in general relativity. *Astriphys. Spa. Sci.* **1987**, *139*, 175. [CrossRef]
58. Deng, Y.; Mannheim, P.D. Acceleration-free spherically symmetric inhomogeneous cosmological model with shear viscosity. *Phys. Rev. D* **1991**, *44*, 1722. [CrossRef]
59. Huang, W.-H. Effects of the shear viscosity on the character of cosmological evolution. *J. Math. Phys.* **1990**, *31*, 659. [CrossRef]
60. Samanta, G.C.; Myrzakulov, R.; Shah, P. Kaluza–Klein Bulk Viscous Fluid Cosmological Models and the Validity of the Second Law of Thermodynamics in f(R,T) Gravity. *Zeits. Naturfor.* **2017**, *72*, 365. [CrossRef]
61. Satish, J.; Venkateswarlu, R. Bulk viscous fluid cosmological models in f(R, T) gravity. *Chin. J. Phys.* **2016**, *54*, 830.
62. Beesham, A. Cosmological Models with a Variable Cosmological Term and Bulk Viscous Models. *Phys. Rev. D* **1993**, *48*, 3539. [CrossRef] [PubMed]
63. Colistete, R., Jr.; Fabris, J.C.; Tossa, J.; Zimdahl, W. Bulk viscous cosmology. *Phys. Rev. D* **2007**, *76*, 103516. [CrossRef]
64. Sadatian, S.D. Effects of viscous content on the modified cosmological F(T) model.*EPL* **2019**, *126*, 30004. [CrossRef]
65. Brevik, I. Viscosity in Modified Gravity. *Entropy* **2012**, *14*, 2302–2310. [CrossRef]
66. Singh, C.P.; Kumar, P. Friedmann model with viscous cosmology in modified $f(R,T)$ gravity theory. *Eur. Phys. J. C* **2014**, *74*, 3070. [CrossRef]
67. Srivastava, M.; Singh, C.P. New holographic dark energy model with constant bulk viscosity in modified $f(R,T)$ gravity theory. *Astrophys. Space Sci.* **2018**, *363*, 117. [CrossRef]
68. Paolis, F.D.; Jamil, M.; Qadir, A. Black holes in bulk viscous cosmology. *Int. J. Theor. Phys.* **2010**, *49*, 621–632. [CrossRef]
69. Brevik, I.; Jamil, M. Black holes in the turbulence phase of viscous rip cosmology. *Int. J. Geom. Meth. Mod. Phys.* **2019**, *16*, 1950030. [CrossRef]
70. Jimenez, R.; Loeb, A. Constraining cosmological parameters based on relative galaxy ages. *ApJ* **2002**, *573*, 37. [CrossRef]
71. Sharov, G.S.; Bhattacharya, S.; Pan, S.; Nunes, R.C.; Chakraborty, S. A new interacting two-fluid model and its consequences. *Mon. Not. R. Astron. Soc.* **2017**, *466*, 3497. [CrossRef]

72. Blake, C.; Kazin, E.A.; Beutler, F.; Davis, T.M.; Parkinson, D.; Brough, S.; Colless, M.; Contreras, C.; Couch, W.; Croom, S.; et al. The WiggleZ Dark Energy Survey: Mapping the distance-redshift relation with baryon acoustic oscillations. *Mon. Not. R. Astron. Soc.* **2011**, *418*, 1707. [CrossRef]
73. Scolnic, D.M.; Jones, D.O.; Rest, A.; Pan, Y.C.; Chornock, R.; Foley, R.J.; Huber, M.E.; Kessler, R.; Narayan, G.; Riess, A.G.; et al. The complete light-curve sample of spectroscopically confirmed SNe Ia from Pan-STARRS1 and cosmological constraints from the combined pantheon sample. *ApJ* **2018**, *859*, 101. [CrossRef]
74. Lazkoz, R.; Lobo, F.S.N.; Ortiz-Banos, M.; Salzano, V. Observational constraints of $f(Q)$ gravity. *Phys. Rev. D* **2019**, *100*, 104027. [CrossRef]
75. Beh, J.T.; Loo, T.H.; De, A. Geodesic Deviation Equation In $f(Q)$ Gravity. *arXiv* **2021**, arXiv:2107.04513.
76. Dimakis, N.; Paliathanasis, A.; Christodoulakis, T. Quantum Cosmology in $f(Q)$ theory. *arXiv* **2021**, arXiv:2108.01970.
77. Maluf, J.W. The teleparallel equivalent of general relativity. *Ann. Phys.* **2013**, *525*, 339. [CrossRef]
78. Hohmann, M. General covariant symmetric teleparallel cosmology. *Phys. Rev. D* **2021**, *104*, 124077. [CrossRef]
79. Vagnozzi, S.; Loeb, A.; Moresco, M. The cosmic chronometers take on spatial curvature and cosmic concordance. *Astrophys. J.* **2021**, *908*, 84.
80. Gusakov, M.E. Bulk viscosity of superfluid neutron stars. *Phys. Rev. D* **2007**, *76*, 083001. [CrossRef]
81. Gusakov, M.E.; Kantor, E.M. Bulk viscosity of superfluid hyperon stars. *Phys. Rev. D* **2008**, *78*, 083006. [CrossRef]
82. Haensel, P.; Levenfish, K.P.; Yakovlev, D.G. Bulk viscosity in superfluid neutron star cores-III. Effects of hyperons. *Astron. Astrophys.* **2002**, *381*, 1080. [CrossRef]
83. Ryden, B. *Introduction to Cosmology*; Addison Wesley: San Francisco, CA, USA, 2003.
84. Odintsov, S.D.; Gomez, D.S.C.; Sharov, G.S. Testing the equation of state for viscous dark energy. *Phys. Rev. D* **2020**, *101*, 044010. [CrossRef]
85. Fabris, J.C.; Goncalves, S.V.B.; Ribeiro, R. Bulk viscosity driving the acceleration of the Universe. *Gen. Rel. Grav.* **2006**, *38*, 495. [CrossRef]
86. Meng, X.-H.; Dou, X. Friedmann cosmology with bulk viscosity: A concrete model for dark energy. *Comm. Theor. Phys.* **2009**, *52*, 377.
87. Brevik, I.; Gorbunova, O. Dark energy and viscous cosmology. *Gen. Rel. Grav.* **2005**, *37*, 2039. [CrossRef]
88. Gron, O. Viscous inflationary universe models. *Astrophys. Space Sci.* **1990**, *173*, 191–225. [CrossRef]
89. Eckart, C. The thermodynamics of irreversible processes. III. Relativistic theory of the simple fluid *Phys. Rev.* **1940**, *58*, 919.
90. Ren, J.; Meng, X.-H. Cosmological model with viscosity media (dark fluid) described by an effective equation of state. *Phys. Lett. B* **2006**, *633*, 1–8. [CrossRef]
91. Harko, T.; Koivisto, T.S.; Lobo, F.S.N.; Olmo, G.J.; Rubiera-Garcia, D. Coupling matter in modified Q gravity. *Phys. Rev. D* **2018**, *98*, 084043. [CrossRef]
92. Mackey, D.F.; Hogg, D.W.; Lang, D.; Goodman, J. emcee: The MCMC hammer. *Publ. Astron. Soc. Pac.* **2013**, *125*, 306. [CrossRef]
93. Sharov, G.S.; Vasilie, V.O. How predictions of cosmological models depend on Hubble parameter data sets. *Math. Model. Geom.* **2018**, *6* 1. [CrossRef]
94. Beutler, F.; Blake, C.; Colless, M.; Jones, D.H.; Staveley-Smith, L.; Campbell, L.; Parker, Q.; Saunders, W.; Watson, F. The 6dF Galaxy Survey: Baryon acoustic oscillations and the local Hubble constant. *Mon. Not. R. Astron. Soc.* **2011**, *416*, 3017. [CrossRef]
95. Jarosik, N.; Bennett, C.L.; Dunkley, J.; Gold, B.; Greason, M.R.; Halpern, M.; Hill, R.S.; Hinshaw, G.; Kogut, A.; Komatsu, E. SEVEN-YEAR WILKINSON MICROWAVE ANISOTROPY PROBE (WMAP) OBSERVATIONS: SKY MAPS, SYSTEMATIC ERRORS, AND BASIC RESULTS. *Astrophys. J. Suppl.* **2011**, *192*, 14. [CrossRef]
96. Giostri, R.; Santos, M.V.d.; Waga, I.; Reis, R.R.R.; Calvao, M.O.; Lago, B.L. From cosmic deceleration to acceleration: New constraints from SN Ia and BAO/CMB. *J. Cosm. Astropart. Phys.* **2012**, *03*, 027. [CrossRef]
97. Mukherjee, P.; Banerjee, N. Non-parametric reconstruction of the cosmological jerk parameter. *Eur. Phys. J. C* **2021**, *81*, 36. [CrossRef]
98. Tripp, R. A two-parameter luminosity correction for Type IA supernovae. *Astron. Astrophys.* **1998**, *331*, 815.
99. Kessler, R.; Scolnic, D. Correcting type Ia supernova distances for selection biases and contamination in photometrically identified samples. *Astrophys. J.* **2017**, *836*, 56. [CrossRef]
100. Anagnostopoulos, F.K.; Basilakos, S.; Saridakis, E.N. Observational constraints on Myrzakulov gravity. *Phys. Rev. D* **2021**, *103*, 104013. [CrossRef]
101. Vagnozzi, S.; Pacucci, F.; Loeb, A. Implications for the Hubble tension from the ages of the oldest astrophysical objects. *JHEAp* **2022, 36**, 27–35. [CrossRef]
102. Raychaudhuri, A. Relativistic cosmology I. *Phys. Rev.* **1955**, *98*, 1123. [CrossRef]
103. Capozziello, S.; Lobo, F.S.N.; Mimoso, J.O. Energy conditions in modified gravity. *Phys. Lett. B* **2014**, *730*, 280–283. [CrossRef]
104. Sahni, V.; Saini, T.D.; Starobinsky, A.A.; Alam, U. Statefinder-a new geometrical diagnostic of dark energy. *JETP Lett.* **2003**, *77*, 201. [CrossRef]
105. CDMS II Collaboration. *Science* **2010**, *327*, 1619.
106. Akerib, D.S.; Araujo, H.M.; Bai, X.; Bailey, A.J.; Balajthy, J.; Bedikian, S.; Bernard, E.; Bernstein, A.; Bolozdynya, A.; Bradley, A.; et al. First Results from the LUX Dark Matter Experiment at the Sanford Underground Research Facility. *Phys. Rev. Lett.* **2014**, *112*, 091303. [CrossRef] [PubMed]

107. Essig, R.; Manalaysay, A.; Mardon, J.; Sorensen, P.; Volansky, T. First direct detection limits on sub-GeV dark matter from XENON10. *Phys. Rev. Lett* **2012**, *109*, 021301. [CrossRef] [PubMed]
108. Bohmer, C.G.; Harko, T.; Lobo, F.S.N. Dark matter as a geometric effect in f(R) gravity. *Astrop. Phys.* **2008**, *29*, 386. [CrossRef]
109. Mannheim, P.D.; O'Brien, J.G. Fitting galactic rotation curves with conformal gravity and a global quadratic potential. *Phys. Rev. D* **2012**, *85*, 124020. [CrossRef]

universe

MDPI

Article

On the Role of Constraints and Degrees of Freedom in the Hamiltonian Formalism

Alexey Golovnev

Centre for Theoretical Physics, The British University in Egypt, El Sherouk City 11837, Egypt; agolovnev@yandex.ru

Abstract: Unfortunately, the Hamiltonian mechanics of degenerate Lagrangian systems is usually presented as a mere recipe of Dirac, with no explanation as to how it works. It then comes to discussing conjectures of whether all primary constraints correspond to gauge symmetries, and it goes all the way to absolutely wrong claims such as the statement that electrodynamics or gravity only have two physical components each, with others being spurious. One has to be very careful because non-dynamical, or constrained, does not mean unphysical. In this article, I give a pedagogical introduction to the degenerate Hamiltonian systems, showing both very simple mechanical examples and general arguments about how it works. For the familiar field theory models, I explain why the gauge freedom there "hits twice" in the sense of producing twice as many first-class constraints as gauge symmetries, and why primary, and only primary, constraints should be put into the total Hamiltonian.

Keywords: Hamiltonian mechanics; field theory; systems with constraints; gauge theories

1. Introduction

Hamiltonian mechanics is a very nice formulation for the whole plethora of dynamical models, both for classical and quantum physics, as well as for mathematics, since it allows to naturally work with a system of first-order differential equations. In case of non-degenerate Lagrangians, we get unconstrained Hamiltonian systems which express the time-derivatives of all their canonical variables as functions of the variables themselves. Therefore, we need one and only one Cauchy datum for every canonical variable in order to solve the equations, at least locally.

In every good textbook on classical mechanics, it is explained how the Hamiltonian formalism of non-degenerate systems works. However, when we face degenerate systems, we have to be more careful and to also introduce some constraints on the values of momenta. These are called primary constraints, and they reduce the number of needed Cauchy data because some combinations of canonical variables get to strictly vanish.

Here comes another advantage of the Hamiltonian analysis. Since all equations are of the first derivative order, in order to find the number of Cauchy data, or the degrees of dynamical freedom, we simply need to know how many combinations of canonical variables are subject to such restrictions. There are some limitations to this statement however. First, there might be a gauge freedom, which makes some variables simply not determined by the equations at all. We call them unphysical and care only about gauge-invariant quantities. Second, the restrictions might be more numerous than the primary ones. The equation of a primary constraint staying zero all the time could demand that some other combinations of variables also vanish, and we call those requirements secondary constraints. Third, Hamiltonian constraints do not always have a clear meaning in the Lagrangian setting. For example, if a variable was totally absent from the Lagrangian, it would have an arbitrary dependence on time, however its momentum would be constrained to be zero. I will discuss all of these issues.

Degenerate systems are very important in modern theoretical physics. Indeed, when we have a gauge symmetry, it certainly means that the Lagrangian does not depend on some combination(s) of the fundamental variables in any way, neither through coordinates nor through velocities. Therefore, the system is degenerate and we have primary constraint(s). Very commonly, and I will explain why, such gauge-related constraint produce another constraint. For example, electrodynamics has one gauge freedom and two constraints, while gravity has four gauge freedoms and eight constraints.

Unfortunately, the procedure of working with such systems is usually explained as just a recipe which, as a matter of fact, works well. It works indeed. However, then comes a strange conjecture from the lectures of Dirac [1] that every "first-class", to be defined below, constraint corresponds to a gauge freedom. It is taken very seriously in most of the literature, even though, if taken literally, it is simply wrong. In a sense, the familiar secondary first-class constraints are indeed the consequences of the gauge symmetry, however they do not increase the number of these symmetries, nor possible gauge choices. The number of gauge freedoms is the number of primary first-class constraints only, not more, and even this amount is not always realized as a usual gauge symmetry, due to some other types of singular behavior of Lagrangians, as I also explain below.

Following the wrong conjecture, as well as putting also the secondary constraints into the Hamiltonian, leads people to claiming that there are only two physical variables in Maxwell electrodynamics, and the same in General Relativity. It is certainly true that those indeed have this number of dynamical modes, but it does not justify calling other non-gauge variables spurious. Doing so they basically say that Coulomb forces and Newtonian potentials are unphysical, which is a bit too much to say. In this paper, I first show how the Hamiltonian analysis works in the simplest mechanical cases, although also mentioning some modern physics issues, then I explain why it works and how it must be used, and finally I review its application to electromagnetism and gravity in a bit more detail.

1.1. A Brief Review of Hamiltonian Mechanics

I will denote the dependence of a Lagrangian on coordinates and velocities as

$$L(\mathfrak{x},\mathfrak{y},\mathfrak{z}) = L(x,\dot{x},y,\dot{y},z,\dot{z})$$

which will mean that L is taken as a Lagrangian for a system of all the mentioned variables, even if it does not depend on some or all of them. For every Lagrangian variable, the momentum is defined as

$$\pi_x \equiv \frac{\partial L}{\partial \dot{x}}. \tag{1}$$

In cases when the matrix $\frac{\partial^2 L}{\partial \dot{x}_i \partial \dot{x}_j}$ is non-degenerate, the Lagrangian system of $L(\mathfrak{x}_i)$ is called non-degenerate, too, and this definition (1) of $\pi = \pi(x,\dot{x})$ can be inverted to $\dot{x} = \dot{x}(\pi, x)$, at least locally.

Then we introduce the Hamiltonian function

$$\mathcal{H}(x_i,\dot{x}_i,\pi_{x_i}) = \sum_i \pi_{x_i}\dot{x}_i - L(\mathfrak{x}_i). \tag{2}$$

For a non-degenerate system, and up to a possible non-uniqueness of solving for velocities in terms of momenta, we can finally define the canonical Hamiltonian as

$$H(x_i,\pi_{x_i}) = \mathcal{H}(x_i,\dot{x}_i(\pi_{x_i},x_i),\pi_{x_i}) = \left(\sum_i \pi_{x_i}\dot{x}_i - L(\mathfrak{x}_i)\right)\Bigg|_{\dot{x}_i:\quad \pi_{x_i} = \frac{\partial L}{\partial \dot{x}_i}}. \tag{3}$$

One can easily find the differential equations satisfied by the Hamiltonian by simply looking at the variation of the Hamiltonian function (2) restricted to the surface defined by imposing the equation for momenta (1):

$$\delta H = \delta\left(\sum_i \pi_{x_i}\dot{x}_i - L(\mathfrak{x}_i)\right)\Bigg|_{\pi_{x_i}=\frac{\partial L}{\partial \dot{x}_i}} = \sum_i\left(\dot{x}_i\delta\pi_{x_i} - \frac{\partial L}{\partial x_i}\delta x_i\right) \tag{4}$$

which implies

$$\frac{\partial H}{\partial \pi_{x_i}} = \dot{x}_i \quad \text{and} \quad \frac{\partial H}{\partial x_i} = -\frac{\partial L}{\partial x_i}.$$

Of course, the first equation simply solves velocities in terms of momenta, while the second one can then be compared with the Lagrangian equation by observing that the latter tells us that

$$\frac{\partial L}{\partial x_i} = \frac{d}{dt}\frac{\partial L}{\partial \dot{x}_i} = \dot{\pi}_{x_i},$$

and therefore the Hamiltonian shape of the equations of motion is

$$\dot{x}_i = \frac{\partial H}{\partial \pi_{x_i}} \quad \text{and} \quad \dot{\pi}_{x_i} = -\frac{\partial H}{\partial x_i}. \tag{5}$$

For a convex Lagrangian, it can be simply viewed as the new Legendre transform, which is involutive, and therefore brings us back to the Lagrangian with the variational principle written in terms of $\int(\pi\dot{x} - H)$.

The equations of motion (5) can be presented in the beautiful language of symplectic geometry,

$$\dot{f}(x,\pi) = \{f, H\},$$

such that the canonical coordinates and momenta are the Darboux coordinates for the (anti-symmetric) Poisson bracket: $\{x_i, \pi_{x_j}\} = \delta_{ij}$, with brackets among coordinates or among momenta being zero. Therefore, we have a natural symplectic geometry on the phase space defined by the Poisson bracket of two functions $f(x,\pi)$ and $g(x,\pi)$:

$$\{f, g\} \equiv \sum_i\left(\frac{\partial f}{\partial x_i}\cdot\frac{\partial g}{\partial \pi_{x_i}} - \frac{\partial f}{\partial \pi_{x_i}}\cdot\frac{\partial g}{\partial x_i}\right). \tag{6}$$

Here I will not go into any more detail on that. There are many nice and rigorous expositions of the Hamiltonian mechanics for non-degenerate systems; see for example the book by Arnold [2].

1.1.1. Non-Quadratic Kinetic Lagrangians–Hamiltonians with Branches

Let me briefly mention that it all works with no surprise if the Lagrangian is a positive-definite quadratic form of velocities. Having higher powers of velocities brings its own issues. Of course, even if the Hessian $\frac{\partial^2 L}{\partial \dot{x}_i \partial \dot{x}_j}$ is non-degenerate, it might have some singular phase space points with a vanishing determinant. On top of that, even in the regular parts, the solution $\dot{x} = \dot{x}(\pi, x)$ is often non-unique.

As a very simple example, consider $L(\mathfrak{x}) = \frac{1}{3}\dot{x}^3$. The momentum is $\pi = \dot{x}^2$, and it is an even function of $\dot{x}$. The Hamiltonian is easily found to be $H = \pm\frac{2}{3}\pi^{3/2}$, and its two branches correspond to positive and negative velocities. This example is very trivial, however there are other options. For instance, one can also play with time crystals [3].

1.1.2. Degenerate Systems—The Dirac's Recipe

Now I will turn to the main topic of this paper: Hamiltonian systems with constraints. We call a Lagrangian degenerate if its Hessian $\frac{\partial^2 L}{\partial \dot{x}_i \partial \dot{x}_j}$ has a vanishing determinant, and therefore it has a non-trivial kernel and cannot be inverted. Such systems cannot be written as $\ddot{x}_i = f_i(x, \dot{x})$.

In this case, the definition (1) of the momenta in terms of velocities is not invertible either. The variations of velocities are mapped into variations of momenta by action of the Hessian matrix. Assuming some smoothness and constancy of rank, the dimension of its kernel gives the number of independent velocity variations, which do not change the momenta at all. On the other hand, this dimension is the codimension of its image in the momentum space, and the equation of the image submanifold is the set of the primary constraints, in the number equal to the dimension of that kernel. Let us define the primary constraints Φ_a as equations

$$\Phi_a(\pi_{x_i}, x_i) = 0$$

to be put on top of the usual Hamiltonian equations, and the index a running from 1 to the dimension of the Hessian's kernel. In other words, the primary constraints are necessary and sufficient (again, at least locally) conditions for a set of momenta π_{x_i} to be in the image of the mapping (1).

We can define the Hamiltonian function (2) the same way again. The sad news is that we cannot find the velocities as functions of momenta in order to substitute them there. However, the calculation of its variation (4) goes the same way as before, the only difference being that the assumption of the Equation (1) put into this variation generically cannot work, for any velocity, unless we always put the momentum onto the primary constraint manifold. However, it is not what we can control when taking derivatives of the Hamiltonian once we have forgotten the initial origin of this function.

Moreover, we cannot even find a unique canonical Hamiltonian with the usual property (4) of the first variation. Indeed, if we change it as $H \longrightarrow H + \sum\limits_a \Phi_a \cdot \mathcal{F}_a(x_i, \pi_{x_i})$ with arbitrary functions $\mathcal{F}_a$, it does not change the Equation (4) because in the case of imposing the condition (1), both on the background and on the variation, we always have $\Phi_a = 0$. Therefore, the next natural step is to take an arbitrary possible choice of the canonical Hamiltonian and define the total Hamiltonian

$$H_T = H + \sum_a \lambda_a \Phi_a \tag{7}$$

with arbitrary functions λ_a called Lagrange multipliers.

The main recipe is that the equations are then

$$\dot{x}_i = \frac{\partial H_T}{\partial \pi_{x_i}} \qquad \text{and} \qquad \dot{\pi}_{x_i} = -\frac{\partial H_T}{\partial x_i} \qquad \text{and the constraints} \qquad \Phi_a = 0. \tag{8}$$

Dirac preferred to call the last equality a weak one and to denote it as $\Phi_a \approx 0$. It is weak in the meaning that it is valid on the constraint surface only, and not when one goes away from it, and in particular the first two equations do get non-trivial terms from differentiating the Φ-s. For me, it is just yet another equation which I do not want to distinguish anyhow. On the contrary, if I want to stress that something is identically zero, I use the sign of $\equiv 0$. After all, it is all but absolutely common that even when a function takes the zero value, its derivative might be non-vanishing. Interestingly enough, Dirac in his paper [4] denoted weak equalities with the usual equality sign, while for strong equalities he used the identical equality notation, even though his definition of a strong equality there was rather that of $\mathcal{O}(\epsilon^2)$ as opposed to $\mathcal{O}(\epsilon)$ type.

My claim about the system of Equation (8) is that, in every reasonable case, this is already the final answer, and nothing else is needed in order to have the equations identical to the Lagrangian ones. One way to look at this is that we are doing the variation (4) always respecting the condition that the momenta are only those which indeed fulfill the condition (1) and therefore do correspond to at least some velocity. I will later give more detailed explanations. However, intuitively one can see it as varying $\int(p\dot{x} - H + \lambda\Phi)$, with the last term both imposing the necessary requirements on the momenta and reproducing every possible canonical Hamiltonian.

Let me stress it once more. The Equation (8) with primary, and only primary, constraints are all that we need. However, when we want to easily and reliably count the degrees of freedom, we need to be more careful.

All in all, when I take $\Phi_a = 0$ as an equation of motion, it immediately implies that $\dot{\Phi}_a = 0$, too. Taken together with other equations, it might or might not entail new constraints, $\tilde{\Phi}_m$, which are called secondary. Using the other equations for that can also be rephrased in terms of Poisson brackets (6):

$$0 = \dot{\Phi}_a = \{\Phi_a, H_T\}$$

which takes the form of a function of coordinates and momenta, with no time-derivatives of those. Then we can do the same with the secondary constraints, to possibly get tertiary ones, and so on. I will call all of them secondary $\tilde{\Phi}_m$. At every step, if the Poisson bracket of a constraint with the total Hamiltonian is not either identically zero or expressed in terms of the constraints themselves (weakly zero in the language of Dirac), it means either a restriction on the Lagrange multipliers or yet another constraint.

Once all the constraints are found, they reduce the number of independent Cauchy data, while those Lagrange multipliers of primary constraints which have not been fixed in this process count unphysical variables in the system. Here comes the classification of the constraints. We can compute the whole matrix of Poisson brackets of all constraints with each other. If every element of this matrix is "weakly zero", i.e., the constraints form a closed algebra with respect to the brackets, we say that they are first-class. If the matrix is, to the contrary, non-degenerate on the constraint surface, then they are second-class. If the matrix is neither of that, one can separate the constraints, maybe after some redefinitions, into first- and second-class ones.

If all the constraints are first-class, then the conditions of $\{\Phi_a, H_T\} = 0$ and $\{\tilde{\Phi}_m, H_T\} = 0$ can not be satisfied by conditions on the Lagrange multipliers. Therefore, in this case every primary constraint goes with an arbitrary λ in the Hamiltonian equations of motion (8), normally representing a gauge freedom. If a primary constraint led to a secondary one with which it is of second-class, then it seems that its Lagrange multiplier gets fixed by preservation of its secondary friend. Since every constraint kills a Cauchy datum, while first-class ones are also associated with gauge symmetries, the common lore says that each first-class constraint and each pair of second-class constraints reduce the number of degrees of freedom by one.

1.1.3. My Comments

In this paper I argue that it is not that simple, and it is better to always check what happens in each particular case, even though the Hamiltonian formalism can sometimes be more transparent than the Lagrangian one, indeed. My main point is that the amount of gauge freedom is normally reflected in the number of only primary first-class constraints. Even this statement requires certain regularity of the Lagrangian. It can happen that some variables enter the Lagrangian only being multiplied by a factor that vanishes together with its first variation on every configuration in which other variables satisfy all the equations. Then the former variables will enjoy the full unphysical freedom of being arbitrary functions of time, at the same time without any off-shell symmetry of such Lagrangian. I will give an example below.

In what concerns the Dirac conjecture, the relation between first-class constraints and gauge symmetries is not trivial. In a sense, all of them are normally related to some gauge symmetry. However, in non-trivial cases when the "gauge hits twice", not even the primary ones are the precise generators of the gauge transformations, with this role being played by some combination of them with the secondary one(s) [5]. If we take the conjecture as a statement that all first-class constraints are independent generators of gauge transformations, then it is simply wrong, in all the cases which have secondary first-class constraints.

There is one more point about that. It is claimed sometimes [6] that it would be better to add all the constraints to the Hamiltonian, not only primary but also secondary ones, in order to get what is called an extended Hamiltonian,

$$H_E = H_T + \sum_m \tilde{\lambda}_m \tilde{\Phi}_m = H + \sum_a \lambda_a \Phi_a + \sum_m \tilde{\lambda}_m \tilde{\Phi}_m. \tag{9}$$

First, it is not needed. The primary constraints are already enough to reproduce the Lagrangian equations of motion in the Hamiltonian language. Second, at least with the first-class constraints, it is simply wrong. In this case, adding a secondary constraint to the Hamiltonian also adds a new arbitrary Lagrange multiplier producing a fake effective gauge symmetry on top of the real one.

2. Very Simple Examples of Systems with Constraints

I will first illustrate the workings of constrained Hamiltonian formalism with very simple examples, then give some general mechanical statements, and finally discuss the usual field theories we have in physics. There are several good expositions of Hamiltonian mechanics with constraints [6–8], however more as a recipe, and the important subtleties are rarely paid enough attention to.

2.1. *Trivial Gauge Symmetry*

Let me start from a very simple example

$$L(x) = 0$$

which has one variable $x(t)$ and a gauge symmetry which kills it. Finally, we have nothing physical, one gauge freedom and zero degrees of freedom. With the same success, I could have taken $L = 1$, for anyway its variation under arbitrary δx is zero.

2.1.1. Failure of Forgetting to Put Primary Constraints into the Hamiltonian

We get the zero momentum $\pi_x = 0$, which is its only (first-class) constraint $\Phi \equiv \pi_x$, and the Hamiltonian function (2), $\mathcal{H} = \pi_x \dot{x} - L$, is equal to zero on the constraint surface for any choice of $\dot{x}$. If we simply take

$$H = 0,$$

then the equations of motion (5) tell us that

$$\dot{x} = 0 \qquad \text{and} \qquad \dot{\pi}_x = 0.$$

It gives then higher freedom of the momentum than in the initial Lagrangian formulation, but at the same time requires more from the variable x which used to be absolutely free. Even adding the equation of $\pi_x = 0$ by hand would not help to restore the arbitrariness of $x(t)$. A primary constraint is not a mere selection of a surface inside a phase space, it genuinely changes the dynamics. Therefore, we must use the primary constraints in the Hamiltonian.

It is also a first sign of the need for being careful when calculating the number of degrees of freedom in the Hamiltonian formulation. If I assume that a zero Hamiltonian $H \equiv 0$ for n variables has no constraints, it means $2n$ required Cauchy data, and works like n degrees of freedom. However, half of these data simply set the momenta to arbitrary constant values which have no influence on the coordinates. For finding all the coordinates we need only n Cauchy data, which set their otherwise arbitrary constant values, and then it goes like n halves degrees of freedom. In particular, as we saw above, in a theory with one variable it gives $\frac{1}{2}$ degree of freedom.

2.1.2. Zero Hamiltonian with a Primary Constraint

Now, let us correctly take the primary constraint into account and construct the total Hamiltonian (7)

$$H_T = H + \lambda\Phi = 0 + \lambda\pi_x$$

with an arbitrary Lagrange multiplier λ. The new Equations (8) are

$$\dot{x} = \lambda \qquad \text{and} \qquad \dot{\pi}_x = 0 \qquad \text{with the constraint} \qquad \pi_x = 0.$$

Now we are totally back to the Lagrangian situation. The momentum is fully fixed by the primary constraint, and the full freedom in the behavior of the coordinate is restored by the Lagrange multiplier being an arbitrary function. The equation $\dot{x} = \lambda$ still requires one Cauchy datum to solve it for x in terms of λ, but it makes no real sense since the Lagrange multiplier is assumed to be an absolutely arbitrary function of time, as no extra condition restricts it; and the same is finally true of x.

Note that we actually could have taken an absolutely different canonical Hamiltonian here. Of course, $H = 0$ is indeed equal to the function $\mathcal{H} = \pi_x\dot{x} - L$ at the surface of $\pi_x = 0$ with $L = 0$. However, any canonical H of the form $H = \pi_x \cdot f(x, \pi_x)$ would do the same. The total Hamiltonian $H_T = \pi_x \cdot f(x, \pi_x) + \lambda\pi_x$ produces then equivalent equations of motion. In other words, the canonical Hamiltonian (3) for a degenerate system is not unique. In fact, if it was not for the primary constraint put into H_T, these choices would have given different equations of motion.

If for the model being trivial, the choice of $H = 0$ looks like the only natural one above, we could also take a somewhat less trivial case of $L(\mathfrak{x}, \mathfrak{y}) = \frac{1}{2}(\dot{y}^2 - y^2)$ with the Hamiltonian $H_T = \frac{1}{2}\left(\pi_y^2 + y^2\right) + \lambda\pi_x$ yielding the usual harmonic oscillator of $y(t)$ and total freedom for $x(t)$ again the same way as before. It is still very natural, to not add any contribution of π_x to the canonical Hamiltonian. A bit more entertaining, however, is to take

$$L(\mathfrak{x}, \mathfrak{y}) = \frac{1}{2}\Big((\dot{x} - \dot{y})^2 - (x - y)^2\Big).$$

In this case, the function $\mathcal{H} = \frac{1}{2}\Big((\dot{x} - \dot{y})^2 + (x - y)^2\Big)$ can be represented as $H = \frac{1}{2}(\pi_x^2 + (x - y)^2)$, or $H = \frac{1}{2}\left(\pi_y^2 + (x - y)^2\right)$, with only symmetry considerations suggesting either $H = \frac{1}{4}\left(\pi_x^2 + \pi_y^2\right) + \frac{1}{2}(x - y)^2$ or $H = -\frac{1}{2}\pi_x\pi_y + \frac{1}{2}(x - y)^2$. Looking at the $\frac{x \pm y}{\sqrt{2}}$ variables offers yet another symmetric choice of $H = \frac{1}{8}\left(\pi_x - \pi_y\right)^2 + \frac{1}{2}(x - y)^2$. Those are all the same at the surface of the primary constraint $\pi_x + \pi_y = 0$, with the same equations of motion produced by $H_T = H + \lambda(\pi_x + \pi_y)$.

2.1.3. Adding Total Time Derivatives

Note that Lagrangians different by only a total time derivative are equivalent, at least at the level of equations of motion. Therefore, a gauge transformation might also be preserving a Lagrangian up to a total time derivative, or a Lagrangian density up to a total divergence, in field theories.

Therefore, another option to take the model with empty physical content is

$$L(\mathfrak{x}) = f(x)\dot{x}.$$

We get then a non-zero momentum $\pi_x = f(x)$, the natural choice of zero canonical Hamiltonian, and $H_T = \lambda(\pi_x - f(x))$. The equations, together with the constraint, then read

$$\dot{x} = \lambda, \qquad \dot{\pi}_x = f' \cdot \lambda, \qquad \pi_x = f.$$

They are obviously satisfied by an arbitrary $x(t)$ and $\pi_x(t) = f(x(t))$.

For the Lagrangian variable $x(t)$, the result is of course the same as it was with the zero Lagrangian. However, the Hamiltonian description goes in a slightly different way, due to the different definition of the momentum, and therefore the constraint $\Phi = \pi - f$ instead of $\Phi = \pi$. It shows that if a gauge symmetry is realized only up to a total derivative term, the Hamiltonian representation of it might well go in a more elaborate style. One good field theory example of it is the Lorentz symmetry [9] of the Teleparallel Equivalent of General Relativity (TEGR).

2.2. A Very Simple Gauge Symmetry Can also Hit Twice

There exists a simple slogan [6,10] that "a gauge symmetry hits twice", meaning that it produces two Hamiltonian constraints. As we have seen in the trivial example, or as can be found in Refs. [6,10] as well, it is not always true. I will come to this later, but let me also mention now that another example is TEGR. If both its diffeomorphism and Lorentz symmetries hit twice, then using the usual naive counting, it would have $16 - 2 \cdot 4 - 2 \cdot 6 = -4$ degrees of freedom in four spacetime dimensions.

However, one can easily construct a simple mechanical model in which a gauge hits twice indeed:

$$L(\mathfrak{x}, \mathfrak{y}) = \frac{1}{2}(\dot{x} + y)^2.$$

We get the momenta $\pi_x = \dot{x} + y$ and $\pi_y = 0$, the latter being the primary constraint $\Phi = \pi_y$. Then the total Hamiltonian (7) can be written as

$$H_T = \frac{1}{2}\pi_x^2 - y\pi_x + \lambda\pi_y,$$

with all the possible additions of π_y to the canonical part if one wants to write the same story in a slightly more complicated form. The equations of motion (8) are

$$\begin{cases} \dot{x} = \pi_x - y \\ \dot{y} = \lambda \end{cases} \quad \text{and} \quad \begin{cases} \dot{\pi}_x = 0 \\ \dot{\pi}_y = \pi_x \end{cases} \quad \text{with the constraint} \quad \pi_y = 0.$$

These equations, or the formal "preservation" of the primary constraint, yield the secondary constraint $\tilde{\Phi} \equiv \pi_x = 0$ which in turn is trivially preserved due to one of the equations, or to the fact of H_T not depending on x. Then what we get is $\dot{x} = -y$ with an arbitrary $y(t)$. It seems we have one pure gauge variable, two first-class constraints, and one Cauchy datum needed, therefore $\frac{1}{2}$ degrees of freedom. Note however that the same system of equations can be interpreted as an arbitrary x and $y = -\dot{x}$ which then means no Cauchy data at all. The very naive counting looked like $\frac{1}{2}$, but in reality there is 0 dynamical degrees of freedom, even though not both variables are pure gauge. The physical combination of variables is $\dot{x} + y$, and it is fully constrained to be equal to zero. The traditional counting of dynamical modes by subtracting the number of first-class constraints works well in this case.

The Lagrangian story is of course the same. The equations of motion are

$$\frac{d}{dt}(\dot{x} + y) = 0 \qquad \text{and} \qquad \dot{x} + y = 0.$$

The first one is a dynamical equation from variation of x, and the second one is a constraint, the only constraint of the Lagrangian formalism. The action has a gauge symmetry under

$$x \longrightarrow x + u(t), \qquad y \longrightarrow y - \dot{u}(t)$$

so that only one combination of the two variables, $\dot{x} + y$, is physical. However, the primary constraint $\Phi = \pi_y$ transforms only y, leaving x intact, which is not a gauge transformation. This is the reason of the secondary constraint appearing. In other words, it is the same as what happens in electrodynamics: the gauge symmetry mixes a velocity (time derivative) with another coordinate (other spatial derivatives), producing two first-class constraints by only one gauge symmetry.

Note that the variation of $y(t)$ in the action has produced the only Lagrangian constraint $\dot{x} + y = 0$, with another combination of the two variables being pure gauge. This constraint corresponds to the secondary Hamiltonian constraint, while the primary one has no relation to the Lagrangian variables, it is simply $\pi_y = 0$, which in itself does not restrict any velocity. In the Lagrangian language, we have one gauge freedom, which strips one combination of variables of physical meaning, while the physical one is subject to a constraint. If we choose a gauge of $x = 0$, the physical equation (not a gauge choice!) tells us that $y = 0$, which means zero Cauchy data. At the same time, the gauge of $y = 0$ has a remaining symmetry of constant u, which is reflected in the solution of an arbitrary constant x, not contributing to the physical gauge-invariant variable $\dot{x} + y$. In the Hamiltonian language, we have both momenta constrained, while for the two coordinates there is one arbitrary Lagrange multiplier, so that naively the system needs one Cauchy datum. The origin of this number is in the fact that the Hamiltonian equations look in such a way as if taking the gauge in terms of y was more natural.

Beware of Secondary Constraints Put into the Hamiltonian

It is often stated, for example in the classic book of Henneaux and Teitelboim [6], that it is more natural to take both primary and secondary constraints on equal footing. This is totally wrong! First, it is not natural even from the basic framework of the Hamiltonian formalism. The primary constraints are the restrictions on possible values of momenta. They might lead to some other restrictions on the variables, called secondary constraints, but those are not new restrictions, as they directly follow from the Hamiltonian equations and the primary constraints (8). Second, adding the secondary constraints to the total Hamiltonian (7) gives an extended Hamiltonian (9) which often changes the physics of the system.

In particular, for our model we get

$$H_E = H_T + \tilde{\lambda}\tilde{\Phi} = \frac{1}{2}\pi_x^2 - y\pi_x + \lambda\pi_y + \tilde{\lambda}\pi_x,$$

with the new equations of motion

$$\begin{cases} \dot{x} = \pi_x - y + \tilde{\lambda} \\ \dot{y} = \lambda \end{cases} \quad \text{and} \quad \begin{cases} \dot{\pi}_x = 0 \\ \dot{\pi}_y = \pi_x \end{cases} \quad \text{with the constraints} \quad \begin{cases} \pi_y = 0 \\ \pi_x = 0 \end{cases}.$$

As also in the correct approach, we have both momenta equal zero; however now the two independent Lagrange multipliers make both Lagrangian variables arbitrary functions of time. In other words, we have spoiled the correspondence with the Lagrange equations. The total Hamiltonian was equivalent to the initial Lagrangian system with one gauge freedom and one physical variable, although a constrained one. At the same time, the extended Hamiltonian introduced one more arbitrary Lagrange multiplier, which resulted in a new model with two unphysical freedoms and not any single physical variable, thus rather corresponding to $L(\mathfrak{x}, \mathfrak{y}) = 0$.

Note also that primary and secondary constraints are not the same, even if the number of Lagrange multipliers is kept correct. If we had put only $\tilde{\Phi}$ in our Hamiltonian, it would have produced constant π_y but not necessarily zero, and constant y with an arbitrary $x(t)$, a totally different system.

2.3. Trivial Second-Class Constraints

The simplest example of a second-class constraint pair would be

$$L(\mathfrak{x}) = -x^2.$$

It immediately has a constraint of $\pi_x = 0$ and the naturally possible canonical Hamiltonian $H = x^2$.

The total Hamiltonian is $H_T = x^2 + \lambda\pi_x$. Its equations $\dot{x} = \lambda$ and $\dot{\pi}_x = x$ together with the primary constraint $\pi_x = 0$ reproduce the Lagrangian equation $x = 0$, which is also the secondary constraint. In this case, the Lagrangian system has one single constraint $x = 0$ which, being of non-derivative character, is enough for reducing the number of Cauchy data to zero. The Hamiltonian representation does the same by having two second-class constraints. At the same time, its Lagrange multiplier is fixed to $\lambda = 0$.

Note that the incorrectness of adding the secondary constraint to the Hamiltonian in the previous example was due to another arbitrary Lagrange multiplier changing the level of predictability. It is not necessarily the case with the second-class constraints since all the multipliers are constrained then. For example, obviously $H_E = x^2 + \lambda\pi_x + \tilde{\lambda}x$ reproduces the same Lagrangian equation of $x = 0$ and demands $\lambda = \tilde{\lambda} = 0$.

It happened because both the coordinate and its momentum are strictly fixed by the pair of constraints, which can not be changed by adding any new terms to the Hamiltonian. Related to that, and as a simple exercise about the non-uniqueness of the Hamiltonian, one can check that changing the canonical Hamiltonian (3) in this case from x^2 to, say, $\pi_x^2 + x^2$ would change neither the H_T nor the H_E results, while taking it as $\pi_x + x^2$ would shift the multiplier λ by 1 with no change to the physics either.

I will not discuss second-class constraints any longer. Let me only mention that introducing Dirac brackets is often taken as a must for the Hamiltonian analysis. This is of course not true. Their purpose is different. If we want to canonically quantize a system with second-class constraints by promoting the Poisson brackets to commutators of operators, then we cannot impose all the constraints in terms of operators, not even as the physical subspace of the Hilbert space belonging to the kernels of all the constraints, unless there is no physics and its subspace is empty. Here come Dirac brackets, as a part of an algorithm for quantitation, which we do not need in this paper.

2.4. First-Class without a Gauge

Now I take another Lagrangian

$$L(\mathfrak{x}, \mathfrak{y}) = \frac{1}{2}y\dot{x}^2$$

which gives the primary constraint of $\pi_y = 0$ and the total Hamiltonian (7)

$$H_T = \frac{\pi_x^2}{2y} + \lambda\pi_y.$$

The Lagrangian equations of motion were

$$y\ddot{x} + \dot{x}\dot{y} = 0 \qquad \text{and} \qquad \dot{x}^2 = 0$$

implying an arbitrary constant value of x and an absolutely arbitrary $y(t)$. The Hamiltonian Equation (8) yields

$$\begin{cases} \dot{x} = \frac{\pi_x}{y} \\ \dot{y} = \lambda \end{cases} \quad \text{and} \quad \begin{cases} \dot{\pi}_x = 0 \\ \dot{\pi}_y = \frac{\pi_x^2}{2y^2} \end{cases} \quad \text{with the constraint} \quad \pi_y = 0$$

with the same result for the Lagrangian variables while requiring both momenta to be zero.

Note that the secondary constraint of $\pi_x = 0$ followed from the primary constraint and the Hamiltonian equation for π_y. Of course, it is nothing but requiring that the primary constraint is preserved in time. In terms of required Cauchy data, it is one half of a degree of freedom, not zero as the traditional counting would suggest. At the same time, the set of constraints,

$$\Phi = \pi_y \qquad \text{and} \qquad \tilde{\Phi} = \pi_x,$$

is a system of two first-class constraints, with no gauge freedom in the Lagrangian model.

This model was also considered in the form of $L = e^y \dot{x}^2$ in the book [6] by Henneaux and Teitelboim. Their claim is that it has a gauge freedom for y but not for x, and therefore is a counterexample to Dirac's conjecture. My point is that the Dirac's conjecture is actually never correct. So, I do not argue against this conclusion. However, in this case there is simply no gauge freedom in the Lagrangian at all. It shows that sometimes a primary first-class constraint might be unrelated to any gauge freedom.

Why they call it a gauge system in the book [6] is because an arbitrary $y(t)$ satisfies the equations of motion. However, in this case, it is not due to a symmetry of the Lagrangian, not even up to a total derivative. It is due to another kind of singularity here: the variable y enters the Lagrangian being multiplied by a square of something which is equal to zero when equations of motion are satisfied, and therefore its contribution to the equations does not identically disappear but rather gets multiplied by something which is zero due to another equation.

If the Lagrangian was of the form $L(\mathfrak{x}, \mathfrak{y}) = f(y) g^2(x, \dot{x})$ with whatever smooth functions f, with $f' \neq 0$, and g, it would be precisely the same story with the equations of motion satisfied by any function $y(t)$ and a function $x(t)$ which is a solution of the equation $g = 0$. Indeed, y is then arbitrary. However, I will not call it a gauge symmetry. It shows that the unphysical nature of a variable, in the sense of not being determined by equations of motion, might sometimes not be coming from a gauge freedom.

On the other hand, this non-gauge freedom is indeed a kind of singularity I mentioned. If we had taken a Lagrangian $L = y\dot{x}$, it would have been a model with two constant in time Lagrangian variables governed by a zero canonical Hamiltonian and a pair of second-class constraints. From the Cauchy data counting, it has one degree of freedom, although it rather looks like two halves. An analogous result would be with $L = y(\dot{x}^2 - x^2)$ when $x \neq 0$, or in a Lorentz-invariant field model such as the mimetic fluid [11,12] $\lambda\left((\partial_\mu \phi)(\partial^\mu \phi) - 1\right)$, for in these cases the extra variable is multiplied in the Lagrangian by something which does have a non-vanishing first variation around the physical trajectory. Needless to say, if one wants to have it singular again, we can always put an extra square as in the fg^2 model mentioned above.

2.4.1. On Bifurcations, New Mimetic, Teleparallel Craze, and All That

Most likely, the main reason why Henneaux and Teitelboim [6] took the previous example with e^y instead of simply y was in order to avoid a point with zero π_x. In fact, in the case of $L = y\dot{x}^2$ the variable y being zero does not change anything in the equations of motion. The Hamiltonian looks singular when $y \to 0$, however everything works fine if we simply ignore the fact that $\frac{1}{y}$ might become infinite. On the other hand, one could try to consider an extra primary constraint of $\pi_x = 0$ when $y = 0$. If applied naively, it gives a zero canonical Hamiltonian with both momenta being primary constraints, and therefore producing an incorrect result of arbitrary $x(t)$.

In fact, the reason of nothing special at $y = 0$ is that the momentum π_x as a function of y had only a simple root, which was not enough for it acting as a fundamental constraint. If we had $L(\mathfrak{x}, \mathfrak{y}) = y^2 \dot{x}^2$, then the locus of zero y would be a special place with a bifurcation: the function $x(t)$ is indeed absolutely arbitrary there. All the $y \neq 0$ regimes can then be described as above with substitution of y by y^2, while the story at $y = 0$ goes like a zero Hamiltonian with constraints for both momenta and y being set to zero.

There is no general understanding of how to work with these bifurcations at the Hamiltonian level. Although usually it happens only at a very specific point, which we are trying to avoid as pathological, for example, like a locus of $f' = 0$ in $f(R)$ gravity. It is more problematic that, even though the bifurcations appearing at the primary level are usually quite obvious right away, they can also come at the later stages, and in a less immediately visible manner. Namely, the nature of constraints can be changing.

As yet another example of a non-gauge singular situation at the Lagrangian level, let us consider a Lagrangian

$$L(\mathfrak{x}, \mathfrak{y}) = \frac{1}{2}\Big(\dot{x}^2 - x^2(y-x)^2\Big).$$

Its total Hamiltonian with the primary constraint $\Phi = \pi_y$

$$H_T = \frac{1}{2}\Big({\pi_x}^2 + x^2(y-x)^2\Big) + \lambda \pi_y$$

produces the secondary constraint $\tilde{\Phi} = x^2(y-x)$. Their Poisson bracket is $\{\tilde{\Phi}, \Phi\} = x^2$. Therefore, at $x \neq 0$ we have a pair of second-class constraints, while at $x = 0$ it turns into two first-class constraints. The very obvious Lagrangian equations show that indeed these regimes are genuinely different. We either have $x = 0$ and totally arbitrary $y(t)$, or $y = x$ and the dynamical equation of $\ddot{x} = 0$.

A more familiar case of that is the Proca theory. When we have a vector field with the Lagrangian density $L(\mathfrak{A}) = -\frac{1}{4}F_{\mu\nu}F^{\mu\nu} + V(A_\mu A^\mu)$ with $F_{\mu\nu} \equiv \partial_\mu A_\nu - \partial_\nu A_\mu$, the usual $U(1)$ gauge symmetry is broken by the potential term V, and therefore all four components are physical with three of them being dynamical and A_0 being constrained by a pair of second-class constraints. However, if there exists a point with $V' = 0$, then the gauge invariance is spontaneously restored for linear perturbations around it, and the constraints get to have their Poisson bracket vanishing. It effectively brings us back to the gauge-invariant case with only three physical components and two of them dynamical. Therefore, the number of degrees of freedom appears to not be well-defined.

Perhaps the most cautious attitude would be to simply avoid such situations. However, sometimes people want to use them on purpose. For example, mimetic gravity [11,12] was built by a metric transformation $\tilde{g}_{\mu\nu} = g^{\alpha\beta}(\partial_\alpha\phi)(\partial_\beta\phi)g_{\mu\nu}$ from the General Relativity (GR). If the physical metric is $\tilde{g}_{\mu\nu}$ while $g_{\mu\nu}$ and ϕ are used as fundamental variables, then the role of conformal mode is played by the scalar field which, due to more derivatives, gives more solutions in the shape of an effective ideal fluid [12]. Very many different modifications have been considered [13,14]. However, one can also stick to the simple initial idea and generalize it to disformal transformations which then produce the same mimetic gravity as long as the metric transformation is not invertible, otherwise it is pure GR with a spurious scalar field on top [15]. Recently, it was noticed [16,17] that in most cases the metric transformation is invertible, except at some singular loci which then produce extra solutions of mimetic type, too. Personally, I am skeptical about using such ill-posed theories [18], but it would be interesting to learn how to work with them.

Another amazing story of recent years is modified teleparallel gravity such as $f(T)$. It breaks the local Lorentz invariance in the space of tetrads (I assume its pure tetrad formulation), with however a rather chaotic zoo of remnant symmetries [19]. The structure of the constraint algebra is so difficult that there is even no consensus on the number of degrees of freedom [20]. What is almost fully clear is that there is some new dynamics on top of the usual two modes of GR [20,21]. However, the trivial Minkowski space is an obvious strong coupling regime with the Lorentz symmetry restoration, and therefore no new modes in the linear perturbations. What is much more surprising is that no new dynamical modes are found around the usual cosmologies either [22,23], even though the Lorentz symmetry is broken there and the new features of cosmological perturbation theory arise, such as non-zero gravitational slip [22]. It might be the case that such theories do require a genuinely novel insight into the Hamiltonian methods.

2.4.2. A Brief Remark on Proca-Nuevo De Nuevo

Let me also mention that, inspired by ghost-free massive gravity [24,25], yet another vector field theory has been proposed [26], with the Lagrangian built in terms of elementary symmetric polynomials of the eigenvalues of a square root of the matrix $\delta^\mu_\nu + \partial^\mu A_\nu + \partial_\nu A^\mu + \partial^\mu A_\alpha \partial_\nu A^\alpha$. It has a primary constraint which sounds promising since dynamical temporal components are usually deadly. Very recently, a new investigation of the model appeared [27], although only in 2D. Lagrangian analysis of constraints [28] was used to claim that the primary constraint is the only one in the story. Then, assuming that there must be the same number of Hamiltonian constraints, and that it cannot be first-class in case of no gauge symmetry, the conclusion was that it is a case of a constraint which is second-class with itself, and therefore the model has $\frac{3}{2}$ degrees of freedom [27].

A self-second-class constraint is a rare and exciting situation, known for example in versions of Hořava gravity. In my opinion, it is a very interesting topic; however, given the subtleties of the dictionary between the Lagrangian and the Hamiltonian formalisms I reviewed above, it would be very important to do an explicit Hamiltonian analysis, as well as to get more Lagrangian details. Of course, the structure of the model is not very simple, but in the case of massive gravity the full Hamiltonian analysis was successfully done by the brute force approach [29,30], and some other methods of working with the square roots of matrices are also known [31,32].

3. Hamiltonian Mechanics with Constraints

Now let me discuss what happens in general, mainly for Lagrgangians quadratic in velocities. My main goal here is to show that in general a total Hamiltonian is enough to reproduce the Lagrangian physics. For demonstrating the fact that insisting on extended Hamiltonians is incorrect, it is enough to have some explicit examples, in the previous and in the following Sections.

Let us take a Lagrangian

$$L(\mathfrak{x},\mathfrak{y}) = \frac{1}{2}\dot{x}^2 + b(x,y)\dot{x} + c(x,y)\dot{y} + d(x,y) \tag{10}$$

which has the momenta $\pi_x = \dot{x} + b$ and $\pi_y = c$. It produces the primary constraint $\Phi = \pi_y - c(x,y)$, and the total Hamiltonian (7) can be chosen as

$$H_T = \frac{1}{2}(\pi_y - b)^2 - d + \lambda(\pi_y - c).$$

It is only for clarity of presentation that I take only two variables. It will not change much of what follows, if I take x as an n-dimensional vector, and y—as an m-dimensional one.

The Lagrangian equations of motion are easily found to be

$$\begin{cases} \ddot{x} = \left(\frac{\partial c}{\partial x} - \frac{\partial b}{\partial y}\right) \cdot \dot{y} + \frac{\partial d}{\partial x} \\ 0 = \left(\frac{\partial b}{\partial y} - \frac{\partial c}{\partial x}\right) \cdot \dot{x} + \frac{\partial d}{\partial y} \end{cases}$$

while the Hamiltonian ones (8) are

$$\begin{cases} \dot{x} = \pi_x - b \\ \dot{y} = \lambda \end{cases} \quad \text{and} \quad \begin{cases} \dot{\pi}_x = (\pi_x - b)\frac{\partial b}{\partial x} + \frac{\partial d}{\partial x} + \lambda\frac{\partial c}{\partial x} \\ \dot{\pi}_y = (\pi_x - b)\frac{\partial b}{\partial y} + \frac{\partial d}{\partial y} + \lambda\frac{\partial c}{\partial y} \end{cases} \quad \text{with the constraint} \quad \pi_y - c = 0.$$

Differentiating $\dot{x} = \pi_x - b$, we get $\ddot{x} = \dot{\pi}_x - \frac{\partial b}{\partial x}\dot{x} - \frac{\partial b}{\partial y}\dot{y}$ which upon substitution of the equation for $\dot{\pi}_x$, and taking into account $\pi_x - b = \dot{x}$ and $\lambda = \dot{y}$, gives the first Lagrangian equation. Analogously, differentiation of the constraint, $\dot{\pi}_y = \frac{\partial c}{\partial x}\dot{x} + \frac{\partial c}{\partial y}\dot{y}$, transforms the equation for $\dot{\pi}_y$ into the second Lagrangian equation. Therefore, the Hamiltonian method works well indeed.

What has happened here is the following. As we discussed in the Introduction, for non-degenerate systems the canonical Hamiltonian (3) has a very nice property (4) of $\frac{\partial H}{\partial \pi} = \dot{x}$

and $\frac{\partial H}{\partial x} = -\frac{\partial L}{\partial x}$, then the former equation means simply that $\pi = \frac{\partial L}{\partial \dot{x}}$ while $\dot{\pi} = -\frac{\partial H}{\partial x}$ comes back to the Lagrangian equation of $\frac{d}{dt}\frac{\partial L}{\partial \dot{x}} = \frac{\partial L}{\partial x}$. Degenerate systems are more intricate. As long as we keep the momenta strictly equal to their Lagrangian definition, the variation of the canonical Hamiltonian is still given by $\delta H = \dot{x}\delta\pi - \frac{\partial L}{\partial \dot{x}}\delta x$. However, once we have found some corresponding expression $H = H(x, \pi)$ and forgotten about $\dot{x}$, it fails to be unique and can contain arbitrary combinations of primary constraints.

In the current example, it corresponds to adding arbitrary combinations of $\pi_y - c$ to the Hamiltonian, which we explicitly took into account by adding the primary constraint term in H_T. Therefore, the equation for $\dot{x}$ has correctly found the velocity of x in terms of its momentum, while $\dot{y}$ remains arbitrary, at least at this stage. The momentum π_y did not depend on $\dot{y}$, which therefore cannot be found from the values of momenta, and the primary constraint term takes care of that by putting a Lagrange multiplier in place of the non-existing $\dot{y}(\pi_y)$. On the other hand, these additions also make the variations $\frac{\partial H}{\partial x}$ and $\frac{\partial H}{\partial y}$ different from $-\frac{\partial L}{\partial x}$ and $-\frac{\partial L}{\partial y}$, by the terms with derivatives of $c(x, y)$, which are taken care of by the primary constraint in H_T, too.

Therefore, different values of λ in H_T correspond to all possible definitions of the canonical Hamiltonian with the correct variation at the primary constraint surface. When we impose also the equation of the primary constraint, $\pi_y - c = 0$ in our case, it comes back to the desired variations. Either a normal to constraint surface variation determines the only correct λ, which is then like the contact force acting on a particle from the surface on which it moves, and the variable y appears to be physical but constrained, or we have it simply arbitrary, and this is then a case of gauge freedom and its induced primary first-class constraint.

Note that we have checked it without ever talking about the secondary constraints. The equations simply tell us everything about the Lagrangian variables, and looking at other consequences is needed only for counting the number of degrees of freedom. In particular, if $\frac{\partial b}{\partial y} \equiv \frac{\partial c}{\partial x}$, so that $b\dot{x} + c\dot{y}$ is a total time derivative, and $\frac{\partial d}{\partial y} \equiv 0$, then $y(t)$ is a purely gauge mode, the first Lagrangian equation is the only one and it is dynamical; at the same time the Hamiltonian formalism has an arbitrary Lagrange multiplier and one first-class constraint, just for the formal definition of the momentum of the unphysical variable.

Otherwise, both variables are physical but we are getting the secondary constraint expressed by the second Lagrangian equation. In this case there is a pair of second-class Hamiltonian constraints. The Lagrangian formalism immediately has a constraint for $\dot{x}$, and substituting it into the equation for $\ddot{x}$ we get another constraint for $\dot{y}$. In the Hamiltonian language, the restriction of $\dot{y}$ comes as a fixed value of the Lagrange multiplier λ required by preservation of the secondary constraint. All in all, we have two Cauchy data and therefore one, or two halves, degree of freedom. A special case would be if $\frac{\partial b}{\partial y} \equiv \frac{\partial c}{\partial x}$ is still satisfied but with $\frac{\partial d}{\partial y}$ not identically zero. Then y is fully constrained to an extremum of d while $x(t)$ is having a fully dynamical equation of motion.

Note that if x and y, as well as b and c, were multidimensional, it could potentially cause a complicated constraint algebra, in terms of Poisson brackets, with many possible secondary constraints and so on. However, the formal demonstration that the Lagrangian and the Hamiltonian (H_T) equations do tell us the same physics would go with no big change.

A Hint of More General Considerations

We have just discussed the Hamiltonian formalism for Lagrangian systems quadratic in velocities. Let me now give some abstract description of how it should work in general. Imagine we have a Lagrangian $L(\mathfrak{x}_i)$ with its equations

$$\frac{\partial^2 L}{\partial \dot{x}_i \partial \dot{x}_j}\ddot{x}_j + \frac{\partial^2 L}{\partial \dot{x}_i \partial x_j}\dot{x}_j - \frac{\partial L}{\partial x_i} = 0$$

and are trying to construct a Hamiltonian (3) for it.

Let us assume that, around a point of the configuration space we are looking at, there exists a matrix Q such that

$$\frac{\partial^2 L}{\partial \dot{x}_i \partial \dot{x}_j} = Q \cdot \begin{pmatrix} \ell & 0 \\ 0 & 0 \end{pmatrix} \cdot Q^{-1}$$

with $\ell \neq 0$ and $Q^T = Q^{-1}$, with ^{T} now meaning "transpose". Only to make it less cumbersome have I taken it as two-dimensional, and it can be thought of as two multidimensional blocks with ℓ then being a non-degenerate matrix.

From the definition of momenta (1), $\pi_{x_i} \equiv \frac{\partial L}{\partial \dot{x}_i}$, we get the following variations of them in terms of the Lagrangian velocities:

$$\overrightarrow{\delta\pi} = Q \begin{pmatrix} \ell & 0 \\ 0 & 0 \end{pmatrix} Q^{-1} \overrightarrow{\delta\dot{x}} + \left(\frac{\partial^2 L}{\partial \dot{x} \partial x} \right) \overrightarrow{\delta x}. \tag{11}$$

If I denote the projector to the second component as $P_2 \equiv \begin{pmatrix} 0 & 0 \\ 0 & 1 \end{pmatrix}$, then the primary constraint in terms of the variations acquires the form of

$$P_2 Q^{-1} \left(\overrightarrow{\delta\pi} - \left(\frac{\partial^2 L}{\partial \dot{x} \partial x} \right) \overrightarrow{\delta x} \right) = 0 \tag{12}$$

showing that QP_2Q^{-1} is the projector to the direction in which the momentum $\vec{\pi}$ can only be changed due to a change of coordinates and does not feel the velocities. At the same time, in the complimentary direction of QP_1Q^{-1}, there is a fixed relation between the momentum and the velocity.

Then, as we have discussed before, even in the degenerate case with the non-invertible relation (11) between variations, we can build a canonical Hamiltonian (3) trying to reproduce the desired variations (4) as

$$\delta H = \sum_i \left(\dot{x}_i \delta\pi_{x_i} - \frac{\partial L}{\partial x_i} \delta x_i \right).$$

What I have swept under the rug is the displeasing question of which precise velocity stays there. Even if we do only a variation allowed by the constraint, it can actually be any velocity that corresponds to the value of $\vec{\pi}$, therefore near our point it is, roughly speaking, something definite in QP_1Q^{-1} subspace, and it can be arbitrary in QP_2Q^{-1}.

What exactly we get, depends on how we construct the canonical Hamiltonian. Recall that it is not unique. Our usual way was to avoid adding any terms proportional to the primary constraint. For example, in the calculations above with the Lagrangian (10) we could have added something proportional to $\pi_y - c$ to the canonical Hamiltonian. However, we did not. This was a way to avoid the unwanted velocity in the term $(\pi_y - c)\dot{y}$ by simply putting it to zero. This way, not only did we expel π_y from the canonical Hamiltonian but also got rid of the function $c(x, y)$ in it.

In other words, we construct the canonical Hamiltonian from only the "QP_1Q^{-1}-part" of momenta which, together with keeping track of the variational relation (12), means that we are losing some of the $\overrightarrow{\delta x}$ variations. Without trying to give a good mathematical meaning to all that, let me say that we basically get only the QP_1Q^{-1} projection of $\dot{x}_i$ from the $\frac{\partial H}{\partial \pi_{x_i}}$ derivative, while in calculating $\overrightarrow{\frac{\partial L}{\partial x}}$ we first subtract a term such as $\overrightarrow{\dot{x}}^T \cdot QP_2Q^{-1} \overrightarrow{\frac{\partial L}{\partial \dot{x}}}$ from the Lagrangian. In our previous example (10), we can check it by putting $\lambda = 0$ for now in its Hamiltonian equation. Indeed, the $\frac{\partial H}{\partial \pi}$ part gave only the non-trivial value of $\dot{x}$ having $\dot{y} = 0$, while the $\frac{\partial H}{\partial x}$ equations lack derivatives of $c(x, y)$ since it had dropped off from the canonical Hamiltonian.

What was removed from the equations by this restriction on the canonical Hamiltonian, is restored by adding the $\lambda\Phi$ term to the total Hamiltonian H_T. Indeed, the variation of the constraint has the form of Equation (12), and then it gives an arbitrary QP_2Q^{-1} component of the velocity and a $QP_2Q^{-1}\frac{\partial^2 L}{\partial\dot{x}\partial x}$ term to the equation for $\dot{\pi}$, therefore taking care of the change in variation of x in the canonical piece. Note that this is precisely what happened with λ-terms in the Hamiltonian equations for the model (10).

The important point is that, precisely as before, it has restored too much. We have got an arbitrary value of the velocity component corresponding to the kernel of the Hessian. It is expected that the multiplier λ either remains arbitrary, which usually shows a case of gauge symmetry, or it will be fixed by equations (and give a secondary second-class constraint).

Needless to say, what I have given here is not a mathematical proof of the Lagrangian–Hamiltonian correspondence for beyond the quadratic-in-velocities case. For example, we would have to worry about the arguments of the coefficients given by partial derivatives of the Lagrangian, and so on. It is an interesting problem, but I only meant to provide a gist of how it should work.

4. Our Usual Field Theories

After having discussed the mechanical cases, which in my opinion are very important for understanding the theoretical foundations and the workings of the method, I turn to the field theories we often use in physics, namely electromagnetism and gravity.

4.1. Electrodynamics

I will consider Maxwell electrodynamics in vacuum. Therefore, the action is very simple,

$$L(\mathfrak{A}) = -\frac{1}{4}F_{\mu\nu}F^{\mu\nu} \qquad \text{with} \qquad F_{\mu\nu} \equiv \partial_\mu A_\nu - \partial_\nu A_\mu \qquad \text{by definition.} \tag{13}$$

The momenta (1) are easily found to be

$$\pi_0 = 0 \qquad \text{and} \qquad \pi_i = F_{0i} = \dot{A}_i - \partial_i A_0,$$

from which we immediately see the primary constraint $\Phi = \pi_0$ and the total Hamiltonian

$$H_T = \frac{1}{2}\pi_i^2 + \pi_i\partial_i A_0 + \frac{1}{4}F_{ij}^2 + \lambda\pi_0.$$

As before, we could have added an arbitrary function of π_0 to the canonical Hamiltonian, but it is anyway taken care of by the primary constraint term.

Note that the Lagrangian equations of motion are

$$\partial_\mu F^{\mu\nu} = 0.$$

At the same time, the Lagrangian has a gauge symmetry of $A_\mu \longrightarrow A_\mu + \partial_\mu\psi$, therefore only three components out of four are physical. And still, not all of them are dynamical because one of the equations, the temporal one, is a constraint, $0 = \partial_i F_{0i} = \partial_i\dot{A}_i - \triangle A_0$. Naively, it looks like reducing the number of degrees of freedom by $\frac{1}{2}$. However, one possible gauge choice is $\partial_i A_i = 0$, which can be achieved by the gauge transformation with ψ such that $\triangle\psi = -\partial_i A_i$, and then the result is a full strength constraint of $\triangle A_0 = 0$. We have two dynamical variables then and one more physical but constrained. Of course, the latter corresponds to the Gauß's law of $\overrightarrow{\partial}\cdot\overrightarrow{E} = 0$ in absence of electric charges, with $\overrightarrow{E}$ being the electric field.

The Hamiltonian equations we get are

$$\begin{cases} \dot{A}_0 = \lambda \\ \dot{A}_i = \pi_i + \partial_i A_0 \end{cases} \qquad \text{and} \qquad \begin{cases} \dot{\pi}_0 = \partial_i\pi_i \\ \dot{\pi}_i = \partial_j F_{ji} \end{cases} \qquad \text{with the constraint} \qquad \pi_0 = 0. \tag{14}$$

Again, it fully reproduces the Lagrangian theory. Indeed, from the primary constraint and the equation for $\dot{\pi}_0$, which is also simply $\{\Phi, H_T\}$, we get the secondary constraint

$$\partial_i \pi_i = 0$$

which, using the equation for $\dot{A}_i$, immediately reproduces the Lagrangian constraint. We easily find the dynamical equations:

$$\partial_j F_{ji} = \dot{\pi}_i = \ddot{A}_i - \partial_i \dot{A}_0 = \dot{F}_{0i}.$$

Incidentally, we see that the full gauge invariance is there: only the combinations of $F_{\mu\nu}$ matter. Again, naively, the equation for A_0 looks like requiring one Cauchy datum. However, this is of course spurious since it makes no sense to worry about an initial value of something whose derivative is given by an arbitrary function. At a deeper level, this is irrelevant because having fixed the value of A_0, we are still allowed to do another gauge transformation with a parameter depending on spatial coordinates only, while if we have fixed the value of $\dot{A}_0$, then we can still have another gauge parameter which is at most linear in time. This symmetry is not so clear from the Hamiltonian representation. However it is also there because the vector potential A_i comes only in combinations of F_{0i} and F_{ij}, even in the Hamiltonian equations.

On Incorrectness of Extended Hamiltonians Again

Let me state again that extended Hamiltonians are not correct descriptions of gauge theories. In this case we would get

$$H_E = \frac{1}{2}\pi_i^2 + \pi_i \partial_i A_0 + \frac{1}{4}F_{ij}^2 + \lambda \pi_0 + \tilde{\lambda}\partial_i \pi_i$$

with an extra, and totally unjustified, arbitrary variable $\tilde{\lambda}$ in the equations. Indeed, the only modification is to one of the equations:

$$\dot{A}_i = \pi_i + \partial_i A_0 + \partial_i \tilde{\lambda}.$$

Now we are allowed to change temporal and longitudinal fields independently, and this is totally wrong. It is not the usual electrodynamics! It is all right for perturbatively quantizing photons in vacuum, but it totally neglects Coulomb forces as if those were unphysical.

Indeed, if we accept both Lagrange multipliers, even the electric field is not physical. We can then produce whatever effective electric charges: $\partial_i F_{0i} = \triangle \tilde{\lambda}$, out of nothing. Note that the case with $\tilde{\lambda} = \tilde{\lambda}(t)$ changes nothing, and this is because of the symmetry of our Hamiltonian: $\tilde{\lambda}(t)\partial_i \pi_i \to \pi_i \partial_i \tilde{\lambda}(t) = 0$. However, allowing here for two fully arbitrary Lagrange multipliers is wrong. It would be as if I could freely do gauge transformations for temporal and spatial components separately $\begin{cases} A_0 \longrightarrow A_0 + \dot{\phi} \\ A_i \longrightarrow A_i + \partial_i \psi \end{cases}$ with independent $\phi = \int dt \int dt\, \lambda$ and $\psi = \int dt \int dt\, \lambda + \int dt\, \tilde{\lambda}$, which is not a symmetry of electrodynamics.

In particular, sometimes researchers state [10,33] that in electrodynamics we must fix two gauges, $A_0 = 0$ and $\partial_i A_i = 0$. No, this is not a gauge choice! As I showed above as well, we cannot arbitrarily change both temporal and longitudinal fields. This choice is possible only in electrodynamics without sources, and it is not only a gauge choice but also a solution of the Lagrangian constraint equation. Indeed, since we have the Gauß's law (for vanishing electric charge density)

$$\partial_i \dot{A}_i - \triangle A_0 = 0,$$

once we have chosen a gauge of $A_0 = 0$, we get $\frac{d}{dt}(\partial_i A_i) = 0$. Then it can be reduced to $\partial_i A_i = 0$ using a gauge parameter ψ depending on spatial coordinates only, which is a totally legitimate remnant symmetry then. By no means is it a new gauge choice. It is simply the fact that $A_0 = 0$ does not fully fix a gauge.

Out of pure curiosity, note also that our total Hamiltonian Equation (14) has reproduced the initial definition of the spatial momenta: $\pi_i = \dot{A}_i - \partial_i A_0$ which is not surprising since the definition of momenta was invertible in its spatial part. This success is totally destroyed by introduction of the extended Hamiltonian. If we demand it by hand however, then the extended Hamiltonian picture comes back to the real physics, and fixes the new Lagrange multiplier to be a function of only time, which makes it practically non-existent together with any incorrect effects it has produced. However, there is no natural reason for doing so on the Hamiltonian side.

4.2. General Relativity

Another important gauge theory we have is General Relativity. Its particularly convenient representation for the analysis is in the ADM variables [34]. Those are given in terms of the metric written as

$$g_{\mu\nu}dx^\mu dx^\nu = (N^2 - N_i N_j \gamma^{ij})dt^2 - 2N_i dt dx^i - \gamma_{ij} dx^i dx^j.$$

Therefore, we use the spatial metric $\gamma_{ij} \equiv -g_{ij}$ and all the spatial indices are operated on by this spatial metric, while all the other metric components are expressed in terms of new variables, the lapse N and shift N_i.

Using the extrinsic curvatures of the constant time slices

$$K_{ij} = \frac{1}{2N}\left(\overset{(3)}{\nabla}_i N_j + \overset{(3)}{\nabla}_j N_i - \dot{\gamma}_{ij}\right),$$

and after some very straightforward but relatively cumbersome algebra [35], one can find the curvature scalar

$$\sqrt{-g}R = \sqrt{\gamma}N\left(\overset{(3)}{R} + K^{ij}K_{ij} - K^i_i K^j_j\right) - 2\partial_0\left(\sqrt{\gamma}K^i_i\right) + 2\sqrt{\gamma}\overset{(3)}{\nabla}_j\left(K^i_i N^j\right) - 2\sqrt{\gamma}\overset{(3)}{\triangle}N$$

in terms of that for the spatial metric. Neglecting the total derivatives, we come to the Lagrangian

$$L(\mathfrak{N}, \mathfrak{N}_i, \mathfrak{g}_{ij}) = \sqrt{\gamma}N\left(\overset{(3)}{R} + K^{ij}K_{ij} - K^i_i K^j_j\right). \tag{15}$$

Then we see that $\pi_N = \pi_{N_i} = 0$ while $\pi^{ij} \equiv \pi_{\gamma_{ij}} = \sqrt{\gamma}\left(K^k_k \gamma^{ij} - K^{ij}\right)$, and the total Hamiltonian is

$$H_T = -\int d^3x \sqrt{\gamma}\left(N\left(\overset{(3)}{R} + \frac{1}{\gamma}\left(\frac{1}{2}\left(\pi^j_j\right)^2 - \pi_{ik}\pi^{ik}\right)\right) + 2N^i \overset{(3)}{\nabla^k} \pi_{ik}\right) + \lambda\pi_N + \lambda_i \pi_{N_i}.$$

The secondary constraints are directly obvious from the Hamiltonian. More work is required to check that all the constraints are first-class, which is quite natural however because all the four pairs again come from a gauge symmetry that mixes temporal and spatial derivatives. In other words, the secondary constraints come about because the gauge transformations are not simply changes of lapse and shift.

It is very common in the literature to omit the λ-terms, and to simply treat this Hamiltonian system as the one with zero canonical Hamiltonian, the secondary constraints in the role of primary ones, and the lapse and shift as Lagrange multipliers. In this case, it is in fact correct since the equations are not changed then, except for renaming some field variables into Lagrange multipliers. Indeed, if I change from $H_T = yf(x, \pi_x) + \lambda\pi_y$ to $H_T = \lambda f(x, \pi_x)$, the only difference then is that the equations for x and π_x instead of the factor of y, which was arbitrary anyway as $\dot{y} = \lambda$, now simply get another arbitrary factor, that of λ itself. The only point to remember then is that the diffeomorphisms become somewhat hidden since they now involve transformations of the new Lagrange multipliers, which is unusual.

As in the case of electrodynamics, the naive counting of dynamical degrees of freedom works well. From the 10 variables, they subtract 8 first-class constraints and get 2 degrees of freedom. What is however wrong is to claim that there are only two physical combinations of the metric components in General Relativity. That would mean to recognize only gravitational waves (GWs) as physical and to neglect all the Newtonian forces. This is not how the world is built.

The best way to see that there are more physical components is to study the cosmological perturbation theory [36]. The coordinate change functions δx^{μ} are all the four independent gauge parameters, and they leave six variables physical. On top of the two graviton polarizations, there are two components of a physical vector and two physical scalars (Bardeen potentials). While the vectors are not very important for cosmology because they are decaying almost inevitably, the scalars play an instrumental role in calculating the CMB observables and the large scale structure. They are not dynamical by themselves, and the acoustic waves appear because of the matter contents of the Universe, but how these waves evolve and lead to structure formation is crucially governed by gravity, and not just in its two GW polarizations.

4.2.1. Gravity with Tetrads—Not Every Gauge Hits Twice

So, as we have seen, the familiar gauge symmetries normally "hit twice" because of mixing different derivatives of different components. At the same time, it is not difficult to construct examples of not hitting twice. In the case of electrodynamics, I could have put $A_{\mu} - B_{\mu}$ everywhere instead of only the simple vector A_{μ}. This is a very trivial symmetry, and it would be associated with four new first-class constraints of $\pi_{A_{\mu}} + \pi_{B_{\mu}} = 0$, which would not "hit twice" and would really be the new gauge generators precisely by themselves.

However, in gravity we have a more appealing way to go. Namely, let us write the GR in terms of tetrads. At the same time, I will not introduce any spin connection. What I have in mind is teleparallel-type generalizations of general relativity rather than coupling fermions. We take

$$L(e^{a}_{\mu}) = -\sqrt{g}R(g) \qquad \text{where} \qquad g_{\mu\nu} \equiv \eta_{ab}e^{a}_{\mu}e^{b}_{\nu}$$

and get the gauge symmetry of Lorentz rotations $e^{a}_{\mu} \longrightarrow \Lambda^{a}_{b}e^{b}_{\mu}$. Since the metric is invariant under such transformations, and with no derivatives of Λ involved in them, it is very easy to see that there appear six Lorentz constraints, which do not hit twice. They simply subtract the six newly introduced variables (from 10 components of the metric to 16 of the tetrad), and do not do anything else.

An interesting point is that I was not entirely honest in the paragraph above. It is that simple only if our Lagrangian is indeed written in terms of the metric. However, there is also the Teleparallel Equivalent of General Relativity (TEGR), which is the basis of modified teleparallel theories. The action of TEGR is not given in terms of the metric. Its Lagrangian is locally Lorentz invariant only up to a surface term (which is its difference from the usual Lagrangians of GR), and therefore the Lorentz behavior of the Hamiltonian formalism becomes much more difficult and interesting [9].

4.2.2. On the Canonicity of ADM

The final topic I want to touch upon is the strange claims that appeared some time ago [37] and continue to be seen in the current literature, including this Topical Collection [38]. The main statement there is that the ADM formalism is not a canonical Hamiltonian treatment of general relativity, and is not equivalent to Dirac's approach to

Hamiltonian relativity. Of course, it simply cannot be true. The only difference between different approaches to Hamiltonian GR is in the boundary terms omitted in order to not have higher derivatives and to not bother with unnecessary complications of the Ostrogradsky procedure, and also in the choice of variables.

The ADM variables of course complicate the way diffeomorphisms look, simply because those are different variables. Nevertheless, all the symmetries are there, since the Lagrangian density is still a scalar density, up to a boundary term. Nothing is done there, except rewriting it in less obviously symmetric variables. Castellani [5], for example, was able to trace the gauge transformations there. It does not matter if we have to parametrize the gauge transformations in a more difficult way for finding them in one procedure or another. One can add any function of canonical variables multiplied by a primary constraint to the canonical Hamiltonian. It will not change any physics, but will reparametrize the Lagrange multiplier. The only real question is whether the symmetry is there—and it is.

However, one point I would like to specifically mention is their claim that ADM variables are not canonical variables for gravity [37]. First of all, there is nothing sacred about being canonical. If I do a non-canonical transformation, it simply means that my Poisson bracket will be changed, and my Hamiltonian equations will look less nice. Moreover, one and the same theory can be written with different choices of canonical variables, simply by having different Hamiltonians. One way to get that is by using boundary terms. For example, Lagrangians $L(\mathfrak{x},\mathfrak{y}) = \frac{1}{2}(\dot{x}^2+\dot{y}^2)$ and $L(\mathfrak{x},\mathfrak{y}) = \frac{1}{2}(\dot{x}^2+\dot{y}^2) + x\dot{y} + y\dot{x}$ are equivalent. However, they have different definitions of momenta. Therefore, canonical variables of one of them are not canonical for another one. The difference has no physical implication however, because the Hamiltonians are also different and the final equations for x and y are equivalent.

Therefore, different Hamiltonian formulations of GR might indeed have different symplectic structures, and it does not mean that only one of them is the correct one. On the other hand, they argue so by saying that the change of variables from metric to ADM is not canonical [37]. This is not true either. If two theories are obtained from each other by only a change of Lagrangian variables, then their canonical variables are related by a conformal transformation.

It is very simple to explain what went wrong with their calculation of $\{N, \pi^{ij}\}$ in which they got a non-zero result [37]. If we go from one set of variables to another, the momenta are also changed, even for those variables that are not changed. In particular, $\dot{g}_{00} = 2N\dot{N} - 2N_i\dot{N}_j\gamma^{ij} + N^iN^j\dot{g}_{ij}$. Therefore, if the new variables are N, N_i and g_{ij}, then

$$\pi^{ij}_{\text{ADM}} = \pi^{ij}_{\text{metric}} + \pi^{00}_{\text{metric}}N^iN^j = \pi^{ij}_{\text{metric}} + \pi^{00}_{\text{metric}}\frac{g^{0i}g^{0j}}{(g^{00})^2},$$

and therefore

$$\{N, \pi^{ij}_{\text{ADM}}\} = \left\{\frac{1}{\sqrt{g^{00}}}\,,\, \pi^{ij}_{\text{metric}} + \pi^{00}_{\text{metric}}\frac{g^{0i}g^{0j}}{(g^{00})^2}\right\} = 0.$$

Of course, $\pi^{00} = 0$ in the Dirac approach, so that it would mean that $\pi^{ij}_{\text{ADM}} = \pi^{ij}_{\text{metric}}$ on shell, or in terms of Lagrangian velocities, so that a direct and naive implementation of the change of variables in the degenerate Lagrangian would not be canonical in this sense. However, barring the possibility of different neglected boundary terms, the transition from metric to ADM variables is a canonical transformation if we employ the transformations such as $\pi^{ij}_{\text{ADM}} = \pi^{ij}_{\text{metric}} + \pi^{00}_{\text{metric}}N^iN^j$, not changing anything except indeed reparametrizing the primary Lagrange multipliers in terms of canonical variables. This is yet another manifestation of the canonical Hamiltonian (3) not being unique for a degenerate system.

5. Conclusions

My main conclusion is that we should start being more serious about the constrained Hamiltonian analysis and stop taking it as simply some sacred prescription that has to be taken with no deep thought. Even the simplest examples can demonstrate very interesting behavior. Moreover, a very important point is that the adjectives of "physical" and "dynamical" are not the same. Both Maxwell electrodynamics and general relativity have only two dynamical modes each, but it is incorrect to say the same about their physical modes.

When it comes to our usual theories, they are well understood and the Hamiltonian formalism indeed works well even in the most naive ways. However, nowadays we are also studying many new and exotic fields theories, up to often classifying all possible models in a certain big class. In these situations, and with much less good physical intuition currently available, we must exercise even more care than ever before.

Funding: This research received no external funding.

Data Availability Statement: There are no new data related to this study.

Conflicts of Interest: The author declares no conflict of interest.

References

1. Dirac, P.A.M. *Lectures on Quantum Mechanics*; Dover Publications: New York, NY, USA, 2001.
2. Arnold, V.I. *Mathematical Methods of Classical Mechanics*; Springer: Berlin/Heidelberg, Germany, 1989.
3. Shapere, A.; Wilczek, F. Classical Time Crystals. *Phys. Rev. Lett.* **2012**, *109*, 160402. [CrossRef]
4. Dirac, P.A.M. Generalized Hamiltonian Dynamics. *Can. J. Math.* **1950**, *2*, 129. [CrossRef]
5. Castellani, L. Symmetries in constrained Hamiltonian systems. *Ann. Phys.* **1982**, *143*, 357. [CrossRef]
6. Henneaux, M.; Teitelboim, C. *Quantization of Gauge Systems*; Princeton University Press: Princeton, NJ, USA, 1994.
7. Gitman, D.M.; Tyutin, I.V. *Quantization of Fields with Constraints*; Springer: Berlin/Heidelberg, Germany, 1990.
8. Rothe, H.J.; Rothe, K.D. *Classical and Quantum Dynamics of Constrained Hamiltonian Systems*; World Scientific: Singapore, 2010.
9. Golovnev, A.; Guzmán, M.J. Lorentz symmetries and primary constraints in covariant teleparallel gravity. *Phys. Rev. D* **2021**, *104*, 124074. [CrossRef]
10. Valenzuela, M. Quantization of a pseudoclassical system with gauge and time-reparametrization invariance. *arXiv* **2022**, arXiv:2212.02414.
11. Chamseddine, A.H.; Mukhanov, V. Mimetic Dark Matter. *J. High Energy Phys. JHEP* **2013**, *135*, 1–5. [CrossRef]
12. Golovnev, A. On the recently proposed Mimetic Dark Matter. *Phys. Lett. B* **2014**, *728*, 39. [CrossRef]
13. Sebastiani, L.; Vagnozzi, S.; Myrzakulov, R. Mimetic gravity: A review of recent developments and applications to cosmology and astrophysics. *Adv. High Energy Phys.* **2017**, *2017*, 3156915. [CrossRef]
14. Chamseddine, A.H.; Mukhanov, V. Resolving Cosmological Singularities. *J. Cosmol. Astropart. Phys.* **2017**, *2007*, 009. [CrossRef]
15. Deruelle, N.; Rua, J. Disformal Transformations, Veiled General Relativity and Mimetic Gravity. *J. Cosmol. Astropart. Phys.* **2014**, *2014*, 002. [CrossRef]
16. Jiroušek, P.; Shimada, K.; Vikman, A.; Yamaguchi, M. Disforming to Conformal Symmetry. *J. Cosmol. Astropart. Phys.* **2022**, *2022*, 019. [CrossRef]
17. Jiroušek, P.; Shimada, K.; Vikman, A.; Yamaguchi, M. New Dynamical Degrees of Freedom from Invertible Transformations. *arXiv* **2022**, arXiv:2208.05951.
18. Golovnev, A. The Variational principle, Conformal and Disformal transformations, and the degrees of freedom. *J. Math. Phys.* **2023**, *64*, 012501. [CrossRef]
19. Ferraro, R.; Fiorini, F. Remnant group of local Lorentz transformations in f(T) theories. *Phys. Rev. D* **2015**, *91*, 064019. [CrossRef]
20. Golovnev, A.; Guzmán, M.J. Foundational issues in f(T) gravity theory. *Int. J. Geom. Methods Mod. Phys.* **2021**, *18*, 2140007. [CrossRef]
21. Golovnev, A.; Guzmán, M.J. Non-trivial Minkowski backgrounds in f(T) gravity. *Phys. Rev. D* **2021**, *103*, 044009. [CrossRef]
22. Golovnev, A.; Koivisto, T. Cosmological perturbations in modified teleparallel gravity models. *J. Cosmol. Astropart. Phys.* **2018**, *2018*, 012. [CrossRef]
23. Bahamonde, S.; Dialektopoulos, K.F.; Hohmann, M.; Said, J.L.; Pfeifer, C.; Saridakis, E.N. Perturbations in Non-Flat Cosmology for f(T) gravity. *arXiv* **2022**, arXiv:2203.00619.
24. de Rham, C.; Gabadadze, G.; Tolley, A.J. Resummation of Massive Gravity. *Phys. Rev. Lett.* **2011**, *106*, 231101. [CrossRef]
25. de Rham, C. Massive gravity. *Living Rev. Relativ.* **2014**, *17*, 7. [CrossRef]
26. de Rham, C.; Pozsgay, V. New class of Proca interactions. *Phys. Rev. D* **2020**, *102*, 083508. [CrossRef]
27. Errasti Díez, V. (Extended) Proca-Nuevo under the two-dimensional loupe. *arXiv* **2022**, arXiv:2212.02549.

28. Errasti Díez, V.; Maier, M.; Méndez-Zavaleta, J.A.; Tehrani, M.T. A Lagrangian constraint analysis of first order classical field theories with an application to gravity. *Phys. Rev. D* **2020**, *102*, 065015. [CrossRef]
29. Hassan, S.F.; Rosen, R.A. Resolving the Ghost Problem in non-Linear Massive Gravity. *Phys. Rev. Lett.* **2012**, *108*, 041101. [CrossRef] [PubMed]
30. Hassan, S.F.; Rosen, R.A. Confirmation of the Secondary Constraint and Absence of Ghost in Massive Gravity and Bimetric Gravity. *J. High Energy Phys.* **2012**, *2012*, 123. [CrossRef]
31. Golovnev, A. On the Hamiltonian analysis of non-linear massive gravity. *Phys. Lett. B* **2012**, *707*, 404. [CrossRef]
32. Golovnev, A.; Smirnov, F. Dealing with ghost-free massive gravity without explicit square roots of matrices. *Phys. Lett. B* **2017**, *770*, 209. [CrossRef]
33. Ferraro, R. Noether's second theorem in teleparallel gravity. *arXiv* **2022**, arXiv:2210.13541.
34. Arnowitt, R.; Deser, S.; Misner, C.W. Republication of: The dynamics of general relativity. *Gen. Relativ. Gravit.* **2008**, *40*, 1997. [CrossRef]
35. Golovnev, A. ADM analysis and massive gravity. In Proceedings of the 7th Mathematical Physics Meeting: Summer School and Conference on Modern Mathematical Physics, Belgrade, Serbia, 9–19 September 2012.
36. Mukhanov, V.F.; Feldman, H.A.; Brandenberger, R.H. Theory of cosmological perturbations. *Phys. Rep.* **1992**, *215*, 203. [CrossRef]
37. Kiriushcheva, N.; Kuzmin, S.V. The Hamiltonian formulation of General Relativity: Myths and reality. *Cent. Eur. J. Phys.* **2011**, *9*, 576. [CrossRef]
38. Frolov, A.M. Metric Gravity in the Hamiltonian Form—Canonical Transformations—Dirac's Modifications of the Hamilton Method and Integral Invariants of the Metric Gravity. *Universe* **2022**, *8*, 533. [CrossRef]

universe

MDPI

Article

Realization of Bounce in a Modified Gravity Framework and Information Theoretic Approach to the Bouncing Point

Sanghati Saha and Surajit Chattopadhyay *

Department of Mathematics, Amity University Kolkata, Major Arterial Road, Action Area II, Newtown, Rajarhat, Kolkata 700135, India

* Correspondence: schattopadhyay1@kol.amity.edu or surajitchatto@outlook.com; Tel.: +91-9830736116 or +91-8240384649

Abstract: In this work, we report a study on bouncing cosmology with modified generalized Chaplygin Gas (mgCG) in a bulk viscosity framework. Reconstruction schemes were demonstrated in Einstein and modified $f(T)$ gravity framework under the purview of viscous cosmological settings. We also took non-viscous cases into account. We studied the equation of state (EoS) parameter under various circumstances and judged the stability of the models through the sign of the squared speed of sound. We observed the mgCG behaving like avoidance of Big Rip in the presence of bulk viscosity at the turnaround point and in non-viscous cases, a phantom-like behavior appears. The turnaround point equation of state parameter crosses the phantom boundary, violating NEC. The role of the mgCG's model parameters was also investigated before and after the bounce. A Hubble flow dynamics was carried out and, it was revealed that mgCG is capable of realizing an inflationary phase as well as an exit from inflation. An $f(T)$ gravitational paradigm was also considered, where the mgCG density was reconstructed in the presence of bulk viscosity. The role of the parameters associated with the bouncing scale factor, describing how fast the bounce takes place, was also studied in this framework. Finally, the reconstructed mgCG turned out to be stable against small perturbations irrespective of the presence of bulk viscosity and modified gravity scenario. Finally, the reconstruction scheme was assessed using statistical analysis, Shannon entropy.

Keywords: modified generelized Chaplygin gas; bulk viscosity; bounce scale factor; Einstein gravity; modified $f(T)$ gravity; null energy condition; Big Rip singularity; Shannon entropy

1. Introduction

Countless observational studies of the cosmic universe, viz. Ia Supernovae (SNe Ia) [1–4], the Cosmic Microwave Background (CMB) [5,6], Wilkinson Microwave Anisotropy Probe(WMAP) [7–11], large scale redshift surveys structure (LSS) [12–17], show that the current universe is in a phase of accelerated expansion [18]. Dark energy (DE) is the term given to a hypothetical fluid with negative pressure that is spearheading the universe's accelerating expansion. This gravitationally repulsive exotic energy exerts a repulsion characterized by the equation of state (EoS) parameter $w = \frac{p}{\rho} < -1/3$. Almost three fourths of total mass-energy density of the universe [19] is made up of this unknown cosmic fluid. Like dark matter, it is invisible. It can be detected from astrophysical observations between cosmic entities. Dark energy still remains an unknown form of energy. DE, the exotic energy, is in the stress–energy–momentum tensor of the field equations in Einstein's theory of general relativity. There is a direct correlation between the matter–energy of the universe and the global or local geometry of the universe in the General Theory of Relativity (GTR) [20–24]. DE is detected by its impact on the rate at which the universe expands and its effect on the rate at which galaxies and gravitationally bound groups of galaxies form through gravitational perturbations. The oldest explanation of dark energy is that it is an innate, fundamental energy density inherent to vacuum energy, which is the key ingredient in Einstein's cosmological constant [25]. In [26], the method of regularization of vacuum energy is observed. Martin et al. briefly described the method of measurement of the cosmological constant as being constrained by experiments with planet orbits in our solar system or atomic spectra in cosmology. In [27], the global nature of the cosmological constant problem is emphasized by highlighting the significance of radiative instability as well as Weinberg's no-go theorem. In this lecture series, the defined sequester is a new representational approach to the cosmological constant problem that incorporates both general relativity and renormalization by cancellation of the standard model vacuum energy.

The simplest candidate of DE is Einstein's cosmological term symbolized as Λ with $\omega_\Lambda = -1$ [28–31], introduced by Albert Einstein to augment his field equations of the General Theory of Relativity. Numerous dark energy models having a time varying EoS parameter supported by observations such as quintessence [32–34], phantom [35–38], quintom [39–42], tachyon field [43–45], hessence [46–49], k-essence [50,51], Chaplygin gas [52–55], holographic dark energy [56–60], etc., have been presented to find out the cause behind the cosmic accelerated expansion. A broader picture of the various facets of DE review is available in the studies [61–70]. Cold dark matter (CDM) [71,72] is another dark component of the universe that presents $\Omega{\sim}0.3$ and is driven by the cosmological interpretation of the rotation curves of galaxies and the formation of a large-scale structure. Of late, a theoretical model has been proposed with the introduction of an unusual, hypothetical fluid, viz. Chaplygin Gas (CG) [73–75]. Chaplygin Gas is one of the many candidates for dark energy and it is capable of unifying CDM and DE. This is also a perfect fluid used in general relativity to model idealized distributions of matter, specified with the equation of state

$$p = -\frac{A}{\rho^\alpha}. \tag{1}$$

In this equation, p denotes pressure, density is denoted by ρ and A is a positive constant. Initially, it was proposed that $\alpha = 1$ [76]. Following that, according to the general conclusion, it was proposed that $\alpha \neq 1$ [77]. This equation of state, introduced by Sergey Chaplygin, serves as an effective mathematical approximation for calculating the aerodynamical lifting force on an aircraft wing. This isentropic fluid only admits a supersymmetric generalization. Chaplygin Gas plays an important role in telling the story of string theory. The foremost major feature of Chaplygin Gas is that it can describe a shift to the current stage of cosmic acceleration from a decelerated cosmic expansion and as a single substance, it can unify dark matter and dark energy, which together constitute the dark sector of the universe. This non-homogeneous Chaplygin Gas provides a broad realistic explanation uniting the disparate roles of DE and DM. This exotic fluid presents the simplest deformations of the most typical ΛCDM models. A. Kamenshchik and M. C. Bento proposed the generalized Chaplygin Gas (CGC) model [78,79] at the start of the 21st century, with $A > 0$ and $0 \leqslant \alpha \leqslant 1$ with equation of state Equation (1). Chaplygin Gas can easily interpolate between a non-relativistic matter phase of vanishing pressure $(p = 0)$ and a negative-pressure era dominated by DE [80]. A modified Chaplygin Gas (mCG) model was proposed by Debnath et al. [81] where isotropous pressure p of the cosmological fluid adheres to

$$p = A\rho - \frac{B}{\rho^\alpha}, \tag{2}$$

and $0 \leqslant A \leqslant 1, 0 \leqslant \alpha \leqslant 1$ and B is a positive constant. Fundamental characteristics of mCG as a DE candidate were studied in [82–84]. Because the behavior of CG is similar to that of dark matter (DM) with vanishing pressure in the early stages of the universe and similar to that of a cosmological constant in later phases, this significant and interesting behavior of the CG model has been widely studied in the literature [85]. Moreover, this model was expanded to several forms [86–88]. These developments were the prime motivation behind our study of the modified generalized Chaplygin Gas (mgCG) model along with the equation of state

$$p = \epsilon\rho - (1+\epsilon)A\rho^{-\gamma}, \tag{3}$$

where ϵ, A and γ are three real parameters.

Based on observational studies, we can say that our current universe is facing an accelerated expansion and it is important to note that this late-time acceleration is by no means the lone accelerated phase of the cosmos. In the preliminary stage, the inflationary phase is another phase of such expansion. Several irregularities of the Big Bang cosmology were accordingly communicated in the works of [89,90] within the framework of the inflationary cosmology exemplary for the early universe. Unification of early inflation and late-time cosmic acceleration was demonstrated in [91,92] and denoted a phantom cosmology uncovering the shift to the inflationary era, bearing in mind the transition to the non-phantom standard cosmology and retrieving the recently observed dark energy era. Furthermore, modern cosmologists continue to debate whether the universe began to expand as a result of a Big Bang singularity or as a result of a non-singular, bouncing scenario. Inflation is one of many cosmological scenarios that provides an explanation for the early universe and finds a possible answer to the horizon and flatness problems. In addition, it predicts an almost scale-invariant curvature perturbation power spectrum. However, tracing back in time, the inflationary scenario explains the consequences of an initial cosmic singularity, well known as the Big Bang singularity, where geodesic incompleteness causes the space-time curvature to diverge at the point of the space-time singularity.

A bouncing cosmology is an alternative to inflation that can produce a curvature power spectrum that is scale invariant and the bouncing scenario leads to non-singular universe evolution [92–102]. In the scenario of matter bouncing, the universe experiences a matter-dominated period with the onset of the contraction. In addition, a bounce occurs without any singularity. To put it another way, it implies that the universe transitioned from an accelerated collapse era to an expanding era over the bounce without exhibiting any singular behavior. Following the bounce, the universe immediately enters a phase of matter-dominated expansion. Beside the Big Bang singularity, there are many other types of finite-time singularities in cosmology studied in [103]. The severest amongst them is the Big Rip or Type-I singularity; however, there are another three similar finite time singularities, namely Type-II, Type-III or Type-IV singularities [104–108]. In the case of Type-I, II or III singularities, the universe's effective energy density and/or effective pressure diverge at the point of singularity. In contrast, in the Type-IV singularity, the effective energy density and the effective pressure of the universe stay finite during the instance of the singularity and consequently the Type-IV singularity is the milder one among the finite-time singularities. Specifically, unlike the Big Rip singularity, the geodesic incompleteness is absent in the Type-IV singularity. As a result, the universe can easily pass through a Type-IV singularity (if such a singularity exists) and the Type-IV has no catastrophic effects on observable quantities. Thus, it is possible that the universe encountered a Type-IV singularity in the past in the course of its evolution. It sailed through it without any geodesic incompleteness. Nevertheless, the existence of a Type-IV singularity might exercise noteworthy influence on the evolution of the universe, and also on the generation era of the primordial perturbation modes, as the singularity can globally influence the Hubble parameter.

Viscous cosmology plays a pivotal role within the scientific community in the evolution of the universe. A few studies and reviews on viscous cosmological analysis are presented in [109–114]. In this context, bulk viscosity is attributed to space isotropy and shear viscosity is linked to space anisotropy. Shear viscosities are not considered because the CMB describes meaningful anisotropies, so bulk viscosity can make a case for the universe's accelerated cosmic expansion. In the early universe, when studying inflationary cosmology, the concept of bulk viscosity was proposed in the literature without any requirement of dark energy. Hence, bulk viscosity could be one of the reasons for the current accelerated expansion of the universe. Nowadays, several authors have tried to explain the late-time acceleration via bulk viscosity without any dark energy constituent. In theory, deviations from local thermodynamic stability can cause bulk viscosity to form, but a detailed mechanism for bulk viscosity formation is still elusive. In cosmology, when the matter content of the universe expands or contracts too fast as a cosmological fluid, then the effective pressure is generated to bring the system back to its thermal stability. The manifestation of such an effective pressure is bulk viscosity. To analyze the late acceleration of the universe, the casual theory, according to which viscosity propagates at a finite speed, has been used. Cataldo et al. [115] proposed a study on late-time acceleration using the casual theory of bulk viscosity. Basically, they used an ansatz for the Hubble parameter (inspired by the Eckart theory) and they have shown the transition of the universe from the Big Rip singularity to the phantom behavior.

In the expansion history of the universe, unifying early-phase inflation and later-phase acceleration can be felt in scalar-tensor gravity. Modified gravity provides a true union of early-time inflation and late-time acceleration in this case. It can also form the basis for explanations of both dark energy and dark matter [116,117]. Inflationary expansion of the universe has been observed in several modified gravity frameworks viz. $f(R)$ gravity [118], scalar curvature corrections to general relativity [119], higher-derivative quantum gravity [120], unimodular $f(T)$ gravity [121], mimetic $f(R)$ gravity [122,123] and $f(R,\phi)$ gravity [124]. An extension of the General Theory of Relativity is that modified gravity causes a transition from deceleration to acceleration in the evolution of the universe. By introducing a scalar field after the conformal transformation, we can use $f(R)$-modified gravity. Without choosing a DE candidate, $f(T)$ gravity, the other modified gravity with torsion T, can explain the late-time acceleration with the thermal expanding history of the universe [125–129]. Torsion's role in gravity has been extensively studied in recent decades, with the main goal of bringing gravity closer to its gauge formulation and incorporating spin in a geometric description. Based on the paradigm of $f(T)$ gravity, where $f(T)$ is an arbitrary function of the torsion scalar, we can incorporate cosmological and astrophysical applications both at the background and perturbation levels, in different eras along the cosmic expansion. The $f(T)$ gravity construction, as an alternative to a cosmological constant, can provide a theoretical interpretation of the late-time universe acceleration and it can easily accommodate the regular thermal expanding history, including the radiation and cold dark matter dominated phases. Furthermore, in the early universe, inflation could be achieved through the $f(T)$ paradigm and the Big Bang singularity could be avoided due to the appearance of non-singular bounces.

In information theory, the uncertainty of a random variable is called the entropy inherent to the variable's possible outcomes. The distribution of a discrete random variable X that accepts values from the alphabet $\mathcal{X}$ is as follows:

$$p : \mathcal{X} \to [0,1], \tag{4}$$

$$H(X) = -\sum_{i=1}^{n} P(x_i) \log P(x_i), \tag{5}$$

where Σ denotes the sum over possible values of the variable. Claude Shannon introduced the concept of information entropy in his 1948 paper "A Mathematical Theory of Communication" and that is why it is also referred to as Shannon entropy [130–135]. Shannon entropy is a measure of uncertainty in statistical thermodynamics theory. In probabilistic information theory, entropy is a measure of data produced uncertainty for some event or variable. The higher value of Shannon entropy reflects the higher level of uncertainty measured in the model. This evaluation is carried out on the basis of maximum entropy. There are several ways in which uncertainty can be found. However, the sources of uncertainty at the bouncing point have not been examined. In this study, we use the probabilistic information theory and the initial conditions to quantify uncertainty. Moreover, we incorporated sources of deterministic uncertainty in this model. Here, we considered two scenarios described as "pre" and "post" bounce scenarios. As this particular non-singular bouncing model favours a late-time cosmic speed-up phenomenon, we wanted to find the change in uncertainty for this speed-up. In this study, it is observed that the uncertainty is significantly higher near the bouncing point due to inflation.

The purpose of the present work is to have an insight into the bouncing behavior of mgCG under the purview of modified theory of gravity. Furthermore, another focus of the study is to implement a probabilistic information theory-based approach to discern how the models behave from the point of view of uncertainty theory. The evolution of uncertainty with changes in the mathematical expectations of the associated random variables is quantified through Shannon entropy, which is optimized by using the Lagrangian optimization method. Details are given in subsequent sections.The plan of the work is the following. In Section 2, we construct the mgCG realization of the bouncing scenario, drawing forth the corresponding analytical solutions. In Sections 3–8, we proceed to the reconstruction of generalized scenarios to explain inflationary expansion of the universe. Finally, Section 9 is devoted to finding the uncertainty towards the turnaround point and Section 10 is devoted to the conclusions.

2. Equation of State of Barotropic Fluid

Considering the bouncing scale factor [136] form of

$$a(t) = a_0\left(1+\frac{3t^2\sigma}{2}\right)^{\frac{1}{3}}. \tag{6}$$

In this case, the scale factor at the turnaround point is a_0 and a positive parameter σ describes the bounce speed. Additionally, we have shown the advantages of taking semi-analytic solutions and this estimation of mathematical solutions demonstrates the existence of bouncing solutions in the framework of matter-dominated contraction and expansion. Therefore, from Equation (6), the derived Hubble parameter is

$$H = \frac{t\sigma}{1+\frac{3t^2\sigma}{2}}. \tag{7}$$

Here, we have taken the background evolution fluid as barotropic mgCG whose equation of state (EoS) [137] is

$$p = \epsilon\rho - (1+\epsilon)A\rho^{-\gamma}, \tag{8}$$

where we have denoted thermodynamic pressure as p, total energy density as ρ and ϵ, A and γ parameterize Equation (8). The constraints on the mgCG barotropic fluid described above are $\epsilon > \frac{1}{3}$, $A > 0$ and $\gamma > 0$. The transition of the vanishing pressure phase can describe the evolution of the universe from a relativistic particle-dominated era ($\epsilon = \frac{1}{3}$) to a dark energy-dominated negative pressure era. Here, we considered the universe to be filled with dark energy and dark matter as a single component and neglected the contribution of other components throughout this paper whereas dark matter with no pressure results in $p_m = 0$. Therefore,

$$\rho = \rho_x + \rho_m, \tag{9}$$

where ρ_x denotes dark energy density, ρ_m denotes dark matter density and ρ denotes total energy density.

In a Friedmann–Robertson–Walker (FRW) universe dominated with bulk viscous fluid, the Friedmann field equations [138] are

$$\frac{6\ddot{a}}{a} = -(\rho + 3(p+\Pi)), \tag{10}$$

and

$$3H^2 = \rho_x + \rho_m, \tag{11}$$

where the universe's total energy density, originating from a scalar field of the background effective fluid is ρ. Here, we assumed that the background fluid is a viscous Chaplygin Gas that drives the early period of the universe's expansion, where Π is the bulk viscous pressure and Π is described by $\Pi = -3H\xi$, where ξ is the viscosity coefficient, which is a function of the Hubble parameter. Again, from the conservation equation of pressureless dark matter, we know that

$$\dot{\rho_m} + 3H\rho_m = 0. \tag{12}$$

$$Or\ \rho_m = \rho_{m0}a(t)^{-3}. \tag{13}$$

From Equations (11) and (13), we obtain

$$\rho_x = 3H^2 - \rho_{m0}a(t)^{-3}. \tag{14}$$

From Equations (8) and (14) we obtain the expression of reconstructed p of mgCG.

$$p = -12^{-\gamma}A(1+\epsilon)\left(\frac{t^2\sigma^2}{(2+3t^2\sigma)^2}\right)^{-\gamma} + \frac{12t^2\epsilon\sigma^2}{(2+3t^2\sigma)^2}. \tag{15}$$

Therefore, the EoS parameter defined by $\omega = \frac{p}{\rho}$, where p is thermodynamic pressure and ρ is total energy density, can be derived from the form

$$\omega = \epsilon - 12^{-1-\gamma}A(1+\epsilon)\left(\frac{t^2\sigma^2}{(2+3t^2\sigma)^2}\right)^{-1-\gamma}. \tag{16}$$

Alternatively, ω can be written as

$$\omega = \epsilon - 12^{-1-\gamma}A(1+\epsilon)\left(\frac{(2+3t^2\sigma)^2}{t^2\sigma^2}\right)^{1+\gamma}. \tag{17}$$

In Equation (17), we can notice, at the bouncing point $t = 0$, the denominator becomes zero. The reconstructed density and EoS parameters are plotted in Figure 1 and Figure 2, respectively. The reason for this divergence in Figure 2 is as follows; in Equation (17), ϵ is a positive quantity and the following term tends to infinity towards the turnaround point for all parameter values.

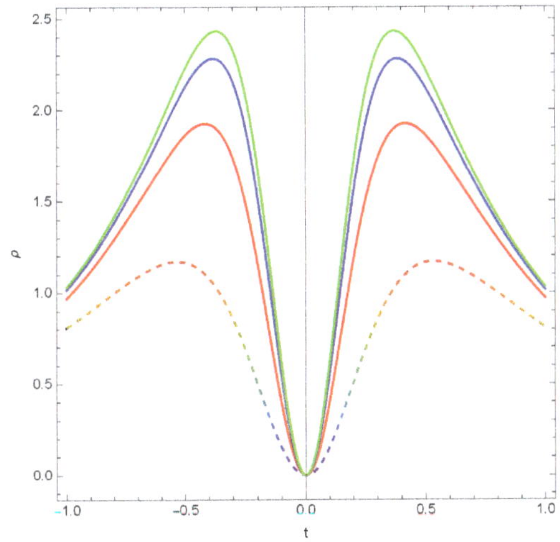

Figure 1. Reconstructed density evolution in the modified generalized Chaplygin Gas scenario with $\rho_{m_0} = 0.32$, $a_0 = 0.2$, $\epsilon = 0.43$, $\gamma = 0.89$ and $A = 0.96$ varying $\sigma = (2.34, 3.85, 4.56, 4.86)$ in units for rainbow-dashed, red-solid curve, blue-solid curve and green-solid curve, respectively, with the bouncing point arising at $t = 0$.

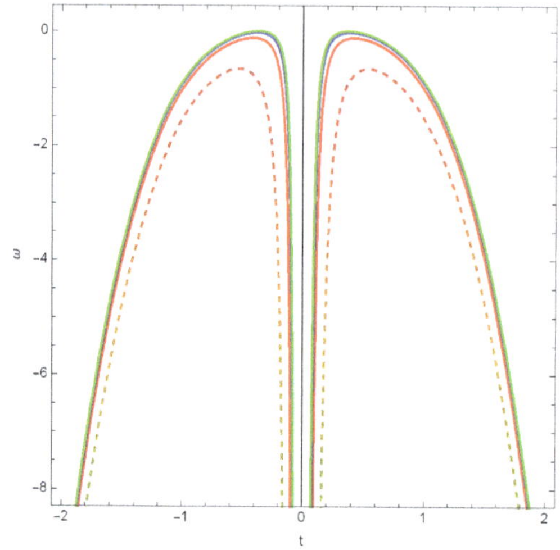

Figure 2. Reconstructed EoS parameter evolution in the modified generalized Chaplygin Gas scenario with $\rho_{m_0} = 0.32$, $a_0 = 0.2$, $\epsilon = 0.43$, $\gamma = 0.89$ and $A = 0.96$ varying $\sigma = (2.34, 3.85, 4.56, 4.86)$ in units for rainbow-dashed, red-solid curve, blue-solid curve and green-solid curve, respectively, with the bouncing point arising at $t = 0$.

From Equations (15) and (10), we obtained the expression of bulk viscous pressure (Π),

$$\Pi = \frac{12^{-\gamma}\left(\frac{t^2\sigma^2}{(2+3t^2\sigma)^2}\right)^{1-\gamma}\left(A(1+\epsilon)(2+3t^2\sigma)^2 - 3^{\gamma}4^{1+\gamma}\sigma\left(\frac{t^2\sigma^2}{(2+3t^2\sigma)^2}\right)^{\gamma}(2+3t^2\epsilon\sigma)\right)}{t^2\sigma^2}. \tag{18}$$

By simplifying Equation (18), we can see that the denominator is zero at the bouncing point $t = 0$. For that reason, the evolution of the reconstructed bulk viscous pressure diverges in Figure 3 and increases infinitely towards the turnaround point.

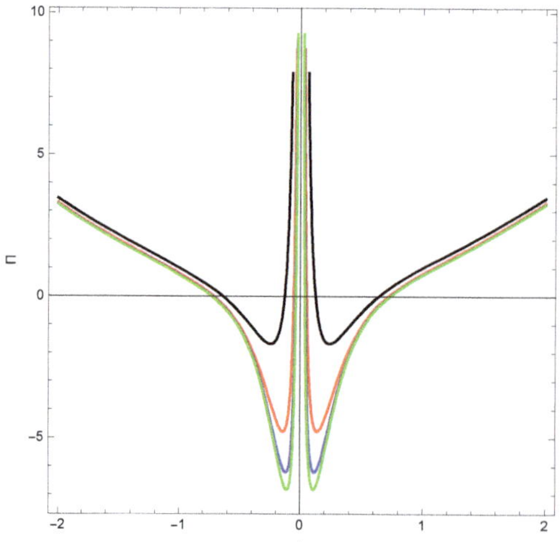

Figure 3. Reconstructed bulk viscous pressure Π evolution in the modified generalized Chaplygin Gas scenario with $\rho_{m_0} = 0.32$, $a_0 = 0.2$, $\epsilon = 0.43$, $\gamma = 0.89$ and $A = 0.96$ varying $\sigma = (2.34, 3.85, 4.56, 4.86)$ in units for black-solid, red-solid curve, blue-solid curve and green-solid curve, respectively.

Here, the bulk viscous pressure (Π) can be written as a function of Hubble parameter $\Pi = -3H\xi$ where ξ is the coefficient of viscosity. Then, reconstructed ξ can be written as:

$$\xi(H) = -\frac{3^{-1-\gamma}4^{-\gamma}\left(1+\frac{3t^2\sigma}{2}\right)\left(\frac{t^2\sigma^2}{(2+3t^2\sigma)^2}\right)^{1-\gamma}\left(A(1+\epsilon)(2+3t^2\sigma)^2-3^{\gamma}4^{1+\gamma}\sigma\left(\frac{t^2\sigma^2}{(2+3t^2\sigma)^2}\right)^{\gamma}(2+3t^2\epsilon\sigma)\right)}{t^3\sigma^3}. \quad (19)$$

Equation (19) can be simplified and towards bouncing point $t = 0$ the denominator tends to zero and that is why the evolution of the reconstructed viscosity coefficient diverges in Figure 4. The effective EoS parameter ω_{eff} can be defined as the ratio of effective pressure $(p + \Pi)$ and total energy density ρ. We can derive the expression of the effective EoS parameter from Equations (14), (15) and (18) which is plotted in Figure 5 with respect to cosmic time t. As bulk viscous pressure itself diverges towards the bouncing point, the effective EoS parameter also diverges.

We plotted the reconstructed mgCG density evolution with respect to cosmic time t in Figure 1 and discovered that mgCG energy density decreases due to contraction in the pre-bounce scenario, is the lowest at the bouncing point at $t = 0$ and increases after bounce due to expansion. This figure supports the inflationary expansion of the universe, as it depicts a rapid increase in density due to inflation, followed by density stabilization in later phases of the universe. The behavior of the background evolution of energy density in Figure 1 is continuous and well behaved. The bounce is clearly visible in Figure 2, the EoS parameter crosses the phantom boundary and there is an avoidance of Big Rip singularity in a viscous mgCG scenario in Figure 5. In Figures 3 and 4, we plotted bulk viscous pressure versus cosmic time t and viscosity coefficient ξ using $\Pi = -3H\xi$.

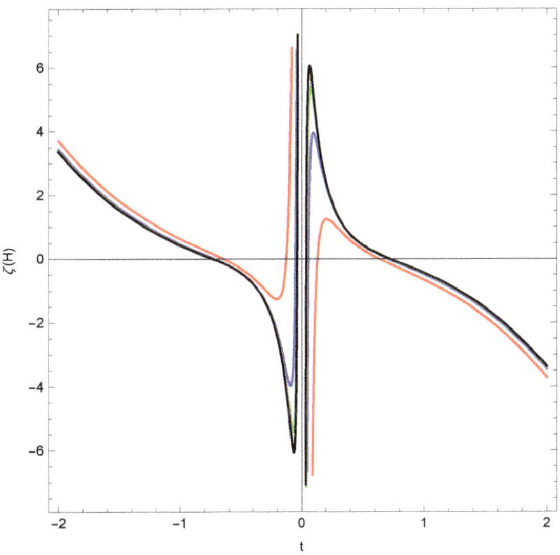

Figure 4. Reconstructed bulk viscosity coefficient ξ evolution in the modified generalized Chaplygin Gas scenario with $\rho_{m_0} = 0.32$, $a_0 = 0.2$, $\epsilon = 0.43$, $\gamma = 0.89$ and $A = 0.96$ varying $\sigma = (2.34, 3.85, 4.56, 4.86)$ in units for black-solid, red-solid curve, blue-solid curve and green-solid curve, respectively.

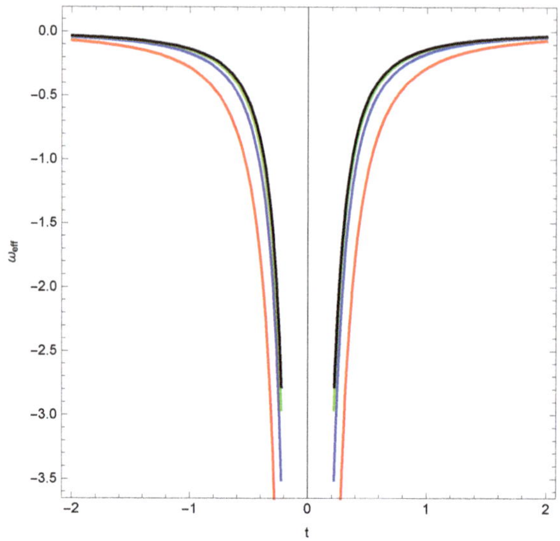

Figure 5. Reconstructed effective EoS parameter evolution in the modified generalized Chaplygin Gas scenario with $\rho_{m_0} = 0.32$, $a_0 = 0.2$, $\epsilon = 0.43$, $\gamma = 0.89$ and $A = 0.96$ varying $\sigma = (2.34, 3.85, 4.56, 4.86)$ in units for black-solid, red-solid curve, blue-solid curve and green-solid curve, respectively, with the bouncing point arising at $t = 0$.

3. Bounce Cosmology with Respect to E-Folding Number

We have demonstrated the theory of perturbations for the framework of mgCG evolution in every single bouncing layout associated with inflationary observables, which is important for us. In particular, the Hubble horizon decays faster than the wavelengths of the generated vacuum state fluctuations and contraction continues before the phase of bounce. The fluctuations after leaving the Hubble radius form bounce and later in the expanding phase, the fluctuations re-enter the inner parts of the horizon. Here, Equation (6) is the form of the bouncing scale factor where a_0 is the scale factor at the bouncing point and the positive parameter σ denotes the speed of the bounce. Therefore, Equation (7) is the derived Hubble parameter. Now, we considered the e-folding time scale N [139] in spite of cosmic time t, associated with the scale factor as shown:

$$e^{-N} = \frac{a_0}{a}. \tag{20}$$

Or $N = \ln(\frac{a}{a_0})$, which implies $H = \dot{N}$.
Therefore,

$$N = \frac{1}{3}\ln\left(1 + \frac{3t^2\sigma}{2}\right). \tag{21}$$

Then, we converted the cosmic time (t) in terms of e-folding number N and found the thermodynamic pressure p, density ρ and EoS parameter (ω) in terms of e-folding time scale.

$$t(N) = \pm\sqrt{\frac{2(-1+e^{3N})}{3\sigma}}. \tag{22}$$

and we obtained

$$p = 2e^{-6N}(-1+e^{3N})\epsilon\sigma - 2^{-\gamma}A(1+\epsilon)\left(e^{-6N}(-1+e^{3N})\sigma\right)^{-\gamma}. \tag{23}$$

$$\rho_{Total} = 2e^{-6N}\left(-1+e^{3N}\right)\sigma. \tag{24}$$

We have shown the evolution of the time derivative of e-fold number $\dot{N}$ versus time t in Figure 6. It has been demonstrated that the scale factor at the bouncing point, i.e., $a = a_0$; we have $N = 0$. In Figure 6, $\dot{N}$ is symmetric about the bouncing time $t = 0$ and it supports the realization of nonsingular bounce. It has been observed that before the bounce $t < 0$, we have $\dot{N} < 0$ due to contraction of the universe as $\dot{a} < 0$ and after the bounce $t > 0$, expanding attribute of $\dot{N}$ occurred due to expansion of the universe.

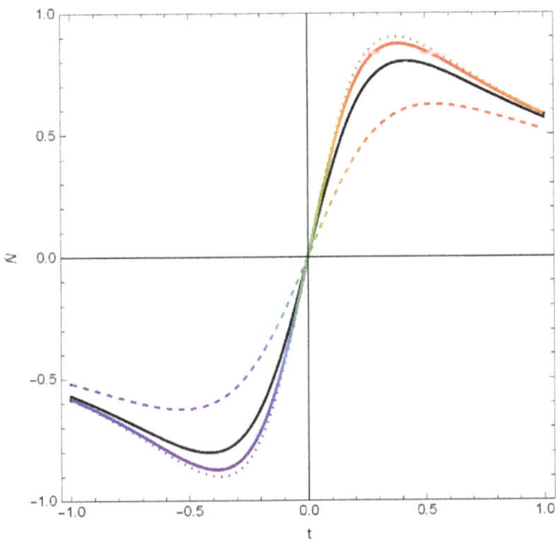

Figure 6. Evolution of expression of time derivative of e-folding Number with respect to cosmic time with $\rho_{m_0} = 0.32$, $a_0 = 0.2$, $\epsilon = 0.43$, $\gamma = 0.89$ and $A = 0.96$ varying $\sigma = (2.34, 3.85, 4.56, 4.86)$ in units for rainbow-dashed curve, rainbow-dotted curve, black-solid curve and rainbow-solid curve, respectively, the bouncing point at $t = 0$.

Here, we plotted the evolution of the expression of ω defined by $\omega = \frac{p}{\rho}$ and ω_{eff} defined by $\omega_{eff} = \frac{p+\Pi}{\rho}$ with respect to time derivative of the e-folding number in Figures 7 and 8.

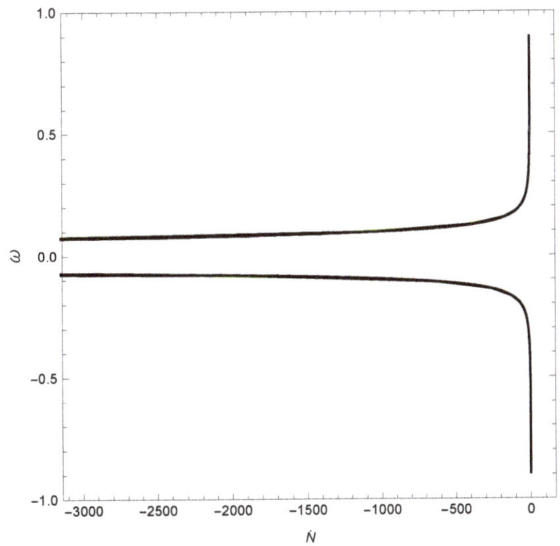

Figure 7. Evolution of expression of ω with respect to time derivative of e-folding number.

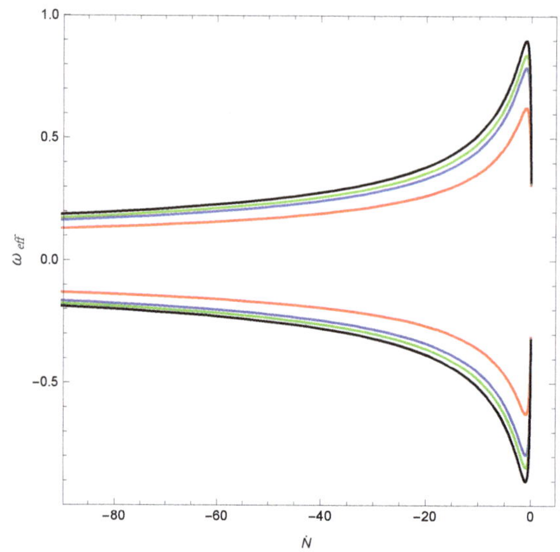

Figure 8. Evolution of expression of ω_{eff} with respect to time derivative of e-folding number.

4. Hubble Flow Dynamics

In the early universe, cosmic inflation is one of the main paradigms for explaining cosmological parameters and contains phase of exponential expansion of the universe and it accomplishes the paths for initiating in homogeneities on cosmological scales. We consider the Hubble flow parameters [140], proposed in the studies of cosmologists Hoffman [141], Turner [142] and Kinney, defined by

$$\epsilon_{i+1} = \frac{dln\epsilon_i}{dN}. \tag{25}$$

where N is the e-folding time scale.

$$\epsilon_1 = 1 - \frac{\ddot{a}}{aH^2} = -\frac{\dot{H}}{H^2}. \tag{26}$$

$$\epsilon_2 = \frac{\dot{\epsilon_1}}{H\epsilon_1}. \tag{27}$$

For inflation, $\ddot{a} > 0$, ϵ_1 has to be less than 1. It is very obvious for vanishing inflationary epoch ϵ_1 has to take the value of unity.

$$\epsilon_1 = \frac{3}{2} - \frac{1}{t^2\sigma}. \tag{28}$$

$$\epsilon_2 = \frac{-2}{t^2\sigma} + \frac{12}{-2+3t^2\sigma}. \tag{29}$$

Here, we plotted Hubble flow parameters in Figure 9 and Figure 10, respectively. From Figure 9 we can see at first ϵ_1 is symmetric about the turnaround point. It shows an increasing pattern before the bounce. Then, after the bounce it has an increasing pattern also. Hence, we understand that across the bounce $\epsilon_1 << 1$ that suits the required condition for inflation. After that, it crosses the boundary of $\epsilon_1 = 1$ and it indicates a nice exit from inflation. Thus, we have demonstrated the capability of realization of the inflationary era and as well as an exit from inflation with mgCG background fluid by Figure 10. In the pre-bounce scenario $H < 0$ and ϵ_1 and $\dot{\epsilon_1} < 0$. We have observed in Figure 10 just before bounce $\epsilon_2 < 0$ which supports the theory and immediately after the bounce $H > 0$ and ϵ_2 is also less than zero by supporting inflation. Then, it becomes positive.

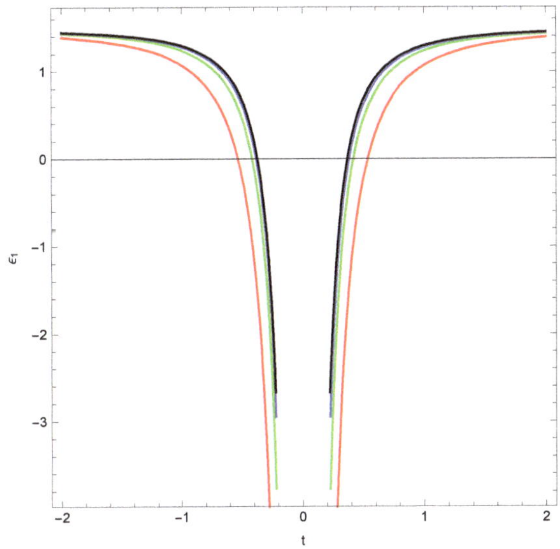

Figure 9. Hubble flow parameter ϵ_1 for the reconstructed mgCG density with time t.

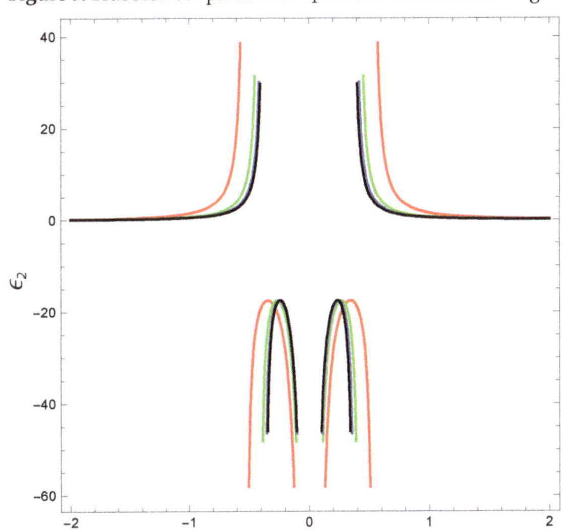

Figure 10. Hubble flow parameter ϵ_2 for the reconstructed mgCG density with time t.

5. Inflation via Scalar Field

In this section, we introduce a scalar field ϕ which is minimally coupled dependent on a potential $V(\phi)$ [143]. Therefore, in a curved space-time the Lagrangian [144] will be

$$\mathcal{L}_\phi = -\frac{1}{2}\mathcal{G}^{\alpha\beta}\partial_\alpha\phi\partial_\beta\phi - V(\phi). \tag{30}$$

Here, $\mathcal{G}^{\alpha\beta}$ is the metric tensor and $\partial_\alpha\phi$, $\partial_\beta\phi$ are covariant derivatives. Therefore, by considering the scalar field ϕ the effective mgCG energy density $\tilde{\rho_c}$ and the isotropic pressure incorporated with bulk viscosity will be

$$\tilde{\rho_c} = \frac{\dot{\phi}^2}{2} + V(\phi). \tag{31}$$

$$\tilde{p_c} = \frac{\dot{\phi}^2}{2} - V(\phi). \tag{32}$$

Now we want to demonstrate inflationary dynamics through the scalar field. In this present section, we considered a flat FRW universe filled with viscous mgCG in the scalar field framework. Taking $\tilde{\rho_c} = \rho$ for our background evolution of viscous mgCG, we have the following from Equations (31) and (32)

$$\dot{\phi}^2 = (1+\epsilon)\left(-12^{-\gamma}A\left(\frac{t^2\sigma^2}{(2+3t^2\sigma)^2}\right)^{-\gamma} + \frac{12t^2\sigma^2}{(2+3t^2\sigma)^2}\right). \tag{33}$$

Hence,

$$\phi(t) = \sqrt{(1+\epsilon)}\left(\frac{2^{-\gamma}3^{-\gamma/2}\sqrt{A}t(1+\frac{3t^2\sigma}{2})^{-\gamma}(\frac{t^2\sigma^2}{(2+3t^2\sigma)^2})^{-\gamma/2}{}_2F_1\left[\frac{1-\gamma}{2}, -\gamma, 1+\frac{1-\gamma}{2}, -\frac{3t^2\sigma}{2}\right]}{-1+\gamma} + \frac{\sqrt{\frac{t^2\sigma^2}{(2+3t^2\sigma)^2}}(2+3t^2\sigma)\log[2+3t^2\sigma]}{\sqrt{3}t\sigma}\right). \tag{34}$$

Then, we can easily find the expression of $V(\phi(t))$. The basic purpose of this section is to demonstrate whether it is possible to have inflationary expansion from the viscous mgCG in scalar field formalism in a post-bounce scenario. Therefore, from Equations (31) and (32), we calculated this particular expression of potential.

$$V(\phi(t)) = \frac{1}{2}\left(12^{-\gamma}A(1+\epsilon)\left(\frac{t^2\sigma^2}{(2+3t^2\sigma)^2}\right)^{-\gamma} - \frac{12t^2(-1+\epsilon)\sigma^2}{(2+3t^2\sigma)^2}\right). \tag{35}$$

We plotted non-singular $\dot{\phi}(t)$ in Figure 11 and the scalar field in Figure 12 respectively. So, the evolution of kinetic energy and the scalar field with cosmic time t is presented in Figure 11 and Figure 12, respectively. From Figure 12 we can show in the early universe, i.e., the pre-bounce scenario, the manner of the scalar field is decaying and rising in the later time of the universe. We have shown in Figure 13 the difference $v - \dot{\phi}^2$ and this figure is consistent with explaining inflationary expansion of the universe as $v > \dot{\phi}^2$ at the turnaround point and after the bounce scenario.

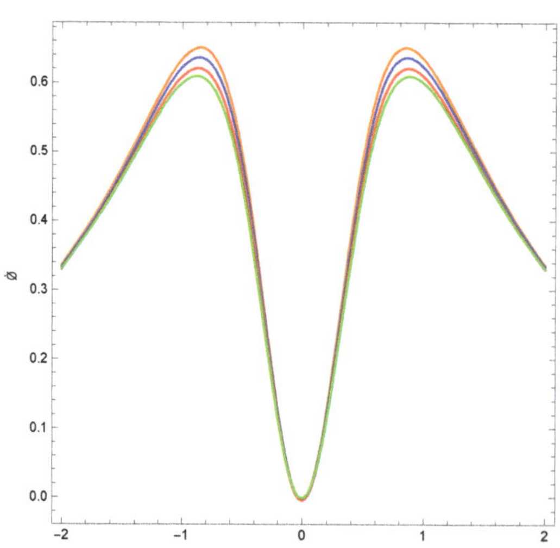

Figure 11. Evolution of kinectic energy with respect to cosmic time t.

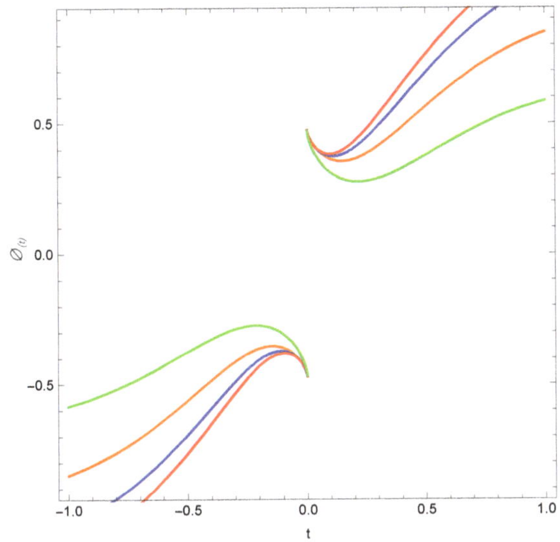

Figure 12. Evolution of scalar field $\phi(t)$ with respect to cosmic time t.

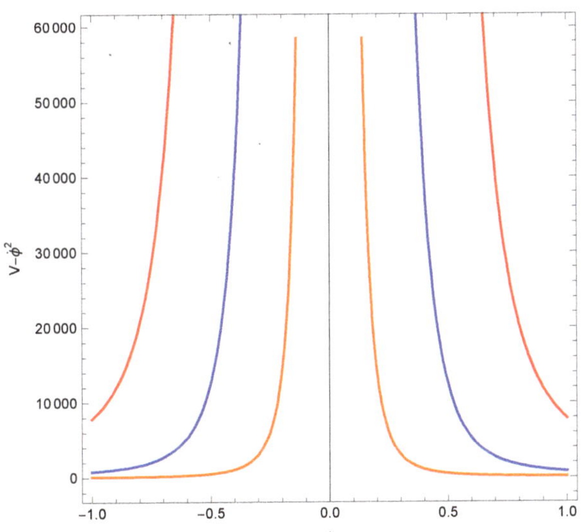

Figure 13. Evolution of $v - \dot{\phi}^2$ with respect to cosmic time t.

6. Fractional Density

In this section, we consider density parameter Ω. The definition of the density parameter is the ratio of the actual (or observed) density to the critical density of the Friedmann universe. Numerous studies explaining the expansion rate of the universe indicate that the universe is very close to reaching the critical density that would cause it to expand indefinitely. To study the expansion of the universe in a different way, we expressed the density as a fraction of the density required for the critical condition. To prove our universe is dominated by dark energy in this expansion era, we have studied fractional densities. The previous history of the Big Bang is viewed as being at first radiation dominated, then matter dominated and now having passed into the era where dark energy is the dominant influence. Let us define the fractional densities [145]:

$$\Omega_{DE} = \frac{\rho_{mgCG}}{3H^2}. \tag{36}$$

$$\Omega_{DM} = \frac{\rho_m}{3H^2}. \tag{37}$$

$$\Omega_{DE} + \Omega_{DM} = 1. \tag{38}$$

Here, reconstructed Ω_{DM} and Ω_{DE} are as follows:

$$\Omega_{DM} = \frac{\rho_{m_0}(2+3t^2\sigma)}{12t^2\sigma^2}. \tag{39}$$

$$\Omega_{DE} = 1 - \frac{\rho_{m_0}(2+3t^2\sigma)}{12t^2\sigma^2}. \tag{40}$$

Here, we plotted the reconstructed fractional densities against cosmic time t in Figure 14 and it is clear that there is a transition from a matter dominated universe to an energy dominated universe. In the above figure, we can see dark matter is decreasing with time and dark energy is increasing with time with mgCG background evolution fluid.

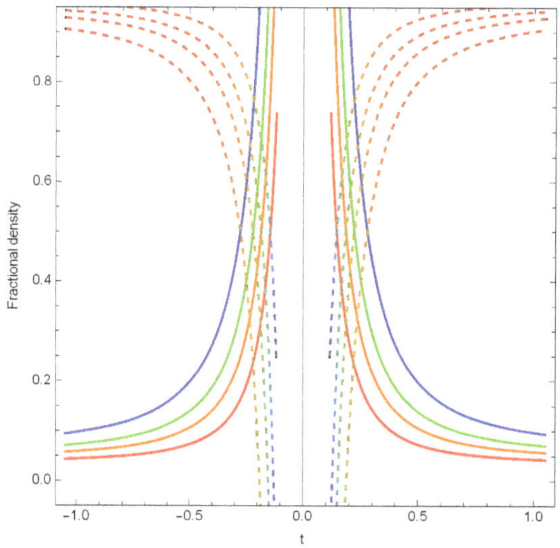

Figure 14. Evolution of fractional density with respect to cosmic time t with decreasing matter density shown by the solid-curve and increasing energy density by the dashed-curve with $\rho_{m_0} = 0.32$, $a_0 = 0.2$, $\epsilon = 0.43$, $\gamma = 0.89$ and $A = 0.96$ varying $\sigma = (2.34, 3.85, 4.56, 4.86)$ in units.

7. MGCG in *f(T)* Gravity

Here, introducing torsion T as a fundamental component [146], we considered "teleparallelism" connected with the Weitzenböck as a theory of gravity apart from Einstein's General Theory of Relativity. There are many cosmological studies (e.g., [147]), considered "teleparallelism" as a gravitational model with Lagrangian density as a function of T $[f(T)]$ to unify the inflationary epoch with the late time cosmological acceleration. In the study of Bamba et al. [148], finite-time future singularities have been studied with their thermodynamics. In the theory of modified teleparallel $f(T)$ gravity [149], the action I is given by

$$I = \frac{1}{16\pi G}\int d^4x\sqrt{-g}[f(T)] + \mathcal{L}_m. \tag{41}$$

Here, the matter Lagrangian density and gravitational constant are denoted by L_m and G, respectively, and g is the determinant of the metric tensor $g_{\mu v}$. In the exponential teleparallel $f(T)$ gravity scenario, by choosing energy density and pressure as ρ_m and p_m, respectively, the modified Friedmann equations are [150]

$$H^2 = \frac{\kappa^2}{3}(\rho_m + \rho_T). \tag{42}$$

$$2\dot{H} + 3H^2 = -\kappa^2(p_m + p_T). \tag{43}$$

Based on cosmological principles, the Big Bang theory is made up and from the Robertson–Walker metric; we can obtain the definition of κ, where κ is the spatial curvature, scaled so as to take the values 0 or ± 1. Here, we have taken the spherical universe and $\kappa = 1$.

Here, we constructed the $f(T)$ theory subject to bouncing scale factor by taking mgCG as the background evolution fluid of the universe.

$$\rho_T = \frac{1}{2\kappa^2}(2Tf' - f - T), \tag{44}$$

and

$$p_T = \frac{-2\dot{H}}{\kappa^2}[2Tf'' + f' - 1] - \rho_T. \tag{45}$$

In the above-stated equations, ρ_T and p_T are the Torsion contributed matter energy density and the pressure counterpart of standard ρ_m and p_m. We can write $f(T)$, f' and f'' in terms of cosmic time t easily. Therefore,

$$f' = \frac{\dot{f}}{\dot{T}}, \tag{46}$$

$$f'' = \frac{\dot{T}\ddot{f} - \ddot{T}\dot{f}}{\dot{T}^3}. \tag{47}$$

Here, the upper dot denotes the cosmic time derivative. For a pressureless dark matter $p_m = 0$ and the continuity equation can be integrated to $\rho_m = \rho_{m0}a^{-3}$.

$$f(t) = \frac{2\rho_{m_0} + 2\sqrt{6}\rho_{m_0}t\sqrt{\sigma}ArcTan\left[\sqrt{\frac{3}{2}}t\sqrt{\sigma}\right] + tC_1}{2 + 3t^2\sigma}. \tag{48}$$

In this case, for background evolution, we chose $\omega_T = \omega_{mgCG}$ in the bulk viscous scenario. Then, from Equation (3), we can write,

$$\omega_{mgCG} = \epsilon - (1+\epsilon)A\rho^{-\gamma-1}. \tag{49}$$

$$\omega_T = \omega_{mgCG}. \tag{50}$$

$$\rho_T = \frac{A(1+\epsilon)}{\epsilon - \omega_T}^{\frac{1}{\gamma+1}} = 12^{1+\gamma}(\frac{t^2\sigma^2}{(2+3t^2\sigma)^2}). \tag{51}$$

In Figure 15, we present the arbitrary torsion contributed function $f(t)$. The evolution of the reconstructed torsion contributed density is presented in Figure 16. where a symmetric pattern is observed around the bouncing point. In the pre-bounce scenario $f(t)$ has a decaying pattern and after bounce in the inflationary scenario $f(t)$ follows an increasing pattern. After the inflationary expanding era $f(t)$ regains its falling behavior. So, we can say this model shows realistic reconstruction. Figure 16 shows increasing energy density in both pre- and post-bounce scenarios and is continuous at $t = 0$ for $f(T)$ gravity also. However, after the inflation rho_T has a falling pattern tendency and then it remains stable for all values of σ. The torsion reconstructed EoS parameter is plotted in Figure 17. In the pre-bounce scenario, at first the EoS parameter avoids Big Rip singularity by transiting from less than -1 to $\frac{-1}{3}$but t tends to the turnaround point again; it crosses the phantom boundary. In the inflationary era, there is a transition from less than -1 to $\frac{-1}{3}$ which is the avoidance of Big Rip singularity in the post-bounce scenario and afterwards again it crosses the phantom boundary in the later universe. However, w_T always has crossed $\frac{-1}{3}$ as required in accelerated expansion of the post-bounce scenario.

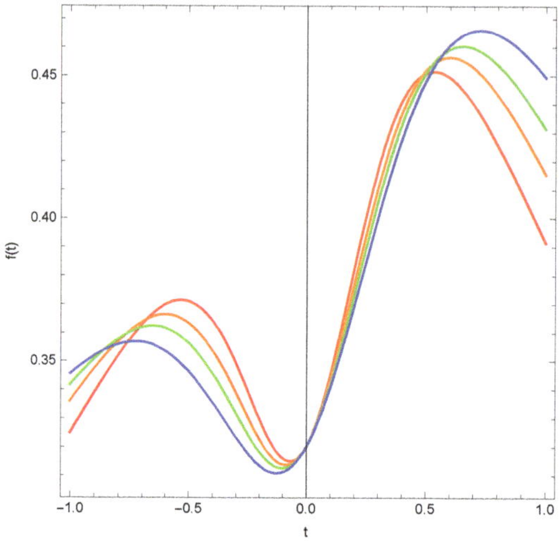

Figure 15. Reconstructed arbitrary torsion contributed function f evolution with respect to cosmic time t in the modified generalized Chaplygin Gas scenario with $\rho_{m_0} = 0.32$, $a_0 = 0.2$, $\epsilon = 0.43$, $\gamma = 0.89$ and $A = 0.96$ varying $\sigma = (2.34, 3.85, 4.56, 4.86)$ in units for orange-solid curve, red-solid curve, blue-solid curve and green-solid curve, respectively, with the bouncing point arising at $t = 0$.

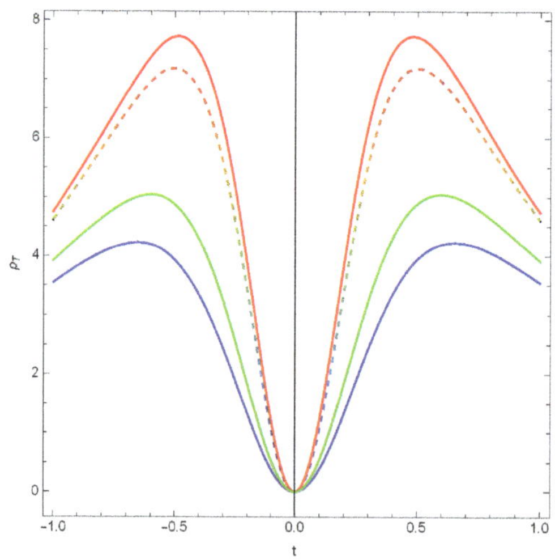

Figure 16. Reconstructed torsion contributed density evolution ρ_T in the modified generalized Chaplygin Gas scenario with $\rho_{m_0} = 0.32$, $a_0 = 0.02$, $\epsilon = 0.64$, $\gamma = 0.68$ and $A = 2.3$ varying $\sigma = (2.65, 2.85, 1.56, 1.86)$ in units for rainbow-dashed curve, red-solid curve, blue-solid curve and green-solid curve, respectively, with the bouncing point arising at $t = 0$.

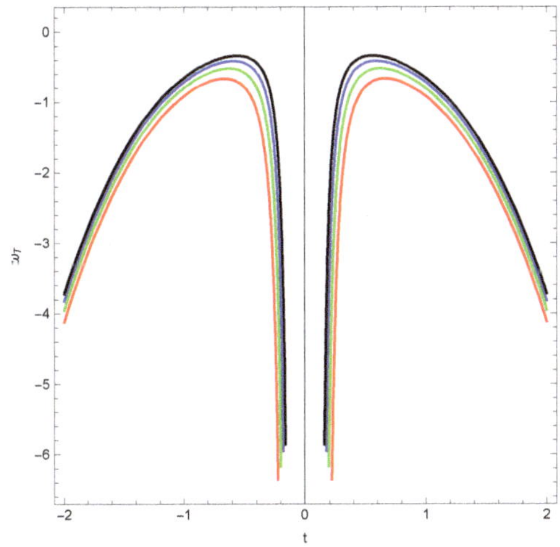

Figure 17. Reconstructed torsion contributed EoS parameter evolution ω_T in the modified generalized Chaplygin Gas scenario with $\rho_{m_0} = 0.32$, $a_0 = 0.2$, $\epsilon = 0.43$, $\gamma = 0.89$ and $A = 0.96$ varying $\sigma = (2.34, 3.85, 4.56, 4.86)$ in units for black-solid curve, red-solid curve, blue-solid curve and green-solid curve, respectively, with the bouncing point arising at $t = 0$.

8. Stability Analysis

Here, we have implemented a study on stability analysis in the framework of Einstein's as well as modified $f(T)$ gravity. Here, we used square speed of sound [151,152] to find the stability of the model.

$$V_s^2 = \frac{dp}{d\rho}. \tag{52}$$

The stability of the model depends on $V_s^2 \geq 0$. If the value is negative it shows instability of the model. Here, we incorporated the squared speed of sound for mgCG reconstruction schemes in Einstein's as well as modified $f(T)$ gravity scenario and plotted V_s^2 versus cosmic time t. In Figure 18, we plotted V_s^2 in Einstein's framework which is continuous and well behaved. In Figure 19, we present the stability analysis by plotting V_s^2 in a modified $f(T)$ gravity scenario. Here, for the $f(T)$ gravity scenario it diverges and does not exist near the turnaround point.

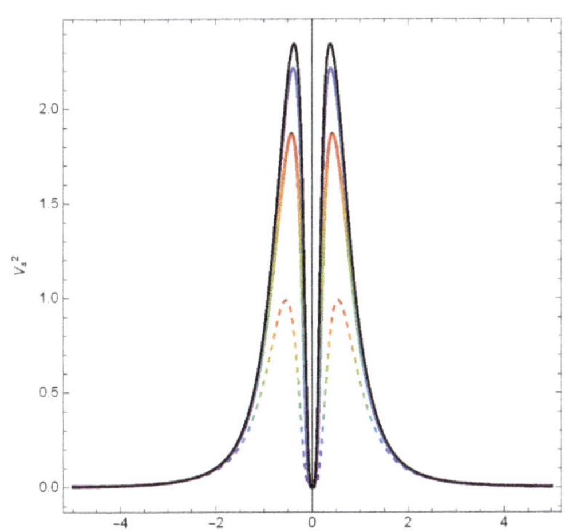

Figure 18. Evolution of square speed of sound $V_s^2 > 0$ for Einstein gravity scenario.

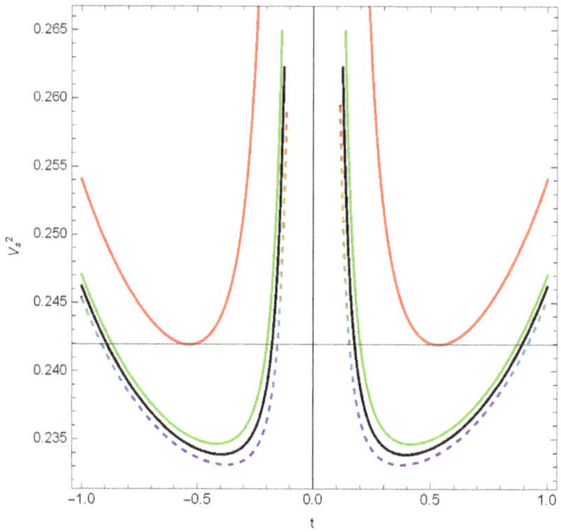

Figure 19. Evolution of square speed of sound $V_s^2 > 0$ for modified $f(T)$ gravity scenario.

The square speed of sound diverges at the bouncing point for the model in the $f(T)$ modified gravity framework although it is continuous in Einstein's gravity framework. This indicates the possible instability of the model at the bouncing point and requires further study.

9. Uncertainty Towards Bouncing Point

Here, we present a detailed study on observational $H(z)$ data from Luminous Red Galaxies of Sloan Digital Sky Survey Data Release Seven (SDSS DR7), which are available at [153]. Initially let us have a glance at the illustrative statistics related to the data set regarding the bouncing scenario. In the above stated dataset, we have Hubble data $H(z)$ with respect to redshift (z) data. We know that

$$\frac{1}{a(t)} = 1 + z, \tag{53}$$

where $a(t)$ refers to scale factor and z refers to the Hubble dataset. We converted the redshift dataset to cosmic time.

In the subsequent section, we present a rigorous study on finding uncertainty towards a bouncing point through a probabilistic information theory using the method of Shannon entropy maximization. Here, we applied the probabilistic information theory to obtain a deterministic picture of the uncertainty correlated with the evolution of the universe. The Shannon entropy quantifies the uncertainty associated with a stochastic variable explained in a given probability space. For this case, an entropy maximizing probability distribution is acquired by construction of Lagrangian [154]

$$\mathcal{L} = -\sum_{x=1}^{n} p_x \ln p_x - \alpha(\sum_{x=1}^{n} p_x - 1) - \beta(\sum_{x=1}^{n} p_x y_x - E(y)), \tag{54}$$

$$\frac{\partial \mathcal{L}}{\partial \alpha} = 1 - \sum_{x=1}^{n} p_x y_x, \tag{55}$$

$$\frac{\partial \mathcal{L}}{\partial \beta} = E(y) - \sum_{x=1}^{n} p_x y_x, \tag{56}$$

$$\frac{\partial \mathcal{L}}{\partial p_x} = -\ln p_x - 1 - \alpha - \beta y_x. \tag{57}$$

Solving the above equations, the entropy maximizing probability distribution becomes

$$p_x = \frac{\exp(-\beta y_x)}{\sum_{x=1}^{n} \exp(-\beta y_m)} \quad \text{where } m = 1, 2, 3, \ldots, n \tag{58}$$

In Equation (58), the probabilities prevail in the manner explained above. It must be noted that β is attained through the Newton–Raphson method. This probability distribution is used to calculate the uncertainty in respect of Shannon entropy $H(p)$ as

$$H(p_1, p_2, p_3, \ldots, p_n) = -\sum_{x=1}^{n} p_x \ln p_x \tag{59}$$

Here, we varied the E[t], where t is the cosmic time. This adaptation is conducted by taking the cosmic time towards bouncing point from the pre-bounce scenario as well as from the post-bounce scenario. Appropriately, the entropy maximizing probability distribution was attained for all expectations. It should be mentioned that the values of β are attained through application of the Newton–Raphson method on the following equation

$$\sum_{x=1}^{n} (t_x - E(t)) \exp(-\beta t_x) = 0 \tag{60}$$

The variation of Shannon entropy with respect to cosmic time is presented in Figure 20. It is clear from Figure 20 that the uncertainty, computed through Shannon entropy, is lower at early times and it is greater near the turnaround point of the universe. After, the bounce uncertainty is higher as the inflationary expansion has occurred. As Shannon entropy is an estimation of randomness in a system, we observe that the uncertainty is reduced as we evolve from Big Bang inflationary to late time universe. Therefore, the reconstruction scheme presented here has the potential to be considered as a unifying model for early inflation and late time acceleration.

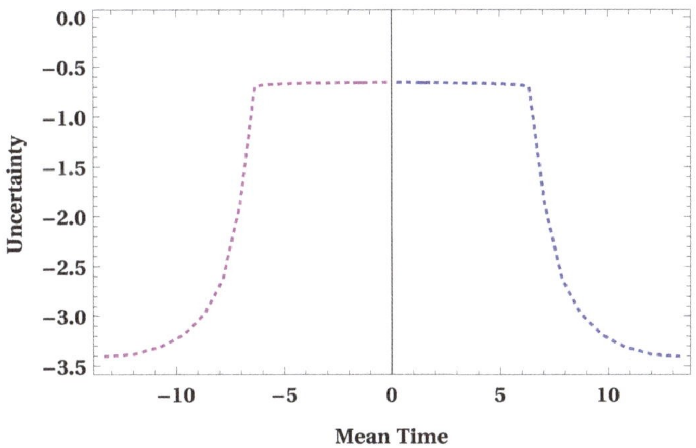

Figure 20. Uncertainty towards turnaround point.

From the Figure 20, we can see, at the very early contracting phase of the universe, uncertainty is very much less and after that uncertainty is increasing when cosmic time leads to the turnaround point. After that, at the inflationary era uncertainty is higher and later the universe uncertainty is decreasing at the post-bounce scenario.

10. Discussion and Concluding Remarks

In the concluding remarks, there are a few important points to be discussed. We started with modified generalized Chaplygin Gas (mgCG) as the background evolution.

The mgCG can interpolate the phases between standard pressure at high energy density fluids and negatively constant pressure with low density energy fluids by varying GCG to the mgCG [155]. The study of Zhang and Zhu [156] is the most relevant study in this framework of the manuscript, in which taking Chaplygin Gas to act as DE interacting with DM results in an EoS parameter (ω) in a bulk viscous scenario capable of crossing $\omega < -1$. Other references [156–160], which are relevant to this work, take a different variant of noninteracting Chaplygin Gas with a $f(T)$ reconstruction scheme and establish the necessary conditions for crossing the phantom boundary.

In this paper, we looked at a bouncing scale factor of the form $a(t) = a_0(\frac{1+3t^2\sigma}{2})^{\frac{1}{3}}$ where a_0 is the scale factor at the bouncing point and σ is a positive parameter describing the speed with which the bounce occurs. The bouncing point appeared at $t = 0$. This model realized big-bounce in the framework of Einstein gravity as well as $f(T)$ modified gravity, which violates the null energy condition (NEC) $p + \rho \geq 0$ for all values of σ. Incorporating the Friedmann equation, we derived the mgCG reconstructed density, the EoS parameter and the effective EoS parameter, respectively, and plotted them against cosmic time t. In the very first section of this manuscript, we plotted the reconstructed mgCG density evolution with respect to cosmic time t in Figure 1 and the figure supports the inflationary expansion of the universe; we can see in the figure that density increases due to inflation after the bounce and then remains stable in the later phase of the universe. Figure 1 shows non-singular bounce, which is a very obvious result of taking nonsingular scale factor and in Figure 2 we can see the EoS parameter crosses the phantom boundary but in Figure 5 avoidance of Big Rip has been demonstrated. Our next purpose was to show inflationary dynamics. In Figure 6, we used the e-folding time scale to calculate the time derivative of the e-fold number $\dot{N}$ versus time t and plotted it against the cosmic t. The scale factor at the bouncing point, $a = a_0$, has been shown to be $N = 0$. In Figure 6, $\dot{N}$ is symmetric about the bouncing time $t = 0$ and allows for nonsingular bounce realization. It is observed that before the bounce $t < 0$, $\dot{N} < 0$ due to universe contraction as a time-varying expansion factor $\dot{a} < 0$ and after the bounce $t > 0$, $\dot{N} > 0$ due to universe expansion. We plotted Hubble flow parameters in Figures 9 and 10. Figure 9 depicts an increasing pattern of ϵ_1 prior to the bounce. Following the bounce, $\epsilon_1 << 1$ meets the necessary condition for inflation. After that, it crosses the boundary of $\epsilon_1 = 1$ and indicates a nice exit from inflation. As a result, Figure 10 depicts the realization of inflation. In Figure 10, just before the bounce, $\epsilon_2 < 0$ is less than zero, supporting the inflationary theory and immediately after the bounce $H > 0$, ϵ_2 is less than zero, also supporting inflationary theory. Then, it becomes positive.

We have shown in Figure 12 the minimally coupled scalar field $\phi(t)$ and demonstrated its behavior. In the pre-bounce case, it behaves in a decaying manner and in a rising manner in the later time of the universe. We have shown in Figure 13 the difference $v - \dot{\phi}^2$ and this figure explains the inflationary expansion of the universe as $v > \dot{\phi}^2$ at the turnaround point and the post-bounce scenario as well. The fractional densities are plotted in Figure 14 and we can clearly understand the transition from DM domination to energy domination. In the next phase, we have shifted to a modified gravity framework which is chosen to be $f(T)$ gravity. The scale factor for this purpose has been chosen in bouncing form $a(t) = a_0(\frac{1+3t^2\sigma}{2})^{\frac{1}{3}}$. Our intention was to demonstrate the cosmology of $f(T)$ gravity where the function f, density ρ_T supports the inflationary cosmology of the universe by considering a correspondence between ω_c and ω_T. The torsion contributed EoS parameter always crosses $\frac{-1}{3}$ as required in accelerated expansion of the post-bounce scenario and it also supports inflationary cosmology. During the bounce, this model is stable in Einstein's and as well as the $f(T)$ modified gravity scenario against small perturbations. We checked the stability analysis of gravitational perturbations through square speed of sound.

Through the application of Shannon entropy, we analyzed the outcomes. The variation of Shannon entropy with cosmic time is presented in Figure 20. It is clear from Figure 20 that the uncertainty, computed through Shannon entropy, is lower at early stages of the universe. It is more significant in the later stage of the universe towards the turnaround point. After the bounce, uncertainty is higher as the inflationary expansion has occurred. Shannon entropy is an estimation of randomness in a system. We observe that the uncertainty is reduced as we evolve from Big Bang inflationary to late time universe. Therefore, the reconstruction scheme presented here has the potential to be considered a unifying model for early inflation and late time acceleration.

While concluding, let us mention the newness of the current work with respect to some related works. First, let us mention the work by us in [161] where generalized Chaplygin Gas was studied in a $f(T)$ gravity framework and the Shannon entropy approach was adopted to optimize the uncertainty associated with the model. Contrary to [161], the current work is extended much further by focusing on the realization of cosmological bounce in a reconstruction scheme for modified gravity with modified generalized Chaplygin Gas as the background fluid. In this context, another relevant work by one of the authors of the current paper, namely "A study on the bouncing behavior of modified Chaplygin Gas in presence of bulk viscosity and its consequences in the modified gravity framework" where modified Chaplygin Gas is studied in $f(T)$ gravity for non-singular bounce, concentrating on the suitability of the bouncing scale factor in inflationary cosmology as well. The current work deviates much from [162] in the sense that here every cosmological consequence was throughly checked for pre-bounce and post-bounce scenarios. Furthermore, the Shannon entropy theory was extensively incorporated to visualize the uncertainty associated with the bouncing point with respect to pre-bounce contraction and post-bounce expansion. We hope to expand this research to include other types of modified gravity, as well as the baro holographic fluid.

Author Contributions: S.S.: Methodology, Formal analysis, Graph plotting, Writing—original draft, Graph plotting. S.C.: Writing—review and editing, Validation, Project administration. All authors have read and agreed to the published version of this manuscript.

Funding: This research received no external funding.

Data Availability Statement: Not applicable.

Acknowledgments: The authors are thankful to the anonymous reviewers for constructive suggestions.

Conflicts of Interest: The authors declare no conflict of interest.

References

1. Perlmutter, S.; Aldering, G.; Valle, M.D.; Deustua, S.; Ellis, R.S.; Fabbro, S.; Fruchter, A.; Goldhaber, G.; Groom, D.E.; Hook, I.M.; Kim, A.G. Discovery of a supernova explosion at half the age of the Universe. *Nature* **1998**, *391*, 51–54. [CrossRef]
2. Perlmutter, S.; Aldering, G.; Goldhaber, G.; Knop, R.A.; Nugent, ; Castro, G.; Deustua, S.; Fabbro, S.; Goobar, A.; Groom, D.E.; Hook, I.M. Measurements of Ω and Λ from 42 high-redshift supernovae. *Astrophys. J.* **1999**, *517*, 565. [CrossRef]
3. Riess, A.G.; Filippenko, A.V.; Challis, ; Clocchiatti, A.; Diercks, A.; Garnavich, M.; Gillil, ; R.L.; Hogan, C.J.; Jha, S.; Kirshner, R.P.; Leibundgut, B. Observational evidence from supernovae for an accelerating universe and a cosmological constant. *Astron. J.* **1998**, *116*, 1009. [CrossRef]
4. Riess, A.G.; Strolger, L.G.; Tonry, J.; Casertano, S.; Ferguson, H.C.; Mobasher, B.; Challis, P.; Filippenko, A.V.; Jha, S.; Li, W.; Chornock, R. Type Ia supernova discoveries at z > 1 from the Hubble Space Telescope: Evidence for past deceleration and constraints on dark energy evolution. *Astrophys. J.* **2004**, *607*, 665. [CrossRef]
5. Miller, A.D.; Caldwell, R.; Devlin, M.J.; Dorwart, W.B.; Herbig, T.; Nolta, M.R.; Page, L.A.; Puchalla, J.; Torbet, E.; Tran, H.T. A Measurement of the Angular Power Spectrum of the Cosmic Microwave Background from l = 100 to 400. *Astrophys. J.* **1999**, *524*, L1. [CrossRef]
6. Boisseau, B.; Esposito-Farese, G.; Polarski, D.; Starobinsky, A.A. Reconstruction of a scalar-tensor theory of gravity in an accelerating universe. *Phys. Rev. Lett.* **2000**, *85*, 2236. [CrossRef] [PubMed]
7. Bennett, C.L.; Larson, D.; Weil, ; J.L.; Jarosik, N.; Hinshaw, G.; Odegard, N.; Smith, K.M.; Hill, R.S.; Gold, B.; Halpern, M.; Komatsu, E. Nine-year Wilkinson Microwave Anisotropy Probe (WMAP) observations: Final maps and results. *Astrophys. J. Suppl. Ser.* **2013**, *208*, 20. [CrossRef]
8. Bennett, C.L.; Hill, R.S.; Hinshaw, G.; Larson, D.; Smith, K.M.; Dunkley, J.; Gold, B.; Halpern, M.; Jarosik, N.; Kogut, A.; Komatsu, E. Seven-year wilkinson microwave anisotropy probe (WMAP*) observations: Are there cosmic microwave background anomalies? *Astrophys. J. Suppl. Ser.* **2011**, *192*, 17. [CrossRef]

9. Larson, D.; Dunkley, J.; Hinshaw, G.; Komatsu, E.; Nolta, M.R.; Bennett, C.L.; Gold, B.; Halpern, M.; Hill, R.S.; Jarosik, N.; Kogut, A. Seven-year wilkinson microwave anisotropy probe (WMAP*) observations: power spectra and WMAP-derived parameters. *Astrophys. J. Suppl. Ser.* **2011**, *192*, 16. [CrossRef]
10. Bridle, S.L.; Lahav, O.; Ostriker, J.P.; Steinhardt, J. Precision cosmology? Not just yet... *Science* **2013**, *299*, 1532–1533. [CrossRef]
11. Spergel, D.N.; Verde, L.; Peiris, H.V.; Komatsu, E.; Nolta, M.R.; Bennett, C.L.; Halpern, M.; Hinshaw, G.; Jarosik, N.; Kogut, A.; Limon, M. First-year Wilkinson Microwave Anisotropy Probe (WMAP)* observations: determination of cosmological parameters. *Astrophys. J. Suppl. Ser.* **2003**, *148*, 175. [CrossRef]
12. Bahcall, N.A.; Ostriker, J.P.; Perlmutter, S.; Steinhardt, J . The cosmic triangle: Revealing the state of the universe. *Science* **1999**, *284*, 1481–1488. [CrossRef]
13. Tegmark, M.; Strauss, M.A.; Blanton, M.R.; Abazajian, K.; Dodelson, S.; Sandvik, H.; Wang, X.; Weinberg, D.H.; Zehavi, I.; Bahcall, N.A.; et al. Cosmological parameters from SDSS and WMAP. *Phys. Rev. D* **2004**, *69*, 103501. [CrossRef]
14. Anand, S.; Chaubal, P.; Mazumdar, A.; Mohanty, S. Cosmic viscosity as a remedy for tension between PLANCK and LSS data. *J. Cosmol. Astropart. Phys.* **2017**, *11*, 5. [CrossRef]
15. Ciarcelluti, P. Cosmology with mirror dark matter II: cosmic microwave background and large scale structure. *Int. J. Mod. Phys.* **2005**, *14*, 223–256. [CrossRef]
16. Willis, J.P.; Pacaud, F.; Valtchanov, I.; Pierre, M.; Ponman, T.; Read, A.; Andreon, S.; Altieri, B.; Quintana, H.; Dos, Santos, S.; Birkinshaw, M. The XMM Large-Scale Structure survey: An initial sample of galaxy groups and clusters to a redshift z < 0.6. *Mon. Not. R. Astron. Soc.* **2005**, *363*, 675–691.
17. Li, H.; Xia, J.Q. Constraints on dark energy parameters from correlations of CMB with LSS. *J. Cosmol. Astropart. Phys.* **2010**, *4*, 026. [CrossRef]
18. Amendola, L.; Tsujikawa, S. *Dark Energy: Theory and Observations*; Cambridge University Press: Cambridge, UK, 2010.
19. Hinshaw, G.; Weil, P.; J.L.; Hill, R.S.; Odegard, N.; Larson, D.; Bennett, C.L.; Dunkley, J.; Gold, B.; Greason, M.R.; Jarosik, N.; et al. Five-year wilkinson microwave anisotropy probe* observations: Data processing, sky maps and basic results. *Astrophys. J. Suppl. Ser.* **2009**, *180*, 225. [CrossRef]
20. Rahaman, F. *The General Theory of Relativity: A Mathematical Approach*; Cambridge University Press, University Printing House: Cambridge, UK, 2021.
21. Anderson, J.L. General Relativity: Gravitation and Cosmology. Principles and Applications of the General Theory of Relativity. Steven Weinberg. Wiley, New York. *Science* **1973**, *179*, 1227–1228. [CrossRef]
22. Weinberg, S. Cosmological production of baryons. *Phys. Rev. Lett.* **1979**, *42*, 850. [CrossRef]
23. Crispino, L.C.; Kennefick, D.J. A hundred years of the first experimental test of general relativity. *Nat. Phys.* **2019**, *15*, 416-419. [CrossRef]
24. Synge, J.L.; Romain, J. Relativity: The general theory. *Phys. Today* **1961** , *14*, 50. [CrossRef]
25. Colistete Jr, R.; Fabris, J.C.; Tossa, J.; Zimdahl, W. Bulk viscous cosmology. *Phys. Rev. D.* **2007**, *76*, 103516. [CrossRef]
26. Martin, J. Everything you always wanted to know about the cosmological constant problem (but were afraid to ask). *C. R. Phys.* **2012**, *13*, 566–665. [CrossRef]
27. Padilla, A. Lectures on the cosmological constant problem. *arXiv* **2015**, arXiv:1502.05296.
28. Weinberg, S. The cosmological constant problem. *Rev. Mod. Phys.* **1989**, *61*, 1. [CrossRef]
29. Zel'dovich, Y.B. The cosmological constant and the theory of elementary particles. *Sov. Phys. USP* **1968**, *11*, 11381–11393
30. Sahni, V.; Krasiński, A. Republication of: The cosmological constant and the theory of elementary particles (By Ya. B. Zeldovich). *Gen. Relativ. Gravit.* **2008**, *40*, 1557–1591. [CrossRef]
31. Padmanabhan, T. Cosmological constant—The weight of the vacuum. *Phys. Rep.* **2003**, *380*, 235–320. [CrossRef]
32. Wu, J.P.; Ma, D.Z.; Ling, Y. Quintessence reconstruction of the new agegraphic dark energy model. *Phys. Lett.* **2008**, *663*, 152–159. [CrossRef]
33. Dutta, S.; Saridakis, E.N.; Scherrer, R.J. Dark energy from a quintessence (phantom) field rolling near a potential minimum (maximum). *Phys. Rev.* **2009**, *79*, 103005. [CrossRef]
34. Cai, Y.F.; Saridakis, E.N.; Setare, M.R.; Xia, J.Q. Quintom cosmology: Theoretical implications and observations. *Phys. Rep.* **2010**, *493*, 1–60. [CrossRef]
35. Debnath, U.; Chattopadhyay, S.; Hussain, I.; Jamil, M.; Myrzakulov, R. Generalized second law of thermodynamics for FRW cosmology with power-law entropy correction. *Eur. Phys. J.* **2012**,*72*, 1–6. [CrossRef]
36. Nojiri, S.I.; Odintsov, S.D.; Tsujikawa, S. Properties of singularities in the (phantom) dark energy universe. *Phys. Rev.* **2005**, *71*, 063004. [CrossRef]
37. Vikman, A. Can dark energy evolve to the phantom? *Phys. Rev. D* **2005**, *71*, 023515. [CrossRef]
38. Nojiri, S.I.; Odintsov, S.D. Quantum de Sitter cosmology and phantom matter. *Phys. Lett.* **2003**, *562*, 147–152. [CrossRef]
39. Elizalde, E.; Nojiri, S.I.; Odintsov, S.D. Late-time cosmology in a (phantom) scalar-tensor theory: Dark energy and the cosmic speed-up. *Phys. Rev.* **2004**, *70*, 043539. [CrossRef]
40. Sami, M.; Toporensky, A.; Tretjakov, V.; Tsujikawa, S. The fate of (phantom) dark energy universe with string curvature corrections. *Phys. Lett.* **2005**, *619*, 193–200. [CrossRef]
41. Nojiri, S.I.; Odintsov, S.D. Final state and thermodynamics of a dark energy universe. *Phys. Rev.* **2004**, *70*, 103522. [CrossRef]

42. Guo, Z.K.; Piao, Y.S.; Zhang, X.; Zhang, Y.Z. Cosmological evolution of a quintom model of dark energy. *Phys. Lett.* **2005**, *608*, 177–182. [CrossRef]
43. Cai, Y.F.; Li, M.; Lu, J.X.; Piao, Y.S.; Qiu, T.; Zhang, X. A string-inspired quintom model of dark energy. *Phys. Lett.* **2007**, *651*, 1–7. [CrossRef]
44. Zhao, G.B.; Xia, J.Q.; Li, M.; Feng, B.; Zhang, X. Perturbations of the quintom models of dark energy and the effects on observations. *Phys. Rev.* **2005**, *72*, 123515. [CrossRef]
45. Feng, B.; Li, M.; Piao, Y.S.; Zhang, X. Oscillating quintom and the recurrent universe. *Phys. Lett.* **2006**, *634*, 101–105. [CrossRef]
46. Bagla, J.S.; Jassal, H.K.; Padmanabhan, T. Cosmology with tachyon field as dark energy. *Phys. Rev.* **2003**, *67*, 063504. [CrossRef]
47. Setare, M.R.; Sadeghi, J.; Amani, A.R. Interacting tachyon dark energy in non-flat universe. *Phys. Lett.* **2009**, *673*, 241–246. [CrossRef]
48. Copel, ; E.J.; Garousi, M.R.; Sami, M.; Tsujikawa, S. What is needed of a tachyon if it is to be the dark energy? *Phys. Rev.* **2005**, *71*, 043003.
49. Wei, H.; Cai, R.G.; Zeng, D.F. Hessence: A new view of quintom dark energy. *Class. Quantum Gravity* **2005**, *22*, 3189. [CrossRef]
50. Alimohammadi, M.; Sadjadi, H.M. Attractor solutions for general hessence dark energy. *Phys. Rev.* **2006**, *73*, 083527. [CrossRef]
51. Chimento, L.P. Extended tachyon field, Chaplygin Gas and solvable k-essence cosmologies. *Phys. Rev.* **2004**, *69*, 123517.
52. Gorini, V.; Kamenshchik, A.; Moschella, U.; Pasquier, V. The Chaplygin Gas as a model for dark energy. In *The Tenth Marcel Grossmann Meeting: On Recent Developments in Theoretical and Experimental General Relativity, Gravitation and Relativistic Field Theories (In 3 Volumes)*; World Scientific Publishing Company: Singapore, 2008; pp. 840–859. .
53. Saha, P.; Debnath, U. Study of anisotropic compact stars with quintessence field and modified chaplygin gas in f (T) gravity. *Eur. Phys. J. C* **2019**, *79*, 1–16. [CrossRef]
54. Setare, M.R. Holographic Chaplygin DGP cosmologies. *Int. J. Mod. Phys.* **2009**, *18*, 419–427. [CrossRef]
55. Avelino, P.; Bolejko, K.; Lewis, G.F. Nonlinear Chaplygin Gas cosmologies. *Phys. Rev.* **2014**, *89*, 103004. [CrossRef]
56. Li, M. A model of holographic dark energy. *Phys. Lett.* **2004**, **603**, 1–5. [CrossRef]
57. Elizalde, E.; Nojiri, S.I.; Odintsov, S.D.; Wang, P. Dark energy: Vacuum fluctuations, the effective phantom phase and holography. *Phys. Rev.* **2005**, *71*, 103504. [CrossRef]
58. Jamil, M.; Saridakis, E.N.; Setare, M.R. Holographic dark energy with varying gravitational constant. *Phys. Lett.* **2009**, *679*, 172–176. [CrossRef]
59. Chattopadhyay, S.; Debnath, U.; Holographic dark energy scenario and variable modified Chaplygin Gas. *Astrophys. Space Sci.* **2009**, *319*, 183–185. [CrossRef]
60. Bamba, K.; Capozziello, S.; Nojiri, S.I.; Odintsov, S.D. Dark energy cosmology: The equivalent description via different theoretical models and cosmography tests. *Astrophys. Space Sci.* **2012**, *342*, 155–228. [CrossRef]
61. Debnath, U.; Chattopadhyay, S. Statefinder and Om diagnostics for interacting new holographic dark energy model and generalized second law of thermodynamics. *Int. J. Theor. Phys.* **2013**, *52*, 1250–1264. [CrossRef]
62. Ghosh, R.; Pasqua, A.; Chattopadhyay, S. Generalized second law of thermodynamics in the emergent universe for some viable models of $f(T)$ gravity. *Eur. Phys. J. Plus* **2013**, *128*, 1–11. [CrossRef]
63. Saridakis, E.N.; Basilakos, S. The generalized second law of thermodynamics with Barrow entropy. *Eur. Phys. J.* **2021**, *81*, 1–6. [CrossRef]
64. Singh, C.P.; Kumar, A. Viscous Ricci dark energy and generalized second law of thermodynamics in modified $f(R,T)$ gravity. *Mod. Phys. Lett.* **2018**, *33*, 1850225. [CrossRef]
65. Benaoum, H. Accelerated universe from modified Chaplygin Gas and tachyonic fluid. *Universe* **2022**, *8*, 340. [CrossRef]
66. Polarski, D. Dark energy. *Int. J. Mod. Phys. D* **2013**, *22*, 1330027. [CrossRef]
67. Straumann, N. Dark energy: Recent developments. *Mod. Phys. Lett. A* **2008**, *21*, 1083–1098. [CrossRef]
68. Yoo, J.; Watanabe, Y. Theoretical models of dark energy. *Int. J. Mod. Phys. D* **2021**, *21*, 1230002. [CrossRef]
69. Sahni, V.; Starobinsky, A. Reconstructing dark energy. *Int. J. Mod. Phys. D* **2006**, *15*, 2105–2132. [CrossRef]
70. Sahni, V.; Shafieloo, A.; Starobinsky, A.A. Model-independent evidence for dark energy evolution from Baryon acoustic oscillations. *Astrophys. J. Lett.* **2014**, *793*, L40. [CrossRef]
71. Spergel, D.N.; Steinhardt, J. Observational evidence for self-interacting cold dark matter. *Phys. Rev. Lett.* **2000**, *84*, 3760. [CrossRef]
72. Liddle, A.R.; Lyth, D.H. The Cold dark matter density perturbation. *Phys. Rep.* **1993**, *231*, 1–105. [CrossRef]
73. Amendola, L.; Finelli, F.; Burigana, C.; Carturan, D. WMAP and the generalized Chaplygin Gas. *J. Cosmol. Astropart. Phys.* **2003**, *2003*, 005. [CrossRef]
74. Setare, M.R. Interacting holographic generalized Chaplygin gas model. *Phys. Lett. B* **2007**, *654*,1–6. [CrossRef]
75. Gorini, V.; Kamenshchik, A.; Moschella, U. Can the Chaplygin Gas be a plausible model for dark energy? *Phys. Rev.* **2003**, *67*, 063509. [CrossRef]
76. Kamenshchik, A.; Moschella, U.; Pasquier, V. An alternative to quintessence. *Phys. Lett. B* **2001**, *511*, 265–268 [CrossRef]
77. Bento, M.C.; Bertolami, O.; Sen, A. Generalized Chaplygin Gas, accelerated expansion and dark Energy-Matter unification. *Phys. Rev. D* **2002**, *66*, 043507. [CrossRef]
78. Bento, M.D.C.; Bertolami, O.; Sen, A.A. Generalized Chaplygin Gas and cosmic microwave background radiation constraints. *Phys. Rev.* **2003**, *67*, 063003. [CrossRef]
79. Sen, A.A.; Scherrer, R.J. Generalizing the generalized Chaplygin Gas. *Phys. Rev.* **2005**, *72*, 063511. [CrossRef]

80. Pun, C.S.J.; Gergely, L.; Mak, M.K.; Kovcs, Z.; Szab, G.M.; Harko, T. Viscous dissipative Chaplygin Gas dominated homogenous and isotropic cosmological models. *Phys. Rev. D* **2008**, *77*, 063528. [CrossRef]
81. Debnath, U.; Banerjee, A.; Chakraborty, S. Role of modified Chaplygin Gas in accelerated universe. *Class. Quantum Grav.* **2004**, *21*, 5609. [CrossRef]
82. Pourhassan, B. Viscous modified cosmic Chaplygin Gas cosmology. *Int. J. Mod. Phys. D* **2013**, *22*, 1350061 [CrossRef]
83. Chattopadhyay, S. Modified Chaplygin Gas equation of state on viscous dissipative extended holographic Ricci dark energy and the cosmological consequences. *Int. J. Mod. Phys. D* **2017**, *26*, 1750042. [CrossRef]
84. Jawad, A.; Ilyas, A.; Rani, S. Dynamics of modified Chaplygin Gas inflation on the Brane with bulk viscous pressure. *Int. J. Mod. Phys. D* **2017**, *26*, 1750031 [CrossRef]
85. Bedran, M.L.; Soares, V.; Araujo, M.E. Temperature evolution of the FRW universe filled with modified Chaplygin Gas. *Phys. Lett.* **2008**, *659*, 462–465. [CrossRef]
86. Panigrahi, D.; Chatterjee, S. Thermodynamics of the variable modified Chaplygin Gas. *J. Cosmol. Astropart. Phys.* **2016**, *2016*, 052. [CrossRef]
87. Salti, M.; Kangal, E.E.; Aydogdu, O. Variable Polytropic gas cosmology. *Ann. Phys.* **2019**, *407*, 166–178. [CrossRef]
88. Kamenshchik, A.Y.; Tronconi, A.; Venturi, G. Reconstruction of scalar potentials in induced gravity and cosmology. *Phys. Lett. B* **2011**, *702*, 191–196. [CrossRef]
89. Gorbunov, D.S.; Rubakov, V.A. *Introduction to the Theory of the Early Universe*; World Scientific: Singapore, 2018.
90. Odintsov, S.D.; Oikonomou, V.K. Deformed matter bounce with dark energy epoch. *Phys. Rev. D* **2016**, *94*, 064022. [CrossRef]
91. Nojiri, S.; Odintsov, S.D. Modified gravity with negative and positive powers of the curvature: Unification of the inflation and of the cosmic acceleration. *Phys. Rev. D* **2003**, *68*, 123512. [CrossRef]
92. Capozziello, S.; Nojiri, S.; Odintsov, S.D. Unified phantom cosmology: Inflation, dark energy and dark matter under the same standard. *Phys. Lett. B* **2006**, *632*, 597–604. [CrossRef]
93. Huang, Q.G.; Li, M. The holographic dark energy in a non-flat universe. *J. Cosmol. Astropart. Phys.* **2004**, *2004*, 013. [CrossRef]
94. Wang, B.; Gong, Y.; Abdalla, E. Transition of the dark energy equation of state in an interacting holographic dark energy model. *Phys. Lett.* **2005**, *624*, 141–146. [CrossRef]
95. Cai, Y.F. Exploring bouncing cosmologies with cosmological surveys. *Sci. China Phys. Mech. Astron.* **2014**, *57*, 1414–1430. [CrossRef]
96. Mukhanov, V.F.; Feldman, H.A.; Brandenberger, R.H. Theory of cosmological perturbations. *Phys. Rep.* **1992**, *215*, 203–333. [CrossRef]
97. Nojiri, S.; Odintsov, S.D.; Oikonomou, V. Modified gravity theories on a nutshell: inflation, bounce and late-time evolution. *Phys. Rep.* **2017**, *692*, 1–104. [CrossRef]
98. Bamba, K.; Makarenko, A.N.; Myagky, A.N.; Nojiri, S.I.; Odintsov, S.D. Bounce cosmology from F (R) gravity and F (R) bigravity. *J. Cosmol. Astropart. Phys.* **2014**, *2014*, 008. [CrossRef]
99. Bamba, K.; Odintsov, S.D. Inflationary cosmology in modified gravity theories. *Symmetry* **2015**, *7*, 220–240. [CrossRef]
100. Nojiri, S.I.; Odintsov, S.D.; Saridakis, E.N. Holographic bounce. *Nucl. Phys.* **2019**, *949*, 114790. [CrossRef]
101. Elizalde, E.; Nojiri, S.; Odintsov, S.D.; Sez-Gmez, D.; Faraoni, V. Reconstructing the universe history, from inflation to acceleration, with phantom and canonical scalar fields. *Phys. Rev. D* **2008**, *77*, 106005 [CrossRef]
102. Odintsov, S.D.; Paul, T. Bounce universe with finite-time singularity. *Universe* **2022**, *8*, 292. [CrossRef]
103. Nojiri, S.I.; Odintsov, S.D.; Paul, T. Towards a smooth unification from an ekpyrotic bounce to the dark energy era. *Phys. Dark Universe* **2022**, *35*, 100984. [CrossRef]
104. Capozziello, S.; De Laurentis, M. Extended theories of gravity. *Phys. Rep.* **2011**, *509*, 167-321. [CrossRef]
105. Barrow, J.D.; Graham, A.A. Singular inflation. *Phys. Rev.* **2015**, *91*, 083513. [CrossRef]
106. Nojiri, S.; Odintsov, S.D.; Oikonomou, V.K. Quantitative analysis of singular inflation with scalar-tensor and modified gravity. *Phys. Rev.* **2015**, *91*, 084059. [CrossRef]
107. Nojiri, S.; Odintsov, S.D.; Oikonomou, V.K. Singular inflation from generalized equation of state fluids. *Phys. Lett.* **2015**, *747*, 310–320. [CrossRef]
108. Barrow, J.D. Sudden future singularities. *Class. Quantum Gravity* **2004**, *21*, L79. [CrossRef]
109. Weinberg, S. Entropy generation and the survival of protogalaxies in an expanding universe. *Astrophys. J.* **1971**, *168*, 175 [CrossRef]
110. Clifton, T.; Barrow, J.D. The power of general relativity. *Phys. Rev. D* **2005**, *72*, 103005 . [CrossRef]
111. Medina, S.B., Nowakowski, M.; Batic, D. Viscous cosmologies. *Class. Quantum Gravity* **2019**, *36*, 215002. [CrossRef]
112. Jaffe, R.L. Casimir effect and the quantum vacuum. *Phys. Rev.* **2005**, *72*, 021301. [CrossRef]
113. Myrzakulov, R.; Sebastiani, L. Bounce solutions in viscous fluid cosmology. *Astrophys. Space Sci.* **2014**, *352*, 281–288. [CrossRef]
114. Brevik, I.; Timoshkin, A.V. Viscous fluid holographic bounce. *Int. J. Geom. Methods Mod. Phys.* **2020**, *17*, 2050023. [CrossRef]
115. Cataldo, M.; Cruz, N.; Lepe, S. Viscous dark energy and phantom evolution. *Phys. Lett.* **2005**, *619*, 5–10. [CrossRef]
116. Hu, M.G.; Meng, X.H. Bulk viscous cosmology: Statefinder and entropy. *Phys. Lett.* **2006**, *635*, 186–194. [CrossRef]
117. Clifton, T.; Ferreira, G.; Padilla, A.; Skordis, C. Modified gravity and cosmology. *Phys. Rep.* **2012**, *513*, 1–189. [CrossRef]
118. Nojiri, S.I.; Odintsov, S.D. Introduction to modified gravity and gravitational alternative for dark energy. *Int. J. Geom. Methods Mod. Phys.* **2007**, *4*, 115–145. [CrossRef]
119. Tsujikawa, S. Modified gravity models of dark energy. In *Lectures on Cosmology*; Springer: Berlin/Heidelberg, Germany, 2010; pp. 99–145.

120. Josset, T., Perez, A. and Sudarsky, D Dark energy from violation of energy conservation. *Phys. Rev. Lett.* **2008**, *118*, 021102. [CrossRef] [PubMed]
121. Myrzakulov, R. Accelerating universe from F (T) gravity. *Eur. Phys. J. C* **2011**, *71*, 1752. [CrossRef]
122. Chattopadhyay, S.; Pasqua, A. Reconstruction of $f(T)$ gravity from the Holographic dark energy. *Astrophys. Space Sci.* **2012**, *344*, 269–274. [CrossRef]
123. Joyce, A.; Lombriser, L.; Schmidt, F. Dark energy vs. modified gravity. *arXiv* **2016**, arXiv:1601.06133.
124. Bertschinger, E.; Zukin, P. Distinguishing modified gravity from dark energy. *Phys. Rev.* **2008**, *78*, 024015. [CrossRef]
125. Wang, Y. Differentiating dark energy and modified gravity with galaxy redshift surveys. *J. Cosmol. Astropart. Phys.* **2008**, *2008*, 021. [CrossRef]
126. Capozziello, S.; Matsumoto, J.; Nojiri, S.I.; Odintsov, S.D. Dark energy from modified gravity with Lagrange multipliers. *Phys. Lett.* **2010**, *693*, 198–208. [CrossRef]
127. Hu, W.; Sawicki, I. Parametrized post-Friedmann framework for modified gravity. *Phys. Rev.* **2007**, *76*, 104043. [CrossRef]
128. Nojiri, S.I.; Odintsov, S.D. Modified gravity and its reconstruction from the universe expansion history. *J. Phys. Conf. Ser.* **2007**, *66*, 012005. [CrossRef]
129. Sanders, R. VModified gravity without dark matter. *The Invisible Universe: Dark Matter and Dark Energy*: Springer: Berlin/Heidelberg, Germany, 2007; pp. 375–402.
130. Klir, G.J.; Folger, T.A. *Fuzzy Sets, Uncertainty and Information*; Prentice-Hall, Inc.: Hoboken, NJ, USA, 1987.
131. Witten, E. A mini-introduction to information theory. *Riv. Del Nuovo C.* **2020**, *43*, 187–227. [CrossRef]
132. Crupi, V.; Nelson, J.D.; Meder, B.; Cevolani, G.; Tentori, K. Generalized information theory meets human cognition: Introducing a unified framework to model uncertainty and information search. *Cogn. Sci.* **2018**, *42*, 1410–1456. [CrossRef] [PubMed]
133. Amani, A.R.Pourhassan, B. Viscous generalized Chaplygin gas with arbitrary. *Int. J. Theor. Phys.* **2013**, *52*, 1309-1313 [CrossRef]
134. Saha, S.; Chattopadhyay, S. Exploring of the summer monsoon rainfall around the Himalayas in time domain through maximization of Shannon entropy. *Theor. Appl. Climatol.* **2020**, *141*, 133–141. [CrossRef]
135. Klir, G.J. Uncertainty and information: foundations of generalized information theory. *Kybernetes* **2006**, *35*, 1297–1299. [CrossRef]
136. Cai, Y.-F.; Chen, S.-H.; Dent, J.B.; Dutta, S.; Saridakis, E.N. Matter bounce cosmology with the $f(T)$ gravity. *Class. Quantum Grav.* **2011**, *28*, 215011. [CrossRef]
137. Maartens, R. Dissipative cosmology. *Class. Quantum Gravity* **1995**, *12*, 1455. [CrossRef]
138. Albrecht, A.; Magueijo, J. Time varying speed of light as a solution to cosmological puzzles. *Phys. Rev.* **1999**, *59*, 043516. [CrossRef]
139. Padmanabhan, T. Emergent perspective of gravity and dark energy. *Res. Astron. Astrophys.* **2012**, *12*, 891. [CrossRef]
140. Coone, D.; Roesta, D.; Venninc, V. The Hubble flow of plateau inflation. *J. Cosmol. Astrophys.* **2015**, *1511*, 010. [CrossRef]
141. Hoffman, M.B.; Turner, M.S. Kinematic constraints to the key inflationary observables. *Phys. Rev. D* **2001**, *64*, 023506 [CrossRef]
142. Kinney, W.H. Inflation: Flow, fixed points and observables to arbitrary order in slow roll. *Phys. Rev. D* **2002**, *66*, 083508. [CrossRef]
143. Nojiri, S.I.; Odintsov, S.D. Unifying phantom inflation with late-time acceleration: Scalar phantom–non-phantom transition model and generalized holographic dark energy. *Gen. Relativ. Gravit.* **2006**, *38*, 1285–1304. [CrossRef]
144. Sahni, V.; Saini, T.D.; Starobinsky, A.A.; Alam, U. Statefinder—A new geometrical diagnostic of dark energy. *J. Exp. Theor. Phys. Lett.* **2003**, *77*, 201–206. [CrossRef]
145. Lam, Y.M.P. Perturbation Lagrangian theory for scalar fields-Ward-Takahashi identity and current algebra. *Phys. Rev.* **1972**, *6*, 2145. [CrossRef]
146. Paliathanasis, A., Barrow, J.D.; Leach, P.G.L. Cosmological solutions of f (T) gravity. *Phys. Rev. D* **2016**, *94*, 023525. [CrossRef]
147. Chattopadhyay, S.; Chakraborty, G. A reconstruction scheme for f (T) gravity: Variable generalized Chaplygin dark energy gas form. *Astron. Nachrichten* **2021**, *342*, 103–109. [CrossRef]
148. Abdalla, M.C.B.; Nojiri, S.I.; Odintsov, S.D. Consistent modified gravity: Dark energy, acceleration and the absence of cosmic doomsday.*Class. Quantum Gravity* **2005**, *22*, L35. [CrossRef]
149. Cai, Y.F., Capozziello, S., De Laurentis, M.; Saridakis, E.N. f (T) teleparallel gravity and cosmology. *Rep. Prog. Phys.* **2016**, *79*, 106901. [CrossRef] [PubMed]
150. Capozziello, S.; Nojiri, S.; Odintsov, S.D. Dark energy: The equation of state description versus scalar-tensor or modified gravity. *Phys. Lett.* **2006**, *634*, 93–100. [CrossRef]
151. Setare, M.R. Interacting holographic dark energy model in non-flat universe. *Phys. Lett.* **2006**, *642*, 1–4. [CrossRef]
152. Hannestad, S. Constraints on the sound speed of dark energy. *Phys. Rev.* **2005**, *71*, 103519. [CrossRef]
153. Zhang, C., Zhang, H., Yuan, S., Liu, S., Zhang, T.J. and Sun, Y.C.; Four new observational H (z) data from luminous red galaxies in the Sloan Digital Sky Survey data release seven. *Res. Astron. Astrophys.* **2014**, *14*, 1221. [CrossRef]
154. Chattopadhyay, S.; Chattopadhyay, G.; Midya, S.K. Shannon entropy maximization supplemented by neurocomputing to study the consequences of a severe weather phenomenon on some surface parameters. *Nat. Hazards* **2018**, *93*, 237–247. [CrossRef]
155. Aziza, A.; Chakraborty, G.; Chattopadhyay, S. Variable generalized Chaplygin gas in f (Q) gravity and the inflationary cosmology. *Int. J. Mod. Phys. D* **2021**, *30*, 2150119. [CrossRef]
156. Jamil, M.; Myrzakulov, Y.; Razina, O.; Myrzakulov, R. Modified Chaplygin Gas and solvable F-essence cosmologies. *Astrophys. Space Sci.* **2011**, *336*, 315–325. [CrossRef]
157. Karami, K.; Abdolmaleki, A. Generalized second law of thermodynamics in $f(T)$ gravity. *J. Cosmol. Astropart. Phys.* **2012**, *2012*, 007. [CrossRef]

158. Lu, J.; Xu, L.; Li, J.; Chang, B.; Gui, Y.; Liu, H. Constraints on modified Chaplygin Gas from recent observations and a comparison of its status with other models. *Phys. Lett.* **2008**, *662*, 87–91. [CrossRef]
159. Karami, K. Chaplygin scalar field reconstruction of the modified ghost dark energy model. *Can. J. Phys.* **2015**, *93*, 855–861. [CrossRef]
160. Yang, H.; Wang, J. Concentration in vanishing pressure limit of solutions to the modified Chaplygin Gas equations. *J. Math. Phys.* **2016**, *57*, 111504. [CrossRef]
161. Saha, S.; Chattopadhyay, S. Viscous generalised Chaplygin Gas under the purview of $f(T)$ gravity and the model assessment through probabilistic information theory. *Phys. Scr.* **2022**, *97*, 045006. [CrossRef]
162. Chattopadhyay, S. A study on the bouncing behavior of modified Chaplygin Gas in presence of bulk viscosity and its consequences in the modified gravity framework. *Int. J. Geom. Methods Mod. Phys.* **2017**, *14*, 1750181. [CrossRef]

universe

MDPI

Article

On Born's Reciprocal Relativity, Algebraic Extensions of the Yang and Quaplectic Algebra, and Noncommutative Curved Phase Spaces

Carlos Castro Perelman

Ronin Institute, 127 Haddon Place, Montclair, NJ 07043, USA; perelmanc@hotmail.com

Abstract: After a brief introduction of Born's reciprocal relativity theory is presented, we review the construction of the *deformed* quaplectic group that is given by the semi-direct product of $U(1,3)$ with the *deformed* (noncommutative) Weyl–Heisenberg group corresponding to *noncommutative* fiber coordinates and momenta $[X_a, X_b] \neq 0$; $[P_a, P_b] \neq 0$. This construction leads to more general algebras given by a two-parameter family of deformations of the quaplectic algebra, and to further algebraic extensions involving antisymmetric tensor coordinates and momenta of higher ranks $[X_{a_1 a_2 \cdots a_n}, X_{b_1 b_2 \cdots b_n}] \neq 0$; $[P_{a_1 a_2 \cdots a_n}, P_{b_1 b_2 \cdots b_n}] \neq 0$. We continue by examining algebraic extensions of the Yang algebra in extended noncommutative phase spaces and compare them with the above extensions of the deformed quaplectic algebra. A solution is found for the exact analytical mapping of the noncommuting x^μ, p^μ operator variables (associated to an 8D curved phase space) to the canonical Y^A, Π^A operator variables of a flat 12D phase space. We explore the geometrical implications of this mapping which provides, in the *classical* limit, the embedding functions $Y^A(x,p), \Pi^A(x,p)$ of an 8D curved phase space into a flat 12D phase space background. The latter embedding functions determine the functional forms of the base spacetime metric $g_{\mu\nu}(x,p)$, the fiber metric of the vertical space $h^{ab}(x,p)$, and the nonlinear connection $N_{a\mu}(x,p)$ associated with the 8D cotangent space of the 4D spacetime. Consequently, we find a direct link between noncommutative curved phase spaces in lower dimensions and commutative flat phase spaces in higher dimensions.

Keywords: Born Reciprocal Relativity; Yang Algebra; Phase Spaces; Finsler Geometry

1. Introduction: Born's Reciprocal Relativity Theory

Most of the work devoted to quantum gravity has been focused on the geometry of spacetime rather than phase space per se. The first indication that phase space should play a role in quantum gravity was raised by [1]. The principle behind Born's reciprocal relativity theory [2–5] was based on the idea proposed long ago by [1] that coordinates and momenta should be unified on the same footing. Consequently, if there is a limiting speed (temporal derivative of the position coordinates) in nature, there should be a maximal force as well, since force is the temporal derivative of the momentum. The principle of maximal acceleration was advocated earlier on by [6–9]. A *maximal* speed limit (speed of light) must be accompanied with a *maximal* proper force (which is also compatible with a *maximal* and *minimal* length duality) [5,10].

We explored in [5,10] some novel consequences of Born's reciprocal relativity theory in a flat phase space and generalized the theory to the curved spacetime scenario. We provided, in particular, some specific results from Born's reciprocal relativity which are *not* present in special relativity. These are: a momentum-dependent time delay in the emission and detection of photons; the relativity of chronology; an energy-dependent notion of locality; a superluminal behavior; the relative rotation of photon trajectories due to the aberration of light; the invariance of area cells in the phase space; and modified dispersion relations.

The generalized velocity and force (acceleration) boosts (rotations) transformations of the *flat* 8D phase-space coordinates, where $X^i, T, E, P^i; i = 1,2,3$ are **c**-valued (classical) variables which are *all* boosted (rotated) into each other, were given by [2–4] based on the group $U(1,3)$, which is the Born version of the Lorentz group $SO(1,3)$. The $U(1,3) = SU(1,3) \times U(1)$ group transformations leave invariant the symplectic two-form $\Omega = -\, dT \wedge dE + \delta_{ij} dX^i \wedge dP^j; i,j = 1,2,3$ and also the following Born–Green line interval in the $flat$ 8D phase space

$$(d\omega)^2 = c^2(dT)^2 - (dX)^2 - (dY)^2 - (dZ)^2 + \\ \frac{1}{b^2}\left((dE)^2 - c^2(dP_x)^2 - c^2(dP_y)^2 - c^2(dP_z)^2\right) \tag{1}$$

The maximal proper force is set to be given by b. The rotations, velocity, and force (acceleration) boosts leaving invariant the symplectic two-form and the line interval in the 8D phase space are rather elaborate; see [2–4] for details.

These transformations can be simplified drastically when the velocity and force (acceleration) boosts are both parallel to the x-direction and leave the transverse directions Y, Z, P_y, P_z intact. There is now a subgroup $U(1,1) = SU(1,1) \times U(1) \subset U(1,3)$ which leaves invariant the following line interval

$$(d\omega)^2 = c^2(dT)^2 - (dX)^2 + \frac{(dE)^2 - c^2(dP)^2}{b^2} = \\ (d\tau)^2\left(1 + \frac{(dE/d\tau)^2 - c^2(dP/d\tau)^2}{b^2}\right) = (d\tau)^2\left(1 - \frac{F^2}{F^2_{max}}\right), P = P_x \tag{2}$$

where one has factored out the proper time infinitesimal $(d\tau)^2 = c^2 dT^2 - dX^2$ in (2). The proper force interval $(dE/d\tau)^2 - c^2(dP/d\tau)^2 = -F^2 < 0$ is "spacelike" when the proper velocity interval $c^2(dT/d\tau)^2 - (dX/d\tau)^2 > 0$ is timelike. The analog of the Lorentz relativistic factor in Equation (2) involves the ratios of two proper *forces*.

One may set the maximal proper force acting on a fundamental particle of Planck mass to be given by $F_{max} = b \equiv m_P c^2/L_P$, where m_P is the Planck mass and L_P is the postulated minimal Planck length. Invoking a minimal/maximal length duality, one can also set $b = M_U c^2/R_H$, where R_H is the Hubble scale and M_U is the observable mass of the universe. Equating both expressions for b leads to $M_U/m_P = R_H/L_P \sim 10^{60}$. The value of b may also be interpreted as the maximal string tension.

The $U(1,1)$ group transformation laws of the phase-space coordinates X, T, P, E which leave the interval (2) invariant are [2–4]

$$T' = T\cosh\xi + \left(\frac{\xi_v\, X}{c^2} + \frac{\xi_a P}{b^2}\right)\frac{\sinh\xi}{\xi} \tag{3a}$$

$$E' = E\cosh\xi + (-\xi_a X + \xi_v P)\frac{\sinh\xi}{\xi} \tag{3b}$$

$$X' = X\cosh\xi + \left(\xi_v T - \frac{\xi_a\, E}{b^2}\right)\frac{\sinh\xi}{\xi} \tag{4a}$$

$$P' = P\cosh\xi + \left(\frac{\xi_v E}{c^2} + \xi_a T\right)\frac{\sinh\xi}{\xi} \tag{4b}$$

where ξ_v is the velocity-boost rapidity parameter, ξ_a is the force (acceleration) boost rapidity parameter, and ξ is the net effective rapidity parameter of the primed-reference frame.

These parameters ξ_a, ξ_v, ξ are defined, respectively, in terms of the velocity $v = dX/dT$ and force $f = dP/dT$ (related to acceleration) as

$$\tanh\left(\frac{\xi_v}{c}\right) = \frac{v}{c}; \quad \tanh\left(\frac{\xi_a}{b}\right) = \frac{F}{F_{max}}, \quad \xi = \sqrt{\left(\frac{\xi_v}{c}\right)^2 + \left(\frac{\xi_a}{b}\right)^2} \tag{5}$$

The $U(1,3)$ generators $Z_{ab} = \frac{1}{2}(L_{[ab]} + M_{(ab)})$ are comprised of the 6 ordinary Lorentz generators $L_{[ab]}$ and 10 force (acceleration) boost/rotation generators $M_{(ab)}$, giving a total of 16 generators.

It is straightforward to verify that the transformations (4a,b) leave invariant the phase-space interval $c^2(dT)^2 - (dX)^2 + ((dE)^2 - c^2(dP)^2)/b^2$ but *do not* leave separately invariant the proper time interval $(d\tau)^2 = c^2 dT^2 - dX^2$, nor the interval in energy–momentum space $\frac{1}{b^2}[(dE)^2 - (dP)^2]$. Only the *combination*

$$(d\omega)^2 = (d\tau)^2\left(1 - \frac{F^2}{F_{max}^2}\right) \tag{6}$$

is truly left invariant under force (acceleration) boosts (4a,b). They also leave invariant the symplectic two-form (phase-space areas) $\Omega = -\, dT \wedge dE + dX \wedge dP$.

Some readers might note that the $U(1,3)$ algebra is usually not used in standard formulations of particle kinematics and as symmetries in the dynamical, mechanical, and field-theoretic models. For instance, Kalman [11] long ago studied the $SU(1,3)$ group (and its discrete representations) as a dynamical group for hadrons. By a dynamical group, one means in general (a noncompact one) which gives the actual energy or mass spectrum of a quantum mechanical system [12].

Low [2–4] has explained in great detail that since $U(1,3)$ is noncompact, the $U(1,3)$ infinite-dimension unitary representations contain discrete series that may be decomposed into *infinite* ladders where the rungs are finite dimensional irreducible unitary $U(3)$ representations. In particular, the rest and null frames yield the groups $SU(3), SU(2)$, and $U(1)$ that appear in the Standard Model, and which is very appealing. If one has a single particle state, under force-boosts (acceleration) transformations, one would expect to transform it into a compound state that decomposes into a sum of single particle states representing the particle interactions of nonuniform velocity frames of reference.

Low [2–4] has argued that one could think of the timelike states as the rungs of the ladder and Poincare transformations transform these rungs into themselves with *no* mixing of states that are on *different* rungs; likewise with the null states. There are no Poincare transformations that take timelike states into null states. However, when one considers noninertial frames the states in different rungs of the ladder can transform into each other, and timelike and null states can mix. The reason is that due to the nonzero rates of change of the momentum, one expects the dynamical symmetry to describe transitions between these states when viewed from the interacting frames.

One should also add that these arguments presented by [2–4] bear a resemblance to the Unruh effect (the Fulling–Davies–Unruh effect) [13] which is a kinematic prediction of quantum field theory that an *accelerating* observer will observe a thermal bath, such as a blackbody radiation, whereas an inertial observer would observe none. In other words, the background appears to be warm from an accelerating reference frame. Heuristically, for a uniformly accelerating observer, the ground state of an inertial observer is seen as a *mixed* state in thermodynamic equilibrium with a nonzero temperature bath of thermal photons and whose temperature is proportional to the acceleration.

Low [2–4] also constructed the eigenvalue equations for the representation of the set of Casimir invariant operators which define the field equations of the system. The applications of the *deformed* quaplectic algebras studied in this work, in particular corresponding to the *deformed* Heisenberg algebras, to theoretical physics models remain to be studied, and in particular, within the context of quantum field theories in noncommutative spacetimes. This is beyond the scope of this work.

After this brief introduction of Born's reciprocal relativity theory, in Section 2 we review the construction of the *deformed* quaplectic group that is given by the semidirect product of $U(1,3)$ with the *deformed* (noncommutative) Weyl–Heisenberg group corresponding to *noncommutative* fiber coordinates and momenta $[X_a, X_b] \neq 0$; $[P_a, P_b] \neq 0$. This construction leads at the end of Section 2 to more general algebras given by a two-parameter family of deformations of the quaplectic algebra and to local gauge theories of gravity based on the latter deformed quaplectic algebras.

We continue in Section 2 by examining the algebraic extensions of the Yang algebra in extended noncommutative phase spaces and compare them with the extensions of the deformed quaplectic algebra involving antisymmetric tensor coordinates and momenta of higher ranks $[X_{a_1 a_2 \cdots a_n}, X_{b_1 b_2 \cdots b_n}] \neq 0$; $[P_{a_1 a_2 \cdots a_n}, P_{b_1 b_2 \cdots b_n}] \neq 0$.

In Section 2, a solution is found for the exact analytical mapping of the noncommuting x^μ, p^μ operator variables (associated with an 8D curved phase space) to the canonical Y^A, Π^A operator variables of a flat 12D phase space. We explore the geometrical implications of this mapping which provides, in the *classical* limit, the embedding functions $Y^A(x,p), \Pi^A(x,p)$ of an 8D curved phase space into a flat 12D phase-space background. The latter embedding functions determine the functional forms of the base spacetime metric $g_{\mu\nu}(x,p)$, the fiber metric of the vertical space $h^{ab}(x,p)$, and the nonlinear connection $N_{a\mu}(x,p)$ associated with the 8D cotangent space of the 4D spacetime. We finalize with some concluding remarks.

2. The Deformed Quaplectic Group and Complex Gravity

To begin this section we review the construction of the *deformed* quaplectic group given by the semidirect product of $U(1,3)$ with the deformed (noncommutative) Weyl–Heisenberg group involving noncommutative coordinates and momenta [14]. Then, we proceed to construct a two-parameter family of deformed quaplectic algebras parametrized by two complex coefficients α, β.

The (undeformed) quaplectic group is given by the semidirect product of $U(1,3)$ with the Weyl–Heisenberg group and was studied in detail by [2–4]. Physically, the quaplectic group is basically the "phase-space" version of the Poincare group (which is given by the semidirect product of the Lorenz group $SO(1,3)$ with the translation group T_4), where the translation group is replaced by the Weyl–Heisenberg group and the Lorentz group is replaced by $U(1,3)$.

The deformed Weyl–Heisenberg algebra involves the generators

$$Z_a = \frac{1}{\sqrt{2}}\left(\frac{X_a}{\lambda_l} - i\frac{P_a}{\lambda_p}\right); \quad Z_a^\dagger = \frac{1}{\sqrt{2}}\left(\frac{X_a}{\lambda_l} + i\frac{P_a}{\lambda_p}\right); \quad a = 1,2,3,4. \tag{7}$$

Notice that we must *not* confuse the *generators* X_a, P_a (associated with the fiber coordinates of the internal space of the fiber bundle) with the ordinary base spacetime coordinates and momenta x_μ, p_μ. The local gauge theory based on the deformed quaplectic algebra was constructed in the fiber bundle over the base manifold which is a 4D curved spacetime with *commuting* coordinates $x^\mu = x^0, x^1, x^2, x^3$ [14]. The (deformed) quaplectic group acts as the automorphism group along the internal fiber coordinates. Therefore, we must *not* confuse the *deformed* complex gravitational theory constructed in [14] with the noncommutative gravity work in the literature where the spacetime coordinates x^μ are not commuting.

The four fundamental length, momentum, temporal, and energy scales are, respectively,

$$\lambda_l = \sqrt{\frac{\hbar c}{b}}; \quad \lambda_p = \sqrt{\frac{\hbar b}{c}}; \quad \lambda_t = \sqrt{\frac{\hbar}{bc}}; \quad \lambda_e = \sqrt{\hbar bc}. \tag{8}$$

where b is the *maximal* proper force associated with the Born's reciprocal relativity theory. In the natural units $\hbar = c = b = 1$, all four scales become *unity*. The gravitational coupling is given by

$$G = \frac{c^4}{\mathcal{F}_{max}} = \frac{c^4}{b}. \tag{9}$$

and the four scales then coincide with the Planck length, momentum, time, and energy, respectively. One may postulate the maximal proper force to be given by

$$\mathcal{F}_{\max} = m_P \frac{c^2}{L_P} \tag{10}$$

where L_P is the Planck scale, the Planck mass m_P is assumed to be the *maximal* mass of a *fundamental* particle, and $\frac{c^2}{L_P}$ is postulated to be its *maximal* proper acceleration. In natural units $\hbar = c = G = 1$, $\mathcal{F}_{max} \to 1$.

The generators of the $U(1,3)$ algebra given by Z_{ab} are Hermitian $(Z_{ab})^\dagger = Z_{ab}$, with $a, b = 1,2,3,4$; while the generators of the *deformed* Weyl–Heisenberg algebra $Z_a, Z_a^\dagger$ are Hermitian-conjugate pairs like $L_+ = L_x + iL_y, L_- = L_x - iL_y$ in the $SO(3)$ algebra. Note that the Hermitian-conjugate pairs of generators $Z_a, Z_a^\dagger$ in Equations (7) are not independent from each other, hence one is *not* doubling the number of physical dimensions. For instance, the complex variables $z^\mu = x^\mu + ip^\mu; \bar{z}^\mu = x^\mu - ip^\mu; \mu = 1,2,\cdots,D$ are not independent but complex-conjugate pairs. The number of physical dimensions of the 2D phase space remains the same.

The standard quaplectic group [2–4] is given by the semidirect product of the $U(1,3)$ group and the unmodified Weyl–Heisenberg $H(1,3)$ group $\mathcal{Q}(1,3) \equiv U(1,3) \otimes_s H(1,3)$ and is defined in terms of the generators $Z_{ab}, Z_a, Z_a^\dagger, \mathcal{I}$ described below with $a, b = 1,2,3,4$.

A careful analysis reveals that the generators $Z_a, Z_a^\dagger$ (comprised of Hermitian *and* anti-Hermitian pieces) of the *deformed* Weyl–Heisenberg algebra can be defined in terms of judicious linear combinations of the Hermitian $U(1,4)$ algebra generators Z_{AB}, where $A, B = 1,2,3,4,5; a, b = 1,2,3,4$ and $\eta_{AB} = diag\ (+,-,-,-,-)$. The linear combination is defined after introducing the following *complex*-valued coefficients as follows:

$$Z_a = (-i)^{1/2}(Z_{a5} - iZ_{5a}); \quad Z_a^\dagger = (i)^{1/2}(Z_{a5} + iZ_{5a}); \quad Z_{55} = \frac{\mathcal{I}}{2} \tag{11}$$

The reason behind this particular choice of the complex coefficients appearing in Equation (11) is explained below in Equation (20a–c). The Hermitian generators of the $U(1,4)$ algebra are given by $Z_{AB} \equiv \mathcal{E}_A^B$ and $Z_{BA} \equiv \mathcal{E}_B^A$; notice that the position of the indices is very relevant because $Z_{AB} \neq Z_{BA}$. The commutators are

$$[\mathcal{E}_a^b,\ \mathcal{E}_c^d] \ = \ -i\,\delta_c^b\,\mathcal{E}_a^d + i\,\delta_a^d\,\mathcal{E}_c^b; \quad [\mathcal{E}_c^d,\ \mathcal{E}_a^5] \ = \ -i\,\delta_a^d\,\mathcal{E}_c^5; \quad [\mathcal{E}_c^d,\ \mathcal{E}_5^a] \ = \ i\,\delta_c^a\,\mathcal{E}_5^d \tag{12}$$

and $[\mathcal{E}_5^5,\ \mathcal{E}_5^a] = -i\,\delta_5^5\,\mathcal{E}_5^a$ .... such that now, $\mathcal{I}(= 2Z_{55})$ *no* longer commutes with $Z_a, Z_a^\dagger$. The generators Z_{ab} of the $U(1,3)$ algebra can be decomposed into the Lorentz subalgebra generators $L_{[ab]}$ and the "shearlike" generators $M_{(ab)}$ as

$$Z_{ab} \equiv \frac{1}{2}(M_{(ab)} + L_{[ab]}) \Rightarrow L_{ab} \equiv L_{[ab]} = (Z_{ab} - Z_{ba}); \quad M_{ab} \equiv M_{(ab)} = (Z_{ab} + Z_{ba}), \tag{13}$$

where the "shearlike" generators $M_{(ab)}$ and the Lorentz generators $L_{[ab]}$ are Hermitian. The explicit commutation relations of the M_{ab}, L_{ab} generators are given by

$$[L_{ab}, L_{cd}] = i(\eta_{bc}L_{ad} - \eta_{ac}L_{bd} - \eta_{bd}L_{ac} + \eta_{ad}L_{bc}). \tag{14a}$$

$$[M_{ab}, M_{cd}] = -i(\eta_{bc}L_{ad} + \eta_{ac}L_{bd} + \eta_{bd}L_{ac} + \eta_{ad}L_{bc}). \tag{14b}$$

$$[L_{ab}, M_{cd}] = i(\eta_{bc}M_{ad} - \eta_{ac}M_{bd} + \eta_{bd}M_{ac} - \eta_{ad}M_{bc}). \tag{14c}$$

Therefore, given $Z_{ab} = \frac{1}{2}(M_{ab} + L_{ab})$, $Z_{cd} = \frac{1}{2}(M_{cd} + L_{cd})$ after straightforward algebra, it leads to the $U(1,3)$ commutators

$$[Z_{ab}, Z_{cd}] = -i(\eta_{bc}Z_{ad} - \eta_{ad}Z_{cb}) \tag{14d}$$

as expected. By extension, the $U(1,4)$ commutators are[1]

$$[Z_{AB}, Z_{CD}] = -i(\eta_{BC}Z_{AD} - \eta_{AD}Z_{CB}). \tag{14e}$$

The commutators of the Lorentz boosts generators L_{ab} with the X_c, P_c generators are

$$[L_{ab}, X_c] = i(\eta_{bc}X_a - \eta_{ac}X_b); \quad [L_{ab}, P_c] = i(\eta_{bc}P_a - \eta_{ac}P_b). \tag{15}$$

The Hermitian M_{ab} generators are the "reciprocal" boosts/rotation transformations which *exchange* X for P, in addition to boosting (rotating) those variables, and one ends up with the commutators of M_{ab} with the X_c, P_c generators given by

$$[M_{ab}, \frac{X_c}{\lambda_l}] = -\frac{i}{\lambda_p}(\eta_{bc}P_a + \eta_{ac}P_b); \quad [M_{ab}, \frac{P_c}{\lambda_p}] = -\frac{i}{\lambda_l}(\eta_{bc}X_a + \eta_{ac}X_b). \tag{16}$$

The commutators in Equation (14d) and the definitions in Equation (11) lead to

$$\begin{aligned} [Z_{ab}, Z_c] &= (-i)^{3/2}(\eta_{bc}Z_{a5} + i\eta_{ac}Z_{5b}) \\ [Z_{ab}, Z_c^\dagger] &= -(i)^{1/2}(i\eta_{bc}Z_{a5} + \eta_{ac}Z_{5b}), \end{aligned} \tag{17}$$

which are consistent with the commutators in Equation (14a–c) and the definitions in Equations (11) and (13). The right-hand side of Equation (17) can be rewritten in terms of $Z_a, Z_a^\dagger, Z_b, Z_b^\dagger$ after the following replacements:

$$Z_{a5} = \frac{1}{2}[(-i)^{1/2}Z_a^\dagger + (i)^{1/2}Z_a], \quad Z_{b5} = \frac{1}{2i}[(-i)^{1/2}Z_a^\dagger - (i)^{1/2}Z_a]. \tag{18}$$

After some algebra one finds

$$\begin{aligned} [Z_{ab}, Z_c] &= -\frac{i}{2}\eta_{bc}Z_a + \frac{i}{2}\eta_{ac}Z_b - \frac{1}{2}\eta_{bc}Z_a^\dagger - \frac{1}{2}\eta_{ac}Z_b^\dagger \\ [Z_{ab}, Z_c^\dagger] &= -\frac{i}{2}\eta_{bc}Z_a^\dagger + \frac{i}{2}\eta_{ac}Z_b^\dagger + \frac{1}{2}\eta_{bc}Z_a + \frac{1}{2}\eta_{ac}Z_b. \end{aligned} \tag{19}$$

The particular *choice* of the complex coefficients appearing in Equation (11) leads to the following *deformed* Weyl–Heisenberg algebra

$$[Z_a, Z_b^\dagger] = -(\eta_{ab}\mathcal{I} + M_{ab}); \quad [Z_a, Z_b] = [Z_a^\dagger, Z_b^\dagger] = -iL_{ab} \tag{20a}$$

$$[Z_a, \mathcal{I}] = 2Z_a^\dagger; \quad [Z_a^\dagger, \mathcal{I}] = -2Z_a; \quad [Z_{ab}, \mathcal{I}] = 0. \quad \mathcal{I} = 2Z_{55} \tag{20b}$$

leading to

$$\left[\frac{X_a}{\lambda_l}, \mathcal{I}\right] = 2i\frac{P_a}{\lambda_p}; \quad \left[\frac{P_a}{\lambda_p}, \mathcal{I}\right] = 2i\frac{X_a}{\lambda_l} \tag{20c}$$

and the metric $\eta_{ab} = (+1, -1, -1, -1)$ is used to raise and lower indices. The Planck constant is given in terms of the length and momentum scales of Equation (8) as $\hbar = \lambda_l \lambda_p$. In $\hbar = 1$ units, $\lambda_l \lambda_p \to 1$.

The deformed quaplectic algebra is given explicitly by Equations (14d), (17), (19), and (20a–c) and obeys the Jacobi identities by virtue of the definitions in Equations (11) and (13). After recurring directly to the definitions in Equation (7), one finds that Equation (20a) explicitly reflects the *deformation* of the Weyl–Heisenberg algebra resulting from the noncommutative algebra of coordinates and momenta given by

$$\left[\frac{X_a}{\lambda_l}, \frac{P_b}{\lambda_p}\right] = i(\eta_{ab}\mathcal{I} + M_{ab}) \tag{21a}$$

$$[X_a, X_b] = -i(\lambda_l)^2 L_{ab}; \quad [P_a, P_b] = i(\lambda_p)^2 L_{ab}. \tag{21b}$$

One could interpret the term $\eta_{ab}\,\mathcal{I} + M_{ab}$ as a matrix-valued Planck constant $\hbar_{ab}$ (in units of $\hbar = 1$). One may also note that the generator $\mathcal{I}$ no longer commutes with $Z_a, Z_a^\dagger$, but it *exchanges* them, as one can see from Equation (20b) resulting from the definition of $\mathcal{I}$ given by $\mathcal{I} \equiv 2Z_{55} = M_{55}$.

One of the salient features of the construction of the deformed quaplectic (Weyl–Heisenberg) algebra is that by varying the values of the following complex coefficients α, β appearing in the linear combinations

$$Z_a = \alpha\, Z_{a5} + \beta Z_{5a}; \quad Z_a^\dagger = \alpha^* Z_{a5} + \beta^* Z_{5a}; \quad Z_{55} = \frac{\mathcal{I}}{2}, \tag{22}$$

it furnishes different commutation relations than the ones described by Equations (20a–c) and (21a,b). The latter commutators are found in the special case when $\alpha = (-i)^{1/2}$, $\beta = (-i)^{3/2}$, as chosen in Equation (11). For instance, if either $\alpha = 0$ or $\beta = 0$ it leads instead to vanishing commutators $[Z_a, Z_b^\dagger] = [Z_a, Z_b] = [Z_a^\dagger, Z_b^\dagger] = 0$ as a result of Equation (14e). In turn, one would have $[X_a, X_b] = [P_a, P_b] = [X_a, P_b] = 0$ instead of Equations (21a,b). Therefore, the introduction of nonvanishing complex coefficients α, β, via Equation (22), yields a two-parameter family of deformed fiber coordinates and momenta algebras parametrized by α, β. In particular, one may explicitly introduce these parameters by writing $Z_a(\alpha, \beta), Z_a^\dagger(\alpha^*, \beta^*)$.

After introducing the complex-valued vierbein $E^a_\mu = e^a_\mu + i f^a_\mu$, it leads to the complex metric

$$g_{\mu\nu} \equiv E^a_\mu (E^b_\nu)^* \eta_{ab} = g_{(\mu\nu)} + i g_{[\mu\nu]} \tag{23a}$$

with

$$g_{(\mu\nu)} = (e^a_\mu e^b_\nu + f^a_\mu f^b_\nu)\eta_{ab}, \quad i g_{[\mu\nu]} = -i(e^a_\mu f^b_\nu - e^b_\nu f^a_\mu)\eta_{ab}. \tag{23b}$$

The 4×4 complex metric $g_{\mu\nu}$ is Hermitian $g^\dagger_{\mu\nu} = g_{\mu\nu}$ as a result of $g_{\nu\mu} = (g_{\mu\nu})^*$. To verify that $g_{[\mu\nu]} = -g_{[\nu\mu]}$, one just needs to relabel the indices $a \leftrightarrow b$ in Equation (23b) and recur to $\eta_{ba} = \eta_{ab}$.

The two-parameter family of $U(1,4)$-valued Hermitian gauge fields is given by

$$\mathbf{A}_\mu = \Omega^{ab}_\mu Z_{ab} + \frac{1}{L}[E^a_\mu Z_a(\alpha, \beta) + (E^a_\mu)^* Z^\dagger_a(\alpha^*, \beta^*)] + \Omega_\mu\, \mathcal{I}, \tag{24}$$

where L is a length scale that is introduced for dimensional reasons since the physical units of $\mathbf{A}_\mu$ are $(length)^{-1}$. $\Omega^{ab}_\mu Z_{ab}$ is given by $\frac{1}{2}(\Omega^{(ab)}_\mu M_{ab} + \Omega^{[ab]}_\mu L_{ab})$, and $Z_a(\alpha, \beta), Z^\dagger_a(\alpha^*, \beta^*)$ are displayed in Equation (22).

One can rewrite the two-parameter family of $U(1,4)$-valued Hermitian gauge fields (24) as

$$\mathbf{A}_\mu = \Omega_\mu^{ab} Z_{ab} + \Omega_\mu^{(a5)} M_{a5} + \Omega_\mu^{[a5]} L_{a5} + \Omega_\mu \mathcal{I}, \quad \Omega_\mu \equiv \Omega_\mu^{55}. \tag{25}$$

After some straightforward algebra, one finds that the real-valued connection components $\Omega_\mu^{a5}, \Omega_\mu^{5a}$ are given by suitable linear combinations of the e_μ^a, f_μ^a components of the complex-valued vierbein as follows

$$\Omega_\mu^{a5} = e_\mu^a \left(\frac{\alpha + \alpha^*}{L} \right) - f_\mu^a \left(\frac{\alpha - \alpha^*}{iL} \right); \quad \Omega_\mu^{5a} = e_\mu^a \left(\frac{\beta + \beta^*}{L} \right) - f_\mu^a \left(\frac{\beta - \beta^*}{iL} \right), \tag{26a}$$

such that

$$\Omega_\mu^{(a5)} \equiv \frac{1}{2} (\Omega_\mu^{a5} + \Omega_\mu^{5a}), \quad \Omega_\mu^{[a5]} \equiv \frac{1}{2} (\Omega_\mu^{a5} - \Omega_\mu^{5a}). \tag{26b}$$

Because $\alpha \neq \beta$, one finds that $\Omega_\mu^{a5} \neq \Omega_\mu^{5a}$; consequently, $\Omega_\mu^{(a5)} \neq 0; \Omega_\mu^{[a5]} \neq 0$. Therefore, the introduction of the two distinct complex coefficients α, β is tantamount to choosing an infinite family of real-valued connection components $\Omega_\mu^{a5}, \Omega_\mu^{5a}$ given by the many different linear combinations of e_μ^a and f_μ^a. The real-valued coefficients of these linear combinations are given by the real and imaginary parts of α and β as displayed in Equation (26a). One should also emphasize that *no* zero torsion conditions were imposed in reaching the relations in Equation (26a,b) between $\Omega_\mu^{a5}, \Omega_\mu^{5a}$ and e_μ^a, f_μ^a.

The Hermitian $U(1,4)$-valued field strength is defined by

$$\mathbf{F}_{\mu\nu} \;=\; \partial_\mu \mathbf{A}_\nu \;-\; \partial_\nu \mathbf{A}_\mu \;+ i\,[\mathbf{A}_\mu,\; \mathbf{A}_\nu], \tag{27}$$

from which one can read the curvature components $R_{\mu\nu}^{(ab)}; R_{\mu\nu}^{[ab]}$, and the other components of the field strength (such as torsion), in terms of the connection components (and their derivatives) of Equation (24) from the following decomposition of the field strength

$$\mathbf{F}_{\mu\nu} = R_{\mu\nu}^{(ab)} M_{ab} + R_{\mu\nu}^{[ab]} L_{ab} + \frac{1}{L} [F_{\mu\nu}^a Z_a(\alpha,\beta) + (F_{\mu\nu}^a)^* Z_a^\dagger(\alpha^*,\beta^*)] + F_{\mu\nu} \mathcal{I}. \tag{28}$$

By proceeding as one did in [14], one may then construct the generalized actions for complex gravity after using the complex metric (vierbein) and its inverse to raise and lower indices. The simplest actions can have terms linear and quadratic in the curvature and also quadratic terms in the torsion. For further details, we refer to [14].

Alternatively, one could instead start with the $U(1,4)$-valued Hermitian gauge field in Equation (25) leading to the field strength

$$\mathbf{F}_{\mu\nu} = R_{\mu\nu}^{(ab)} M_{ab} + R_{\mu\nu}^{[ab]} L_{ab} + R_\mu^{(a5)} M_{a5} + R_\mu^{[a5]} L_{a5} + F_{\mu\nu} \mathcal{I} \tag{29}$$

and expressed in terms of $\Omega_\mu^{(ab)}, \Omega_\mu^{[ab]}, e_\mu^a, f_\mu^a, \Omega_\mu^{55} = \Omega_\mu$, and their derivatives. Note that $U(1,4)$ has 25 generators, whereas the metric affine group in 4D, given by the semidirect product of $GL(4,R)$ with the translation group T_4, has 20 generators. Therefore, the complex gravitational theory based on $U(1,4)$ and inspired from Born's reciprocal relativity theory, has more degrees of freedom than the metric affine theory of gravity in 4D. This is not surprising since one is dealing with gravity in curved phase spaces. There is also torsion in our construction.

A curved phase-space action associated with the geometry of the cotangent bundle of spacetime and based on Lagrange–Finsler and Hamilton–Cartan geometry [15–18] can be found in [19–21]. To conclude this section, there are two different approaches to construct generalized gravitational theories in curved phase spaces: (i) via the $U(1,4)$ local gauge theory construction presented here, or (ii) via Finsler's geometric methods.

3. The Yang Algebra versus the Deformed Quaplectic Algebra

This section is devoted to an extensive analysis of the Yang and the deformed quaplectic algebras associated with noncommutative phase spaces. Secondly, we present extensions of such algebras involving antisymmetric tensor coordinates and momenta of different ranks.

3.1. The Yang Algebra and Its Extension via Generalized Angular Momentum Operators in Higher Dimensions

Given a flat 6D spacetime with coordinates $Y^M = \{Y^1, Y^2, Y^3, Y^4, Y^5, Y^6\}$ and a metric $\eta_{MN} = diag(-1, +1, +1, \ldots, +1)^2$, the Yang algebra [22,23], which is an extension of the Snyder algebra [24], can be derived in terms of the $SO(5,1)$ Lorentz algebra generators described by the angular momentum/boost operators[3]

$$J^{MN} = -(Y^M \Pi^N - Y^N \Pi^M) = iY^M \frac{\partial}{\partial Y_N} - iY^N \frac{\partial}{\partial Y_M}, \tag{30}$$

where $\Pi^M = -i(\partial/\partial Y_A)$ is the canonical conjugate momentum variable to Y^M. Their commutators are

$$[Y^M, Y^N] = 0, [\Pi^M, \Pi^N] = 0, \quad [Y^M, \Pi^N] = i\eta^{MN}, \quad M, N = 1,2,3,4,5,6. \tag{31}$$

The coordinates Y^M commute. The momenta Π^M also commute, and the canonical conjugate variables Y^M, Π^N obey the Weyl–Heisenberg algebra in 6D.

Adopting the units $\hbar = c = 1$, the correspondence among the noncommuting 4D spacetime coordinates x^μ, the noncommuting momenta p^μ, and the Lorentz $SO(5,1)$ algebra generators leading to the Yang algebra [22,23] is given by

$$x^\mu \leftrightarrow L_P J^{\mu 5} = -L_P(Y^\mu \Pi^5 - Y^5 \Pi^\mu) \tag{32a}$$

$$p^\mu \leftrightarrow \frac{1}{\mathcal{L}} J^{\mu 6} = -\frac{1}{\mathcal{L}}(Y^\mu \Pi^6 - Y^6 \Pi^\mu), \quad \mu, \nu = 1,2,3,4, \tag{32b}$$

which requires the introduction of an ultraviolet cutoff scale L_P given by the Planck scale, and an infrared cutoff scale $\mathcal{L}$ that can be set equal to the Hubble scale R_H (which determines the cosmological constant). It is very important to emphasize that despite the introduction of two length scales $L_P, \mathcal{L}$, the Lorentz symmetry is not lost. This is one of the most salient features of the Snyder [24] and Yang [22,23] algebras[4].

The other generators are given by

$$\mathcal{N} \equiv J^{56} = -(Y^5\, \Pi^6 - Y^6\, \Pi^5), \; J^{\mu\nu} = -(Y^\mu\, \Pi^\nu - Y^\nu\, \Pi^\mu), \; \mu, \nu = 1,2,3,4 \tag{33}$$

One can then verify that the Yang algebra is recovered after imposing the correspondence in Equations (32a,b) and (33)

$$[x^\mu, x^\nu] = -iL_P^2 J^{\mu\nu}, \quad [p^\mu, p^\nu] = -i(\frac{1}{\mathcal{L}})^2 J^{\mu\nu}, \quad \eta^{55} = \eta^{66} = 1 \tag{34}$$

$$[x^\mu, J^{\nu\rho}] = i(\eta^{\mu\rho} x^\nu - \eta^{\mu\nu} x^\rho) \tag{35}$$

$$[p^\mu, J^{\nu\rho}] = i(\eta^{\mu\rho} p^\nu - \eta^{\mu\nu} p^\rho) \tag{36}$$

$$[x^\mu, p^\nu] = -i\eta^{\mu\nu} \frac{L_P}{\mathcal{L}} \mathcal{N}, \quad [J^{\mu\nu}, \mathcal{N}] = 0 \tag{37}$$

$$[x^\mu, \mathcal{N}] = iL_P \mathcal{L} p^\mu, \quad [p^\mu, \mathcal{N}] = -i\frac{1}{L_P \mathcal{L}} x^\mu \tag{38}$$

where the $[J^{\mu\nu}, J^{\rho\sigma}]$ commutators are the same as in the $SO(3,1)$ Lorentz algebra in 4D. They are of the form

$$[J^{\mu_1\mu_2}, J^{\nu_1\nu_2}] = -i\eta^{\mu_1\nu_1} J^{\mu_2\nu_2} + i\eta^{\mu_1\nu_2} J^{\mu_2\nu_1} + \\ i\eta^{\mu_2\nu_1} J^{\mu_1\nu_2} - i\eta^{\mu_2\nu_2} J^{\mu_1\nu_1}, \quad \hbar = c = 1 \tag{39}$$

The generators are assigned to be Hermitian so there are i factors in the right-hand side of Equation (39) since the commutator of two Hermitian operators is anti-Hermitian. The 4D spacetime metric is $\eta_{\mu\nu} = diag(-1,1,1,1)$.

Before continuing, it is important to point out the differences/similarities between the $U(1,4)$ algebra and the Yang algebra which is based on $SO(4,2)$ (or $SO(5,1)$). Firstly, $U(1,4)$ has 25 generators while $SO(4,2)$ has 15. Secondly, the modified Weyl–Heisenberg algebra in Equation (37) *differs* from the one displayed by Equation (21a). Equation (34) is similar to Equation (21b); Equation (38) is similar to Equation (20c); and Equations (35) and (36) are trivially similar to Equation (15). Thirdly, there is *no* analog in the Yang algebra of the Hermitian M_{ab} generators which act as the "reciprocal" boosts/rotation transformations which *exchange* X for P, in addition to boosting (rotating) those variables, and leading to the commutators of M_{ab} with the X_c, P_c generators given by Equation (16).

Another difference between the Yang and the deformed quaplectic algebra is that in the Yang algebra case, one adds two additional coordinates and momenta Y^5, Y^6, Π^5, and Π^6 in order to construct the $SO(4,2), SO(5,1)$ algebras with 15 generators. Whereas in the (deformed) quaplectic algebra case, one adds one additional coordinate and momentum Y^5, Π^5, and the extra generators $M_{ab}, M_{a5}, M_{55} = \mathcal{I}$ in order to construct the $U(1,4)$ algebra with 25 generators. Furthermore, the construction of the Yang algebra requires the two length scales $L_P, \mathcal{L}$, whereas in the (deformed) quaplectic algebra, one has the length scale λ_l and the momentum scale λ_p.

One may also clarify that quantum phase spaces can be described by real or complex phase space coordinates. A typical example of the use of complex coordinates is in the description of the coherent state $|z\rangle$ that is defined to be the unique eigenstate of the (bosonic) annihilation operator $\hat{a}|z\rangle = z|z\rangle$ [25]. The formal solution of this eigenvalue equation is the vacuum state displaced to a location z in phase space, and it is obtained by letting the unitary displacement operator $D(z)$ operate on the vacuum $|z\rangle = e^{z\hat{a}^\dagger - z^*\hat{a}}|0\rangle = D(z)|0\rangle$, where the annihilation operator $\hat{a} = \hat{X} + i\hat{P}$ and creation operator $\hat{a}^\dagger = \hat{X} - i\hat{P}$ are expressed in terms of the phase-space coordinates associated with the quantum harmonic oscillator.

Using the representation of the coherent state in the basis of Fock states, one finds $|z\rangle = e^{-\frac{|z|^2}{2}} \sum_{n=0}^{\infty} \frac{z^n}{\sqrt{n!}} |n\rangle = e^{-\frac{|z|^2}{2}} e^{z\hat{a}^\dagger} e^{-z^*\hat{a}}|0\rangle$, where $|n\rangle$ are the energy (number) eigenvectors of the quantum harmonic oscillator Hamiltonian $H = \hbar\omega(\hat{a}^\dagger\hat{a} + \frac{1}{2})$ [25].

Pertaining the role of the $U(1,4)$ symmetry, one should point that there is a standard procedure to obtain the $U(N)$ generators $E_{jk} = a_j^\dagger a_k$ in terms of the complex $Cl(2N, C)$ algebra generators via the creation and annihilation *fermionic* oscillators defined as follows: $a_j = \frac{1}{2}(\Gamma_{2j} + i\,\Gamma_{2j-1})$; $a_j^\dagger = \frac{1}{2}(\Gamma_{2j} - i\,\Gamma_{2j-1})$; $j = 1, 2, \cdots, N$. One can verify that the following *anticommutators* $\{a_j, a_k^\dagger\} = \delta_{jk}$; $\{a_j, a_k\} = 0$; $\{a_j^\dagger, a_k^\dagger\} = 0$ lead to the $U(N)$ commutation relations $[E_{jk}, E_{lm}] = \delta_{kl}E_{jm} - \delta_{jm}E_{lk}$. This construction is just a reflection of the fact that $U(N) \subset SO(2N)$. In particular, $U(4) \subset SO(8)$.

After this detour, given the above correspondence (9), we can *extend* it further to the higher-grade polyvector-valued coordinates and momenta operators in noncommutative Clifford phase spaces [26,27]. Given a Clifford algebra $\{\gamma^\mu, \gamma^\nu\} = 2\eta^{\mu\nu}\mathbf{1}$, a polyvector-valued coordinate is defined as $\mathbf{X} = X_M \Gamma^M$ and admits the following expansion in terms of the Clifford algebra generators in D-dimensions, $\mathbf{1}, \gamma^\mu, \gamma^{\mu_1} \wedge \gamma^{\mu_2}, \cdots, \gamma^{\mu_1} \wedge \gamma^{\mu_2} \wedge \cdots \wedge \gamma^{\mu_D}$, as follows:

$$\mathbf{X} = X\mathbf{1} + X_\mu \gamma^\mu + X_{\mu_1\mu_2}\gamma^{\mu_1} \wedge \gamma^{\mu_2} + X_{\mu_1\mu_2\mu_3}\gamma^{\mu_1} \wedge \gamma^{\mu_2} \wedge \gamma^{\mu_3} + ...+ \\ X_{\mu_1\mu_2\mu_3......\mu_D}\gamma^{\mu_1} \wedge \gamma^{\mu_2} \wedge \gamma^{\mu_3} .... \wedge \gamma^{\mu_D}. \tag{40a}$$

The numerical combinatorial factors can be omitted by imposing the ordering prescription $\mu_1 < \mu_2 < \mu_3 \cdots < \mu_D$. In order to match physical units in each term of (17), a length scale parameter must be suitably introduced in the expansion in Equation (17). In [28,29], we introduced the Planck scale as the expansion parameter in (17), which was set to unity, when one adopted the units $\hbar = c = G = 1$.

Similarly, the polyvector-valued momentum $\mathbf{P} = P_M \Gamma^M$ admits the following expansion in terms of the Clifford algebra generators in D-dimensions

$$\mathbf{P} = P\mathbf{1} + P_\mu \gamma^\mu + P_{\mu_1\mu_2}\gamma^{\mu_1} \wedge \gamma^{\mu_2} + P_{\mu_1\mu_2\mu_3}\gamma^{\mu_1} \wedge \gamma^{\mu_2} \wedge \gamma^{\mu_3} + ...+ \\ P_{\mu_1\mu_2\mu_3......\mu_D}\gamma^{\mu_1} \wedge \gamma^{\mu_2} \wedge \gamma^{\mu_3} .... \wedge \gamma^{\mu_D} \tag{40b}$$

The scalar, vectorial, antisymmetric tensorial coordinates $X, X_\mu, X_{\mu_1\mu_2} = -X_{\mu_2\mu_1}, \cdots, X_{\mu_1\mu_2\cdots\mu_D}$ are the scalar, vector, bivector, trivector, etc., components of the polyvector-valued coordinates. The $X_{\mu_1\mu_2}$ bivector (antisymmetric tensor of rank two) corresponds to an oriented area element. The trivector $X_{\mu_1\mu_2\mu_3}$ (antisymmetric tensor of rank three) corresponds to an oriented volume element, and so forth.

Similarly, the scalar, vectorial, antisymmetric tensorial coordinates $P, P_\mu, P_{\mu_1\mu_2} = -P_{\mu_2\mu_1}, \cdots, P_{\mu_1\mu_2\cdots\mu_D}$ are the scalar, vector, bivector, trivector, etc., components of the polyvector-valued momentum coordinates. The $P_{\mu_1\mu_2}$ bivector (antisymmetric tensor of rank two) corresponds to an oriented areal-momentum element. The trivector $P_{\mu_1\mu_2\mu_3}$ (antisymmetric tensor of rank three) corresponds to an oriented volume–momentum element, and so forth.

We constructed in [26,27] the corresponding nonvanishing commutators among the *noncommutative* antisymmetric tensors $X^{\mu_1\mu_2}, X^{\mu_1\mu_2\mu_3}, \ldots; P^{\mu_1\mu_2}, P^{\mu_1\mu_2\mu_3}, \ldots$ of different ranks. We coined such *extension* of the Yang algebra the Clifford–Yang algebra, since it involves polyvector-valued coordinates and momenta associated with a Clifford algebra. The *noncommuting* bivector coordinates obey

$$[X^{\mu_1\mu_2}, X^{\nu_1\nu_2}] \sim iL_P^4 \eta^{55} J^{\mu_1\mu_2|\nu_1\nu_2}, \quad J^{\mu_1\mu_2|\nu_1\nu_2} \equiv -(Y^{\mu_1\mu_2}\Pi^{\nu_1\nu_2} - Y^{\nu_1\nu_2}\Pi^{\mu_1\mu_2} \tag{41a}$$

$Y^{\mu_1\mu_2}$ is a bivector coordinate associated with the $Cl(5,1)$ algebra of the 6D flat spacetime. $\Pi^{\mu_1\mu_2} = -i(\partial/\partial Y_{\mu_1\mu_2})$ is the corresponding bivector canonical momentum conjugate. Their commutators are

$$[Y^{\mu_1\mu_2}, Y^{\nu_1\nu_2}] = 0, \quad [\Pi^{\mu_1\mu_2}, \Pi^{\nu_1\nu_2}] = 0, \quad [Y^{\mu_1\mu_2}, P^{\nu_1\nu_2}] = i\eta^{\mu_1\mu_2|\nu_1\nu_2}, \tag{41b}$$

where the generalized metric involving bivector indices is defined as

$$\eta^{\mu_1\mu_2|\nu_1\nu_2} = \eta^{\nu_1\nu_2|\mu_1\mu_2} = \eta^{\mu_1\nu_1}\eta^{\mu_2\nu_2} - \eta^{\mu_1\nu_2}\eta^{\mu_2\nu_1} \tag{41c}$$

The *noncommuting* bivector momenta obey

$$[P^{\mu_1\mu_2}, P^{\nu_1\nu_2}] \sim i\mathcal{L}^{-4}\eta^{66} J^{\mu_1\mu_2|\nu_1\nu_2}, \tag{41d}$$

and so forth. All the commutators have the same structural form of a generalized angular momentum algebra as follows

$$[J^{A(r_1)|B(r_2)}, J^{C(s_1)|D(s_2)}] = -i\eta^{A(r_1)|C(s_1)} J^{B(r_2)|D(s_2)} + i\eta^{A(r_1)|D(s_2)} J^{B(r_2)|C(s_1)} + i\eta^{B(r_2)|C(s_1)} J^{A(r_1)|D(s_2)} - i\eta^{B(r_2)|D(s_2)} J^{A(r_1)|C(s_1)}, \quad \hbar = c = 1, \tag{41e}$$

where the grades of the polyvector indices $A(r_1)B(r_2), C(s_1)$, and $D(s_2)$ appearing in the generators are r_1, r_2, s_1, and s_2, respectively. The shorthand notation for $J^{a_1 a_2 \cdots a_{r_1}|b_1 b_2 \cdots b_{r_2}}$ is $J^{A(r_1)|B(r_2)}, \cdots$. The generalized metric tensor $\eta^{A|C} = 0$ if the grade of A is *not* equal to the grade of C. Similarly, $\eta^{A|D} = 0$ if the grade of A is *not* equal to the grade of $D, \cdots$. Moreover, $\eta^{\mu 5} = \eta^{\mu 6} = 0$ since the 6D metric is diagonal. The commutators (41e) ensure that the Jacobi identities are satisfied. In addition, we found the spectrum of the quantum harmonic oscillator in noncommutative spaces in terms of the eigenvalues of the generalized angular momentum operators in higher dimensions and discussed how to extend these results to higher-grade polyvector-valued coordinates and momenta. For full details, we refer the reader to [26,27].

3.2. Realization of the Deformed Quaplectic Algebra and its Extensions

We saw above how the *noncommutative* coordinates and momenta of the Yang algebra in 4D can be realized in terms of the angular momentum operators in 6D, which, in turn, are expressed in terms of the canonical-conjugate variables Y^M, Π^N in 6D shown in Equations (32a,b) and (33) and obeying the standard commutation relations displayed in Equations (31). Inspired by this procedure, next, we find a realization of the *deformed* quaplectic algebra generators in terms of the canonical coordinate and momentum variables Y_a, Π_b, Y_5, Π_5 as follows:

$$M_{ab} = M_{ba} = \frac{1}{2}(Y_a\Pi_b + \Pi_b Y_a) + \frac{1}{2}(Y_b\Pi_a + \Pi_a Y_b) \tag{42a}$$

$$M_{a5} = M_{5a} = \frac{1}{2}(Y_a\Pi_5 + \Pi_5 Y_a) + \frac{1}{2}(Y_5\Pi_a + \Pi_a Y_5), \quad M_{55} = (Y_5\Pi_5 + \Pi_5 Y_5) \tag{42b}$$

$$L_{ab} = -L_{ba} = \frac{1}{2}(Y_a\Pi_b + \Pi_b Y_a) - \frac{1}{2}(Y_b\Pi_a + \Pi_a Y_b) \tag{42c}$$

$$L_{a5} = -L_{5a} = \frac{1}{2}(Y_a\Pi_5 + \Pi_5 Y_a) - \frac{1}{2}(Y_5\Pi_a + \Pi_a Y_5) \tag{42d}$$

From Equations (41a-e) and (42a–d), one then finds an explicit realization of the generators $Z_{AB} = \frac{1}{2}(M_{AB} + L_{AB})$ of the deformed quaplectic algebra, with $A, B = 1,2,3,4,5$, directly in terms of the canonical coordinate and momentum variables Y_a, Π_b, Y_5, Π_5, and obeying the following commutation relations:

$$[Y_a, Y_b] = 0, \quad [Y_a, Y_5] = 0, \quad [\Pi_a, \Pi_b] = 0 \tag{43a}$$

$$[\Pi_a, \Pi_5] = 0, \quad [Y_a, \Pi_b] = i\eta_{ab}, \quad [Y_5, \Pi_5] = i\eta_{55}. \tag{43b}$$

From Equation (43a,b), one learns that when $a \neq b$, the generator M_{ab} reduces to $Y_a\Pi_b + Y_b\Pi_a$, and when $a = b$, $M_{aa} = Y_a\Pi_a + \Pi_a Y_a$, while the generator $L_{ab} = Y_a\Pi_b - Y_b\Pi_a$. Similarly, M_{a5} reduces to $Y_a\Pi_5 + Y_5\Pi_a$, $M_{55} = Y_5\Pi_5 + \Pi_5 Y_5$, and $L_{a5} = Y_a\Pi_5 - Y_5\Pi_a$

The antisymmetric rank-two tensor coordinates and momenta operators' extensions of the expressions in Equations (41a–e) and (42a–d) are given by:

$$M_{a_1a_2|b_1b_2} = \frac{1}{2}(Y_{a_1a_2}\Pi_{b_1b_2} + \Pi_{b_1b_2}Y_{a_1a_2}) + \frac{1}{2}(Y_{b_1b_2}\Pi_{a_1a_2} + \Pi_{a_1a_2}Y_{b_1b_2}) \tag{44a}$$

$$L_{a_1a_2|b_1b_2} = \frac{1}{2}(Y_{a_1a_2}\Pi_{b_1b_2} + \Pi_{b_1b_2}Y_{a_1a_2}) - \frac{1}{2}(Y_{b_1b_2}\Pi_{a_1a_2} + \Pi_{a_1a_2}Y_{b_1b_2}) \tag{44b}$$

where

$$M_{a_1a_2|b_1b_2} = -M_{a_2a_1|b_1b_2} = -M_{a_1a_2|b_2b_1} = M_{b_1b_2|a_1a_2} \tag{45a}$$

$$L_{a_1a_2|b_1b_2} = -L_{a_2a_1|b_1b_2} = -L_{a_1a_2|b_2b_1} = -L_{b_1b_2|a_1a_2} \tag{45b}$$

Given $M_{a_1a_2|b_1b_2}, L_{a_1a_2|b_1b_2}$ the generalization of the operator Z_{ab} is

$$Z_{a_1a_2|b_1b_2} \equiv \frac{1}{2}(M_{a_1a_2|b_1b_2} + L_{a_1a_2|b_1b_2}) \tag{45c}$$

The generalization of the commutators in Equations (14a–c) corresponding to the $M_{a_1a_2|b_1b_2}, L_{a_1a_2|b_1b_2}$ generators is given by

$$\begin{aligned}[L_{a_1a_2|b_1b_2}, L_{c_1c_2|d_1d_2}] = i\eta_{b_1b_2|c_1c_2}L_{a_1a_2|d_1d_2} - i\eta_{a_1a_2|c_1c_2}L_{b_1b_2|d_1d_2} - \\ i\eta_{b_1b_2|d_1d_2}L_{a_1a_2|c_1c_2} + i\eta_{a_1a_2|d_1d_2}L_{b_1b_2|c_1c_2}\end{aligned} \tag{46}$$

$$\begin{aligned}[M_{ab}, M_{cd}] = -i\eta_{b_1b_2|c_1c_2}L_{a_1a_2|d_1d_2} - i\eta_{a_1a_2|c_1c_2}L_{b_1b_2|d_1d_2} - \\ i\eta_{b_1b_2|d_1d_2}L_{a_1a_2|c_1c_2} - i\eta_{a_1a_2|d_1d_2}L_{b_1b_2|c_1c_2}\end{aligned} \tag{47}$$

$$\begin{aligned}[L_{ab}, M_{cd}] = i\eta_{b_1b_2|c_1c_2}M_{a_1a_2|d_1d_2} - i\eta_{a_1a_2|c_1c_2}M_{b_1b_2|d_1d_2} + \\ i\eta_{b_1b_2|d_1d_2}M_{a_1a_2|c_1c_2} - i\eta_{a_1a_2|d_1d_2}M_{b_1b_2|c_1c_2}\end{aligned} \tag{48}$$

where

$$\eta^{a_1a_2|b_1b_2} \equiv \eta^{a_1b_1}\eta^{a_2b_2} - \eta^{a_1b_2}\eta^{a_2b_1} \tag{49}$$

From Equations (45c) and (46)–(49), one finds that

$$[Z_{a_1a_2|b_1b_2}, Z_{c_1c_2|d_1d_2}] = -i(\eta_{b_1b_2|c_1c_2}Z_{a_1a_2|d_1d_2} - \eta_{a_1a_2|d_1d_2}Z_{c_1c_2|b_1b_2}). \tag{50}$$

This is a result of the canonical antisymmetric rank-two tensor coordinates and momenta variables $Y_{a_1a_2}, \Pi_{b_1b_2}$ obeying the following commutation relations (the generalization of Equation (43a,b)

$$[Y_{a_1a_2}, Y_{b_1b_2}] = 0, \quad [\Pi_{a_1a_2}, \Pi_{b_1b_2}] = 0, \quad [Y_{a_1a_2}, \Pi_{b_1b_2}] = i\eta_{a_1a_2|b_1b_2} \tag{51}$$

The other *dimensionless* generators are[5]

$$\begin{aligned}M_{a_1a_2|5} &= \frac{Y_{a_1a_2}}{\lambda_l^2}\frac{\Pi_5}{\lambda_p} + \frac{Y_5}{\lambda_l}\frac{\Pi_{a_1a_2}}{\lambda_p^2}, \\ M_{5|a_1a_2} &= \frac{Y_5}{\lambda_l}\frac{\Pi_{a_1a_2}}{\lambda_p^2} + \frac{Y_{a_1a_2}}{\lambda_l^2}\frac{\Pi_5}{\lambda_p} \\ L_{a_1a_2|5} &= \frac{Y_{a_1a_2}}{\lambda_l^2}\frac{\Pi_5}{\lambda_p} - \frac{Y_5}{\lambda_l}\frac{\Pi_{a_1a_2}}{\lambda_p^2},\end{aligned} \tag{52}$$

$$L_{5|a_1a_2} = \frac{Y_5}{\lambda_l}\frac{\Pi_{a_1a_2}}{\lambda_p^2} - \frac{Y_{a_1a_2}}{\lambda_l^2}\frac{\Pi_5}{\lambda_p} \tag{53}$$

such that

$$Z_{a_1a_2|5} = \frac{1}{2}(M_{a_1a_2|5} + L_{a_1a_2|5}), \quad Z_{5|a_1a_2} = \frac{1}{2}(M_{5|a_1a_2} + L_{5|a_1a_2}) \tag{54}$$

and leading to the following generators

$$Z_{[a_1a_2]} \equiv \frac{1}{\sqrt{2}}\left(\frac{X_{a_1a_2}}{\lambda_l^2} - i\frac{P_{a_1a_2}}{\lambda_p^2}\right) = \alpha Z_{a_1a_2|5} + \beta Z_{5|a_1a_2}, \tag{55a}$$

$$Z^\dagger_{[a_1a_2]} \equiv \frac{1}{\sqrt{2}}\left(\frac{X_{a_1a_2}}{\lambda_l^2} + i\frac{P_{a_1a_2}}{\lambda_p^2}\right) = \alpha^* Z_{a_1a_2|5} + \beta^* Z_{5|a_1a_2} \tag{55b}$$

where α, β are suitable complex-valued coefficients chosen so that[6]

$$[Z_{[a_1a_2]}, Z^\dagger_{[b_1b_2]}] = -(\eta_{a_1a_2|b_1b_2}\mathcal{I} + M_{a_1a_2|b_1b_2}) \tag{56}$$

$$[Z_{[a_1a_2]}, Z_{[b_1b_2]}] = [Z^\dagger_{[a_1a_2]}, Z^\dagger_{[b_1b_2]}] = -iL_{a_1a_2|b_1b_2}. \tag{57}$$

Finally, from Equations (55a,b)–(57), one arrives at the desired result

$$\left[\frac{X_{a_1a_2}}{\lambda_l^2}, \frac{P_{b_1b_2}}{\lambda_p^2}\right] = i(\eta_{a_1a_2|b_1b_2}\mathcal{I} + M_{a_1a_2|b_1b_2}) \tag{58}$$

$$[X_{a_1a_2}, X_{b_1b_2}] = -i(\lambda_l)^4 L_{a_1a_2|b_1b_2}; \quad [P_{a_1a_2}, P_{b_1b_2}] = i(\lambda_p)^4 L_{a_1a_2|b_1b_2}; \tag{59}$$

The above construction can be *extended* to higher-rank antisymmetric tensor coordinates and momenta $Y_{a_1a_2,a_3}, \Pi_{a_1a_2a_3}, \cdots$ leading to the generators $Z_{a_1a_2a_3|b_1b_2b_3}; Z_{a_1a_2a_3|5}; Z_{5|a_1a_2a_3}, \ldots,$ and whose commutators are the extensions of the equations above. The end result is

$$\left[\frac{X_{a_1a_2\cdots a_n}}{\lambda_l^n}, \frac{P_{b_1b_2\cdots b_n}}{\lambda_p^n}\right] = i(\eta_{a_1a_2\cdots a_n|b_1b_2\cdots b_n}\mathcal{I} + M_{a_1a_2\cdots a_n|b_1b_2\cdots b_n}) \tag{60}$$

$$[X_{a_1a_2\cdots a_n}, X_{b_1b_2\cdots b_n}] = -i(\lambda_l)^{2n} L_{a_1a_2\cdots a_n|b_1b_2\cdots b_n} \tag{61a}$$

$$[P_{a_1a_2\cdots a_n}, P_{b_1b_2\cdots b_n}] = i(\lambda_p)^{2n} L_{a_1a_2\cdots a_n|b_1b_2\cdots b_n}, \tag{61b}$$

where $\eta_{a_1a_2\cdots a_n|b_1b_2\cdots b_n}$ can be written as the determinant of the $n \times n$ matrix whose entries are $\eta^{a_ib_j}$ with $i,j = 1,2,\cdots,n$. The same occurs with $\delta^{a_1a_2\cdots a_n}_{b_1b_2\cdots b_n}$ where the entries are $\delta^{a_i}_{b_j}$. One finds that Equations (60) and (61a,b) do not differ too much from those corresponding equations of the Clifford–Yang algebra [26,27]. In the latter algebra, $\mathcal{I} = 2Z_{55} = M_{55}$ is replaced by $\mathcal{N} \equiv J^{56}$; there are no $M_{a_1a_2\cdots a_n|b_1b_2\cdots b_n}$ terms, and λ_l, λ_p are replaced by $L_P, \mathcal{L}^{-1}$, respectively, where $L_P, \mathcal{L}$ are the lower and upper length scales.

To sum up, all the commutation relations can be obtained from

$$[Z_{a_1a_2\cdots a_n|b_1b_2\cdots b_n}, Z_{c_1c_2\cdots c_n|d_1d_2\cdots d_n}] = \\ -i(\eta_{b_1b_2\cdots b_n|c_1c_2\cdots c_n} Z_{a_1a_2\cdots a_n|d_1d_2\cdots d_n} - \eta_{a_1a_2\cdots a_n|d_1d_2\cdots d_n} Z_{c_1c_2\cdots c_n|b_1b_2\cdots b_n}). \tag{62a}$$

$$[Z_{a_1a_2\cdots a_n|5}, Z_{5|b_1b_2\cdots b_n}] = -i(\eta_{55} Z_{a_1a_2\cdots a_n|b_1b_2\cdots b_n} - \eta_{a_1a_2\cdots a_n|b_1b_2\cdots b_n} Z_{55}), \quad \cdots \tag{62b}$$

We finalize this section by pointing out that Meljanac and collaborators introduced also the tensorial canonical Heisenberg algebras as a tool to provide the solution, e.g., of the Snyder models describing noncommutative quantum spacetime coordinates. In particular, the Yang model and its generalizations were discussed very recently [30] .

4. Curved Phase Space Due to Noncommutative Coordinates and Momenta

Noncommuting momentum operators are a reflection of the spacetime curvature after invoking the QM prescription $p_\mu \leftrightarrow -i\hbar\nabla_\mu$. By Born's reciprocity, noncommuting coordinates are a reflection of the momentum space curvature after invoking $x_\mu \leftrightarrow i\hbar\tilde{\nabla}_\mu$, where the tilde derivatives represent derivatives with respect to the momentum variables.

Having reviewed the basics of the Yang algebra of noncommutative phase spaces, Born's reciprocal relativity, and the extended Yang and (deformed) quaplectic algebras, in this section, we provide a solution for the exact analytical mapping of the noncommuting x^μ, p^μ operator variables (associated to an 8D curved phase space) into the canonical Y^A, Π^A operator variables of a flat 12D phase space. We explore the geometrical implications of this mapping which provides, in the *classical* limit, the embedding functions $Y^A(x,p), \Pi^A(x,p)$ of an 8D curved phase space into a flat 12D phase space background. The latter embedding functions determine the functional forms of the base spacetime metric $g_{\mu\nu}(x,p)$, the fiber metric of the vertical space $h^{ab}(x,p)$, and the nonlinear connection $N_{a\mu}(x,p)$ associated with the 8D cotangent space of the 4D spacetime.

Instead of working with the above *canonical* coordinates Y^A and momenta Π^A in a flat 12D phase space ($A = 1,2,\cdots,5,6$), the authors [31] were interested in finding Hermitian realizations of the above Yang algebra in an 8D phase space, and given in terms of the *canonical* variables $\tilde{x}_\mu, \tilde{p}_\mu$ satisfying $[\tilde{x}_\mu, \tilde{x}_\nu] = [\tilde{p}_\mu, \tilde{p}_\nu] = 0$, and $[\tilde{x}_\mu, \tilde{p}_\nu] = i\eta_{\mu\nu}$, with $\mu, \nu = 1,2,3,4$.

The Yang model studied by [31] was characterized by the choice of the commutator $[x_\mu, p_\nu] = i\gamma_{\mu\nu}(x,p)$, and where the rank-two tensor $\gamma_{\mu\nu}(x,p)$ is of the form

$$\gamma_{\mu\nu} = h(x^2, p^2, x \cdot p + p \cdot x)\eta_{\mu\nu} \tag{63}$$

with h a judicious function of the Lorentz scalars $x^2, p^2, x \cdot p + p \cdot x$, which is determined by solving the Jacobi identities. The rank-two tensor $\gamma_{\mu\nu}(x,p)$ is what leads to the generalized uncertainty relations. The triple special relativity model [32], an extension of [33,34], was characterized by a different choice of $\gamma_{\mu\nu}(x,p)$. The Lorentz generators were represented as

$$\mathcal{J}_{\mu\nu} = \frac{1}{2}(x_\mu p_\nu - x_\nu p_\mu + p_\nu x_\mu - p_\mu x_\nu) \tag{64}$$

In particular, the authors [31] looked for representations where the generators $\mathcal{J}_{\mu\nu}$ and the tensor $\gamma_{\mu\nu}$ could be written in terms of the canonical variables $\tilde{x}_\mu$ and $\tilde{p}_\nu$. This required the arduous task of finding the nontrivial map among the *noncanonical* variables x_μ, p_μ and the canonical ones $\tilde{x}_\mu, \tilde{p}_\mu : x_\mu = x_\mu(\tilde{x}, \tilde{p}); p_\mu = p_\mu(\tilde{x}, \tilde{p})$. The map was found iteratively in powers of $\tilde{x}, \tilde{p}$. The explicit technical details of this map can be found in [31].

4.1. *Mapping of x^μ, p^μ to the Y^A, Π^A Variables in Flat Phase Space*

The Y^5, Y^6, Π^5, and Π^6 canonical coordinates and momenta (operators) in the flat 12D phase space are scalars from the point of view of the 8D curved phase space parametrized by the noncanonical coordinates x^μ and momenta p^μ. Therefore, Y^5, Y^6, Π^5, and Π^6 must be functions of the Lorentz scalars

$$x^2 = \eta_{\mu\nu}x^\mu x^\nu, \quad p^2 = \eta_{\mu\nu}p^\mu p^\nu, \quad x \cdot p = \eta_{\mu\nu}x^\mu p^\mu, \quad p \cdot x = \eta_{\mu\nu}p^\mu x^\nu, \quad \mu,\nu = 1,2,3,4 \tag{65}$$

Setting $\alpha = \mathcal{L}^{-1}, \beta = L_P$, due to the Born reciprocity principle, one must have functions $f(z_1, z_2, z_3)$ of the arguments z_1, z_2, and z_3 given by the following combination of Hermitian variables (operators)

$$z_1 \equiv (\alpha^2 x^2 + \beta^2 p^2), \quad z_2 \equiv (x \cdot p + p \cdot x), \quad z_3 \equiv i(x \cdot p - p \cdot x), \quad \alpha = \mathcal{L}^{-1}, \quad \beta = L_P \tag{66}$$

The arguments z_1, z_2, and z_3 are invariant under $\alpha \leftrightarrow \beta$, $x \leftrightarrow p$, and $i \leftrightarrow -i$, if one wishes to implement Born's reciprocity symmetry. Therefore, one must have functions of the form

$$Y^5 = Y^5(z_1, z_2, z_3), \quad Y^6 = Y^6(z_1, z_2, z_3), \quad \Pi^5 = \Pi^5(z_1, z_2, z_3), \quad \Pi^6 = \Pi^6(z_1, z_2, z_3) \tag{67}$$

For instance, one could have functions linear in z_1, z_2, and z_3 defined as follows

$$Y^5(x,p) = a_1(\alpha^2 x^2 + \beta^2 p^2) + b_1(x \cdot p) + b_1^*(p \cdot x) + c_1 \tag{68a}$$

$$Y^6(x,p) = a_2(\alpha^2 x^2 + \beta^2 p^2) + b_2(x \cdot p) + b_2^*(p \cdot x) + c_2 \tag{68b}$$

$$\Pi^5(x,p) = a_3(\alpha^2 x^2 + \beta^2 p^2) + b_3(x \cdot p) + b_3^*(p \cdot x) + c_3 \tag{68c}$$

$$\Pi^6(x,p) = a_4(\alpha^2 x^2 + \beta^2 p^2) + b_4(x \cdot p) + b_4^*(p \cdot x) + c_4. \tag{68d}$$

where a_i, b_i, c_i ($i = 1,2,3,4$) are judicious numerical (dimensionful) coefficients. The units of the coefficients in Equations (68a,b) are those of length, while those in Equations (68c,d) are those of mass. Note that the b_i coefficients in Equations (68a–e) are complex-valued: $b_i = \gamma_i + i\delta_i$. The reason is that the combination

$$b_i(x \cdot p) + b_i^*(p \cdot x) = \gamma_i(x \cdot p + p \cdot x) + i\delta_i(x \cdot p - p \cdot x) = \gamma_i z_2 + \delta_i z_3, \quad i = 1,2,3,4 \tag{68e}$$

ensures that Equation (68e) is Hermitian by construction. Equation (68e) is also invariant under Born's reciprocity $x \leftrightarrow p$ and $i \leftrightarrow -i$. We show that Equations (68a–e) should, in principle, provide satisfactory solutions to the embedding problem defined below.

The $[x^\mu, p^\nu]$ commutator is defined as

$$[x^\mu, p^\nu] \;=\; x^\mu\, p^\nu \;-\; p^\nu\, x^\mu \;=\; i\,\gamma^{\mu\nu}(x,p), \tag{69}$$

where $\gamma^{\mu\nu}(x,p)$ is a second-rank tensor, not necessarily symmetric, that we refrain from identifying as a metric tensor. The above commutator can also be expressed in terms of the 6D angular momenta variables displayed by Equations (32a,b) and (33) as

$$\begin{gathered}[x^\mu, p^\nu] = i\gamma^{\mu\nu}(x,p) = -i\alpha\beta J^{56}(x,p)\eta^{\mu\nu} = \\ i\alpha\beta[Y^5(x,p)\Pi^6(x,p) - Y^6(x,p)\Pi^5(x,p)]\eta^{\mu\nu}, \quad \alpha = \mathcal{L}^{-1}, \beta = L_P\end{gathered} \tag{70}$$

Therefore, from Equations (69) and (70), one arrives at the following relation, after contracting both equations with $\eta_{\mu\nu}$,

$$\begin{gathered}\tfrac{1}{4i}\eta_{\mu\nu}(x^\mu p^\nu - p^\nu x^\mu) = \tfrac{1}{4i}(x \cdot p - p \cdot x) = \\ \alpha\beta(Y^5(x,p)\Pi^6(x,p) - Y^6(x,p)\Pi^5(x,p))) = -\alpha\beta\mathcal{N}\end{gathered} \tag{71}$$

Therefore, in this particular case, one finds that the tensor is symmetric $\gamma^{\mu\nu}(x,p) = \Phi(x,p)\eta^{\mu\nu}$ and such that the conformal factor $\Phi(x,p)$ is Hermitian and given by the left-hand side of Equation (71). The right-hand side of (71) is Hermitian because J^{56} is Hermitian due to the canonical and Hermiticity nature of the 6D variables: $(Y^5\Pi^6)^\dagger = \Pi^6 Y^5 = Y^5\Pi^6$, and $(Y^6\Pi^5)^\dagger = \Pi^5 Y^6 = Y^6\Pi^5$ resulting from the commutators of the 6D canonical variables given by Equation (31).

From Equations (32a,b), one learns that the 4D operators x^μ, p^μ admitted a 6D angular momentum realization of the form

$$x^\mu = \beta J^{\mu 5} = -\beta(Y^\mu \Pi^5 - Y^5 \Pi^\mu), \quad \beta = L_P \tag{72}$$

$$p^\mu = \alpha J^{\mu 6} = -\alpha(Y^\mu \Pi^6 - Y^6 \Pi^\mu), \quad \alpha = \mathcal{L}^{-1} \tag{73}$$

From Equations (72) and (73), one can deduce the relation

$$\mathcal{J}^{\mu\nu} = x^\mu p^\nu - x^\nu p^\mu = \alpha\beta J^{56}(Y^\mu \Pi^\nu - Y^\nu \Pi^\mu), \tag{74}$$

where $J^{56} \equiv \mathcal{N}$ and $J^{\mu\nu}$ are given by Equation (33) explicitly in terms of the 6D canonical variables Y^A, Π^B.

One can *invert* the relations in Equations (72) and (73) as follows. After multiplying Equations (72) and (73) on the *right* by Π^6 and Π^5, respectively, and subtracting the top equation from the bottom one, it yields

$$\beta^{-1} x^\mu \Pi^6 - \alpha^{-1} p^\mu \Pi^5 = \Pi^\mu \mathcal{N} = \mathcal{N} \Pi^\mu \tag{75a}$$

due to the canonical nature of the 6D variables Y^A and Π^A described by the commutators in Equation (31) and which allows us to reorder the relevant factors due to the commutativity.

Moreover, multiplying Equations (72) and (73) on the *right* by Y^6 and Y^5, respectively, and subtracting the top equation from the bottom one yield

$$\beta^{-1} x^\mu Y^6 - \alpha^{-1} p^\mu Y^5 = Y^\mu \mathcal{N} = \mathcal{N} Y^\mu \tag{75b}$$

Next, we see that the functional forms of $Y^5(x,p), Y^6(x,p), \Pi^5(x,p)$, and $\Pi^6(x,p)$ provided by Equations (68a–e) lead to solutions to Equation (71), which, in turn, yields automatically the solutions to Equation (75a,b). In doing so, one finds the solutions to the embedding problem $Y^\mu = Y^\mu(x,p); \Pi^\mu = \Pi^\mu(x,p)$, with $\mathcal{N}(x,p) \equiv J^{56}(x,p) = -(Y^5\Pi^6 - Y^6\Pi^5)(x,p)$, where $[\mathcal{N}, Y^\mu] = [\mathcal{N}, \Pi^\mu] = 0$. The operator $\mathcal{N}$ appearing in the right-hand side of Equation (75a,b) can be moved to the left-hand side via the inverse $\mathcal{N}^{-1}$ operator, and that can be defined as a formal power series as follows: $[1-(1-\mathcal{N})]^{-1} = 1 + (1-\mathcal{N}) + (1-\mathcal{N})^2 + \cdots$.

Thus, from Equations (71) and (75a,b) one can then construct the maps from the x^μ, p^μ noncanonical (operator) variables in 4D to the canonical (operator) variables Y^A, Π^A in 6D. After a laborious but straightforward procedure we find the following family of solutions

$$Y^5(x,p) = \kappa_1 \beta z_1 + \kappa_2 \beta z_2 + \kappa_3 \beta z_3 + \kappa_4 \beta \tag{76a}$$

$$Y^6(x,p) = \kappa_1 \beta z_1 + \kappa_2 \beta z_2 + \kappa_3 \beta z_3 + (\kappa_4 + 1)\beta \tag{76b}$$

$$\Pi^5(x,p) = \kappa_1 \beta^{-1} z_1 + \kappa_2 \beta^{-1} z_2 + \frac{5}{4}\kappa_3 \beta^{-1} z_3 + \kappa_4 \beta^{-1} \tag{76c}$$

$$\Pi^6(x,p) = \kappa_1 \beta^{-1} z_1 + \kappa_2 \beta^{-1} z_2 + \frac{5}{4}\kappa_3 \beta^{-1} z_3 + (\kappa_4 + 1)\beta^{-1} \tag{76d}$$

where $\kappa_3 = (\alpha\beta)^{-1}$ and κ_1, κ_2, and κ_4 are three arbitrary parameters. This is due to the nonlinearity of the equations that one is solving. These solutions (76a–d) have the form $Y^6 = Y^5 + \beta; \Pi^5 = \Pi^6 - \beta^{-1}$ such that $\alpha\beta\ Y^{[5}\ \Pi^{6]} = -\frac{z_3}{4} = -\alpha\beta\mathcal{N}$ as required by Equation (71).

When one takes the classical limit, upon restoring $\hbar$ which was set to unity in the terms $\gamma_i z_2 \to \frac{\gamma_i}{\hbar} z_2$ of Equations (68e), in order to match units, one can see that these terms are *singular* in the $\hbar \to 0$ limit, whereas the terms $\frac{\delta_i}{\hbar} z_3 \to -4\delta_i$ are well-behaved and yield constants.

For these reasons, we just adhere to the following prescription when finding the *classical* limit of the embedding functions $Y^A(x,p), \Pi^A(x,p)$. We could simply drop the *singular* $\frac{1}{\hbar} z_2$ terms in Equation (76a–d) by setting the arbitrary constant κ_2 to zero $\kappa_2 = 0$ and set the $\frac{1}{\hbar} z_3$ terms to constants that can be reabsorbed into a redefinition of the κ_4 parameter in the explicit solutions for Y^5, Y^6, Π^5, and Π^6 given by Equation (76a–d). In doing so, one ends up with the following expressions in the *classical* limit

$$Y^5(z_1) = \kappa_1 \beta z_1 + \beta(\kappa_4 - 4(\alpha\beta)^{-1}) \tag{77a}$$

$$Y^6(z_1) = \kappa_1 \beta z_1 + \beta(\kappa_4 + 1 - 4(\alpha\beta)^{-1}) \tag{77b}$$

$$\Pi^5(z_1) = \kappa_1 \beta^{-1} z_1 + \beta^{-1}(\kappa_4 - 5(\alpha\beta)^{-1}) \tag{77c}$$

$$\Pi^6(z_1) = \kappa_1 \beta^{-1} z_1 + \beta^{-1}(\kappa_4 + 1 - 5(\alpha\beta)^{-1}) \tag{77d}$$

To conclude, one can finally obtain the explicit solutions for $Y^\mu, (z_1, x^\mu, p^\mu); \Pi^\mu(z_1, x^\mu, p^\mu)$, in the classical limit, given in terms of the functions $Y^5(z_1), Y^6(z_1), \Pi^5(z_1)$, and $\Pi^6(z_1)$ in Equation (77a–d) (and x^μ, p^μ) as follows:

$$\alpha x^\mu \Pi^6(z_1) - \beta p^\mu \Pi^5(z_1) = -\Pi^\mu(z_1, x^\mu, p^\mu) \tag{78a}$$

$$\alpha x^\mu Y^6(z_1) - \beta p^\mu Y^5(z_1) = -Y^\mu(z_1, x^\mu, p^\mu) \tag{78b}$$

where $z_1 \equiv \alpha^2 x^2 + \beta^2 p^2$, $\alpha = \mathcal{L}^{-1}; \beta = L_P$. Next, we study the geometrical implications of the (classical) embedding solutions found in this section and provided by Equations (77a–d) and (78a,b).

4.2. Embedding an 8D Curved Phase Space into a 12D Flat Phase Space

The previous section involved the use of coordinates and momenta *operators*. In this section, we shall deal with *classical* variables (c-numbers) x, p. A more rigorous notation in the previous section would have been to assign "hats" to operators $\hat{x}^\mu, \hat{p}^\mu; \hat{Y}^A, \hat{\Pi}^A$. For the sake of simplicity, we avoided it. The geometry of the cotangent bundle of spacetime (phase space) can be best-explored within the context of Lagrange–Finsler, Hamilton–Cartan geometry [15–18]. The line element in the 8D curved phase space is

$$(ds)^2 = g_{\mu\nu}(x,p) dx^\mu dx^\nu + h^{ab}(x,p)(dp_a + N_{a\mu}(x,p)dx^\mu)(dp_b + N_{b\nu}(x,p)dx^\nu) \tag{79}$$

where $g_{\mu\nu}(x,p), h^{ab}(x,p)$ are the base spacetime and internal space metrics, respectively, with $a,b = 1,2,3,4$, $\mu,\nu = 1,2,3,4$, and $N_{a\mu}(x,p)$ is the nonlinear connection.

One should note that the metric tensor $g_{\mu\nu}$ is *not* the vertical Hessian of the square of a Finsler function, and h^{ab} is *not* the inverse of $g_{\mu\nu}$. h^{ab} represents, physically, the cotangent bundle's internal-space metric tensor which is independent from the base-spacetime metric tensor $g_{\mu\nu}$. The number of total components of $g_{\mu\nu}, h^{ab}, N_{a\mu}$ is $10 + 10 + 16 = 36 = (8 \times 9)/2)$.

The generalized (vacuum) gravitational field equations associated with the geometry of the 8D cotangent bundle differ considerably from the standard (vacuum) Einstein field equations in 8D based on Riemannian geometry. Thus, for instance, by using a base-spacetime $g_{\mu\nu}$ metric to be *independent* from the internal-space metric h_{ab} and a nonlinear connection $N_{\mu a}$, it might avoid the reduction of the solutions of the generalized gravitational field equations to the standard Schwarzschild (Tangherlini) solutions when radial symmetry is imposed.

For example, in [19] we further studied a scalar-gravity model in curved phase spaces proposed by [20,21]. After a very laborious procedure, the variation of the action S with respect to the fundamental fields

$$\frac{\delta S}{\delta g_{\mu\nu}} = 0, \quad \frac{\delta S}{\delta h_{ab}} = 0, \quad \frac{\delta S}{\delta N_{\mu a}} = 0, \quad \frac{\delta S}{\delta \Phi} = 0 \tag{80}$$

led to very *complicated* field equations which differed considerably from the Einstein field equations. Exact nontrivial analytical solutions for the base-spacetime $g_{\mu\nu}$, the internal-space metric h_{ab} components, the nonlinear connection N_{ia}, and the scalar field Φ were found that obeyed the generalized gravitational field equations, in addition to satisfying the *zero*-torsion conditions for *all* of the torsion components. See [19] for details.

The embedding of the 8D curved phase space into the 12D flat phase space is described by equating the 8D line interval ds^2 in (79) with the 12D one $ds^2 = \eta_{AB} dZ^A dZ^B$. After doing so, given $Z^A \equiv (Y^A, \Pi^A)$ one learns that

$$g_{\mu\nu} + h^{ab} N_{a\mu} N_{b\nu} = \eta_{AB} \frac{\partial Z^A}{\partial x^\mu} \frac{\partial Z^A}{\partial x^\nu} \tag{81}$$

$$h^{ab} = \eta_{AB} \frac{\partial Z^A}{\partial p_a} \frac{\partial Z^A}{\partial p_b} \tag{82}$$

$$h^{ab} N_{b\nu} = \eta_{AB} \frac{\partial Z^A}{\partial p_a} \frac{\partial Z^A}{\partial x^\nu} \quad A, B = 1, 2, \cdots, 5, 6 \tag{83}$$

Equations (81)–(83) *determine* the functional form of $g_{\mu\nu}, h^{ab}, N_{a\mu}$ after one inserts the functional forms of the embedding functions $Z^A(x,p) = Y^A(x,p), \Pi^A(x,p)$ found in the previous section. However, there is a subtlety: to match indices with the ones appearing in Equations (77a–d) and (78a,b) it is necessary to make the following key *replacements* (index adjustments) $p_a \to p^\sigma, p_b \to p^\tau, h^{ab} \to h_{\sigma\tau}, N_{a\mu} \to N^\sigma_\mu, N_{b\nu} \to N^\tau_\nu$ in Equations (79) and (81)–(83).

To sum up, the (classical) embedding functions $Z^A(x,p) = Y^A(x,p), \Pi^A(x,p)$ obtained in the previous section in Equations (77a–d) and (78a,b) determine the functional form of $g_{\mu\nu}, h^{ab}, N_{a\mu}$ in Equations (81)–(83), after adjusting the indices. The key question is whether or not the solutions found for $g_{\mu\nu}, h^{ab}, N_{a\mu}$ *also* solve the vacuum field equations. If not, can one find the appropriate field/matter sources which are consistent with these solutions? It is natural to assume that quantum matter/fields could be the source of the noncommutativity of the spacetime coordinates and momenta. After all, quantum fields live in spacetime. If this were not the case, what then is the source of this phase-space noncommutativity? Is it spacetime foam, dark matter, dark energy? If one expects to have a space–time–matter unification in the quantum gravity program, then, if matter curves spacetime, spacetime, in turn, could backreact on matter curving momentum space, "curving matter". To conclude, to find solutions of Equations (81)–(83) for $g_{\mu\nu}, h^{ab}, N_{a\mu}$ (after adjusting indices) is a highly nontrivial task, and so is to verify that they also solve the field equations in [19–21].

5. Concluding Remarks

After a review of Born's reciprocal relativity and its physical implications, this work was mainly devoted to the Yang and the deformed quaplectic algebras associated with noncommutative phase spaces, and to their extensions involving antisymmetric tensor coordinates and momenta of different ranks. Our approach to construct extended Yang algebras *differs* from the study by [35]. We finalized with an analysis of the embedding of an 8D curved phase space into a 12D flat phase space which provided a direct link between noncommutative curved phase spaces in lower dimensions to commutative flat phase spaces in higher dimensions. Left from our discussion was the role of quantum groups, Hopf algebras, κ-deformed Poincare algebras, and of the deformed special relativity [32–34,36–40]. This will be the subject of future investigations.

Funding: This research received no external funding.

Acknowledgments: We are indebted to M. Bowers for assistance, to the referees for their suggestions to improve the manuscript, and to Stavrinos and Saridakis for their very kind invitation to contribute to this topical collection on Modified Theories of Gravity and Cosmological Applications.

Conflicts of Interest: The author declares no conflict of interest.

Notes

1 Strictly speaking, $U(1,4)$ is a pseudo-unitary group. After performing the Weyl unitary "trick" via an analytical continuation $U(1,4) \to U(5)$, one obtains the unitary group $U(5)$ comprised of 5×5 unitary matrices obeying $U^\dagger = U^{-1}$. A unitary matrix can be written as $U = e^A$, where A is an anti-Hermitian matrix $A^\dagger = -A$, and any anti-Hermitian matrix A can be written as $A = \pm iH$, where H is Hermitian; therefore, all group elements can be written in the form $U = e^{\pm i\theta^{AB} Z_{AB}}$, where θ^{AB} are the corresponding parameters associated to every generator.

2 We choose a different signature than the one in the Introduction.

3 Our choice differs by a minus sign from the conventional definition.

4 A simple inspection reveals that a correspondence of the form $\frac{x^\mu}{L_P} = a_1 J^{\mu 5} + b_1 J^{\mu 6}; \mathcal{L}p^\mu = a_2 J^{\mu 5} + b_2 J^{\mu 6}$ will automatically lead to $b_1 = 0, a_2 = 0$; or $b_2 = 0, a_1 = 0$ resulting from the antisymmetry of the commutators $[x^\mu, x^\nu], [p^\mu, p^\nu]$

5 Since $\lambda_l \lambda_p = 1$, in units of $\hbar = 1$, the powers of λ_l, λ_p decouple explicitly from Equation (44a,b)

6 Note that one must not confuse $Z_{ab} \equiv \frac{1}{2}(M_{ab} + L_{ab})$ with $Z_{[a_1 a_2]}$ defined by Equation (55a)

References

1. Born, M. Elementary Particles and the Principle of Reciprocity. *Nature* **1949**, *163*, 207. [CrossRef] [PubMed]
2. Low, S. $U(3,1)$ Transformations with Invariant Symplectic and Orthogonal Metrics. *Il Nuovo Cim. B* **1993**, *108.*, 841. [CrossRef]
3. Low, S. Representations of the canonical group, (the semi-direct product of the unitary and Weyl-Heisenberg groups), acting as a dynamical group on noncommutative extended phase space. *J. Phys. A Math. Gen.* **2002**, *35*, 5711. [CrossRef]
4. Low, S. Canonically relativistic quantum mechanics: Casimir field equations of the Quaplectic group. *arXiv* **2004**, arXiv:math-ph/0404077.
5. Castro, C. Is Dark Matter and Black-Hole Cosmology an Effect of Born's Reciprocal Relativity Theory? *Can. J. Phys.* **2019**, *97*, 198–209. [CrossRef]
6. Cainiello, E.R. Is there a Maximal Acceleration. *Il Nuovo Cim.* **1981**, *32*, 65. [CrossRef]
7. Brandt, H. Finslerian Fields in the Spacetime Tangent Bundle. *Chaos Solitons Fractals* **1999**, *10*, 267. [CrossRef]
8. Brandt, H. Maximal proper acceleration relative to the vacuum. *Lett. Nuovo Cimento* **1983**, *38*, 522. [CrossRef]
9. Toller, M. The geometry of maximal acceleration. *Int. J. Theor. Phys.* **1990**, *29*, 963. [CrossRef]
10. Castro, C. Some consequences of Born's Reciprocal Relativity in Phase Spaces. *Found. Phys.* **2005**, *35*, 971. [CrossRef]
11. Kalman, C.S. $SU(1,3)$ as a Dynamical Group: Analysis of All the Discrete Representations. *Can. J. Phys.* **1973**, *51*, 1573. [CrossRef]
12. Barut, A.O.; Bohm, A. Dynamical Groups and Mass Formula. *Phys. Rev.* **1965**, *139*, B1107. [CrossRef]
13. Unruh, W. Notes on black-hole evaporation. *Phys. Rev. D* **1976**, *14*, 870. [CrossRef]
14. Castro, C. On Born's Deformed Reciprocal Complex Gravitational Theory and Noncommutative Gravity. *Phys. Lett. B* **2008**, *668*, 442–446. [CrossRef]
15. Vacaru, S.; Stavrinos, P.; Gaburov, E.; Gonta, D. *Clifford and Riemann-Finsler Structures in Geometric Mechanics and Gravity*; Geometry Balkan Press: Bucharest, Romania; 2005; 693p.
16. Rund, H. *The Differential Geometry of Finsler Spaces*; Springer: Berlin/Heidelberg, Germany, 1959.
17. Bucataru, I.; Miron, R. *Finsler-Lagrange Geometry*; Editura Academiei Romane: Bucharest, Romania, 2007.
18. Miron, R.; Hrimiuc, D.; Shimada, H.; Sabau, S. *The Geometry of Hamilton and Lagrange Spaces*; Kluwer Academic Publishers: Dordrecht, The Netherlands; Boston, MA, USA, 2001.
19. Castro Perelman, C. Born's Reciprocal Relativity Theory, Curved Phase Space, Finsler Geometry and the Cosmological Constant. *Ann. Phys.* **2020**, *416*, 168143. [CrossRef]
20. Minas, G.; Saridakis, E.N.; Stavrinos, P.C.; Triantafyllopoulos, A. Bounce cosmology in generalized modified gravities. *arXiv* **2019**, arXiv:1902.06558v2.
21. Ikeda, S.; Saridakis, E.N.; Stavrinos, P.C.; Triantafyllopoulos, A. Cosmology of Lorentz fiber-bundle induced scalar-tensor theories. *Phys. Rev. D* **2019**, *100*, 124035. [CrossRef]
22. Yang, C. On Quantized Spacetime. *Phys. Rev.* **1947**, *72*, 874. [CrossRef]

23. Tanaka, S. A Short Essay on Quantum Black Holes and Underlying Noncommutative Quantized Space-Time. *Class. Quantum Grav.* **2017**, *34*, 015007. [CrossRef]
24. Snyder, H. Quantized space-time. *Phys. Rev.* **1947**, *67*, 38. [CrossRef]
25. Klauder, J.R.; Skagerstam, B. *Coherent States*; World Scientific: Singapore, 1985.
26. Castro Perelman, C. The Clifford-Yang Algebra, Noncommutative Clifford Phase Spaces and the Deformed Quantum Oscillator. *Int. J. Geom. Methods Mod. Phys.* **2023**. [CrossRef]
27. Castro Perelman, C. A Noncommutative Spacetime Realization of Quantum Black Holes, Regge Trajectories and Holography. *J. Geom. Phys.* **2022**, *173*, 104435. [CrossRef]
28. Castro, C.; Pavsic, M. The Extended Relativity Theory in Clifford spaces. *Prog. Phys.* 2005, *1*, 31.
29. Castro, C.; Pavsic, M. On Clifford algebras of spacetime and the Conformal Group. *Int. J. Theor. Phys.* **2003**, *42*, 1693. [CrossRef]
30. Meljanac, S.; Mignemi, S. Noncommutative Yang model and its generalizations. *J. Math. Phys.* **2023**, *64*, 023505. [CrossRef]
31. Meljanac, S.; Mignemi, S. Generalizations of Snyder model to curved spaces. *Phys. Lett. B* **2022**, *833*, 137289. [CrossRef]
32. Magueijo, J.; Smolin, L. Lorentz Invariance with an Invariant Energy Scale. *Phys. Rev. Lett.* **2002**, *88*, 190403. [CrossRef]
33. Amelino-Camelia, G.; Freidel, L.; Kowalski-Glikman, J.; Smolin, L. The principle of relative locality. *Phys. Rev. D* **2011**, *84*, 084010. [CrossRef]
34. Amelino-Camelia, G.; Freidel, L.; Kowalski-Glikman, J.; Smolin, L. Relative locality: A deepening of the relativity principle. *Gen. Rel. Grav.* **2011**, *43*, 2547. [CrossRef]
35. Lukierski, J.; Meljanac, S.; Mignemi, S.; Pachol, A. Quantum perturbative solutions of extended Snyder and Yang models with spontaneous symmetry breaking. *arXiv* **2022**, arXiv:2212.02361.
36. Gutierrez-Sagredo, I.; Ballesteros, A.; Gubitosi, G.; Herranz, F.J. Quantum groups, non-commutative Lorentzian spacetimes and curved momentum spaces. In *Spacetime Physics 1907–2017*; Duston, C., Holman, M., Eds.; Minkowski Institute Press: Montreal, QC, Canada, 2019; pp. 261–290.
37. Ballesteros, A.; Gubitosi, G.; Gutierrez-Sagredo, I.; Herranz, F.J. Curved momentum spaces from quantum (Anti-)de Sitter groups in (3+1) dimensions. *Phys. Rev. D* **2018**, *97*, 106024. [CrossRef]
38. Ballesteros, A.; Gutierrez-Sagredo, I.; Herranz, F.J. The κ-(A)dS noncommutative spacetime. *Phys. Lett. B* 2019, *796*, 93. [CrossRef]
39. Arzano, M.; Kowalski, J. Quantum particles in non-commutative space-time: An identity crisis. *arXiv* **2022**, arXiv:2212.03703.
40. Trzesniewski, T. 3*D* gravity, point particles and deformed symmetries. *arXiv* **2022**, arXiv:2212.14031.

universe

MDPI

Review

Geometric Outlines of the Gravitational Lensing and Its Astronomic Applications

Bin Shen *,† and Mingyang Yu †

School of Mathmatics, Southeast University, Nanjing 211189, China
* Correspondence: shenbin@seu.edu.cn; Tel.: +86-025-52090590
† These authors contributed equally to this work.

Abstract: Gravitational lensing is a topic of great application value in the field of astronomy. The properties and research methods of gravitational lensing are closely related to the geometric and relativistic characteristics of the background universe. This review focuses on the theoretical research and application of strong lenses and weak lenses. We first introduce the basic principles of gravitational lensing, focusing on the geometric basis of geometric lensing, the representation of deflection angles, and the curvature relationship in different geometric spaces. In addition, we summarize the wide range of applications of gravitational lensing, including the application of strong gravitational lensing in Schwarzschild black holes, time delay, the cosmic shearing based on weak lensing, the applications in signal extraction, dark matter, and dark energy. In astronomy, through the use of advanced astronomical instruments and computers, analyzing gravitational lensing effects to understand the structure of galaxies in the universe is an important topic at present. It is foreseeable that gravitational lensing will continue to play an important role in the study of cosmology and will enrich our understanding of the universe.

Keywords: gravitational lensing; deflection angle; Einstein ring; Schwarzschild black hole; dark matter and dark energy; time delay; cosmic shear

MSC: 85-06

1. Introduction

Light rays are deflected in a gravitational field. The light of the background galaxy is focused by the gravitational field of the foreground galaxy. Such a light deflecting phenomenon, which is similar to that observed in an optical lens, is called gravitational lensing. Einstein's general theory of relativity first predicted and explained the phenomenon of gravitational lensing [1]. The first lens system, QSO 0957+561A, B, was discovered in 1979 [2]. Since then, increasing attention has been paid to the research of gravitational lensing. Currently, gravitational lensing can be divided into three categories (i.e., strong gravitational lensing, weak gravitational lensing, and microlensing) based on the geometric configuration among the background galaxy, the foreground galaxy, and the observer, as well as the mass–energy distribution of the foreground galaxy. Strong gravitational lensing can produce multiple and severely distorted images [3]. However, it can greatly brighten the lights from distant galaxies, which is quite spectacular. Thus, strong gravitational lensing is often used to study distant galaxies [4]. Unlike strong gravitational lensing, weak gravitational lensing is more common and produces slightly distorted images [5]. It has more applications, for example, as the information from weak lensing is necessary to reveal the whole mass distribution of the cluster.

Gravitational lensing has valuable applications, especially in the detection of dark matter and dark energy, which do not emit electromagnetic radiation but constitute 95% contents of the universe. However, dark matter is a major component of matters with gravitational effects. In addition, considering the relationship between gravitational lensing, dark energy, and the large-scale structure evolution of the universe, gravitational lensing can also be one of the detection tools of dark energy. Furthermore, we can use gravitational lensing to study the cosmological parameter, the Hubble constant, and the age and density of the universe [6]. Recently, Jee I. et al. proposed a method to determine the Hubble constant by using the angular diameter distance to strong gravitational lenses as a suitable calibrator [7].

Moreover, gravitational lensing was also adopted to investigate the black holes and wormholes. Recently, Bohn et al. analyzed a binary black hole merger [8]. Lu and Xie studied an Extended Uncertainty Principle black hole [9]. Sharif et al. numerically evaluated the deflection angles in the strong field limit to find the locations of wormholes [10]. Ono T. et al. studied the deflection angle of light for an observer and source at a finite distance from a rotating Teo wormhole, and the correction of the deflection angle due to the limited distance from the rotating wormhole was obtained [11]. Takahashi et al. discussed the negative-mass compact objects and the number density of Ellis wormholes [12]. Ovgun applied the Gauss–Bonnet theorem to Damour–Solodukhin wormhole spacetimes to probe weak gravitational lensing [13]. Recently, researchers focus on the black hole shadow, which is the dark region in the center of a black hole image. It appears because the light rays are close enough to be captured by the black hole itself, thereby leaving a black shadow in the observer's visual field. Perlick et al. proposed the derivation of the angular magnitude of shadows in static sphere symmetric spacetime. They calculated the black hole shadows in the expanding universe and also considered the effect of plasma on black hole shadows [14]. Akiyama et al. assembled the Event Horizon Telescope. The images they observed were consistent with the prediction of general relativity concerning the shadow of the Kerr black hole [15]. In addition, the first observation of Sagittarius A* (Sgr A*) using the Event Horizon Telescope (EHT) can be found in [16]. The gravitational lensing effect is also of great value for gravitational wave detection. For example, Cao et al. analyzed the influence of lenses on gravitational wave signal parameters [17]. Oguri used gravitational lensing magnification to analyze the distribution of gravitational waves generated by binary star mergers [18]. Today, various emerging technologies have also been used to explore gravitational lensing, such as convolutional neural networks [19] and 3D printing technology [20]. With the further development of science and technology, it is foreseeable that the study and application of gravitational lensing will be improved, which would help us understand the vastness of universe on a deeper level.

In this manuscript, we first briefly introduce the fundamental principle of the gravitational lensing in Section 2. Then, we discuss the progress of the strong lensing of a Schwarzschild black hole and the time delay in Section 3. Subsequently, cosmic shear, signal extraction, dark matter, and dark energy detection based on weak lensing are introduced in Section 4. For a better understanding of gravitational lensing, we recommend the following references [21–24].

2. Fundamental Principle

2.1. Deflection Angle

When a photon passes by a massive object, such as a galaxy or a galaxy cluster, the deflection of its trajectory is described by the deflection angle $\hat{\alpha}$. Assuming the effective refractive index $n = 1 - 2\Phi/c^2$, one can obtain $\hat{\alpha}$ from the Fermat's principle that

$$\hat{\vec{\alpha}} = \frac{2}{c^2} \int \vec{\nabla}_\perp \Phi \mathrm{d}s, \tag{1}$$

where Φ is the scalar Newtonian potential that obeys the Poisson's equation. Here, we use $\hat{\vec{\alpha}}$ to denote the angle with a direction since the right-hand side of (1) is an integral of a vector. However, we will denote the angle without the arrow in this manuscript since we are more concerned about the magnitude of the angle than its direction.

For a point lens, in Newtonian mechanics, the deflection angle $\hat{\alpha}$ is deduced from (1) to be

$$\hat{\alpha} = \frac{2GM}{bc^2}.$$

where M is the mass of lens and b is the closest distance of the photon to the lens, while, in General Relativity, the deflection angle $\hat{\alpha}$ of a point lens with gravitational potential $\Phi(r) = -GM/r$ is deduced to be

$$\hat{\alpha} = \frac{4GM}{bc^2},$$

which is twice of that in Newtonian mechanics. However, the deflection angle of light is small and generally on the order of arc–sec.

Takizawa et al. [25] defined the deflection angle of light (α_K) in an integral form to eliminate the asymptotically flatness of space–time. Namely,

$$\alpha_K = \iint_{\Omega_R+\Omega_S} K dS + \int_{P_R}^{P_S} \kappa_g d\ell + \phi_{RS}, \tag{2}$$

where Ω_R is a trilateral specified by the points R, P_0, and P_R, and Ω_S is a trilateral specified by the points S, P_0, and P_S. The right-hand side of (2) contains the radial coordinate $r \in [r_0, r_R] \cup [r_0, r_S]$, where r_0 means the closest approach of light. Ishihara et al. [26] defined the deflection angle of light (α_I) under the assumption of asymptotic flatness as

$$\alpha_I \equiv \Psi_R - \Psi_S + \phi_{RS}, \tag{3}$$

in which Ψ_R and Ψ_R are the angles between the radial direction and the light ray at the source position and at the receiver position, respectively. ϕ_{RS} is a coordinate angle between the receiver and source. Without the assumption of asymptotic flatness, Takizawa et al. [25] proved the consistency of the above two definitions on a static and spherically symmetric spacetime, i.e.,

$$\alpha_I = \alpha_K. \tag{4}$$

In [27], Takizawa et al. also defined a deflection angle α_G as

$$\alpha_G \equiv \theta - \Psi_S + \phi_{RS}, \tag{5}$$

in which θ denotes the angular direction of the lensed image with respect to the lens direction. It follows from (3) and (5) that

$$\alpha_I = \alpha_G, \tag{6}$$

where we use $\Psi_R = \theta$. One can deduce from (4) and (6) the equivalence of the above three definitions.

The gravitational lensing deflection angle can reflect the overall geometry of the foreground universe. In 2008, Gibbon and Werner first proposed a mathematical method to measure the asymptotic deflection angle $\hat{\alpha}$ of the gravitational lenses adopting the Gauss–Bonnet method [28]. In 2012, Werner discovered that for a stationary observer in the Kerr spacetime, a spatial light ray is a geodesic in a certain Randers metric space [29]. Given that a light ray is in a geodesically complete surface whose Gaussian curvature in the optical geometry is K. It can be shown that its asymptotic deflection angle $\hat{\alpha}$ is

$$\hat{\alpha} = -\iint_{S_\infty} K dS, \tag{7}$$

where the integral is used over the infinite region S_∞ of the surface bounded by the light ray excluding the lens. Based on the same approach, in 2020, Halla and Perlick used the Gauss–Bonnet theorem to provide a mathematical representation of the deflection angle of gravitational lenses in the NUT metric space [30].

As a generalization of traditional geometric backgrounds, Finsler background space-time, especially Randers spacetime, has a wide range of physical applications. There is also a corresponding deflection angle definition on it. We refer to [31] for more details of the underlying geometry of the SFR gravitational model, as well as the field equations for the SFR metric. Kapsabelis et al. [31] derived the deflection angle in SFR spacetime as

$$\alpha_{SFR} = \frac{4GM}{b}\frac{1-a}{\sqrt{1-2a}}. \tag{8}$$

Expanding (8) in powers of $a = \frac{\tilde{A}_0 b}{J}$, the deflection angle can be written as follows:

$$\alpha_{SFR} \approx \left(1 + \frac{a^2}{2}\right)\frac{4GM}{b}, \tag{9}$$

where J is the angular momentum, and $\tilde{A}_0$ is a constant obtained when solving field equations.

Compared to the deflection angle $\alpha_{GR} = \frac{4GM}{b}$ in General relativity, α_{SFR} includes a small additional Randers contribution term s, which yields a slight deviation from α_{GR}. Considering by $|\tilde{A}_0| \ll 1$, it follows that

$$\lim_{\tilde{A}_0 \to 0} \alpha_{SFR} = \alpha_{GR}. \tag{10}$$

The difference between the deflection angle in SFR spacetime and α_{GR} may be due to Lorentz violations [32], or the small amount of energy that is added to the gravitational potential of SFR.

Recently, the first author of this work introduced, for the first time, the asymptotically Minkowskian flat properties on the general Finsler metric manifold. Moreover, an integral representation of the integral symmetric normal deflection angle γ_∞ of the gravitational lens on an asymptotic symmetric Minkowskian flat Finsler manifold in a large range was provided [33]. Concretely,

$$\gamma_\infty = -L\iint_{D_\infty} \frac{1+\mathcal{T}_N(T)}{L(x)}[X^*(K\omega^1 \wedge \omega^2 - J\omega^1 \wedge \omega^3) + \log \mathcal{T}_N(T) \wedge X^*(\omega^3)], \tag{11}$$

where D_∞ is a region bounded by the deflection light, L and $L(x)$ are the lengths of the indicatrixes at the infinity and at x, respectively, $\mathcal{T}_N(T)$ is the T-curvature, K is the Gauss curvature, and J is the Landsberg curvature. Formula (11) is a generalization of (7) and can be adopted to illustrate a new geometric explanation of dark matter and dark energy, as well as their effects.

2.2. Lens Equation

In general, the scale of the lens is much smaller than both the distance from the light source to the lens D_{LS} and the distance from the lens to the observer D_L. Therefore, the lens is often reduced to fall on a flat surface. For lenses with a mass distribution of ρ, we use ξ to represent the position vector on the plane of the lens. Its surface mass density is provided by $\Sigma(\xi) = \int \rho(\xi, s)ds$. Then, the deflection angle at ξ is

$$\hat{\alpha}(\xi) = \frac{4GM(\xi)}{c^2\xi} = \frac{4G}{c^2}\int \frac{(\xi - \xi')\,\Sigma(\xi')}{|\xi - \xi'|^2} \mathrm{d}^2\xi'^2, \tag{12}$$

in which we call $R_{sc} = \frac{2GM(\xi)}{c^2}$ the Schwarzschild radius.

The line-of-sight dominates the plane of the light source, the plane of the lens, and the optical axis of the observer. Without a lens, the observer will see the light from the light source at an angle of β to the optical axis. Under the gravitation of massive objects, the observer will see the angle θ. Thus, using geometric relations, under the setting $\xi = D_L\theta$, we can write it as

$$\beta = \theta - \alpha(\theta), \tag{13}$$

$$\alpha(\theta) = -\frac{D_{LS}}{D_S}\hat{\alpha}(\theta), \tag{14}$$

where (13) is known as the lens equation and has further expression and applications [34,35]. A schematic geometry of the lens equation can be seen in Figure 1.

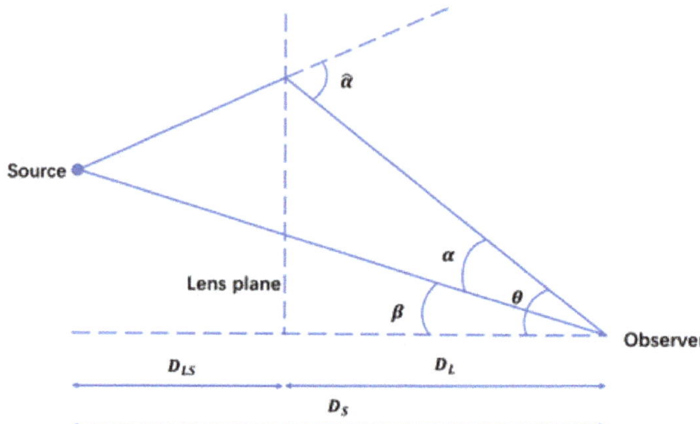

Figure 1. A schematic geometry of the lens equation.

Equations (12)–(14) are classic lens equations in gravitational lensing. Nowadays, the study of lens equations has been further advanced and different forms have been proposed. We introduce more definitions of the deflection angle, as well as a class of exact gravitational lens equations at finite distances.

Takizawa et al. deduced the exact gravitational lens equation for α_G (in Equation (5)) as

$$\alpha_G - \theta - \arcsin\left(\frac{D_L}{\sqrt{(D_{LS})^2 + (D_S)^2\tan^2\beta}}\sin\theta\right) + \arctan\left(\frac{D_S}{D_{LS}}\tan\beta\right) = 0. \tag{15}$$

They also considered a static and spherically symmetric solution in Weyl conformal gravity, and provided iterative solutions for the finite-distance lens equation up to the third order.

Furthermore, Takizawa et al. [36] showed that the hyperbolic (spherical) trigonometry can be used in the exact lens equation for a de-Sitter (anti-dS) background and discussed possible effects by the curvature of the dS/AdS background. Moreover, they provided a unified form of Lens equations in Euclidean, hyperbolic, and spherical geometry by

$$\alpha - \theta = \arcsin\left(\sqrt{\frac{1 + K\hat{D}_S^2\tan^2\beta}{\hat{D}_{LS}^2 + \hat{D}_S^2\tan^2\beta}}\hat{D}_L\sin\theta\right) - \arctan\left(\frac{\hat{D}_S}{\hat{D}_{LS}}\tan\beta\right), \tag{16}$$

where the parameters with corner markers are normalized defined quantities, and K denotes 1, 0, and -1 for spherical, flat, and hyperbolic geometry, respectively. Based on the curvatures of the dS/AdS background, they presented the deflection angle in dS/AdS background by

$$\alpha = \frac{2r_g}{D_L\theta} - \frac{r_g\theta}{2D_L}\left[1 + \left(\frac{D_L}{D_{LS}}\right)^2\right] + \frac{r_g\Lambda D_L\theta}{6} + O\left(r_g^2, r_g\theta^3, r_g\Lambda D\theta^3, r_g\Lambda^2 D^3\right), \tag{17}$$

in which $r_g \equiv 2m$ for the mass m. Although dS and AdS are geometrically different, their deflection angles are formally the same.

In small angle approximations, the difference in the form among the dS/AdS lens equations and the exact lens equation in the Minkowski background begins at the third order. The angular separation of the lensed images is decreased by the third-order deviation in the dS lens equation, while it is increased in AdS.

An Einstein ring occurs when the source, lens, and observer are aligned well. That is, using $\beta = 0$ in (13), we can obtain the angular radius of the ring, namely, the Einstein radius, as

$$\theta_E = \sqrt{\frac{4GM(\theta_E)}{c^2}\frac{D_{LS}}{D_L D_S}} = 16.5\left[\frac{M(\theta_E)}{10^{14}h^{-1}M_\odot}\right]^{\frac{1}{2}}\left(\frac{D_{LS}}{D_L D_S}\right)^{\frac{1}{2}}. \tag{18}$$

Combining (12)–(14), the graviational lensing equation can be rewritten as

$$\alpha = \frac{2R_{sc}D_{LS}}{\theta D_L D_S} = \theta - \beta. \tag{19}$$

The property of the Einstein ring is special when there is a point source. In this case, Equations (12), (18), and (19) provide that

$$\theta^2 - \beta\theta - \theta_E^2 = 0, \tag{20}$$

which can be solved by

$$\theta_{A,B} = \frac{1}{2}\beta \pm \sqrt{\left(\frac{1}{2}\beta\right)^2 + \theta_E^2}. \tag{21}$$

Therefore, we can obtain that the angular radius of Einstein's ring is $\theta_{A,B} = \pm\theta_E$ by setting $\beta = 0$ in (21).

The magnification of an Einstein ring is

$$\mu_i = \frac{d\omega_i}{d\omega_s} = \left|\det\left(\frac{\partial\beta}{\partial\theta_i}\right)\right|^{-1}, \tag{22}$$

where $d\omega_i$ is the solid angle covered by the lensed image, and $d\omega_s$ denotes the angle in the absence of the lens.

For Einstein rings, noticing the geometric relationship between the source, the lens, and the observer (i.e., the axisymmetric property), one can simplify (22) as

$$\mu_i = \frac{\theta_i d\theta_i}{\beta d\beta}. \tag{23}$$

Hence, we can deduce the magnification of an Einstein ring from (23) as

$$\mu_E = \frac{2\theta_E}{d\beta}, \tag{24}$$

where $d\beta$ represents the true angular radius of the source.

In the case of the imperfect alignment of sources, lenses and observers, i.e., $\beta \neq 0$, the angular separation between the two images is deduced from (23) and (21) as

$$\Delta\theta = \sqrt{\beta^2 + 4\theta_E^2}.$$

Hence, the magnification is

$$\mu_{A,B} = \frac{1}{4}\left(\frac{\Delta\theta}{\beta} + \frac{\beta}{\Delta\theta} \pm 2\right). \tag{25}$$

Moreover, the total magnification is

$$\mu_T = \frac{1}{2}\left(\frac{\Delta\theta}{\beta} + \frac{\beta}{\Delta\theta}\right). \tag{26}$$

The figure of an Einstein ring as a smiley in the following link, taken by the NASA/ESA Hubble Space Telescope: https://images.nasa.gov/details-GSFC_20171208_Archive_e000791 (accessed on 14 January 2023).

2.3. Convergence and Shear

Lens Equation (13) could be considered as a mapping from the source plane to the lens plane. It can especially be linearized to obtain local information of the mapping. The Jacobian matrix of (13) has component $A_{ij} = \frac{\partial\beta_i}{\partial\theta_j} = \delta_{ij} - \alpha_{ij}$, which could be decomposed into two parts as

$$A(\theta) = \begin{pmatrix} 1-\kappa-\gamma_1 & -\gamma_2 \\ -\gamma_2 & 1-\kappa+\gamma_1 \end{pmatrix} = (1-\kappa)\begin{pmatrix} 1 & 0 \\ 0 & 1 \end{pmatrix} - \begin{pmatrix} \gamma_1 & \gamma_2 \\ \gamma_2 & -\gamma_1 \end{pmatrix}, \tag{27}$$

where κ and $\gamma(\gamma_1, \gamma_2)$ are the convergence and shear parameters, respectively, defined by

$$\kappa = \frac{\Sigma(\theta)}{\Sigma_{crit}} = \frac{1}{2}\nabla_\theta^2\Psi, \tag{28}$$

$$\gamma_1 = |\gamma|\cos(2\phi) = \frac{1}{2}(\psi_{,11} - \psi_{,22}), \tag{29}$$

$$\gamma_2 = |\gamma|\sin(2\phi) = \psi_{,ij}, \tag{30}$$

where Ψ is the lensing potential, whose expression is

$$\Psi = \frac{2}{c^2}\frac{D_{LS}}{D_L D_S}\int \Phi ds = \frac{4GM}{c^2}\frac{D_{LS}}{D_L D_S}\ln|\theta|. \tag{31}$$

Furthermore, $\alpha = \frac{\partial\Psi}{\partial\theta}$, so that (13) could be rewritten as

$$\beta = \theta - \nabla\Psi.$$

Σ_{crit} in (28) is named as the critical surface mass density, and is defined by $\Sigma_{crit} = \frac{c^2 D_s}{4\pi G D_{LS} D_L}$. κ is a parameter of homothetic transformation, indicating that the light source is isotropically amplified or reduced. Meanwhile, γ adds anisotropic effects to the imaging. The distortion of the image can be described by the inverse matrix of the above Jacobian, which is denoted by $M = A^{-1}$.

By the eigenvalues $\lambda_\pm$ of the Jacabian matrix A, one can determine the radial elongation factor to be $\frac{1}{\lambda_-}$, and the tangential factor to be $\frac{1}{\lambda_+}$. It follows for $\hat{\theta} > 0$ that

$$\lambda_+ = \frac{\hat{\beta}}{\hat{\theta}} = 1 - \frac{1}{\hat{\theta}^{n+1}}, \tag{32}$$

$$\lambda_- = \frac{d\hat{\beta}}{d\hat{\theta}} = 1 + \frac{n}{\hat{\theta}^{n+1}}. \tag{33}$$

These imply that the convergence and the shear are

$$\kappa = 1 - \frac{\lambda_+ + \lambda_-}{2} = \frac{1-n}{2}\frac{1}{\hat{\theta}^{n+1}}, \tag{34}$$

$$\gamma = \frac{\lambda_+ - \lambda_-}{2} = -\frac{1+n}{2}\frac{1}{\hat{\theta}^{n+1}}. \tag{35}$$

For $\varepsilon < 0$, the lensing shear can be similarly obtained as

$$\gamma = \frac{\lambda_+ - \lambda_-}{2} = \frac{1+n}{2}\frac{1}{\hat{\theta}^{n+1}}, \tag{36}$$

For cases where the position of the lens is unknown, the origin of the two-dimensional coordinates can be selected as the center of the lens object. For a pair of radially elongated images ($\varepsilon < 0$), they are aligned with each other, while, for a pair of tangentially elongated images, they are parallel to each other. Thus, radial and tangential elongations can be distinguished by measuring such image alignment in observations.

In the weak field and thin lens approximation, the image is tangentially elongated due to the gravitational action of the lens model on the light, while the gravitational repulsion of light by the other models always radially distorts. This feature of the shape of the lens image can be used to search for (or confine) local exotic matter or energy. For more details and developments, we recommend [37].

In fact, cosmic shear is based on measurements of the shape of galaxies, so what is really observed is not the shear γ but the reduced shear $g = \gamma/(1-\kappa)$, which has the same spin-two transformation properties as the shear.

In addition, the magnification of the image, denoted by μ, can be expressed by the inverse of the determinant of the Jacobian matrix, i.e.,

$$\mu = \frac{1}{\det(A)} = \left((1-\kappa)^2 - \gamma_1^2 - \gamma_2^2\right)^{-1}. \tag{37}$$

Furthermore, according to (37), the image is magnified, i.e., $\mu > 1$, so it looks brighter than the source.

Curves with infinity magnification, namely, $\det(A) = 0$, in the image plane are called critical curves. The corresponding curves on the source plane are called caustic curves. If a source is located near a caustic curve, it will be highly amplified and distorted. For example, it may create a huge arc in a star cluster. However, the magnification is not infinite, and a source will only produce one image, provided that the magnification of a gravitational lens is limited anywhere.

3. Strong Gravitational Lensing

In addition to the Einstein ring mentioned above, which has a tremendous visual aesthetic, we can also observe other important strong lensing phenomena. The observation and application of these phenomena have helped us further promote the development of cosmology and the advancement of modern technology.

For strong gravitational lensing, there are many observable values, such as the relative position of the image, the relative flux of the image, and the time delay of the change in luminosity between the lens images. These observables depend not only on the mass distribution of the source but also on cosmological parameters [23]. A mass model of the light source can be built to solve the lens equation with the help of light trace quality (LTM) or non-LTM methods. Today, based on these two methods, there is a lot of software code for strong lens models that can be applied.

Strong lenses can also be used to obtain the magnification of images [38]. Although we do not know the brightness of the light source, the ratio of the luminosity between images is observable. For the observed images A and B, we have

$$\delta\theta_B = M(\theta_B)\delta\beta = M(\theta_B)M^{-1}(\theta_A)\delta\theta_A = M_{BA}\delta\theta_A, \tag{38}$$

where M is the magnification matrix. Formula (38) can be considered as a linear transformation between two lens image spaces. On the other hand, the flux ratio is also an important constraint of the mass model of the lens.

In this section, to demonstrate the effective application of the strong lensing, we introduce the ongoing field of it, including Schwarzschild black hole and time delay.

3.1. Schwarzschild Black Hole

The deflection angle has already been briefly introduced in the last section. In this section we describe the deflection angle in the Schwarzschild's black hole and the considerable measurements in it. The strong field limit coefficients and the deflection angle of the black holes can be obtained by performing the strong field limit method, which is a basic approach of this topic. For instance, Chagoya et al. [39] adopted a hybrid analytic–numerical approximation to determine the deflection angle in Schwarzschild spacetimes. Similar research on deflection angles and related applications can be found in [40–42]. The general Schwarzschild metric is defined by

$$ds^2 = -A(r)dt^2 + B(r)dr^2 + C(r)\left(d\theta^2 + \sin^2\theta d\phi^2\right), \tag{39}$$

where $A(r)$, $B(r)$, and $C(r)$ are functions of r. The corresponding photosphere equation is

$$\frac{C'(r)}{C(r)} = \frac{A'(r)}{A(r)}. \tag{40}$$

The maximum root of (40), denoted by r_m, is called the radius of the photosphere.

For photons from the infinity, when approaching a black hole, there exist deviations from their original orbits. Let u nr the collision parameter, and r_0 be the minimum radial distance of a photon from the black hole during the traveling. The relationship between u and r_0 obeys the conservation of the angular momentum as

$$u = \sqrt{\frac{C(r_0)}{A(r_0)}}. \tag{41}$$

We may choose $\theta = \frac{\pi}{2}$ in (39) for the spherical symmetry of space–time. Therefore, from the radial equation of the photon

$$\frac{1}{2}\dot{r}^2 + \frac{L^2}{2r^2}\left(1 - \frac{2M}{r}\right) = \frac{1}{2}E^2,$$

the expression of the deflection angle becomes [43]

$$\alpha(r_0) = I(r_0) - \pi = 2\int_{r_0}^{\infty} \frac{\sqrt{B}dr}{\sqrt{C}\sqrt{\frac{C}{C_0}\frac{A_0}{A} - 1}} - \pi, \tag{42}$$

in which A_0 and C_0 are the values of $A(r)$ and $C(r)$ at r_0, respectively. The deflection angle $\alpha(r_0)$ depends on r_0. More precisely, $\alpha(r_0)$ increases as r_0 reduces. It requires that $r_0 \leq r_m$ for the photons to be fully absorbed. According to the bijective correspondence of θ and r_0, the deflection angle can be represented by θ as

$$\alpha(\theta) = -\bar{a}\ln\left(\frac{\theta D_{OL}}{u_m} - 1\right) + \bar{b} = -\bar{a}\ln\left(\theta D_{OL}\sqrt{\frac{A(r_m)}{C(r_m)}} - 1\right) + \bar{b}, \tag{43}$$

which is the deflection angle equation under strong field approximation. The two strong field limit coefficients $\bar{a}$ and $\bar{b}$ in (43) could be deduced as

$$\bar{a} = \frac{R(0, r_m)}{2\sqrt{q_m}},$$
$$\bar{b} = I_R(r_m) + \bar{a}\ln\frac{2q_m}{y_m} - \pi,$$

where the quantities marked with the subscript m indicate the values of them at $r_0 = r_m$. The specific measurement parameters may be clarified by observing the strong field effect. Based on it, the structure of the lens should be detected.

Moreover, the relation between the image position and magnification can be obtained by substituting (43) into the lens Equation (13).

3.2. Time Delay

Light travels along geodesics in curved space. The effect of gravity will increase the distance and cause a difference in time compared to the situation without gravity, which is called the time delay. In this case, we will only describe the propagation of light from sources at a great distance from the lens.

When light travels far from the lens, its path could be regarded as the geodesic in the Friedmann–Robertson–Walker gauge, which is an exact solution of Einstein's field equations

$$ds^2 = -dt^2 + R^2(t)\left(\frac{dr^2}{1-kr^2} + r^2(d\theta^2 + \sin^2\theta d\phi^2)\right), \tag{44}$$

where $R(t)$ is the cosmic scale factor, which determines the geometric scale of the space. Denoting the closest point on the light path to the lens by P, the intrinsic length of the light path from the source to P and from P to the observer is

$$\int_{t_S}^{t_O} \frac{dt}{R(t)} = \sigma_{SP} + \sigma_{PO}, \tag{45}$$

in which t_S and t_O are the moments when the photon leaves the source and reaches the observer, respectively. However, the time it requires for a photon to reach the observer without the lens is

$$\int_{t_S}^{t'_O} \frac{dt}{R(t)} = \sigma_{SO}, \tag{46}$$

in which σ_{SO} is the intrinsic distance along the geodesic from the source to the observer, and t'_O is the moment when the photon arrives at the observer. Therefore, comparing (45) and (46), the time delay is

$$\Delta t_f = t_O - t'_O = R(t_O)(\sigma_{SP} + \sigma_{PO} - \sigma_{SO}). \tag{47}$$

Another expression of (47) is the following, with the assistance of the geodesic equation, that is

$$\Delta t_f = \frac{R(t_O) r_O(L) r_O(S) \beta^2}{2 r_L(S)}, \tag{48}$$

where $r_O(S)$ and $r_O(L)$ are the normal radial coordinates of the source S and the point P, respectively, and *beta* is the angle between the observer-to-source connection and the observer-to-P connection.

Accounting for the cosmic redshift effect, that is, $R(t_O) = (1 + z_L)R(t_L)$ and using $d_A(OL) = R(t_L) r_O(L)$, (48) could be modified to be

$$\Delta t_f = \frac{(1 + z_L) d_A(OL) d_A(OS) \beta^2}{2 d_A(LS)}. \tag{49}$$

Time delay has been employed to determine the Hubble constant, but it is imprecise due to insufficient astronomical observations.

A smoothness condition is imposed on the gravitationally deformed paths followed by the photons from the source to the observer. Maggiore, N., et al. [44] considered an alternative formula for time delay, which was generalized to the arbitrary angles of the standard one. It can be applied to investigate the discrepancy between the various estimations of the Hubble constant.

Inspired by its various effective applications, the research of time delay has been developed rapidly. It provides a powerful cosmological detector through the time delay distance.

However, it is not easy to achieve the ideal effect, because the images have the measurement gap, noise, and no prior function model, which causes the robust time delay estimation to be challenging. In fact, Hojjati et al. [45] used Gaussian process techniques and was successfully demonstrated in an accurate blind reconstruction of time delays and a reduction in uncertainties for real data. The results will be more accurate in the future when more information is accumulated.

4. Weak Gravitational Lensing

Strong gravitational lensing causes a large signal and produces phenomena that look extremely spectacular, such as Einstein rings and Einstein crosses. However, such phenomena rarely occur.

Conversely, there are a large number of weak gravitational lensing signals in the universe because the universe is filled with various structures, such as dark matter halos, fibers, etc. Especially in the past two decades, research on weak lenses has flourished. Moreover, weak gravitational lensing plays an important role in the field of astronomical detection [46–48]. Theoretically, if we can accurately measure information such as the deformation and redshift of galaxies in the universe, it is possible to reconstruct the distribution of matter in the whole universe, because the lens signal can reflect the information of matter in the path. The combination of lens signals with different redshifts provides ways to accurately determine the fundamental parameters in cosmology, such as Ω_m, Ω_λ, σ_8.

As related to the Schwarzchild black hole, a static and spherically symmetric modified spacetime metric depends on the inverse distance to the power of positive n ($n = 1$ for Schwarzschild metric, and $n = 2$ for an Ellis wormhole) in the weak-field approximation. Izumi, K., et al. obtained the following deflection angle in such space–time [37].

$$\alpha = \frac{\varepsilon}{b^n} \int_0^{\frac{\pi}{2}} \cos^n \psi d\psi + O\left(\varepsilon^2\right), \tag{50}$$

where ε is the book-keeping parameter. Under the thin lens approximation, the modified lens equation becomes

$$\beta = \frac{b}{D_L} - \frac{D_{LS}}{D_S}\alpha(b). \tag{51}$$

For the case of $\varepsilon > 0$, the matter (and energy) need to be exotic if $n > 1$; (51) could be rewritten as the following in the vectorial form.

$$\hat{\beta} = \hat{\theta} - \frac{\hat{\theta}}{\hat{\theta}^{n+1}} \quad (\hat{\theta} > 0), \tag{52}$$

$$\hat{\beta} = \hat{\theta} - \frac{\hat{\theta}}{\left(-\hat{\theta}\right)^{n+1}} \quad (\hat{\theta} < 0), \tag{53}$$

where we normalize $\hat{\beta} \equiv \frac{\beta}{\theta_E}$ and $\hat{\theta} \equiv \frac{\theta}{\theta_E}$ for the angular position of the image $\theta = \frac{b}{D_L}$, $\hat{\beta}$, and $\hat{\theta}$ denote the corresponding vectors.

4.1. Cosmic Shear and E/B Modes

At present, cosmic shear is extremely important in the physical study of weak lensing, since its signal can reflect the distribution of matter and limit the cosmological parameters [49–51].

Nicola et al. [52] puts predictive constraints on the cosmological and mass calibration parameters for a combination of LSST cosmic shear and Simons Observatory tSZ (thermal Sunyaev-Zel'dovich) cluster counts. They found competitive constraints on cluster cosmology. Mancini et al. [53] studied 3D cosmic shear and Minkowski functionals, and demonstrated parameter inference from Minkowski functionals in a cosmic shear survey. Their work is the first step toward the use of Minkowski functionals as a probe of cosmology beyond the regular method of two-point statistics.

Convergence and shear are briefly described in Section 2.3. In fact, utilizing the Fourier transform, they can be connected as

$$\langle \hat{\gamma}(\mathrm{k})\hat{\gamma}(\mathrm{k})\rangle = \langle \hat{\kappa}(\mathrm{k})\hat{\kappa}(\mathrm{k}')\rangle = (2\pi)^2 C(k)\delta_D(\mathrm{k}-\mathrm{k}'), \tag{54}$$

where the quantities marked with ^ indicate the corresponding values in the Fourier transformation.

Cosmic shear could be adopted to measure the coherent distortion of background galaxies, determine the matter–power spectrum, and illustrate the properties of dark matter. Technically, we represent the distance prefactor as

$$\frac{D_{LS}}{D_L D_S} \to \frac{\chi_S - \chi}{\chi_S \chi}, \tag{55}$$

so that (31) is equal to

$$\Psi(\theta) = \frac{2}{c^2}\int_0^{\chi_S} d\chi \frac{\chi_S - \chi}{\chi_S \chi}\Phi(\chi^\theta, \chi), \tag{56}$$

where χ is the co-moving angular–diameter distance.

From (28) and (56), considering the extended lens, we can deduce the convergence κ to be

$$\kappa = \frac{4\pi G}{c^2}\int_0^{\chi_S} d\chi \frac{(\chi_S - \chi)\chi}{\chi_S} a^2 \rho(\chi), \tag{57}$$

where ρ is the mass density.

Equation (57) means that κ is the geometrically weighted line-of-sight integral of ρ. Noticing that ρ and its cosmological mean value $\bar{\rho}$ are provided by

$$\rho = \bar{\rho}\delta, \tag{58}$$

$$\bar{\rho} = \frac{3H_0^2}{8\pi G}\Omega_{m0} a^{-3}, \tag{59}$$

where δ is the dimensionless density contrast, H_0 is the Hubble constant, and Ω_{m0} is the dimensionless matter–density parameter. Considering (57)–(59), the convergence κ is

$$\kappa = \frac{3}{2}\frac{H_0^2}{c^2}\Omega_{m0}\int_0^{\chi_S} d\chi \frac{\chi(\chi_S - \chi)}{\chi_S}\frac{\delta(\chi)}{a}. \tag{60}$$

It follows from (60) that cosmological parameters are closely related to the evolution of density contrast δ, which depends on local gravity and global geometry. Thus, cosmic shear can be used to explain the accelerating expansion of the universe, that is, whether it should be the existence of dark energy or the correction of gravitational theory.

By the linear growth factor $D_+(a)$ and the shape function $\mathcal{P}(k)$, the matter power spectrum $P_\delta(k)$ is $P_\delta(k) = \sigma_8^2 D_+^2(a)\mathcal{P}(k)$. It can be deduced that the power spectrum of the shear (convergence) $C_k(l)$ has the form that

$$C_\kappa(l) = \frac{9}{4}\left(\frac{H_0}{c}\right)^4 \Omega_{m0}^2 \sigma_8^2 \int_0^{\chi_S} d\chi \left[\frac{D_+(a)}{a}\frac{\chi(\chi_S - \chi)}{\chi_S}\right]^2 \mathcal{P}\left(\frac{l}{\chi}\right).$$

The convergence can be observed by considering the dark matter density contrast δ_m with the number density of galaxies n_g.

We can sketch the shape of a galaxy by the sheared two-point correlation function, which is expressed as

$$\xi_\pm(\theta) = \langle \gamma_t(x_i)\gamma_t(x_j)\rangle \pm \langle \gamma_r(x_i)\gamma_r(x_j)\rangle, \tag{61}$$

where $\theta = |x_i - x_j|$ is the angular distance between two galaxies. However, there are many observational effects, such as the image distortion caused by the camera's optical system, which would affect the final analysis and the determination of the galaxy shape. In the specific data analysis, multiple steps of correction are required. Therefore, it is necessary to understand the impact of systematic errors on the final result.

E/B modules are a common way to assess systematic errors. Observable signals fall into two categories, E-mode without rotation and B-mode sensitive to shear field rotation. By setting the two-dimensional vector field u as

$$u \equiv \nabla\kappa = \begin{pmatrix} \gamma_{1,1} + \gamma_{2,2} \\ \gamma_{2,1} - \gamma_{1,2} \end{pmatrix}, \tag{62}$$

the E-mode and B-mode can be defined by

$$\nabla^2\kappa^E \equiv \nabla \cdot u, \tag{63}$$

$$\nabla^2\kappa^B \equiv \nabla \times u = u_{2,1} - u_{1,2}, \tag{64}$$

$$\nabla^2\phi^{E,B} = 2\kappa^{E,B}. \tag{65}$$

where (65) provides both E-mode and B-mode potentials.

B-mode can be used to determine the system residual values. There are many factors that contribute to the creation of B-molds, such as physical factors and artificial factors. For example, the PSF correction with some sort of approximation will result in a B-mode. Another typical cause is the existence of intrinsic alignment, classified into II correlation and GI correlation, which will produce a B-mode. The B-mode may decrease as the number and density of the background galaxies increases. The following link shows the E/B modes: https://www.cosmostat.org/wp-content/uploads/2017/04/EBmode2.png (accessed on 15 January 2023).

4.2. Lensing Signals

The lensing signal can be used to estimate the cosmological parameters and mass distribution of galaxies [54]. It is a focus of current applications. Hamana et al. [55] studied the dilution effect of galaxy cluster members and foreground galaxies on weak lensing signals from galaxy clusters in HSC surveys. Lombardi et al. [56] detected clear weak lens signals in i (F775W) and z (F850LP) filters for the first time at $z > 1$ using the Hubble Space Telescope (HST).

It is not easy to accurately extract the lens signal, and a lot of information is often discarded at the measuring and preprocessing stage. It is often said that we only observe what we want to observe, and we agree. Here, we briefly describe the basic mathematical methods for measuring signals.

After a series of preprocessing on the collected images, such as background removal, source detection, cosmic rays, and bad pixel identification, we collect shear measurements of the galaxy. Theli integrates the functions of multiple image processing softwares (e.g., sextractor and scmap) to find galaxies and stars on CCDs, identify cosmic rays, measure astrometry and magnitude, and so on. Theli's strategy for better background removal is to model the background twice. The first time is to use the median value of the pixel reading on the CCD to perform the overall background removal. The second time is to find all the sources on the CCD and establish a background model again after subtracting these sources. Once the detection of the source and background removal are complete, astrometry is required. By matching the stars and reference stars in the CCD, the linear term and nonlinear term coefficients required to transition from CCD coordinates to celestial coordinates can be determined.

We need to separate the stars and galaxies once the sources on the CCD are ascertained. According to the PSF effect, the ellipticity of the source is generally not 0. Although we deconvolve the detected source, there still may be a residual PSF effect on the image.

This effect statistically attenuates the galaxy's gravitational lensing signal. The Fourier quadrupole moment method can be adopted to solve this problem. Due to the different statistical methods employed to calculate the shear signal, the presence of a star has no theoretical effect on the weak lens signal. See Zhang et al. for details [57].

It is time to measure the shear after the separation of the star from the galaxy. Several methods have been proposed, such as KSB [58], Lensfit [59], and the classical method proposed by U. Seljak and M. Zaldarriaga [60]. Their idea is to provide a shear estimate evolving the partial derivative of the galaxy's brightness to the position. In Fourier space, considering the PSF effect, the two components of the shear signal g_1 and g_2 can be provided by three shear quantities: G_1, G_2, and N. Precisely,

$$\frac{\langle G_1 \rangle}{\langle N \rangle} = -g_1 + O\left(g_{1,2}^3\right), \tag{66}$$

$$\frac{\langle G_2 \rangle}{\langle N \rangle} = -g_2 + O\left(g_{1,2}^3\right). \tag{67}$$

More information can be referred to [61].

Perturbated by various factors, including geometric and physical ones, the telescope twists the original shape of an object during the imaging. The image field distortion effect is involved in the change of the source at the CCD position. If the detected source is a point and if it is an extended source, the image field distortion is also reflected in the change in the shape of the source. In addition, the weak lens formula can be used to calculate image field distortion [62], and the image field distortion can also be used to check the accuracy of galaxy deformation measurements.

4.3. Dark Energy and Dark Matter

In 1998, observations of Type Ia supernovae found that the brightness of the more distant supernova was dimmer [63], indicating that the expansion of the universe is accelerating. There are still two popular explanations for this phenomenon, namely, the existence of dark energy with negative pressure and the modified theory of gravity. Since the introduction of dark energy, a large number of dark energy models have been proposed, such as RDE [64] and HDE [65]. The ΛCDM model is widely popular [66–68], and its corresponding dark energy equation of state is

$$w(a) \equiv \frac{p_{de}}{\rho_{de}} = -\frac{1}{3}\frac{d\ln\rho_e}{d\ln a} - 1. \tag{68}$$

We can regard weak gravitational lensing signals near low-density regions as cosmological probes to characterize the dark energy equation of state. This is because the low-density region is less affected by the nonlinear evolution of matter and the behavior of the baryon matter. Hence, it is an ideal region to test the dark energy models. According to the different definitions of low-density areas, the corresponding measurement methods are also distinct. Using the article Gruen et al. [69] as an example, they smoothed the density distribution of the foreground galaxies based on the photometric redshift of bright red galaxies (redMaGiC), and obtained lensing signal measurements with high signal-to-noise ratios.

When performing numerical simulations, it is normal to limit the parameters using CMB measurements. Plenty of modified cosmological models involving more variable parameters can be proposed.

Since Zwicky proposed the dark matter, several observations have supported this hypothesis, such as V. Rubin et al.'s measurements of the rotation curves of the Spiral Galaxy [70], and observations of the Bullet Cluster [71]. The density perturbation in the universe enters a nonlinear evolution stage at a redshift of about 10, in which the density disturbance intensity is higher than the critical linear density contrast δ_c and the surrounding matter collapses under gravity to form a dark matter halo. The dark halo mass function is a good cosmological probe. This is because, under the deduction of linear theory, a region collapses to form a self-binding gravitational system when the density perturbation of it reaches $delta_c = 1.69$. Therefore, considering the distribution of matter in the universe to be a density perturbation field, we can predict the number of halos that this density field will produce at any time, as long as the initial density field is deduced linearly and the density of perturbations greater than $delta_c$ is counted [72]. The number of halos with a mass of $[M, M + dM]$ and a redshift of $[z, z + dz]$ can be calculated as

$$\frac{d^2N}{d\Omega dz} = \frac{r^2(z)}{H(z)} \frac{dn(M,z)}{dM} dM, \tag{69}$$

where $n(M,z)$ is number density, which could be calibrated from numerical simulations.

The use of weak gravitational lensing signals to reconstruct the mass function of dark halos stimulates gravitational lensing to be an important tool in the research of dark matter halos. Assuming that the halo has a spherically symmetric density profile, NFW could be utilized to fit the gravitational lensing signal around the dark halo. The mass function of the halo can also be measured using an N-body numerical simulation. The accurate measurement of the halo mass function is crucial for cosmological research, and may provide more abundant and more accurate cosmological models.

Nowadays, the detection of dark matter and dark energy is developing, such as in the detection of dark matter CDMS [73], AMS [74], and the research of dark energy PLANCK [75] and BIGBOSS [76]. Astronomical observations can determine dark matter from small- to large-scale structures. The evolution of the dark energy equation of state can be investigated once the data have been accumulated and analyzed sufficiently. However, there are still specifics to improve to overcome the limitations of detection such as redshift information and supernova data. This will be a field of great value in the future.

Funding: This research was funded by National Natural Science Foundation of China grant number 12001099 and 12271093.

Data Availability Statement: No new data were created or analyzed in this study. Data sharing is not applicable to this article.

Conflicts of Interest: The authors declare no conflict of interest.

References

1. Refsdal, S.; Surdej, J. Gravitational lenses. *Rep. Prog. Phys.* **1994**, *57*, 117–185. [CrossRef]
2. Walsh, D.; Carswell, R.F.; Weymann, R.J. 0957+561 A, B: Twin quasistellar objects or gravitational lens? *Nature* **1979**, *279*, 381–384. [CrossRef]
3. Moller, O.; Blain, A.W. Strong gravitational lensing by multiple galaxies. *Mon. Not. R. Astron. Soc.* **2001**, *327*, 339–349. [CrossRef]
4. Treu, T. Strong Lensing by Galaxies. *Annu. Rev. Astron. Astrophys.* **2010**, *48*, 87–125. [CrossRef]
5. Castro, P.G.; Heavens, A.F.; Kitching, T.D. Weak lensing analysis in three dimensions. *Phys. Rev. D* **2005**, *72*, 023516. [CrossRef]
6. Watanabe, K.; Sasaki, M.; Tomita, K. On distances and the Hubble parameter determination in gravitational lenses. *Astrophys. J.* **1992**, *394*, 38–50. [CrossRef]
7. Jee, I.; Suyu, S.H.; Komatsu, E.; Fassnacht, C.D.; Hilbert, S.; Koopmans, L.V.E. A measurement of the Hubble constant from angular diameter distances to two gravitational lenses. *Science* **2019**, *365*, 1134–1138. [CrossRef]
8. Bohn, A.; Throwe, W.; Herbert, F.; Henriksson, K.; Bunandar, D.; Taylor, N.W.; Scheel, M.A. What does a binary black hole merger look like? *Class. Quantum Gravity* **2015**, *32*, 065002. [CrossRef]
9. Lu, X.; Xie, Y. Probing an Extended Uncertainty Principle black hole with gravitational lensings. *Mod. Phys. Lett. A* **2019**, *34*, 1950152. [CrossRef]
10. Sharif, M.; Iftikhar, S. Strong gravitational lensing in non-commutative wormholes. *Astrophys. Space Sci.* **2015**, *357*, 85. [CrossRef]
11. Ono, T.; Ishihara, A.; Asada, H. Deflection angle of light for an observer and source at finite distance from a rotating wormhole. *Phys. Rev. D* **2018**, *98*, 044047. [CrossRef]
12. Takahashi, R.; Asada, H. Observational upper bound on the cosmic abundances of negative-mass compact objects and ellis wormholes from the Sloan Digital Sky Survey Quasar Lens Search. *Astrophys. J. Lett.* **2013**, *78*, L16. [CrossRef]
13. Ovgun, A. Light deflection by Damour-Solodukhin wormholes and Gauss–Bonnet theorem. *Phys. Rev. D* **2018**, *98*, 044033. [CrossRef]

14. Perlick, V.; Tsupko, O. Calculating black hole shadows: Review of analytical studies. *Phys. Rep.-Rev. Sec. Phys. Lett.* **2022**, *947*, 1–39. [CrossRef]
15. Akiyama, K.; Alberdi, A.; Alef, W.; Asada, K.; Azulay, R.; Baczko, A.-K.; Ball, D.; Balokovic, M.; Barrett, J. First M87 event Horizon Telescope Results. I. the shadow of the supermassive black hole. *Astrophys. J. Lett.* **2019**, *875*, L1.
16. Akiyama, K.; Alberdi, A.; Alef, W.; Algaba, J.C.; Anantua, R.; Asada, K.; Azulay, R.; Bach, U.; Baczko, A.-K. First Sagittarius A* Event Horizon Telescope Results. I. The Shadow of the Supermassive Black Hole in the Center of the Milky Way. *Astrophys. J. Lett.* **2022**, *930*, L12.
17. Cao, Z.J.; Li, L.F.; Wang, Y. Gravitational lensing effects on parameter estimation in gravitational wave detection with advanced detectors. *Phys. Rev. D* **2014**, *90*, 062003. [CrossRef]
18. Oguri, M. Effect of gravitational lensing on the distribution of gravitational waves from distant binary black hole mergers. *Mon Not. R. Astron. Soc.* **2018**, *480*, 3842–3855. [CrossRef]
19. Petrillo, C.E.; Tortora, C.; Chatterjee, S.; Vernardos, G.; Koopmans, L.V.E.; Verdoes Kleijn, G.; Napolitano, N.R.; Covone, G.; Kelvin, L.S.; Hopkins, A.M. Testing convolutional neural networks for finding strong gravitational lenses in KiDS. *Mon. Not. R. Astron. Soc.* **2019**, *482*, 807–820. [CrossRef]
20. Su, J.; Wang, W.G.; Wang, X.; Song, F. Simulation of the Gravitational Lensing Effect of Galactic Dark Matter Halos Using 3D Printing Technology. *Phys. Teach.* **2019**, *57*, 590–593. [CrossRef]
21. Hanson, D.; Challinor, A.; Lewis, A. Weak lensing of the CMB. *Gen. Relativ. Gravit.* **2010**, *42*, 2197–2218. [CrossRef]
22. Bartelmann, M. Gravitational lensing. *Class. Quantum Gravity* **2010**, *27*, 233001. [CrossRef]
23. Futamase, T. Gravitational lensing in cosmology. *Int. J. Mod. Phys. D* **2015**, *24*, 1530011. [CrossRef]
24. Cunha, P.V.P.; Herdeiro, C.A.R. Shadows and strong gravitational lensing: A brief review. *Gen. Relativ. Gravit.* **2018**, *50*, 42. [CrossRef]
25. Takizawa, K.; Ono, T.; Asada, H. Gravitational deflection angle of light: Definition by an observer and its application to an asymptotically nonflat spacetime. *Phys. Rev. D* **2020**, *101*, 104032. [CrossRef]
26. Ishihara, A.; Suzuki, Y.; Ono, T.; Kitamura, T.; Asada, H. Gravitational bending angle of light for finite distance and the Gauss–Bonnet theorem. *Phys. Rev. D* **2016**, *94*, 084015. [CrossRef]
27. Takizawa, K.; Ono, T.; Asada, H. Gravitational lens without asymptotic flatness: Its application to the Weyl gravity. *Phys. Rev. D* **2020**, *102*, 064060. [CrossRef]
28. Gibbons, G.W.; Werner, M.C. Applications of the Gauss–Bonnet theorem to gravitational lensing. *Class. Quantum Gravity* **2008**, *25*, 235009. [CrossRef]
29. Werner, M.C. Gravitational lensing in the Kerr-Randers optical geometry. *Gen. Relativ. Gravit.* **2012**, *44*, 3047–3057. [CrossRef]
30. Halla, M.; Perlick, V. Application of the Gauss–Bonnet theorem to lensing in the NUT metric. *Gen. Relativ. Gravit.* **2020**, *52*, 112. [CrossRef]
31. Kapsabelis, E.; Kevrekidis, P.G.; Stavrinos, P.C.; Triantafyllopoulos, A. Schwarzschild-Finsler-Randers spacetime: Geodesics, dynamical analysis and deflection angle. *Eur. Phys. J. C* **2022**, *82*, 1098. [CrossRef]
32. Kostelecky, V.A. Riemann-Finsler geometry and Lorentz-violating kinematics. *Phys. Lett. B* **2011**, *701*, 137–143. [CrossRef]
33. Shen, B. Gravitational Lensing Effect in the Universe with a Finsler Background. *arXiv* 2023, *submitted*.
34. Asada, H. Images for a binary gravitational lens from a single real algebraic equation. *Astron. Astrophys.* **2002**, *390*, L11. [CrossRef]
35. Bozza, V. Comparison of approximate gravitational lens equations and a proposal for an improved new one. *Phys. Rev. D* **2008**, *78*, 103005. [CrossRef]
36. Takizawa, K.; Asada, H. Gravitational lens on de Sitter background. *Phys. Rev. D* **2022**, *105*, 084022. [CrossRef]
37. Izumi, K.; Hagiwara, C.; Nakajima, K.; Kitamura, T.; Asada, H. Gravitational lensing shear by an exotic lens object with negative convergence or negative mass. *Phys. Rev. D* **2013**, *88*, 024049. [CrossRef]
38. Hezaveh, Y.D.; Holder, G.P. Effects of strong gravitational lensing on millimeter-wave galaxy number counts. *Astrophys. J.* **2011**, *734*, 52. [CrossRef]
39. Chagoya, J.; Ortiz, C.; Rodriguez, B.; Roque, A.A. Strong gravitational lensing by DHOST black holes. *Class. Quantum Gravity* **2021**, *38*, 075026. [CrossRef]
40. Wei, S.-W.; Liu, Y.-X.; Fu, C.-E.; Yang, K. Strong field limit analysis of gravitational lensing in Kerr-Taub-NUT spacetime. *J. Cosmol. Astropart. Phys.* **2012**, *10*, 053. [CrossRef]
41. Jin, X.H.; Gao, Y.X.; Liu, D.J. Strong gravitational lensing of a 4-dimensional Einstein–Gauss–Bonnet black hole in homogeneous plasma. *Int. J. Mod. Phys. D* **2020**, *29*, 2050065. [CrossRef]
42. Bozza, V.; Scarpetta, G. Strong deflection limit of black hole gravitational lensing with arbitrary source distances. *Phys. Rev. D* **2007**, *76*, 083008. [CrossRef]
43. Bozza, V. Gravitational lensing in the strong field limit. *Phys. Rev. D* **2002**, *66*, 103001. [CrossRef]
44. Alchera, N.; Bonici, M.; Cardinale, R.; Domi, A.; Maggiore, N.; Righi, C.; Tosi, S. Analysis of the Angular Dependence of Time Delay in Gravitational Lensing. *Symmetry* **2018**, *10*, 246. [CrossRef]
45. Hojjati, A.; Kim, A.G.; Linder, E.V. Robust strong lensing time delay estimation. *Phys. Rev. D* **2013**, *87*, 123512. [CrossRef]
46. Mandelbaum, R. Instrumental systematics and weak gravitational lensing. *J. Instrum.* **2015**, *10*, C05017. [CrossRef]
47. Ghosh, B.; Durrer, R.; Schafer, B.M. Intrinsic and extrinsic correlations of galaxy shapes and sizes in weak lensing data. *Mon. Not. R. Astron. Soc.* **2021**, *505*, 2594–2609. [CrossRef]

48. Fleury, P.; Larena, J.; Uzan, J.P. Weak Gravitational Lensing of Finite Beams. *Phys. Rev. Lett.* **2017**, *119*, 191101. [CrossRef] [PubMed]
49. Van Waerbeke, L. Shear and magnification: Cosmic complementarity. *Mon. Not. R. Astron. Soc.* **2010**, *401*, 2093–2100. [CrossRef]
50. Heydenreich, S.; Schneider, P.; Hildebrandt, H.; Asgari, M.; Heymans, C.; Joachimi, B.; Kuijken, K.; Lin, C.-A.; Troster, T.; van den Busch, J.L. The effects of varying depth in cosmic shear surveys. *Astron. Astrophys.* **2020**, *634*, A104. [CrossRef]
51. Taylor, P.L.; Bernardeau, F.; Huff, E. x-cut Cosmic shear: Optimally removing sensitivity to baryonic and nonlinear physics with an application to the Dark Energy Survey year 1 shear data. *Phys. Rev. D* **2021**, *103*, 043531. [CrossRef]
52. Nicola, A.; Dunkley, J.; Spergel, D.N. Joint cosmology and mass calibration from thermal Sunyaev-Zel'dovich cluster counts and cosmic shear. *Phys. Rev. D* **2020**, *102*, 083505. [CrossRef]
53. Mancini, A.S.; Taylor, P.L.; Reischke, R.; Kitching, T.; Pettorino, V.; Schafer, B.M.; Zieser, B.; Merket, M. 3D cosmic shear: Numerical challenges, 3D lensing random fields generation, and Minkowski functionals for cosmological inference. *Phys. Rev. D* **2018**, *98*, 103507. [CrossRef]
54. Lewis, A.; Challinor, A. Weak gravitational lensing of the CMB. *Phys. Rep.-Rev. Sec. Phys. Lett.* **2006**, *429*, 1–65. [CrossRef]
55. Hamana, T.; Shirasaki, M.; Lin, Y.T. Weak-lensing clusters from HSC survey first-year data: Mitigating the dilution effect of foreground and cluster-member galaxies. *Publ. Astron. Soc. Jpn.* **2020**, *72*, 78. [CrossRef]
56. Lombardi, M.; Rosati, P.; Blakeslee, J.P.; Ettori, S.; Demarco, R.; Ford, H.C.; Illingworth, G.D.; Clampin, M.; Hartig, G.F.; Benítez, N. Hubble Space Telescope ACS weak-lensing analysis of the galaxy cluster RDCS 1252.9-2927 at z=1.24. *Astron. J.* **2005**, *623*, 42. [CrossRef]
57. Lu, T.; Zhang, J.; Dong, F.; Li, Y.; Liu, D.; Fu, L.; Li, G.; Fan, Z. Testing PSF Interpolation in Weak Lensing with Real Data. *Astron. J.* **2017**, *153*, 197. [CrossRef]
58. Erben, T.; Waerbeke, L.; Bertin, E.; Mellier, Y.; Schneider, P. How accurately can we measure weak gravitational shear? *Astron. Astrophys.* **2001**, *366*, 717–735. [CrossRef]
59. Kannawadi, A.; Hoekstra, H.; Miller, L.; Viola, M.; Conti, I.F.; Herbonnet, R.; Erben, T.; Heymans, C.; Hildebrandt, H.; Kuijken, K. et al. Towards emulating cosmic shear data: Revisiting the calibration of the shear measurements for the Kilo-Degree Survey. *Astron. Astrophys.* **2019**, *624*, A92. [CrossRef]
60. Seljak, U.; Zaldarriaga, M. Measuring dark matter power spectrum from cosmic microwave background. *Phys. Rev. Lett.* **1999**, *82*, 2636–2639. [CrossRef]
61. Zhang, J.; Zhang, P.J.; Luo, W.T. Approaching the CRAMER-RAO bound in weak lensing with PDF symmetrization. *Astron. J.* **2017**, *834*, 8. [CrossRef]
62. Zhang, J.; Dong, F.; Li, H.; Li, X.; Li, Y.; Liu, D.; Luo, W.; Fu, L.; Li, G.; Fan, Z. Testing Shear Recovery with Field Distortion. *Astron. J.* **2019**, *875*, 48. [CrossRef]
63. Gomez, G.; Lopez, R. The Canarias Type Ia Supernova Archive. II. A standard spectral evolution sequence. *Astron. J.* **1998**, *115*, 1096. [CrossRef]
64. Kim, K.Y.; Lee, H.W.; Myung, Y.S. On the Ricci dark energy model. *Gen. Relativ. Gravit.* **2011**, *43*, 1095–1101. [CrossRef]
65. Saadat, H.; Saadat, A.M. Time-Dependent Dark Energy Density and Holographic DE Model with Interaction. *Int. J. Theor. Phys.* **2011**, *50*, 1358–1366. [CrossRef]
66. Mukhopadhyay, U.; Ray, S.; Choudhury, S.B.D. Lambda-CDM universe: A phenomenological approach with many possibilities. *Int. J. Mod. Phys. D* **2008**, *17*, 301–309. [CrossRef]
67. Stachowski, A.; Szydlowski, M. Dynamical system approach to running Lambda cosmological models. *Eur. Phys. J. C* **2016**, *76*, 606. [CrossRef]
68. Yang, R.J.; Zhang, S.N. The age problem in the ΛCDM model. *Mon. Not. R. Astron. Soc.* **2010**, *407*, 1835–1841. [CrossRef]
69. Gruen, D.; Friedrich, O.; Amara, A.; Bacon, D.; Bonnett, C.; Harley, W.; Jain, B.; Jarvis, M.; Kacprzak, T.; Krause, E.; et al. Weak lensing by galaxy troughs in DES Science Verification data. *Mon. Not. R. Astron. Soc.* **2016**, *455*, 3367–3380. [CrossRef]
70. Rubin, V.C.; Ford, W.K. Rotation of andromeda nebula from a spectroscopic survey of emission regions. *Astrophys. J.* **1970**, *159*, 379. [CrossRef]
71. Thompson, R.; Dave, R.; Nagamine, K. The rise and fall of a challenger: The Bullet Cluster in Lambda cold dark matter simulations. *Mon. Not. R. Astron. Soc.* **2015**, *452*, 3030–3037. [CrossRef]
72. Press, W.H.; Schechter, P. Formation of galaxies and clusters of galaxies by self-similar gravitational condensation. *Astron. J.* **1974**, *187*, 425–438. [CrossRef]
73. Farina, M.; Pappadopulo, D.; Strumia, A. CDMS stands for Constrained Dark Matter Singlet. *Phys. Lett. B* **2010**, *688*, 329–331. [CrossRef]
74. Di Mauro, M.; Donato, F.; Fornengo, N.; Vittino, A. Dark matter vs. astrophysics in the interpretation of AMS-02 electron and positron data. *J. Cosmol. Astropart. Phys.* **2016**, *20116*, 031. [CrossRef]

75. Salvatelli, V.; Marchini, A.; Lopez-Honorez, L.; Mena, O. New constraints on coupled dark energy from the Planck satellite experiment. *Phys. Rev. D* **2013**, *88*, 023531. [CrossRef]
76. Stril, A.; Cahn, R.N.; Linder, E.V. Testing standard cosmology with large-scale structure. *Mon. Not. R. Astron. Soc.* **2010**, *404*, 239–246. [CrossRef]

universe

MDPI

Article

Resolution of Cosmological Singularity in Hořava–Lifshitz Cosmology

Ewa Czuchry

National Centre for Nuclear Research, Pasteura 7, 02-093 Warsaw, Poland; ewa.czuchry@ncbj.gov.pl

Abstract: The standard ΛCDM model, despite its agreement with observational data, still has some issues unaddressed, such as the problem of initial singularity. Solving that problem usually requires modifications of general relativity. However, there appeared the Hořava–Lifshitz (HL) theory of gravity, in which equations governing cosmological evolution include a new term scaling similarly as the dark radiation term in the Friedmann equations, enabling a bounce of the universe instead of initial singularity. This review describes past works on the stability of such a bounce in different formulations of HL theory, an initial detailed balance scenario, and further projectable versions containing higher than quadratic terms to the original action.

Keywords: modified gravity; initial singularity; cosmology

1. Introduction

Classical general relativity (GR), apart from its simple beauty and symmetry, has also been strongly confirmed in several experimental tests. However, it does not explain many issues such as dark matter or space-time singularities, including the initial one in cosmology and those inside black holes. In order to answer these issues, there have been many attempts to modify GR both at the classical and quantum level. Specifically, the quantisation of GR cosmology was supposed to resolve the initial singularity problem.

Attempts to quantise gravity could be divided into two categories. One category involved assuming the classical theory of gravity and quantising it in various manners, with the first attempts performed via the covariant quantum gravity. In that classical approach, one repeats the method that was successful in quantising electrodynamics, namely considering the path integral of the Hilbert–Einstein action and then calculating the perturbation of the metric around a background one. The obtained equations, unlike in electrodynamics, are non-renormalizable in higher energies. Canonical quantum gravity considers the ADM $(3+1)$-decomposition of space-time and the quantisation of the constraints obtained from the Hamiltonian. Other attempts have included sophisticated theories such as string theory and loop quantum gravity. These theories manage to solve some problems (such as the cosmological singularity [1]), but they are difficult to phenomenologically test [2,3]. The author of the present paper and colleagues used a combination of coherent states and Weyl quantisation in order to resolve an initial singularity problem; however, at this moment, the obtained models are difficult to validate by observational data [4].

Although there is still no full theory of quantum gravity that has been developed, it is supposed to manifest beyond a characteristic energy scale for quantum gravity $E_{Pl} = \sqrt{\hbar c^5/G}$ built in terms of the speed of light c, the gravitational constant G, and Planck's constant $\hbar$. Therefore, there is the second research direction, which aims to construct a modified version of GR with an improved UV behaviour. General relativity, after many tests performed, seems to be consistent with all current observations. This makes it a very good IR limit of a potential quantum gravity model. Some proposals have been made for UV completions of general relativity in the past [5,6]. They have one thing in common, namely the existence of some cutoff energy scale beyond which quantum effects could be detected, specially for a cutoff energy in the range of TeV. The widely discussed recent proposal is Hořava gravity, which is a proposal of a UV complete theory of gravity. It seems to be renormalizable at high energies, which makes it a candidate for a quantum gravity model [7,8]. The action of this theory contains additional higher order spatial derivatives, and therefore, the theory loses the full diffeomorphism invariance, keeping the (1 + 3) foliation preserving diffeomorphism. Moreover, there is a UV fixed point in this gravity model where there is an anisotropic Lifshitz scaling between time and space. Therefore, the resulting theory is called Hořava–Lifshitz (HL) gravity.

Significant work has been carried out on this theory examining different aspects and properties [9–22]. Many studies were devoted to cosmological solutions [12,19,23], including quantum cosmological ones [1,24–29], braneworlds, and dark radiation [12,21]. Hořava–Lifshitz cosmology obtained a novel feature enabling the existence of bounce instead of the initial singularity predicted by classical GR. There has also been other research focused on finding specific solutions, including black holes and their properties, and many works devoted to phenomenological aspects both astrophysical and concerning dark matter.

A derivation of Hořava–Lifshitz cosmology [12,19,23] made via varying-action written Friedmann–Robertson–Walker space-time metrics resulted in equations analogous to the standard Friedmann ones. These equations contain a new term that scales similarly as dark radiation [12,19,21], i.e., $\sim 1/a^4$ (where a is a scale factor) and provides a negative contribution to energy density. This feature enables obtaining non-singular cosmological evolution, resolving the initial singularity problem [14,21,30]. Such a possibility not only results in avoiding the initial singularity but may have other consequences for potential histories of the universe, such as the scenario of contraction from an infinite size connected by a bounce to expansion to infinite size again or eternal cycles of a similar scenario.

Despite many promises made by this modified theory of gravity, it seems that it contains instabilities and pathologies in different formulations (see, e.g., [22,31–33]). The original Hořava formulation suffers from many problems: the existence of ghost instabilities and strong coupling at IR [10,34], the appearance of a term that violates parity [22], very large values and negative signs of the cosmological constant [35,36], and issues with power-counting renormalisation of the propagation of the scalar mode [13,37]. Some of these problems might be solved by performing an analytic continuation of the parameters of the theory [20].

In the original Hořava formulation, it is assumed via the so-called detailed balance condition that a potential part of the action is derived from the so-called superpotential, which limits the big number of its terms and corresponding independent couplings. Another imposed condition is the demand of projectability, used in a standard cosmology. It requires that lapse function N depends only on time $N = N(t)$. It might seem that this condition is too strict, but on the other hand, it seems that the non-projectable version of Hořava gravity results in a serious strong coupling problem ([34]) and does not possess a valid GR limit at IR. However, some authors [35,38] claim the opposite, proposing adding additional terms to the superpotential (not to the action, thus still keeping detailed balance or eventually softly breaking it) and relaxing projectability. Nonetheless, subsequent works demonstrated that it caused problems with the scalar mode power-counting renormalizability.

One of the simplest models with the detailed balance condition relaxed is the Sotiriou–Visser–Weinfurtner (SVW) generalisation [22]. This version of HL gravity assumes a gravitational action containing terms not only quadratic in curvature but also cubic, as was suggested already in [12,19]. This model still maintains the projectability condition. Generalised Friedmann equations obtained from varying such an action contain not only a dark radiation term $\sim 1/a^4$ but also terms scaling with the $\sim 1/a^6$ term. These new terms, although negligible at large values of a, become dominating at small ones and might modify or cancel bounce solutions. Specifically, as it has the opposite sign as the $1/a^4$ term, it may compensate for the dark radiation term at small scales and result in singular solutions. A similar scenario arrives in the HL gravity with the softly broken detailed balance condition and negative spatial curvature [39].

Nonetheless, the issue of the initial singularity still remains one of the key questions of early universe cosmology. The possibility that it might be avoided in a modified gravity and replaced by a bounce is a very promising feature. In this review, we are going to present the result of the research [14,15] performed via phase portrait techniques on the occurrence and stability of the bounce in the two simplest formulations of HL cosmology: the original one with the imposed detailed balance (DB) condition and the beyond detailed balance (BDB) formulation relaxing this condition. As additional terms in analogs of Friedman equations are proportional to the curvature parameter $K = \{-1, 0, 1\}$, only non-flat cosmologies with $K = \pm 1$ allow the existence of a bounce and the existence of non-singular solutions.

The paper [14] was our first attempt at analysing the possibilities of a cosmological bounce in the original formulation of the theory. The matter sector was described there in terms of a scalar field with a potential given by a quadratic power of that field. However, that work included only a simplified version of the theory with either a vanishing cosmological constant Λ or an HL universe with a non-zero Λ in the region of a small scale factor a.

A more general approach that is easier to fit with observational data is the hydrodynamical approach used in [15], where the matter sector is described in terms of the density ρ and pressure p. In the latter work, it was assumed that w, providing the relation between density and pressure in the equation of state, is constant, which is at some level an idealisation and simplification. At the moment, we do not have the history of the HL universe constructed in a similar way as in the standard ΛCDM model, where we have phases and epochs containing different matter or radiation sectors. Therefore, as we still have a limited understanding on the physical aspects of the theory and its parameters, current research rather describes different analytical possibilities, not some exact physical solutions.

It is important to mention that currently, HL theory and its extensions are not ruled out observationally (although there are tight bounds on some parameters [40]); thus, further observational constraints could shed new light on different specific scenarios or on the whole model, providing a better justification for deeper theoretical research. Several papers have placed bounds on different regions of the Hořava–Lifshitz framework, for example, using cosmological data [16], binary pulsars [41,42], and in the context of dark energy [43]. This has also occurred [18] in the effective field theory formalism of the the extended version of Hořava theory [31].

In [44,45], we investigated the same two basic Hořava–Lifshitz scenarios described earlier in the Introduction as a background theory for further numerical calculations. These calculations were devoted to fit observational data to two analogs of the Friedmann equations that arise in the mentioned HL scenarios. We did not limit our considerations to a flat model but left the curvature density parameter as a free parameter. In those papers, we provided improved observational constraints based on the recent cosmological data set from the cosmic microwave background (Planck CMB) [46], expansion rates of elliptical and lenticular galaxies [47], JLA compilation (joint light-curve analysis) data for Type Ia supernovae (SneIa) [48], Baryon acoustic pscillations (BAO) [49–51], and priors on the Hubble parameter [52].

Consequently, this review, on the one hand, presents our analytical studies of an HL universe aimed towards the possible existence of Big Bounce. On the other hand, we discuss the analytically obtained conditions on HL cosmology parameters that lead to the bounce and its corresponding observational constraints arrived at in our mentioned papers [44,45].

This paper is organised as follows: we first give a brief overview of HL cosmology in both scenarios under consideration in Section 2. In Section 3, the possibility of bounce in both formulations is discussed. Section 4 contains a derivation and description of the phase portraits of the HL cosmology with the imposed condition of detailed balance, while in Section 5, this condition is released. Section 6 contains a summary of results on the possibility of a bounce in HL cosmology. In Section 7, we discuss the possibility of a bounce, taking into account observational constraints, and present the limitation of the underlying theory.

2. Hořava–Lifshitz Cosmology

The main obstacle in quantising gravity is that general relativity in its classical formulation is non-renormalisable. This might be visualised by expanding some quantity $\mathcal{F}$ with respect to the gravitational constant [33] as follows:

$$\mathcal{F} = \sum_{n=0}^{\infty} a_n \left(G_N E^2\right)^n. \tag{1}$$

Here, E is the energy of the system, a_n denotes a numerical coefficient, and G_N is the gravitational coupling constant. Therefore, $E^2 \geq G^{-1}$ and the expansion above diverges. Consequently, as demonstrated, general relativity is not perturbatively renormalisable in the high energy regimes.

There have been many studies pointing out that the ultraviolet behaviour of general relativity might be improved by including higher-order derivatives in the standard gravitational metric. The latter is the Einstein–Hilbert action:

$$S = \int d^4x \sqrt{g} R, \tag{2}$$

where d^4x denotes the volume element of space-time, g is its metric matrix's determinant, and R is a scalar curvature. Including-higher order terms of the derivatives of the metric provides the following action:

$$S = \int d^4x \sqrt{g}(R + f(R_{\mu\nu}R^{\mu\nu})). \tag{3}$$

The additional terms, containing different derivatives of R, $R_{\mu\nu}$, etc., change the graviton propagator from $1/k^2$ into $1/(k^2 - G_N k^4)$ [7,8]. The propagator part proportional to k^{-4} cancels the ultraviolet divergence. However, the resulting theory has time derivatives of $\mathcal{O} > 2$ and is therefore non-unitary. Moreover, it possesses a spin-2 ghost with a non-zero mass [33] and derived form for which action field equations are of the fourth order.

The novel idea of Hořava [7] was to construct a higher-order theory of gravity breaking the Lorentz invariance in the ultraviolet range. In his theory, only the *spatial* derivatives are of $\mathcal{O} > 2$, which evaded the ghost. However, it is necessary that any theory of gravity should be consistent with all current experiments that have not detected any significant violation of Lorentz invariance. Thus, it is necessary to restore the Lorentz invariance in the infrared limit. In order to overcome this problem, Hořava proposed an anisotropic scaling of space and time at high-UV energies, which is known as Lifshitz scaling. In a 4-dimensional space-time, this scaling takes the form:

$$t \to b^{-z} t,\; x^i \to b^{-1} x^i, \tag{4}$$

where $i = 1, 2, 3$, and z is a critical exponent. Lorentz invariance is restored when $z = 1$, but the power-counting renormalizability demands $z \geq 3$ [33], so usually $z = 3$ is assumed. Therefore, the resulting theory is called Hořava–Lifshitz (HL) gravity. Lorentz symmetry is here broken down to transformations $t \to \xi_0(t)$, $x^i \to \xi^i(t, x^k)$, preserving the spatial diffeomorphisms unlike the full space-time diffeomorphism invariance of GR. Thus, such a theory acquires a symmetry, preserving a space-time foliation [7,33], where on each constant time hypersurface, there are allowed arbitrary changes of the spatial coordinates.

Preservation of a space-time foliation and anisotropic scaling between time and space and time introduces the ADM (1 + 3)decomposition of the space-time. The standard ADM metrics in a preferred foliation and with a $(-+++)$ signature are as follows:

$$ds^2 = -N^2 dt^2 + g_{ij}(dx^i + N^i dt)(dx^j + N^j dt). \tag{5}$$

The dynamics are now described in terms of the lapse function N, the shift vector N^i, and the spatial metric g_{ij} ($i, j = 1, 2, 3$). The most general action for such theory can be written as:

$$S = \int d^3x dt\, N\sqrt{g}\left[K^{ij}K_{ij} - \lambda K^2 - \mathcal{V}(g_{ij})\right]. \tag{6}$$

Here, g denotes the determinant of the spatial metric g_{ij}, λ is a dimensionless running coupling constant, $\mathcal{V}$ is a potential term, and K is a trace of the extrinsic curvature of the spatial 3-dimensional hypersurface K_{ij}:

$$K_{ij} = \frac{1}{2N}(\dot{g}_{ij} - \nabla_i N_j - \nabla_j N_i). \tag{7}$$

An overdot denotes a derivative with respect to the time coordinate t. The trace of K_{ij} is K. The potential $\mathcal{V}$ is invariant only under three-dimensional diffeomorphisms [31] and depends only on the spatial metric and its (spatial) derivatives. Thus, it contains only operators constructed from the spatial metric g_{ij} and of dimension 4 and 6.

2.1. Detailed Balance

As the action (6) is very complicated, Hořava [7,32,35] proposed to impose an additional condition, the so-called detailed balance. It assumes that $\mathcal{V}$ could be derived from a superpotential W [7,32,35]:

$$\mathcal{V} = E^{ij}\mathcal{G}_{ijkl}E^{kl}, \quad E^{ij} = \frac{1}{\sqrt{g}}\frac{\delta W}{\delta g_{ij}}, \tag{8}$$

and

$$\mathcal{G}^{ijkl} = \frac{1}{2}\left(g^{ik}g^{jl} + g^{il}g^{jk}\right) - \lambda g^{ij}g^{kl}. \tag{9}$$

By carrying out an analytic continuation (e.g., [20]) of the two constant parameters ω and μ, we obtain the action for Hořava–Lifshitz gravity in the detailed balance condition [32], which reads as

$$\begin{aligned} S_{db} = \int dt\, d^3x \sqrt{g} N \Bigg[& \frac{2}{\kappa^2}\left(K_{ij}K^{ij} - \lambda K^2\right) + \frac{\kappa^2}{2\omega^4}C_{ij}C^{ij} - \frac{\kappa^2\mu}{2\omega^2}\frac{\epsilon^{ijk}}{\sqrt{g}}R_{il}\nabla_j R^l_k \\ & + \frac{\kappa^2\mu^2}{8}R_{ij}R^{ij} + \frac{\kappa^2\mu^2}{8(1-3\lambda)}\left(\frac{1-4\lambda}{4}R^2 + \Lambda R - 3\Lambda^2\right)\Bigg], \end{aligned} \tag{10}$$

where C^{ij} is the Cotton tensor:

$$C^{ij} = \epsilon^{ikl}\nabla_k\left(R^j{}_l - \frac{1}{4}R\delta^j_l\right), \tag{11}$$

ϵ^{ikl} denotes the totally antisymmetric tensor. The parameters κ, ω, and μ arriving in the theory have mass dimensions of, respectively, -1, 0, and 1. The analytic continuation mentioned above reads as $\mu \mapsto i\mu$ and $\omega^2 \mapsto -i\omega^2$, and it enables obtaining the positive values of the cosmological constant Λ as predicted by current observational results in the low-energy regime.

It is expected that action (10) reduces to the Einstein–Hilbert one in the IR limit of the theory. This is possible if the speed of light c and gravitational constant G correspond to HL parameters as follows:

$$G = \frac{\kappa^2}{32\pi c}, \quad c = \frac{\kappa^4\mu^2\Lambda}{8(3\lambda - 1)^2}. \tag{12}$$

The coupling constant λ present in the action (10) is dimensionless. It runs with energy and flows to the three infrared (IR) fixed points ([7]): $\lambda = 1/3$, $\lambda = 1$, or $\lambda = \infty$. However, some of those values seem unphysical. In the region $1 > \lambda > 1/3$, ghost instabilities appear in the IR limit of the theory [53]. The attempt to solve this problem [20] resulted in instabilities re-emerging at the other energy region, in the UV. Thus, the most physically interesting case is the regime $\lambda \geq 1$ that allows for a possible flow towards GR, where $\lambda = 1$. Region $\lambda \leq 1/3$, on the other hand, is disconnected from $\lambda = 1$ and therefore cannot be included in realistic physical considerations.

In order to obtain a cosmological model, it is necessary to populate the universe with matter (and radiation). The simplest method would be to model the matter sector by assuming it is described by a scalar field φ with a quadratic potential $V(\varphi) = \frac{1}{2}m^2\varphi^2$. However, a more realistic approach is to apply a hydrodynamic approximation where matter is described by two quantities p and ρ, which are, respectively, pressure and energy density and fulfil the continuity equation $\dot{\rho} + 3H(\rho + p) = 0$.

To derive equations of HL cosmology, one uses the projectability condition $N = N(t)$ [7], with the spatial part of the metrics being the standard FLRW line element: $g_{ij} = a^2(t)\gamma_{ij}$, $N_i = 0$, where γ_{ij} denotes a maximally symmetric metric with constant curvature:

$$\gamma_{ij}\mathrm{d}x^i\mathrm{d}x^j = \frac{\mathrm{d}r^2}{1-Kr^2} + r^2(\mathrm{d}\theta^2 + \sin^2\theta\mathrm{d}\varphi^2), \tag{13}$$

where values $K = \{-1, 0, 1\}$ correspond, respectively, to closed, flat, and open universes. This background metric implies that

$$C_{ij} = 0, \qquad R_{ij} = \frac{2K}{a^2}g_{ij}, \qquad K_{ij} = \frac{H}{N}g_{ij}, \tag{14}$$

where $H \equiv \dot{a}/a$ denotes the Hubble parameter.

On this background, the gravitational action (10) takes the following form :

$$S_{\mathrm{FRW}} = \int dt\, d^3x\, Na^3 \left\{ \frac{3(1-3\lambda)}{2\kappa^2}\frac{H^2}{N^2} + \frac{3\kappa^2\mu^2\Lambda}{4(1-3\lambda)}\left(\frac{K}{a^2} - \frac{\Lambda}{3}\right) - \frac{\kappa^2\mu^2}{8(1-3\lambda)}\frac{K^2}{a^4} \right\}. \tag{15}$$

In order to obtain equations of motion on a cosmological background, one needs to vary the action (15) with respect to N and a. Only after that can the lapse be set to one, $N = 1$, and terms with density ρ and pressure p are added. This procedure provides the analogs to the Friedmann equations for the projectable Hořava–Lifshitz cosmology with the imposed detailed-balance condition:

$$H^2 = \frac{\kappa^2\rho}{6(3\lambda-1)} \pm \frac{\kappa^4\mu^2}{8(3\lambda-1)^2}\left(\frac{K\Lambda}{a^2} - \frac{\Lambda^2}{2} - \frac{K^2}{2a^4}\right), \tag{16}$$

$$\dot{H} = -\frac{\kappa^2(\rho+p)}{4(3\lambda-1)} \mp \frac{\kappa^4\mu^2}{8(3\lambda-1)^2}\left(\frac{K\Lambda}{a^2} + \frac{K^2}{4a^4}\right), \tag{17}$$

together with the continuity equation:

$$\dot{\rho} + 3H(\rho + p) = 0. \tag{18}$$

In the equations above, there are two signs before the terms with Λ; namely, the upper one corresponds to the $\Lambda < 0$ case, while the lower one describes the analytic continuation $\mu \mapsto i\mu$ providing a positive Λ.

Some terms in the above equations which scale as a^{-4} are similar to the dark energy expressions; therefore, the parameters of energy density ρ_{de} and pressure density p_{de} are interpreted as dark energy parameters:

$$\rho_{de}\Big|_{db} := \frac{3\kappa^2\mu^2K^2}{8(3\lambda-1)a^4} + \frac{3\kappa^2\mu^2\Lambda^2}{8(3\lambda-1)}, \tag{19}$$

$$p_{de}\Big|_{db} := \frac{\kappa^2\mu^2K^2}{8(3\lambda-1)a^4} - \frac{3\kappa^2\mu^2\Lambda^2}{8(3\lambda-1)}. \tag{20}$$

We require that Equations (16) and (17) coincide with the standard Friedmann equations. Thus, we can identify the following:

$$c = \frac{\kappa^2\mu}{4}\sqrt{\frac{\Lambda}{1-3\lambda}}, \quad G = \frac{\kappa^2 c}{32\pi}, \quad \Lambda_E = -\frac{3\kappa^4\mu^2}{3\lambda-1}\frac{\Lambda^2}{32} = \frac{3c^2}{2}\Lambda, \tag{21}$$

respectively, as well as $\mu^2\Lambda = 1/32\pi^2G^2$ and $\lambda = 1$ (which is an IR fixed point). We demand a real value of the speed of light, c; therefore, the cosmological constant Λ has to be negative for $\lambda > 1/3$ and positive for $\lambda < 1/3$. In order to obtain a positive cosmological constant Λ, as suggested by observations, it is necessary to perform in (10) an analytic complex continuation of constant parameters μ and ω as follows $\mu \mapsto i\mu$ and $\omega^2 \mapsto -i\omega^2$. On the level of equations for Hořava–Lifshitz cosmology, varying λ-parameter in the range $[1,\infty)$ results in the running of the speed of light but does not change the structure of Equations (16) and (17).

When we substitute the equation of state $p = w\rho$ and the above expressions linking physical constants and HL parameters to (16) and (17), we obtain the following equations:

$$H^2 = \frac{2}{3\lambda-1}\left[\frac{\rho}{3} \pm \left(\frac{\Lambda_E}{3} - \frac{K}{a^2} + \frac{3}{4\Lambda_E}\frac{K^2}{a^4}\right)\right] \tag{22}$$

$$\dot{H} = \frac{2}{3\lambda-1}\left[-\frac{(1+w)}{2}\rho \pm \left(\frac{K}{a^2} - \frac{3}{2\Lambda_E}\frac{K^2}{a^4}\right)\right]. \tag{23}$$

2.2. Beyond Detailed Balance

The gravitational action (10) contains terms up to quadratic in the curvature. However, a more general renormalizable theory could also contain cubic terms, and there is not an *a priori* reason to keep only quadratic terms ([12,19,33]). Thus, Sotiriou, Visser, and Weinfurtner ([32]) built a projectable theory as the original Hořava theory but without imposing the detailed balance condition in the action.

This formulation led to Friedmann equations with an additional term $\sim 1/a^6$, moreover with additional and uncoupled coefficients:

$$H^2 = \frac{2}{(3\lambda-1)}\left(\frac{\rho}{3} + \sigma_1 + \sigma_2\frac{K}{a^2} + \sigma_3\frac{K^2}{a^4} + \sigma_4\frac{K}{a^6}\right), \tag{24}$$

$$\dot{H} = \frac{2}{(3\lambda-1)}\left(-\frac{p}{2} - \frac{\rho}{2} - \sigma_2\frac{K}{a^2} - 2\sigma_3\frac{K^2}{a^4} - 3\sigma_4\frac{K}{a^6}\right). \tag{25}$$

In order to coincide with the Friedmann equations in the IR limit $\lambda = 1$ and for large a, when terms proportional to $1/a^4$ and to $1/a^6$ become negligibly small, one has to set $\sigma_1 = \Lambda_E/3$ and $\sigma_2 = -1$. However, values of the constants σ_3 and σ_4 are, at this stage, arbitrary. In this way, we obtain the following equations:

$$H^2 = \frac{2}{(3\lambda-1)}\left(\frac{\rho}{3} + \frac{\Lambda_E}{3} - \frac{K}{a^2} + \sigma_3\frac{K^2}{a^4} + \sigma_4\frac{K}{a^6}\right), \tag{26}$$

$$\dot{H} = \frac{2}{(3\lambda-1)}\left(-\frac{\rho(1+w)}{2} + \frac{K}{a^2} - 2\sigma_3\frac{K^2}{a^4} - 3\sigma_4\frac{K}{a^6}\right), \tag{27}$$

We can observe new terms in the above analogs of Friedmann equations proportional to $1/a^6$. They mimic stiff matter, such that $\rho = p$ ($w = 1$), which scales similarly to $\rho_{\text{stiff}} \sim 1/a^6$. These terms are negligibly small at large scales but may play a significant role at small values of a scale parameter, thus changing the dynamics of the universe around initial singularity or a bounce.

3. Existence of Bounce

Hořava–Lifshitz cosmological equations contain additional terms proportional to a^{-4} (DB) and to a^{-6} (BDB) that introduce the possibility of a cosmological bounce, namely, a scenario in which contraction of the universe stops and reverses to expansion (or in the opposite direction). In a DB scenario, from the form of Equation (16), it follows that it is possible that $H = 0$. When this condition is fulfilled at some moment of time, the realisation of the bounce is possible (but not necessary; for that, we also need $\dot{H} \neq 0$. In the case $\lambda = 1$ [12], the bounce may happen in a non-empty Universe equipped with matter at the critical time t_*, $a = a_*$, when the critical energy density reaches the following value:

$$\rho = \rho_* = \frac{12}{\kappa^2}\left(\frac{K}{a_*^2} + \frac{\Lambda_E}{3} + \frac{\kappa^4\mu^2}{64}\frac{K^2}{a_*^4}\right), \tag{28}$$

This value is determined by the values of couplings κ and μ.

Additionally, a continuity equation implies that at the bounce, $\dot{H} > 0$. Therefore, when the condition $H = 0$ is fulfilled, we also have a sufficient condition for the existence of a bounce $\dot{H} \neq 0$. $\dot{H} > 0$ is only possible as a transition from a contracting to an expanding phase, but not the reverse. Moreover, there is another condition for a realisation bounce [30] that requires that $(\frac{\rho}{12} - p) > 0$ and the energy density of regular matter scales less quickly than dark matter terms.

Near the bounce—for small values of the parameter a—the dominating terms in the Friedmann Equations (22) and (23) are the terms scaling as a^{-4}, while other terms become insignificant. Particularly, H^2, $\dot{H}$, and ρ scale as $a^{-3(1+w)}$, where w is a constant parameter in the equation of state $p = w\rho$. Subsequently, if $w > -\frac{1}{3}$, the density term dominates over the curvature term $\sim 1/a^2$.

In the BDB scenario, bounce might happen at the critical density:

$$\rho_* = -\Lambda_E + 3\frac{K}{a_*^2} - 3\frac{\sigma_3 K^2}{a_*^4} - \frac{3\sigma_4 K}{a_*^6}. \tag{29}$$

For a flat universe and positive cosmological constant, bounce is not positive, as the resulting critical density becomes negative.

4. Bounce Stability in the Detailed Balance Formulation

We are mainly interested in the possibility of the appearance of a bounce that could be given by dynamics of variables a and H. From Equation (22), we might determine ρ and then insert its formula into (23). This way, we obtain two systems, with one containing the formula for density and its derivative via the continuity Equation (22) but still dependent on a and H. The second system is independent and consists of two equations describing the evolution of a and H.

Specifically, Equation (22) provides the following expression for ρ:

$$\rho = \frac{3(3\lambda - 1)}{2}H^2 \mp \left(\Lambda_E - 3\frac{K}{a^2} + \frac{9}{4\Lambda_E}\frac{K^2}{a^4}\right). \tag{30}$$

This expression substituted in (23) results in

$$\dot{H} = \frac{\pm 1}{3\lambda - 1}\left[(1+w)\Lambda_E - (3w+1)\frac{K}{a^2} + \frac{3(3w-1)}{4\Lambda_E}\frac{K^2}{a^4}\right] - \frac{3}{2}(1+w)H^2. \tag{31}$$

Adding the the definition of the Hubble parameter:

$$\dot{a} = aH, \tag{32}$$

we have a two dimensional dynamical system.

The set of Equations (31) and (32) is difficult to solve analytically. However, we are interested not in detailed solutions but in the qualitative analysis. For this purpose, we use the method of the phase portraits, where we search for critical points and analyse their character. These points are locations where the derivatives of all the dynamic variables, in our case when the r.h.s. of (31) and (32), vanish. What we obtain are the only points where phase trajectories could start, end, or intersect. Moreover, they can also appear in infinity. In this case, a suitable coordinate transformation, the so-called Poincaré projection, is used to project the complete phase space onto a compact region. The nature of these points, both finite and infinite, is given by the properties of the Jacobian matrix of the linearised equations at those points. All that information provides a qualitative analysis of the dynamical system.

The method of finding critical points consists of setting all right-hand sides of dynamical equations to zero, thus finding points where the derivatives of dynamical variables vanish. In the case of two equations, (31) and (32), the corresponding solutions are the two following P_1 and P_2 in the phase-space (a, H):

$$P_1 : a^2 = \frac{3K}{2\Lambda_E}, \quad H = 0, \tag{33}$$

$$P_2 : a^2 = \frac{(3w-1)K}{(1+w)2\Lambda_E}, \quad H = 0. \tag{34}$$

These two points exist when the values of a obtained via the square root of the expression on the right-hand side of the above equations are real and nonnegative. Thus the point P_1 exists if $K/\Lambda_E > 0$ if we assume a positive cosmological constant, therefore, only for $K > 0$. Point P_2 exists when $w > 1/3$ and $K/\Lambda_E > 0$ or $w < 1/3$ and $K/\Lambda_E < 0$. Thus, we have two critical points existing if the equation of state parameter $w > 1/3$. Moreover, they are both finite, unless $w = -1$, when P_2 blows to infinity. As mentioned above, due to $\dot{H} > 0$, both points represent a bouncing solution.

In order to complete the analysis, the stability properties of the critical points are needed (rather informally, a stable critical point $\mathbf{x}_0$ is one where, given any small distance ϵ to $\mathbf{x}_0$ and any initial condition within a radius $\rho < \epsilon$ around $\mathbf{x}_0$, the trajectory of the system shall not go further away from $\mathbf{x}_0$ than ϵ. An unstable critical point is the point which is not stable). They are determined by the eigenvalues of the Jacobian A of the system (31) and (32). Eigenvalues of A with non-zero real parts indicate hyperbolic points. They include sources (unstable) with positive real parts, saddles for real parts of opposite signs, and sinks (stable) corresponding to negative real parts. Critical points at which all the eigenvalues have real parts different from zero are called hyperbolic. Among them, one can distinguish sources (unstable) with positive real parts, saddles with real parts of different signs, and sinks (stable) for negative real parts. If at least one eigenvalue has a real part equal to zero it is then called a non-hyperbolic critical point. For such points, it is not possible to obtain conclusive information about the stability from the Jacobian matrix and other tools; e.g., numerical simulation [54] should be then used.

In the case of (31) and (32), the eigenvalues of A at P_1 are both imaginary, and it is a center for all the values of the parameters. The character of P_2 is more complicated and depends on the values of Λ_E, K and w. Thus, P_2 is a center when $K/\Lambda_E < 0$ and $-1 \leq w < 1/3$, with a special subcase for $w = -1$ that being so it becomes a linear center, therefore becoming a center with only one eigenvector. Otherwise, it becomes a saddle and is thus without a bouncing possibility.

To obtain a full picture of the dynamics of the universe, the information about critical points that occur at infinity is also necessary. For this purpose, the so-called Poincaré projection [55,56] is used. It projects the whole infinite phase space (a, H) onto a compact region. Specifically, we introduce the new coordinates $(\tilde{a}, \tilde{H})$, which are written in polar coordinates r, ϕ: $\tilde{a} = r\cos\phi$ and $\tilde{H} = r\sin\phi$. Moreover:

$$a = \frac{r}{1-r}\cos\phi, \tag{35}$$

$$H = \frac{r}{1-r}\sin\phi, \tag{36}$$

It is also necessary to rescale the time parameter t, which takes infinite values, by introducing the new time parameter T in a similar way, i.e. $dT = dt/(1-r)$. In such coordinates, the phase space is now compactified to a sphere of radius one and its interior. Here, infinity corresponds to $r = 1$. We have to keep in mind that a scale factor a may take only nonnegative values, thus actually being a semi-sphere (there is some small difference in interpreting phase trajectories in the original infinite phase space and in the Poincaré projected phase space. In the latter, there could be regions in which the parameter $\tilde{a}$ increases, although the parameter $\tilde{H}$ is negative, whereas this behaviour is absent in the former. This is due to the fact that parameters $(\tilde{a}, \tilde{H})$ are geometric objects without the same physical meaning as the scale factor a and the Hubble parameter H).

This procedure provides the dynamical equations in terms of r, ϕ, and their derivatives with respect to the new time T. At the surface of the sphere, if we limit $r = 1$, there are 3 solutions $P_3 = (1,0)$, $P_4 = (1,\pi/2)$, $P_6 = (1,-\pi/2)$, written in polar coordinates (r,ϕ). These critical points are now hyperbolic unless $w = -1$, resulting in P_4 and P_6 being, respectively, a repelling and an attracting node. For $w = -1$ points, P_4 and P_6 are non-hyperbolic, and numerical simulations provide that they are saddles and also ends of a separatrice.

The numerical phase portraits are presented in Figure 1, which contains the deformed phase space scaled to fit on the compactified sphere. We observe that bounce scenarios are only possible when one of the critical points P_1 and P_2 exist and is a center. Then, we have closed orbits around them, and the Universe might go through eternal cycles of expansion and collapse, connected by a bounce of a finite size, expansion, etc. However, the point P_1 describes a less physical bouncing solution with a density $\rho = 0$. A more interesting case is when P_2 is a center, as the density ρ is non-zero at that point. The special case of $w = -1$ provides a third bounce scenario around the linear center P_2 located now at ∞. In this case, the universe begins in a static infinite state as $H = 0$ $a = \infty$, then contracts to a finite size and rebounces to a static infinite universe.

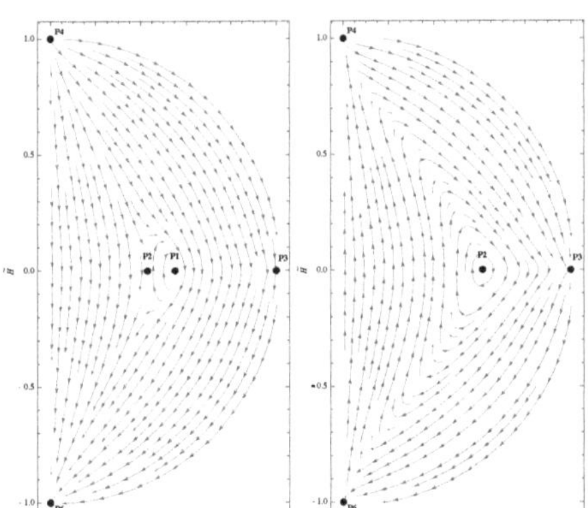

(**a**) Two bouncing solutions (**b**) One bouncing solution

Figure 1. Projected phase space of HL universe [15] in DB condition. Figure (**a**) is the case of $K/\Lambda_E > 0$ and $w > 1/3$. Figure (**b**), $K/\Lambda_E < 0$, and $-1 < w < 1/3$.

5. Bounce Stability in the beyond Detailed Balance Formulation

In the Sotiriou, Visser, and Weinfurtner formulation, the generalised Friedmann Equations (26) and (27) contain additional terms $\sim 1/a^6$ and uncoupled coefficients.

Solving Equation (26) for ρ provides:

$$\rho = 3\frac{(3\lambda - 1)}{2}H^2 - \Lambda_E - 3\frac{K}{a^2} - 3\frac{\sigma_3 K^2}{a^4} - \frac{3\sigma_4 K}{a^6}. \tag{37}$$

Substituting this expression on ρ into (27) and using the equation of state $p = w\rho$ results in

$$\dot{H} = \frac{2}{3\lambda - 1}\left(\frac{\Lambda_E(1+w)}{2} - \frac{K(1+3w)}{2a^2} + \frac{\sigma_3(-1+3w)K^2}{2a^4} + \frac{3\sigma_4(1+w)K}{2a^6}\right) - \frac{3(1+w)}{2}H^2. \quad (38)$$

As in the DB case, supplementing the above equation with the definition of the Hubble parameter provides the two-dimensional dynamical system.

Again, we search for critical points where $\dot{a}$ and $\dot{H}$. These points fulfil $H = 0$ and obtain the following condition:

$$\Lambda_E(1+w)a^6 - K(1+3w)a^4 + \sigma_3(-1+3w)K^2a^2 + 3\sigma_4(-1+w)K = 0 \quad (39)$$

This is a bicubic equation, which, in general, possesses quite complicated solutions but might be simplified in two special cases: namely when $w = -1$, describing the equation of state of the cosmological constant, and in the case of radiation described by $w = 1/3$. Outside of these two cases, critical points of the system (32) and (38) have the following coordinates: $(a_x, 0)$. Here, a_x^2 is a root of the cubic equation:

$$\Lambda_E(1+w)x^3 - K(1+3w)x^2 + \sigma_3(-1+3w)K^2x + 3\sigma_4(-1+w)K = 0. \quad (40)$$

Such an equation might have zero, one, two, or three real solutions depending on the sign of its discriminant. Moreover, if they exist, they are either always stable or always unstable depending on the sign of $K/(3\lambda - 1)$. Their character depends on the values of a_x, Λ_E, σ_3, and σ_4. The most significant feature of oscillating (and bouncing) solutions in the SVW formulation is the existence of two centres with a saddle between them (three finite critical points) for some values of parameters. In a more realistic situation, that includes a dynamical change of state parameter, so it would be possible to go from one oscillating bouncing solution to another.

In order to study the stability properties of infinite critical points, one again has to perform the Poincarè transformation. This leads to similar results as in the detailed balance scenario. Points at infinity are transformed to the sphere $r = 1$. Two points at $\phi = \pi/2$ and at $-\pi/2$ are, respectively, the repelling and attracting node, respectively. The point at $\phi = 0$ is non-hyperbolic.

Figure 2 shows the example of the phase space of a system with three finite critical points. Here, points S_1 and S_3 are centres, and a point S_2 is a saddle.

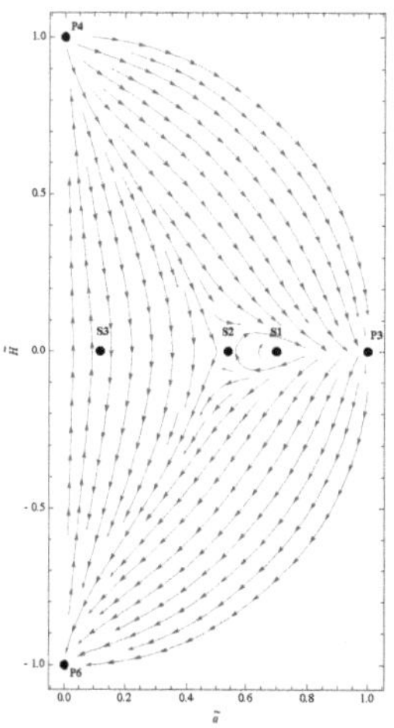

Figure 2. Projected phase space of the HL universe in beyond detailed balance formulation with 3 critical points existing [15].

6. Discussion

This paper reviews the research performed on the occurrence and the properties of the cosmological bounce in different formulations of projectable versions of Hořava–Lifshitz gravity, with and without a detailed balance condition. The analogs of the Friedmann equations in both these models contain a term scaling as $1/a^4$ and similar to dark radiation. That additional term enables that the Hubble parameter might be $H = 0$ at some moment of time. This is a necessary condition for the realisation of the bounce, while an additional condition $\dot{H} \neq 0$ makes it sufficient. In the Sotiriou, Visser, and Weinfurtner model, there is an additional term $1/a^6$ in the analogs of Friedmann equations. This term is of arbitrary sign, so it can enhance the possibility of a bounce or cancel it.

The biggest difference between the detailed balance theory and its breaking arrives for small values of a scale parameter a as the SVW gravity term $1/a^6$ plays role only for small values of a and becomes insignificant for bigger values. This difference is visible in the phase portraits of both theories and the number of potential bouncing solutions. In the original Hořava formulation, there exists one bouncing solution for all values of parameters, but it corresponds to a density of $\rho = 0$. For a non-zero $\rho = 0$, there might be a bouncing solution if $K/\Lambda_E < 0$ and $-1 \leq w < 1/3$, but for other values of these parameters, a bounce is not possible.

The SVW HL cosmology is a bit more complicated as there are additional terms in the analogs of Friedmann equations. There exist bouncing solutions for some values of parameters of the theory; however, the range of parameters that lead only to singular solutions is wider than in the detailed balance scenario. One very interesting special case includes two centres with a saddle between them (corresponding to three finite critical points). If one takes into account the dynamical change of state parameter, which is a much more realistic scenario, it might be possible to go from one oscillating solution to another bouncing solution. The problem is that the existence of such solutions depends on the values of coupling constants σ_3 and σ_4, and their physical interpretation still remains an open question.

Moreover, in both these formulations, bouncing non-singular solutions exist only in the case of a non-flat universe $K \neq 0$. In a DB scenario, from the form of Equations (33) and (34), it follows that for flat cosmologies with $K = 0$, the bounce happens at $a = 0$ unless $w = -1$ (stiff matter) or $\Lambda_E = 0$. In a zero-size universe, the metric becomes singular, and we cannot avoid an initial singularity. A similar situation happens in NDB. When we substitute $K = 0$ into Equation (39), we then obtain Equation $\Lambda_E(1 + w)a^6 = 0$, providing the values of a for which the bounce happens. Again, unless $w = -1$ or $\Lambda_E = 0$, the only solution is $a = 0$. Therefore, for most realistic values of w and Λ_E, the only non-singular bouncing solutions arrive in non-flat cosmologies.

A specific subcase of described solutions is presented in [14] where, for the purpose of illustration, it was assumed that matter in the pre-bounce epoch is described by a scalar field φ with a quadratic potential. That paper describes the dynamics of a Hořava–Lifshitz universe in two simplified models: one with a vanishing cosmological constant Λ_E and the other as an HL universe with a non-zero Λ_E but in the region of a small scale factor a. These two limitations result from performing some simplifications in the equation of motion, which are valid only in the regime of a small a or in the case of $\Lambda = 0$.

However, even in those simplified settings, it was possible to answer the question of possible scenarios realising a bounce and whether it is generic for the theory or not. We have found stable solutions leading to a Big Crunch or starting at a Big Bounce, both staying within the regime of a small a. Compared to results with the standard cosmology [56], we observed that in the HL formulation, there were additional critical points allowing the existence of a bounce and enabling new possible families of trajectories.

7. Conclusions

The obtained cosmological results presented here are promising and suggest there is a possibility to replace the initial cosmological singularity of GR by finite bouncing solutions. However, one must also consider that there are many problems and contradicting statements in the different formulations and extensions of HL-type theories. Aside from that aspect, there are also observational bounds on the existence of the Hořava–Lifshitz gravity and the values of its constants and parameters.

At present, HL-type theories, including the original one and its extensions, are not yet ruled out by observational data. However, there now are tight bounds on some parameters of the theory [40] from the binary neutron star merger GW170817 [57]. Therefore, it is possible that further observational data might either rule out some specific scenarios or the whole model. It is also possible that some agreement with observations could provide a better justification for additional theoretical research, as it is still hoped that HL gravity could offer a promising cosmological scenario without an initial singularity and solve some shortcomings of classical GR, such as non-renormalizability and thus problems with quantisation.

There are several observational bounds on different regions of the Hořava–Lifshitz framework, e.g., using data from binary pulsars [41,42], using general cosmological data [16], and also bounds in the context of dark energy [43]. In the context of dark matter and dark energy, there are also bounds on general Lorentz violations [9,58]. There is also quite recent research performed in the effective field theory formalism [18] of the extension HL gravity [31]. However, this analysis is reduced to a flat background space-time, which limits the overall number of parameters.

As was presented above, bouncing non-singular solutions in two basic HL cosmological scenarios exist only in the case of a non-flat universe $K \neq 0$. Therefore, it is interesting to compare that condition to observational constraints on the curvature parameter. Particularly, in our papers [44,45], we have placed new bounds on the parameters of Hořava–Lifshitz cosmology in its projectable version with and without imposing detailed balance conditions. We found very interesting results on spatial curvature. Namely, in [44], the original HL model was well fitted with a non-zero spatial curvature with accuracy to more than 3σ, whereas when we relaxed the detailed balance condition, we again obtained a positive non-zero spatial curvature at 1σ accuracy. In [45]. we used a larger and more updated dataset with additional high-energy sources such as quasars and gamma-ray bursts. We obtained that the HL universe both in the DB scenario and the NDB scenario is fitted with a negative curvature parameter at 1σ. On the other hand, similar calculations were made for ΛCDM model, such as in [59], where the negative curvature parameter was obtained from Planck data, weak galaxy lensing, and SH_0ES (supernovae and H_0 for the equation of state) local cosmic distance ladder measurements of the expansion rate.

As all those calculations included BAOs, therefore, there is a need for a further investigation of the curvature parameter, which could possibly finally exclude some of the HL models or zero-curvature ΛCDM models. Regardless, those results seem to be fascinating in view of future observations and also somehow demonstrate why an analysis limited to zero spatial curvature is somehow limited. Still, non-singular bouncing solutions in an HL universe appear only for non-zero spatial curvatures, so these two topics are related.

We have to take into account that most obtained bounds on the parameters of the HL cosmology are similar to those in the ΛCDM model. Of course, the ΛCDM model still has fewer parameters, and from this point of view should be preferred; it also fits the data well. However, one has to also bear in mind the theoretical aspects of Hořava gravity, which make it a good candidate for an ultraviolet complete theory of gravity. There are also several implications such as the possible resolution of the initial cosmological singularity, so there are still many reasons to keep investigating this model and its extensions.

Funding: This research received no external funding.

Conflicts of Interest: The author declares no conflict of interest.

Abbreviations

The following abbreviations are used in this manuscript:

GR	General Relativity
SVW	Sotiriou-Visser-Weinfurtner
HL	Hořava-Lifshitz
ΛCDM	Lambda cold dark matter
ADM	Arnowitt, Deser and Misner
IR	Infrared
UV	Ultraviolet

References

1. Bojowald, M.; Date, G. Quantum Suppression of the Generic Chaotic Behavior Close to Cosmological Singularities. *Phys. Rev. Lett.* **2004**, *92*, 071302. [CrossRef]
2. Quevedo, F. Is String Phenomenology an Oxymoron? 2016. Available online: http://xxx.lanl.gov/abs/1612.01569 (accessed on 23 March 2023).
3. Girelli, F.; Hinterleitner, F.; Major, S. Loop Quantum Gravity Phenomenology: Linking Loops to Observational Physics. *SIGMA Symmetry Integrability Geom. Methods Appl.* **2012**, *8*, 098. [CrossRef]
4. Bergeron, H.; Czuchry, E.; Gazeau, J.P.; Małkiewicz, P.; Piechocki, W. Singularity avoidance in a quantum model of the Mixmaster universe. *Phys. Rev. D* **2015**, *92*, 124018. [CrossRef]
5. Arkani-Hamed, N.; Dimopoulos, S.; Dvali, G.R. The Hierarchy problem and new dimensions at a millimeter. *Phys. Lett. B* **1998**, *429*, 263–272. [CrossRef]
6. Dvali, G. Black Holes and Large N Species Solution to the Hierarchy Problem. *Fortsch. Phys.* **2010**, *58*, 528–536. [CrossRef]
7. Hořava, P. Quantum Gravity at a Lifshitz Point. *Phys. Rev. D* **2009**, *79*, 084008. [CrossRef]
8. Hořava, P.; Melby-Thompson, C.M. General Covariance in Quantum Gravity at a Lifshitz Point. *Phys. Rev. D* **2010**, *82*, 064027. [CrossRef]
9. Audren, B.; Blas, D.; Ivanov, M.M.; Lesgourgues, J.; Sibiryakov, S. Cosmological constraints on deviations from Lorentz invariance in gravity and dark matter. *J. Cosmol. Astropart. Phys.* **2015**, *1503*, 016. [CrossRef]
10. Blas, D.; Pujolas, O.; Sibiryakov, S. On the Extra Mode and Inconsistency of Horava Gravity. *J. High Energy Phys.* **2009**, *10*, 029. [CrossRef]
11. Blas, D.; Sanctuary, H. Gravitational Radiation in Hořava Gravity. *Phys. Rev. D* **2011**, *84*, 064004. [CrossRef]
12. Calcagni, G. Cosmology of the Lifshitz universe. *J. High Energy Phys.* **2009**, *9*, 112. [CrossRef]
13. Colombo, M.; Gümrükçüoğlu, A.E.; Sotiriou, T.P. Hořava gravity with mixed derivative terms: Power counting renormalizability with lower order dispersions. *Phys. Rev. D* **2015**, *92*, 064037. [CrossRef]
14. Czuchry, E. The Phase portrait of a matter bounce in Hořava-Lifshitz cosmology. *Class. Quant. Grav.* **2011**, *28*, 085011. [CrossRef]
15. Czuchry, E. Bounce scenarios in the Sotiriou-Visser-Weinfurtner generalization of the projectable Hořava-Lifshitz gravity. *Class. Quant. Grav.* **2011**, *28*, 125013. [CrossRef]
16. Dutta, S.; Saridakis, E.N. Observational constraints on Hořava-Lifshitz cosmology. *J. Cosmol. Astropart. Phys.* **2010**, *1001*, 013. [CrossRef]
17. Dutta, S.; Saridakis, E.N. Overall observational constraints on the running parameter λ of Hořava-Lifshitz gravity. *J. Cosmol. Astropart. Phys.* **2010**, *1005*, 013. [CrossRef]
18. Frusciante, N.; Raveri, M.; Vernieri, D.; Hu, B.; Silvestri, A. Hořava Gravity in the Effective Field Theory formalism: From cosmology to observational constraints. *Phys. Dark Univ.* **2016**, *13*, 7–24. [CrossRef]
19. Kiritsis, E.; Kofinas, G. Hořava-Lifshitz Cosmology. *Nucl. Phys. B* **2009**, *821*, 467–480. [CrossRef]
20. Lü, H.; Mei, J.; Pope, C.N. Solutions to Hořava Gravity. *Phys. Rev. Lett.* **2009**, *103*, 091301. [CrossRef]
21. Saridakis, E.N. Hořava-Lifshitz Dark Energy. *Eur. Phys. J. C* **2010**, *67*, 229–235. [CrossRef]
22. Sotiriou, T.P.; Visser, M.; Weinfurtner, S. Quantum gravity without Lorentz invariance. *J. High Energy Phys.* **2009**, *10*, 033. [CrossRef]
23. Mukohyama, S. Horava-Lifshitz Cosmology: A Review. *Class. Quant. Grav.* **2010**, *27*, 223101. [CrossRef]
24. Bertolami, O.; Zarro, C.A.D. Hořava-Lifshitz quantum cosmology. *Phys. Rev. D* **2011**, *84*, 044042. [CrossRef]
25. Pitelli, J.P.M.; Saa, A. Quantum Singularities in Horava-Lifshitz Cosmology. *Phys. Rev. D* **2012**, *86*, 063506. [CrossRef]
26. Vakili, B.; Kord, V. Classical and quantum Hořava-Lifshitz cosmology in a minisuperspace perspective. *Gen. Rel. Grav.* **2013**, *45*, 1313–1331. [CrossRef]
27. Oliveira-Neto, G.; Martins, L.G.; Monerat, G.A.; Corrêa Silva, E.V. De Broglie–Bohm interpretation of a Hořava–Lifshitz quantum cosmology model. *Mod. Phys. Lett. A* **2018**, *33*, 1850014. [CrossRef]
28. Oliveira-Neto, G.; Martins, L.G.; Monerat, G.A.; Corrêa Silva, E.V. Quantum cosmology of a Hořava–Lifshitz model coupled to radiation. *Int. J. Mod. Phys. D* **2019**, *28*, 1950130. [CrossRef]
29. Tavakoli, F.; Vakili, B.; Ardehali, H. Hořava-Lifshitz Scalar Field Cosmology: Classical and Quantum Viewpoints. *Adv. High Energy Phys.* **2021**, *2021*, 6617910. [CrossRef]

30. Brandenberger, R. Matter bounce in Hořava-Lifshitz cosmology. *Phys. Rev. D* **2009**, *80*, 043516. [CrossRef]
31. Blas, D.; Pujolas, O.; Sibiryakov, S. Consistent Extension of Hořava Gravity. *Phys. Rev. Lett.* **2010**, *104*, 181302. [CrossRef]
32. Sotiriou, T.P. Hořava-Lifshitz gravity: A status report. *J. Phys. Conf. Ser.* **2011**, *283*, 012034. [CrossRef]
33. Wang, A. Hořava gravity at a Lifshitz point: A progress report. *Int. J. Mod. Phys. D* **2017**, *26*, 1730014. [CrossRef]
34. Charmousis, C.; Niz, G.; Padilla, A.; Saffin, P.M. Strong coupling in Hořava gravity. *J. High Energy Phys.* **2009**, *8*, 070. [CrossRef]
35. Vernieri, D.; Sotiriou, T.P. Hořava-Lifshitz Gravity: Detailed Balance Revisited. *Phys. Rev. D* **2012**, *85*, 064003. [CrossRef]
36. Appignani, C.; Casadio, R.; Shankaranarayanan, S. The Cosmological Constant and Horava-Lifshitz Gravity. *J. Cosmol. Astropart. Phys.* **2010**, *4*, 006. [CrossRef]
37. Vernieri, D. On power-counting renormalizability of Hořava gravity with detailed balance. *Phys. Rev. D* **2015**, *91*, 124029. [CrossRef]
38. Pospelov, M.; Shang, Y. On Lorentz violation in Horava-Lifshitz type theories. *Phys. Rev. D* **2012**, *85*, 105001. [CrossRef]
39. Son, E.J.; Kim, W. Note on nonsingular cyclic universes in the deformed Horava-Lifshitz gravity. *Mod. Phys. Lett. A* **2011**, *26*, 719–725. [CrossRef]
40. Emir Gumrukcuoglu, A.; Saravani, M.; Sotiriou, T.P. Hořava gravity after GW170817. *Phys. Rev. D* **2018**, *97*, 024032. [CrossRef]
41. Yagi, K.; Blas, D.; Barausse, E.; Yunes, N. Constraints on Einstein-Æther theory and Hořava gravity from binary pulsar observations. *Phys. Rev. D* **2014**, *89*, 084067; Erratum in *Phys. Rev. D* **2014**, *90*, 069902; Erratum in *Phys. Rev. D* **2014**, *90*, 069901. [CrossRef]
42. Yagi, K.; Blas, D.; Yunes, N.; Barausse, E. Strong Binary Pulsar Constraints on Lorentz Violation in Gravity. *Phys. Rev. Lett.* **2014**, *112*, 161101. [CrossRef]
43. Park, M.i. A Test of Horava Gravity: The Dark Energy. *J. Cosmol. Astropart. Phys.* **2010**, *1*, 001. [CrossRef]
44. Nilsson, N.A.; Czuchry, E. Hořava-Lifshitz cosmology in light of new data. *Phys. Dark Univ.* **2019**, *23*, 100253. [CrossRef]
45. Czuchry, E.; Nilsson, N.A. On the energy flow of λ in Hořava-Lifshitz cosmology. *arXiv* **2023**, arXiv:2304.09766. [CrossRef]
46. Ade, P.A.R.; Aghanim, N.; Arnaud, M.; Ashdown, N.; Aumont, J.; Baccigalupi, C.; Banday, A.J.; Barreiro, R.B.; Bartlett, J.G.; Bartolo, N.; et al. Planck 2015 results. XIII. Cosmological parameters. *Astron. Astrophys.* **2016**, *594*, A13. [CrossRef]
47. Moresco, M. Raising the bar: New constraints on the Hubble parameter with cosmic chronometers at $z \sim 2$. *Mon. Not. R. Astron. Soc.* **2015**, *450*, L16–L20. [CrossRef]
48. Betoule, M.; Kessler, R.; Guy, J.; Mosher, J.; Hardin, D.; Biswas, R.; Astier, A.; El-Hage, P.; Konig, M.; Kuhlmann, S.; et al. Improved cosmological constraints from a joint analysis of the SDSS-II and SNLS supernova samples. *Astron. Astrophys.* **2014**, *568*, A22. [CrossRef]
49. Blake, C.; Brough, S.; Colless, M.; Contreras, C.; Couch, W.; Croom, S.; Croton, D.; Davis, T.M.; Drinkwater, M.J.; Forster, K.; et al. The WiggleZ Dark Energy Survey: Joint measurements of the expansion and growth history at $z < 1$. *Mon. Not. Roy. Astron. Soc.* **2012**, *425*, 405–414. [CrossRef]
50. Alam, S.; Ata, M.; Bailey, S.; Beutler, F.; Bizyaev, D.; Blazek, J.A.; Bolton, A.S.; Brownstein, J.R.; Burden, A.; Chuang, C.H.; et al. The clustering of galaxies in the completed SDSS-III Baryon Oscillation Spectroscopic Survey: Cosmological analysis of the DR12 galaxy sample. *Mon. Not. Roy. Astron. Soc.* **2017**, *470*, 2617–2652. [CrossRef]
51. Font-Ribera, A.; Kirkby, D.; Miralda-Escudé, J.; Ross, N.P.; Slosar, A.; Rich, J.; Aubourg, É.; Bailey, S.; Bhardwaj, V.; Bautista, J.; et al. Quasar-Lyman α Forest Cross-Correlation from BOSS DR11: Baryon Acoustic Oscillations. *J. Cosmol. Astropart. Phys.* **2014**, *5*, 027. [CrossRef]
52. Bennett, C.L.; Larson, D.; Weiland, J.L.; Hinshaw, G. The 1% Concordance Hubble Constant. *Astrophys. J.* **2014**, *794*, 135. [CrossRef]
53. Bogdanos, C.; Saridakis, E.N. Perturbative instabilities in Horava gravity. *Class. Quant. Grav.* **2010**, *27*, 075005. [CrossRef]
54. Arrowsmith, D.K.; Place, C.M.; Place, C. *An Introduction to Dynamical Systems*; Cambridge University Press: Cambridge, UK, 1990.
55. Felder, G.N.; Frolov, A.; Kofman, L. Warped geometry of brane worlds. *Class. Quantum Gravity* **2002**, *19*, 2983. [CrossRef]
56. Felder, G.; Frolov, A.; Kofman, L.; Linde, A. Cosmology with negative potentials. *Phys. Rev. D* **2002**, *66*, 023507. [CrossRef]
57. Abbott, B.P.; Abbott, R.; Abbott, T.D.; Acernese, F.; Ackley, K.; Adams, C.; Adams, T.; Addesso, P.; Adhikari, R.X.; Adya, V.B.; et al. Gravitational Waves and Gamma-rays from a Binary Neutron Star Merger: GW170817 and GRB 170817A. *Astrophys. J. Lett.* **2017**, *848*, L13. [CrossRef]
58. Audren, B.; Blas, D.; Lesgourgues, J.; Sibiryakov, S. Cosmological constraints on Lorentz violating dark energy. *J. Cosmol. Astropart. Phys.* **2013**, *8*, 039. [CrossRef]
59. Handley, W. Curvature tension: Evidence for a closed universe. *Phys. Rev. D* **2021**, *103*, L041301. [CrossRef]

universe

MDPI

Article

The Finsler Spacetime Condition for (α,β)-Metrics and Their Isometries

Nicoleta Voicu [1,*,†], Annamária Friedl-Szász [1,†], Elena Popovici-Popescu [1,†] and Christian Pfeifer [2]

[1] Faculty of Mathematics and Computer Science, Transilvania University, Iuliu Maniu Str. 50, 500091 Brasov, Romania; szasz.anma@gmail.com (A.F.-S.); elena.c.popovici@gmail.com (E.P.-P.)

[2] ZARM, University of Bremen, 28359 Bremen, Germany; christian.pfeifer@zarm.uni-bremen.de

* Correspondence: nico.voicu@unitbv.ro

† These authors contributed equally to this work.

Abstract: For the general class of pseudo-Finsler spaces with (α,β)-metrics, we establish necessary and sufficient conditions such that these admit a Finsler spacetime structure. This means that the fundamental tensor has a Lorentzian signature on a conic subbundle of the tangent bundle and thus the existence of a cone of future-pointing time-like vectors is ensured. The identified (α,β)-Finsler spacetimes are candidates for applications in gravitational physics. Moreover, we completely determine the relation between the isometries of an (α,β)-metric and the isometries of the underlying pseudo-Riemannian metric a; in particular, we list all (α,β)-metrics which admit isometries that are not isometries of a.

Keywords: Finsler geometry; (α,β)-metric; isometry

1. Introduction

Finsler geometry, which is the geometry of a manifold described by a general geometric length measure for curves, has numerous applications in physics [1]. In the context of gravitational physics, it is the perfect mathematical framework to describe the gravitational field of a kinetic gas [2,3], it models the propagation of particles subject to deformed/doubly special relativity symmetries employed in quantum gravity phenomenology [4–6], and it emerges naturally in the context of theories based on broken/deformed Lorentz invariance, such as the standard model extension (SME) [7–9] or very special relativity (VSR) [10].

In general, Finsler geometry allows for a huge variety of structures, which are much more varied than those of pseudo-Riemannian structures on manifolds. Therefore, it is important to classify Finsler geometries in order to identify the best models for specific applications.

Among all Finsler structures, the class of (α,β)-metrics, obtained by constructing a geometric length measure for curves from a (pseudo)-Riemannian metric a and a 1-form b, are the easiest to construct and the most used in practice. Notorious examples include Bogoslovsky–Kropina (or m-Kropina) metrics, which represent the framework for VSR and its generalization, very general relativity (VGR) [10–14]—also used for dark energy models [15]—and Randers metrics, used, for instance, in the description of the propagation of light in stationary spacetimes [16–19], for the motion of an electrically charged particle in an electromagnetic field, in the study of Finsler gravitational waves [20,21], as an extended background for field theories [22] and black holes [23], or in Zermelo's navigation problem [16,24].

All the above-mentioned applications require Finsler metrics of Lorentzian signature. While there exists a rich literature on positive definite Finsler metrics (and in particular, on (α,β)-ones [25–30]), Lorentzian Finsler geometry is by far least understood and investigated [31]. In this paper, we study for the first time in full generality two questions about

Lorentzian (α, β)-Finsler structures, which have been only partially tackled in the literature (mostly only for very particular cases):

1. The necessary and sufficient conditions for an (α, β)-metric to define a Finsler spacetime structure;
2. Determining the isometries of general (α, β)-metrics.

Regarding point 1. above, the very definition of a Finsler spacetime is actually still a matter of debate [32–41]. Yet, in recent years, though the various definitions may differ in their minute details, they all converge to the following understanding: at each point of a Finslerian spacetime, there should exist a convex cone with null boundary—interpreted as the cone of future-pointing time-like vectors—on which the Finsler metric tensor must be well defined, smooth (maybe with the exception of one singular direction [38]), and with Lorentzian signature.

Starting from this understanding, we determine the conditions for a general (α, β)-metric, with a completely arbitrary 1-form, to be smooth and to have Lorentzian signature inside such a cone. Furthermore, we present concrete examples that are interesting for applications such as Randers and Bogoslovsky–Kropina metrics (extending previous studies [12,42] for the case of non-space-like 1-forms) as well as generalized m-Kropina and exponential metrics.

For the second problem, isometries of (α, β)-metrics, to the best of our knowledge, the only cases when these were known are Bogoslovsky–Kropina deformations of Minkowski metrics [28,43], as well as Randers and Kropina metrics [27]. Here, we determine infinitesimal isometries of *general* (α, β)-metrics.

The structure of this paper is as follows. Section 2 reviews the necessary notions of Finsler spacetimes for our later construction. Section 3 consists of the investigation of the conditions for an (α, β)-metric to define a spacetime structure, and presents our main theorem, Theorem 1. A complete classification, using simple conditions, is then given for the most used classes in Section 4. In Section 5, we determine the infinitesimal isometries of general (α, β)-metrics. Section 6 briefly presents our conclusions. In Appendix A, we display the proof of our formula for the determinant of the fundamental tensor of a general (α, β)-metric and of its inverse.

2. Preliminaries

We begin by recalling the concept of a Finsler spacetime in this section.

There are numerous attempts to find a suitable definition of a Finsler spacetime, i.e., of pseudo-Finsler geometry with a Finsler metric of Lorentzian signature, where among the first are those of Beem [40] and Asanov [41]. However, it quickly turned out that these definitions given are too restrictive to cover numerous interesting physical examples, such as m-th root metrics, Randers metrics, or m-Kropina metrics. Since then, several definitions of Finsler spacetimes have been developed [32–39], all agreeing that the Finsler metric tensor should be of Lorentzian signature on some (conic) subset of the tangent bundle, but differing in the precise details of where it must be smooth or only continuous. The origin of these fine differences lies in the various applications and examples on which the authors focused when formulating their definitions; thorough discussions of the differences between the distinct approaches to indefinite Finsler spacetime geometry can be found, e.g., in [34,39].

In the following, we will use the notion of Finsler spacetime as defined by two of us in [39], as it is the most permissive one which still allows for well-defined curvature-related quantities on the entire future-pointing time-like domain. Yet, as we will point out below, our approach can be applied with a minimal modification to the (even more permissive) definition by Caponio and co-workers [37,38].

Prior to introducing the notion of Finsler spacetime, we briefly introduce the manifolds we are working on and the preliminary notions of conic subbundle and pseudo-Finsler structure.

For the whole article, let M be a four-dimensional, connected, orientable, smooth manifold, let (TM, π, M) be its tangent bundle, and $\overset{\circ}{TM} = TM \backslash \{0\}$ be the tangent bundle without its zero section. We will denote the coordinates of a point $x \in U \subset M$ in a local chart (U, φ) by $(x^i)_{i=\overline{0,3}}$, and the naturally induced local coordinates of points $(x, \dot{x}) \in \pi^{-1}(U)$ by $(x^i, \dot{x}^i)$. Commas $_{,i}$ denote partial differentiation with respect to the coordinates x^i and dots $_{\cdot i}$ denote partial differentiation of coordinates $\dot{x}^i$. Furthermore, whenever there is no risk of confusion, we will omit for simplicity the indices of the coordinates.

A *conic subbundle* of TM is a non-empty open submanifold $\mathcal{Q} \subset \overset{\circ}{TM}$, which projects by π on the entire manifold, i.e., $\pi(\mathcal{Q}) = M$, and possesses the so-called *conic property:*

$$(x, \dot{x}) \in \mathcal{Q} \ \Rightarrow \ (x, \lambda\dot{x}) \in \mathcal{Q}, \quad \forall\lambda > 0.$$

Furthermore, a *pseudo-Finsler structure* on M, see [44], is a smooth function $L : \mathcal{A} \to \mathbb{R}$ defined on a conic subbundle $\mathcal{A} \subset \overset{\circ}{TM}$, obeying the following conditions:

1. Positive two-homogeneity: $L(x, \alpha\dot{x}) = \alpha^2 L(x, \dot{x})$, $\forall\alpha > 0, \forall(x, \dot{x}) \in \mathcal{A}$.
2. At any $(x, \dot{x}) \in \mathcal{A}$ and in one (and then, in any) local chart around $(x, \dot{x})$, the Hessian:

$$g_{ij} = \frac{1}{2}\frac{\partial^2 L}{\partial \dot{x}^i \partial \dot{x}^j} = \frac{1}{2}L_{\cdot i \cdot j} \tag{1}$$

is nondegenerate.

We note that, in general, the functions $g_{ij} = g_{ij}(x, \dot{x})$ have a nontrivial $\dot{x}$-dependence; more precisely, they define a mapping

$$g : \mathcal{A} \to T^0_2 M, \quad (x, \dot{x}) \mapsto g_{(x,\dot{x})} = g_{ij}(x, \dot{x}) dx^i \otimes dx^j, \tag{2}$$

called the *Finslerian metric tensor.* This is generally not a tensor field on M (as it depends on vectors $\dot{x} \in T_x M$), but it plays a largely similar role to the one of the metric tensor in pseudo-Riemannian geometry. The particular case when $g = g_{(x)}$ only (that is, $L(x, \dot{x}) = a_{ij}(x)\dot{x}^i\dot{x}^j$ is *quadratic* in $\dot{x}$) corresponds to pseudo-Riemannian geometry, see, for example, [45].

Definition 1 (Finsler spacetime, following [39]). *A Finsler spacetime (M, L), is a four-dimensional manifold M with a pseudo-Finsler structure $L : \mathcal{A} \to \mathbb{R}$ (where M is connected and orientable), obeying the additional third condition:*

3. There exists a connected conic subbundle $\mathcal{T} \subset \mathcal{A}$ with connected fibers $\mathcal{T}_x = \mathcal{T} \cap T_x M$, $x \in M$, such that on each $\mathcal{T}_x : L > 0$, g has a Lorentzian signature $(+, -, -, -)$ and L can be continuously extended as 0 to the boundary $\partial\mathcal{T}_x$.

Physically, the *Finsler spacetime function L* is interpreted as the interval $ds^2 = L(x, dx)$.

On a Finsler spacetime, there exists the following important subsets of TM:

1. The conic subbundle $\mathcal{A}$ where L is defined, smooth, and contains nondegenerate Hessians is called the set of *admissible vectors;* we will typically understand $\mathcal{A}$ as the *maximal* subset of $\overset{\circ}{TM}$ with these properties.
2. The conic subbundle $\mathcal{T}$, where the signature of g and the sign of L agree, will be interpreted as the set of *future-pointing time-like vectors.*

Note. From the above definition, it follows that all the fibers $\mathcal{T}_x$ of $\mathcal{T}$ are actually *convex* cones, see [39].

The definition in [34] can be recovered by setting $\mathcal{A} = \mathcal{T}$ and imposing that L extends *smoothly* to the boundary $\partial\mathcal{T}$, whereas the definition in [37,38] can be recovered by allowing L to be of class $\mathcal{C}^1$ only along one direction in each cone $\mathcal{T}_x$.

In a Finsler spacetime, the arc length of a curve $\gamma : [a,b] \to M,\ t \mapsto \gamma(t)$:

$$l_\gamma = \int_a^b \sqrt{|L(\gamma(t), \dot{\gamma}(t))|}dt$$

is well defined (independent of the parametrization), by virtue of the two-homogeneity of L. Moreover, for future-pointing time-like curves (defined by the fact that $\dot{\gamma}(t) = \frac{d\gamma(t)}{dt}$ belongs to the cone$\mathcal{T}_{\gamma(t)}$ for all t—interpreted as worldlines of massive particles), it gives the *proper time* along the respective worldline.

Having clarified our notion of Finsler spacetimes, we can proceed and present general conditions for (α, β)-Finsler metrics, so that they indeed define Finsler spacetimes.

3. Spacetime Conditions for (α, β)-Metrics

Finsler metrics of (α, β)-type can nicely be classified and studied due to their fairly simple building blocks, which are a pseudo-Riemannian metric a and a 1-form b on M.

More precisely, consider a Lorentzian metric $a = a_{ij}dx^i \otimes dx^j$ (we use the signature convention $(+,-,-,-)$) and a 1-form $b = b_i dx^i$ on M. Let us denote, in any local chart:

$$B = b(\dot{x}) = b_i\dot{x}^i, \qquad A(x,\dot{x}) = a_x(\dot{x},\dot{x}) = a_{ij}\dot{x}^i\dot{x}^j; \tag{3}$$

in comparison to the usual notations α, β in the literature on positive definite Finsler spaces, this is:

$$\alpha = \sqrt{|A|}, \qquad \beta = B. \tag{4}$$

Throughout the paper, for simplicity, we will also refer to the Lorentzian metric a as A.

An (α, β)-*metric* on M is, by definition, a pseudo-Finsler structure $L : \mathcal{A} \to \mathbb{R}$ (where $\mathcal{A} \subset \overset{\circ}{T}M$ is a conic subbundle) of the form:

$$L = A\Psi(s), \quad s = \frac{B^2}{A}, \tag{5}$$

where $\Psi = \Psi(s)$ is smooth on the set $s(\mathcal{A})$ of values $s \in \mathbb{R}$ corresponding to vectors in $\mathcal{A}$.

The metric tensor components $g_{ij} = g_{ij}(x,\dot{x})$ are easily found as:

$$g_{ij} = a_{ij}(\Psi - s\Psi') + \Psi' b_i b_j + \frac{1}{2}A^2\Psi'' s_{\cdot i} s_{\cdot j}, \tag{6}$$

with inverse (A10) given in Appendix A, see also [13].

In the following, we will investigate the conditions such that functions L as in (5) define a Finsler spacetime structure on M. In other words, we will investigate the existence of a conic subbundle $\mathcal{T} \subset \mathcal{A}$ satisfying the requirement of Definition 1. With this aim, let us make the following assumption:

Assumption 1. *At each $x \in M$, the cone $\mathcal{T}_x$ of L and the future-pointing time-like cone $\mathcal{T}_x^a$ of the metric a have a non-empty intersection.*
In other words, subluminal speeds of particles, as measured by a Finslerian observer, are not all superluminal from the point of view of an observer using a (pseudo-)Riemannian arc length.

Fix, in the following, an arbitrary point $x \in M$ and an arbitrary local chart with local coordinates $(x^i, \dot{x}^i)$ in a neighborhood of $\pi^{-1}(x)$. In the following, we will omit the point x from the writing of $(x,\dot{x})$-dependent quantities; that is, we will write simply $L = L(\dot{x})$, $A = A(\dot{x})$, $g_{\dot{x}} = g_{(x,\dot{x})}$, etc.

We denote by $\langle\ ,\ \rangle$ the scalar product with respect to a_x, that is:

$$A := \langle \dot{x}, \dot{x} \rangle, \quad \langle b, b \rangle = a^{ij} b_i b_j = a_{ij} \tilde{b}^i \tilde{b}^j, \quad \langle \tilde{b}, \dot{x} \rangle = b(\dot{x}) = B, \tag{7}$$

where tildes are used to indicate the raising/lowering of indices by a, e.g., $\tilde{b}^i = a^{ij} b_j$ (in contrast, we will not use tildes when raising/lowering indices with g; that is, $b^i = g^{ij} b_j$, etc.).

Remark 1.

1. *Using the above assumption $\mathcal{T}_x \cap \mathcal{T}_x^a \neq \varnothing$ and $L_{|\mathcal{T}_x} > 0$, it follows that:*

$$\forall \dot{x} \in \mathcal{T}_x : \quad A > 0, \quad \Psi > 0; \tag{8}$$

   in particular, $\mathcal{T}_x$ must be completely contained in the future-pointing cone of a.
   Indeed, according to the mentioned assumption, for any $x \in M$, there exists at least one vector y in the intersection $\mathcal{T}_x \cap \mathcal{T}_x^a$ which thus must satisfy $A(y) > 0, \Psi(y) > 0$. As both a A and Ψ are assumed to be smooth (and hence, continuous) on $\mathcal{T}_x$, in order to change sign, they should pass through 0. But the vanishing of either A or Ψ entails $L = 0$, which is in contradiction with the positiveness axiom for L inside $\mathcal{T}$; therefore, (8) *must hold throughout $\mathcal{T}_x$.*
2. *On the boundary $\partial\mathcal{T}_x$, L can be continuously prolonged as to satisfy:*

$$A = 0 \quad or \quad \Psi = 0. \tag{9}$$

Lemma 1 (***The domain of*** s). *In any (α, β)-metric Finsler spacetime, the set of values s such that $s(\dot{x}) \in \mathcal{T}_x$ is an interval*

$$I \subset [s_0, \infty), \qquad s_0 := \max\{\langle b, b \rangle, 0\}. \tag{10}$$

Proof. Let us first show that the set $I = \{s(\dot{x}) |\ \dot{x} \in \mathcal{T}_x\}$ is contained in $[s_0, \infty)$, that is:

$$s(\dot{x}) \geq s_0, \qquad \forall \dot{x} \in \mathcal{T}_x.$$

If b is a-time-like, i.e., $\langle b, b \rangle > 0$, then, taking into account that $\dot{x}$ is by definition also a-time-like, the reverse Cauchy–Schwarz inequality tells us that: $\langle b, b \rangle \langle \dot{x}, \dot{x} \rangle \leq \langle b, \dot{x} \rangle^2$, which, using (7), is nothing but:

$$\langle b, b \rangle A \leq B^2;$$

that is,

$$s \geq \langle b, b \rangle \quad \Rightarrow \quad s \geq \max\{\langle b, b \rangle, 0\} = s_0.$$

In the case when $\langle b, b \rangle \leq 0$, the statement is trivially satisfied, as $s = \dfrac{B^2}{A}$ is the ratio of two quantities that are non-negative on $\mathcal{T}_x$, and hence $s \geq 0 = s_0$.

Furthermore, since on $\mathcal{T}_x$, $A \neq 0$, the rational function $s = s(\dot{x}) : \mathcal{T}_x \to \mathbb{R}$ is continuous; using the connectedness axiom for $\mathcal{T}_x$, it follows that $I = s(\mathcal{T}_x) \subset \mathbb{R}$ must also be connected; namely, it is an interval. □

In Figure 1 below, we represent various relative positions of the future-pointing cone $\mathcal{T}_x^a$ of A (in orange) and the conic set $\Psi > 0$ (in blue). It is important to note that the cone $\mathcal{T}_x$ is always the *intersection* of these two sets.

(a) (b) (c) (d)

Figure 1. The relative positions of the spacetime cones. **(a)** $\partial\mathcal{T}_x : \{A = 0\}$, **(b)** $\partial\mathcal{T}_x : \{\Psi = 0\}$, **(c)** $\partial\mathcal{T}_x : \{A = 0\}or\{\Psi = 0\}$, **(d)** $\partial\mathcal{T}_x \cap \partial\mathcal{T}_x^a = \ell$.

Remark 2 (sharpness (or non-sharpness) of the bounds for s). *To precisely establish the interval I, we first need the critical points of the function $s = s(\dot{x})$. These turn out to be situated:*

- In the plane $B = 0$; in this case, the corresponding critical value is $s = 0$.
- On the ray directed by $\dot{x} = \tilde{b}$ emanating from the origin; this yields the critical value $s = \langle b, b\rangle$.

The above statement is justified as follows. Differentiating the expression $s = \frac{B^2}{A}$, we find that the condition $\dot{\partial}_i s = 0$ is equivalent to:

$$B(b_i A - B a_{ij}\dot{x}^j) = 0. \tag{11}$$

The vanishing of the first factor means precisely $B = 0$, whereas the second one gives, after raising indices with the help of a^{jk}, that $\dot{x}$ is proportional to $\tilde{b}$.
Taking into account the above Lemma, we note that the corresponding critical values are—if attained inside $\mathcal{T}_x$—*minimal* values for s. This way, we find:

1. The lower bound $s = s_0$ is attained for $\dot{x} \in \mathcal{T}_x$ in two situations:
   i. When $b \in \mathcal{T}_x$ is L-time-like (which implies that, in particular, it is also a-time-like, meaning that $\langle b, b\rangle > 0$) and $\dot{x}$ is collinear to $\tilde{b}$;
   ii. When the critical hyperplane $B = 0$ for s intersects $\mathcal{T}_x$ (a necessary condition for this is that b is a-space-like).
2. For the upper bound, we have two possibilities:
   i. If the boundary $\partial\mathcal{T}_x$ contains points where $A = 0, B \neq 0$, (as in Figure 1a,c,d), then, approaching these points, we will have $s \to \infty$; therefore, in this case, $s \in [s_0, \infty)$, where the upper bound is sharp.
   ii. If A does not vanish anywhere on $\partial\mathcal{T}_x$ (situation (b) in Figure 1, when $\partial\mathcal{T}_x$ consists only of points where $\Psi = 0$, $A > 0$), then—since obviously we cannot have $A = 0$ *inside* $\mathcal{T}_x$ either—it follows that s has a finite supremum on $\mathcal{T}_x$, in other words, we can safely write $s \in [s_0, s_1)$ for some finite value $s_1 > 0$.

Having determined the domain of s, the next step for determining the precise conditions relating Ψ and $\langle b, b\rangle$ such that L defines a spacetime structure is to find the sign of the coefficient $\Psi - s\Psi'$ in (6).

Lemma 2. *In any (α, β)-metric Finsler spacetime and at any $\dot{x} \in \mathcal{T}_x$, it holds that:*

$$\Psi - s\Psi' > 0.$$

Proof. Let $I = s(\mathcal{T}_x) \subset [s_0, \infty)$ be the interval defined above. We will proceed in three steps:

Step 1. There exists at least one $\dot{x}$ on the boundary $\partial\mathcal{T}_x$ such that $B(\dot{x}) \neq 0$:

Pick an arbitrary $v \in \partial\mathcal{T}_x$. If v does not belong to the hyperplane $H := \{v \in T_xM \mid B(v) = 0\}$, then the statement is proven with $\dot{x} = v$.

Hence, in the following, let us assume that $v \in \partial\mathcal{T}_x \cap H$.

The cone $\mathcal{T}_x$ is, by definition, open (in the topology of $T_xM \simeq \mathbb{R}^4$); therefore, it cannot be entirely contained in the hyperplane H. This way, there exists $u \in \mathcal{T}_x$ such that $B(u) \neq 0$. We will show that, going along the line:

$$\ell := \{u - \lambda v \mid \lambda \in \mathbb{R}\}$$

we must encounter a boundary point which is not in H:

- We first note that ℓ has no common points with H. Indeed, for any $\lambda \in \mathbb{R}$, we have $B(u - \lambda v) = B(u) - \lambda B(v) = B(u) \neq 0$.
- Second, $\ell \cap \overline{\mathcal{T}}_x$ is a non-empty, connected set. Indeed, ℓ contains an interior point of $\mathcal{T}_x$, which is u. Moreover, since $\mathcal{T}_x$ is convex, this means that its closure—which is also

necessarily convex—must intersect ℓ by a segment, a half-line, or the whole ℓ. In any case, $\ell \cap \overline{\mathcal{T}}_x$ is connected.

- Third, we show that the set $\overline{\mathcal{T}}_x \cap \ell$ intersects the cone $\mathcal{T}_x^a : A = 0$. To this aim, let us build the function $f : \mathbb{R} \to \mathbb{R}$,

$$f(\lambda) := A(u - \lambda v) = \langle u, u \rangle - 2\lambda \langle u, v \rangle + \lambda^2 \langle v, v \rangle.$$

This function always has at least one root, λ_0, as follows. *(i)* If $\langle v, v \rangle = 0$, i.e., $v \in \partial \mathcal{T}_x^a$, then, since u is a-time-like, it cannot be a-orthogonal to v, which means $\langle u, v \rangle \neq 0$. Hence, in this case f is of first degree in λ and thus has one zero. *(ii)* If $\langle v, v \rangle > 0$, i.e., $v \in \mathcal{T}_x^a$, then f is quadratic with a halved discriminant $\Delta = \langle u, v \rangle^2 - \langle u, u \rangle \langle v, v \rangle \geq 0$ by virtue of the reverse Cauchy–Schwarz inequality for a. Hence, again, f has real roots.

These roots λ_0 correspond to points $\dot{x} = u - \lambda_0 v \in \partial \mathcal{T}_x^a$.

- Finally, taking into account the connectedness of $\overline{\mathcal{T}}_x \cap \ell$, we find that, moving away from u along the line ℓ, we stay in $\mathcal{T}_x$ and, at some point, we must hit the boundary $\partial \mathcal{T}_x$. If, in a worst case scenario, we do not hit first a point where $\Psi = 0$, then we must anyway reach a root λ_0 of f, which will thus give a boundary point $\dot{x}$ for $\mathcal{T}_x \subset \mathcal{T}_x^a$. Moreover, since $\ell \cap H = \varnothing$, at this point we always have $B(\dot{x}) \neq 0$.

Step 2. There exists a subset $I_0 \subset I$ on which $\Psi - s\Psi' > 0$:

Pick $\dot{x} \in \partial \mathcal{T}_x$ such that $B(\dot{x}) \neq 0$; then, on a small enough neighborhood $V \subset \mathcal{T}_x$ of $\dot{x}$, B is still nonvanishing, i.e., $s \neq 0$. Then, on the set $s(V)$, we can write:

$$L = A\Psi = B^2 \frac{\Psi}{s}, \tag{12}$$

where the function $s \mapsto \dfrac{\Psi(s)}{s}$ is well defined, smooth, and strictly positive on $s(V)$. On the other hand, as we approach $\dot{x} \in \partial \mathcal{T}_x$, the function L must tend to 0. Since $B(\dot{x})$ cannot vanish, the one which has to vanish is $\lim\limits_{s \to s(\dot{x})} \dfrac{\Psi(s)}{s}$, meaning that $\dfrac{\Psi}{s}$ must be strictly decreasing on some interval $I_0 \subset I$. The statement then follows from noticing that:

$$\Psi - s\Psi' = -s^2 \frac{d}{ds}\left(\frac{\Psi}{s}\right) > 0, \quad \forall s \in I_0.$$

Step 3. $\Psi - s\Psi' \neq 0$ must have a constant sign on the entire I:

Assuming that there exists some $s \in I$ such that $\Psi - s\Psi' = 0$, then, at the corresponding vectors $\dot{x} \in \mathcal{T}_x$, we would have, by (6), $g_{ij}(\dot{x}) = \Psi' b_i b_j + \frac{1}{2} A^2 \Psi'' s_{\cdot i} s_{\cdot j}$. However, by direct, brute force computation of $det(g)$, we can see that in four dimensions, such a matrix is degenerate, which is not acceptable. Thus, $\Psi - s\Psi' \neq 0$ on I, which, together with the connectedness of I and $\Psi - s\Psi' > 0$ on $I_0 \subset I$, yields the statement. □

The above Lemma allows us to state a simple condition such that, for all $\dot{x} \in \mathcal{T}_x$, $g_{\dot{x}}$ has a Lorentzian signature $(+, -, -, -)$ (thus agreeing with the sign of L).

Lemma 3. *Assume $L = A\Psi : \mathcal{A} \to \mathbb{R}$ is a pseudo-Finsler (α, β)-metric structure, $x \in M$ is an arbitrarily fixed point, and $\mathcal{T}_x \subset \mathcal{A}_x$ is a conic set satisfying (8) and (9). Then, at any $\dot{x} \in \mathcal{T}_x$, the following statements are equivalent:*

(i) $g_{\dot{x}}$ *has $(+, -, -, -)$ signature and is negative definite on the $g_{\dot{x}}$-orthogonal complement of $\dot{x}$;*

(ii) $\det g_{\dot{x}} < 0$.

Proof. $(i) \to (ii)$ is obvious.

(ii) $\to$ *(i)* : Assuming $\det(g_{\dot{x}}) < 0$, the signature of $g_{\dot{x}}$ can be either $(+,-,-,-)$ or $(-,+,+,+)$. Using the $(+,-,-,-)$ signature of a, we will show that, actually, the latter situation is not possible, as it would entail that in any diagonal form, $g_{\dot{x}}$ must have at least two minus signs.

Fix an arbitrary L-time-like vector $\dot{x} \in \mathcal{T}_x$, which is not collinear to $\tilde{b}$. Since $\mathcal{T}_x \subset \mathcal{T}_x^a$, the vector $\dot{x}$ must then be also time-like with respect to a. Let us construct a $g_{\dot{x}}$-orthogonal basis $\mathcal{B} = \{e_0, e_1, e_2, e_3\}$ as follows. Pick $e_0 := \dot{x}$. Then,

$$g_{\dot{x}}(e_0, e_0) = g_{\dot{x}}(\dot{x}, \dot{x}) = L > 0.$$

As e_1, e_2, we will choose any two mutually perpendicular vectors in the (two-dimensional) a-orthogonal complement of $Span\{\tilde{b}, \dot{x}\}$. This means:

$$\langle \tilde{b}, v\rangle = 0, \ \ \langle \dot{x}, v\rangle = 0, \qquad v \in \{e_1, e_2\}. \tag{13}$$

With this choice, we obtain:

$$\begin{aligned} g_{\dot{x}}(\dot{x}, v) &= g_{ij}(\dot{x})\dot{x}^i v^j = \frac{1}{2}L_{\cdot i}(\dot{x})v^i = \frac{1}{2}(A_{\cdot i}\Psi + A\Psi' s_{\cdot i})v^i \\ &= \langle \dot{x}, v\rangle \Psi + \frac{1}{2}A\Psi'\left(s_{\cdot i}v^i\right), \end{aligned}$$

where we have used $A_{\cdot i}v^i = 2a_{ik}\dot{x}^k v^i = 2\langle \dot{x}, v\rangle$. Furthermore, using (5), we obtain $s_{\cdot i} = \frac{1}{A}(2Bb_i - sA_{\cdot i})$, which, using (13), gives:

$$s_{\cdot i}v^i = \frac{1}{A}(2B\langle \tilde{b}, v\rangle - 2s\langle \dot{x}, v\rangle) = 0$$

and finally,

$$g_{\dot{x}}(\dot{x}, v) = 0,$$

that is, e_1 and e_2 are indeed $g_{\dot{x}}$-orthogonal to $\dot{x}$.

It remains to check the sign of $g_{\dot{x}}(v, v) = g_{ij}(\dot{x})v^i v^j$ for $v \in \{e_1, e_2\}$. Substituting g_{ij} from (6) and taking into account that $b_i v^i = 0$, $s_{\cdot i}v^i = 0$, we find:

$$g_{\dot{x}}(v, v) = (\Psi - s\Psi')\langle v, v\rangle.$$

On the one hand, the assumption $\langle \dot{x}, v\rangle = 0$ implies that v must be a-space-like, i.e., $\langle v, v\rangle < 0$. On the other hand, using Lemma 2, the first factor above is strictly positive. All in all, we obtain:

$$g_{\dot{x}}(v, v) < 0, \quad v \in \{e_1, e_2\}$$

and $g_{\dot{x}}(\dot{x}, \dot{x}) > 0$. Then, for any choice of the fourth basis vector e_3 in the $g_{\dot{x}}$-orthogonal complement of e_0, e_1, e_2, using the hypothesis that $\det(g_{\dot{x}}) < 0$, we find

$$g_{\dot{x}}(e_3, e_3) < 0,$$

which proves *(i)*. □

Furthermore, let us introduce the function $\sigma = \sigma(s) : I \to \mathbb{R}$ (with I as in Lemma 1) as:

$$\sigma := \frac{(\Psi - s\Psi')^2}{\Psi}. \tag{14}$$

A direct computation (see Appendix A) then proves the following Proposition.

Proposition 1. *For any pseudo-Finsler function* $L = A\Psi(s)$, $s = \frac{B^2}{A}$, *the determinant of its Finslerian metric tensor* $g_{\dot{x}}$ *is:*

$$\det(g_{\dot{x}}) = \Psi^2(\Psi - s\Psi')\det(a)\frac{d}{ds}[(s - \langle b,b\rangle)\sigma]. \tag{15}$$

Using the expression of $\det(g)$ and the above Lemmas, we are now able to prove the main result of this section.

Theorem 1 (*The spacetime conditions*). *Let* M *be a four-dimensional, connected, orientable manifold. An* (α,β)*-metric function* $L : \mathcal{A} \to \mathbb{R}$, $L = A\Psi(s)$, $s = \frac{B^2}{A}$, *with the underlying pseudo-Riemannian metric* A *of Lorentzian signature* $(+,-,-,-)$, *defines a Finsler spacetime structure if and only if there exists a conic subbundle* $\mathcal{T} \subset \mathcal{A}$ *with connected fibers* $\mathcal{T}_x$ *obeying the following conditions:*

(i) $A > 0, \Psi > 0$ *on* $\mathcal{T}_x$ *and* $\lim_{\dot{x}\to\partial\mathcal{T}_x}(A\Psi) = 0$.

(ii) *For all values of* s *corresponding to vectors* $\dot{x} \in \mathcal{T}_x$:

$$\Psi - s\Psi' > 0, \qquad (s - \langle b,b\rangle)\frac{d}{ds}\ln\sigma > -1. \tag{16}$$

Proof. $\to$: Assuming that (M,L) is a Finsler spacetime, then, by definition, on each of its future-pointing time-like cones $\mathcal{T}_x$, $x \in M$, (i) must hold. The first inequality (ii) follows from Lemma 2. Then, using this inequality, together with $\Psi > 0$, $\det g_{\dot{x}} < 0$, and $\det a < 0$, in expression (15) for $\det g_{\dot{x}}$, we obtain:

$$\frac{d}{ds}[(s - \langle b,b\rangle)\sigma] =: \frac{d}{ds}(\rho\sigma) > 0,$$

where $\rho(s) := s - \langle b,b\rangle$ is known by Lemma 1 to obey $\rho \geq 0$. Besides, $\frac{d\rho}{ds} = 1$, which means that the above inequality can be re-expressed as:

$$\frac{d}{ds}(\rho\sigma) = \rho'\sigma + \rho\sigma' = \sigma + \rho\sigma' > 0.$$

Using $\sigma > 0$, this is in turn equivalent to:

$$\rho\frac{\sigma'}{\sigma} > -1,$$

which is precisely the second inequality (16).
$\leftarrow$: Assume now that there exists a conic subbundle $\mathcal{T} \subset \mathcal{A}$ with connected fibers and obeying conditions (i) and (ii). Then, in order to prove that (M,L) is a Finsler spacetime, it is sufficient to show that $g_{\dot{x}}$ has $(+,-,-,-)$ signature for all $\dot{x} \in \mathcal{T}_x$, which, using Lemma 3, is the same as $\det g_{\dot{x}} < 0$. With this aim, we note that, using $\Psi > 0$, $\Psi - s\Psi' > 0$, this reduces to:

$$\frac{d}{ds}[(s - \langle b,b\rangle)\sigma] > 0.$$

The latter is equivalent to $\sigma + (s - \langle b,b\rangle)\sigma' > 0$, which, by virtue of $\sigma = \frac{(\Psi - s\Psi')^2}{\Psi} > 0$, is nothing but the second inequality (16) and is thus satisfied by the hypothesis. □

Remark 3.

1. *Since $s - \langle b, b\rangle \geq 0$, then any strictly positive and monotonically increasing function σ will obey our condition, as, in this case, we will have the stronger inequality: $(s - \langle b, b\rangle)\frac{d}{ds}\ln\sigma \geq 0$.*
2. *The above theorem also works if one allows Ψ to be of class C^1 only along one direction in each $\mathcal{T}_x$, as in [37,38], with the only mention that the second condition* (16) *(which involves second derivatives of Ψ) will only hold outside that direction.*

4. Examples

4.1. Lorentzian Metrics $L = \kappa A$

These are given by

$$\Psi = \kappa, \tag{17}$$

where $\kappa > 0$ is a constant and they are always Finsler spacetimes in the sense of the above definition. Indeed, in this case, we obtain $\Psi - s\Psi' = \kappa > 0, \quad \sigma = \kappa > 0$, and $(s - \langle b, b\rangle)\frac{\sigma'}{\sigma} = 0$. Hence, the conditions in (16) are trivially satisfied; the light cones of L are obviously the light cones of A.

4.2. Randers Metrics $L = \epsilon(\sqrt{A} + B)^2$, $\epsilon = sign(\sqrt{A} + B)$

Before determining the corresponding function Ψ, some preliminary remarks on the future-pointing time-like cones $\mathcal{T}_x, x \in M$ of a Randers spacetime are made.

- The cone $\mathcal{T}_x$ is a connected component of the conic set $\sqrt{A} + B > 0$, more precisely, the connected component of the cone

$$(a_{ij} - b_i b_j)\dot{x}^i\dot{x}^j > 0 \tag{18}$$

  lying in the future-pointing cone $\mathcal{T}_x^a$. This is seen as $A = 0$, which is a singularity for L, cannot occur within the cone (18). Hence, the boundary $L = 0$ must be a "sharper" cone than $A = 0$, see also [42].
- On the boundary $\partial\mathcal{T}_x$, we must have $L = 0$, which implies:

$$B = -\sqrt{A} \leq 0. \tag{19}$$

  This leads to the even more interesting conclusion below.
- At any point $x \in M$ of a Randers spacetime, *the whole cone* $\mathcal{T}_x$ lies in the half-space $B < 0$:

$$B(\dot{x}) < 0, \quad \forall \dot{x} \in \mathcal{T}_x. \tag{20}$$

  In particular, the hyperplane $B = 0$ cannot intersect the interior of $\mathcal{T}_x$.
  To justify the above, let us first show that $B = 0$ cannot occur inside $\mathcal{T}_x$. Indeed, if this were the case, then, fixing an arbitrary $u \in \mathcal{T}_x \cap \{B = 0\}$, the fact that $\mathcal{T}_x$ is open in the topology of $T_xM \simeq \mathbb{R}^4$ guarantees that there must be an entire Euclidean ball centered at u which remains in $\mathcal{T}_x$. However, since u is in the hyperplane $B = 0$, half of this ball will be contained in the hyperspace $B > 0$; in other words, the intersection $\mathcal{C} := \mathcal{T}_x \cap \{B > 0\}$ is non-empty. By its definition, $\mathcal{C}$ is an open, conic, and convex set and, moreover, by continuity, on the boundary $\partial\mathcal{C}$, we have $B \geq 0$. On the other hand, proceeding as in the first step of the proof of Lemma 2, we find that there must exist a point $\dot{x} \in \partial\mathcal{C}$ where $B \neq 0$. Such a point is, thus, a point of $\partial\mathcal{T}_x$, where, in addition, $B > 0$, which is in contradiction with the above remark. Therefore, $B = 0$ cannot happen in the interior of $\mathcal{T}_x$ (but only possibly on its boundary). An immediate consequence of this is that B must have a constant nonzero sign on $\mathcal{T}_x$; moreover, this sign is negative.

The last remark points out that inside the future-pointing time-like domain for Randers spacetimes, we always have $\sqrt{s} = -\frac{B}{\sqrt{A}}$. In other words, the restriction $\Psi : \mathcal{T} \to \mathbb{R}$ (which is relevant to our work) has the expression:

$$\Psi = (1 - \sqrt{s})^2.$$

The quantities in Theorem 1 are then immediately obtained as:

$$\Psi - s\Psi' = 1 - \sqrt{s}, \qquad \sigma = 1. \tag{21}$$

We are now able to prove the following result.

Proposition 2. *A Randers-type deformation of a four-dimensional Lorentzian metric a defines a Finsler spacetime structure if and only if*

$$0 \le \langle b, b \rangle < 1.$$

If this is the case, then each of its future-pointing time-like cones $\mathcal{T}_x$ is the intersection of the cone $A - B^2 > 0$ with the future-pointing cone of a at x.

Proof. $\to$: Assume that L defines a spacetime structure. The fact that, in this case, the cones $\mathcal{T}_x$ are the intersections of the cones $\sqrt{A} > -B$ (which is the same as $A > B^2$, taking into account that $B < 0$) with $\mathcal{T}_x^a$ was already discussed in the first remark above. Furthermore, according to Theorem 1, on its future-pointing time-like cones, we must have $\Psi - s\Psi' = 1 - \sqrt{s} > 0$, which is equivalent to $s < 1$. Using Lemma 1, this gives

$$\langle b, b \rangle \le s < 1.$$

To prove the first inequality $\langle b, b \rangle \ge 0$, let us assume that this is not the case, i.e., $\langle b, b \rangle < 0$. We will show that, in this case, the hyperplane $B = 0$ must intersect $\mathcal{T}_x$. To this aim, fix an arbitrary point $x \in M$ and an a-orthonormal basis $\{e_0, e_1, e_2, e_3\}$, with e_0 being a-time-like and e_1 being collinear to $\tilde{b}$. It follows that $\tilde{b} = \lambda e_1$ for some $\lambda \in \mathbb{R}^*$. Accordingly, $A = 0$ becomes equivalent to $(\dot{x}^0)^2 - (\dot{x}^1)^2 - (\dot{x}^2)^2 - (\dot{x}^3)^2 = 0$ and the hyperplane $B = 0$ is described by the equation $\dot{x}^1 = 0$. Taking into account (18), it follows that the intersection between this hyperplane and $\mathcal{T}_x$ is the region of the three-dimensional cone:

$$(\dot{x}^0)^2 - (\dot{x}^2)^2 - (\dot{x}^3)^2 > 0, \tag{22}$$

lying in $\mathcal{T}_x^a$ (that is, with $\dot{x}^0 > 0$), which is by far non-empty. This is in contradiction with the fact that the hyperplane $B = 0$ cannot intersect $\mathcal{T}_x$, as shown above. We conclude that our assumption was false, that is, $\langle b, b \rangle \ge 0$.

$\leftarrow$: Assume now that $0 \le \langle b, b \rangle < 1$ and let us show that L satisfies the conditions of Theorem 1. Define $\mathcal{T}$ as the intersection of the future-pointing cone bundle $\mathcal{T}^a$ with the set $A - B^2 > 0$, i.e.,

$$(a_{ij} - b_i b_j)\dot{x}^i \dot{x}^j > 0; \tag{23}$$

in particular, this entails $s \in (0, 1)$. The latter inequality makes it clear that, at each point $x \in M$, the boundary $\partial \mathcal{T}_x$ is a quadric. Using the hypothesis $\langle b, b \rangle < 1$ and the Lorentzian signature of a, we find that:

$$det(a_{ij} - b_i b_j) = det(a_{ij})(1 - \langle b, b \rangle) < 0.$$

This way, the matrix $(a_{ij} - b_i b_j)$ can only have $(+, -, -, -)$ or $(-, +, +, +)$ signatures. Moreover, regarding its entries as continuous functions of b, we find that, for $b = 0$, the signature of our matrix coincides with that of a, as the matrices themselves coincide. Since the signature cannot change without passing through a zero of the determinant, we find that for all b in the given domain, we still have a $(+,-,-,-)$ signature. In other words, each set $\mathcal{T}_x$ is the interior of a convex (and hence, connected) cone.

Moreover, on each fiber $\mathcal{T}_x, x \in M$, we have:

- The functions A and Ψ are, by definition, positive inside $\mathcal{T}$ and, when approaching the boundary of $\mathcal{T}$ (that is, for $s \to 1$), we find that $\Psi \to 0$ and hence, $L \to 0$. The smoothness of Ψ follows as the value $s = 0$ (which would correspond to $B = 0$) cannot appear inside $\mathcal{T}_x$. This happens since b is, by hypothesis, non-space-like with respect to a; that is, $B = 0$ cannot occur within $\mathcal{T}_x^a$ and accordingly cannot occur in $\mathcal{T}_x$.
- The conditions in (ii) of Theorem 1 follow immediately using (21).

Thus, $L = A\Psi$ defines a Finsler spacetime structure. □

In Figure 2, the possible relative positions of hyperplane $B = 0$ with respect to the time-like cones $\mathcal{T}_x$ and $\mathcal{T}_x^a$ are displayed, with the same color meaning as in Figure 1.

(**a**) b light-like (**b**) b time-like

Figure 2. Possible positions of the hyperplane $B = 0$ with respect to the spacetime cones.

Remark 4.

1. *The result above extends those in [42] by taking into consideration the case when b is a-space-like and actually proving that this never defines a Randers spacetime.*
2. *A very recent article*[1]*, [21], proposed a modification of the standard Randers metric. With our above notations and signature convention $(+,-,-,-)$, on the set of interest $A > 0$, this reads:*

$$L = \epsilon(\sqrt{A} - |B|)^2. \tag{24}$$

This is a very interesting extension, as it removes the request that B should be strictly negative inside $\mathcal{T}_x$. Thus, for modified Randers metrics (24)*, the restriction $\langle b, b\rangle \geq 0$ (which was motivated precisely by the fact that once crossing the hyperplane $B = 0$, B changes sign) is no longer necessary. Moreover, on regions where $L > 0$ and $A > 0$ (in which the cones $\mathcal{T}_x$ must be contained), we have $\Psi = (1 - \sqrt{s})^2$ just as above (with the only difference that, in the definition of s, we no longer have $|B| = -B$). Thus, a completely similar reasoning shows that* (24) *leads to a well-defined spacetime structure if and only if $\langle b, b\rangle < 1$. This is consistent with the result in [21].*

4.3. Bogoslovsky–Kropina Metrics $L = A^{1-q}B^{2q}$

We will leave aside the case when $q = 0$, as it is trivial.

Proposition 3. *A Bogoslovsky–Kropina metric $L = A^{1-q}B^{2q}$ with $q \neq 0$ defines a spacetime structure if and only if one of the following occurs:*

(i) $\langle b, b\rangle > 0$ *and* $q \in [-1, 1)$*. In this case, the future-pointing cones of L coincide with those of A.*

(ii) $\langle b, b\rangle = 0$ *and* $q \in (-1, 1)$*. In this case, the future-pointing cones of L coincide with those of A.*

(iii) $\langle b, b\rangle < 0$ *and* $q \in (0, 1)$*. In this case, the future-pointing cones of L are obtained by intersecting the future-pointing cones of A with the half-space $B > 0$.*

Proof. The Finsler function is expressed as $L = As^q$, which means that Ψ has the expression:

$$\Psi(s) := s^q. \tag{25}$$

$\rightarrow$: Assume that L defines a spacetime structure and let us first identify the future-pointing cones $\mathcal{T}_x$ of L, together with the domain of the definition of Ψ.

Fix $x \in M$. As we have seen above, for any (α, β)-metric, the cone $\mathcal{T}_x$ must be contained in $\mathcal{T}_x^a$. Moreover, in our specific case, boundary points must be either on the cone $A = 0$ or in the plane $B = 0$. Thus:

1. If b is non-space-like with respect to a, then $B = 0$ cannot happen inside $\mathcal{T}_x^a$ (it can, in the worst case, when b is a-light-like, happen on its boundary). In this case, we thus have $\mathcal{T}_x = \mathcal{T}_x^a$; moreover, the reverse Cauchy–Schwarz inequality tells us that $s \geq \langle b, b \rangle$, with equality for $\dot{x} \sim \tilde{b}$. In other words, the minimum value $s_0 = \langle b, b \rangle$ is always attained on the closure of $\mathcal{T}_x$.
2. If b is a-space-like, then points with $B = 0$ will always exist inside $\mathcal{T}_x^a$. We note that, in order to have a finite limit for $L = A^{1-q}B^q$ as we approach the hyperplane $B = 0$, we must necessarily have:
$$q > 0. \tag{26}$$
In this case, the cone $\mathcal{T}_x$ will be the region of the cone $\mathcal{T}_x^a$ situated in the half-space $B > 0$ and s will tend to zero as we approach the boundary points with $B = 0$.
3. On the other hand, as shown above, the boundary of each cone $\mathcal{T}_x$ of any (α, β)-metric spacetime must contain at least one point where $B \neq 0$. In our case, this means that there will always be at least one boundary point satisfying $A = 0$, i.e., we necessarily have in $\mathcal{T}_x$, values $s \rightarrow \infty$.

Briefly: in any case, Ψ is defined on the entire interval:

$$s \in (s_0, \infty), \ \ s_0 := max\{\langle b, b \rangle, 0\}. \tag{27}$$

Having identified the function Ψ, we are now ready to rewrite, in our case, the conditions in Theorem 1.

- Since, inside the above-identified cones $\mathcal{T}_x$, we obviously have $A > 0$ and $\Psi > 0$, where Ψ is smooth, in order to check these conditions, we must impose that when approaching any point of the boundary $\mathcal{T}_x$, we should have $L = A^{1-q}B^q \rightarrow 0$. This immediately implies:
$$q < 1. \tag{28}$$
- The first inequality $\Psi - s\Psi' > 0$ becomes $s^q(1-q) > 0$, i.e., it is equivalent to the same inequality $q < 1$.

Noting that $\sigma := s^q(q-1)^2$, the second inequality (16) reads $(s - \langle b, b \rangle)\frac{q}{s} > -1$, which, taking into account that $s > 0$, is equivalent to

$$(q+1)s > \langle b, b \rangle q. \tag{29}$$

To solve this inequality, we note that its left hand side is linear in s; hence, it is necessary and sufficient to check its non-strict version at the endpoints of the interval (s_0, ∞) (with just a bit of extra attention to $q = -1$, for which this linear function identically vanishes). Thus:

- If $\langle b, b \rangle > 0$, then at the lower bound $s_0 = \langle b, b \rangle > 0$, the (non-strict) inequality (29) is identically satisfied. Imposing it for $s \rightarrow \infty$, we find:
$$q \geq -1, \tag{30}$$
which proves the necessity of condition (i) in our statement.

- If $\langle b,b\rangle = 0$, then for $s_0 = 0$, the (non-strict) inequality (29) is identically satisfied. Imposing it for $s \to \infty$, we find:
$$q \geq -1. \tag{31}$$
Yet, we note that $q = -1$ cannot actually happen, as it would lead to equality in (29) for all s, which proves the necessity of condition (ii) in our statement.
- If $\langle b,b\rangle < 0$, then for $s \to \infty$, we find $q > -1$, whereas the lower bound $s_0 = 0$ gives $q \geq 0$, proving the necessity of condition (iii).

$\leftarrow$: Assuming one of the situations (i)–(iii) happens, then the conditions in Theorem 1 are immediately satisfied on the said cones by $\Psi : (s_0, \infty) \to \mathbb{R}$, $\Psi(s) = s^q$. Hence, L defines a spacetime structure. □

4.4. Generalized m-Kropina (VGR) Metrics

These are given by:

$$L(x,\dot{x}) = As^{-p}(k+ms)^{p+1} = B^{-2p}\left(kA+mB^2\right)^{p+1}, \tag{32}$$

where $p,k,m \in \mathbb{R}$ are constants. As $k = 0$ gives a degenerate metric and $m = 0$ provides the already discussed Bogoslovsky–Kropina class, we will assume here that $k,m \neq 0$. The corresponding function Ψ is given by:

$$\Psi(s) = \frac{(k+ms)^{p+1}}{s^p}.$$

Proposition 4. *The generalized m-Kropina metric $L(x,\dot{x}) = As^{-p}(k+ms)^{p+1}$ defines a Finsler spacetime metric if and only if $k > 0$, $m < 0$, and one of the following situations occurs:*

(i) $\langle b,b\rangle \geq 0$ *and* $p \in (-1,1]$.
In this case, the future-pointing time-like cones of L are the connected components of the cones $kA + mB^2 > 0$ contained in the future-pointing time-like cones $\mathcal{T}^a_x$.

(ii) $\langle b,b\rangle < 0$ *and* $p \in (-1,0)$.
In this case, the future-pointing time-like cones of L are regions of the the cones $kA + mB^2 > 0$ lying both in the future-pointing time-like cones $\mathcal{T}^a_x$ and in the half-space $B > 0$.

Proof. $\rightarrow$: Assume L defines a Finsler spacetime structure. Let us first make some general remarks on the future-pointing time-like cones $\mathcal{T}_x$, that hold regardless of the causal character of b. To this aim, fix an arbitrary $x \in M$.

The latter equality in (32) reveals that the only boundary points $\dot{x} \in \partial\mathcal{T}_x$ where $B \neq 0$ are in the set $kA + mB^2 = 0$, which corresponds to $s = -\frac{k}{m}$. Since such boundary points must exist and lead to the finite value $L = 0$ and, on the other hand, they must correspond to strictly positive values of s, we obtain, respectively:

$$p+1 > 0, \quad -\frac{k}{m} > 0. \tag{33}$$

The latter relation points out that k and m must have opposite signs. To see precisely what these signs are, we evaluate the first condition in Theorem 1 (ii):

$$\Psi - s\Psi' > 0 \quad \Leftrightarrow \quad k(p+1)(k+ms)^p s^{-p} > 0, \tag{34}$$

which shows that only viable variant is:

$$k > 0, \quad m < 0, \tag{35}$$

as claimed. This means that the inequality $k + ms > 0$ is equivalent to $s < -\dfrac{k}{m}$, which gives, for $\dot{x} \in \mathcal{T}_x$:

$$s \in I \subset \left(0, -\frac{k}{m}\right),$$

where the upper bound $-\dfrac{k}{m}$ of the interval I is sharp. The lower bound, however, depends on the a-causal character of b.

The second condition in Theorem 1 (ii) becomes:

$$\rho\frac{\sigma'}{\sigma} > -1 \quad \Leftrightarrow \quad (\langle b,b\rangle m - kp + k)s + \langle b,b\rangle kp > 0. \tag{36}$$

This holds for every s corresponding to $\dot{x} \in \mathcal{T}_x$ if and only if it holds (non-strictly) at the bounds of the interval I.

At the upper bound $s \to -\dfrac{k}{m}$, this gives:

$$(p-1)(\langle b,b\rangle m + k) \leq 0.$$

Taking into account $s \geq \langle b,b\rangle$ together with $m < 0$, it follows that

$$k + m\langle b,b\rangle \geq k + ms > 0. \tag{37}$$

Therefore, we must have $p - 1 \leq 0$ and, all in all,

$$p \in (-1, 1]. \tag{38}$$

Another important consequence of (37) is that it ensures that the cones $kA + mB^2 > 0$ are, indeed, convex cones. This is seen as the matrix $(ka_{ij} + mb_ib_j)$ has the determinant $k^4 \det(a)(1 + \frac{m}{k}\langle b,b\rangle) < 0$. A quick similar reasoning to the one in the Randers subsection shows that its signature is the same as the one of a, i.e., $(+,-,-,-)$. In other words, $kA + mB^2 > 0$ is the interior of a convex cone. Moreover, as $kA > kA + mB^2$, this cone is completely contained in the cone $A > 0$.

It remains to evaluate (36) at the lower bound for s. Since this depends on the sign of $\langle b,b\rangle$, we distinguish two major cases:

1. Assume $\langle b,b\rangle \geq 0$. Then, $B = 0$ cannot happen inside the future-pointing time-like cones of a, and hence cannot occur in the Finslerian ones $\mathcal{T}_x$. Thus, in this case, each of the cones $\mathcal{T}_x$ is the connected component of the convex cone:

$$kA + mB^2 > 0 \tag{39}$$

contained in $\mathcal{T}_x^a$. In this case, a quick check shows that:

$$(ka_{ij} + mb_ib_j)\tilde{b}^i\tilde{b}^j = \langle b,b\rangle(k + m\langle b,b\rangle) \geq 0, \tag{40}$$

which means that one of the vectors $\tilde{b}$ or $-\tilde{b}$ belongs to $\overline{\mathcal{T}}_x$. In either case, we find that the corresponding value $s_0 = \langle b,b\rangle$ is attained on the closure of $\mathcal{T}_x$ and it represents the lower bound for $s \in I$.

At this lower bound, inequality (36) reduces to:

$$\langle b,b\rangle(k + m\langle b,b\rangle) \geq 0$$

and is identically satisfied for any value of $p \in (-1, 1]$.

2. Suppose $\langle b,b\rangle < 0$. Then, the hyperplane $B = 0$ has common points with the cone $kA + mB^2 > 0$. However, points with $B = 0$ are either singular points for L (if

$p > 0$), or null cone points if $p < 0$. Obviously, the only viable situation is the second one, namely:

$$p \in (-1, 0). \tag{41}$$

Knowing this, evaluation of (36) at the lower bound of the interval for s, which is in this case $s = 0$, gives:

$$\langle b, b \rangle kp \geq 0 \tag{42}$$

which always happens, as $\langle b, b \rangle < 0, p < 0, k > 0$.
The future-pointing time-like cones of L are then the intersections of the cone $kA + mB^2 > 0$ with the future-pointing cones $\mathcal{T}_x^a$ and the half-space $B > 0$.

$\Leftarrow$: Assuming $k > 0$ and $m < 0$, then the matrix $(ka_{ij} + mb_ib_j)$ has a Lorentzian signature $(+, -, -, -)$, meaning that the set $kA + mB^2 > 0$ is a convex cone, which is, moreover, contained in the cone $A > 0$. Then, assuming one of the situations (i) or (ii) holds, one can immediately check the conditions of Theorem (1) for the specified $\mathcal{T}_x$ (note that statement (ii) in the theorem is equivalent to (34) and (36)). □

4.5. Exponential Metrics $L = Ae^{P(s)}$

Here, $P = P(s)$ denotes an arbitrary smooth function for all s in the interval $I = (\max\{0, \langle b, b \rangle\}, \infty)$. Since $e^{P(s)}$ has no zeros, the cone $\mathcal{T}_x$ is the entire cone of a, and, accordingly, the domain of s is the entire interval I above.

Proposition 5. *The exponential metric $L = Ae^{P(s)}$ defines a Finsler spacetime structure if and only if:*

$$\lim_{s \to \infty} \frac{e^{P(s)}}{s} = 0, \tag{43}$$

$$1 - sP' > 0, \qquad 1 - sP' > (s - \langle b, b \rangle)\left(s(P')^2 + 2sP'' + P'\right), \quad \forall s \in I.$$

If this is the case, then the future-pointing time-like cones of L coincide with those of the Lorentzian metric A.

Proof. To justify the above statement, we note that, in this case, L and A always have the same sign, which means that the future-pointing time-like cones of L and A must be the same. Moreover,

$$\Psi(s) = e^{P(s)};$$

therefore, the first relation (43) is equivalent to $\lim\limits_{\dot{x} \to \partial \mathcal{T}_x} (A\Psi) = 0$ and the second one is equivalent to $\Psi - s\Psi' > 0$, whereas a brief calculation using $\sigma = e^P(1 - sP')^2$ reveals that the latter one is just $\rho \dfrac{\sigma'}{\sigma} > -1$. □

Example 1. *A concrete example, which resembles the so-called Maxwell–Boltzmann distribution function in the kinetic theory of gases, is obtained by considering an arbitrary metric a on M, together with a time-like 1-form $b \in \Omega_1(M)$ and*

$$\Psi : [\langle b, b \rangle, \infty) \to \mathbb{R}_+, \quad \Psi(s) = e^{k - \frac{\langle b, b \rangle^2}{2s^2}},$$

where $k > 0$ is a constant. The first two conditions in (43) are immediately checked. Besides, a brief calculation shows that the third one is equivalent to

$$s^4 + \langle b, b \rangle s^3 + 5\langle b, b \rangle^2 s^2 - \langle b, b \rangle^4 > 0.$$

The latter is trivially satisfied at $s = \langle b, b \rangle > 0$ *and the left hand side is a monotonically increasing function on* $(0, \infty)$*; hence, the inequality holds for all* $s \in [\langle b, b \rangle, \infty)$.

5. Isometries of (α, β)-Metric Spacetimes

An *isometry* of a pseudo-Finsler space is, [28], a diffeomorphism $\varphi : M \to M$ whose natural lift $T\varphi : TM \to TM$ leaves L invariant. Leaving aside possible discrete symmetries, we will turn our attention to one-parameter Lie groups of isometries. Their generators, known as *Finslerian Killing vector fields* $\xi = \xi^i \partial_i \in \mathcal{X}(M)$, are given by the equation:

$$\xi^C(L) = 0, \tag{44}$$

where $\xi^C = \xi^i \partial_i + \xi^i_{,j} \dot{x}^j \dot{\partial}_i$ is the complete lift of ξ to TM.

In particular, for (α, β)-metric functions $L = A\Psi(s)$, with $s = \frac{B^2}{A}$, the Killing vector condition reads $\xi^C(A)\Psi + A\Psi' \xi^C(s) = 0$, equivalently:

$$\xi^C(A)(\Psi - s\Psi') + 2\Psi' B \xi^C(B) = 0, \tag{45}$$

where

$$\xi^C(A) = (\mathfrak{L}_\xi a_{ij}) \dot{x}^i \dot{x}^j, \quad \xi^C(B) = (\mathfrak{L}_\xi b_i) \dot{x}^i$$

are polynomial expressions of degree two and one, respectively, in $\dot{x}$.

Particular case: trivial symmetries. Assume L is non-Riemannian, that is, $\Psi'(s) \neq 0$ at least on some interval $I_0 \subset I$. If ξ is a Killing vector field of L, then

$$\xi^C(A) = 0 \quad \Leftrightarrow \quad \xi^C(B) = 0,$$

which, differentiating with respect to $\dot{x}$, gives:

$$\mathfrak{L}_\xi a = 0 \quad \Leftrightarrow \quad \mathfrak{L}_\xi b = 0.$$

In other words, if ξ is a Killing vector field of a, then it is a Killing vector field of any (α, β)-metric $L = A\Psi$ with 1-form $b \in \Omega_1(M)$ invariant under the flow of ξ, regardless of the form of Ψ.

We will call Killing vectors of L obeying any of the equivalent conditions $\mathfrak{L}_\xi a = 0$ or $\mathfrak{L}_\xi b = 0$ *trivial symmetries of* L.

Nontrivial symmetries. In the following, let us explore which (α, β)-metrics admit nontrivial symmetries. We find the following result, which is valid for all pseudo-Finsler (α, β) metrics, independently of their signature.

Theorem 2. *A non-Riemannian* (α, β) *pseudo-Finsler function* $L = A\Psi$ *admits nontrivial Killing vector fields if and only if:*

$$\Psi = \begin{cases} c s^{\frac{\mu_1}{\mu_1 - 2\lambda_2}} |\mu_2 s + \mu_1 - 2\lambda_2|^{-\frac{2\lambda_2}{\mu_1 - 2\lambda_2}}, & \text{if } \mu_1 \neq 2\lambda_2 \\ c s e^{-\frac{2\lambda_2}{\mu_2 s}}, & \text{if } \mu_1 = 2\lambda_2,\ \mu_2 \neq 0 \end{cases} \tag{46}$$

for some smooth functions $c, \lambda_2, \mu_1, \mu_2$ *of* x *only, and there exist solutions* $\xi \in \mathcal{X}(M)$ *of the equations*

$$\xi^C(B) = \kappa \lambda_2 B, \qquad \xi^C(A) = \kappa \left(\mu_1 A + \mu_2 B^2 \right), \tag{47}$$

for some $\kappa = \kappa(x)$ *which does not identically vanish. If this is the case, the nontrivial Killing vectors of* L *are precisely the solutions of* (47).

Proof. $\to$: Assume that L admits a nontrivial Killing vector field $\xi \in \mathcal{X}(M)$, i.e., $\xi^C(L) = 0$, but $\xi^C(A)$ does not identically vanish.

Fix $x \in M$ and let us restrict our attention to an open, conic subset of $\mathcal{T}_x$ where $\xi^C(A) \neq 0$. For these, we can rewrite the Killing Equation (45) as:

$$\frac{B\xi^C(B)}{\xi^C(A)} = \frac{s\Psi' - \Psi}{2\Psi'} = \frac{1}{2}\left(s - \frac{\Psi}{\Psi'}\right). \tag{48}$$

In particular, the latter expression must be equal to a ratio of two homogeneous polynomial expressions of degree two in $\dot{x}$. However, on the other hand, it is a function of B^2 and A; the only such possibility is:

$$\frac{s\Psi' - \Psi}{2\Psi'} = \frac{\lambda_1 A + \lambda_2 B^2}{\mu_1 A + \mu_2 B^2}, \tag{49}$$

for some $\lambda_1, \lambda_2, \mu_1, \mu_2$ depending on x only. Depending on whether λ_1, μ_1 vanish or not, we distinguish four situations:

1. $\lambda_1, \mu_1 = 0$. In this case, $\dfrac{s\Psi' - \Psi}{2\Psi'} = \dfrac{\lambda_2}{\mu_2} =: \kappa(x)$ only. Integration of this equation gives:

$$\Psi = c(s - 2\kappa), \quad c = c(x),$$

which means that $L = A\Psi = c(B^2 - 2\kappa A)$ is actually pseudo-Riemannian.

2. $\lambda_1 = 0, \mu_1 \neq 0$. In this situation, the first equality (48) gives, after simplification by B,

$$\frac{\xi^C(B)}{\xi^C(A)} = \frac{\lambda_2 B}{\mu_1 A + \mu_2 B^2}.$$

We note that, on the right hand side, on the one hand, we must have $\lambda_2 \neq 0$ (otherwise we only get trivial symmetries $\xi^C(B) = 0$) and, on the other hand, the numerator and the denominator must be given by relatively prime polynomials; in the contrary case, A would admit B as a factor, which is not possible, since its $\dot{x}$-Hessian $(2a_{ij})$ is nondegenerate. However, then B must divide $\xi^C(B)$, that is,

$$\xi^C(B) = \kappa\lambda_2 B,$$

where, given that $\deg B = \deg \xi^C(B) = 1$, we must have $\kappa = \kappa(x)$ only. Accordingly,

$$\xi^C(A) = \kappa\left(\mu_1 A + \mu_2 B^2\right).$$

The functions Ψ corresponding to such symmetries are then obtained from (49), which in our case reads

$$s - \frac{\Psi}{\Psi'} = \frac{2\lambda_2 s}{\mu_1 + \mu_2 s} \tag{50}$$

and can be directly integrated to give

$$\Psi = c\exp\left(\int \frac{(\mu_1 + s\mu_2)ds}{\mu_2 s^2 + (\mu_1 - 2\lambda_2)s}\right). \tag{51}$$

Calculation of the integral then leads to (46). We note that the situation $\mu_1 - 2\lambda_2 = 0$, $\mu_2 = 0$ is not possible, as it would lead in (50) to $\frac{\Psi}{\Psi'} = 0$.

3. $\lambda_1 \neq 0, \mu_1 = 0$. This gives $\dfrac{B\xi^C(B)}{\xi^C(A)} = \dfrac{\lambda_1 A + \lambda_2 B^2}{\mu_2 B^2}$ and, equivalently,

$$B^3\mu_2\xi^C(B) = \xi^C(A)\left(\lambda_1 A + \lambda_2 B^2\right). \tag{52}$$

This implies that B must divide $\lambda_1 A + \lambda_2 B^2$, since $\xi^C(A)$ is of degree two and hence it cannot "swallow" more than a factor of B^2 of the B^3 from the left hand side. This in turn implies that B divides A, thus leading to a degenerate $\dot{x}$-Hessian for A, which is impossible.

4. $\lambda_1, \mu_1 \neq 0$. In this case, we have:

$$\frac{B\xi^C(B)}{\xi^C(A)} = \frac{\lambda_1 A + \lambda_2 B^2}{\mu_1 A + \mu_2 B^2}. \tag{53}$$

The ratio on the right hand side is either irreducible or a function of x only. This is derived as follows. Assuming that it can be simplified by a first degree factor, then both $\lambda_1 A + \lambda_2 B^2$ and $\mu_1 A + \mu_2 B^2$ must be decomposable; in particular, they must have degenerate $\dot{x}$-Hessians, i.e.,

$$\det(\lambda_1 a_{ij} + \lambda_2 b_i b_j) = \det(\mu_1 a_{ij} + \mu_2 b_i b_j) = 0.$$

Using Lemma A1 (see Appendix A), that is $\lambda_1^4 \det a \left(1 + \frac{\lambda_2}{\lambda_1}\langle b, b\rangle\right) = \mu_1^4 \det a \left(1 + \frac{\mu_2}{\mu_1}\langle b, b\rangle\right) = 0$, leads to

$$\frac{\lambda_2}{\lambda_1} = \frac{\mu_2}{\mu_1}.$$

The latter actually means that the ratio $\frac{\lambda_1 A + \lambda_2 B^2}{\mu_1 A + \mu_2 B^2} = \frac{\lambda_1}{\mu_1}$ depends on x only (i.e., it is actually simplified by a second degree factor, not a first degree one, as assumed). Thus, we only have two possibilities:

(a) $\frac{\lambda_1 A + \lambda_2 B^2}{\mu_1 A + \mu_2 B^2} = \kappa(x)$, that is, $\frac{s\Psi' - \Psi}{2\Psi'} = \kappa$, which, by a similar reasoning to Case 1, entails that L is actually pseudo-Riemannian.

(b) $\frac{\lambda_1 A + \lambda_2 B^2}{\mu_1 A + \mu_2 B^2}$ is irreducible. In this case, from (53), we find that B must either divide $\xi^C(A)$ (but, then, $\frac{\lambda_1 A + \lambda_2 B^2}{\mu_1 A + \mu_2 B^2} = \frac{B\xi^C(B)}{\xi^C(A)}$ would equal a ratio of first degree polynomials, which contradicts the irreducibility assumption) or it must divide $\lambda_1 A + \lambda_2 B^2$, which, taking into account that A and B are always relatively prime, leads to $\lambda_1 = 0$, in contradiction with the hypothesis $\lambda_1 \neq 0$. Therefore, there are no properly Finslerian functions L with $\lambda_1, \mu_1 \neq 0$.

We conclude that the only valid possibility is Case 2, thus giving the statement of the theorem.

$\leftarrow$: Conversely, assuming that ξ and Ψ satisfy (47) and (46), a direct computation shows that $\xi^C(L) = \xi^C(A)\Psi + A\Psi'\xi^C(s) = 0$, and hence ξ is a Killing vector for $L = A\Psi$.

If L is non-Riemannian, these symmetries are always nontrivial. This is seen by the fact that $\xi^C(A) = 0$ is only possible when $\mu_1 = \mu_2 = 0$, which is, $\Psi = const.$, corresponding to a Lorentzian metric. □

The above result gives necessary and sufficient conditions for an (α, β)-metric to admit nontrivial symmetries. There remains, of course, the question whether such metrics really exist. To show that the answer is affirmative, we present below a concrete example.

Example 2. ***Nontrivial symmetries of a Bogoslovsky–Kropina spacetime metric.*** *Consider, on $M = \mathbb{R}^4$, a conformal deformation a of the Minkowski metric $\eta = diag(1,-1,-1,-1)$ and a light-like 1-form b, as follows:*

$$a = e^{2qx^0}\eta, \quad b = e^{(q-1)x^0}\left(dx^0 + dx^1\right),$$

where $q \in (0,1)$ is arbitrarily fixed. With these data, the Bogoslovsky-type metric:

$$L = As^q, \quad s = \frac{B^2}{A} = \frac{(b(\dot{x}))^2}{a(\dot{x},\dot{x})}$$

is a nontrivial Finsler spacetime function, whose $\Psi(s) = s^q$, fits into class (46) (for $\lambda_2 = q - 1, \ \mu_1 = 2q, \ \mu_2 = 0, \ \kappa = c = 1$).

We take ξ as the time translation generator $\xi = \partial_0$, which gives:

$$\xi^C = \partial_0.$$

Then, using $A = e^{2qx^0}\eta_{ij}\dot{x}^i\dot{x}^j$, $B = e^{(q-1)x^0}\left(\dot{x}^0 + \dot{x}^1\right)$, we immediately find:

$$\xi^C(A) = 2qA \neq 0, \quad \xi^C(B) = (q-1)B,$$

which are precisely the conditions in (47). Therefore, ξ is a nontrivial Killing vector field for L.

6. Conclusions

Among all possible Finsler metrics, the class of (α,β)-metrics is of particular interest, since it allows for numerous applications and explicit calculations. Up to now, it was not known how to identify (α,β)-Finsler spacetimes in general. They have potential applications in physics, and in particular in gravitational physics.

In this article, we identified the necessary and sufficient conditions such that an (α,β)-Finsler metric defines a Finsler spacetime in Theorem 1; these are the conditions that ensure a Lorentzian signature inside a convex cone with null boundary, in each tangent space. As we demonstrated afterwards, this enables us to find the best candidates for physical applications belonging to different subclasses of (α,β)-metrics. For Randers metrics, we found that the defining 1-form must be non-space-like and with bounded norms with respect to the pseudo-Riemannian metric (Proposition 2). For Bogoslovsky–Kropina metrics, the value of the power parameter is constrained (Proposition 3). For generalized m-Kropina and exponential type metrics, we provided simple necessary and sufficient Finsler spacetime constraints in Propositions 4 and Proposition 5, respectively.

Moreover, we investigated the existence of isometries of (α,β)-metrics. Surprisingly, it turned out that there can exist isometries of a Finsler metric which are not isometries of the metric[2] a, as we pointed out in Theorem 2 and, subsequently, in a concrete example. The physical consequences of the existence of these isometries still must be determined. This paper is meant as a starting point for—we hope—numerous research works. Here are just some ideas we want to pursue in the near future:

1. To use (α,β)-metrics with given symmetries as ansatzes for solving the Finslerian gravitational field equation in [3,32,42].
2. To calculate and understand the precise physical interpretation of the various non-Riemannian notions such as Berwald or Landsberg curvature (see, e.g., [47]), with a special focus on spacetimes with (α,β)-metrics. We conjecture that these non-Riemannian quantities could account for at least a part of the observed dark energy phenomenology.
3. In light of the above, it will be important to construct some non-trivial Ricci flat Finsler (α,β)-metric spacetimes with some nonzero non-Riemannian quantities. Some

interesting Ricci flat Finsler spacetimes have been constructed, although they are not in an (α, β)-form, see, e.g., [48].

4. To construct the most general spatially spherically symmetric and the most general cosmologically symmetric spacetimes of (α, β)-metric type, whose underlying Lorentzian metrics do not possess such symmetries.
5. To completely classify (α, β)-metrics which lead to well-defined Finsler spacetimes, for which the underlying pseudo-Riemannian metric a has a non-Lorentzian signature.

Author Contributions: The authors all contributed substantially to the derivation of the presented results as well as analysis, drafting, review, and finalization of the manuscript. All authors have read and agreed to the published version of the manuscript.

Funding: C.P. is funded by the Deutsche Forschungsgemeinschaft (DFG, German Research Foundation) —Project Number 420243324 and acknowledges the excellence cluster Quantum Frontiers funded by the Deutsche Forschungsgemeinschaft (DFG, German Research Foundation) under Germany's Excellence Strategy—EXC-2123 QuantumFrontiers—390837967.

Acknowledgments: This article is based upon work from COST Action CA21136 (Addressing observational tensions in cosmology with systematics and fundamental physics—CosmoVerse) and the authors would like to acknowledge networking support by the COST Action CA18108 (Quantum Gravity Phenomenology in the Multi-Messenger Approach), supported by COST (European Cooperation in Science and Technology). The authors are grateful to Andrea Fuster and Sjors Heefer for numerous useful discussions and their insight.

Conflicts of Interest: The authors declare no conflict of interest.

Appendix A. Calculation of $\det(g_{\dot{x}})$

Consider, in the following, $\dim M = n$ and a pseudo-Finsler (α, β)-metric $L = A\Psi : \mathcal{A} \to \mathbb{R}$. We will prove in the following that:

$$\det(g_{\dot{x}}) = \Psi^2 (\Psi - s\Psi')^{n-3} \det(a) \frac{\partial}{\partial s} \left((s - \langle b, b \rangle) \frac{(\Psi - s\Psi')^2}{\Psi} \right), \tag{A1}$$

where primes denote differentiation with respect to s and $\langle b, b \rangle := a^{ij} b_i b_j$.

To this aim, the first step is to see that, in any local chart:

$$g_{ij} = (\Psi - s\Psi') a_{ij} + \Psi' b_i b_j + \frac{1}{2} A \Psi'' s_{\cdot i} s_{\cdot j} \tag{A2}$$

This is easily obtained by differentiating L with respect to $\dot{x}^i$, $\dot{x}^j$ and using the relations $A_{\cdot ij} = 2a_{ij}$ and

$$A s_{\cdot ij} + s_{\cdot i} A_{\cdot j} + s_{\cdot j} A_{\cdot i} = 2 b_i b_j - 2 s a_{ij} \ ;$$

the latter follows by differentiation of the identity $B^2 = sA$ twice, with $\dot{x}^i$ and $\dot{x}^j$.

The above formula for g_{ij} suggests that, in order to calculate $\det(g_{ij})$, we can apply the following known result from linear algebra twice:

Lemma A1 ([45], pp. 287–288). *If $Q \in \mathcal{M}_n(\mathbb{C})$ has $\det(Q_{ij}) \neq 0$, then:*

1. $\det(Q_{ij} + \delta C_i C_j) = \det(Q_{ij}) \left(1 + \delta C^k C_k\right)$ *for all n-vectors $C = (C_i) \in \mathbb{C}^n$, $\delta \in \mathbb{R}$, where the indices of C_i have been raised by means of the inverse (Q^{ij}).*
2. *If $1 + \delta C^k C_k \neq 0$, the inverse of the matrix $(Q_{ij} + \delta C_i C_j)$ exists and has the entries:*

$$Q^{ij} - \frac{\delta C^i C^j}{1 + \delta C^k C_k}. \tag{A3}$$

Note. Some other versions of this result, e.g., [49], require the matrix Q to be symmetric. Yet, as stated in [45], this restriction (though it applies to all the examples in this article), is not necessary, so we preferred to present the results in full generality.

Indeed, applying twice the above Lemma, we easily find:

Corollary A1. *If $Q \in \mathcal{M}_n(\mathbb{R})$ has $\det(Q_{ij}) \neq 0$, $\delta, \mu \in \mathbb{R}$ and the components of the n-vectors $B_k, C_k \in \mathbb{R}$ are such that at least one of the quantities $1 + \delta B^k B_k$, $1 + \mu C^k C_k$ is nonzero, then:*

$$\det(Q_{ij} + \delta B_i B_j + \mu C_i C_j) = \det(Q_{ij}) \left[\left(1 + \delta B^k B_k\right)\left(1 + \mu C^k C_k\right) - \delta\mu \left(B^k C_k\right)^2 \right], \quad \text{(A4)}$$

where $B^k = Q^{kj} B_j$, $C^k = Q^{kj} C_j$.

Moreover, if $\left(1 + \delta B^k B_k\right)\left(1 + \mu C^k C_k\right) - \delta\mu \left(B^k C_k\right)^2 \neq 0$, the inverse of the matrix $(Q_{ij} + \delta B_i B_j + \mu C_i C_j)$ exists and has the entries:

$$Q^{ij} - \frac{\mu\left(1 + B^k B_k\right) B^i C^j - \mu\delta B^k C_k (B^i C^j + B^j C^i) + \delta\left(1 + \mu C^k C_k\right) B^i B^j}{(1 + \delta B^k B_k)(1 + \mu C^k C_k) - \delta\mu (B^k C_k)^2} \quad \text{(A5)}$$

Proof. Denote $\hat{Q}_{ij} = Q_{ij} + \delta B_i B_j$. Assuming that $1 + \delta B^k B_k \neq 0$, from Lemma A1, we first find:

$$\det(\hat{Q}_{ij}) = \det(Q_{ij})\left(1 + \delta B^k B_k\right) \neq 0$$

Therefore $(\hat{Q}_{ij})$ is invertible, with an inverse as in (A3). The result then follows by applying Lemma A1 again on the matrix with entries $\hat{Q}_{ij} + \mu C_i C_j$. The case when $1 + \mu C^k C_k \neq 0$ is completely similar.

The formula for the inverse matrix can be obtained similarly, by applying point 2 of the above Lemma twice. □

We are now ready to calculate $\det(g_{\dot{x}})$. To this aim, let us fix an arbitrary $x \in M$ and an arbitrary local chart on $\pi^{-1}(x) \subset TM$. Taking into account Lemma 2, there is no loss of generality if we limit our attention to conic subsets of TM where $\Psi - s\Psi' \neq 0$. On such subsets, we can write:

$$g_{ij} = (\Psi - s\Psi')(a_{ij} + \delta b_i b_j + \mu s_{\cdot i} s_{\cdot j}); \quad \text{(A6)}$$

hence, we can apply the above corollary for:

$$Q_{ij} = a_{ij}, \quad B_k := b_k, \quad C_k = s_{\cdot k}, \quad \delta = \frac{\Psi'}{\Psi - s\Psi'}, \quad \mu = \frac{1}{2}\frac{A\Psi''}{\Psi - s\Psi'}. \quad \text{(A7)}$$

- To calculate the blocks appearing in (A4), we use $A_{\cdot i} = 2a_{ik}\dot{x}^k$, which yields: $a^{ij} A_{\cdot i} = 2\dot{x}^j$, $a^{ij} A_{\cdot i} b_j = 2B$, $s_{\cdot i} = \frac{2}{A}(Bb_i - s\tilde{x}_i)$ (where $\tilde{x}_i = a_{ij}\dot{x}^j$) and finally:

$$B^k B_k = \langle b, b \rangle, \qquad C^k C_k = \frac{4s}{A}(\langle b, b \rangle - s), \qquad \left(B^k C_k\right)^2 = \frac{4s}{A}(\langle b, b \rangle - s)^2. \quad \text{(A8)}$$

- On subsets where $1 + \delta b^k b_k \neq 0$, we can apply the corollary, which gives, after a brief computation:

$$\det(g_{ij}) = (\Psi - s\Psi')^{n-2} \det(a_{ij}) \left[\Psi(\Psi - s\Psi') + (\langle b, b \rangle - s)\left(\Psi\Psi' + 2s\Psi\Psi'' - s\Psi'^2\right) \right].$$

The square bracket can be rewritten as:

$$\frac{\Psi^2}{(\Psi - s\Psi')}\frac{\partial}{\partial s}\left((s - \langle b,b\rangle)\frac{(\Psi - s\Psi')^2}{\Psi}\right),$$

which gives the desired result (A1).

- The result can then be prolonged by continuity also at points where $1 + \delta b^k b_k = 0$. To see this, we note that this equality cannot happen on any entire interval $I_0 \subset I$, as this would entail that, for $s \in I_0$,

$$1 + \frac{\Psi'\langle b,b\rangle}{\Psi - s\Psi'} = 0 \qquad \Leftrightarrow \qquad \frac{\Psi'}{\Psi} = \frac{1}{s - \langle b,b\rangle}.$$

That is, $\Psi = \kappa(s - \langle b,b\rangle)$, where $\kappa = \kappa(x)$ only. However, on the (open) subset $s^{-1}(I_0) \subset \mathcal{A}$, we would then have $L = A\Psi = \kappa\left(B^2 - \langle b,b\rangle A\right)$, which has degenerate Hessian $g_{ij} = \kappa\left(b_i b_j - \langle b,b\rangle a_{ij}\right)$ and hence does not represent a pseudo-Finsler function.

In particular, for $n = \dim M = 4$, we find:

$$\det(g_{\dot{x}}) = \Psi^2(\Psi - s\Psi')\det(a)\frac{\partial}{\partial s}\left((s - \langle b,b\rangle)\frac{(\Psi - s\Psi')^2}{\Psi}\right) \tag{A9}$$

Now, if the determinant is nonzero, i.e., $(s - \langle b,b\rangle)\frac{(\Psi - s\Psi')^2}{\Psi} \neq$ *const.*, we can apply again the result (A5) from the above corollary for the metric tensor (A6). Using the blocks (A8), after some calculations we obtain the inverse

$$\begin{aligned} g^{ij} = {} & \frac{1}{\Psi - s\Psi'}a^{ij} + \frac{s^2}{\nu A}\Psi''[\Psi'(s - \langle b,b\rangle) - \Psi]\dot{x}^i\dot{x}^j \\ & + \frac{2Bs}{\nu A}\Psi\Psi''(b^i\dot{x}^j + b^j\dot{x}^i) + \frac{1}{\nu}[\Psi'(\Psi - s\Psi') - 2s\Psi\Psi'']b^i b^j, \end{aligned} \tag{A10}$$

where

$$\nu = (\Psi - s\Psi')\left[\Psi(\Psi - s\Psi') + (\langle b,b\rangle - s)\left(\Psi\Psi' + 2s\Psi\Psi'' - s\Psi'^2\right)\right].$$

The same results can be also obtained using [13].
Note that ν defined above can be related to σ and ρ from Theorem 1 as

$$\nu = \Psi^2\frac{d}{ds}(\rho\sigma).$$

Notes

1 This article appeared during the production process of the present paper.

2 A contrary statement appears in the arxiv version of the paper [46]; yet, in the published version [28], this statement does not appear anymore.

References

1. Pfeifer, C. Finsler spacetime geometry in Physics. *Int. J. Geom. Meth. Mod. Phys.* **2019**, *16*, 1941004. [CrossRef]
2. Saridakis, E.N.; Lazkoz, R.; Salzano, V.; Moniz, P.V.; Capozziello, S.; Jiménez, J.B.; De Laurentis, M.; Olmo, G.J. *Modified Gravity and Cosmology: An Update by the CANTATA Network*; Springer: Cham, Switzerland, 2021.
3. Hohmann, M.; Pfeifer, C.; Voicu, N. Relativistic kinetic gases as direct sources of gravity. *Phys. Rev.* **2020**, *D101*, 024062. [CrossRef]
4. Addazi, A.; Alvarez-Muniz, J.; Batista, R.A.; Amelino-Camelia, G.; Antonelli, V.; Arzano, M.; Asorey, M.; Atteia, J.-L.; Bahamonde, S.; Bajardi, F.; et al. Quantum gravity phenomenology at the dawn of the multi-messenger era—A review. *Prog. Part. Nucl. Phys.* **2022**, *125*, 103948. [CrossRef]
5. Lobo, I.P.; Pfeifer, C. Reaching the Planck scale with muon lifetime measurements. *Phys. Rev. D* **2021**, *103*, 106025. [CrossRef]
6. Amelino-Camelia, G.; Barcaroli, L.; Gubitosi, G.; Liberati, S.; Loret, N. Realization of doubly special relativistic symmetries in Finsler geometries. *Phys. Rev. D* **2014**, *90*, 125030. [CrossRef]

7. Kostelecký, A. Riemann-Finsler geometry and Lorentz-violating kinematics. *Phys. Lett. B* **2011**, *701*, 137–143. [CrossRef]
8. Shreck, M. Classical Lagrangians and Finsler structures for the nonminimal fermion sector of the Standard-Model Extension. *Phys. Rev.* **2016**, *D93*, 105017. [CrossRef]
9. Kostelecký, V.A.; Russell, N.; Tso, R. Bipartite Riemann–Finsler geometry and Lorentz violation. *Phys. Lett.* **2012**, *B716*, 470–474. [CrossRef]
10. Gibbons, G.; Gomis, J.; Pope, C. General very special relativity is Finsler geometry. *Phys. Rev. D* **2007**, *76*, 081701. [CrossRef]
11. Cohen, A.G.; Glashow, S.L. Very special relativity. *Phys. Rev. Lett.* **2006**, *97*, 021601. [CrossRef]
12. Fuster, A.; Pabst, C. Finsler pp-waves. *Phys. Rev.* **2016**, *D94, 104072*. [CrossRef]
13. Fuster, A.; Pabst, C.; Pfeifer, C. Berwald spacetimes and very special relativity. *Phys. Rev.* **2018**, *D98*, 084062. [CrossRef]
14. Elbistan, M.; Zhang, P.M.; Dimakis, N.; Gibbons, G.W.; Horvathy, P.A. Geodesic motion in Bogoslovsky-Finsler spacetimes. *Phys. Rev. D* **2020**, *102*, 024014. [CrossRef]
15. Bouali, A.; Chaudhary, H.; Hama, R.; Harko, T.; Sabau, S.V.; Martín, M.S. Cosmological tests of the osculating Barthel–Kropina dark energy model. *Eur. Phys. J. C* **2023**, *83*, 121. [CrossRef]
16. Gibbons, G.W.; Herdeiro, C.A.R.; Warnick, C.M.; Werner, M.C. Stationary Metrics and Optical Zermelo-Randers-Finsler Geometry. *Phys. Rev. D* **2009**, *79*, 044022. [CrossRef]
17. Werner, M.C. Gravitational lensing in the Kerr-Randers optical geometry. *Gen. Rel. Grav.* **2012**, *44*, 3047–3057. [CrossRef]
18. Caponio, E.; Javaloyes, M.Á.; Masiello, A. On the energy functional on Finsler manifolds and applications to stationary spacetimes. *Math. Ann.* **2011**, *351*, 365–392. [CrossRef]
19. Caponio, E.; Javaloyes, M.A.; Caja, M.S. On the interplay between Lorentzian Causality and Finsler metrics of Randers type. *Rev. Mat. Iberoam.* **2011**, *27*, 919–952. [CrossRef]
20. Heefer, S.; Pfeifer, C.; Fuster, A. Randers pp-waves. *Phys. Rev. D* **2021**, *104*, 024007. [CrossRef]
21. Heefer, S.; Fuster, A. Finsler gravitational waves of (α, β)-type and their observational signature. *arXiv* **2023**, arXiv:2302.08334.
22. Silva, J.E.G. A field theory in Randers-Finsler spacetime. *EPL* **2021**, *133*, 21002. [CrossRef]
23. Kapsabelis, E.; Triantafyllopoulos, A.; Basilakos, S.; Stavrinos, P.C. Applications of the Schwarzschild—Finsler—Randers model. *Eur. Phys. J. C* **2021**, *81*, 990. [CrossRef]
24. Caponio, E.; Javaloyes, M.A.; Sánchez, M. Wind Finslerian structures: From Zermelo's navigation to the causality of spacetimes. *arXiv* **2014**, arXiv:1407.5494.
25. Bacso, S.; Cheng, S.; Shen, Z. Curvature properties of (α, β)-metrics. *Adv. Stud. Pure Math.* **2005**, *48*, 73–110.
26. Sabau, S.; Shimada, H. Classes of Finsler spaces with (α, β)-metrics. *Rep. Math. Phys.* **2001**, *47*, 31–48. [CrossRef]
27. Matsumoto, M. Theory of Finsler spaces with (α, β)-metric. *Rep. Math. Phys.* **1992**, *31*, 43–83. [CrossRef]
28. Li, X.; Chang, Z.; Mo, X. Symmetries in a very special relativity and isometric group of Finsler space. *Chin. Phys. C* **2011**, *35*, 535. [CrossRef]
29. Elgendi, S.G.; Kozma, L. (α, β)-Metrics Satisfying the T-Condition or the σT-Condition. *J. Geom. Anal.* **2020**, *31*, 7866–7884. [CrossRef]
30. Crampin, M. Isometries and Geodesic Invariants of Finsler Spaces of (α, β) Type. Preprint on *Research Gate*. 2022. Available online: https://www.researchgate.net/publication/360335742_Isometries_and_geodesic_invariants_of_Finsler_spaces_of_a_b_type (accessed on 12 April 2023). [CrossRef]
31. Javaloyes, M.A.; Pendás-Recondo, E.; Sánchez, M. An account on links between Finsler and Lorentz Geometries for Riemannian Geometers. *arXiv* **2022**, arXiv:2203.13391.
32. Pfeifer, C.; Wohlfarth, M. Causal structure and electrodynamics on Finsler spacetimes. *Phys. Rev.* **2011**, *D84*, 044039. [CrossRef]
33. Lammerzahl, C.; Perlick, V.; Hasse, W. Observable effects in a class of spherically symmetric static Finsler spacetimes. *Phys. Rev.* **2012**, *D86*, 104042. [CrossRef]
34. Javaloyes, M.A.; Sánchez, M. On the definition and examples of cones and Finsler spacetimes. *RACSAM* **2020**, *114*, 30. [CrossRef]
35. Hasse, W.; Perlick, V. Redshift in Finsler spacetimes. *Phys. Rev.* **2019**, *D100*, 024033. [CrossRef]
36. Bernal, A.; Javaloyes, M.A.; Sánchez, M. Foundations of Finsler Spacetimes from the Observers' Viewpoint. *Universe* **2020**, *6*, 55. [CrossRef]
37. Caponio, E.; Masiello, A. On the analyticity of static solutions of a field equation in Finsler gravity. *Universe* **2020**, *6*, 59. [CrossRef]
38. Caponio, E.; Stancarone, G. Standard static Finsler spacetimes. *Int. J. Geom. Meth. Mod. Phys.* **2016**, *13*, 1650040. [CrossRef]
39. Hohmann, M.; Pfeifer, C.; Voicu, N. Mathematical foundations for field theories on Finsler spacetimes. *J. Math. Phys.* **2022**, *63*, 032503. [CrossRef]
40. Beem, J.K. Indefinite Finsler spaces and timelike spaces. *Can. J. Math.* **1970**, *22*, 1035. [CrossRef]
41. Asanov, G.S. *Finsler Geometry, Relativity and Gauge Theories*; D. Reidel Publishing Company: Dordrecht, The Netherlands, 1985.
42. Hohmann, M.; Pfeifer, C.; Voicu, N. Finsler gravity action from variational completion. *Phys. Rev.* **2019**, *D100*, 064035. [CrossRef]
43. Bogoslovsky, G. A special-relativistic theory of the locally anisotropic space-time. *Il Nuovo C. B Ser. 11* **1977**, *40*, 99. [CrossRef]
44. Bejancu, A.; Farran, H. *Geometry of Pseudo-Finsler Submanifolds*; Springer: Berlin/Heidelberg, Germany, 2000.
45. Bao, D.; Chern, S.S.; Shen, Z. *An Introduction to Finsler-Riemann Geometry*; Springer: New York, NY, USA, 2000.
46. Li, X.; Chang, Z.; Mo, X. Isometric group of (α, β)-type Finsler space and the symmetry of Very Special Relativity. *arXiv* **2010**, arXiv:1001.2667.
47. Shen, Z. *Differential Geometry of Spray and Finsler Spaces*; Springer: Dordrecht, The Netherlands, 2001.

48. Marcal, P.; Shen, Z. Ricci flat Finsler metrics by warped product. *Proc. Am. Math. Soc.* **2023**, *151*, 2169–2183.
49. Chern, S.S.; Shen, Z. *Riemann-Finsler Geometry*; Nankai Tracts in Mathematics: Volume 6; World Scientific: Singapore, 2005. [CrossRef]

universe

MDPI

Article

Newtonian Fractional-Dimension Gravity and Galaxies without Dark Matter

Gabriele U. Varieschi

Department of Physics, Loyola Marymount University, Los Angeles, CA 90045, USA; gabriele.varieschi@lmu.edu

Abstract: We apply Newtonian fractional-dimension gravity (NFDG), an alternative gravitational model, to some notable cases of galaxies with little or no dark matter. In the case of the ultra-diffuse galaxy AGC 114905, we show that NFDG methods can effectively reproduce the observed rotation curve using a variable fractional dimension $D(R)$, as was performed for other galaxies in previous studies. For AGC 114905, we obtain a variable dimension in the range $D \approx 2.2$–3.2, but our fixed $D = 3$ curve can still fit all the experimental data within their error bars. This confirms other studies indicating that the dynamics of this galaxy can be described almost entirely by the baryonic mass distribution alone. In the case of NGC 1052-DF2, we use an argument based on the NFDG extension of the virial theorem applied to the velocity dispersion of globular clusters showing that, in general, discrepancies between observed and predicted velocity dispersions can be attributed to an overall fractal dimension $D < 3$ of the astrophysical structure considered, and not to the presence of dark matter. For NGC 1052-DF2, we estimate $D \approx 2.9$, thus confirming that this galaxy almost follows standard Newtonian behavior. We also consider the case of the Bullet Cluster merger (1E0657-56), assumed to be one of the strongest proofs of dark matter existence. A simplified but effective NFDG model of the collision shows that the observed infall velocity of this merger can be explained by a fractional dimension of the system in the range $D \simeq 2.4$–2.5, again, without using any dark matter.

Keywords: Newtonian fractional-dimension gravity; modified gravity; modified Newtonian dynamics; dark matter; galaxies

1. Introduction

Newtonian fractional-dimension gravity (NFDG) is an alternative gravitational model introduced in 2020–2021 through a series of papers ([1–3], hereafter, referred to as papers I–III, respectively), with the goal of modeling galactic dynamics without using any dark matter (DM) component. The main assumption of NFDG is to consider galaxies, and possibly other astrophysical objects, as fractal structures characterized by a fractional-dimension function $D(R)$, where D represents a variable Hausdorff-type fractal dimension, which typically depends on the radial coordinate of the structure being studied and can assume any positive real value (usually, $1 < D \lesssim 3$).[1]

An intuitive understanding of the concept of fractal dimension is based on fractal geometry, as originally introduced by B. Mandelbrot [5], where a fractal object containing a detailed structure at small scales is characterized by a fractal dimension that can differ from the standard topological dimension (i.e., in contrast to the standard integer dimension $D = 1, 2, 3$). Fractal phenomena, displaying self-similarity and fractal dimensions, are being uncovered in stellar systems, galaxies, and other astrophysical structures [6], thus leading to the applications of fractals to astrophysics and cosmology.

The main motivation for considering a theory of gravity based on a fractional dimension $1 < D \lesssim 3$ is that it leads to a gravitational potential of the form $\sim 1/r^{D-2}$ (or $\sim \ln r$ for $D = 2$), as in Equation (1) below. This gravitational potential interpolates naturally between the standard Newtonian one ($\sim 1/r$, for $D = 3$) and those yielding flat galactic rotation curves, without the need to introduce any DM component. We also note that

NFDG should be considered [1–3] as a pure modification of the laws of gravity, as briefly outlined in Section 2. Therefore, NFDG will not affect other fundamental interactions such as electromagnetism, or the way electromagnetic waves propagate in the universe, etc.

In papers I–III, this model was shown to work effectively for general structures characterized by spherical or axial symmetries, as well as for three notable cases of rotationally supported galaxies (NGC 6503, NGC 7814, and NGC 3741), by fitting their rotation curves without using any DM contribution. In a subsequent paper IV [7], a relativistic version of the model was presented, while in our paper V [8], four more galaxies were analyzed (NGC 5033, NGC 6674, NGC 5055, NGC 1090) in relation to the so-called external field effect (EFE), and accurate fits to their rotation velocity data were obtained also in these cases.

In all these previous papers, we discussed possible points of contact with other existing alternative gravities (modified Newtonian dynamics—MOND [9–17]; conformal Gravity—CG [18–23]; modified gravity—MOG [24–26]; etc.), as well as the links with fractional gravity theories in multi-scale spacetimes [27–34]. In addition to these alternative theories of gravity, many other ideas have recently been explored, such as gravitational confinement [35], graviton–graviton interactions [36], and unexpected relativistic corrections [37], including gravitomagnetic effects [38,39], just to name a few.

In this short paper, we will apply our NFDG methods to the case of galaxies which appear to have little or no dark matter. In Section 2, we will consider the case of the ultra-diffuse galaxy (UDG) AGC 114905 [40,41], while in Section 3, we will study another UDG, NGC 1052-DF2 [42–44], which is another remarkable case of a galaxy lacking dark matter. In Section 4, we will also briefly analyze the Bullet Cluster merger (1E0657-56), which is usually referred to as empirical proof of the existence of dark matter [45,46].

Our NFDG model will prove to be effective in all these cases: for the first two, it will confirm that these galaxies can be considered to be virtually free of DM, while in the case of the Bullet Cluster, we will show that its dynamics can also be explained by NFDG without including DM contributions.

2. NFDG and the Ultra-Diffuse Galaxy AGC 114905

NFDG is based on a heuristic extension of Gauss's law for gravitation to a lower-dimensional spacetime $D+1$, where $D \leq 3$ can become a non-integer space dimension. In our previous papers, we introduced a fractional-dimension Poisson equation, $\nabla_D^2 \Phi = \frac{4\pi G}{l_0}\rho$, where ∇_D^2 is a generalized D-dimensional Laplacian (see Ref. [1], or the more detailed discussion in Ref. [7]), and ρ here represents the rescaled mass density. The solution to this Poisson equation, for a point-like mass m placed at the origin, yields the following NFDG gravitational potential [8]:

$$\Phi_{NFDG}(r) = -\frac{2\pi^{1-D/2}\Gamma(D/2)\, Gm}{(D-2)l_0 r^{D-2}}\ ;\ D \neq 2 \tag{1}$$
$$\Phi_{NFDG}(r) = \frac{2\, Gm}{l_0}\ln r\ ;\ D = 2,$$

where G is the gravitational constant, the radial coordinate r is considered to be a rescaled dimensionless coordinate, i.e., $r \to r/l_0$, and l_0 is an appropriate scale length, which is required for dimensional correctness in fractional gravity models.

This potential is then generalized to extended source mass distributions, particularly spherically symmetric and axially symmetric distributions, and used to model the three main components of galactic baryonic matter: the spherical bulge mass distribution (if present) and the cylindrical stellar disk and gas distributions. It is assumed that the space dimension D can be a function of the field point coordinates, but we usually neglect its space derivatives in the computation of the NFDG gravitational field:

$$\mathbf{g}_{NFDG}(R) = -\frac{1}{l_0}\frac{d\Phi_{NFDG}(R)}{dR}\widehat{\mathbf{R}}. \tag{2}$$

This field is usually computed just in the galactic disk plane and it is then expressed as a function of the (dimensionless) radial coordinate R in the same plane, with the scale length l_0 included for dimensional correctness. Similarly, the variable dimension function $D = D(R)$ is considered a function of the same radial variable and will characterize each particular galaxy studied with NFDG methods.

Finally, the circular velocity, for stars rotating in the main galactic plane, is simply obtained from Equation (2) as:

$$v_{circ}(R) = \sqrt{l_0 R |\mathbf{g}_{NFDG}(R)|}, \tag{3}$$

where the NFDG field is computed by using the variable dimension function $D = D(R)$ mentioned above. The observed rotation velocity data for each galaxy are then fitted by plotting these circular velocities as a function of the radial distance from the galactic center and the best $D(R)$ for each galaxy is obtained by matching the experimental data with the NFDG fits. An alternative method for computing $D(R)$ directly from the galactic mass distributions will be introduced later in this section, but this second method is, at the moment, less effective than the one described above. Full details about the NFDG model can be found in our papers I–V, with the latest version of our computations described in paper V [8].

It should be noted that NFDG does not imply any external field effect [8], so that the solar system gravity is essentially due to the Sun's gravitational field, with other nearby sources at most producing minor tidal forces. At the moment, we have not fully analyzed the role of spherically symmetric sources in NFDG, in order to explain why standard gravity ($D = 3$) is present at the solar system level, or for any other similar star systems. This will be performed in a future publication which will also study other spherically symmetric sources, such as globular clusters or galaxies with spherical symmetry.

However, in our previous papers dealing with axially symmetric sources, we showed how the dynamics of stars orbiting in a disk galaxy depends on the galactic mass distribution, which might be better described by a fractal structure, and thus by an effective NFDG dimension $D \leq 3$, also depending on position. This is the main NFDG hypothesis, supported by how the galactic rotation curves can be effectively described by NFDG methods [2,3,8]. The three situations analyzed in this manuscript are meant to provide additional support for this hypothesis. In the following, we will first apply NFDG to the case of AGC 114905.

The gas-rich ultra-diffuse galaxy AGC 114905 has always been considered an outlier of the baryonic Tully–Fisher relation, but the latest observations and analyses of this galaxy [40] have shown that its rotation curve can be almost entirely explained by the baryonic mass distribution alone, hence the possibility of the existence of dark-matter-free galaxies. This was not the first galaxy to be shown to be lacking dark matter, but since the NFDG methods outlined above can be easily applied to this case, we will consider AGC 114905 as our first and main example.

We note that AGC 114905 represents a challenge for other alternative theories of gravity, such as MOND, as shown in the original paper by Mancera Piña et al. [40]. However, a subsequent work by Banik et al. [47] has shown that the MOND analysis can be consistent with the experimental data, by invoking a much lower inclination angle for this galaxy than that assumed in Ref. [40].

The data and properties of AGC 114905 were obtained from Refs. [40,48], beginning with the mass modeling for this galaxy. In Figure 1, we show the mass data for the two main baryonic components (stars data—black circles; gas data—gray circles) as computed by Mancera Piña et al. [40,48] from the original luminosity data. This figure is equivalent to the right panel of Figure 1 in Ref. [40], although we use different units for the surface mass density Σ.

These data can then be interpolated with a spline function of the appropriate order (solid-orange curve for stars, and solid-blue curve for the gas component), or we can use the density profiles as in Ref. [40] (dashed-black curve for stars and dashed-gray curve for the gas component). These profiles for the stellar disk and gas component (HI plus helium) surface mass densities are defined as:

$$\Sigma_{disk}(R) = \left(\frac{M_*}{2\pi R_d^2}\right) e^{-R/R_d} \tag{4}$$
$$\Sigma_{gas}(R) = \Sigma_{0,gas} e^{-R/R_1}(1 + R/R_2)^{\alpha},$$

where $M_* = 1.3 \times 10^8 M_\odot$ and $R_d = 1.79\ kpc$ for the exponential stellar disk profile, while $\Sigma_{0,gas} = 3.2\ M_\odot/pc^2$, $R_1 = 1.1\ kpc$, $R_2 = 16.5\ kpc$ and $\alpha = 18$ for the gas profile [40]. The first profile in Equation (4) is a simple exponential function typically used to model stellar disks, while the second profile in the same equation is the best analytical fit for the gas component introduced in Ref. [40].

As in [40], we adopted a sech^2 profile along the vertical direction with a constant thickness of $z_d = 0.196\ R_d^{0.633} \approx 280\ pc$ [49] for the stellar disk, and a Gaussian profile with a constant vertical scale height $z_d = 250\ pc$ for the gas disk. These vertical profiles, together with the radial ones in Equation (4), were used as the main input for our NFDG mass distributions.[2]

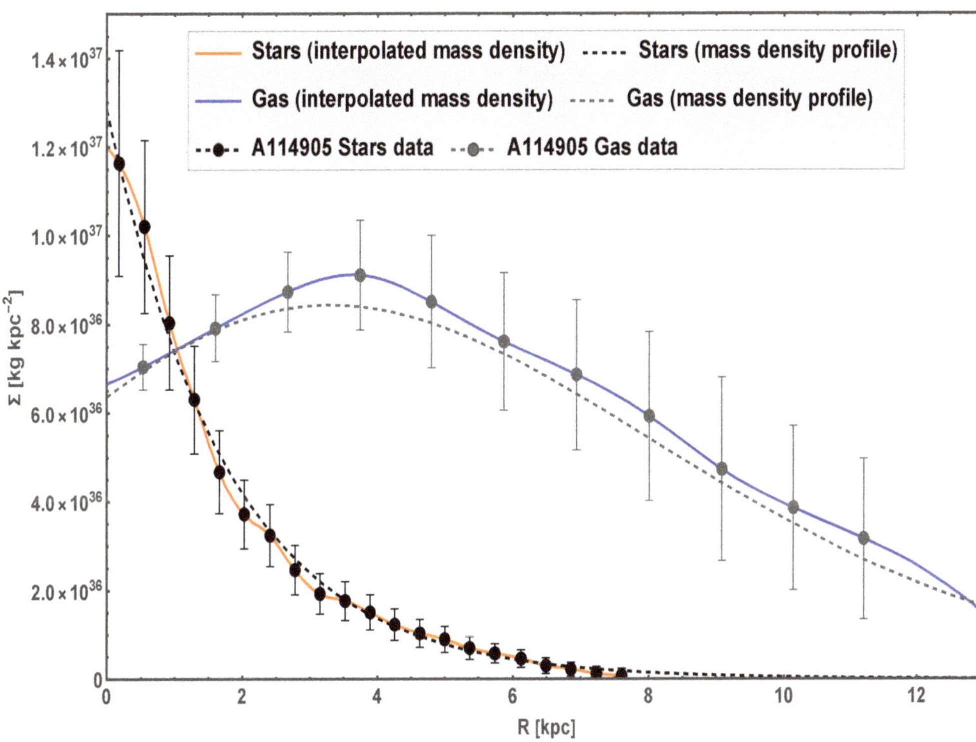

Figure 1. Surface mass densities for AGC 114905. The data for the stellar and gas components, obtained from Mancera Piña et al. [40,48], are interpolated with spline functions (solid-orange and solid-blue curves), or with the mass density profiles in Equation (4) (dashed curves).

We then used the latest version of the NFDG computation, as detailed in Appendix A of our paper V [8], in order to obtain the variable dimension function $D(R)$, which yields the best NFDG fit to the observed circular velocity data for AGC 114905. These experimental data points were obtained from the same sources in Refs. [40,48], and are shown in Figure 2 (bottom panel, black circles with error bars), together with the original contributions [40,48] expected for the stars (orange—dot-dashed line), gas (cyan—dot-dashed line), and total baryonic mass (stars plus gas—magenta dot-dashed line). These data and related contributions are the same as those shown in Figure 4 of Ref. [40].

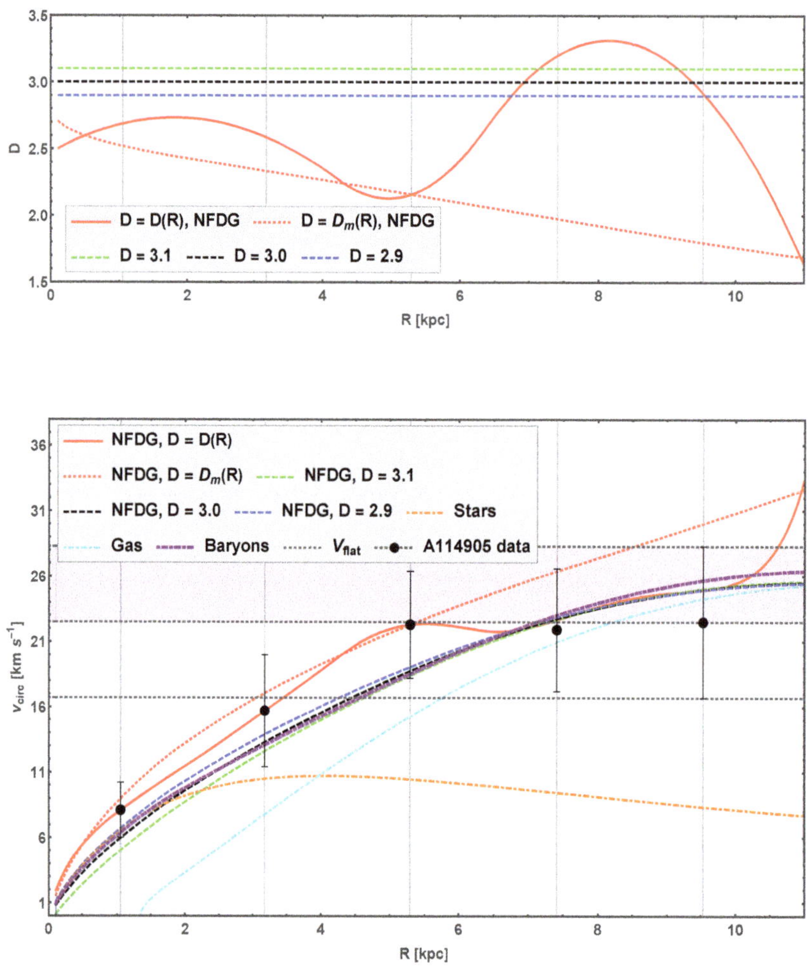

Figure 2. NFDG results for AGC 114905. Top panel: NFDG variable dimension $D(R)$ based directly on the galaxy data (red-solid curve) and NFDG variable dimension $D_m(R)$ based on the mass dimension of a fractal system (red-dotted curve), are compared with fixed values $D = 3.1$, $D = 3.0$, and $D = 2.9$ (dashed lines). Bottom panel: NFDG rotation curves (circular velocity vs. radial distance) compared to the original data [40,48] (black circles with error bars). The NFDG best fit for the variable dimension $D(R)$ is shown by the red-solid line; the NFDG fit for $D_m(R)$ is shown by the red-dotted line; while NFDG fits for fixed values $D = 3.1$, $D = 3.0$, and $D = 2.9$ are shown, respectively, by the green-dashed, black-dashed, and blue-dashed lines. The original computation by Mancera Piña et al. [40,48] of the contributions expected from the stars (orange dot-dashed line), from gas (cyan dot-dashed line), and total baryonic mass (magenta, dot-dashed line) is also shown. The asymptotic flat velocity band (horizontal gray band) is instead based on the data/errors of the last experimental point.

The additional curves shown in both panels of Figure 2 are instead related to our NFDG computations. In the top panel, we show our best NFDG variable dimension function $D(R)$ as a red-solid line, together with three fixed dimension values $D = 3.1,\ 3.0,\ 2.9$, related to the green, black, and blue dotted lines, respectively. The corresponding NFDG fits to the rotation data are shown in the bottom panel using the lines of the same color/type.

The main NFDG fit is represented by the red-solid line in the bottom panel of the figure. It should be noted that this fit is based on only five experimental data points available for AGC 114905, as opposed to the many experimental points which were available for each of the galaxies studied in our previous papers (based on the SPARC data catalog [50]). Therefore, the NFDG computations of the variable dimension function $D(R)$ and related rotation velocity (red-solid curves in both panels) were only carried out for the five radial distances of the known experimental data (shown by the vertical gray-thin lines in both panels of Figure 2), while interpolation was used in between these radial distances, making our predictions somewhat less effective than the results obtained in previous papers.

However, our main NFDG $D(R)$ fit in the bottom panel can effectively match the experimental data: the first three data points are perfectly matched, while for the last two, our red-solid curve is close to the data points and well within their respective error bars. The other NFDG $D = 3.1,\ 3.0,\ 2.9$ fits in the bottom panel (in green, black, and blue dashed curves, respectively) were computed in order to compare these fixed-dimension curves with the experimental data, as well as with the original baryonic curve (magenta, dot-dashed). Our NFDG $D = 3.0$ curve (black-dashed) is close to the original baryonic curve (magenta, dot-dashed, from Ref. [40]), with minor differences probably due to the different methods used to compute these purely Newtonian components.

Instead considering the NFDG $D = 3.1,\ 2.9$ curves (green and blue, dashed lines) over the radial range, we observe that the spread between these curves becomes progressively less pronounced with the increasing radial distances, with all the NFDG fits becoming almost equivalent in the radial region between the last two experimental points. Again, this shows some limitations of our NFDG fits for the last two data points, probably due to the lack of more precise experimental data in this outer radial region.

An alternative NFDG computation is also shown by the red-dotted curves in both panels of Figure 2. This alternative analysis was based on the discussion in Appendix C of Ref. [8], where the variable dimension function is considered as a field depending on the radial coordinate R and denoted by $D_m(R)$ in the following. In this approach, $D_m(R)$ is now based on the concept of mass dimension of a fractal system, i.e., the mass dimension of a disk-shaped source is defined as $M(R) \approx M_0(R/R_0)^{D_m(R)-1}$, with $M(R)$ being the total baryonic mass within radius R, and R_0 representing a scale length of the system with related mass M_0.

Following these definitions, it was shown in the same Appendix C of Ref. [8] that this new field $D_m(R)$ can be easily computed as:

$$D_m(R) \approx 1 + \ln\left[\frac{M(R)}{M(R_0)}\right] / \ln\left[\frac{R}{R_0}\right], \tag{5}$$

where $M(R) = 2\pi \int_0^R \Sigma_{tot}(R')R'dR'$ and $M_0 \equiv M(R_0) = 2\pi \int_0^{R_0} \Sigma_{tot}(R')R'dR'$. As such, the field $D_m(R)$ is directly predicted from the baryonic matter distribution alone (described by the total surface mass density $\Sigma_{tot}(R')$) and not based on the observed rotational velocity data points, as was performed instead using the previous method for $D = D(R)$ (red-solid curves).

As also discussed in Ref. [8], the value of R_0 (and related mass M_0) can be left as a free parameter, or simply chosen by matching one of the experimental data points. We opted for this simpler solution and just arranged this free parameter to match the third data point in the bottom panel of Figure 2. The resulting circular velocity fit (red-dotted curve) in the bottom panel seems to effectively match the first three data points, as well as the fourth one within its error bar, while the fifth experimental point is not well fitted by this curve.

The corresponding red-dotted $D_m(R)$ curve in the top panel shows a smooth variation in the variable dimension, as opposed to the more pronounced changes in the red-solid $D(R)$ curve, computed with the original NFDG method. We would like to remark that this new method of direct computation of the field $D_m(R)$ is still a work in progress and will be improved in future publications.

Apart from these limitations of our NFDG fits in the outer radial region, we conclude that our methods also work reasonably well for the case of AGC 114905, as analyzed in this section. We confirm that an NFDG $D = 3.0$ curve, i.e., purely baryonic, can fit all five experimental data within their error bars, as in the original analysis by Mancera Piña et al. [40], since our black-dashed curve is close to the magenta dot-dashed curve in the bottom panel of Figure 2. However, an improved fit to the experimental points is obtained by our NFDG variable $D(R)$ curve, with values for the variable dimension at the radial distances of the experimental points detailed in Table 1.

Table 1. AGC 114905 circular velocity data (with errors) and related radial distances [40,48], together with the corresponding values of the NFDG variable dimension $D(R)$.

R (kpc)	v_{circ} **(km s^{-1})**	**D (R)**
1.05	8.1 ± 2.1	2.69
3.18	15.7 ± 4.3	2.58
5.29	22.3 ± 4.1	2.16
7.41	21.9 ± 4.7	3.20
9.53	22.5 ± 5.8	2.92

From this table and from the top panel of Figure 2 (red-solid curve), we see that this galaxy is best described in NFDG as having a fractional dimension $D \approx 2.5$–2.7 at lower radial distances around the first two experimental points, then decreasing towards $D \approx 2.2$ around the region of the third experimental point, and finally increasing towards $D \approx 2.9$–3.2 in the outer radial region (the last two experimental points). However, the NFDG fits to these last two points are not perfect, as already mentioned above. We conclude that NFDG can also effectively model this galaxy, without any DM contribution, although the quality of our fit is limited by the low number of experimental data points available.

3. NFDG and the Dynamics of NGC 1052-DF2

Another notable case of a galaxy apparently lacking dark matter is the ultra-diffuse galaxy NGC 1052-DF2, as recently studied by van Dokkum et al. [42–44]. By measuring the radial velocities of ten globular-cluster objects within this galaxy, it was shown [43] that the intrinsic velocity dispersion of these clusters is $\sigma_{gc} = 5.6^{+5.2}_{-3.8}$ km s^{-1}, leading to a 90% confidence upper limit of $\sigma_{gc} < 12.4$ km s^{-1}, with the likelihood level instead giving $\sigma_{gc} = 7.8^{+5.2}_{-2.2}$ km s^{-1} ($\sigma_{gc} < 14.6$ km s^{-1}). These globular-cluster velocity dispersions are very close to the one expected from stars alone, $\sigma_{stars} = 7.0^{+1.6}_{-1.3}$ km s^{-1}, thus implying that this galaxy is very deficient in DM and might be the candidate of a purely baryonic galaxy.

Later papers [44,51–53] improved upon the original analysis of NGC 1052-DF2, while confirming the lack of dark matter in this galaxy. In particular, in Ref. [52], the globular-cluster velocity dispersion was updated as $\sigma_{gc} = 10.6^{+3.9}_{-2.3}$ km s^{-1}, in Ref. [53], the stellar velocity dispersion was revised as $\sigma_{stars} = 8.5^{+2.3}_{-3.1}$ km s^{-1}, and in Ref. [44], the ultra-diffuse galaxy DF2 was seen as part of a group of similar dark-matter free galaxies, probably originated from a bullet–dwarf collision.

It should be noted that the original analysis [42] of this UDG seemed to be inconsistent with MOND predictions. However, the correct MOND predictions should not only include the internal field of the galaxy, but also the external field due to the proximity to the giant parent galaxy NGC 1052. Including this external field effect into the MOND analysis, it was shown [54] that MOND predictions ($\sigma_{MOND} = 13.4^{+4.8}_{-3.7}$ km s^{-1}) are in agreement with the observed velocity dispersions reported above.[3]

In NFDG, the EFE is not present [8], but the non-integer variable dimension D postulated by our model will affect the theoretical computation of the velocity dispersion of NGC 1052-DF2. In this paper, we limit our analysis to a simple argument based on the NFDG version of the virial theorem, while we will leave a more comprehensive analysis of this topic to future work.

In mechanics, the virial theorem [56] links the average over time of the total kinetic energy $\langle T \rangle$ of a stable system of particles with the average over time of the total potential energy of the system $\langle U \rangle$. If the potential energy of a two-particle system is of the form $U(r) = \alpha r^n$, i.e., proportional to some power n of the inter-particle distance r, it is easy to show that the virial theorem implies $2\langle T \rangle = n\langle U \rangle$. For the case of standard inverse-square-distance forces, such as Newtonian gravity, we have $n = -1$ and $2\langle T \rangle + \langle U \rangle = 0$; however, in NFDG, we instead have $n = -(D-2)$, in view of Equation (1), and the NFDG virial theorem becomes:

$$2\langle T \rangle + (D-2)\langle U \rangle = 0. \tag{6}$$

A simple application of the virial theorem to a system of N astrophysical objects, each of mass m', uniformly distributed inside a sphere of radius R and total mass $M = Nm'$, yields $\langle T \rangle = \frac{1}{2}Nm'\langle v^2 \rangle = \frac{1}{2}M\langle v^2 \rangle$ and $\langle U \rangle = -\frac{3}{5}\frac{GM^2}{R}$. Using the standard ($D = 3$) form of the virial theorem and assuming $\langle v^2 \rangle \approx 3\sigma^2$, the "Newtonian" velocity dispersion σ_{Newt} can be estimated from the total mass M within the radius R as:

$$\sigma_{Newt} \approx \sqrt{\frac{GM}{5R}}. \tag{7}$$

Using the NFDG gravitational potential in Equation (1) for a constant dimension $D \neq 2$ and with a rescaled radial coordinate instead of the dimensionless one, i.e., $r \rightarrow r/l_0$ (l_0 is the NFDG scale length), it is straightforward to recompute the time-averaged potential energy of the system as $\langle U \rangle = -\frac{2D\,\pi^{1-D/2}\Gamma(D/2)}{(D-2)(D+2)}\frac{GM^2}{l_0(R/l_0)^{D-2}}$, which can be used in the generalized form of the virial theorem in Equation (6) to obtain the following NFDG velocity dispersion:

$$\sigma_{NFDG} \approx \sqrt{\frac{2D\,\pi^{1-D/2}\Gamma(D/2)GM}{3(D+2)l_0(R/l_0)^{D-2}}}, \tag{8}$$

which reduces to Equation (7) for $D = 3$. Details of the computations leading to Equations (7) and (8) are discussed in Appendix A.

In order to use Equation (8), we need to specify the value of the NFDG scale length l_0, currently undefined. In view of the connections of NFDG with MOND, as described in all our previous papers I–V, we can assume that $l_0 = \sqrt{GM_{ref}/a_0}$, where M_{ref} is an appropriate reference mass and $a_0 = 1.2 \times 10^{-10}$m s^{-2} is the MOND acceleration constant. In this section, we simply assume that $M_{ref} = M$, the total mass of the spherical system within radius R.

It should be noted that the choice of the value for the scale length l_0 did not affect the calculations of the rotation curves of AGC 114905 in Section 2, as well as other rotation curves computed in previous papers, since this quantity is canceled out in the computations. However, in this section and in the following one, we need to use an explicit value for l_0 based on the connection with the MOND acceleration a_0, as outlined above. Since our first paper [1], we have heuristically argued that $a_0 \approx GM/l_0^2$, as this connection allowed NFDG to recover the MOND flat rotational speed $V_f = \sqrt[4]{GMa_0}$, the baryonic Tully–Fisher relation, and other fundamental MOND results. In addition, similar connections were used in other fractional gravity models, such as that proposed by Giusti et al. [57–59], where the scale length was explicitly set as $l_0 = (2/\pi)\sqrt{GM/a_0}$, which is practically equivalent to the

NFDG relation reported above.[4] Following these considerations, we feel confident that our NFDG relation for l_0 can be used in our current analysis, at least as a first approximation for our study of globular clusters in this section and of the Bullet Cluster in Section 4.

Although Equations (7) and (8) are simplified estimates of the velocity dispersion of a system and should be employed just to determine the order of magnitude of these quantities, we used them in the context of NGC 1052-DF2 to show how σ might depend on the effective fractal dimension D of this galaxy. For example, we considered the original data in Ref. [43] assuming that the Newtonian velocity dispersion in Equation (7) corresponds to the one due to stars alone, i.e., $\sigma_{Newt} \approx \sigma_{stars} = 7.0^{+1.6}_{-1.3}$ km s^{-1}, as well as used the estimated total stellar mass as the total mass $M = M_{stars} = 2.0^{+1.0}_{-0.7} \times 10^8 M_{\odot}$.

For each of these values of σ_{Newt} and M (central value and errors), we derived the corresponding radius R from Equation (7), which was then used in the NFDG velocity dispersion in Equation (8), together with the appropriate scale length l_0. This last equation was then numerically solved for the variable dimension D, assuming that the NFDG velocity dispersion is now set equal to the globular-cluster velocity dispersion, i.e., $\sigma_{NFDG} \approx \sigma_{gc} = 7.8^{+5.2}_{-2.2}$ km s^{-1} [43].

The result of this computation is[5]:

$$D = 2.91^{-0.28}_{+0.10}, \tag{9}$$

showing that the increase in the globular-cluster velocity dispersion, compared to the purely baryonic stellar estimate, can be due to an effective decrease in the fractional dimension below the standard $D = 3$ value. However, our NFDG estimate in Equation (9) is still very close to the standard $D \approx 3$, indicating that this galaxy has very little mass discrepancy, and thus confirming the results in the literature.

Figure 3 provides a more graphical illustration of our argument. In this figure, we show the dependence of the NFDG velocity dispersion σ_{NFDG} as a function of the variable dimension D, following Equation (8) and the procedure described above. In particular, the stellar velocity dispersion $\sigma_{stars} = 7.0^{+1.6}_{-1.3}$ km s^{-1} is shown by the black data point corresponding to a standard $D = 3$ spatial dimension and by the horizontal black-dashed line, while the globular-cluster velocity dispersion $\sigma_{gc} = 7.8^{+5.2}_{-2.2}$ km s^{-1} is shown by the red/orange data points (central value/error points, respectively, as also shown with horizontal gray-dotted lines), corresponding to the values of D reported in Equation (9): $D = 2.91$ (red data point), $D = 2.63,\ 3.01$ (orange error data points).

The gray vertical gridlines correspond to all these values of the dimension D, while the main NFDG red-solid curve and the related orange error band clearly show how an increase in the observed globular-cluster velocity dispersion, compared to the purely baryonic (stellar) one, could be due to a lower fractional dimension of the galaxy, instead of being attributed to the presence of dark matter. In other words, this NFDG analysis based on the extension of the virial theorem in Equations (6)–(8) can be used for other galaxies where the discrepancy between baryonic and observed globular-cluster velocity dispersions is usually explained in terms of a DM component. NFDG might be able to explain these discrepancies in terms of a $D \neq 3$ value of the fractional dimension of the galaxy being studied.

It should again be noted that the NFDG approach to globular-clusters dispersion velocity discussed in this section is only a limited, preliminary study. A more comprehensive analysis should be performed in NFDG by the following methods based on the spherical Jeans equation (see [60] and the references therein), or others. We will leave this work to future papers on the subject.

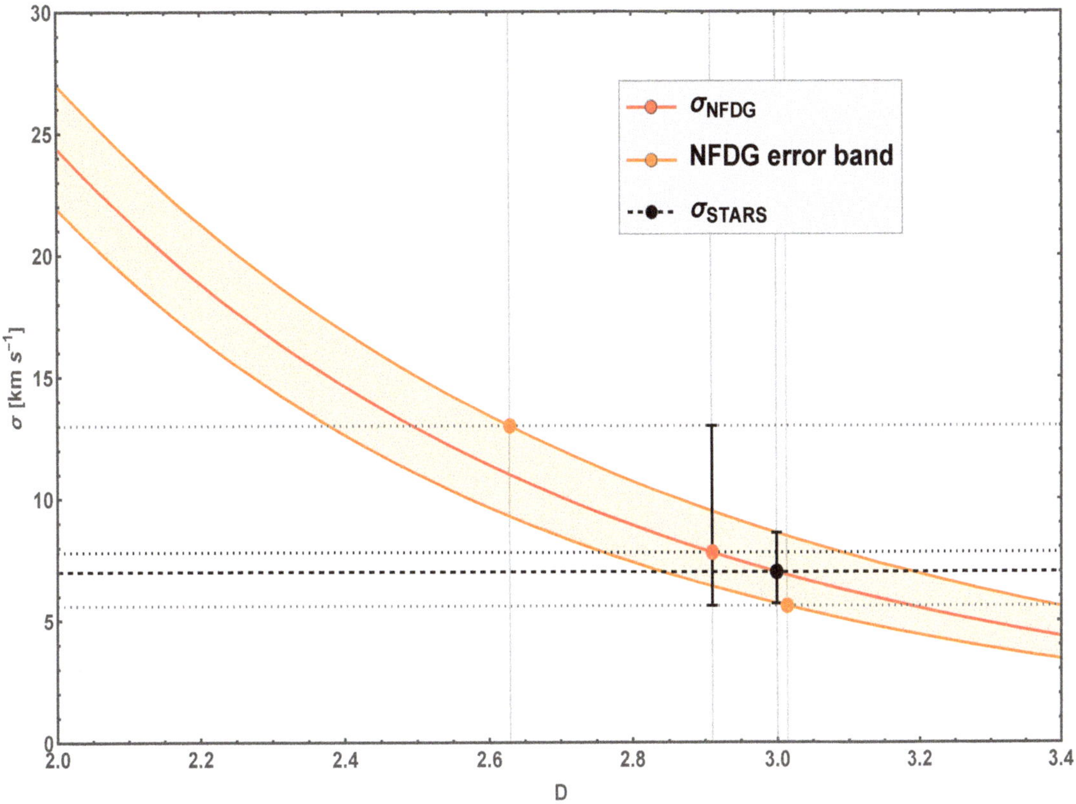

Figure 3. NFDG velocity dispersion as a function of the variable dimension, following Equation (8) (red curve and related orange error band). The black-circle data point and related errors are assumed to correspond the stars' velocity dispersion for $D = 3$, while the red-circle data point and orange-circle error points are assumed to be related to the globular-cluster velocity dispersion with a variable dimension D.

4. NFDG and the Bullet Cluster

The Bullet Cluster (BC) merger (1E0657-06) is considered one of the strongest proofs of the existence of DM [45,46]. This merger is usually described in terms of the gas components of the two colliding clusters (revealed by the X-ray emissions) which, together with standard luminous matter (stars), represent the baryonic content of the cluster pair. In addition, a strong DM component is revealed by gravitational lensing, pointing towards a direct, empirical evidence of dark matter in this collision [45].

A simplified but effective way of modeling this collision [25,34] can be achieved by considering the two clusters as point masses subject to their reciprocal gravitational attraction.[6] Using the updated values in Ref. [46], the main cluster is assumed to have a dynamical mass $M_m = 2 \times 10^{15} M_\odot$, while the bullet cluster dynamical mass is estimated as $M_b = 5 \times 10^{14} M_\odot$. These "dynamical" masses include both baryonic and DM contributions, while purely baryonic masses can be obtained using the estimated baryonic-to-total mass ratio, $M_{bar}/M_{total} = 0.14$ [46], i.e., $M_m^{bar} = 2.8 \times 10^{14} M_\odot$ and $M_b^{bar} = 7.0 \times 10^{13} M_\odot$.

The main objective of this simplified analysis is to obtain the initial infall velocity of the two clusters at a certain separation distance, which was estimated in Ref. [46] as 2900 km s^{-1} at a distance of 2.5 Mpc. Treating the system of the two point-mass clusters as isolated and with the clusters at relative rest at infinity, the infall velocity computation follows the standard calculation of the escape velocity, by setting to zero both the total energy and the total momentum of the system:

$$\frac{1}{2}\left(M_m v_m^2 + M_b v_b^2\right) - G\frac{M_m M_b}{r} = 0$$
$$M_m v_m + M_b v_b = 0. \tag{10}$$

These equations can be easily solved to obtain the standard, Newtonian, infall velocity $(v_m - v_b)_{Newt}$ at the separation distance r:

$$(v_m - v_b)_{Newt} = \sqrt{\frac{2G}{r}(M_m + M_b)}. \tag{11}$$

Using the total cluster masses reported above (including DM) and the separation distance $r = 2.5$ Mpc, one obtains $(v_m - v_b)_{Newt} = 2933$ km s^{-1}, in line with the expected results [46], while using the purely baryonic masses for the two clusters, also reported above, one would instead obtain $(v_m - v_b)_{Newt} = 1097$ km s^{-1}, clearly showing that baryonic masses alone, combined with standard Newtonian gravity, are unable to effectively describe the cluster merger.

However, the same argument can now be used in our NFDG model, by considering the appropriate extension of the gravitational potential energy and using only baryonic masses, since NFDG does not include DM components:

$$U_{NFDG} = -\frac{2\pi^{1-D/2}\Gamma(D/2)GM_m^{bar}M_b^{bar}}{(D-2)l_0 r^{D-2}}, \tag{12}$$

which follows from the NFDG gravitational potential for $D \neq 2$ in Equation (1). Using this NFDG potential energy instead of the Newtonian one in Equation (10) and solving for the NFDG equivalent infall velocity, we obtain:

$$(v_m - v_b)_{NFDG} = \sqrt{\frac{4\pi^{1-D/2}\Gamma(D/2)G}{(D-2)l_0(r/l_0)^{D-2}}\left(M_m^{bar} + M_b^{bar}\right)}. \tag{13}$$

In this last equation, we introduced the rescaled radial coordinate instead of the dimensionless one, i.e., $r \to r/l_0$, with l_0 representing as usual the NFDG scale length. It is easy to check that Equation (13) is reduced to Equation (11) for $D = 3$.

We can now assume that the infall velocity is the same as the one obtained with Equation (11), using the total dynamical masses (including DM) of the two clusters, i.e., set $(v_m - v_b)_{NFDG} = 2933$ km s^{-1} and solving Equation (13) for the value of the fractional dimension D which would yield the above infall velocity at the same separation distance $r = 2.5$ Mpc. Obtaining a reasonable value for D (e.g., $1 \leq D \leq 3$) would show that NFDG can explain the BC dynamics as a fractional-dimension effect, at least within the limits of this simplified model.

In order to numerically solve Equation (13), we again need to specify the value of the NFDG scale length l_0, currently undefined. As performed in the previous section, we set $l_0 = \sqrt{GM_{ref}/a_0}$, where M_{ref} is an appropriate reference baryonic mass and $a_0 = 1.2 \times 10^{-10}$ m s^{-1} is the MOND acceleration constant. The reference mass, although arbitrary, should be related to the baryonic masses of the two clusters, with some possible choices being the total baryonic mass $M_{ref} = M_m^{bar} + M_b^{bar} = 3.5 \times 10^{14} M_\odot$, the individual baryonic masses of each cluster ($M_{ref} = M_m^{bar} = 2.8 \times 10^{14} M_\odot$, or $M_{ref} = M_b^{bar} = 7.0 \times 10^{13} M_\odot$), the reduced mass of the cluster system ($M_{ref} = M_m^{bar} M_b^{bar} / \left(M_m^{bar} + M_b^{bar}\right) = 5.6 \times 10^{13} M_\odot$), etc.

Table 2 shows the values of the fractional dimension D which would yield the infall velocity of $(v_m - v_b)_{NFDG} = 2933$ km s^{-1} in NFDG for different choices of the reference mass M_{ref}, as outlined above. We can see that all these values for D are within a reasonable range of $D \simeq 2.4$–2.5, similar to the values obtained in previous NFDG studies of the dynamics of individual galaxies. Although our analysis of the BC is based on a simplified model of this merger, we conclude that NFDG is potentially able to describe the nature of this phenomenon without using any DM component.

Table 2. NFDG fractional dimension D for the Bullet Cluster system for different choices of the reference mass M_{ref}.

M_{ref}	D
$M_m^{bar} + M_b^{bar}$	2.43
M_m^{bar}	2.45
M_b^{bar}	2.53
$\frac{M_m^{bar} M_b^{bar}}{M_m^{bar} + M_b^{bar}}$	2.54

5. Conclusions

In this work, we expanded our NFDG model to include galaxies with little or no dark matter (AGC 114905 and NGC 1052-DF2), as well as the case of the Bullet Cluster merger (1E0657-06), considered one of the strongest proofs for the existence of dark matter. We used different NFDG methods for the three cases: our standard NFDG procedures, employed in previous studies, were applied to AGC 114905 and produced an effective fit to the rotational velocity data points without any DM.

For NGC 1052-DF2, we extended the virial theorem to cases with dimension $D < 3$, and we were able to show that an increase in the observed velocity dispersion can be the result of a lower overall fractional dimension of the galactic structure. However, for both galaxies, our methods yielded $D \lesssim 3$, indicating that Newtonian dynamics can model these galaxies reasonably well and any residual discrepancy can be explained by a decrease in the dimension D below the standard $D = 3$ value. In other words, NFDG provides a simple explanation of the almost Newtonian behavior of galaxies such as AGC 114905 and NGC 1052-DF2, in terms of a slight change in the fractional dimension from standard $D = 3$, without invoking the EFE and a possible violation of the strong equivalence principle, as it was performed by MOND theories when analyzing the same two galaxies.

Finally, a simplified model of the Bullet Cluster merger (1E0657-06) was used to show that the observed infall velocity could be due to a fractional dimension $D \simeq 2.4$–2.5 of the system, without requiring any DM presence. Although the NFDG methods employed in the last two cases (NGC 1052-DF2 and 1E0657-06) will need to be improved and extended, we conclude that our model has once again shown its effectiveness in describing all these astrophysical structures without resorting to the dark matter paradigm.

Therefore, these results constitute another step towards proving that dark matter might not be an exotic component of the Universe, but rather a problem with the standard gravitational theories. However, future NFDG studies are still needed to show that this model can yield other predictions besides galactic rotation curves. Future directions will possibly include a more detailed study of galactic clusters, gravitational lensing, and an improved analysis of the relativistic version of NFDG [7], for possible applications in astrophysics and cosmology.

Funding: This research received no external funding.

Data Availability Statement: The datasets generated during and/or analyzed in the present study are available from the corresponding author upon reasonable request.

Acknowledgments: This work was supported by the Seaver College of Science and Engineering, Loyola Marymount University—Los Angeles. The author wishes to thank Mancera Piña for sharing AGC 114905 galactic data files and other useful information, and also acknowledges the anonymous reviewers for useful comments and suggestions.

Conflicts of Interest: The authors declare no conflict of interest.

Appendix A. NFDG Virial Theorem and Spherical Structures

In this section, we will detail the computations leading to Equations (7) and (8) in Section 3. Let us consider a system of N astrophysical objects, each of mass m', uniformly distributed inside a sphere of radius R and total mass $M = Nm'$. The time-averaged total kinetic energy is $\langle T \rangle = \frac{1}{2}Nm'\langle v^2 \rangle = \frac{1}{2}M\langle v^2 \rangle \approx \frac{3}{2}M\sigma^2$, and is directly related to the velocity dispersion σ.

The NFDG computation of the time-averaged potential energy $\langle U \rangle$ follows from $d\langle U \rangle = \phi_{NFDG}(r)dm$ with the NFDG potential from Equation (1), $\phi_{NFDG}(r) = -\frac{2\pi^{1-D/2}\Gamma(D/2)}{(D-2)}\frac{Gm}{l_0(r/l_0)^{D-2}}$. NFDG [1,7] is based on the integral of a spherically symmetric function $f = f(r)$ over a D-dimensional metric space χ [62], $\int_\chi f d\mu_H = \frac{2\pi^{D/2}}{\Gamma(D/2)}\int_0^\infty f(r)r^{D-1}dr$, where μ_H denotes an appropriate Hausdorff measure over the space and the fractional dimension D is assumed to be constant. In view of this integral, the total mass $m(r)$, within a sphere of radius r is:

$$m(r) = \frac{2\pi^{D/2}}{\Gamma(D/2)}\int_0^r \rho(r')l_0^3(r'/l_0)^{D-1}d(r'/l_0), \tag{A1}$$

where the NFDG scale length l_0 has been added for dimensional correctness, and the standard formula $m(r) = 4\pi\int_0^r \rho(r')r'^2dr'$ is recovered for $D = 3$.

If the total mass M is uniformly distributed inside the sphere of radius R, i.e., for a constant volume density ρ, from the previous equation, we have $M = m(R) = \frac{2\pi^{D/2}}{\Gamma(D/2)}\rho l_0^{3-D}\frac{R^D}{D}$, and therefore, the constant density ρ can be expressed as $\rho = \frac{\Gamma(D/2)}{2\pi^{D/2}}\frac{DM}{l_0^{3-D}R^D}$ ($\rho = \frac{3M}{4\pi R^3}$ for $D = 3$). From these equations, it also follows that $m(r) = M(\frac{r}{R})^D$, i.e., the mass within a sphere of radius r obviously scales as r^D in NFDG.

Combining all these formulas into the expression for $d\langle U \rangle$ yields:

$$d\langle U \rangle = \phi_{NFDG}(r)dm = -\frac{2\pi^{1-D/2}\Gamma(D/2)}{(D-2)}\frac{DGM^2}{R^{2D}l_0^{3-D}}r^{D+1}dr, \tag{A2}$$

which becomes $d\langle U \rangle = -\frac{3GM^2}{R^6}r^4dr$ for $D = 3$. Integrating Equation (A2) over the whole sphere, we obtain the time-averaged potential energy already reported in Section 3:

$$\langle U \rangle = \int_{sphere}\phi_{NFDG}(r)dm = -\frac{2D\,\pi^{1-D/2}\Gamma(D/2)}{(D-2)(D+2)}\frac{GM^2}{l_0(R/l_0)^{D-2}}, \tag{A3}$$

which yields the standard result for $D = 3$, $\langle U \rangle = -\frac{3GM^2}{5R}$.

Finally, using the NFDG virial theorem in Equation (6) together with $\langle T \rangle \approx \frac{3}{2}M\sigma^2$ and $\langle U \rangle$ from Equation (A3), we obtain the final expression in Equation (8) for the NFDG velocity dispersion σ_{NFDG}.

Notes

1 For a general overview of Newtonian fractional-dimension gravity, also see the NFDG website [4].

2 The alternative radial profiles given by the direct interpolation of the mass densities (solid-orange and solid-blue curves in Figure 1) were also tested, but were found to yield results which were very similar to those based on the analytical profiles in Equation (4), so these alternative results will be omitted here.

3 For a more comprehensive analysis of NGC 1052-DF2 within the context of four popular alternative theories of gravity (MOND, CG, MOG, and Verlinde's emergent gravity), see Ref. [55].

4 This model by Giusti et al. [57–59] is based on a different way of extending the Poisson equation, $\Delta\Phi(\mathbf{r}) = 4\pi G\rho(\mathbf{r})$ into a fractional Poisson equation, $(-\Delta)^s\Phi(\mathbf{r}) = -4\pi G l_0^{2-2s}\rho(\mathbf{r})$, where $(-\Delta)^s$ is the fractional Laplacian and $s \in [1, 3/2]$ is the fractional index. The Newtonian case is recovered for $s = 1$. This fractional Laplacian is defined in terms of fractional operators, while our NFDG is based on standard operators acting on a metric space of fractional dimension D [1,7].

5 It should be noted that the positive/negative error values for σ_{gc} correspond to the negative/positive error values for the dimension D, as shown in Equation (9).

[6] This simplified method discussed in Ref. [25] was then linked to the MOG model. A more detailed analysis of the BC merger in MOG can be found in Ref. [61].

References

1. Varieschi, G.U. Newtonian Fractional-Dimension Gravity and MOND. *Found. Phys.* **2020**, *50*, 1608–1644, Erratum in *Found. Phys.* **2021**, *51*, 41. [CrossRef]
2. Varieschi, G.U. Newtonian Fractional-Dimension Gravity and Disk Galaxies. *Eur. Phys. J. Plus* **2021**, *136*, 183. [CrossRef]
3. Varieschi, G.U. Newtonian Fractional-Dimension Gravity and Rotationally Supported Galaxies. *Mon. Not. R. Astron. Soc.* **2021**, *503*, 1915–1931. [CrossRef]
4. Varieschi, G.U. Newtonian Fractional-Dimension Gravity (NFDG). 2020. Available online: https://gvarieschi.lmu.build/NFDG2020.html (accessed on 1 April 2023).
5. Mandelbrot, B.B. *The Fractal Geometry of Nature*, 3rd ed.; W. H. Freeman and Company: New York, NY, USA, 1983.
6. Baryshev, Y.; Teerikorpi, P. *Discovery of Cosmic Fractals*; World Scientific: River Edge, NJ, USA, 2002; 373p.
7. Varieschi, G.U. Relativistic Fractional-Dimension Gravity. *Universe* **2021**, *7*, 387. [CrossRef]
8. Varieschi, G.U. Newtonian fractional-dimension gravity and the external field effect. *Eur. Phys. J. Plus* **2022**, *137*, 1217. [CrossRef]
9. Milgrom, M. A Modification of the Newtonian dynamics as a possible alternative to the hidden mass hypothesis. *Astrophys. J.* **1983**, *270*, 365–370. [CrossRef]
10. Milgrom, M. A Modification of the Newtonian dynamics: Implications for galaxies. *Astrophys. J.* **1983**, *270*, 371–383. [CrossRef]
11. Milgrom, M. A modification of the Newtonian dynamics: Implications for galaxy systems. *Astrophys. J.* **1983**, *270*, 384–389. [CrossRef]
12. Bekenstein, J.; Milgrom, M. Does the missing mass problem signal the breakdown of Newtonian gravity? *Astrophys. J.* **1984**, *286*, 7–14. [CrossRef]
13. Milgrom, M. MOND: A pedagogical review. *Acta Phys. Polon.* **2001**, *B32*, 3613. [CrossRef]
14. Sanders, R.H.; McGaugh, S.S. Modified Newtonian Dynamics as an Alternative to Dark Matter. *Annu. Rev. Astron. Astr.* **2002**, *40*, 263–317. [CrossRef]
15. Famaey, B.; McGaugh, S. Modified Newtonian Dynamics (MOND): Observational Phenomenology and Relativistic Extensions. *Living Rev. Rel.* **2012**, *15*, 10. [CrossRef] [PubMed]
16. Skordis, C.; Zlosnik, T. New Relativistic Theory for Modified Newtonian Dynamics. *Phys. Rev. Lett.* **2021**, *127*, 161302. [CrossRef] [PubMed]
17. Merritt, D. *A Philosophical Approach to MOND*; Cambridge University Press: Cambridge, MA, USA, 2020. [CrossRef]
18. Mannheim, P.D.; Kazanas, D. Exact Vacuum Solution to Conformal Weyl Gravity and Galactic Rotation Curves. *Astrophys. J.* **1989**, *342*, 635–638. [CrossRef]
19. Mannheim, P.D.; Kazanas, D. Newtonian limit of conformal gravity and the lack of necessity of the second order Poisson equation. *Gen. Rel. Grav.* **1994**, *26*, 337–361. [CrossRef]
20. Mannheim, P.D. Alternatives to dark matter and dark energy. *Prog. Part. Nucl. Phys.* **2006**, *56*, 340–445. [CrossRef]
21. Mannheim, P.D.; O'Brien, J.G. Impact of a global quadratic potential on galactic rotation curves. *Phys. Rev. Lett.* **2011**, *106*, 121101. [CrossRef] [PubMed]
22. Mannheim, P.D.; O'Brien, J.G. Fitting galactic rotation curves with conformal gravity and a global quadratic potential. *Phys. Rev. D* **2012**, *85*, 124020. [CrossRef]
23. Mannheim, P.D. Making the Case for Conformal Gravity. *Found. Phys.* **2012**, *42*, 388–420. [CrossRef]
24. Moffat, J.W. Scalar-tensor-vector gravity theory. *JCAP* **2006**, *3*, 4. [CrossRef]
25. Moffat, J.W.; Toth, V.T. Can Modified Gravity (MOG) explain the speeding Bullet (Cluster)? *arXiv* **2010**, arXiv:1005.2685.
26. Green, M.A.; Moffat, J.W. Modified Gravity (MOG) fits to observed radial acceleration of SPARC galaxies. *Phys. Dark Univ.* **2019**, *25*, 100323. [CrossRef]
27. Calcagni, G. Geometry and field theory in multi-fractional spacetime. *JHEP* **2012**, *1*, 65. [CrossRef]
28. Calcagni, G. Fractal universe and quantum gravity. *Phys. Rev. Lett.* **2010**, *104*, 251301. [CrossRef]
29. Calcagni, G. Quantum field theory, gravity and cosmology in a fractal universe. *JHEP* **2010**, *3*, 120. [CrossRef]
30. Calcagni, G. Multi-scale gravity and cosmology. *JCAP* **2013**, *12*, 41. [CrossRef]
31. Calcagni, G.; De Felice, A. Dark energy in multifractional spacetimes. *Phys. Rev. D* **2020**, *102*, 103529. [CrossRef]
32. Calcagni, G. Multifractional theories: An updated review. *Mod. Phys. Lett. A* **2021**, *36*, 2140006. [CrossRef]
33. Calcagni, G. Classical and quantum gravity with fractional operators. *Class. Quant. Grav.* **2021**, *38*, 165005, Erratum in *Class. Quant. Grav.* **2021**, *38*, 169601. [CrossRef]
34. Calcagni, G.; Varieschi, G.U. Gravitational potential and galaxy rotation curves in multi-fractional spacetimes. *JHEP* **2022**, *8*, 24. [CrossRef]
35. Barker, W.E.V.; Hobson, M.P.; Lasenby, A.N. Does gravitational confinement sustain flat galactic rotation curves without dark matter? *arXiv* **2023**, arXiv:2303.11094.
36. Deur, A. Implications of Graviton-Graviton Interaction to Dark Matter. *Phys. Lett. B* **2009**, *676*, 21–24. [CrossRef]
37. Deur, A. Relativistic corrections to the rotation curves of disk galaxies. *Eur. Phys. J. C* **2021**, *81*, 213. [CrossRef]

38. Cooperstock, F.I.; Tieu, S. Galactic Dynamics via General Relativity: A Compilation and New Developments. *Int. J. Mod. Phys. A* **2007**, *22*, 2293–2325. [CrossRef]
39. Ludwig, G.O. Galactic rotation curve and dark matter according to gravitomagnetism. *Eur. Phys. J. C* **2021**, *81*, 186. [CrossRef]
40. Piña Mancera, P.E.; Fraternali, F.; Oosterloo, T.; Adams, E.A.K.; Oman, K.A.; Leisman, L. No need for dark matter: Resolved kinematics of the ultra-diffuse galaxy AGC 114905. *Mon. Not. R. Astron. Soc.* **2022**, *512*, 3230–3242. [CrossRef]
41. Kong, D.; Kaplinghat, M.; Yu, H.B.; Fraternali, F.; Piña Mancera, P.E. The Odd Dark Matter Halos of Isolated Gas-rich Ultradiffuse Galaxies. *Astrophys. J.* **2022**, *936*, 166. [CrossRef]
42. Van Dokkum, P.; Danieli, S.; Cohen, Y.; Merritt, A.; Romanowsky, A.J.; Abraham, R.; Brodie, J.; Conroy, C.; Lokhorst, D.; Mowla, L. et al. A galaxy lacking dark matter. *Nature* **2018**, *555*, 629–632. [CrossRef]
43. Van Dokkum, P.; Cohen, Y.; Danieli, S.; Romanowsky, A.; Abraham, R.; Brodie, J.; Conroy, C.; Kruijssen, J.; Lokhorst, D.; Mowla, L. et al. A Revised Velocity for the Globular Cluster GC-98 in the Ultra Diffuse Galaxy NGC 1052-DF2. *Res. Notes Am. Astron. Soc.* **2018**, *2*, 54. [CrossRef]
44. Van Dokkum, P.; Shen, Z.; Keim, M.A.; Trujillo-Gomez, S.; Danieli, S.; Dutta Chowdhury, D.; Abraham R.; Conroy, C.; Kruijssen, J.; Nagai, D.; et al. A trail of dark-matter-free galaxies from a bullet-dwarf collision. *Nature* **2022**, *605*, 435–439. [CrossRef]
45. Clowe, D.; Bradac, M.; Gonzalez, A.H.; Markevitch, M.; Randall, S.W.; Jones, C.; Zaritsky, D. A direct empirical proof of the existence of dark matter. *Astrophys. J. Lett.* **2006**, *648*, L109–L113. [CrossRef]
46. Lage, C.; Farrar, G.R. The Bullet Cluster is not a Cosmological Anomaly. *JCAP* **2015**, *2*, 38. [CrossRef]
47. Banik, I.; Nagesh, S.T.; Haghi, H.; Kroupa, P.; Zhao, H. Overestimated inclinations of Milgromian disc galaxies: The case of the ultradiffuse galaxy AGC 114905. *Mon. Not. R. Astron. Soc.* **2022**, *513*, 3541–3548. [CrossRef]
48. Piña Mancera, P.E. (Leiden Observatory, Leiden, The Netherlands). Personal communication, 2022.
49. Bershady, M.A.; Verheijen, M.A.W.; Westfall, K.B.; Andersen, D.R.; Swaters, R.A.; Martinsson, T. The DiskMass Survey. II. Error Budget. *Astrophys. J.* **2010**, *716*, 234–268. [CrossRef]
50. Lelli, F.; McGaugh, S.S.; Schombert, J.M. SPARC: Mass Models for 175 Disk Galaxies with Spitzer Photometry and Accurate Rotation Curves. *Astron. J.* **2016**, *152*, 157. [CrossRef]
51. Wasserman, A.; Romanowsky, A.J.; Brodie, J.; van Dokkum, P.; Conroy, C.; Abraham, R.; Cohen, Y.; Danieli, S. A Deficit of Dark Matter from Jeans Modeling of the Ultra-diffuse Galaxy NGC 1052-DF2. *Astrophys. J. Lett.* **2018**, *863*, L15. [CrossRef]
52. Emsellem, E.; van der Burg, R.F.J.; Fensch, J.; Jerábková, T.; Zanella, A.; Agnello, A.; Hilker, M.; Müller, O.; Rejkuba, M.; Duc, P.-A.; et al. The ultra-diffuse galaxy NGC 1052-DF2 with MUSE—I. Kinematics of the stellar body. *Astron. Astrophys.* **2019**, *625*, A76. [CrossRef]
53. Danieli, S.; van Dokkum, P.; Conroy, C.; Abraham, R.; Romanowsky, A.J. Still Missing Dark Matter: KCWI High-Resolution Stellar Kinematics of NGC1052-DF2. *Astrophys. J. Lett.* **2019**, *874*, L12. [CrossRef]
54. Famaey, B.; McGaugh, S.; Milgrom, M. MOND and the dynamics of NGC 1052-DF2. *Mon. Not. R. Astron. Soc.* **2018**, *480*, 473–476. [CrossRef]
55. Islam, T.; Dutta, K. Modified Gravity Theories in Light of the Anomalous Velocity Dispersion of NGC1052-DF2. *Phys. Rev. D* **2019**, *100*, 104049. [CrossRef]
56. Clausius, R. XVI. On a mechanical theorem applicable to heat. *Lond. Edinb. Dublin Philos. Mag. J. Sci.* **1870**, *40*, 122–127. [CrossRef]
57. Giusti, A. MOND-like Fractional Laplacian Theory. *Phys. Rev. D* **2020**, *101*, 124029. [CrossRef]
58. Giusti, A.; Garrappa, R.; Vachon, G. On the Kuzmin model in fractional Newtonian gravity. *Eur. Phys. J. Plus* **2020**, *135*, 798. [CrossRef]
59. Benetti, F.; Lapi, A.; Gandolfi, G.; Salucci, P.; Danese, L. Dark Matter in Fractional Gravity I: Astrophysical Tests on Galactic Scales. *arXiv* **2023**, arXiv:2303.15767.
60. Wolf, J.; Martinez, G.D.; Bullock, J.S.; Kaplinghat, M.; Geha, M.; Munoz, R.R.; Simon, J.D.; Avedo, F.F. Accurate Masses for Dispersion-supported Galaxies. *Mon. Not. R. Astron. Soc.* **2010**, *406*, 1220. [CrossRef]
61. Moffat, J.W.; Rahvar, S.; Toth, V.T. Applying MOG to lensing: Einstein rings, Abell 520 and the Bullet Cluster. *Galaxies* **2018**, *6*, 43. [CrossRef]
62. Svozil, K. Quantum field theory on fractal spacetime: A new regularisation method. *J. Phys. A Math. Gen.* **1987**, *20*, 3861–3875. [CrossRef]

universe

MDPI

Communication

Production of Primordial Black Holes in Improved E-Models of Inflation

Daniel Frolovsky [1] **and Sergei V. Ketov** [1,2,3,*]

[1] Interdisciplinary Research Laboratory, Tomsk State University, Tomsk 634050, Russia; daniylfrolovsky@yandex.kz
[2] Department of Physics, Tokyo Metropolitan University, Tokyo 192-0397, Japan
[3] Kavli Institute for the Physics and Mathematics of the Universe (WPI), The University of Tokyo Institutes for Advanced Study, Chiba 277-8583, Japan
* Correspondence: ketov@tmu.ac.jp

Abstract: E-type α-attractor models of single-field inflation were generalized further in order to accommodate production of primordial black holes (PBHs) via adding a near-inflection point to the inflaton scalar potential at smaller scales, in good agreement with measurements of cosmic microwave background (CMB) radiation. A minimal number of new parameters were used but their fine-tuning was maximized in order to increase the possible masses of PBHs formed during an ultra-slow-roll phase, leading to a large enhancement in the power spectrum of scalar (curvature) perturbations by 6 or 7 orders of magnitude against the power spectrum of perturbations observed in CMB. It was found that extreme fine-tuning of the parameters in our models can lead to the formation of moon-sized PBHs, with masses of up to 10^{26} g, still in agreement with CMB observations. Quantum corrections are known to lead to the perturbative upper bound on the amplitude of large scalar perturbations responsible for PBH production. The quantum (one-loop) corrections in our models were found to be suppressed by one order of magnitude for PBHs with masses of approximately 10^{19} g, which may form the whole dark matter in the Universe.

Keywords: inflation; primordial black holes

1. Introduction

The paradigm of cosmological inflation proposed by Guth and Linde [1,2] is a possible answer to internal problems of standard (Einstein–Friedmann) cosmology. Inflation as the amplifier of curvature perturbations is also a possible origin of seeds of structure formation in the early Universe [3,4]. The existence of inflation is supported by measurements of cosmic microwave background (CMB) radiation [5–7].

The CMB measurements highlight a small window into high energy physics in the form of the power spectrum of perturbations on limited (CMB) scales and lead to important restrictions on viable inflationary models, but do not allow a reconstruction of the unique underlying inflationary model. There are uncertainties even among the simplest single-field (quintessence) models of inflation in the form of the (canonical) inflaton scalar potential. These uncertainties can be exploited to win more room for the single-field inflationary models by extending the flexibility of their predictions about inflation, while simultaneously describing the formation of primordial black holes (PBHs) from large scalar perturbations at lower (than CMB) scales [8,9]. A standard mechanism of PBH formation is given by adding a near-inflection point to the inflaton potential [10–13]. The observational constraints on PBHs are also subject to uncertainties [14,15] even in the limited context of the single-field models of inflatons with a near-inflection point [16]. Therefore, it makes sense to investigate the amount of flexibility of fine-tuning in the well-motivated models of inflation and PBH production on a case-by-case basis.

In this paper, we revisit the *α-attractor* (single-field) inflationary models introduced in refs. [17,18] and propose new generalizations of them. The simplest attractor model with $\alpha = 1$ is given by the famous Starobinsky model [19] motivated by gravitational interactions, with an inflaton as the Nambu–Goldstone boson associated with the spontaneous breaking of scale invariance [20]. After taking into account the CMB constraints alone, the Starobinsky model has no free parameters and leads to a sharp prediction for the tensor power spectrum tilt of the gravitational waves induced by inflation. The (canonical) inflaton scalar potential in the Starobinsky model reads

$$V(\phi) = \frac{3}{4}M_{\rm Pl}^2 M^2\left[1-\exp\left(-\sqrt{\tfrac{2}{3}}\phi/M_{\rm Pl}\right)\right]^2 \equiv V_0(1-2y)+\mathcal{O}(y^2),$$
$$y = \exp\left(-\sqrt{\tfrac{2}{3}}\phi/M_{\rm Pl}\right), \tag{1}$$

where $M \sim 10^{-5}M_{\rm Pl}$ is the inflaton mass and $M_{\rm Pl}$ is the reduced Planck mass. This leads to the Mukhanov–Chibisov formula [21] for the tilt of scalar perturbations, $n_s \approx 1-2/N_e$, and the tensor-to-scalar ratio $r \approx 12/N_e^2$, where N_e is the number of e-folds measuring the duration of inflation, in very good agreement with the CMB measurements [5–7],

$$n_s = 0.9649 \pm 0.0042 \quad (68\%\ \text{C.L.})\ , \quad r < 0.032 \quad (95\%\ \text{C.L.}), \tag{2}$$

for $N_e = 55 \pm 10$. Actually, only the leading exponent in Equation (1), i.e., the term linear in y, is relevant for the tilts during slow-roll inflation predicted by the Starobinsky model. This simple observation allows one to extend the Starobinsky model by a single new parameter $\alpha > 0$ entering the new variable y as [17,18]

$$y = \exp\left(-\sqrt{\tfrac{2}{3\alpha}}\phi/M_{\rm Pl}\right), \tag{3}$$

without changing the Mukhanov–Chibisov formula for n_s. However, the tensor-to-scalar ratio changes via

$$r \approx \frac{12\alpha}{N_e^2}, \tag{4}$$

thus adding more flexibility against future measurements of r.

The scalar potentials of the simplest α-attractor models fall into two classes: the simplest *E-models* have the scalar potential shown in Equation (1) but with the variable y defined by Equation (3), whereas the simplest *T-models* have the scalar potential

$$V_{\rm T}(\phi) = V_0\tanh^2\frac{\phi/M_{\rm Pl}}{\sqrt{6\alpha}} \equiv V_0\tilde{r}^2\ , \quad \tilde{r} = \tanh\frac{\phi/M_{\rm Pl}}{\sqrt{6\alpha}} \equiv \tanh\varphi\ , \tag{5}$$

where the scale of inflation is given by V_0. Further generalizations are also possible, while keeping the predictions for the cosmological tilts n_s and r during inflation. For example, generalized T-models with the scalar potential $V_{\alpha,f}(\varphi) = f^2(\tilde{r})$ and a monotonically increasing (during slow roll) function $f(\tilde{r})$ were used in refs. [22–24] for engineering a near-inflection point in the potential and PBH production, see also ref. [25]. The T-model potentials can also be made periodic by changing their global shape after replacing the function $\tilde{r}(\varphi)$ in Equation (5) by the periodic (Jacobi) elliptic function $\mathrm{sn}(\varphi|k)$ with the elliptic modulus $0 < k^2 < 1$, in the limit

$$\mathrm{sn}(\varphi|k) \approx \tanh\varphi \quad \text{for} \quad k^2 \to 1\ , \tag{6}$$

thus combining theoretically attractive features of chaotic inflation and natural inflation.

The power spectrum of scalar perturbations in some generalized T-models with a near-inflection point for PBH production and the related spectrum of induced gravitational waves were derived in ref. [26]. Similar results for the E-models were obtained in ref. [27].

The scalar potential used in ref. [27] has the form

$$V(\phi) = \frac{3}{4} M_{Pl}^2 M^2 \left[1 - y + y^2(\beta - \gamma y)\right]^2 , \tag{7}$$

with y defined by Equation (3) and the new (positive) parameters β and γ needed for engineering a near-inflection point. The ansatz (7) was motivated in ref. [26] by small values of y during slow-roll inflation, keeping agreement with CMB measurements [28] and having a near-inflection point at lower scales for certain (fine-tuned) values of the new parameters, which was needed for viable inflation and PBH production.[1]

As was demonstrated in ref. [27], the potential (7) leads to PBH production, with the PBH masses corresponding to asteroid-sized black holes in the small mass window, where those PBHs may form the whole dark matter in the Universe. However, in ref. [27], it was achieved at the price of having the agreement with the CMB value of n_s outside 1σ (though within 2σ), with fine tuning the parameters and ignoring quantum corrections. Actually, the higher the masses of the PBH produced, the lower (in red) the scalar tilt n_s becomes. Therefore, the following questions arise:

- Is it possible to reach the perfect (within 1σ) agreement with the CMB value of n_s in E-type models of inflation and PBH formation?
- Is it possible to increase the PBH masses beyond asteroid size?
- Is it possible so suppress (one-loop) quantum corrections?

In this communication, we improve our earlier findings about E-models in order to achieve a perfect agreement (inside 1σ) with the observed values of CMB tilts, reconsider PBH formation and comment on (one-loop) quantum corrections in our new E-type models.

2. More General E-Models

At least two extra parameters are needed, as in Equation (7), for engineering a near-inflection point. However, the parameters in Equation (7) do not have clear meaning, so it is useful to replace them by other (related) parameters as [23,27]

$$\beta = \frac{1}{1-\xi^2} \exp\left[\sqrt{\frac{2}{3\alpha}} \frac{\phi_i}{M_{Pl}}\right], \quad \gamma = \frac{1}{3(1-\xi^2)} \exp\left[2\sqrt{\frac{2}{3\alpha}} \frac{\phi_i}{M_{Pl}}\right]. \tag{8}$$

The two real parameters (ϕ_i, ξ) have a clear meaning; when $\xi = 0$, the potential has an inflection point at $\phi = \phi_i$ only and when $0 < \xi \ll 1$, the potential also has a local minimum (dip) and a local maximum (bump) at $y^{\pm}_{\text{ext.}}$, respectively, while both extrema are equally separated from the inflection point,

$$y^{\pm}_{\text{ext.}} = y_i(1 \pm \xi) , \quad y_i = \exp\left[-\sqrt{\frac{2}{3\alpha}} \frac{\phi_i}{M_{Pl}}\right] . \tag{9}$$

The analyticity of the inflaton potential with respect to y (near $y = 0$), responsible for the infinite plateau at $y \to +\infty$, can be relaxed by adding *negative* powers of y. This is motivated by supergravity cosmology [30] and string cosmology [31], where such terms naturally appear. For our purposes, we use this resource in the minimalistic manner by adding merely a single term with a negative power to the E-model potential (7). We find adding a term proportional to y^{-2} is more efficient than y^{-1}, so we only consider the case of

$$V(\phi) = \frac{3}{4} M_{Pl}^2 M^2 \left[1 - y - \theta y^{-2} + y^2(\beta - \gamma y)\right]^2 , \tag{10}$$

with the new (third) real parameter θ that is supposed to be small enough in order to keep agreement with CMB. An impact of the new parameter on the shape of the scalar potential is illustrated on Figure 1. The shape needed for inflation and PBH production is not generic, and is found after scanning the parameter space and fine-tuning the parameter values.

As is clear from Figure 1 that a negative θ brings the potential up, and a positive θ brings the potential down, against the plateau for $\theta = 0$ for large values of ϕ/M_{Pl}.

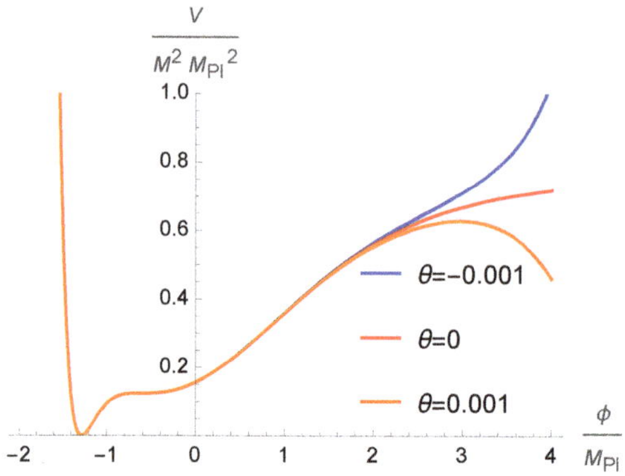

Figure 1. The potential (10) with $\alpha = 0.74265$, $\xi = 0.01213$ and $\phi_i/M_{Pl} = -0.61115$ for the selected values of θ.

The impact of the new parameter θ on the slow-roll (SR) parameters

$$\epsilon_{sr}(\phi) = \frac{M_{Pl}^2}{2}\left[\frac{V'(\phi)}{V(\phi)}\right]^2, \qquad \eta_{sr}(\phi) = M_{Pl}^2 \frac{V''(\phi)}{V(\phi)}, \tag{11}$$

is displayed in Figure 2, which shows that the SR phase is in agreement with CMB. The end of this SR phase of inflation is determined by the condition $|\eta| = 1$. The standard pivot scale of the horizon exit, when scalar perturbations leave the horizon, is $k_* = 0.05\ \text{Mpc}^{-1}$.

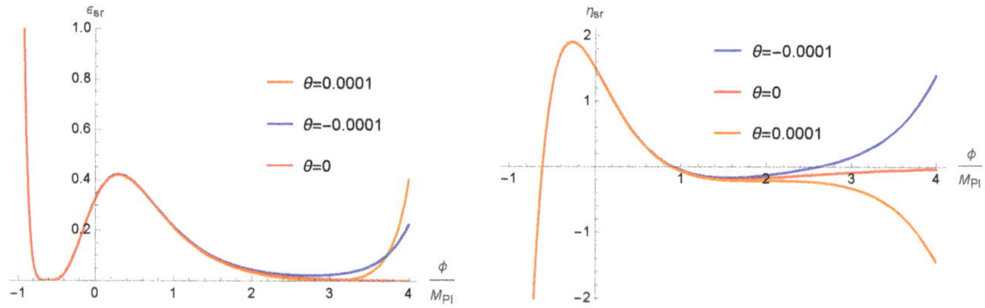

Figure 2. The ϕ-dependence of the slow-roll parameters at $\alpha = 0.74265$, $\xi = 0.01213$ and $\phi_i/M_{Pl} = -0.61115$ for the selected values of θ.

The raising tail of the profile of $\epsilon_{sr}(\phi)$ near the horizon exit on the left plot in Figure 2 increases the tensor-to-scalar ratio r, while the need to increase n_s implies the negative sign of θ, as is clear from the right plot in Figure 2. To increase the PBH masses (see the next section), we can go up to the maximal value of $r_{max.} = 0.032$ according to Equation (2).

Our numerical solutions of the standard equations of motion for the inflaton $\phi(t)$ as a function of time t and the Hubble function $H(t)$ with the same values of the parameters (α, ξ, ϕ_i) for the selected values of $\theta < 0$ are shown in Figure 3. It shows the duration of inflation, defined by the length of the plateau, for different values of θ.

The Hubble flow functions are defined by

$$\epsilon(t) = -\frac{\dot{H}}{H^2}, \qquad \eta(t) = \frac{\dot{\epsilon}}{H\epsilon}. \tag{12}$$

The (numerically derived) dynamics of these function are shown in Figure 4.

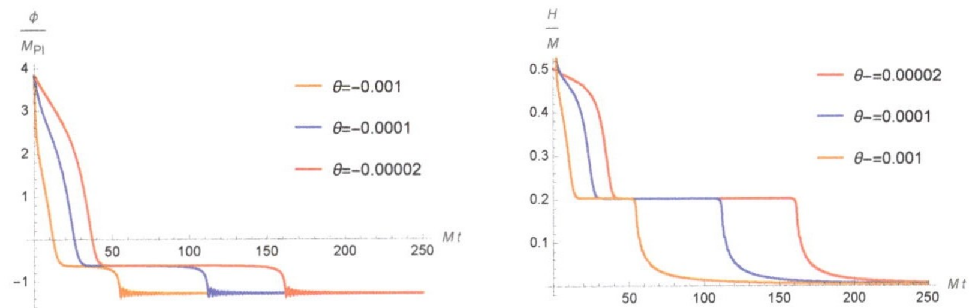

Figure 3. The time evolution of the inflaton $\phi(t)$ and the Hubble function $H(t) = \dot{a}/a$ (where $a(t)$ is the cosmic factor in the metric of a spatially flat Friedmann universe) with the parameters $\alpha = 0.74265$, $\xi = 0.01213$ and $\phi_i/M_{\rm Pl} = -0.61115$, and the horizon exit at $\phi_{\rm in}/M_{\rm Pl} = 3.886$.

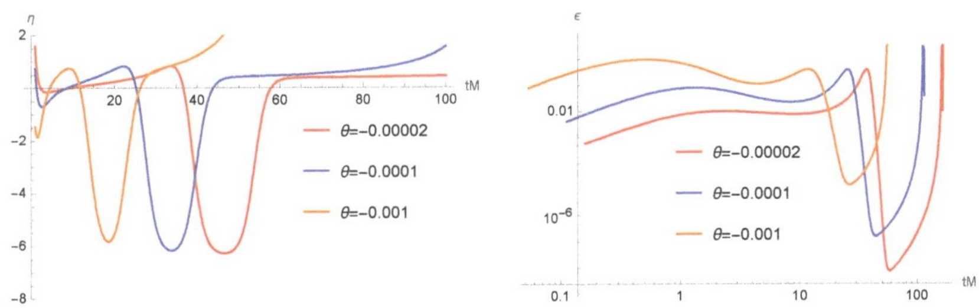

Figure 4. The Hubble flow functions $\eta(t)$ and $\epsilon(t)$ with the parameters $\alpha = 0.74265$, $\xi = 0.01213$ and $\phi_i/M_{\rm Pl} = -0.61115$.

Figure 4 shows the existence of the ultra-slow-roll (USR) phase [13,32], where the first derivative of the inflaton potential and the Hubble flow function $\epsilon(t)$ drop to very low values. During the USR phase, the Hubble flow function $\eta(t)$ also drops from near zero to

$$\Delta\eta \approx -6. \tag{13}$$

Our findings for the CMB scalar tilt n_s with the fine-tuned parameters (α, ξ, ϕ_i) are displayed in Figure 5, where we also show the dependence of n_s upon $\phi_{\rm in}$ and θ.

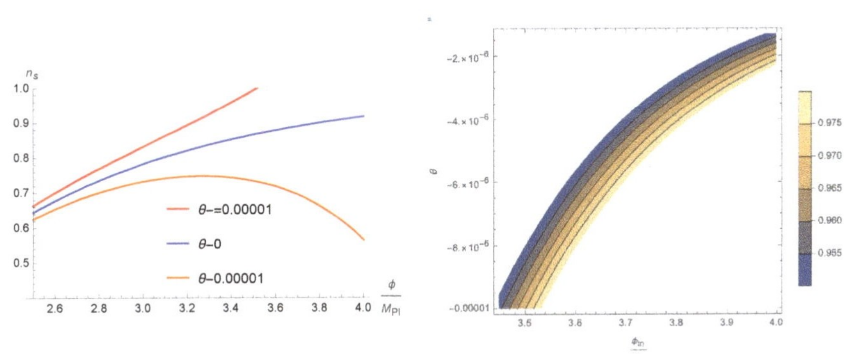

Figure 5. The dependence of the CMB scalar tilt n_s upon the inflaton field ϕ (on the left) and θ and $\phi_{\rm in}$ at the horizon exit (on the right). The other parameters are $\alpha = 0.74265$, $\xi = 0.01213$ and $\phi_i = -0.61115$.

With the exit value of $\phi_{\rm in}/M_{\rm Pl} = 3.886$ and the parameter $\theta = -0.000002597$, we find

$$n_s = 0.96498 \quad \text{and} \quad r = 0.03196\,, \tag{14}$$

which are fully consistent with the CMB measurements (2). This value of r saturates the current CMB upper bound on the tensor-to-scalar ratio and maximizes the PBH masses (see the next section).

3. Power Spectrum and PBH Formation

A standard procedure of (numerically) computing the power spectrum $P_R(k)$ of scalar (curvature) perturbations depending upon scale k is based on the Mukhanov–Sasaki (MS) equation [33,34]. There is a simple analytic formula for P_R in the SR approximation, see, e.g., ref. [12], which reads

$$P_R = \frac{H^2}{8M_{\rm Pl}^2\pi^2\epsilon} \; . \tag{15}$$

We used both approaches in our models and found that the difference between the exact results from numerically solving the MS equation and those derived from the SR Formula (15) is small (see Figure 6 as an example), and is irrelevant for our purposes.

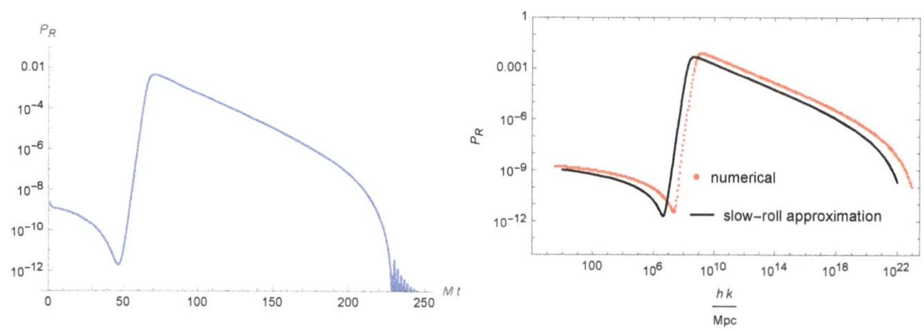

Figure 6. The power spectrum of scalar perturbations with the parameters $\alpha = 0.74265$, $\xi = 0.01213$, $\phi_i = -0.61115$, $\theta = -0.000002597$ and $\phi_{\rm in}/M_{\rm Pl} = 3.886$. The peak location is at $k_{\rm peak} = 10^9\ {\rm Mpc}^{-1}$.

Knowing the location (k) of the peak in the power spectrum allows us to estimate the masses $M(k)$ of the produced PBH as [14,15]

$$\frac{M(k)}{M_\odot} \simeq 10^{-16}\left(\frac{k}{10^{14}{\rm Mpc}^{-1}}\right)^{-2} , \tag{16}$$

where $M_\odot$ is the mass of the sun, $M_\odot = 1.998 \cdot 10^{33} g$. For instance, with the peak value $k_{peak} \approx 10^9\ {\rm Mpc}^{-1}$, we obtain

$$M_{\rm PBH} = 1.74 \cdot 10^{26}\ g \approx 10^{-7} M_\odot \ , \tag{17}$$

which is just between the mass of the moon, $M_{\rm Moon} = 7.34 \cdot 10^{25}$ g, and the mass of the Earth, $M_\oplus = 5.97 \cdot 10^{27}$ g. The PBH mass (17) should be understood as the *maximal* PBH mass possible in our classical E-models after extreme fine-tuning of the parameters, in agreement with CMB measurements.

In regard to other possible values of the CMB tilts n_s and r, the squared amplitude $\Delta^2_{\rm peak}$ of curvature perturbations at the horizon exit $r_* = 2\pi/k_*$, and the related PBH masses according to Equation (16), derived from our models, are summarized in Table 1.

The plots of the Hubble flow functions (12) in the particular model with the parameters $\alpha = 0.74265$, $\xi = 0.01213$, $\phi_i = -0.61115$ and $\theta = -0.000002597$ are given in Figure 7. Figure 7 shows the existence of a USR phase with $|\eta| > 1$ between two SR phases, where the USR begins at $Mt_s \approx 46$ and ends at $Mt_e \approx 70$ with the USR duration of $M(t_e - t_s) \approx 24$. Our plots in Figure 7 are similar to those in the literature for the inflationary models with a USR phase, see, e.g., Figure 1 in ref. [35] for comparison.

Table 1. The values of the CMB tilts n_s and r, the squared amplitude $\Delta^2_{\rm peak}$ of curvature perturbations in the peak and the PBH masses (16) in our single-field inflationary E-models with the potential (10) and the tuned values of the parameters.

n_s	r	α	θ	ϕ_{in}	$\Delta^2_{\rm peak}$	M_{PBH}
0.96498	0.03196	0.74256	−0.000002597	3.886	0.008	$1.7 \cdot 10^{26}$ g
0.96494	0.03098	0.74260	−0.000002472	3.9	0.007	$8 \cdot 10^{25}$ g
0.96496	0.01569	0.74250	−0.000000820	4.2	0.003	$5 \cdot 10^{19}$ g

At the end of this section, we briefly comment on expected one-loop quantum corrections in our E-models by using the results and discussions in refs. [36–41].

The Hubble flow function $\eta(t)$ can be approximated by a step function during the USR phase, which leads to the (one-loop) perturbative bound on validity of the classical results in the form [36,38]

$$\tfrac{1}{4}(\Delta\eta)^2\left(1.1+\log\tfrac{k_e}{k_s}\right)\Delta^2_{\rm peak} \ll 1 \ , \tag{18}$$

where the amplitude squared of scalar curvature perturbations was taken from refs. [36,38] and is of the order $\Delta^2_{\rm peak} \sim \mathcal{O}(10^{-2})$. Given $(\Delta\eta)^2 \approx 36$, Equation (18) implies the perturbative bound on the power spectrum as [36,38]

$$\Delta^2_{\rm peak} \ll 0.055 \sim \mathcal{O}(10^{-2}) \tag{19}$$

which might rule out classical results about PBH production in any single-field model of inflation with a near-inflection point. In our models, we have the smooth function $\eta(t)$ and $\Delta\eta \approx -6.3$ that implies a stronger bound,

$$\Delta^2_{\rm peak} \ll 0.025 \sim \mathcal{O}(10^{-2}) \ . \tag{20}$$

However, the values of $\Delta^2_{\rm peak}$ given in Table 1 are smaller (though not much smaller) than the bound (20). There is an uncertainty in the value of $\Delta^2_{\rm peak} \sim \mathcal{O}(10^{-2})$ used in refs. [36,38] for PBH production, which is related to an uncertainty in the smoothed density contrast constant and relaxes the peak value (needed for PBH formation) to $\Delta^2_{\rm peak} \sim \mathcal{O}(10^{-3})$ [39]. Therefore, our classical results obtained in the fine-tuned E-models for PBH production are still not ruled out by quantum corrections, at least in regard to the bottom line in Table 1.

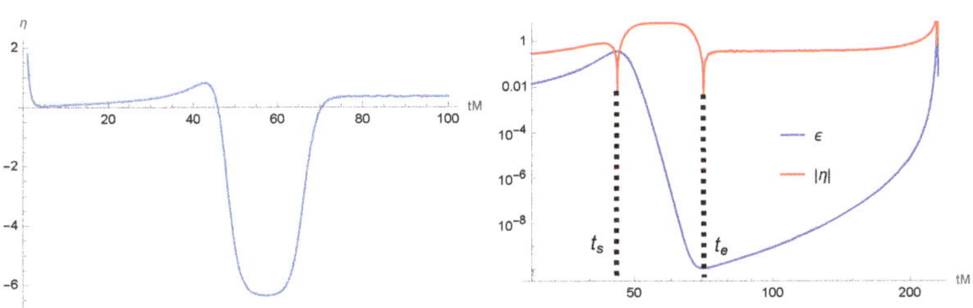

Figure 7. The Hubble flow functions (12) for the parameters $\alpha = 0.74256$, $\xi = 0.01213$, $\phi_i = -0.61115$ and $\theta = -0.000002597$. The other related values are $\phi_{in}/M_{Pl} = 3.886$, $k_s \approx 4.4 \cdot 10^6$ Mpc^{-1}, $k_e \approx 3 \cdot 10^8$ Mpc^{-1} and $\Delta\eta = -6.3$. Both plots show the USR phase for $|\eta| > 1$. The plot (right) is the superposition of the Hubble flows for $\epsilon(t)$ and $|\eta|(t)$.

4. Conclusions

Production of PBHs with masses beyond the Hawking (black hole) radiation limit of 10^{15} g in the early Universe is not a generic feature of single-field models of inflation with a near-inflection point because it requires fine-tuning of the parameters defining the inflaton potential and depends upon the duration of the SR and USR phases. We adopted maximal fine-tuning with the minimal number of parameters in the (generalized) E-type α-attractor models of inflation, while keeping agreement with CMB measurements.

We numerically scanned the parameter space of our models in order to find the parameter values leading to a significant enhancement (by 10^6 to 10^7 times) in the power spectrum of scalar (curvature) perturbations. We confirmed the high sensitivity of our results with a choice of the parameters, in agreement with general expectations [42].

Compared to ref. [27], we added the new parameter θ in the new term (proportional to y^{-2}) in the inflaton scalar potential in order to reach perfect agreement with the observed CMB values of the cosmological tilts n_s and r.

Fine-tuning of the parameters in our models was needed not only to match the CMB measurements (inflation is robust in the E-models) but also to generate PBH production at lower scales during the USR phase, which would lead to a sizable portion of PBHs in the current dark matter. The equations for PBH mass in the literature [14–16] are sensitive to a detailed shape of the peak in the power spectrum and the USR phase duration, so in this paper, we only used the rough estimate in Equation (16). Taking into account these uncertainties may change the PBH masses by 1–2 orders of magnitude.

Though production of PBHs with masses of approximately 10^{26} g was found to be possible in classical theory (see the first row in Table 1), those PBHs may be ruled out by quantum corrections. In regard to the asteroid-sized PBHs with masses of approximately 10^{19} g, which are possible candidates for the whole dark matter in the Universe [14,15,27], the quantum (one-loop) corrections were shown to be suppressed by one order of magnitude according to the estimate in Equation (20) and the third row of Table 1. Hence, our generalized E-models of inflation and PBH production are still not ruled out.

Author Contributions: All authors contributed equally. All authors have read and agreed to the published version of the manuscript.

Funding: This research was partially supported by Tomsk State University under the development program Priority-2030. SVK was also supported by Tokyo Metropolitan University, the Japanese Society for Promotion of Science under grant no. 22K03624 and the World Premier International Research Center Initiative, MEXT, Japan.

Data Availability Statement: No new data was created.

Conflicts of Interest: The authors declare no conflict of interest.

Note

1 Similar models were proposed and studied in ref. [29].

References

1. Guth, A.H. The Inflationary Universe: A Possible Solution to the Horizon and Flatness Problems. *Phys. Rev. D* **1981**, *23*, 347–356. [CrossRef]
2. Linde, A.D. A New Inflationary Universe Scenario: A Possible Solution of the Horizon, Flatness, Homogeneity, Isotropy and Primordial Monopole Problems. *Phys. Lett. B* **1982**, *108*, 389–393. [CrossRef]
3. Mukhanov, V. *Physical Foundations of Cosmology*; Cambridge University Press: Oxford, UK, 2005. [CrossRef]
4. Liddle, A.R.; Lyth, D.H. *Cosmological Inflation and Large Scale Structure*; Cambridge University Press: Cambridge, UK, 2000.
5. Akrami, Y. et al. [Planck Collaboration] Planck 2018 results. X. Constraints on inflation. *Astron. Astrophys.* **2020**, *641*, A10. [CrossRef]
6. Ade, P.A.R. et al. [BICEP, Keck Collaboration] Improved Constraints on Primordial Gravitational Waves using Planck, WMAP, and BICEP/Keck Observations through the 2018 Observing Season. *Phys. Rev. Lett.* **2021**, *127*, 151301. [CrossRef]

7. Tristram, M.; Tristram, M.; Banday, J.; Górski, M.; Keskitalo, R.; Lawrence, R.; Andersen, J.; Barreiro, B.; Borrill, J.; Colombo, L.; Eriksen, K.; et al. Improved limits on the tensor-to-scalar ratio using BICEP and Planck data. *Phys. Rev. D* **2022**, *105*, 083524. [CrossRef]
8. Novikov, I.; Zeldovic, Y. Cosmology. *Ann. Rev. Astron. Astrophys.* **1967**, *5*, 627–649. [CrossRef]
9. Hawking, S. Gravitationally collapsed objects of very low mass. *Mon. Not. R. Astron. Soc.* **1971**, *152*, 75. [CrossRef]
10. Ivanov, P.; Naselsky, P.; Novikov, I. Inflation and primordial black holes as dark matter. *Phys. Rev. D* **1994**, *50*, 7173–7178. [CrossRef] [PubMed]
11. Garcia-Bellido, J.; Linde, A.D.; Wands, D. Density perturbations and black hole formation in hybrid inflation. *Phys. Rev. D* **1996**, *54*, 6040–6058. [CrossRef] [PubMed]
12. Garcia-Bellido, J.; Ruiz Morales, E. Primordial black holes from single field models of inflation. *Phys. Dark Univ.* **2017**, *18*, 47–54. [CrossRef]
13. Germani, C.; Prokopec, T. On primordial black holes from an inflection point. *Phys. Dark Univ.* **2017**, *18*, 6–10. [CrossRef]
14. Carr, B.; Kohri, K.; Sendouda, Y.; Yokoyama, J. Constraints on primordial black holes. *Rep. Prog. Phys.* **2021**, *84*, 116902. [CrossRef] [PubMed]
15. Escrivà, A.; Kuhnel, F.; Tada, Y. Primordial Black Holes. *arXiv* **2022**, arXiv:2211.05767.
16. Karam, A.; Koivunen, N.; Tomberg, E.; Vaskonen, V.; Veermäe, H. Anatomy of single-field inflationary models for primordial black holes. *arXiv* **2023**, arXiv:2205.13540.
17. Kallosh, R.; Linde, A. Universality Class in Conformal Inflation. *J. Cosmol. Astropart. Phys.* **2013**, *7*, 002. [CrossRef]
18. Galante, M.; Kallosh, R.; Linde, A.; Roest, D. Unity of Cosmological Inflation Attractors. *Phys. Rev. Lett.* **2015**, *114*, 141302. [CrossRef] [PubMed]
19. Starobinsky, A.A. A new type of isotropic cosmological models without singularity. *Phys. Lett. B* **1980**, *91*, 99–102. [CrossRef]
20. Ketov, S.V.; Starobinsky, A.A. Inflation and non-minimal scalar-curvature coupling in gravity and supergravity. *J. Cosmol. Astropart. Phys.* **2012**, *8*, 22. [CrossRef]
21. Mukhanov, V.F.; Chibisov, G.V. Quantum Fluctuations and a Nonsingular Universe. *JETP Lett.* **1981**, *33*, 532–535.
22. Dalianis, I.; Kehagias, A.; Tringas, G. Primordial black holes from α-attractors. *J. Cosmol. Astropart. Phys.* **2019**, *1*, 037. [CrossRef]
23. Iacconi, L.; Assadullahi, H.; Fasiello, M.; Wands, D. Revisiting small-scale fluctuations in α-attractor models of inflation. *J. Cosmol. Astropart. Phys.* **2022**, *6*, 7. [CrossRef]
24. Braglia, M.; Linde, A.; Kallosh, R.; Finelli, F. Hybrid α-attractors, primordial black holes and gravitational wave backgrounds. *J. Cosmol. Astropart. Phys.* **2023**, *2023*, 033. [CrossRef]
25. Frolovsky, D.; Ketov, S.V. Fitting power spectrum of scalar perturbations for primordial black hole production during inflation. *Astronomy* **2023**, *2*, 47–57. [CrossRef]
26. Frolovsky, D.; Ketov, S.V.; Saburov, S. Formation of primordial black holes after Starobinsky inflation. *Mod. Phys. Lett. A* **2022**, *37*, 2250135. [CrossRef]
27. Frolovsky, D.; Ketov, S.V.; Saburov, S. E-models of inflation and primordial black holes. *Front. Phys.* **2022**, *10*, 1005333. [CrossRef]
28. Ivanov, V.R.; Ketov, S.V.; Pozdeeva, E.O.; Vernov, S.Y. Analytic extensions of Starobinsky model of inflation. *J. Cosmol. Astropart. Phys.* **2022**, *3*, 058. [CrossRef]
29. Mishra, S.S.; Sahni, V. Primordial Black Holes from a tiny bump/dip in the Inflaton potential. *J. Cosmol. Astropart. Phys.* **2020**, *4*, 007. [CrossRef]
30. Aldabergenov, Y.; Ketov, S.V. Primordial black holes from Volkov-Akulov-Starobinsky supergravity. *Fortsch. Phys.* **2023**, *2023*, 2300039. [CrossRef]
31. Cicoli, M.; Conlon, J.P.; Maharana, A.; Parameswaran, S.; Quevedo, F.; Zavala, I. String Cosmology: From the Early Universe to Today. *arXiv* **2023**, arXiv:2303.04819.
32. Dimopoulos, K. Ultra slow-roll inflation demystified. *Phys. Lett. B* **2017**, *775*, 262–265. [CrossRef]
33. Mukhanov, V.F. Gravitational Instability of the Universe Filled with a Scalar Field. *JETP Lett.* **1985**, *41*, 493–496.
34. Sasaki, M. Large Scale Quantum Fluctuations in the Inflationary Universe. *Prog. Theor. Phys.* **1986**, *76*, 1036. [CrossRef]
35. Liu, J.; Guo, Z.-K.; Cai, R.-G. Analytical approximation of the scalar spectrum in the ultraslow-roll inflationary models. *Phys. Rev. D* **2020**, *101*, 083535. [CrossRef]
36. Kristiano, J.; Yokoyama, J. Ruling Out Primordial Black Hole Formation from Single-Field Inflation. *arXiv* **2022**, arXiv:2211.03395.
37. Riotto, A. The Primordial Black Hole Formation from Single-Field Inflation is Not Ruled Out. *arXiv* **2023**, arXiv:2301.00599.
38. Kristiano, J.; Yokoyama, J. Response to criticism on "Ruling Out Primordial Black Hole Formation From Single-Field Inflation": A note on bispectrum and one-loop correction in single-field inflation with primordial black hole formation. *arXiv* **2023**, arXiv:2303.00341.
39. Riotto, A. The Primordial Black Hole Formation from Single-Field Inflation is Still Not Ruled Out. *arXiv* **2023**, arXiv:2303.01727.
40. Choudhury, S.; Panda, S.; Sami, M. Quantum loop effects on the power spectrum and constraints on primordial black holes. *arXiv* **2023**, arXiv:2303.06066.

41. Firouzjahi, H.; Riotto, A. Primordial Black Holes and Loops in Single-Field Inflation. *arXiv* **2023**, arXiv:2304.07801.
42. Cole, P.S.; Gow, A.D.; Byrnes, C.T.; Patil, S.P. Primordial black holes from single-field inflation: A fine-tuning audit. *arXiv* **2023**, arXiv:2304.01997.

universe MDPI

Article

Fab-Four Cosmography to Tackle the Hubble Tension

Celia Escamilla-Rivera [1,*,†], José María de Albornoz-Caratozzolo [2,†] and Sebastián Nájera [1]

1 Instituto de Ciencias Nucleares, Universidad Nacional Autónoma de México, Circuito Exterior C.U., A.P. 70-543, México D.F., Ciudad de México 04510, Mexico; sebastian.najera@correo.nucleares.unam.mx
2 Departamento de Fisica y Matemáticas, Universidad Iberoamericana Ciudad de México, Prolongación Paseo de la Reforma 880, Ciudad de México 01219, Mexico; jose.albornoz@correo.uia.mx
* Correspondence: celia.escamilla@nucleares.unam.mx
† These authors contributed equally to this work.

Abstract: In the context of the Fab-Four theory of gravity in a Friedmann-Lemaître-Robertson-Walker background, in this work we use the cosmography approach to study a particular self-tuning filter solution focused on a zero-curvature fixed point to study the H_0 tension. In this scheme, the equations restrict the universe's evolution to certain scenarios, including radiation-like expansion, matter-like expansion, and late-time acceleration. Furthermore, we build the cosmographic series of the Fab-Four theory to obtain the kinematic parameters as the Hubble constant H_0 and the deceleration parameter q_0 for all the scenarios mentioned. Finally, we compare our results to find that it is possible to alleviate the current discrepancy on H_0 by considering specific requirements on the free parameters of the Fab-Four theory through a self-tuning filter.

Keywords: fab-four; horndeski gravity

1. Introduction

The concordance model of cosmology, Λ Cold Dark Matter (ΛCDM), is thought to work well at explaining several observations of the Universe, including the observed late cosmic acceleration [1,2], the structure of the cosmic microwave background (CMB), the baryon acoustic oscillations (BAOs), among others [3]. In recent years, there have been several issues at hand [4], e.g., CMB anisotropy anomalies [5], cosmic dipoles [4], and the most notable issue: the Hubble constant tension, reaching a statistical discrepancy of $\sim 5\sigma$ [6–9] between early and late measurements.

Direct measurements of the current Hubble constant H_0, through late-universe observations, have yielded a value of $H_0 = 73.3 \pm 1.04$ km s^{-1} Mpc^{-1} [8], while early-universe observations of the CMB give a lower result of $H_0 = 67.4 \pm 0.5$ km s^{-1} Mpc^{-1} [6], differing by around 8–10%. This discrepancy has positioned itself as an important crisis for the standard ΛCDM model. Consequently, several theories have been proposed to solve, alleviate or give insights into the Hubble tension issue. In the same line of thought, some proposals focus on the nature of dark energy, since if the cosmological constant Λ played a bigger role than previously thought in the early stages of the expansion of the Universe, it could alter the calculation for the early-universe Hubble constant [10], as a consequence, this would bring it closer to the locally measured H_0 value. On one hand, changing the equation-of-state of dark energy by making it time-dependent, has also been proposed as a possible explanation [11]. Other theories include adding new interactions, such as interacting dark energy [12,13], interacting dark matter [14] and interacting holographic dark energy [15]; including sterile neutrinos [16], axions [17], or other theorized particles to increase the relativistic degrees of freedom; unifying dark energy and dark matter as a single fluid [18]; introducing cosmological piecewise functions [19]; using a statistical lookback time approach [20], considering the black hole shadows as standard rulers calibrators [21], among many others [22–24]. See, for example, [25], for a summary and comparison of several proposals that involve interesting solutions to the H_0 tension.

On the other hand, alternative gravity theories also attempt to solve the Hubble tension and other problems within standard cosmology. These theories change General Relativity's (GR) Einstein-Hilbert action to introduce different dynamics in the resulting equations of motion, which can explain different phases of expansion of the universe and could explain the Hubble tension [26,27]. Allowing higher order equations of motion, there are theories such as $f(\mathcal{R})$ theories, which introduce a function of the Ricci scalar, $\mathcal{R}$, to act as the action. In the context of Teleparallel Gravity [28], it can be considered $f(\mathcal{T})$ theories, which instead use functions of the torsion scalar. Another way to extend GR is by allowing the Lagrangian density to depend not only on the metric but also on another degree of freedom such as a scalar field ϕ, these are the so-called scalar-tensor theories. If we maintain the requirement of working in four dimensions, and having second-order equations of motion—to avoid Ostrogradsky instabilities [26]—the most general of these theories is Horndeski theory [29,30].

In this work, we focus on a special case of Horndeski's theory first proposed in [29] called the *Fab Four* theory. As a first approach, in [29] a *self-tuning* filter was applied on the original Horndeski action that partially de-couples the curvature from the cosmological constant and reduces the action to only four base Lagrangians, each one a function dependent on combinations of the Riemann tensor and an arbitrary function of the scalar field ϕ. The self-tuning filter consisted in requiring the following [29]:

1. The theory admits a patch of Minkowski vacuum for any value of the cosmological constant Λ, and for instantaneous changes in its value;
2. The theory allows non-trivial cosmologies by introducing additional dynamical terms to the standard one.

According to these two requirements, in [31], an initial outlook was provided on the types of cosmology and its dynamics derived from the Fab-Four theory. Although it is a fairly recent proposal, the Fab-Four theory has been explored extensively and in different contexts. Specifically, there have been several studies involving the $\mathcal{L}_{john}$ term in isolation, since it is the one that mainly drives the self-tuning mechanism [32] and it contains a derivative coupling to the scalar field, e.g., in [33] the $\mathcal{L}_{john}$ term (see Section 2) is studied in detail, specifically its generalization as the $\mathcal{L}_5$ term in Horndeski's theory which adds a dependency on the field kinetic term $X = \nabla_\mu\phi\nabla^\mu\phi$, finding that the resulting *Fab 5* theory can be ghost-free, or stable, or self-tuning, but not all these characteristics at the same time. *Fab 5* cosmology is studied in more detail in [34–36]. After the extension of the original Horndeski Lagrangian [37], adding two more terms in Horndeski's action, it was shown in [38] that these new terms can be included in a generalized Fab-Four theory where the arbitrary functions of the field $V_j(\phi)$ and $V_p(\phi)$ become arbitrary functions of the field and its kinetic term $V_j(\phi, X)$ and $V_p(\phi, X)$. Additionally, it was shown that these new terms also present self-tuning for screening the cosmological constant. In [39], it was argued that the measurement of the speed of gravitational waves (GWs) strongly constrained *John*-like terms in Horndeski's action, along with several other scalar-tensor theories and dark energy (DE) models. However, in [32], it is shown that working in an Effective Field Theory (EFT) framework, the resulting effective action is not constrained as such, and so it still admits *John*-like terms, therefore allowing a self-tuning mechanism in the EFT regime and the possibility to be consistent with the observational data [39,40]. In order to explore GW constraints, see [41]. Furthermore, as any alternative theory of gravity could, the Fab Four proposal could be excluded if we demand that the speed of gravitational waves c_s is exactly the same as the speed of light c along the evolution of the Universe. Therefore, if we consider small deviations from c, then several models can become consistent with the observable data (and their constrictions). New self-tuning models known as *well-tempered* have been developed and deal with the GW constraints by construction [42], and also prevent the scalar field from screening not only the cosmological constant but also all other matter sources in the system. A generalized class of *well-tempered* models and their self-tuning mechanisms has been discussed in [43]. In a quantum gravity context, the study of the quantization of the Fab-Four *John* term results in a Hamiltonian with fractional powers of the momenta, leading to the use of a conformal fractional derivative approach to get a wide range of solutions for the scale factor and the scalar field [44].

In this work, we follow closely the derivations performed for the Fab-Four cosmology [31] to obtain the corresponding equations of motion for a Friedmann-Lemaître-Robertson-Walker (FLRW) background and to apply a dynamical systems approach to solving the dynamical equations. Under this scheme and considering all four components of Fab-Four theory, we also extend this analysis by using a cosmographic approach, which will be useful to obtain kinematic quantities, such as H_0 and q_0. Finally, with these quantities at hand, we will be capable of setting general requirements on the free parameters to relax the Hubble tension and give a likelihood landscape for future works and analyses of this theory. The core idea behind this process is to facilitate the search for a stable cosmological solution through dynamical systems. The advantage of this kind of solution is the possibility to explore those that fulfill the cosmological constraint equation of interest for the analysis scheme. Once this constraint is satisfied, the analytical solution at hand can be tested with observations to establish a cutoff for the cosmological parameters of interest, namely H_0 and q_0. Although there may exist more analytical solutions for the set of equations, our cosmographic approach up to second order allows us to narrow down the options and select the ones that address the late-time cosmic acceleration.

This paper is divided as follows: In Section 2, we present a brief introduction of the results discussed in [31], where the full equations of motion of the system are given for the Fab-Four theory in a FLRW background. Also, we describe the dynamical system approach for this theory to find the dynamical parameters and the corresponding fixed points. We rewrite the equations in the zero-curvature fixed point of interest and solve them to find viable cosmological scenarios. In Section 3, we construct the cosmographic series for the Fab-Four theory and find the expressions for the Hubble constant H_0, and the deceleration parameter q_0 for each cosmological scenario described. Also, we will restrict our analysis to a couple of cases due to the solutions found in the dynamical system approach. In Section 4, we discuss how our solutions from the Fab-Four theory are able to relax the H_0 tension by following a set of requirements on the free parameters of the theory. Finally, we present our conclusions and considerations for future work in Section 5.

2. Fab-Four Cosmology as a Dynamical System

The Fab-Four model comes from a restriction set upon Horndeski's theory, which simplifies it greatly so that the Lagrangian density is now just the sum of four base Lagrangians [29]:

$$\mathcal{L}_{\text{Fab-Four}} = \mathcal{L}_{\text{john}} + \mathcal{L}_{\text{paul}} + \mathcal{L}_{\text{george}} + \mathcal{L}_{\text{ringo}}, \tag{1}$$

where each Lagrangian is defined as a combination of derivatives of the scalar field ϕ, combinations of the metric tensor, and an arbitrary function of the field $V_i(\phi)$, with $i = j, p, g, r$, namely:

$$\mathcal{L}_{john} = \sqrt{-g}V_{john}G^{\mu\nu}\nabla_\mu\phi\nabla_\nu\phi, \tag{2}$$

$$\mathcal{L}_{paul} = \sqrt{-g}V_{paul}P^{\mu\nu\alpha\beta}\nabla_\mu\phi\nabla_\alpha\phi\nabla_\nu\phi\nabla_\beta\phi, \tag{3}$$

$$\mathcal{L}_{george} = \sqrt{-g}V_{george}R, \tag{4}$$

$$\mathcal{L}_{ringo} = \sqrt{-g}V_{ringo}\hat{\mathcal{G}}, \tag{5}$$

where $G_{\mu\nu}$ is the Einstein tensor, $P_{\mu\nu\alpha\beta}$ is the double dual of the Riemann tensor, R is the Ricci scalar and $\hat{\mathcal{G}}$ is the gauss-Bonnet combination [31]. Given this notation and considering an FLRW metric whose line element is given by

$$ds^2 = -dt^2 + a^2(t)\left(\frac{dr^2}{1-kr^2} + r^2 d\Omega^2\right), \tag{6}$$

where $a(t)$ is the scale factor, which denotes the relative size of the co-moving, constant-time hypersurfaces of space-time, $d\Omega^2 = d\theta^2 + \sin^2\theta d\phi^2$ is the element of solid angle and k is the spatial curvature constant. Here, $k = 0$ gives a flat space, while $k > 0$ ($k < 0$) corresponds to positive (negative) curvature. We can express the Fab-Four Lagrangians as follows [31]:

$$\mathcal{L}_j = 3a^3 V_j(\phi)\dot{\phi}^2\left(H^2 + \frac{k}{a^2}\right), \tag{7}$$

$$\mathcal{L}_p = -3a^3 V_p(\phi)\dot{\phi}^3 H\left(H^2 + \frac{k}{a^2}\right), \tag{8}$$

$$\mathcal{L}_g = -6a^3 V_g(\phi)\left(H^2 - \frac{k}{a^2}\right) - 6a^3 V_{g,\phi}(\phi)H\dot{\phi}, \tag{9}$$

$$\mathcal{L}_r = -8a^3 V_{r,\phi}(\phi)\dot{\phi}H\left(H^2 + 3\frac{k}{a^2}\right). \tag{10}$$

Here, $V_{i,\phi}$ denotes the derivative of V_i with respect to the field, $dV_i/d\phi$, and $\dot{\phi} = U^a\nabla_a\phi$, with $U^a = \delta^a{}_t$ the 4-velocity. Following [31], we find that in the presence of standard matter and a cosmological constant Λ, the Hamiltonian density is given by the following:

$$\mathcal{H} = \mathcal{H}_j + \mathcal{H}_p + \mathcal{H}_g + \mathcal{H}_r + \rho_\Lambda = 0, \tag{11}$$

where, for convenience, we can express each Hamiltonian in conformal time by considering $N = \ln a$, where A' denotes differentiation with respect to N, such that $A' = dA/dN = \dot{A}(1/H)$:

$$\mathcal{H}_j = 3V_j\phi'^2 H^2\left(3H^2 + \frac{k}{a^2}\right), \tag{12}$$

$$\mathcal{H}_p = -3V_p\phi'^3 H^4\left(5H^2 + 3\frac{k}{a^2}\right), \tag{13}$$

$$\mathcal{H}_g = -6V_g\left[\left(H^2 + \frac{k}{a^2}\right) + H^2\frac{V_{g,\phi}\phi'}{V_g}\right], \tag{14}$$

$$\mathcal{H}_r = -24V_{r,\phi}\phi' H^2\left(H^2 + \frac{k}{a^2}\right). \tag{15}$$

The equations for the scalar field ϕ can be obtained through varying

$$E_i^{(\phi)} = -\frac{d}{dt}\left(\frac{\partial\mathcal{L}_i}{\partial\dot{\phi}}\right) + \frac{\partial\mathcal{L}_i}{\partial\phi}, \tag{16}$$

which gives the following expressions [31]:

$$E_j^{(\phi)}\dot{\phi}H^{-1} = \frac{1}{a^3\Delta_2}\left(\frac{\mathcal{H}_j a^6\Delta_2^2}{3\Delta_2 - 2\frac{k}{a^2}}\right)', \tag{17}$$

$$E_p^{(\phi)}\dot{\phi}H^{-1} = \frac{2}{a^{3/2}(H\Delta_2)^{1/2}}\left(\frac{\mathcal{H}_p a^{9/2}(H\Delta_2)^{1/2}\Delta_2}{5\Delta_2 - 2\frac{k}{a^2}}\right)', \tag{18}$$

$$E_g^{(\phi)}\dot{\phi}H^{-1} = -3\frac{V_g'}{a}(\Delta_2 a^4)', \tag{19}$$

$$E_r^{(\phi)}\dot{\phi}H^{-1} = \frac{\mathcal{H}_r}{H\Delta_2}\left[a^3\left(\frac{k}{a^2}\Delta_1 + \frac{1}{3}\Delta_3\right)\right]', \tag{20}$$

with $\Delta_n = H^n - (\sqrt{-k}/a)^n$. Notice that this function goes to zero as the latter equations approach a self-tuning solution corresponding to a Ricci flat scenario and the full system of equations that arises is coupled with second-order differential equations, which are extremely difficult to solve analytically. Therefore, a viable path to reach a set of possible solutions is through a dynamical systems approach[31]. In order to implement this procedure, we consider the following dynamical variables:

$$h = \frac{H'}{H}, \qquad \sigma = \frac{\sqrt{-k}}{Ha}, \tag{21}$$

notice that if h =const. we obtain that $H = H_0 a^h \Rightarrow a \propto t^{-\frac{1}{h}}$, where a matter-like expansion would arise from $h = -3/2$, radiation-like expansion from $h = -2$, and an inflationary expansion coming from $-1 < h < 0$. To build the dynamical system according to the latter variables, we define the following expressions [31]:

$$x = H^\alpha \phi', \tag{22}$$

$$y_j = H^{4-2\alpha} V_j, \quad y_p = H^{6-3\alpha} V_p, \quad y_g = H^2 V_g, \quad y_r = H^{4-\alpha} V_{r,\phi}, \tag{23}$$

$$\lambda_j = H^{-\alpha} \frac{V_{j,\phi}}{V_j}, \quad \lambda_p = H^{-\alpha} \frac{V_{p,\phi}}{V_p}, \quad \lambda_g = H^{-\alpha} \frac{V_{g,\phi}}{V_g}, \quad \lambda_r = H^{-\alpha} \frac{V_{r,\phi\phi}}{V_{r,\phi}}, \tag{24}$$

$$\mu_j = \frac{V_j V_{j,\phi\phi}}{(V_{j,\phi})^2}, \quad \mu_p = \frac{V_p V_{p,\phi\phi}}{(V_{p,\phi})^2}, \quad \mu_g = \frac{V_g V_{g,\phi\phi}}{(V_{g,\phi})^2}, \quad \mu_r = \frac{V_{r,\phi} V_{r,\phi\phi\phi}}{(V_{r,\phi\phi})^2}, \tag{25}$$

where α is an arbitrary constant that provides freedom in re-defining the scalar field while maintaining equations of motion that do not explicitly depend on H. This freedom will later be fixed to a particular value in Section 3.

The evolution equations for these variables are directly obtained by differentiating Equations (21)–(25) with respect to N. Finally, we obtain:

$$\sigma' = \sqrt{-k}\frac{1}{H^2a^2}(H'a + Ha) = -\sigma(h+1), \tag{26}$$

$$x' = \alpha H^{\alpha-1} H' \phi' + H^\alpha \phi'' = x\left(\alpha h + \frac{\phi''}{\phi'}\right), \tag{27}$$

$$y_i' = \beta_i H^{\beta_i} h V_i + H^{\beta_i} V_{i,\phi} \phi' = y_i(\beta_i h + \lambda_i x), \tag{28}$$

$$\lambda_i' = -\alpha h H^{-\alpha} \frac{V_{i.\phi}}{V_i} + H^{-\alpha} \phi' \frac{V_{i,\phi\phi}}{V_i} - H^{-\alpha} \frac{V_{i,\phi}^2}{V_i^2} \phi' = \lambda_i(-\alpha h + \lambda_i x(\mu_i - 1)), \tag{29}$$

where $i = j, p, g, r$ and β_i denotes the exponents of H in Equation (23). Knowing how these variables evolve, we can easily find their fixed points. Considering constant μ_i allows for potentials in a power law form. For the y_i and the λ_i, these fixed points are those where

$$\beta_i h + \lambda_i x = 0, \tag{30}$$

$$\lambda_i x(\mu_i - 1) = \alpha h. \tag{31}$$

In this fixed point, combining Equations (30) and (31) yields:

$$\lambda_i\left(-\frac{\beta_i h}{\lambda_i}\right)(\mu_i - 1) = \alpha h \Rightarrow (1 - \mu_i) = \frac{\alpha}{\beta_i}, \tag{32}$$

which allows us to directly find the exponents for all four potentials, and they become

$$V_j(\phi) = c_j \phi^{\frac{4}{\alpha}-2}, \tag{33}$$

$$V_p(\phi) = c_p \phi^{\frac{6}{\alpha}-3}, \tag{34}$$

$$V_g(\phi) = c_g \phi^{\frac{2}{\alpha}}, \tag{35}$$

$$V_{r,\phi}(\phi) = c_r \phi^{\frac{4}{\alpha}-1}. \tag{36}$$

Furthermore, the fixed point for σ corresponds to $\sigma(h+1) = 0$, i.e., fixed points solutions exist for $\sigma = 0$ and $h = -1$. On this line of thought, we are interested in the case where $\sigma = 0$, which refers to vanishing spatial curvature, as it is consistent with observations [6]. In this selected fixed point, by also using Equations (30) and (31), the full scalar field equation Equation (20) can be written as follows:

$$2xy_j(h+3) - 3x^2y_p(h+3) - 2\lambda_g y_g(h+2) - 8y_r(h+1) = 0. \tag{37}$$

Similarly, the Hamiltonian constraint Equation (11) can be written as

$$9x^2y_j - 15x^3y_p - 6y_g(1-2h) - 24xy_r = -\rho_\Lambda. \tag{38}$$

Following the arguments in [31], because the self-tuning filter demands that the solutions hold for any value of the cosmological constant, the parameters in the potentials (α, c_i) should be all independent of ρ_Λ. This assumption constrains the possible values of the terms present in Equations (37) and (38) while satisfying the following:

$$(3+h)c_p(-\alpha h)^{2-\frac{2}{\alpha}} = 0, \tag{39}$$
$$-2c_j\alpha h(3+h) - 8c_r(h+1) = 0, \tag{40}$$
$$h(2+h)c_g(-\alpha h)^{-1+\frac{2}{\alpha}} = 0. \tag{41}$$

We can distinguish several values for h that fulfill these requirements, which are associated with different behaviors of a, due to $H = H_0a^h$. Two cases of interest in this article are arbitrary for $h \neq 0$ and radiation for $h = -2$.

Under the arbitrary case, we focus on three cases: Firstly, the case of $h = -1$ corresponds to an evolution where $H^2 \propto a^{-2}$, indicating an accelerated expansion. Secondly, the case of $h = -3/2$ exhibits an evolution with $H^2 \propto a^{-3}$, resembling matter-like expansion. Lastly, we consider the radiation case, which is applicable for studying the standard cosmological model. This case encompasses an inflationary period followed by a radiation-dominated epoch, then a matter-like expansion facilitating structure formation, and finally, a late-time inflationary expansion that aligns with our current observations [3]. The applicability of such cases is due to the fact that we are looking for a viable cosmological scheme that can reproduce the current cosmic dynamics. Of course, in this case, we are considering solely a particular scalar field dynamic minimally coupled to gravity; so, the possibility that this scalar field can imitate dark matter and the effects of dark energy could set up a *toy* contemporary universe model.

3. Fab Four Cosmography

In what follows, we proceed with the standard cosmographic methodology [45–47] to derive the kinematic parameters, which will be evaluated at current cosmic times. The advantage of considering this approach is the geometry construction behind it. By assuming the structure of the Cosmological Principle, we can factorize the scale factor around the present time. However, we should take into account that we must relax the issues of such an approach: (i) the convergence of the series, which can be attended by using observables in order to restrain the order of it. (ii) the cosmic systematics that could be propagated on the cosmographic parameters. In such a case, we require a control of the numerical fit. Notice that for this analysis, we must select a fixed stable point and two possible cosmological scenarios.

Staying in the fixed point $\sigma = 0$, we can start by constructing cosmographic series based on the Taylor series expansion context of the scale factor a, defined as the following:

$$a = 1 + H_0\Delta t - \frac{1}{2}q_0H_0^2\Delta t^2 + \mathcal{O}(\Delta t^3), \tag{42}$$

where q is defined as

$$q = -\frac{\ddot{a}a}{\dot{a}^2} = -\frac{\dot{H}}{H^2} - 1 = -\frac{H'}{H} - 1, \tag{43}$$

and q_0 is the current deceleration parameter. To obtain H_0, we first choose an appropriate scenario as described in Section 2, substitute it into the Hamiltonian constraint Equation (11), and solve the system for the Hubble parameter H. Furthermore, with H we can find q using Equation (43). Finally, we can then evaluate both expressions at the current time to obtain H_0 and q_0, respectively.

Specifically, we chose to build the cosmographical series because it is a purely kinematic expression where other parameters such as matter density or curvature do not play any role. Therefore, we can focus solely on the kinematic terms without specifying other parameters of the model.

At this point, we consider the radiation-dominant behaviour case (for the early-universe calculation of H_0), and the arbitrary case (for the late-universe calculation of H_0). On one hand, for the arbitrary case, we have only two potential functions different from zero (as can be seen in Equations (39)–(41) by letting h be arbitrary):

$$V_j(\phi) = c_{\text{arb}}(h+1)\phi^{\frac{4}{\alpha}-2}, \tag{44}$$

$$V_{r,\phi}(\phi) = -\frac{\alpha}{4}h(3+h)c_{\text{arb}}\phi^{\frac{4}{\alpha}-1}, \tag{45}$$

where the subindex *arb* denotes arbitrary. Therefore, the only Hamiltonian parts considered would be $\mathcal{H}_j$ and $\mathcal{H}_r$; thus, we can write Equation (11) as follows:

$$9c_{\text{arb}}(1+h)H^4\phi'^2\phi^{\frac{4}{\alpha}-2} + 6c_{\text{arb}}\alpha h(3+h)H^4\phi'\phi^{\frac{4}{\alpha}-1} + \rho_\Lambda = 0, \tag{46}$$

$$-3c_{\text{arb}}H^4\phi'\phi^{\frac{4}{\alpha}-1}\Big[3(h+1)\phi'\phi^{-1} + 2\alpha h(3+h)\Big] = \rho_\Lambda. \tag{47}$$

From the latter, we can solve for H:

$$H^{\text{arb}} = \frac{\sqrt[4]{\rho_\Lambda}\phi^{-\frac{1}{\alpha}+\frac{1}{4}}}{\sqrt[4]{3c_{\text{arb}}\phi'}\sqrt[4]{-3(h+1)\phi'\phi^{-1} - 2\alpha h(3+h)}}, \tag{48}$$

which, evaluating at current times, gives us the following:

$$H_0^{\text{arb}} = \frac{\sqrt[4]{\rho_{\Lambda 0}}\phi_0^{-\frac{1}{\alpha}+\frac{1}{4}}}{\sqrt[4]{3c_{\text{arb}}\phi_0'}\sqrt[4]{-3(h+1)\phi_0'\phi_0^{-1} - 2\alpha h(3+h)}}, \tag{49}$$

where we have discarded the negative signs of the fourth roots to ensure a real, positive value of H_0. Substituting Equation (48) in Equation (43), and evaluating at the current time gives us the following:

$$q_0^{\text{arb}} = -1 - \frac{\rho'_{\Lambda 0}}{4\rho_{\Lambda 0}} - \frac{h\alpha\phi_0\phi_0'^2(3+h)(\alpha-4) + 3\phi_0'^3(h+1)(\alpha-2) - h\alpha^2\phi_0^2\phi_0''(3+h)}{2\alpha\phi_0\phi_0'\left[2h\alpha\phi_0(3+h) + 3\phi_0'(h+1)\right]} - \frac{3\alpha\phi_0\phi_0'\phi_0''(h+1)}{2\alpha\phi_0\phi_0'\left[2h\alpha\phi_0(3+h) + 3\phi_0'(h+1)\right]}. \tag{50}$$

On the other hand, for the radiation scenario, the potentials to be considered are as follows:

$$V_j(\phi) = c_1\phi^{\frac{4}{\alpha}-2}, \tag{51}$$

$$V_g(\phi) = c_2\phi^{\frac{2}{\alpha}}, \tag{52}$$

$$V_{r,\phi}(\phi) = -\frac{\alpha}{2}c_1\phi^{\frac{4}{\alpha}-1}, \tag{53}$$

which come from considering $h = -2$ in Equations (39)–(41); with the corresponding Hamiltonian parts $\mathcal{H}_j$, $\mathcal{H}_g$ and $\mathcal{H}_r$. The full Hamiltonian constraint for this case is:

$$9c_1H^4\phi'^2\phi^{\frac{4}{\alpha}-2} - 6c_2H^2\phi^{\frac{2}{\alpha}}\left(1+\frac{2}{\alpha}\phi^{-1}\phi'\right) + 12c_1\alpha H^4\phi'\phi^{\frac{4}{\alpha}-1} + \rho_\Lambda = 0, \tag{54}$$

where solving for H we obtain:

$$H_\pm^{\text{rad}} = \sqrt{\frac{\alpha\rho_\Lambda\phi^{1-\frac{2}{\alpha}}}{3c_2(\alpha\phi+2\phi')\pm\alpha\gamma(\alpha)}}. \tag{55}$$

Here, we have defined the quantity:

$$\gamma(\alpha) = \frac{1}{\alpha}\sqrt{9c_2^2\alpha^2\phi^2 - 12\alpha\phi\phi'(-3c_2^2+c_1\alpha^2\rho_\Lambda) + 9\phi'^2(4c_2^2 - c_1\alpha^2\rho_\Lambda)}, \tag{56}$$

which will appear frequently in our results. From Equation (55) and using Equation (56) we can obtain values for H_0 and q_0 for the radiation scenario:

$$H_{0\pm}^{\text{rad}} = \sqrt{\frac{\alpha\rho_{\Lambda0}\phi_0^{1-\frac{2}{\alpha}}}{3c_2(\alpha\phi_0+2\phi_0')\pm\alpha\gamma_0(\alpha)}}, \tag{57}$$

$$\begin{aligned} q_0^{\text{rad}} = {} & -1+\frac{1}{2}\Bigg[-\frac{\rho_{\Lambda0}'}{\rho_{\Lambda0}} - \frac{(\alpha-2)\phi_0'}{\alpha\phi_0} + \frac{3\alpha\phi_0'[2c_2(3\alpha c_2\phi_0+6c_2\phi_0'\pm\alpha\gamma)]}{2\alpha\gamma(3\alpha c_2\phi_0+6c_2\phi_0'\pm\alpha\gamma)} \\ & + \frac{6\phi_0''[2c_2(3\alpha c_2\phi_0+6c_2\phi_0'\pm\alpha\gamma) - \alpha^2c_1\rho_{\Lambda0}(2\alpha\phi_0+3\phi_0')]}{2\alpha\gamma(3\alpha c_2\phi_0+6c_2\phi_0'\pm\alpha\gamma)} \\ & - \frac{3\alpha^2c_1\phi_0'(\rho_{\Lambda0}'(4\alpha\phi_0+3\phi_0')+4\alpha\rho_{\Lambda0}\phi_0')}{2\alpha\gamma(3\alpha c_2\phi_0+6c_2\phi_0'\pm\alpha\gamma)}\Bigg]. \end{aligned} \tag{58}$$

Now, in the calculated values for H_0 and q_0 in Equations (49), (50) and (57)–(59) there are several free parameters, namely: c_{arb}, c_1, c_2, α, h and the scalar field and its derivative at current time ϕ_0 and ϕ_0'. We can choose an arbitrary value of α to fix the freedom in redefining the scalar field. Following [31], we can select α such that $\alpha h = -1$. Alternatively, we can choose values of α to ensure we can obtain power law potentials of the form $V \propto \phi^2$, which gives an inflationary behaviour. Only the cases where $\alpha = 1$ and $\alpha = 2$, give potentials of this kind, while $\alpha = 1$ is preferred, as it also fulfills $\alpha h = -1$ when $h = -1$ (inflation).

3.1. Case for $\alpha = 1$

For this case, we consider such $\alpha = 1$ value in Equations (49) and (50), which yields:

$$H_0^{\text{arb}} = \frac{\sqrt[4]{\rho_{\Lambda0}}}{\phi_0^{\frac{3}{4}}\sqrt[4]{3c_{\text{arb}}\phi_0'}\sqrt[4]{-3(h+1)\phi_0'\phi_0^{-1} - 2h(3+h)}}, \tag{59}$$

and

$$q_0^{\text{arb}} = -1 - \frac{\rho_{\Lambda0}'}{4\rho_{\Lambda0}} + \frac{3h\phi_0\phi_0'^2(3+h) + 3\phi_0'^3(h+1) + h\phi_0^2\phi_0''(3+h) + 3\phi_0\phi_0'\phi_0''(h+1)}{2\phi_0\phi_0'[2h\phi_0(3+h) + 3\phi_0'(h+1)]}. \tag{60}$$

Now, choosing a value for h determines how the Hubble parameter depends on a according to Equation (21), and thus specifies the type of expansion. We can study the cases for $h = -3/2$ and $h = -1$, which correspond to matter-dominated expansion and inflation, respectively. Substituting these values of h in Equations (59) and (60) gives:

$$H_0^{\text{matter}} = \frac{\sqrt[4]{2\rho_{\Lambda 0}}}{\sqrt{3}\phi_0^{\frac{3}{4}}\sqrt[4]{c_{\text{arb}}\phi_0'\left(3+\phi_0'\phi_0^{-1}\right)}}, \tag{61}$$

$$q_0^{\text{matter}} = -1 - \frac{\rho_{\Lambda 0}'}{4\rho_{\Lambda 0}} + \frac{\phi_0'^2(9\phi_0 + 2\phi_0') + \phi_0\phi_0''(3\phi_0 + 2\phi_0')}{4\phi_0\phi_0'(3\phi_0 + \phi_0')}, \tag{62}$$

which correspond to $h = -3/2$, and

$$H_0^{\text{inf}} = \frac{\sqrt[4]{\rho_{\Lambda 0}}}{\sqrt{2}\phi_0^{\frac{3}{4}}\sqrt[4]{3c_{arb}\phi_0'}}, \tag{63}$$

$$q_0^{\text{inf}} = -1 - \frac{\rho_{\Lambda 0}'}{4\rho_{\Lambda 0}} + \frac{3\phi_0'}{4\phi_0} + \frac{\phi_0''}{4\phi_0'}, \tag{64}$$

which corresponds to the inflationary case $h = -1$.

Now, the case for radiation-like expansion is given by substituting $\alpha = 1$ in $\gamma(\alpha)$ as in Equations (57)–(59):

$$H_{0\pm}^{\text{rad}} = \sqrt{\frac{\rho_{\Lambda 0}\phi_0^{-1}}{3c_2(\phi_0 + 2\phi_0') \pm \gamma_0(1)}}, \tag{65}$$

$$\begin{aligned} q_{0\pm}^{\text{rad}} = & -1 - \frac{\rho_{\Lambda 0}'}{2\rho_{\Lambda 0}} + \frac{\phi_0'}{2\phi_0} + \frac{3\phi_0'(2c_2(3c_2\phi_0 + 6c_2\phi_0' \pm \gamma(1)) - c_1(\rho_{\Lambda 0}'(4\phi_0 + 3\phi_0') + 4\rho_{\Lambda 0}\phi_0'))}{4\gamma(1)(3c_2\phi_0 + 6c_2\phi_0' \pm \gamma(1))} \\ & + \frac{6\phi_0''(2c_2(3c_2\phi_0 + 6c_2\phi_0' \pm \gamma(1)) - c_1\rho_{\Lambda 0}(2\phi_0 + 3\phi_0'))}{4\gamma(1)(3c_2\phi_0 + 6c_2\phi_0' \pm \gamma(1))}. \end{aligned} \tag{66}$$

3.2. *Case for $\alpha = 2$*

Another way to obtain potentials of the form ϕ^2 is with $\alpha = 2$. Substituting this value on Equations (49), (50), and (57)–(59), and following a similar procedure as in the latter case, we obtain the following:

$$H_0^{\text{arb}} = \frac{\sqrt[4]{\rho_{\Lambda 0}}}{\sqrt[4]{3c_{\text{arb}}\phi_0\phi_0'}\sqrt[4]{-3(h+1)\phi_0'\phi_0^{-1} - 4h(3+h)}}, \tag{67}$$

$$q_0^{\text{arb}} = -1 - \frac{\rho_{\Lambda 0}'}{4\rho_{\Lambda 0}} + \frac{2h(h+3)(\phi_0'^2 + \phi_0\phi_0'') + 3(h+1)\phi_0'\phi_0''}{2\phi_0'(4h\phi_0(3+h) + 3\phi_0'(h+1))}, \tag{68}$$

and for the matter case,

$$H_0^{\text{matter}} = \frac{\sqrt[4]{\rho_{\Lambda 0}}}{\sqrt[4]{3c_{\text{arb}}\phi_0\phi_0'}\sqrt[4]{9 + \frac{3\phi_0'}{2\phi_0}}}, \tag{69}$$

$$q_0^{\text{matter}} = -1 - \frac{\rho_{\Lambda 0}'}{4\rho_{\Lambda 0}} + \frac{3\phi_0'^2 + \phi_0''(\phi_0 + \phi_0')}{2\phi_0'(6\phi_0 + \phi_0')}. \tag{70}$$

The inflation case gives us the following:

$$H_0^{\rm inf} = \frac{\sqrt[4]{\rho_{\Lambda 0}}}{2^{\frac{3}{4}}\sqrt[4]{3c_{\rm arb}\phi_0\phi_0'}}, \tag{71}$$

$$q_0^{\rm inf} = -1 - \frac{\rho_{\Lambda 0}'}{4\rho_{\Lambda 0}} + \frac{\phi_0'}{4\phi_0} + \frac{\phi_0''}{4\phi_0'}, \tag{72}$$

$$H_{0\pm}^{\rm rad} = \sqrt{\frac{\rho_{\Lambda 0}}{3c_2(\phi_0+\phi_0')\pm\gamma(2)}}, \tag{73}$$

and for the radiation case,

$$q_{0\pm}^{\rm rad} = -1 - \frac{\rho_{\Lambda 0}'}{2\rho_{\Lambda 0}} + \frac{\pm 6c_2\phi_0'\gamma(2) \pm 6c_2\phi_0''\gamma(2) + 18c_2^2(\phi_0+\phi_0')(\phi_0'+\phi_0'')}{4\gamma(2)(3c_2(\phi_0+\phi_0')+\gamma(2))} - \frac{3c_1(8\rho_{\Lambda 0}\phi_0'^2 + \rho_{\Lambda 0}'\phi_0'(8\phi_0+3\phi_0') + 2\rho_{\Lambda 0}\phi_0''(4\phi_0+3\phi_0'))}{4\gamma(2)(3c_2(\phi_0+\phi_0')+\gamma(2))}. \tag{74}$$

For both values of α, we notice similarities in all forms of H_0, which are all proportional to $\propto \sqrt[4]{\rho_\Lambda}$. Additionally, we notice similarities in the expressions for q_0, where all of them can be written in the same functional form:

$$q_0 = -1 - A\left(\frac{\rho_\Lambda'}{\rho_\Lambda}\right) + B\left(\frac{\phi'}{\phi}\right) + C\left(\frac{\phi''}{\phi'}\right), \tag{75}$$

where A, B and C are, in general, functions of the parameters ρ_Λ, ϕ, ϕ', ϕ'' and the constants $c_{\rm arb}$, c_1 and c_2. Considering standard GR with a cosmological constant and a scalar field as the source in an FLRW background, the deceleration parameter has the same functional form.

4. Fab-Four on the H_0 Tension

According to the latter results, we can now discuss the possible requirements to relax the value of H_0 in the cosmological scenarios discussed.

With the expressions obtained for $H_0^{\rm inf}$ and $H_0^{\rm rad}$ we can now calculate the predicted discrepancy between both values. The most recent measurements [6,8] suggest a discrepancy of around $\sim$10% between the values for the Hubble constant in the early universe and those in the late universe. As such, we expect that

$$\delta_{H_0} \equiv 1 - \frac{H_0^{\rm rad}}{H_0^{\rm inf}} \approx 0.1 \tag{76}$$

For $\alpha = 1$, we can use Equations (63)–(65) to get:

$$\delta_{H_0,\alpha=1} = 1 - \frac{\sqrt{2}\sqrt[4]{3c_{\rm arb}\rho_{\Lambda 0}\phi_0\phi_0'}}{\sqrt{3c_2(\phi_0+2\phi_0')\pm\gamma(1)}}. \tag{77}$$

On the other hand, for $\alpha = 2$, using Equations (71)–(73), we obtain the following

$$\delta_{H_0,\alpha=2} = 1 - \frac{\sqrt[4]{24c_{\rm arb}\rho_{\Lambda 0}\phi_0\phi_0'}}{\sqrt{3c_2(\phi_0+\phi_0')\pm\gamma(2)}}. \tag{78}$$

Now, before proceeding, several requirements arise in the process of obtaining expressions for H_0, q_0, and δ_{H_0} to ensure that the solutions are real. These are summarized in Table 1.

Table 1. Requirements on the Fab-Four derived constants in the results for H_0, q_0, and δ_{H_0}.

Case		Requirements
General		$\rho_{\Lambda 0} > 0$, $c_{arb}\phi_0' > 0$, $-3(h+1)\phi_0'\phi_0^{-1} - 2\alpha h(3+h) > 0$, $\phi_0 > 0$
$\alpha = 1$	General	$9c_2^2\phi_0^2 - 12\phi_0\phi_0'(-3c_2^2 + c_1\rho_{\Lambda 0}) + 9{\phi_0'}^2(4c_2^2 - c_1\rho_{\Lambda 0}) > 0$, $3c_2(\phi_0 + 2\phi_0') \pm \gamma_0(1) > 0$
	$h = -3/2$ (Matter)	$3 + \phi_0'\phi_0^{-1} > 0$
	$h = -1$ (Inflation)	No extra requirements
$\alpha = 2$	General	$9c_2^2\phi_0^2 + 6\phi_0\phi_0'(3c_2^2 - 4c_1\rho_{\Lambda 0}) + 9{\phi_0'}^2(c_2^2 - c_1\rho_{\Lambda 0}) > 0$ $3c_2(\phi_0 + \phi_0') \pm \gamma_0(2) > 0$
	$h = -3/2$ (Matter)	$6 + \phi_0'\phi_0^{-1} > 0$
	$h = -1$ (Inflation)	No extra requirements

Taking into account the restrictions given by Table 1, it is possible to get values for $\delta_{H_0} = 0.1$, by considering the requirements on the five free parameters that appear in the expressions: ϕ_0, ϕ_0', c_{arb}, c_1 and c_2. We also have another two parameters, ρ_Λ' and ϕ'', which only appear in the expressions for q_0.

However, the behaviour of ρ_Λ comes into play. The Fab-Four model was introduced to handle the cosmological constant fine-tuning problem [29,31]. As such, instead of the cosmological constant being constant, it is assumed that it is a piecewise constant function, with several phase transitions that change its value. Therefore, ρ_Λ is discontinuous at these phase transitions. This means that the *r.h.s* of Equation (11) is discontinuous, but the *l.h.s* of this equation is composed of the Hamiltonian parts, which depend only on the functions a, ϕ, and ϕ'. Since a is continuous, and ϕ as well - by construction - it follows that ϕ' must be discontinuous to maintain equality at the phase transitions. Consequently, both ρ_Λ' and ϕ'' contain delta-functions localized at the transition times. These same arguments are used to initially restrict Horndeski's full theory of gravity [29].

At these phase-transitions, the self-tuning nature of the Fab-Four theory is active to still allow a standard cosmology to develop [29]. However, by construction, the self-tuning filter can only work when the curvature is non-zero. Since the latest observations assert a negligible curvature [6], we assume that we are not presently in a phase transition. For this work, we are only taking the present-day values of these parameters ($\rho_{\Lambda 0}'$, ϕ_0''), which we will assume to be zero since we are not currently in a phase-transition, where there would be a delta-function.

Therefore, substituting $\rho_{\Lambda 0}' = 0$ and $\phi_0'' = 0$, from Equations (64)–(66) and (72)–(74), we obtain:

$$q_{0\,\alpha=1}^{\text{inf}} = -1 + \frac{3\phi_0'}{4\phi_0}, \tag{79}$$

$$q_{0\,\alpha=1}^{\text{rad}} = -1 + \frac{\phi_0'}{2\phi_0} + \frac{3\phi_0'(2c_2(3c_2\phi_0 + 6c_2\phi_0' \pm \gamma(1)) - 4c_1\rho_{\Lambda 0}\phi_0')}{4\gamma(1)(3c_2\phi_0 + 6c_2\phi_0' \pm \gamma(1))}, \tag{80}$$

$$q_{0\,\alpha=2}^{\text{inf}} = -1 + \frac{\phi_0'}{4\phi_0}, \tag{81}$$

$$q_{0\,\alpha=2}^{\text{rad}} = -1 + \frac{\pm 3c_2\phi_0'\gamma(2) + 9c_2^2\phi_0'(\phi_0 + \phi_0') - 12c_1\rho_{\Lambda 0}{\phi_0'}^2}{2\gamma(2)(3c_2(\phi_0 + \phi_0') + \gamma(2))}, \tag{82}$$

for each $\alpha = 1, 2$ case, respectively. Now, given a value of q_0 through the latest observations [8] ($q_0 \approx -0.5$), we can either force the deceleration parameters (evaluated at the current time) of both the early and late universe to be the same, or we can leave them to be different.

In this work, we assume $q_0^{\text{inf}} = q_0^{\text{rad}} \approx -0.5$ for simplicity, and we leave the other case for future work. Given this restriction, with the expressions for q_0^{inf} Equations (79) and (81), we will get a fixed value for ϕ_0'/ϕ_0 (depending on the choice of α). Then, using the expressions for $q_0^{\rm rad}$ Equations (80) and (82), we can constrain the value of one more parameter, say, for example, c_1.

Finally, we are only left with two free parameters: $c_{\rm arb}$ and either c_1 or c_2. These are enough to get $\delta_{H_0} \approx 0.1$, and reducing the number of free parameters is important to enhance the likelihood of this model in Bayesian statistic comparisons [48].

5. Conclusions

In this work, we consider the possibility of tackling the Hubble tension using the Fab-Four theory. This scalar-tensor theory of gravity applies a self-tuning filter on the more general Horndeski theory, to solve the cosmological constant fine-tuning problem. This leaves a much more manageable theory, consisting only of four base Lagrangians, given by Equations (7)–(10). We follow the procedure discussed in [31] to obtain a simpler form of the theory, using a dynamical system approach to find the fixed points of the theory, and we concentrate on the $\sigma = 0$ fixed-point, which denotes zero curvature (as supported by observations [6]). In this fixed point, the equation of motion of the scalar field Equation (37) and the Hamiltonian restriction Equation (38) of the theory are such that we have certain allowed types of cosmological scenarios, among which are the inflationary expansion, radiation-like expansion, and a matter-like phase. This shows the utility of the Fab-Four theory, as *turning on-and-off* certain potentials lead naturally to different types of cosmological behaviors, and the full dynamics of the theory allow for a smooth, continuous shift between these types of cosmological evolution.

Furthermore, we construct the Fab-Four cosmography in the zero-curvature fixed point. In particular, we derive analytical expressions for the current Hubble constant H_0, and the deceleration parameter q_0, for the matter, inflation, and radiation scenarios, as well as for different choices of the arbitrary constant α. To do this, we take into account the forms of the potentials, substitute them into the Hamiltonian constraint, which results in quartic equations of the Hubble parameter, and then solve for H. These equations yield four solutions: *(i)* in the arbitrary case, we discard two complex roots and a negative one; *(ii)* in the radiation case, we can only discard two negative roots, leaving both signs of the expression for $H_\pm^{\rm rad}$. Examining Table 1, we also notice that the requirements do not restrict the choice of signs in the expressions for $H_\pm^{\rm rad}$ and $q_\pm^{\rm rad}$, and both signs can give $\delta_{H_0} = 0.1$, which is why we have left them written explicitly in all expressions. Afterward, we differentiate the results for H according to Equation (43) to obtain expressions for the deceleration parameter, which all ultimately take the functional form of Equation (75). Using these expressions evaluated at current time, we can compare the Hubble constant in the late universe with the Hubble constant in the early universe (radiation-like expansion).

Finally, we show that it is possible to get the desired discrepancy between the early-universe and the late-universe Hubble constants, of around $\sim$10%, showing that the Hubble constant could, in reality, be dynamical. Therefore, we have succeeded in our objective to show that it is indeed possible to get a disagreement between the observed Hubble constant if measured through local means, and the one obtained from observations of the CMB, due to the different dynamics put forward by the Fab-Four theory.

The final results contain five arbitrary parameters which are constrained by the requirements in Table 1. However, using the expressions for the deceleration parameter, we can reduce the number of free parameters in these results to only two constants. In future work, we will propose performing an analysis with observational data to obtain confidence limits on the free variables of the theory. For this, extracting other observables from the Fab Four theory, e.g., expressions for the luminosity distance, and even perturbation theory results such as the angular power spectrum from the CMB would be convenient. Afterward, a possible course of action would be to make a Bayesian statistics comparison of this model's prediction against the standard ΛCDM to test its plausibility, as well as to further constrain the possible values of the free parameters to match observations [48]. Furthermore, the possibility to explore beyond Horndeski theories [26] could be a path in order to tackle the cosmological tensions at early cosmic times.

Author Contributions: All authors discussed the results and contributed to the final manuscript. All authors have read and agreed to the published version of the manuscript.

Funding: C.E.-R. is supported by PAPIIT UNAM Project TA100122. S.N. acknowledges financial support from SEP–CONACYT postgraduate grants program.

Data Availability Statement: Not applicable.

Acknowledgments: C.E.-R. acknowledges the Royal Astronomical Society as FRAS 10147 and is supported by PAPIIT UNAM Project TA100122. S.N. acknowledges financial support from SEP—CONACYT postgraduate grants program. This article is based upon work from COST Action CA21136 Addressing observational tensions in cosmology with systematics and fundamental physics (CosmoVerse) supported by COST (European Cooperation in Science and Technology).

Conflicts of Interest: The authors declare no conflict of interest.

Abbreviations

The following abbreviations are used in this manuscript:

FLRW	Friedmann-Lemaître-Robertson-Walker
CMB	Cosmic Microwave Background
BAO	Baryon Acoustic Oscillations
GR	General Relativity
GW	Gravitational Wave

References

1. Riess, A.G.; Filippenko, A.V.; Challis, P.; Clocchiatti, A.; Diercks, A.; Garnavich, P.M.; Gilliland, R.L.; Hogan, C.J.; Jha, S.; Kirshner, R.P. Observational evidence from supernovae for an accelerating universe and a cosmological constant. *Astron. J.* **1998**, *116*, 1009. [CrossRef]
2. Perlmutter, S.; Aldering, G.; Goldhaber, G.; Knop, R.A.; Nugent, P.; Castro, P.G.; Deustua, S.; Fabbro, S.; Goobar, A.; Groom, D.E.; et al. Measurements of Ω and Λ from 42 High-Redshift Supernovae. *Astrophys. J.* **1999**, *517*, 565–586.
3. Dodelson, S.; Schmidt, F. *Modern Cosmology*, 2nd ed.; Academic Press: Cambridge, MA, USA, 2021.
4. Perivolaropoulos, L.; Skara, F. Challenges for ΛCDM: An update. *New Astron. Rev.* **2022**, *95*, 101659. [CrossRef]
5. Schwarz, D.J.; Copi, C.J.; Huterer, D.; Starkman, G.D. CMB anomalies after Planck. *Class. Quantum Gravity* **2016**, *33*, 184001. [CrossRef]
6. Aghanim, N. et al. [Plank Collaboration] *Planck* 2018 results. *Astron. Astrophys.* **2020**, *641*, A6. [CrossRef]
7. Kamionkowski, M.; Riess, A.G. The Hubble Tension and Early Dark Energy. *arXiv* **2022**, arXiv:2211.04492.
8. Riess, A.G.; Yuan, W.; Macri, L.M.; Scolnic, D.; Brout, D.; Casertano, S.; Jones, D.O.; Murakami, Y.; Anand, G.S.; Breuval, L.; et al. A Comprehensive Measurement of the Local Value of the Hubble Constant with 1 km s^{-1} Mpc^{-1} Uncertainty from the Hubble Space Telescope and the SH0ES Team. *Astrophys. J. Lett.* **2022**, *934*, L7. [CrossRef]
9. Di Valentino, E.; Anchordoqui, L.A.; Akarsu, Ó; Ali-Haimoud, Y.; Amendola, L.; Arendse, N.; Asgari, M.; Ballardini, M.; Basilakos, S.; Battistelli, E.; et al. Snowmass2021—Letter of interest cosmology intertwined II: The hubble constant tension. *Astropart. Phys.* **2021**, *131*, 102605.
10. Karwal, T.; Kamionkowski, M. Dark energy at early times, the Hubble parameter, and the string axiverse. *Phys. Rev. D* **2016**, *94*, 103523. [CrossRef]
11. Zhao, G.B.; Raveri, M.; Pogosian, L.; Wang, Y.; Crittenden, R.G.; Handley, W.J.; Percival, W.J.; Beutler, F.; Brinkmann, J.; Chuang, C.H.; et al. Dynamical dark energy in light of the latest observations. *Nat. Astron.* **2017**, *1*, 627–632. [CrossRef]
12. Aich, A. Interacting Dark Energy: New parametrization and observational constraints. *arXiv* **2022**, arXiv:2207.09079.
13. Kim, R.G.; Jong, S.; Hyok O.G. Cosmic coincidence problem in interacting holographic dark energy models. *New Astron.* **2023**, *101*, 102015. [CrossRef]
14. Cao, S.; Zhu, Z.H.; Liang, N. Observational constraints on interacting dark matter model without dark energy. *Astron. Astrophys.* **2011**, *529*, A61. [CrossRef]
15. Escamilla-Rivera, C.; Gamboa, A. A new parameterized interacting holographic dark energy. *Eur. Phys. J. Plus* **2022**, *137*, 254.
16. Salvio, A.; Scollo, S. Axion Sterile Neutrino Dark Matter. *Universe* **2021**, *7*, 354.
17. Berlin, A.; Hooper, D. Axion-Assisted Production of Sterile Neutrino Dark Matter. *Phys. Rev. D* **2017**, *95*, 075017.
18. Farnes, J.S. A unifying theory of dark energy and dark matter: Negative masses and matter creation within a modified ΛCDM framework. *Astron. Astrophys.* **2018**, *620*, A92.

19. Sandoval-Orozco, R.; Escamilla-Rivera, C. Cosmological piecewise functions to treat the local Hubble tension. *Eur. Phys. J. Plus* **2022**, *137*, 819.
20. Escamilla-Rivera, C.; Nájera, J.A. Comment on: "Questioning the H_0 tension via the look-back time". *arXiv* **2023**, arXiv:2303.02140.
21. Escamilla-Rivera, C.; Torres Castillejos, R. H0 Tension on the Light of Supermassive Black Hole Shadows Data. *Universe* **2023**, *9*, 14.
22. Di Valentino, E.; Mena, O.; Pan, S.; Visinelli, L.; Yang, W.; Melchiorri, A.; Mota, D.F.; Riess, A.G.; Silk, J. In the realm of the Hubble tension—A review of solutions. *Class. Quantum Gravity* **2021**, *38*, 153001. [CrossRef]
23. Adhikari, R.X.; Anchordoqui, L.A.; Fang, K.; Sathyaprakash, B.S.; Tollefson, K.; Lewis, T.R.; Engel, K.; Aboubrahim, A.; Akarsu, O.; Akrami, Y.; et al. Report of the Topical Group on Cosmic Probes of Fundamental Physics for for Snowmass 2021. *arXiv* **2022**, arXiv:2209.11726.
24. Abdalla, E.; Abellán, G.F.; Aboubrahim, A.; Agnello, A.; Akarsu, Ó; Akrami, Y.; Alestas, G.; Aloni, D.; Amendola, L.; Anchordoqui, L.A.; et al. Cosmology intertwined: A review of the particle physics, astrophysics, and cosmology associated with the cosmological tensions and anomalies. *J. High Energy Astrophys.* **2022**, *34*, 49–211.
25. Schöneberg, N.; Franco Abellán, G.; Pérez Sánchez, A.; Witte, S.J.; Poulin, V.; Lesgourgues, J. The H0 Olympics: A fair ranking of proposed models. *Phys. Rept.* **2022**, *984*, 1–55.
26. Kobayashi, T. Horndeski theory and beyond: a review. *Rep. Prog. Phys.* **2019**, *82*, 086901. [CrossRef]
27. Clifton, T.; Ferreira, P.G.; Padilla, A.; Skordis, C. Modified Gravity and Cosmology. *Phys. Rept.* **2012**, *513*, 1–189.
28. Bahamonde, S.; Dialektopoulos, K.F.; Escamilla-Rivera, C.; Farrugia, G.; Gakis, V.; Hendry, M.; Hohmann, M.; Levi Said, J.; Mifsud, J.; Di Valentino, E. Teleparallel gravity: from theory to cosmology. *Rept. Prog. Phys.* **2023**, *86*, 026901.
29. Charmousis, C.; Copeland, E.J.; Padilla, A.; Saffin, P.M. Self-tuning and the derivation of a class of scalar-tensor theories. *Phys. Rev. D* **2012**, *85*, 104040. [CrossRef]
30. Horndeski, G.W. Second-order scalar-tensor field equations in a four-dimensional space. *Int. J. Theor. Phys.* **1974**, *10*, 363–384. [CrossRef]
31. Copeland, E.J.; Padilla, A.; Saffin, P.M. The cosmology of the Fab-Four. *J. Cosmol. Astropart. Phys.* **2012**, *12*, 026.
32. Latosh, B. Fab Four Effective Field Theory Treatment. *Eur. Phys. J. C* **2018**, *78*, 991.
33. Linder, E.V. How Fabulous Is Fab 5 Cosmology? *J. Cosmol. Astropart. Phys.* **2013**, *12*, 032.
34. Starobinsky, A.A.; Sushkov, S.V.; Volkov, M.S. The screening Horndeski cosmologies. *J. Cosmol. Astropart. Phys.* **2016**, *2016*, 007. [CrossRef]
35. Starobinsky, A.A.; Sushkov, S.V.; Volkov, M.S. Anisotropy screening in Horndeski cosmologies. *Phys. Rev. D* **2020**, *101*, 064039. [CrossRef]
36. Galeev, R.; Muharlyamov, R.; Starobinsky, A.A.; Sushkov, S.V.; Volkov, M.S. Anisotropic cosmological models in Horndeski gravity. *Phys. Rev. D* **2021**, *103*, 104015. [CrossRef]
37. Gleyzes, J.; Langlois, D.; Piazza, F.; Vernizzi, F. New Class of Consistent Scalar-Tensor Theories. *Phys. Rev. Lett.* **2015**, *114*, 211101. [CrossRef]
38. Babichev, E.; Charmousis, C.; Langlois, D.; Saito, R. Beyond Fab Four. *Class. Quant. Grav.* **2015**, *32*, 242001.
39. Ezquiaga, J.M.; Zumalacárregui, M. Dark Energy After GW170817: Dead Ends and the Road Ahead. *Phys. Rev. Lett.* **2017**, *119*, 251304. [CrossRef]
40. Copeland, E.J.; Kopp, M.; Padilla, A.; Saffin, P.M.; Skordis, C. Dark energy after GW170817 revisited. *Phys. Rev. Lett.* **2019**, *122*, 061301.
41. Bordin, L.; Copeland, E.J.; Padilla, A. Dark energy loopholes some time after GW170817. *J. Cosmol. Astropart. Phys.* **2020**, *2020*, 063. [CrossRef]
42. Appleby, S.; Linder, E.V. The well-tempered cosmological constant. *J. Cosmol. Astropart. Phys.* **2018**, *2018*, 034. [CrossRef]
43. Emond, W.T.; Li, C.; Saffin, P.M.; Zhou, S.Y. Well-Tempered Cosmology. *J. Cosmol. Astropart. Phys.* **2019**, *05*, 038.
44. Torres, I.; Fabris, J.C.; Piattella, O.F.; Batista, A.B. Quantum Cosmology of Fab Four John Theory with Conformable Fractional Derivative. *Universe* **2020**, *6*, 50.
45. Bamba, K.; Capozziello, S.; Nojiri, S.; Odintsov, S.D. Dark energy cosmology: the equivalent description via different theoretical models and cosmography tests. *Astrophys. Space Sci.* **2012**, *342*, 155–228.
46. Capozziello, S.; D'Agostino, R.; Luongo, O. Extended Gravity Cosmography. *Int. J. Mod. Phys. D* **2019**, *28*, 1930016.
47. Escamilla-Rivera, C.; Capozziello, S. Unveiling cosmography from the dark energy equation of state. *Int. J. Mod. Phys. D* **2019**, *28*, 1950154.
48. Escamilla-Rivera, C.; de Albornoz-Caratozzolo, J.; Najera, S. In preparation. 2023.

universe MDPI

Article

Reconstructing Modified and Alternative Theories of Gravity

Dalia Saha [1,2], Manas Chakrabortty [3] and Abhik Kumar Sanyal [1,*]

[1] Department of Physics, Jangipur College, Murshidabad 742213, West Bengal, India; daliasahamandal1983@gmail.com
[2] Department of Physics, University of Kalyani, Nadia 741235, West Bengal, India
[3] Department of Physics, University of Bankura, Bankura 722155, West Bengal, India; manas.chakrabortty001@gmail.com
* Correspondence: sanyal_ak@yahoo.com

Abstract: A viable radiation-dominated era in the early universe is best described by the standard (FLRW) model of cosmology. In this short review, we demonstrate reconstruction of the forms of $F(R)$ in the modified theory of gravity and the metric compatible $F(\mathrm{T})$ together with the symmetric $F(Q)$ in alternative teleparallel theories of gravity, from different perspectives, primarily rendering emphasis on a viable FLRW radiation era. Inflation has also been studied for a particular choice of the scalar potential. The inflationary parameters are found to agree appreciably with the recently released observational data.

Keywords: modified theories and alternative theories of gravity; reconstruction from radiation era; inflation

1. Introduction

Luminosity versus red-shift data curve associated with SN1a standard candles requires a modification of the FLRW universe at the late stage of cosmic evolution. On the other hand, the early vacuum-dominated era requires to incorporate a stage of very rapid growth of the universe (Inflation) in order to alleviate the flatness, the horizon and the structure formation problems associated with the standard model of cosmology. However, the radiation-dominated era and the early matter-dominated era are best described by the standard model of cosmology. In this connection, let us mention that the FLRW universe, or the FLRW model, stands for the standard model of cosmology and must not be confused with the 'FLRW metric'. In fact, based on the cosmological principle (homogeneity and isotropy), Friedman in 1922 and Lemaître in 1927 independently showed that a universe filled with perfect barotropic fluid undergoes decelerated expansion ($a \propto \sqrt{t}$ in the radiation era and $a \propto t^{(2/3)}$ in the matter-dominated era, where $a(t)$ is the scale factor). Later, Robertson (1935, 1936) and Walker (1936) independently demonstrated that the metric used by Friedmann and Lemaître provides the most general Riemannian geometry compatible with homogeneity and isotropy. After Hubble's discovery in 1930 along with Eddington's proof (1931) that the static universe with a positive cosmological constant (Λ) is unstable, Einstein withdrew Λ and agreed that the universe is expanding. Since then, it is dubbed as the standard model of cosmology or the FLRW model. However, for the reason stated above, the metric which was commonly known as Robertson–Walker metric, currently called as FLRW metric. Note that the metric and the model are different. In the background of the FLRW metric, if the energy momentum tensor is modified by considering any other type of fluid (dissipative, scalar field or even a cosmological constant), it is different from the FLRW model.

Since on the one hand, the theoretical estimation of the cosmological constant (the sum of zero point energy of all quantum fields existed in the early universe) is 10^{120} orders of magnitude larger than the value required to solve the cosmic puzzle (late-stage of accelerated expansion), and on the other, no trace of dark energy is observed after performing precession experiments [1], the bulk of recent research is therefore oriented in search of alternatives. Consequently, a host of modified theories of gravity have been proposed in recent years as alternatives to the dark energy. However, considering the fact that, apart from the violation of superposition principle, the modified $F(R)$ theories, being associated with higher derivative terms, sometimes run from Ostrogradsky's instability [2]; alternative theories of gravity, commonly known as teleparallel theories, are currently in the limelight.

Curvature is the trait of (and attributed to) the 'general theory of relativity' (GTR), where the torsion-less Levi-Civita connection is used. The minimal modified version of GTR is the so called $F(R)$ theory of gravity. Alternatively, it is possible to ascribe gravity to torsion, by using the curvature-less ($R^{\alpha}{}_{\beta\mu\nu} = 0$) Weitzenböck connection. Since $\nabla_{\mu}g_{\alpha\beta} = 0$, where ∇ is the covariant derivative with respect to the affine connection instead of the Levi-Civita connection, it is often referred to as metric (compatible) teleparallel gravity. Its extended version is the $F(\mathrm{T})$ theory of gravity, where T is the quadratic torsion scalar constructed from the torsion tensor. Yet another class of gravity theory exists, by the name symmetric teleparallel gravity. It is constructed from the curvature-less ($R^{\alpha}{}_{\beta\mu\nu} = 0$) and torsion-less ($\mathrm{T}^{\lambda}{}_{\mu\nu} = 0$) general affine connection, which is symmetric in lower indices. It is a generalized version of the $F(Q)$ gravity theory, where Q stands for the quadratic non-metricity scalar constructed from the non-metricity tensor ($Q_{\lambda\mu\nu} = \nabla_{\lambda}g_{\mu\nu}$). Clearly, in these so-called teleparallel theories of gravity, the affine connection is independent of the metric tensor, and so these are essentially the metric-affine theories, where both the metric and the connection act as dynamical variables. In the first order, both the teleparallel theories end up with GTR apart from a total derivative term and are dubbed as 'Teleparallel Equivalent of General Relativity' (TEGR) and 'Symmetric Teleparallel Equivalent of General Relativity' (STEGR), respectively. However, their extended versions deviate considerably from GTR and its modified versions $F(R)$ or else, particularly giving second-order differential field equations, thus avoiding Ostrogradsky's instability. However, the $F(\mathrm{T})$ gravity theory is endured with a serious pathology of non-invariance under local Lorentz transformations of the tetrad field. It may be stated that Lorentz symmetry is the only known fundamental physical symmetry. A lack of Lorentz invariance requires preferred frames to parallelize the spacetime, implying the existence of extra degrees of freedom (dof) compared to GTR. The canonical formulation of the singular action of $F(\mathrm{T})$ gravity theory requires constraint analysis. The extra dof appear due to the six first-class constraints associated with the theory [3,4]. Recent investigation suggests that the number of extra dofs is one [5–7]. Next, is the problem endured with strong coupling issue. This appears due to the fact that some of the genuine physical degrees of freedom lose their kinetic term at the quadratic order and, consequently, the standard perturbation theory breaks down on these backgrounds [8,9]. Under these circumstances, the extended version of symmetric teleparallel $F(Q)$ gravity theory, apparently being free from such problems [10], attracted lot of attention in the recent years. Nonetheless, it is to be mentioned that, as long as the cosmic puzzle is not being resolved unambiguously, focus on different modified and alternative theories of gravity will persist.

Whatever the theory might be, whenever an arbitrary functional form like $F(R)$, $F(\mathrm{T})$ or $F(Q)$ is retained in the action, it becomes mandatory to expatiate such functional forms in view of some additional physically viable conditions. Two possibilities have largely been explored so far. One is to invoke the presence of dynamical (Noether) symmetry, while the other is to follow the reconstruction program. Both have come out with certain specific functional forms in all the cases. Additionally, some forms are also chosen by hand (on some physical ground). Most of these claim to associate early inflation in the vacuum era together with late-time accelerated expansion in the current matter-dominated era. The important point to be noted is that the cosmic evolution in the radiation-dominated era is not studied in general, as if the universe jumped from the early inflationary stage to the late stage of the matter-dominated era. On the contrary, a viable radiation-dominated era which can explain the standard Big-Bang Nucleosynthesis (BBN), accounts for the observed structure formation and the formation of CMBR, and is best described by the standard (FLRW) model of cosmology. This manuscript is devoted to reconstructing all three different forms for the gravity theories with curvature $[F(R)]$ and torsion $[F(\mathrm{T})]$, as well as non-metricity $[F(Q)]$ of different extended theories of gravity in view of a viable radiation-dominated era.

In Section 2, we briefly review the three above-mentioned pillars of gravity theory. In Section 3, we follow the reconstruction program as mentioned, to explore the forms of $F(R)$, and $F(\mathrm{T})$ gravity theories, primarily in view of a viable Friedmann-like radiation-dominated era. The form of $F(\mathrm{T})$ is also known from the Noether symmetry consideration, which is different from the form obtained here. We, therefore, have studied the energy conditions and inflation, taking into account different forms of $F(\mathrm{T})$ so obtained. Since $F(Q)$ gravity in a coincident gauge is no different from $F(\mathrm{T})$ in the present context, all the results obtained for the $F(\mathrm{T})$ theory hold as well in the $F(Q)$ theory of gravity. Yet another largely developed theory of gravity, the so called Palatini formalism, has been discussed briefly at the end of Section 3. Finally, we conclude in Section 4.

2. The Three Pillars of Gravitation

The building block of gravitation is the general affine connection $\Gamma^{\alpha}{}_{\mu\nu}$, which is constructed from the Levi-Civita connection ($\{_{\mu}{}^{\alpha}{}_{\nu}\}$), the contorsion tensor ($K^{\alpha}{}_{\mu\nu}$) and the disformation tensor ($L^{\alpha}{}_{\mu\nu}$) as

$$\Gamma^{\alpha}{}_{\mu\nu} = \{_{\mu}{}^{\alpha}{}_{\nu}\} + K^{\alpha}{}_{\mu\nu} + L^{\alpha}{}_{\mu\nu}, \tag{1}$$

where

$$\begin{aligned} \{_{\mu}{}^{\alpha}{}_{\nu}\} &= \frac{1}{2}g^{\alpha\lambda}\left(g_{\mu\lambda,\nu} + g_{\nu\lambda,\mu} - g_{\mu\nu,\lambda}\right), \\ K^{\alpha}{}_{\mu\nu} &= \frac{1}{2}g^{\alpha\lambda}\left(\mathrm{T}_{\lambda\mu\nu} - \mathrm{T}_{\mu\lambda\nu} - \mathrm{T}_{\nu\lambda\mu}\right), \\ L^{\alpha}{}_{\mu\nu} &= \frac{1}{2}g^{\alpha\lambda}\left(Q_{\lambda\mu\nu} - Q_{\mu\lambda\nu} - Q_{\nu\lambda\mu}\right). \end{aligned} \tag{2}$$

while the Levi-Civita connection is formed from the derivatives of the metric, and the contorsion tensor, being the difference between the connections with and without torsion, is again equivalent to the difference between the Weitzenböck ($\Gamma^{\lambda}{}_{a\mu} = e^{a\lambda}\partial_{\mu}e_{a\nu}$, where $e^{a}{}_{\mu}$ are the components of the orthonormal basis vectors of the tetrad field, commonly called the vielbein (or vierbein) field, to be more specific. Here, Latin indices stands for Lorentz indices, while Greek indices stand for space time. Since both run from 0 to 3, we shall omit using Latin indices in the following.)[1] and the Levi-Civita connections, which are formed from the torsion tensor as

$$\mathrm{T}^{\alpha}{}_{\mu\nu} = \partial_{\mu}e^{\alpha}{}_{\nu} - \partial_{\nu}e^{\alpha}{}_{\mu} = \Gamma^{\alpha}{}_{[\mu,\nu]} = \Gamma^{\alpha}{}_{\mu\nu} - \Gamma^{\alpha}{}_{\nu\mu}, \tag{3}$$

and, finally, the disformation tensor is constructed from the non-metricity tensor $Q_{\alpha\mu\nu} = \nabla_{\alpha}g_{\mu\nu} \neq 0$ as

$$Q_{\alpha\mu\nu} = \nabla_{\alpha}g_{\mu\nu} = g_{\mu\nu,\alpha} - \Gamma^{\lambda}{}_{\mu\alpha}g_{\lambda\nu} - \Gamma^{\lambda}{}_{\nu\alpha}g_{\mu\lambda}. \tag{4}$$

Einstein's GTR is built from the non-vanishing Riemann tensor formed from the Levi-Civita connection,

$$R^{\alpha}{}_{\beta\mu\nu} = \partial_{\mu}\{_{\nu}{}^{\alpha}{}_{\beta}\} - \partial_{\nu}\{_{\mu}{}^{\alpha}{}_{\beta}\} + \{_{\mu}{}^{\alpha}{}_{\lambda}\}\{_{\nu}{}^{\lambda}{}_{\beta}\} - \{_{\nu}{}^{\alpha}{}_{\lambda}\}\{_{\mu}{}^{\lambda}{}_{\beta}\}, \tag{5}$$

the simple reason being that the Levi-Civita connections are the unique affine torsion-free connections on the tangent bundle of a manifold that preserves the Riemannian metric $g^{\mu\nu}{}_{;\nu} = 0$. So, once a metric is specified, the unique non-vanishing components of the Riemann tensor are found from the Levi-Civita connections or Christoffel symbols, to be specific[2] (79). The above relation (79) is contracted to find the Ricci tensor ($R_{\mu\nu}$) and, under further contraction, results in the Ricci scalar (R),

while the Einstein–Hilbert action for GTR is constructed from the Ricci scalar, the so called modified theory of gravity is constructed from $F(R)$ as

$$A = \int \left[F(R)\sqrt{-g}\, d^4x \right] + \mathcal{S}_m, \tag{6}$$

with $\mathcal{S}_m$ being the matter action. The field equations corresponding to the action (6) may be expressed as

$$F'(R)R_{\mu\nu} - \frac{1}{2}F(R)g_{\mu\nu} - \nabla_\mu \nabla_\nu F'(R) + g_{\mu\nu}\Box F'(R) = \kappa T_{\mu\nu}. \tag{7}$$

In the above, $\Box F'(R) = F^{;\mu}{}_{;\mu} = \frac{1}{\sqrt{-g}}\partial_\mu(\sqrt{-g}\, g^{\mu\nu}\partial_\nu F')$, where the semicolon stands for the covariant derivative, and $\kappa = 8\pi G$. It is noteworthy that the $F(R)$ theory of gravity, under weak field approximation, leads to the 'GTR' [11,12]. However, once flatness is assumed, the curvature vanishes and the orientation of vectors remain unaltered under parallel transport along a curve, and the geometry is teleparallel. The condition for torsion-based teleparallelism is $\bar{R}^\alpha_{\beta\mu\nu} = 0 = \tilde{R}^\alpha_{\beta\mu\nu}$, where over-bar and over-tilde stand for the Riemann curvature formed from the Weitzenböck connections and connections associated with the non-metricity tensor, respectively. The role of R is played by the torsion scalar T and the action for generalized gravity theory with torsion [13] may be expressed as

$$A = \int |e| F(\mathrm{T}) d^4x + \mathcal{S}_m, \tag{8}$$

where $|e| = \det e^i_\mu = \sqrt{-g}$, e^i_μ are the components of the vierbien field $\mathbf{e_i}(x^\mu) = e^\mu_i \partial\mu$. The metric tensor is obtained from the dual vierbein as $g_{\mu\nu}(x) = \eta_{ij} e^i_\mu(x) e^j_\nu(x)$. The role of the Ricci scalar (R) in the Einstein–Hilbert action is played in the metric teleparallel theory by the torsion scalar (T), which is found from the torsion tensor

$$\mathrm{T}^\rho{}_{\mu\nu} \equiv e^\rho{}_i [\partial_\mu e^i{}_\nu - \partial_\nu e^i{}_\mu], \tag{9}$$

as

$$\mathrm{T} = S_\rho{}^{\mu\nu} \mathrm{T}^\rho{}_{\mu\nu}, \quad \text{where,} \quad S_\rho{}^{\mu\nu} = \frac{1}{2}\left(K_\rho{}^{\mu\nu} + \partial^\mu_\rho \mathrm{T}_\alpha{}^{\alpha\nu} - \partial^\nu_\rho \mathrm{T}_\alpha{}^{\alpha\mu} \right), \tag{10}$$

which is called the super-potential, and the affine connection takes the Weitzenböck form,

$$\Gamma^\alpha{}_{\mu\nu} = e^\alpha{}_\lambda \partial_\nu e^\lambda{}_\mu. \tag{11}$$

The above action (8) leads to the following field equations for $F(\mathrm{T})$ gravity:

$$e^{-1}\partial\mu(e S_i^{\mu\nu}) F_{,\mathrm{T}} - e_i^\lambda \mathrm{T}^\rho_{\mu\lambda} S_\rho^{\nu\mu} F_{,\mathrm{T}} + S_i^{\mu\nu}\partial\mu(F(\mathrm{T})) F_{,\mathrm{TT}} + \frac{1}{4} e_i^\nu F(\mathrm{T}) = \frac{1}{4} e_i^\rho T_\rho^\nu, \tag{12}$$

where $S^{\mu\nu}{}_i = e^\rho{}_i S^{\mu\nu}{}_\rho$ and $F_{,\mathrm{T}}$ denotes differentiation with respect to T.

In symmetric teleparallel gravity, on the contrary, the role of Ricci scalar is played by the non-metricity scalar (Q), and again the generalized action is given by

$$A = \int F(Q)\sqrt{-g} d^4x + \mathcal{S}_m, \tag{13}$$

where

$$\begin{aligned} Q_{\alpha\mu\nu} &= \frac{\partial g_{\mu\nu}}{\partial x^\alpha} - \Gamma^\beta{}_{\alpha\mu} g_{\beta\nu} - \Gamma^\beta{}_{\alpha\nu} g_{\mu\beta}; \\ Q_\alpha &= g^{\mu\nu} Q_{\alpha\mu\nu} = Q_{\alpha\nu}{}^\nu; \qquad \hat{Q}_\alpha = g^{\mu\nu} Q_{\mu\alpha\nu} = Q_{\nu\alpha}{}^\nu; \\ Q &= -\frac{1}{4} Q_{\alpha\mu\nu} Q^{\alpha\mu\nu} + \frac{1}{2} Q_{\alpha\mu\nu} Q^{\mu\nu\alpha} + \frac{1}{4} Q_\mu Q^\mu - \frac{1}{2} Q_\mu \hat{Q}^\mu. \end{aligned} \tag{14}$$

Varying the above action (13) with respect to the metric tensor $g^{\mu\nu}$ and the connection, the following field equations are obtained:

$$\frac{2}{\sqrt{-g}} \nabla_\lambda (\sqrt{-g} F_Q P^\lambda{}_{\mu\nu}) + \frac{1}{2} F g_{\mu\nu} + F_Q (P_{\nu\rho\sigma} Q_\mu{}^{\rho\sigma} - 2 P_{\rho\sigma\mu} Q^{\rho\sigma}{}_\nu) = -\kappa T_{\mu\nu}, \tag{15}$$

$$\nabla_{\mu}\nabla_{\nu}(\sqrt{-g}F_{,Q}P^{\mu\nu}{}_{\lambda}) = 0, \tag{16}$$

where

$$P^{\alpha}{}_{\mu\nu} = \frac{1}{2}Q_{(\mu\ \nu)}^{\ \alpha} - \frac{1}{4}Q^{\alpha}{}_{\mu\nu} + \frac{1}{4}(Q^{\alpha} - \hat{Q}^{\alpha})g_{\mu\nu} - \frac{1}{4}\delta^{\alpha}{}_{(\mu}Q_{\nu)} \tag{17}$$

is called the non-metricity conjugate tensor. In the above, we have used the first bracket in the suffix to denote symmetrization of indices. The covariant formulation [14] of the field Equation (15) is

$$F_{,Q}G_{\mu\nu} + \frac{1}{2}g_{\mu\nu}(QF_{,Q} - F(Q)) + 2F_{,QQ}\nabla_{\lambda}QP^{\lambda}{}_{\mu\nu} = -\kappa T_{\mu\nu} \tag{18}$$

where $G_{\mu\nu} = R_{\mu\nu} - \frac{1}{2}g_{\mu\nu}R$, which corresponds to the Levi-Civita connection, and $F_{,Q}$ stands for the derivative of $F(Q)$ with respect to Q.

Having briefly discussed the three different ways attributed to the gravity, sometimes dubbed as the 'geometrical trinity of gravity', we now proceed in the following section to find the specific forms of $F(R)$, $F(\mathrm{T})$ and $F(Q)$.

3. Reconstruction from the Radiation Era

As mentioned, the standard FLRW model of cosmology fits best in the radiation- and early matter-dominated eras. In particular, a Friedmann-like radiation-dominated era expedites the standard Big-Bang-Nucleosynthesis (BBN), accounts for the observed structure formation (once the seeds of structure are given) and the Cosmic Microwave Background Radiation (CMBR). Therefore, exploring suitable forms of different theories of gravity in view of early Friedmann-like radiation era is supposed to be the best option. It may be mentioned that the modified or generalized alternative theories of gravity might admit Friedmann-like solutions, depending on their forms. For example, the first-order Palatini theory of gravity and also the two telleparallel theories, viz. TEGR and STEGR, are equivalent to GTR. As a result, Friedmann solutions automatically follows. We also remind the reader that if modified gravity includes terms like $F(R) = \alpha R + \beta R^2 + \gamma R^n$ or others, so that R dominates in the middle, then Friedmann-like solutions result in the radiation- as well as in the early matter-dominated eras. Nonetheless, modified and alternative theories should be treated as viable only if these theories admit Friedmann-like solutions in the radiation- and early matter-dominated eras. Therefore, in this article, we restrict ourselves to the cosmological principle, considering the isotropic and homogeneous Robertson–Walker (RW) space-time,

$$ds^2 = -dt^2 + a^2(t)\left[\frac{dr^2}{1-kr^2} + r^2(d\theta^2 + r^2 sin^2\theta d\phi^2)\right], \tag{19}$$

and explore the relevant forms of modified gravity theories comprised with the curvature [$F(R)$], torsion [$F(\mathrm{T})$], as well as non-metricity [$F(Q)$].

3.1. $F(R)$ Gravity

The Ricci scalar corresponding to the RW metric (19) is

$$R = 6\left(\frac{\ddot{a}}{a} + \frac{\dot{a}^2}{a^2} + \frac{k}{a^2}\right). \tag{20}$$

The point Lagrangian corresponding to the action (6) is given by [15]

$$L_R = -6a\dot{a}^2F' - 6a^2\dot{a}\dot{R}F'' + a^3(F - RF') + 6kaF' + L_m, \tag{21}$$

with L_m being the matter Lagrangian, comprised a barotropic fluid associated with the thermodynamic pressure (p) and energy density (ρ) along with the cold dark matter (CDM). The field equations are

$$\left(2\frac{\ddot{a}}{a}+\frac{\dot{a}^2}{a^2}+\frac{k}{a^2}\right)F'+\left(\ddot{R}+2\frac{\dot{a}}{a}\dot{R}\right)F''+\dot{R}^2F'''+\frac{1}{2}(F-RF')=-p,$$
$$\left(\frac{\dot{a}^2}{a^2}+\frac{k}{a^2}\right)F'+\frac{\dot{a}}{a}\dot{R}F''+\frac{1}{6}(F-RF')=\frac{\rho}{3}, \tag{22}$$

Adding the above pair of equations, one obtains,

$$2\left(\frac{\ddot{a}}{a}+\frac{\dot{a}^2}{a^2}+\frac{k}{a^2}\right)F'+\left(3\frac{\dot{a}}{a}\dot{R}+\ddot{R}\right)F''++\dot{R}^2F'''+\frac{2}{3}(F-RF')=\frac{\rho}{3}-p. \tag{23}$$

Clearly, the radiation era ($p=\frac{1}{3}\rho$) does not evolve like standard (FLRW) model ($a\propto\sqrt{t}$) for the spatially flat ($k=0$) universe and, as such, a viable form of $F(R)$ remains obscure. For $k\neq 0$, on the other hand, the differential equation cannot be solved, either in the radiation or in the vacuum-dominated ($\rho=0=p$) era, other than the fact that, for $k=0$, the vacuum era admits a de-Sitter solution. On the contrary, it may be mentioned that different forms of $F(R)$ emerge in different eras, in view of the Noether symmetry [16].

Nonetheless, interesting results emerge from a generalized four-dimensional string effective action associated with higher-order curvature invariant terms, being expressed in the following form [17]:

$$A=\int\left[f(\phi)R+BF(R)-\frac{\omega(\phi)}{\phi}\phi_{,\mu}\phi^{,\mu}-V(\phi)-\kappa\mathcal{L}_m\right]\sqrt{-g}d^4x, \tag{24}$$

where B is the coupling constant, $f(\phi)$ is the coupling parameter, $\omega(\phi)$ is the variable Brans–Dicke parameter, and $\mathcal{L}_m$ is the matter Lagrangian density. It had been revealed [17] that the above action admits a conserved current $J^\mu=(3f'^2+2f\frac{\omega}{\phi})^{\frac{1}{2}}$ under the condition

$$B\left[RF_{,R}+3(F_{,R})^{;\alpha}_{;\alpha}-2f\right]=\frac{\kappa}{2}T^\mu_\mu,\quad\text{provided}\quad V=\lambda f^2, \tag{25}$$

where λ is a constant. To perceive the very important role of such a conserved current, a particular case was also studied [17]. For example, considering $F(R)=R^2$, the condition for the existence of conserved current (25) reads as

$$R^{;\mu}{}_{;\mu}=\Box R=\left(\ddot{R}+3\frac{\dot{a}}{a}R\right)=\frac{\kappa}{12B}T^\mu_\mu. \tag{26}$$

Now, if the scalar field (ϕ) is completely used up in the process of driving inflation and reheating under particle creation then, since in the radiation-dominated era $T^\mu{}_\mu=T=\rho-3p=0$, in view of (26), $\Box R=(\ddot{R}+3\frac{\dot{a}}{a}R)=0$. Further, since in the very early vacuum-dominated era, sufficient inflation makes the universe spatially flat, i.e., $k=0$, the above Equation (26) admits a Friedmann-like solution, $a=a_0\sqrt{t}$ in the radiation-dominated era as demonstrated in [17]. This is a unique result, since even in the presence of a higher-order curvature, invariant-term Friedmann-like radiation era is admissible.

Finally, we consider pure curvature-induced gravity theory, a yet another generalized form of $F(R)$ action, considered in [18] being expressed as

$$A=\int\left[\alpha R+BR^{\frac{3}{2}}+\gamma R^2-\mathcal{L}_m\right]\sqrt{-g}d^4x. \tag{27}$$

The above action was found more suitable to explain cosmic evolution right from the very early stage to date, since it satisfies all the strong conditions necessary for a viable $F(R)$ theory of gravity. It may be mentioned that the $R^{\frac{3}{2}}$ term appeared as a consequence of the Noether symmetry in the R-W metric, both in vacuum- as well as in the matter-dominated eras [19–23]. At the initial stage, the R^2 term dominates and a de-Sitter solution is realizable. This leads to the inflationary epoch and reheating following the mechanism of particle production via scalaron decay, exploiting gravity only without invoking phase transition [24,25]. After the reheating is over, the universe evolves as Friedmann-like radiation ($a \propto \sqrt{t}$) and early matter ($a \propto t^{\frac{2}{3}}$)-dominated eras, and finally accelerated expansion is realized due to the presence of the combination of linear term R and non-linear term $R^{\frac{3}{2}}$ [21]. Further, following numerical analysis [18] taking deceleration parameter q as a function of the red-shift parameter z, three distinct cases were analyzed to establish the fact that, within a particular range of values for β and γ, the q versus z plot depicts that the universe was in the pure radiation era at $z > 3200$. Thereafter, the deceleration parameter falls off from the matter–radiation equality epoch to the decoupling epoch. It falls even more sharply thereafter, and a Friedmann-type ($q \approx 0.5$) matter-dominated era is reached at around $z \approx 200$. The deceleration parameter then starts increasing slowly, and it is peaked to ($q = 1$), explaining re-ionization of the inter-galactic medium (IGM). Subsequently, late time accelerated expansion is initiated and the phantom divide line is crossed, to make a second transition out of it at $z \approx 0.5$.

In a nutshell, higher-order gravity theory is a viable option to explain cosmological evolution, from the early vacuum-dominated era to date, via a Friedmann-like radiation- and early matter-dominated eras. Additionally, re-ionization of the IGM may also be explained in the process. Inflation in the Starobinski model [24,25], with the R^2 term, is widely explored in the literature and, therefore, we leave it here.

3.2. Torsion-Based Metric $F(\mathrm{T})$ Teleparallel Gravity

Although the modified $F(R)$ theory of gravity seems to produce a viable theory of gravity, the fact that the field equations are fourth-order means it is vulnerable to Ostrogradsky's instability and also suffers from the violation of the superposition principle. This led us to consider alternative telleparallel theories of gravity, as already discussed. The main virtue of telleparallel theory of gravity is that the field equations are second-order as in the case of Lanczos–Lovelock gravity. Yet another possible fundamental motivation for teleparallelism is that it was thought to account for the otherwise anomalous dimension and, hence, non-renormalisability of the GTR action, providing a totally different approach towards a ultra-violet completion of gravity. Here, we shall consider generalized metric telleparallel gravity theory. In recent years, a generalized version of the 'teleparallel gravity' with torsion, namely the $F(\mathrm{T})$ theory of gravity (where T stands for the torsion scalar), also dubbed as generalized 'gravity with torsion', has been proposed as an alternative to both the dark energy theories and the modified theories of gravity. Primarily, the $F(\mathrm{T})$ theory of gravity was proposed to drive inflation. Later, it was applied to drive the current accelerated expansion of the present universe without considering dark energy [26,27]. It is worth mentioning that 'teleparallel equivalent of general relativity' (TEGR) is established for $F(\mathrm{T}) \propto \mathrm{T}$, since dynamically it leads to GTR. Further, as demonstrated in [18], when the trace of the electro-magnetic radiation field tensor ($T = \rho - 3p$) is zero, the contracted GTR equation $R \propto T$ enforces the Ricci scalar to vanish ($R = 0$) and the Friedmann-like decelerated expansion $[a(t) \propto \sqrt{t}]$ results automatically in the radiation-dominated era. This is true also in the case of modified theory of gravity, as demonstrated above. However, in the case of torsion, although $F(\mathrm{T}) \propto \mathrm{T}$ leads to GTR, the trace of the energy–momentum tensor $T = 0$ does not lead to the static solution $\mathrm{T} = 0$. Therefore, even though all the results of GTR hold, the pathology of discontinuous evolution of the Ricci scalar (large initially, vanishing in the middle and small at present) is averted. However, unlike GTR and $F(R)$ theories, gravity with torsion is not generally covariant by default. Nonetheless, it has been argued that it may be made so by introducing a new variable, viz. a spin-connection [13,28]. The reason being that the spin-connection enters the teleparallel action only as a surface term, and does not contribute to the field equations.

Now, the components of the vierbein field for the RW metric (19) are expressed in terms of the cosmological scale factor $a(t)$ as

$$e^i_\mu = \mathrm{diag}(1, a(t), a(t), a(t)), \tag{28}$$

and the torsion scalar reads as [29]

$$\mathrm{T} = -6H^2 + \frac{6k}{a^2}, \tag{29}$$

where $H = \frac{\dot{a}}{a}$ is the Hubble parameter. Thus, the field Equations (12) for RW metric are

$$12H^2 F_{,\mathrm{T}} + F(\mathrm{T}) = \rho, \tag{30}$$

$$4\left(\frac{k}{a^2} + \dot{H}\right)\left[12H^2 F_{,\mathrm{TT}} + F_{,\mathrm{T}}\right] - F(\mathrm{T}) - 4F_{,\mathrm{T}}[2\dot{H} + 3H^2] = p, \tag{31}$$

where ρ and p are the energy density and thermodynamic pressure of a barotropic fluid (inclusive of dark matter component), respectively, and $F_{,\mathrm{T}}$ stands for derivative of $F(\mathrm{T})$ with respect to T. Bianchi identity does not hold naturally in teleparallel gravity theories. However, in the RW metric under consideration, it holds even for extended models such as $F(\mathrm{T})$ [29]. For example, taking the time derivative of the first Equation (30), and also adding the the two (30) and (31), one obtains

$$\begin{aligned}
\dot{\rho} &= 12H^2\dot{\mathrm{T}}F_{,\mathrm{TT}} + 24H\dot{H}F_{,\mathrm{T}} + \dot{\mathrm{T}}F_{,\mathrm{T}} \\
&= -144H^3\dot{H}F_{,\mathrm{TT}} - \frac{144kH^3F_{,\mathrm{TT}}}{a^2} + 12H\dot{H}F_{,\mathrm{T}} - \frac{12kHF_{,\mathrm{T}}}{a^2}, \\
\text{and } (\rho + p) &= 48\dot{H}H^2F_{,\mathrm{TT}} - 4\dot{H}F_{,\mathrm{T}} + \frac{48k}{a^2}H^2F_{,\mathrm{TT}} + \frac{4k}{a^2}F_{,\mathrm{T}}.
\end{aligned} \tag{32}$$

Combining the above two, one finds

$$\dot{\rho} + 3H(\rho + p) = 0, \tag{33}$$

which is the energy–momentum conservation law, an outcome Bianchi identity. Thus, one may also write

$$\rho = \rho_0 a^{-3(1+\omega)}, \tag{34}$$

for the barotropic equation of state $p = \omega\rho$, as usual, where ω is the state parameter.

Let us now proceed to find the form of $F(\mathrm{T})$ for flat space ($k = 0$), already obtained earlier in view of cosmological evolution [30]. First, combining Equations (30) and (31) for flat space $k = 0$, one finds for the vacuum era ($p = 0 = \rho$),

$$\dot{H}\left[12H^2F_{,\mathrm{TT}} - F_{,\mathrm{T}}\right] = 0. \tag{35}$$

Clearly, two possibilities emerge from Equation (35): (i) $F = F_0\sqrt{\mathrm{T}}$, where F_0 is a constant of integration. In view of this form of $F(\mathrm{T})$, either of the field Equations (30) or (31) yields simply the definition and, hence, no dynamic results. Indeed, it is expected, since the above form of $F(\mathrm{T})$ only results in a divergent term in the action. (ii) $H = \lambda$ (where λ is a constant), leading to the de-Sitter solution $a = a_0\exp(\lambda t)$, while $F(\mathrm{T}) = F_0\exp(-\frac{t}{12\Lambda^2})$ emerges as an exponentially decaying function, which is not very promising.

Since the vacuum-dominated era does not yield a reasonably viable form of $F(\mathrm{T})$, let us now advance further to study the radiation-dominated era. Starting from the action (8), if a solution in the form $a = a_0t^n$ is sought in the radiation-dominated era ($p = \frac{1}{3}\rho$), for which the energy–momentum conservation law yields $\rho a^4 = \rho_{r0}$, where a_0, n and ρ_{r0} are constants, then the following form of $F(\mathrm{T})$ is found [16]:

$$F(\mathrm{T})_r = 2f_1\sqrt{-6\mathrm{T}} + \frac{\rho_{r0}}{6a_0^4(4n-1)n^{4n}(-6)^{2n-1}}\mathrm{T}^{2n}, \tag{36}$$

where f_1 is a constant and the suffix r stands for the radiation-dominated era. Note that the first term of the above equation is essentially a divergent term in the RW metric under consideration. Thus, only the second term is left and, hence,

$$F(\mathrm{T})_r = \frac{\rho_{r0}}{6a_0^4(4n-1)n^{4n}(-6)^{2n-1}}\mathrm{T}^{2n}. \tag{37}$$

It is quite apparent that under the choice $n = \frac{1}{2}$, the radiation era evolves exactly like the standard (FLRW) model; however, the action too reduces to that of GTR, since $[F(T) \propto \mathrm{T}]$ (apart from a total derivative term), which is TEGR, as already mentioned. Of course, $n = \frac{3}{4}$ also leads to decelerated expansion, but such a slow deceleration suggests the formation of CMBR at latter times than observed. Let us next focus on the matter-dominated era ($p = 0$), for which the energy–momentum conservation law yields $\rho a^3 = \rho_{m0}$, where ρ_{m0} is a constant, and suffix m stands for the matter-dominated era. Proceeding in the similar manner as before, one finds

$$F(\mathrm{T})_m = \frac{\rho_{m0}}{3a_0^3(3n-1)n^{3n}(-6)^{\frac{3n}{2}-1}}\mathrm{T}^{\frac{3n}{2}}, \tag{38}$$

where the suffix m stands for the matter-dominated era. Thus, for $n = \frac{2}{3}$, the matter-dominated era evolves ($a \propto t^{\frac{2}{3}}$) in the same manner as the standard (FLRW) model of cosmology, and GTR is recovered through TEGR. However, it was also discussed in [16] that when torsion is attributed to gravity, usually a form such as $F(\mathrm{T}) = f_0\mathrm{T} + f_1\mathrm{T}^2 + \cdots$ is chosen to combat early deceleration in the Friedmann form $[a(t) \propto t^{\frac{2}{3}}]$, followed by a late-time cosmic acceleration in the matter (pressure-less dust)-dominated era. In view of (36) and (38), it is clear that $F(\mathrm{T}) \propto \mathrm{T}$ gives an exactly Friedmann-like radiation-dominated era $[a(t) \propto t^{\frac{1}{2}}]$, and early pressure-less dust-dominated era $[a(t) \propto t^{\frac{2}{3}}]$, respectively. Thus, following generalized form of $F(\mathrm{T})$,

$$F(\mathrm{T}) = f_0\mathrm{T} + f_1(-\mathrm{T})^{\frac{3}{2}} + f_2(\mathrm{T}^2) + \cdots \tag{39}$$

might be useful to study cosmic evolution. On the contrary, Noether symmetry analysis [16] demands that instead of T^2, one should associate T^3 and higher odd integral powers in the action. That is, a viable form that might explain the cosmic evolutionary history may be in the form $F(\mathrm{T}) = f_0\mathrm{T} + f_1\mathrm{T}^3 + \cdots$. It may be mentioned that the current analysis also reveals the fact that pure $F(\mathrm{T})$ gravity in vacuum ($\rho = p = 0$) does not give rise to any dynamics. Therefore, to drive inflation, and also to avert the pathological behavior of pure $F(\mathrm{T})$ gravity in the very early vacuum-dominated era, either unimodular $F(\mathrm{T})$ gravity has to be considered [31], or a scalar field should be associated with $F(\mathrm{T})$ gravity theory [30] and the action may be proposed as

$$A = \int \left[f_0\mathrm{T} + f_1\mathrm{T}^3 + \cdots - \frac{1}{2}\phi_{,\mu}\phi^{,\mu} - V(\phi)\right]\sqrt{-g}d^4x, \tag{40}$$

where the scalar field drives inflation at the very early stage and decays to an insignificant value in the process of particle creation. Therefore, once the inflation is over, the universe enters the radiation- and, thereafter, the matter-dominated eras, whence T dominates to envisage the standard model. Now, since $\mathrm{T} = -6H^2 < 0$ by definition, therefore, as the Hubble parameter decreases further, the odd-integral higher degree terms start dominating and become responsible for late-time cosmic acceleration. The great conceptual advantage over modified theories of gravity is that, unlike the Ricci scalar, $\mathrm{T} \neq 0$, at any stage of cosmic evolution in the middle.

3.2.1. Energy Conditions

Before we proceed further, it is necessary to fix the signature of the coefficients (f_0, f_1 and f_2) associated with the two different forms of $F(\mathrm{T})$ obtained above, viz. (i) $F(\mathrm{T}) = f_0\mathrm{T} + f_1(-\mathrm{T})^{\frac{3}{2}} + f_2(\mathrm{T}^2) + \cdots$, and (ii) $F(\mathrm{T}) = f_0\mathrm{T} + f_1\mathrm{T}^3 + f_2\mathrm{T}^5 + \cdots$, in view of the energy conditions. For perfect fluid $T_{\mu\nu} = (\rho + p)u_\mu u_\nu + pg_{\mu\nu}$, the energy conditions are:

1. Null energy condition: $\rho + p \geq 0$.
2. Weak energy condition: $\rho + p \geq 0$ and $\rho \geq 0$.
3. Dominant energy condition: $\rho \geq |p|$.
4. Strong energy condition: $\rho + p \geq 0$ and $\rho + 3p \geq 0$.

Note that there is no restriction on the thermodynamic pressure, i.e., $p < 0$ is allowed; nonetheless, if both $\rho > 0$ and $p > 0$, all energy conditions are satisfied simultaneously.

Form 1. *Fixing coefficients of* $F(\mathrm{T}) = f_0\mathrm{T} + f_1(-\mathrm{T})^{\frac{3}{2}} + f_2(\mathrm{T}^2) + \cdots$.

To expatiate the signature associated with the coefficients of $F(\mathrm{T})$, let us analyze the form of for each individual term separately.

1. For, $F(\mathrm{T}) = f_0\mathrm{T}$, the field Equations (30) and (31) read as
$$6f_0H^2 = \rho, \quad \text{and} \quad -6f_0H^2 - 4f_0\dot{H} = p, \tag{41}$$
respectively. Therefore, f_0 has to be positive so that $\rho > 0$. Consequently, $\rho + p = -4f_0\dot{H} > 0$ is also satisfied, because $\dot{H} < 0$, in the expanding model.
2. Again for, $F(\mathrm{T}) = f_1(-\mathrm{T})^{\frac{3}{2}}$, the field Equation (30) is
$$24f_1\sqrt{6}H^3 = \rho, \quad \text{and} \quad -\frac{9}{2}f_1\sqrt{6}\dot{H}H - 24f_1\sqrt{6}H^3 = p, \tag{42}$$
so, to keep ρ positive, f_1 has to be positive and $\rho + p = -\frac{9}{2}f_1\sqrt{6}\dot{H}H > 0$ is also satisfied, since $\dot{H} < 0$.
3. Finally, for $F(\mathrm{T}) = f_2\mathrm{T}^2$, the field Equations (30) and (31) may be cast as
$$-108f_2H^4 = \rho, \quad \text{and} \quad 144f_2\dot{H}H^2 + 108f_2H^4 = p, \tag{43}$$
respectively. Therefore, $\rho > 0$ is ensured, provided that $f_2 < 0$ and, finally, to satisfy the weak energy condition, $\rho + p = 144f_2\dot{H}H^2 > 0$ also ensures $f_2 < 0$, since $\dot{H} < 0$ in the expanding model.

Thus, all the energy conditions are satisfied, provided that $F(T)$ has the form
$$F(\mathrm{T}) = f_0\mathrm{T} + f_1(-\mathrm{T})^{\frac{3}{2}} - f_2(\mathrm{T}^2), \tag{44}$$
in which all the coefficients are positive, i.e., $f_0 > 0$, $f_1 > 0$, and $f_2 > 0$, and we terminate after the third term.

Form 2. *Fixing the coefficients of* $F(\mathrm{T}) = f_0\mathrm{T} + f_1(\mathrm{T}^3) + f_2(\mathrm{T}^5) + \cdots$.

Likewise, let us fix the signature of the coefficients (f_0, f_1 and f_2) appearing in the second form of $F(\mathrm{T})$, considering each term separately.

1. For $F(\mathrm{T}) = f_0\mathrm{T}$, the field Equations (30) and (31) are expressed as
$$6f_0H^2 = \rho, \quad \text{and} \quad -6f_0H^2 - 4f_0\dot{H} = p, \tag{45}$$
respectively. Therefore, to ensure $\rho > 0$, f_0 must be positive and, thus, $\rho + p = -4f_0\dot{H} > 0$ is also satisfied as $\dot{H} < 0$ in an expanding model.
2. Again, for $F(\mathrm{T}) = f_1(\mathrm{T})^3$, the field Equations (30) and (31) are
$$1512f_1H^6 = \rho, \quad \text{and} \quad -2160f_1\dot{H}H^4 - 1512f_1H^6 = p. \tag{46}$$
Therefore, $\rho > 0$ is ensured provided $f_1 > 0$, and $\rho + p = -2160f_1\dot{H}H^4 > 0$ is also ensured if $f_1 > 0$ as $\dot{H} < 0$, and as a result, the weak energy condition is also satisfied.
3. Finally, for $F(\mathrm{T}) = f_2\mathrm{T}^5$, the field Equation (30) takes the form
$$69984f_2H^{10} = \rho, \quad \text{and} \quad 181440f_2\dot{H}H^8 - 69984f_2H^{10} = p. \tag{47}$$
Hence, $\rho > 0$ together with $\rho + p = 181440f_2\dot{H}H^8 > 0$, provided $f_2 > 0$ as $\dot{H} < 0$, and so the weak energy condition is also satisfied.

Thus, all the energy condition are satisfied, provided the form of $F(\mathrm{T})$ is
$$F(\mathrm{T}) = f_0\mathrm{T} + f_1(\mathrm{T})^3 + f_2(\mathrm{T}^5), \tag{48}$$
where all the coefficients f_0, f_1, f_2 are positive, considering no additional terms.

3.2.2. Slow Roll Inflation

We have seen that both the forms $F(\mathrm{T})$ of generalized metric teleparallel gravity theory presented in (44) and (48) admit FLRW-type radiation- and early matter-dominated eras, and can trigger accelerated expansion in the late stage of cosmological evolution. Further, both the forms are validated by energy conditions. To explore their behavior in the very early universe, let us consider slow roll inflation for both the models to find the inflationary parameters and to compare with the currently observed constraints.

Cosmological inflation, that occurred sometimes between 10^{-42} s and $10^{-32\pm6}$ s, not only can solve the horizon, flatness and monopole problems, but also generates the seeds of perturbation required to trigger the structure formation at a latter epoch. Although it is a quantum theory of perturbation, where gravity is treated as classical, while all other fields remain quantized, classical field equations are well suited to study inflation. The recently released data-sets [32,33] imposed tighter constraints on the inflationary parameters n_s ($0.9631 \leq n_s \leq 0.9705$), as well as on the tensor-to-scalar ratio ($r < 0.055$). More recently, combination of Planck PR4 data with ground-based experiments, such as BICEP/Keck 2018 (BK18), BAO and CMB lensing data, tightens the tensor to scalar ratio even further to $r < 0.032$ [34]. Nonetheless, in recent years, r has been constrained starting from $r < 0.14$ to the above-mentioned value and, therefore, we presume that r might be restricted to even less value in more precise future experiments, such as polarized CMB space missions (including LiteBIRD) [35]. To study inflation, we incorporate a scalar field ϕ along with a potential $V(\phi)$ in the action, which drives the inflation, as already mentioned in (40). As an example, let us choose a special form of potential $V = V_0 - \frac{V_1}{\phi}$, such that when ϕ becomes large enough, then $V \approx V_0$, representing a flat potential.

Form-1:

$F(\mathrm{T}) = f_0\mathrm{T} + f_1(-\mathrm{T})^{\frac{3}{2}} - f_2(\mathrm{T}^2) + \cdots,$

As the energy condition ensures all the coefficients are positive, so in the vacuum era, the field Equation (30), and the ϕ variation equation are expressed as

$$6f_0H^2 + 24\sqrt{6}f_1H^3 + 108f_2H^4 = V(\phi) + \frac{1}{2}\dot{\phi}^2, \qquad \ddot{\phi} + 3H|\dot{\phi}| + V' = 0. \tag{49}$$

Let us now consider the standard slow-roll conditions $\dot{\phi}^2 \ll V(\phi)$ and $|\ddot{\phi}| \ll 3H|\dot{\phi}|$, on the pair of Equation (49), which therefore finally reduce to

$$108f_2H^4 + 24\sqrt{6}f_1H^3 + 6f_0H^2 - V(\phi) = 0, \qquad 3H\dot{\phi} + V' = 0. \tag{50}$$

Clearly, the first quartic Equation (50) has four roots, each of which are exorbitantly large and complicated, including cubic roots. Hence, these three terms together in $F(T)$ cannot be handled. Therefore, in the following, we consider the combination of (i) the first term (T) and the third term (T^2), and (ii) the first two terms, i.e., T and $(-\mathrm{T})^{\frac{3}{2}}$.

Case 1. $F(\mathrm{T}) = f_0\mathrm{T} - f_2\mathrm{T}^2$:

Earlier, slow roll inflation considering the form $F(\mathrm{T}) = f_0\mathrm{T} + f_2\mathrm{T}^2$, where both $f_0 > 0$ and $f_2 > 0$, has been studied extensively and wonderful agreement with the observed data was found [30]. Nonetheless, the energy condition suggests that the T^2 term must appear with a negative sign. Hence, it is required to look over if the inflationary parameters are still on par with the observational data. The field equations are now

$$6f_0H^2 + 108f_2H^4 = \frac{1}{2}\dot{\phi}^2 + V(\phi), \qquad \ddot{\phi} + 3H|\dot{\phi}| + V'(\phi) = 0, \tag{51}$$

upon which we apply the standard slow-roll conditions $\dot{\phi}^2 \ll V(\phi)$ and $|\ddot{\phi}| \ll 3H|\dot{\phi}|$ to finally obtain

$$\gamma H^4 + 6f_0H^2 - V(\phi) = 0, \qquad 3H\dot{\phi} + V'(\phi) = 0, \tag{52}$$

where $\gamma = 108f_2$. Now for the above choice of the potential $V = V_0 - \frac{V_1}{\phi}$, the slow roll parameters are expressed as

$$\epsilon = \frac{\gamma^2 {V_1}^2}{12\phi^4\left(\sqrt{9f_0^2+\gamma V_1\left(\frac{V_0}{V_1}-\frac{1}{\phi}\right)}\right)\left[-3f_0+\sqrt{9f_0^2+\gamma V_1\left(\frac{V_0}{V_1}-\frac{1}{\phi}\right)}\right]^2}; \qquad \eta = -\frac{2M_P^2}{\phi^2(\frac{V_0}{V_1}\phi - 1)},$$
$$N = \int_{\phi_f}^{\phi_i}\frac{3\phi^2\left[-3f_0+\sqrt{9f_0^2+\gamma V_1\left(\frac{V_0}{V_1}-\frac{1}{\phi}\right)}\right]}{\gamma V_1}d\phi, \tag{53}$$

where N is the number of e-folds. In yet another paper (under preparation), we have found excellent agreement of the inflationary parameters with the observational data. While $N = 50$ is sufficient to solve the horizon and flatness problems, the oscillatory behavior of the scalar field has also been explored. As an example, we present a set of data in Table 1.

Table 1. Data-set for the inflationary parameters for $F(\mathrm{T}) = f_0\mathrm{T} - f_2\mathrm{T}^2$.

$\phi_i\ M_P$	$\frac{V_0}{V_1}\ M_P^{-1}$	$\gamma V_1\ M_P^5$	$f_0\ M_P^2$	$\phi_f\ M_P$	n_s	r	N
3.0	6.5	0.1	0.5	0.69599	0.96425	0.03128	50

It is also important to mention that one may fix the value of γ, which is arbitrary, to set the energy scale of inflation H_* at the sub-Planckian scale.

Case 2. $F(\mathrm{T}) = f_0\mathrm{T} + f_1(-\mathrm{T})^{\frac{3}{2}}$:

This case has never been studied earlier. For the above form of $F(\mathrm{T})$, we apply the standard slow-roll conditions $\dot{\phi}^2 \ll V(\phi)$ and $|\ddot{\phi}| \ll 3H|\dot{\phi}|$ and, as such, the field equations reduce to

$$6f_0H^2 + 24\sqrt{6}f_1H^3 - V(\phi) = 0, \quad 3H\dot{\phi} + V' = 0. \tag{54}$$

Now, the first cubic equation of (54) can be solved for H as

$$H = \frac{1}{\gamma}\left[-2f_0 + \frac{6.35f_0^2 + \left(V(\phi)\gamma^2 - 16f_0^3 + \sqrt{V(\phi)^2\gamma^4 - 32V(\phi)\gamma^2 f_0^3}\right)^{\frac{2}{3}}}{1.26\left(V\gamma^2 - 16f_0^3 + \sqrt{V(\phi)^2\gamma^4 - 32V(\phi)\gamma^2 f_0^3}\right)^{\frac{1}{3}}}\right], \tag{55}$$

where $\gamma = 24\sqrt{6}f_1$. Combining the pair of Equation (54), one can identify the 'potential slow roll parameter' ϵ with the 'Hubble slow roll parameter' (ϵ_1), and can also express η and the number of e-folds, respectively, as

$$\epsilon \equiv \frac{\dot{H}}{H^2} = \frac{V'(\phi)^2}{3H^4[12f_0 + 3\gamma H]}, \qquad \eta = 2f_0\left[\frac{V''(\phi)}{V(\phi)}\right]$$
$$N(\phi) \simeq \int_{t_i}^{t_f} Hdt = \int_{\phi_i}^{\phi_f}\frac{H}{\dot{\phi}}d\phi = \int_{\phi_f}^{\phi_i} 3\frac{\left[-2f_0 + \frac{6.35f_0^2 + \left(V(\phi)\gamma^2 - 16f_0^3 + \sqrt{V(\phi)^2\gamma^4 - 32V(\phi)\gamma^2 f_0^3}\right)^{\frac{2}{3}}}{1.26\left(V\gamma^2 - 16f_0^3 + \sqrt{V(\phi)^2\gamma^4 - 32V(\phi)\gamma^2 f_0^3}\right)^{\frac{1}{3}}}\right]^2}{[\gamma^2 V']}d\phi. \tag{56}$$

Now, for this potential $V = V_0 - \frac{V_1}{\phi}$ under consideration, the expression for ϵ takes a extortionate form, such as

$$\epsilon = \frac{3.18V_1^2\gamma^4\left[V_1\gamma^2(\frac{V_0}{V_1}-\frac{1}{\phi})-16f_0^3+\sqrt{V_1^2(\frac{V_0}{V_1}-\frac{1}{\phi})^2\gamma^4-32V_1(\frac{V_0}{V_1}-\frac{1}{\phi})\gamma^2f_0^3}\right]^{\frac{5}{3}}}{9\phi^4\left[-2.52f_0[V_1\gamma^2(\frac{V_0}{V_1}-\frac{1}{\phi})-16f_0^3+\sqrt{V_1^2(\frac{V_0}{V_1}-\frac{1}{\phi})^2\gamma^4-32V_1(\frac{V_0}{V_1}-\frac{1}{\phi})\gamma^2f_0^3}]^{\frac{1}{3}}+6.35f_0^2+[V_1\gamma^2(\frac{V_0}{V_1}-\frac{1}{\phi})-16f_0^3+\sqrt{V_1^2(\frac{V_0}{V_1}-\frac{1}{\phi})^2\gamma^4-32V_1(\frac{V_0}{V_1}-\frac{1}{\phi})\gamma^2f_0^3}]^{\frac{2}{3}}\right]^4}$$
$$\times\frac{1}{\left[2.52f_0[V_1\gamma^2(\frac{V_0}{V_1}-\frac{1}{\phi})-16f_0^3+\sqrt{V_1^2(\frac{V_0}{V_1}-\frac{1}{\phi})^2\gamma^4-32V_1(\frac{V_0}{V_1}-\frac{1}{\phi})\gamma^2f_0^3}]^{\frac{1}{3}}+6.35f_0^2+[V_1\gamma^2(\frac{V_0}{V_1}-\frac{1}{\phi})-16f_0^3+\sqrt{V_1^2(\frac{V_0}{V_1}-\frac{1}{\phi})^2\gamma^4-32V_1(\frac{V_0}{V_1}-\frac{1}{\phi})\gamma^2f_0^3}]^{\frac{2}{3}}\right]}, \tag{57}$$

while η and N may be expressed, respectively, as

$$\eta = -\frac{4f_0}{\phi^2(\frac{V_0}{V_1}\phi-1)}; \qquad N(\phi) \simeq \int_{t_i}^{t_f} Hdt = \int_{\phi_i}^{\phi_f}\frac{H}{\dot{\phi}}d\phi$$
$$= \int_{\phi_f}^{\phi_i}\left[\frac{3\phi^2}{1.588V_1\gamma^2\left(\sqrt{V_1^2\gamma^4\left(\frac{V_0}{V_1}-\frac{1}{\phi}\right)^2-32V_1\gamma^2f_0^3\left(\frac{V_0}{V_1}-\frac{1}{\phi}\right)}+V_1\gamma^2\left(\frac{V_0}{V_1}-\frac{1}{\phi}\right)-16f^3\right)^{2/3}}\times\right.$$
$$\left\{-2.52f_0\left[\sqrt{V_1^2\gamma^4\left(\frac{V_0}{V_1}-\frac{1}{\phi}\right)^2-32V_1\gamma^2f_0^3\left(\frac{V_0}{V_1}-\frac{1}{\phi}\right)}+V_1\gamma^2\left(\frac{V_0}{V_1}-\frac{1}{\phi}\right)-16f_0^3\right]^{\frac{1}{3}}\right.$$
$$\left.\left.+\left(\sqrt{V_1^2\gamma^4\left(\frac{V_0}{V_1}-\frac{1}{\phi}\right)^2-32V_1\gamma^2f_0^3\left(\frac{V_0}{V_1}-\frac{1}{\phi}\right)}+V_1\gamma^2\left(\frac{V_0}{V_1}-\frac{1}{\phi}\right)-16f_0^3\right)^{2/3}+6.35f_0^2\right\}^2\right]d\phi. \tag{58}$$

Despite such huge structures of the parameters, it is still possible to handle these expressions, and we present a table of the data-set for the expressions (57) and (58) in Table 2. In this table, we have varied ϕ_i within the range $2.80\,M_P \leq \phi_i \leq 3.40\,M_P$, so that r and n_s lie more-or-less within the experimental limit. However, restrictions on n_s and r, restricts the number of e-folds within the range $24 \leq N \leq 44$, which still might solve the horizon and flatness problems.

Table 2. Data-set for the inflationary parameters in the case $F(\mathrm{T}) = f_0\mathrm{T} + f_1(-\mathrm{T})^{\frac{3}{2}}$, taking into account $\frac{V_0}{V_1} = 6\,M_P^{-1},\ V_1\gamma^2 = 3.4\,M_P^7,\ f_0 = 0.5\,M_P^2$ and varying ϕ within the range $2.8\,M_P < \phi_i < 3.4\,M_P$, so that $\phi_f \approx 0.460794\,M_P$ at the end of inflation.

ϕ_i in M_P	n_s	r	N
2.80	0.9656	0.00566	24
2.85	0.9674	0.00520	26
2.90	0.9692	0.00490	27
2.95	0.9707	0.00456	28
3.00	0.9723	0.00426	30
3.10	0.9750	0.00372	33
3.20	0.9773	0.00327	37
3.30	0.9794	0.00288	40
3.40	0.9812	0.00255	44

Let us now compute the energy scale of inflation in view of the relation (55), considering the data: ($N = 40$, for which $\phi_i = 3.3\,M_P$, $f_0 = 0.5\,M_p^2$, $\frac{V_0}{V_1} = 6.0\,M_P^{-1}$, $V_1\gamma^2 = 3.4\,M_P^7$), as depicted in Table 2. Correspondingly we find,

$$H_* = \frac{8.114}{\gamma}. \tag{59}$$

Now, the energy scale of inflation in a single scalar field model in GTR [36] is given by the following expression:

$$H_* = 8 \times 10^{13} \sqrt{\frac{r}{0.2}} GeV = 0.96 \times 10^{13} GeV \approx 3.918 \times 10^{-6} M_P, \tag{60}$$

whose numerical value is computed, taking into account the value of the tensor-to-scalar ratio $r = 0.00288$ from the data-set of Table 2. Thus, in order to match the scale of inflation (59) with the single field scale of inflation (60), we are required to constrain γ, such as $\gamma \approx 2.074 \times 10^6 M_P$. The requirement of the sub-Planckian scale for inflation is the physical ground upon which the parameter γ has been constrained and, consequently, the values of V_1 and V_0 are fixed as well:

$$V_1 = 7.90 \times 10^{-11}\, M_P^5, \qquad V_0 = 6V_1 \approx 4.74 \times 10^{-10}\, M_P^4. \tag{61}$$

Finally, to handle the issue of graceful exit from inflation, we recall the first equation of (54) which, in view, of the above form of the potential, $V(\phi) = V_0 - \frac{V_1}{\phi}$, is expressed as

$$-\frac{\gamma H^3}{V_1} - \frac{6 f_0 H^2}{V_1} + \left[\frac{\dot{\phi}^2}{2V_1} + \left(\frac{V_0}{V_1} - \frac{1}{\phi}\right)\right] = 0. \tag{62}$$

During inflation, H^2 and V_1 are of the same order of magnitude, while the Hubble parameter varies slowly. But, at the end of inflation, the Hubble rate usually decreases sharply, and γH^3 falls much below V_1. Hence, one can neglect both the terms $\frac{\gamma H^3}{V_1}$ and $\frac{f_0 H^2}{V_1}$ without any loss of generality. In the process, one finds

$$\dot{\phi}^2 = -2\left[V_0 - \frac{V_1}{\phi}\right]. \tag{63}$$

Taking into account $\phi_f = 0.460794\, M_P$, $V_0 = 4.74 \times 10^{-10}\, M_P^4$ and $V_1 = 7.90 \times 10^{-11}\, M_P^5$ from computation, the above equation clearly exhibits the oscillatory behavior of the scalar field ϕ

$$\phi = \exp(i\omega t), \tag{64}$$

provided that $\omega \approx 5.335 \times 10^{-5} M_P$.

In a nutshell, although we have not been able to handle all the three terms together, both the pairs nonetheless exhibit excellent agreement with the observational data.

Form-2

$F(\mathrm{T}) = f_0 \mathrm{T} + f_1 (\mathrm{T})^3 + f_2 (\mathrm{T}^5) + \cdots,$

Case 3. $F(\mathrm{T}) = f_0 \mathrm{T} + f_1 (\mathrm{T})^3$:

Here, too it is impossible to handle all three terms together; we therefore only consider the first two terms $F(\mathrm{T}) = f_0 \mathrm{T} + f_1 (\mathrm{T})^3$ and focus to study the slow roll inflation. The field Equation (30) and the ϕ variation equation are now expressed as

$$6 f_0 H^2 + 1080 f_1 H^6 = V(\phi) + \frac{1}{2}\dot{\phi}^2, \qquad \ddot{\phi} + 3H|\dot{\phi}| + V' = 0. \tag{65}$$

Applying the standard slow-roll conditions $\dot{\phi}^2 \ll V(\phi)$ and $|\ddot{\phi}| \ll 3H|\dot{\phi}|$, the pair of Equation (65) finally reduce to

$$\gamma H^6 + 6 f_0 H^2 - V = 0, \qquad 3H\dot{\phi} + V' = 0, \tag{66}$$

where $1080 f_1 = \gamma$. Solving for H^2 in view of (66), we readily obtain

$$H^2 = \frac{\left[\left(V\gamma^2 + \sqrt{V^2\gamma^4 + 32\gamma^3 f_0^3}\right)^{\frac{2}{3}} - 3.175\gamma f_0\right]}{1.26\gamma\left(V\gamma^2 + \sqrt{V^2\gamma^4 + 32\gamma^3 f_0^3}\right)^{\frac{1}{3}}}. \tag{67}$$

Further, combining Equation (66), one can show that the 'potential slow roll parameter' ϵ is equal to the 'Hubble slow roll parameter' (ϵ_1) under the condition

$$\epsilon \equiv -\frac{\dot{H}}{H^2} = \frac{2.52\gamma^4 V'^2\left(V\gamma^2+\sqrt{V^2\gamma^4+32\gamma^3f_0^3}\right)^{\frac{4}{3}}}{18\left[\left(V\gamma^2+\sqrt{V^2\gamma^4+32\gamma^3f_0^3}\right)^{\frac{2}{3}}-3.175\gamma f_0\right]^2}$$
$$\times\frac{1}{\left[\gamma\left[\left(V\gamma^2+\sqrt{V^2\gamma^4+32\gamma^3f_0^3}\right)^{\frac{2}{3}}-3.175\gamma f_0\right]^2+3.175\gamma^2 f_0\left(V\gamma^2+\sqrt{V^2\gamma^4+32\gamma^3f_0^3}\right)^{\frac{2}{3}}\right]}; \tag{68}$$
$$\eta = 2f_0\left(\frac{V''(\phi)}{V(\phi)}\right)$$

Further, one can compute the number of e-folds as

$$N(\phi) \simeq \int_{t_i}^{t_f} Hdt = \int_{\phi_i}^{\phi_f}\frac{H}{\dot{\phi}}d\phi = \int_{\phi_f}^{\phi_i} 3\frac{\left[\left(V\gamma^2+\sqrt{V^2\gamma^4+32\gamma^3f_0^3}\right)^{\frac{2}{3}}-3.175\gamma f_0\right]}{\left[1.26\gamma V'\left(V\gamma^2+\sqrt{V^2\gamma^4+32\gamma^3f_0^3}\right)^{\frac{1}{3}}\right]}d\phi. \tag{69}$$

Now, under the same choice of the potential as above, viz. $V(\phi) = V_0 - \frac{V_1}{\phi}$, the expressions of ϵ, η (68) and N (69) are found as

$$\epsilon = \frac{2.52\gamma^4 V_1^2\left(V_1\gamma^2(\frac{V_0}{V_1}-\frac{1}{\phi})+\sqrt{V_1^2\gamma^4(\frac{V_0}{V_1}-\frac{1}{\phi})^2+32\gamma^3f_0^3}\right)^{\frac{4}{3}}}{18\phi^2\left[\left(V_1\gamma^2(\frac{V_0}{V_1}-\frac{1}{\phi})+\sqrt{V_1^2\gamma^4(\frac{V_0}{V_1}-\frac{1}{\phi})^2+32\gamma^3f_0^3}\right)^{\frac{2}{3}}-3.175\gamma f_0\right]^2}$$
$$\times\frac{1}{\left[\gamma\left[\left(V_1\gamma^2(\frac{V_0}{V_1}-\frac{1}{\phi})+\sqrt{V_1^2\gamma^4(\frac{V_0}{V_1}-\frac{1}{\phi})^2+32\gamma^3f_0^3}\right)^{\frac{2}{3}}-3.175\gamma f_0\right]^2+3.175\gamma^2 f_0\left\{V_1\gamma^2(\frac{V_0}{V_1}-\frac{1}{\phi})+\sqrt{V_1^2\gamma^4(\frac{V_0}{V_1}-\frac{1}{\phi})^2+32\gamma^3f_0^3}\right\}^{\frac{2}{3}}\right]} \tag{70}$$
$$\eta = -\frac{4f_0}{\phi^2(\frac{V_0}{V_1}\phi-1)}, \qquad N(\phi) = \int_{\phi_f}^{\phi_i} 3\phi^2\frac{\left[\left(V_1\gamma^2(\frac{V_0}{V_1}-\frac{1}{\phi})+\sqrt{V_1^2\gamma^4(\frac{V_0}{V_1}-\frac{1}{\phi})^2+32\gamma^3f_0^3}\right)^{\frac{2}{3}}-3.175\gamma f_0\right]}{1.26\gamma V_1\left[V_1\gamma^2(\frac{V_0}{V_1}-\frac{1}{\phi})+\sqrt{V_1^2\gamma^4(\frac{V_0}{V_1}-\frac{1}{\phi})^2+32\gamma^3f_0^3}\right]^{\frac{1}{3}}}d\phi.$$

Again, despite such huge structures of equations, we are able to present a table of the data-set for the expressions (70). In Table 3, we have varied V_1 between $4.5\ M_P^5 \leq V_1 \leq 5.0\ M_P^5$, so that r and n_s lie within the experimental limit. Further, the number of e-folds for Table 3 is found to vary within the range $40 \leq N \leq 45$, which is more-or-less sufficient to solve the horizon and flatness problems. Clearly, the agreement with the observational data is outstanding, since the tensor-to-scalar ratio is able to sustain further constraints, which might appear from future analysis.

The energy scale of inflation (H_*) is shown in the last column of Table 3, which is of the order $10^{-3}\ M_P$, i.e., in the sub-Planckian scale, while the energy scale of inflation in a single scalar field model corresponding to GTR [36] is given by the following expression:

$$H_* = 8\times 10^{13}\sqrt{\frac{r}{0.2}}GeV = 1.28\times 10^{13}GeV \approx 5.25\times 10^{-6}M_P, \tag{71}$$

whose numerical value is computed taking into account the value of the tensor-to-scalar ratio $r = 0.005149$ from the data-set of Table 3.

Table 3. Data-set for the inflationary parameters in the case of $F(\mathrm{T}) = f_0\mathrm{T} + f_1(\mathrm{T})^3$, taking into account $\phi_i = 4.0\ M_P$, $V_0 = 2.0\ M_P^4$, $\gamma = 5 \times 10^7\ M_P^{-2}$, $f_0 = 0.5\ M_P^2$ varying V_1 between 4.5×10^{-6} to $5.0 \times 10^{-6}\ M_P^5$.

ϕ_f in M_P	V_1 in $\times 10^{-6}\ M_P^5$	n_s	r	N	$H_*{}^2$ in $\times 10^{-6}\ M_P^2$
0.75450	4.5	0.96656	0.005836	40	1.33
0.74478	4.6	0.96742	0.005593	41	1.37
0.73540	4.7	0.96823	0.005364	42	1.40
0.72637	4.8	0.96899	0.005149	43	1.43
0.71750	4.9	0.96974	0.004947	44	1.47
0.70900	5.0	0.97044	0.004756	45	1.50

Finally, to handle the issue of graceful exit from inflation, we recall Equation (65) which, in view of the above form of the potential $V(\phi) = V_0 - \frac{V_1}{\phi}$, is expressed as

$$\frac{\gamma H^6}{V_1} + \frac{6f_0H^2}{V_1} + \left[\frac{\dot{\phi}^2}{2V_1} - \left(\frac{V_0}{V_1} - \frac{1}{\phi}\right)\right] = 0. \tag{72}$$

During inflation, H^2 and V_1 are of the same order of magnitude, while the Hubble parameter varies slowly. But, at the end of inflation, the Hubble rate decreases sharply, and γH^6 falls much below V_1. Hence, one can neglect both the terms $\frac{\gamma H^6}{V_1}$ and $\frac{f_0 H^2}{V_1}$ without any loss of generality. In the process, one obtains

$$\dot{\phi}^2 = -2\left[V_0 - \frac{V_1}{\phi}\right]. \tag{73}$$

Taking into account the values: $\phi_f = 0.70900 M_P$, $V_0 = 5.0 \times 10^{-6} {M_P}^4$ and $V_1 = 2.0 \times 10^{-6} {M_P}^5$ from Table 3, it is possible to show that the above equation exhibits following oscillatory behavior,

$$\phi = \exp(i\omega t), \tag{74}$$

provided that $\omega \approx 1.476 \times 10^{-3} M_P$. Hence, a graceful exit is also exhibited. In a nutshell, both the forms $F(\mathrm{T})$ of the generalized metric teleparallel gravity are found quite capable of explaining the cosmological evolutionary history, right from the very early inflationary era up to date.

3.3. Symmetric Teleparallel $F(Q)$ Theory of Gravity

Although everything appears to be nice, non-trivial $F(\mathrm{T})$ theories are not locally Lorentz invariant, the most fundamental symmetry in physics, resulting in extra degrees of freedom not present in GTR. In fact, the lack of local Lorentz invariance requires us to deal with 16 equations, instead of the 10 associated with GTR. Einstein suggested to identify these extra degrees of freedom with the electromagnetic fields. However, he failed to find a consistent tensor-like description of the electromagnetic field equations in this approach. Further, while studying perturbations on the top of RW background, models based on $F(\mathrm{T})$ gravity appear to suffer from strong coupling problems. In particular, if the action is expressed in the Einstein frame, and perturbation is studied around the flat Minkwoski space-time, the spectrum of the 'general theory of relativity' reappears, being free from all additional degrees of freedom. That is, the genuine physical degrees of freedom lose their kinetic term at the quadratic order, resulting in the breakdown of the standard perturbation theory. Since no new scalar mode is present in linear and second-order perturbations in $F(\mathrm{T})$ gravity, a strong coupling problem appears. Additionally, $F(\mathrm{T})$ theory gravity cannot make any prediction for 'cosmic microwave background radiation' and 'large scale structure' within the standard RW model, due to strong coupling [9]. Therefore, $F(Q)$ gravity is currently in the lime-light, since these problems do not arise in $F(Q)$ theory of gravity, as already mentioned. This led people to consider generalized symmetric teleparallel $F(Q)$ gravity theory. Recently $F(Q)$ theory has been studied largely in different perspectives [37–45]. Further, the field equations in view of the variational principle have also been derived [14]. In the $F(Q)$ theory, one may use the special FLRW metric in Cartesian coordinates and the coincident gauge in this setting, which make the calculation easier by reducing the covariant derivative to merely a partial derivative. In the coincidence gauge, all connections are made to vanish globally in a set of local Lorentz frame. As a result, the affine connection vanishes everywhere in a manifold, and also, both the Riemann-tensor $\tilde{R}^{\alpha}_{\beta\mu\nu}$ and the torsion-tensor $\mathrm{T}_{\alpha\mu\nu}$ vanish globally. In the coincidence gauge, the non metricity scalar reads as $Q = -\frac{6\dot{a}^2}{a^2N^2} = -6H^2$ under the gauge choice $N = 1$. Consequently, the Friedmann equations become identical to those of the $F(\mathrm{T})$ gravity theory, and all the results found in connection with $F(\mathrm{T})$ gravity theory holds, averting the problems associated with $F(\mathrm{T})$ gravity theory.

Unfortunately, as soon as one considers a nonlinear generalization of STEGR, i.e., $F(Q)$ theory of gravity, it is also found to run from some serious pathologies. Covariant formulation of the theory requires to introduce Stückelberg fields, in view of which the presence of ghosts induces Ostrogradski's instability. In the recent years, it was shown that there is a strong coupling problem with the scalar perturbations around maximally symmetric (De-Sitter and Minkowski) backgrounds, and there might also be a potentially strong coupling problem in the vector sector for flat cosmology [10]. The strong coupling issue in the maximally symmetric Minkowski background mostly occurs at very small scales, such as when applying it to the black-holes in the asymptotically Minkowski region. Further, the lack of diffeomorphic invariance in the self-interaction of graviton might possibly give rise to the Boulware–Deser ghost [46]. Particularly, while studying cosmological perturbation around the spatially flat $F(Q)$ gravity theory, seven degrees of freedoms are found propagating in the gravitational sector, out of which at least one is a ghost [47]. Nonetheless, the isotropic and homogeneous Robertson–Walker metric (19) under present consideration also admits three non-trivial connections [14,48], apart from the coincidence gauge. The study of the three other connections have been initiated recently [49,50]. More recently, it has been shown that all these three connections can also explain the cosmic evolutionary history, for a linear form of $F(Q) = f_0 Q$, without further modification [51]. Since no generalization to non-linearity is required, these might be free from pathologies.

3.4. Palatini Formalism

For the sake of completeness, let us add a few lines in the context of yet another theory of gravity presented by Palatini long ago, which again admits late-stage of cosmic acceleration upon suitable modification [52]. In fact, while the metric $g_{\mu\nu}$ measures the distances in manifold and angles in tangent space, the connection $\Gamma^{\alpha}{}_{\mu\nu}$ actually maps between different tangent spaces that define parallel transport. Once the physical consideration that the principle of equivalence and causality should be related to each other is demanded, one enters into the realm of GTR, and $\Gamma^{\alpha}{}_{\mu\nu} = \{{}_{\mu}{}^{\alpha}{}_{\nu}\}$, the Levi-Civita. On the contrary, Palatini formalism treats the metric and the connection as independent quantities. Hence, mathematically Palatini formalism appears to be more rigorous, and enriches the geometric structure of the spacetime. The generalized action being a function of both the metric and the connection is given by

$$A(g,\Gamma) = \frac{1}{16\pi G}\int \sqrt{g}d^4x F(\mathfrak{R}) + S_m, \tag{75}$$

where $\mathfrak{R}$ stands for Ricci scalar in Palatini formalism and S_m is the matter action. Note that $\delta\Gamma^{\alpha}{}_{\mu\nu;\alpha} - \delta\Gamma^{\alpha}{}_{\mu\alpha;\nu}$ and, since there is no way to find the connections, the Ricci scalar $\mathfrak{R}$ remains unknown. The field equations are

$$F'(\mathfrak{R})\mathfrak{R}_{\mu\nu} - \frac{1}{2}F(\mathfrak{R})g_{\mu\nu} = \kappa T_{\mu\nu}, \tag{76}$$

whose trace yields

$$F(\mathfrak{R})\mathfrak{R} - 2F(\mathfrak{R}) = \kappa T. \tag{77}$$

One can immediately notice that for traceless fields $T = 0$, such as in the vacuum ($p = \rho = 0$) and in the radiation ($p = \frac{1}{3}\rho$) eras; in the context of cosmology, Equation (77) yields $F(\mathfrak{R}) \propto \mathfrak{R}^2$. As $\mathfrak{R}$ is obscure, so one does not have any idea regarding the evolution. Nonetheless, the evolution, whatever it might be, is identical in both eras, which is physically unrealistic. On the other hand, for $F(\mathfrak{R}) = \alpha\mathfrak{R} + \beta\mathfrak{R}^2$, Equation (77) for traceless fields results in $\mathfrak{R}^{n-1} = \frac{\alpha}{\beta(n-2)}$ for $n \neq 1, 2$. Thus, $\mathfrak{R}$ is a constant, and a de(anti)Sitter solution emerges (we shall make it clear in the following) in both the vacuum and radiation eras, which is unrealistic, as already mentioned. Clearly, one requires considering additional fields, such as a scalar field, to obtain a physically reasonable outcome. But, prior to that, it is somehow required to relate $\mathfrak{R}$ with R, so that one can study cosmological evolution. This is possible under the conformal transformation, $g_{\mu\nu} = F'(\mathfrak{R})\tilde{g}_{\mu\nu}$, whence the action takes the following form [53]:

$$A_P = \frac{1}{2\kappa}\int\sqrt{-\tilde{g}}d^4x[\Phi R - \frac{3}{2\Phi}\partial_\mu\Phi\partial^\mu\Phi - U(\Phi)] + S_m, \tag{78}$$

where

$$\mathfrak{R} = R - \frac{3}{2\Phi^2}\partial_\mu\Phi\partial^\mu\Phi - \frac{3}{\Phi}\Box\Phi,\ \Phi = F'(\mathfrak{R}),\ U(\Phi) = \Phi\chi(\Phi) - F(\chi) \tag{79}$$

and the new field $\chi = \mathfrak{R}$. In the linear theory $F(\mathfrak{R}) \propto \mathfrak{R}$, Φ becomes a constant, so is $U(\Phi)$; as a result, $\mathfrak{R} = R$, in view of (79) and GTR, is retrieved. Phase-space structure of the scalar-tensor equivalence of the Palatini action has been formulated and quantized, and it was found that the semi-classical wavefunction is classically forbidden, which is a serious pathology of the formalism. Next, it was observed that either the scale factor evolves exponentially both in the early vacuum-dominated and radiation-dominated eras, which is unrealistic, or it admits the Friedman-like solution $a \propto \sqrt{t}$ in the radiation era, but unfortunately, a power law inflation is administered in the vacuum era without a graceful exit. The fact that the Palatini formalism runs into contradiction with the standard model of cosmology was also unleashed earlier [54]. Further, in an attempt to analyze the stellar structure, it was revealed that the formalism exhibits a singular behavior, giving rise to infinite tidal forces on the surface [55]. Thus, the Palatini formalism lacks novelty.

4. Concluding Remarks

In this article, we have reviewed some earlier works with suitable modification focusing on the reconstruction program of the forms of modified and generalized teleparallel gravity theories. The importance of the work is that all the forms presented here admit a viable radiation- and early matter-dominated eras, which are on par with the standard (FLRW) model of cosmology. Further, these models can also explain the late-time cosmic accelerated expansion. Since inflation in the context of the modified $F(R)$ theory of gravity has been studied widely, we have not considered it here. However, for the two possible forms of the teleparallel $F(\mathrm{T})$ gravity, we have studied inflation and found that the inflatioanry parameters are quite in agreement with the recently released observational data. Also, since symmetric telepallel $F(Q)$ gravity theory in a coincidence gauge produces a field equation identical to those for $F(\mathrm{T})$ gravity theory, all the results obtained for $F(\mathrm{T})$ gravity also hold for coincidence general relativity.

While the results obtained here are quite encouraging, nonetheless, all these theories suffer from some sort of unavoidable pathologies, as discussed in the literature. In the mean time, linear and second-order scalar perturbations and the strong coupling issues of $F(\mathrm{T})$ gravity have been reinvestigated by applying the effective field theory [51]. Of course, no new scalar mode is found, suggesting a strong coupling issue. Nonetheless, the authors have also shown that this can be avoided at some scale comparable with the cutoff scale of the applicability of the theory [51]. Clearly, nothing has been settled as yet. In this context, it may be mentioned that in the RW metric (19) under consideration, symmetric teleparallel gravity theory admits four different connections. Apart from the coincidence gauge, the other three involve an additional non-vanishing function of time $\gamma(t)$ and, as a result, some difficulty arises in studying these connections. Nonetheless, very recently, we have prepared a detailed study with the three other connections (communicated), which exhibit cosmic evolution at the linear level without any modification. It should also be mentioned that Bianchi identity, which is an outcome of the invariance of curvature by isometries of the metric tensor, holds automatically for GTR ($G^{\mu\nu}{}_{;\mu} = 0$), and is also true for $F(R)$ gravity, resulting in the classical conservation of the energy momentum tensor. Nonetheless, this does not hold in general for teleparallel gravity theories. Even so, in the background of the isotropic and homogeneous Robertson–Walker metric, the Bianchi identity holds for $F(\mathrm{T})$ gravity uniquely. In the case of the $F(Q)$ theory of gravity, the coincidence gauge ($Q = -6H^2$) is usually taken into consideration, in which case the Bianchi identity also holds automatically. Nevertheless, for other connections, it was not known with certainty. It has been exhibited [56] that, apart from one single situation, it holds in general for all the connections. We have also briefly discussed the Palatini formalism, which appears to be a non-viable alternative to GTR.

Author Contributions: The problem was suggested by A.K.S., the computation was carried out by D.S. and checked by M.C. extensively. The draft was written by D.S., checked and modified by M.C., A.K.S. prepared the final version. All authors have read and agreed to the published version of the manuscript.

Funding: This research received no external funding.

Data Availability Statement: Data are contained within the article.

Conflicts of Interest: The authors declare no conflicts of interest.

Notes

1 It may be mentioned that the curvature tensor constructed out of Weitzenböck connection vanishes identically, while a theory of gravity constructed in view of tetrad fields admits both the Weitzenböck and the Riemannian geometry and therefore is more general.

2 Under the choice of local coordinates and coordinate basis x^μ and ∂_μ, respectively, the affine Levi-Civita connection (∇) is called the Christoffel symbol $\{{}_{\mu}{}^{\alpha}{}_{\nu}\}$, so that $\nabla_\mu \partial_\nu = \{{}_{\mu}{}^{\alpha}{}_{\nu}\}$.

References

1. Sabulsky, D.O.; Dutta, I.; Hinds, E.A.; Elder, B.; Burrage, C.; Copel, E.J. Experiment to detect dark energy forces using atom interferometry. *Phys. Rev. Lett.* **2019**, *123*, 061102. [CrossRef] [PubMed]
2. Motohashi, H.; Suyama, T. Third order equations of motion and the Ostrogradsky instability. *Phys. Rev. D* **2015**, *91*, 085009. [CrossRef]
3. Maluf, J.W.; Rocha-Neto, J.F.D. Hamiltonian formulation of general relativity in the teleparallel geometry. *Phys. Rev. D* **2001**, *64*, 084014. [CrossRef]
4. Li, M.; Miao, R.-X.; Miao, Y.-G. Degrees of freedom of f(T) gravity. *J. High Energy Phys.* **2011**, *1107*, 108. [CrossRef]
5. Ferraro, R.; Guzmán, M.J. Hamiltonian formalism for f(T) gravity. *Phys. Rev. D* **2018**, *97*, 104028. [CrossRef]
6. Ferraro, R.; Guzmán, M.J. Quest for the extra degree of freedom in f(T) gravity. *Phys. Rev. D* **2018**, *98*, 124037. [CrossRef]
7. Ferraro, R.; Guzmán, M.J. Pseudoinvariance and the extra degree of freedom in f(T) gravity. *Phys. Rev. D* **2020**, *101*, 084017. [CrossRef]
8. Izumi, K.; Ong, Y.C. Cosmological perturbation in f(T) gravity revisited. *J. Cosmol. Astropart. Phys.* **2013**, *6*, 29. [CrossRef]
9. Golovnev, A.; Koivisto, T. Cosmological perturbations in modified teleparallel gravity models. *J. Cosmol. Astropart. Phys.* **2018**, *2018*, 012. [CrossRef]
10. Jiménez, J.B.; Heisenberg, L.; Koivisto, T.; Pekar, S. Cosmology in f(Q) geometry. *Phys. Rev. D* **2020**, *101*, 103507. [CrossRef]
11. Capozziello, S.; De Laurentis, M. Extended theories of gravity. *Phys. Rep.* **2011**, *509*, 167. [CrossRef]
12. Nojiri, S.; Odintsov, S.D. Unified cosmic history in modified gravity:from $F(R)$ theory to Lorentz non-invariant models. *Phys. Rep.* **2011**, *505*, 59. [CrossRef]
13. Cai, Y.-F.; Capozziello, S.; De Laurentis, M.; Saridakis, E.N. f(T) teleparallel gravity and cosmology. *Rep. Prog. Phys.* **2016**, *76*, 106901. [CrossRef] [PubMed]
14. Zhao, D. Covariant formulation of f(Q) theory. *Eur. Phys. J. C* **2022**, *82*, 303. [CrossRef]
15. Sarkar, K.; Sk, N.; Ruz, S.; Debnath, S.; Sanyal, A.K. Why Noether Symmetry of F(R) Theory Yields Three-Half Power Law? *Int. J. Theor. Phys.* **2013**, *52*, 1515. [CrossRef]
16. Chakrabortty, M.S.K.N.; Sanyal, A.K. A viable form of teleparallel F(T) theory of gravity. *Eur. Phys. J. C* **2023**, *83*, 557. [CrossRef]
17. Sanyal, A.K. Study of symmetry in $F(R)$ theory of gravity. *Mod. Phys. Lett. A* **2010**, *25*, 2667–2668. [CrossRef]

18. Modak, B.; Sarkar, K.; Sanyal, A.K. Modified theory of gravity and the history of cosmic evolution. *Astrophys. Space Sci.* **2014**, *353*, 707. [CrossRef]
19. Capozziello, S.; Francaviglia, M. Extended theories of gravity and their cosmological and astrophysical applications. *Gen. Relativ. Gravit.* **2008**, *40*, 357. [CrossRef]
20. Vakili, B. Noether symmetric $f(R)$ quantum cosmology and its classical correlations. *Phys. Lett. B* **2008**, *669*, 206. [CrossRef]
21. Sarkar, K.; Sk, N.; Debnath, S.; Sanyal, A.K. Viability of Noether Symmetry of $F(R)$ theory of Gravity. *Int. J. Theor. Phys.* **2013**, *52*, 1194. [CrossRef]
22. Sk, N.; Sanyal, A.K. Revisiting Noether gauge symmetry for $F(R)$ theory of gravity. *Astrophys. Space Sci.* **2012**, *342*, 549. [CrossRef]
23. Sk, N.; Sanyal, A.K. Field Independent Cosmic Evolution. *J. Astrophys.* **2013**, *2013*, 590171. [CrossRef]
24. Starobinsky, A.A. A new type of isotropic cosmological models without singularity. *Phys. Lett. B* **1980**, *91*, 99. [CrossRef]
25. Maeda, K. Inflation as a transient attractor in R^2 cosmology. *Phys. Rev. D* **1988**, *37*, 858. [CrossRef]
26. Bengochea, G.R.; Ferraro, R. Dark torsion as the cosmic speed-up. *Phys. Rev. D* **2009**, *79*, 124019. [CrossRef]
27. Linder, E.V. Einstein's other gravity and the acceleration of the universe. *Phys. Rev. D* **2010**, *81*, 127301. [CrossRef]
28. Krššák, M.; Saridakis, E.N. The covariant formulation of f(T) gravity. *Class. Quant. Grav.* **2016**, *33*, 115009.
29. Ferraro, R.; Fiorini, F. Non-trivial frames for $f(T)$ theories of gravity and beyond. *Phys. Lett. B* **2011**, *702*, 75. [CrossRef]
30. Chakrabortty, M.; Sk, N.; Sanyal, S.; Sanyal, A.K. Inflation with F(T) teleparallel gravity. *Eur. Phys. J. Plus* **2021**, *136*, 1213. [CrossRef]
31. Bamba, K.; Odintsov, S.D.; Saridakis, E.N. Inflationary cosmology in unimodular F(T) gravity. *Mod. Phys. Lett. A* **2017**, *32*, 1750114. [CrossRef]
32. Akrami, Y. et al. [Planck Collaboration]. Planck 2018 results. X. Constraints on inflation. *Astron. Astrophys.* **2020**, *641*, A10.
33. Aghanim, N. et al. [Planck Collaboration]. Planck 2018 Results. VI. Cosmological Parameters. *Astron. Astrophys.* **2020**, *641*, A6.
34. Tristram, M.; Banday, A.; Górski, K.; Keskitalo, R.; Lawrence, C.; Andersen, K.; Barreiro, R.; Borrill, J.; Colombo, L.; Eriksen, H.; et al. Improved limits on the tensor-to-scalar ratio using BICEP and Planck. *Phys. Rev. D* **2022**, *105*, 083524. [CrossRef]
35. Hazumi, M.; Ade, P.A.R.; Adler, A.; Allys, E.; Arnold, K.; Auguste, D.; Aumont, J.; Aurlien, R.; Austermann, J.; Baccigalupi, C.; et al. LiteBIRD: JAXA's new strategic L-class mission for all-sky surveys of cosmic microwave background polarization. *Proc. SPIE* **2020**, *11443*, 114432F.
36. Enqvist, K.; Hardwick, R.J.; Tenkanen, T.; Venninb, V.; Wands, D. A novel way to determine the scale of inflation. *JCAP* **2018**, *2*, 6. [CrossRef]
37. Ambrosio, F.D.; Fell, S.D.B.; Heisenberg, L.; Kuhn, S. Black holes in f(Q) gravity. *Phys. Rev. D* **2022**, *105*, 024042. [CrossRef]
38. Chen, S.; Gibbons, G.W.; Li, Y.; Yang, Y. Friedmann's equations and Chebyshev's theorem. *J. Cosmol. Astropart. Phys.* **2014**, *12*, 35. [CrossRef]
39. D' Ambrosio, F.; Heisenberg, L.; Kuhn, S. Revisiting Cosmologies in Teleparallelism. *Class. Quantum Grav.* **2022**, *39*, 025013. [CrossRef]
40. Capozziello, S.; D'Agostino, R. Model-independent reconstruction of f(Q) non-metric gravity. *Phys. Lett.* **2022**, *832*, 137229. [CrossRef]
41. Atayde, L.; Frusciante, N. Can f(Q) gravity challenge ΛCDM? *Phys. Rev. D* **2021**, *104*, 064052. [CrossRef]
42. Esposito, F.; Carloni, S.; Cianci, R.; Vignolo, S. Reconstructing isotropic and anisotropic f($\mathcal{Q}$) cosmologies. *Phys. Rev. D* **2022**, *105*, 084061. [CrossRef]
43. Nester, J.M.; Yo, H.-J. Symmetric teleparallel general relativity. *Chin. J. Phys.* **1999**, *37*, 113.
44. Jimenez, J.B.; Heisenberg, L.; Koivisto, T. Coincident general relativity. *Phys. Rev. D* **2018**, *98*, 044048. [CrossRef]
45. Bahamonde, S.; Järv, L. Coincident gauge for static spherical field configurations in symmetric teleparallel gravity. *Eur. Phys. J. C* **2022**, *82*, 963. [CrossRef]
46. Jiménez, J.B.; Koivisto, T.S. Accidental gauge symmetries of Minkowski spacetime in Teleparallel theories. *Universe* **2021**, *7*, 143. [CrossRef]
47. Gomes, D.A.; Jiménez, J.B.; Cano, A.J.; Koivisto, T.S. On the pathological character of modifications to Coincident General Relativity: Cosmological strong coupling and ghosts in f(Q) theories. *arXiv* **2023**, arXiv:2311.04201.
48. Hohmann, M. General covariant symmetric teleparallel cosmology. *Phys. Rev. D* **2021**, *104*, 124077. [CrossRef]
49. Dimakis, N.; Paliathanasis, A.; Roumeliotis, M.; Christodoulakis, T. FLRW solutions in f(Q) theory: The effect of using different connections. *Phys. Rev. D* **2022**, *106*, 043509. [CrossRef]
50. De, A.; Loo, T.-H. On the viability of f(Q) gravity models. *Class. Quan. Grav.* **2023**, *40*, 115007. [CrossRef]
51. Hu, Y.-M.; Zhao, Y.; Ren, X.; Wang, B.; Sadakis, E.N.; Cai, Y.-F. The effective field theory approach to the strong coupling issue in f(T) gravity. *J. Cosmol. Astropart. Phys.* **2023**, *7*, 60. [CrossRef]
52. Allemandi, G.; Borowiec, A.; Francaviglia, M.; Odintsov, S.D. Dark energy dominance and cosmic acceleration in first order formalism. *Phys. Rev. D* **2005**, *72*, 063505. [CrossRef]
53. Chakrabortty, M.; Sk, N.; Sanyal, A.K. Some aspects of modified theory of gravity in Palatini formalism unveiled. *Mod. Phys. Lett. A* **2020**, *35*, 2050162. [CrossRef]
54. Sotiriou, T.; Faraoni, V. f(R) theories of gravity. *Rev. Mod. Phys.* **2010**, *82*, 451. [CrossRef]
55. Barausse, E.; Sotiriou, T.; Miller, J. A no-go theorem for polytropic spheres in Palatini f(R) gravity. *Class. Quantum Grav.* **2008**, *25*, 062001. [CrossRef]

56. Saha, D. (1. Department of Physics, Jangipur College, Jangipur, India; 2. Department of Physics, University of Kalyani, Nadia, India); Sanyal, A.K. (Department of Physics, Jangipur College, Jangipur, India). Cosmological evolution for all the connections in symmetric teleparallel gravity theory. Personal communication, 2023.

universe

MDPI

Article

Bose and Fermi Gases in Metric-Affine Gravity and Linear Generalized Uncertainty Principle

Aneta Wojnar [1,*,†] and Débora Aguiar Gomes [2,†]

[1] Department of Theoretical Physics & IPARCOS, Complutense University of Madrid, 28040 Madrid, Spain
[2] Laboratory of Theoretical Physics, Institute of Physics, University of Tartu, W. Ostwaldi 1, 50411 Tartu, Estonia; debora.aguiar.gomes@ut.ee
* Correspondence: awojnar@ucm.es
† These authors contributed equally to this work.

Abstract: Palatini-like theories of gravity have a remarkable connection to models incorporating linear generalized uncertainty principles. Considering this, we delve into the thermodynamics of systems comprising both Bose and Fermi gases. Our analysis encompasses the equations of state for various systems, including general Fermi gases, degenerate Fermi gases, Boltzmann gases, and Bose gases such as phonons and photons, as well as Bose–Einstein condensates and liquid helium.

Keywords: Bose gas; Fermi gas; modified gravity; liquid helium; Bose–Einstein condensate; finite temperature corrections; linear GUP

1. Introduction

The consistency of equations within Modified Gravity (MG) is brought into question by numerous indications from prior investigations. For instance, the dependence of chemical potential on gravity suggests that modifications in the description of gravitational fields would have an impact [1]. Modified Gravity has been demonstrated to alter the geodesic deviation equation on the surface of stars, resembling Hook's law and introducing corrections to the polytropic equation of state [2]. Moreover, microscopic quantities such as opacity exhibit modifications, implying a need for an effective treatment [3]. Thermodynamic laws, stellar stability criteria, and properties of Fermi gases also display corrections stemming from gravitational proposals [4–7].

Theoretical descriptions of thermonuclear processes within stars' interiors undergo changes under modified gravity, thereby affecting computations of energy generation rates [8–12]. Some gravity theories introduce a dependence of elementary particle interactions on local energy-momentum distributions [13]. Additionally, specific heats, Debye temperatures, and crystallization processes in white dwarfs are influenced by the gravity model [14,15]. Furthermore, chemical reaction rates, which are influenced by gravity [16], are expected to change with modifications to this interaction.

Relativistic effects in equations of state, when neglected in Tolman–Oppenheimer–Volkoff equations derived from General Relativity (GR), result in an underestimation of compact star limiting masses. Equations of state in curved spacetime for degenerate stars explicitly depend on metric components, leading to alterations in chemical potentials and temperatures [17–20]. Moreover, thermodynamic quantities and equations of state are further modified when (pseudo-)scalar fields, such as axions, are considered [21].

The emerging link between MG and the Generalized Uncertainty Principle (GUP), as shown in [22,23], opens up avenues for testing gravitational theories within Earth-based laboratories. Additionally, techniques advanced by either community can be utilized to evaluate proposals concerning MG or the GUP interchangeably [24]. Notice that such a correspondence is also expected; it is believed that MG is an effective theory of *some*

Quantum Gravity. Many proposals emphasize integrating the quantum structure of space-time and deforming associated quantum phase spaces, leading to the generalization of the Heisenberg uncertainty principle. This emphasis is underscored for the potential measurable effects it offers [25–27]. In theories of quantum gravity, the GUP accounts for possible quantum corrections (see, for instance, [28–33]). Such effects are expected to manifest at small scales, which are of the order of the Planck scale $L_P \sim \sqrt{\frac{\hbar G}{c^3}}$, as detailed in [34–36].

The GUP models introduce modifications to equations of state and microscopic variables, stemming from the interplay between special relativity and gravity. This proposition incorporates a dispersion relation involving energy, mass, and momentum within Heisenberg's Uncertainty Principle, incorporating constants such as the speed of light and the gravitational constant [33,37–41].

Below, we will delve into a review of previous findings regarding ideal gases, derived from the connection between Ricci-based gravity and linear GUP, based on [7,22,24]. Before proceeding, we will provide a brief recap of this correspondence. Additionally, we introduce, for the first time, a theoretical framework for ideal and Boltzmann gases, incorporating Fermi statistics in both idealized and realistic scenarios (with the temperature corrections), along with a simplified portrayal of Bose gases, such as photon and phonon gases, within the context of Ricci-based and linear GUP models.

2. Metric-Affine Gravity and Linear GUP Correspondence

The class of metric-affine gravity theories under consideration is defined by the following action:

$$\mathcal{S} = \int d^4x\sqrt{-g}\mathcal{L}_G(g_{\mu\nu}, R_{\mu\nu}) + \mathcal{S}_m(g_{\mu\nu}, \psi_m)\,. \tag{1}$$

Here, g denotes the determinant of the space-time metric $g_{\mu\nu}$, and $R_{\mu\nu}$ represents the symmetric Ricci tensor, which relies solely on the affine connection $\Gamma \equiv \Gamma^{\lambda}_{\mu\nu}$. To construct the gravitational Lagrangian $\mathcal{L}_G$ as a scalar function using powers of traces of $M^{\mu}{}_{\nu} \equiv g^{\mu\alpha}R_{\alpha\nu}$, the object $M^{\mu}{}_{\nu}$ is introduced.

The matter action is described by

$$\mathcal{S}_m = \int d^4x\sqrt{-g}\mathcal{L}_m(g_{\mu\nu}, \psi_m). \tag{2}$$

In this framework, the matter action couples minimally to the metric, neglecting the torsion (the antisymmetric part of the connection), akin to minimally coupled bosonic fields. Such an assumption also applies to fermionic particles and allows degenerate matter to be simply modeled as a perfect fluid [42]. Likewise, regarding only the symmetric piece, the Ricci tensor has the added benefit of preventing ghostlike instabilities [43–46]. This framework encompasses various gravity theories, including GR, Palatini $f(R)$ gravity, Eddington-inspired Born–Infeld gravity [47], and its extensions [48].

Before proceeding further, we need to introduce the tensor $q_{\mu\nu}$ which is related to the space-time metric $g_{\mu\nu}$ through

$$q_{\mu\nu} = g_{\mu\alpha}\Omega^{\alpha}{}_{\nu}. \tag{3}$$

The deformation matrix $\Omega^{\alpha}{}_{\nu}$ is theory-dependent and determined by the gravitational Lagrangian $\mathcal{L}_G$. For instance, for the considered models (Palatini and EiBI), it is an outcome of the fields Equation (7), and they are, respectively,

$$q_{\mu\nu} = f'(R)g_{\mu\nu} \tag{4}$$

$$q_{\mu\nu} = g_{\mu\nu} + \epsilon\mathcal{R}_{\mu\nu}, \tag{5}$$

where $\epsilon = 1/M_{BI}$, and M_{BI} is the Born–Infeld mass (the theory parameter). For more details, see [48].

The gravitational action encapsulates theories that, despite complex field equations, can be conveniently reformulated [48]:

$$G^{\mu}{}_{\nu}(q) = \frac{\kappa}{|\hat{\Omega}|^{1/2}}\left(T^{\mu}{}_{\nu} - \delta^{\mu}_{\nu}\left(\mathcal{L}_G + \frac{T}{2}\right)\right). \quad (6)$$

Here, $|\hat{\Omega}|$ denotes the determinant of the deformation matrix, and T represents the trace of the energy-momentum tensor of matter fields. The Einstein tensor $G^{\mu}{}_{\nu}(q)$ is linked to a tensor $q_{\mu\nu}$, where the connection Γ adopts the Levi-Civita connection of $q_{\mu\nu}$:

$$\nabla^{\Gamma}_{\mu}(\sqrt{-q}q^{\alpha\beta}) = 0. \quad (7)$$

These theories yield second-order field equations, reducing to GR counterparts in vacuum ($T_{\mu}{}^{\nu} = 0$), implying no extra degrees of freedom propagate in these theories beyond the usual two polarizations of the gravitational field.

The nonrelativistic limit of the field Equations (6) is particularly intriguing. Notice that on the right-hand part of the equations, we are dealing with functions of the energy-momentum tensor and its trace. As the Poisson equation is of the second-order, the only component of the energy-momentum tensor is $T_{00} \approx -T \approx \rho$ (see more details in [49] for the second-order corrections). Regarding the curvature impact, it was also shown there that only R^2 terms have an influence in the second-order approximation. In the most popular models such as Palatini $f(R)$ [49] and EiBI [50,51] gravities, the Poisson equation (more complicated Lagrangians would require a detailed analysis; however, it seems that their non-relativistic limits will not differ significantly from the form studied here as only R^2 terms provide corrections to the second order equations) takes the following particular form:

$$\nabla^2\phi = \frac{\kappa}{2}\left(\rho + \bar{\alpha}\nabla^2\rho\right), \quad (8)$$

where ϕ represents the gravitational potential, $\kappa = 8\pi G$, and $\bar{\alpha}$ stands for the theory parameter. In Palatini $f(R)$, $\bar{\alpha} = 2\bar{\beta}$, with $\bar{\beta}$ accompanying the quadratic term, while in EiBI, $\bar{\alpha} = \epsilon/2$, where $\epsilon = 1/M_{BI}$ and M_{BI} denotes the Born–Infeld mass. The resemblance in the Poisson equation between these two gravity proposals is not coincidental; the EiBI gravity, in the first-order approximation, reduces to Palatini gravity with the quadratic term [52]. Furthermore, only the quadratic term R^2 influences the non-relativistic equations, as higher curvature scalar terms enter the equations at the sixth order [49].

As indicated in [22], the augmentation in the Poisson Equation (8) can be interpreted as a modification to the Fermi gas at a finite temperature. This modification arises when considering a deformation of the phase space represented by the following integral:

$$\frac{1}{(2\pi\hbar)^3}\int \frac{d^3x d^3p}{(1-\sigma p)^d}, \quad (9)$$

where $d = 1$ corresponds to Palatini-like gravity theories. The connection between the deformation parameter σ and the Palatini parameter $\bar{\beta}$ is established as follows:

$$\sigma = \frac{4\pi G}{K_2}\bar{\beta} \quad \text{and} \quad K_2 = \frac{3}{\pi}\frac{h^3 N_A^2}{m_e \mu_e^2}, \quad (10)$$

with m_e being the electron mass, N_A is the Avogadro constant, μ_e is the mean molecular weight per electron defined in terms of the mass fractions of hydrogen X and helium Y as $\frac{1}{\mu_e} = X + \frac{Y}{2}$, and other constants hold their usual definitions.

This correspondence enables the expression of a general partition function in three dimensions within a large volume

$$\ln Z = \frac{V}{(2\pi\hbar)^3}\frac{g}{a}\int \ln\left[1 + aze^{-E/k_BT}\right]\frac{d^3p}{(1-\sigma p)^d}, \quad (11)$$

where $V := \int d^3x$ denotes the volume of the cell in configuration space, and setting $a = 1$ ($a = -1$) corresponds to a system of fermionic (bosonic) particles with energy states E_p. The fugacity is represented by $z = e^{\mu/k_BT}$, with μ as the chemical potential, and g indicating the spin of a particle.

As we can see from (11), the deformation in the phase space introduced here is parameterized by σ. Hence, the procedure described in this work can be paralleled to the GUP theories containing linear corrections in p [53–58]. In the context of the GUP, the deformation is derived through the utilization of the Liouville theorem [59]. Consequently, the effective $\hbar$ depends on the momentum p in the generalized uncertainty relation, leading to a momentum-dependent size of the unit cell for each quantum state in a phase space.

With this modified partition function, one can straightforwardly derive thermodynamic variables for the requisite statistics, focusing primarily on pressure, number of particles, internal energy, and specific heat, which are, respectively, given by

$$P = k_BT\frac{\partial}{\partial V}\ln Z, \tag{12}$$

$$n = k_BT\frac{\partial}{\partial \mu}\ln Z\,|_{T,V}, \tag{13}$$

$$U = k_BT^2\frac{\partial}{\partial T}\ln Z\,|_{z,V}, \tag{14}$$

$$C_V = \frac{\partial U}{\partial T}\,|_V\,. \tag{15}$$

3. Boltzmann Gas

In the case of high temperature, thermodynamic systems described by Bose or Fermi statistics behave as the Boltzmann one. The Maxwell–Boltzmann statistics corresponds to the limit $a \to 0$, from which the general partition function (11) acquires the form:

$$\ln Z = \frac{V}{(2\pi\hbar)^3}\int z e^{-E/k_BT}\frac{d^3p}{(1-\sigma p)^d}\,. \tag{16}$$

Moreover, the distribution function for this statistics is given by

$$f_{\rm MB}(E) = z e^{-E/k_BT}, \tag{17}$$

corresponding to the regime where Fermi–Dirac, Bose–Einstein and Maxwell–Boltzmann statistics become identical, i.e., $(E-\mu)/k_BT >> 1$.

Taking the partition function (16), we can easily obtain

$$P = \frac{1}{2\pi^2\hbar^3}\int \frac{1}{3}p^3\,{}_2F_1(3,d,4,p\sigma)f_{\rm MB}(E)\frac{c^2p}{E}dp, \tag{18}$$

$$n = \frac{V}{2\pi^2\hbar^3}\int f_{\rm MB}(E)\frac{p^2dp}{(1-\sigma p)^d}, \tag{19}$$

$$U = \frac{V}{2\pi^2\hbar^3}\int E f_{\rm MB}(E)\frac{p^2dp}{(1-\sigma p)^d}, \tag{20}$$

where ${}_2F_1$ represents the hypergeometric function.

4. Fermi Statistics

In this extended scenario, utilizing (11) and (12) with $a = 1$ for fermions and $g = 2$ for electrons, we derive the microphysical description of the system employing the Fermi–Dirac distribution $f(E)$:

$$f(E) = \left(1 + z^{-1}e^{E/k_BT}\right)^{-1}, \tag{21}$$

resulting in the expression for pressure as

$$P = \frac{1}{\pi^2\hbar^3}\int \frac{1}{3}p^3\,{}_2F_1(3,d,4,p\sigma)f(E)\frac{c^2p}{E}dp, \tag{22}$$

Meanwhile, the particle number density and internal energy are given by

$$n = \frac{V}{\pi^2\hbar^3}\int f(E)\frac{p^2dp}{(1-\sigma p)^d} \tag{23}$$

$$U = \frac{V}{\pi^2\hbar^3}\int Ef(E)\frac{p^2dp}{(1-\sigma p)^d}. \tag{24}$$

For the case when $|\sigma p| < 1$, the general form of the pressure (22) can be written as the series

$$P = \frac{1}{\pi^2\hbar^3}\int \frac{p^3}{3}\left(\sum_{k=0}^{\infty}\frac{(d)_k\left(\frac{3}{2}\right)_k(\sigma p)^k}{\left(\frac{5}{2}\right)_k k!}\right)f(E)\frac{c^2p}{E}dp. \tag{25}$$

While taking into account only the first two terms of the series (as we consider the gravity deformation only up to linear terms in σ), we have

$$P = \frac{1}{\pi^2\hbar^3}\int \frac{1}{3}p^3\left(1+\frac{3d}{4}\sigma p\right)f(E)\frac{c^2p}{E}dp. \tag{26}$$

We will now discuss its form in more detail.

4.1. Non-Relativistic and Relativistic Degenerate Fermi Gas

We will now shift our focus to a specific form of matter that holds significant importance in stellar physics: degenerate gases. These gases are utilized to characterize the dense cores of stellar and substellar objects, as well as degenerate matter in compact stars. In the toy model that we are employing here, all states with energies lower than the Fermi energy level are occupied, while all states with higher energies remain empty at absolute zero temperature ($T \to 0$). Consequently, the chemical potential μ takes on the value of the Fermi energy E_F. In such scenarios, the Fermi–Dirac distribution takes the form of a step function:

$$f(E) = \begin{cases} 1 & \text{if } E \leq E_F \\ 0 & \text{otherwise.} \end{cases}$$

Consequently, in Equation (26), integration is carried out up to the Fermi energy E_F:

$$P = \frac{1}{\pi^2\hbar^3}\left[\frac{(2m_eE_F)^{5/2}}{6m_e} + \frac{\sigma d}{3}\frac{(2m_eE_F)^3}{8m_e}\right], \tag{27}$$

which corresponds to the effective pressure in Palatini theories when $d = 1$ (see [22]). Using the definition of the electron degeneracy

$$\psi = (\beta E_F)^{-1} = \frac{k_BT}{A}\left(\frac{\mu_e}{\rho}\right)^{2/3}, \tag{28}$$

we can rewrite it as a function of the density such that one deals with the non-relativistic polytrope EoS. Let us write it in a more convenient form for the further analysis:

$$P = K\rho^{\frac{5}{3}} + \sigma K_2\rho^2, \tag{29}$$

where K_2 is given by (10), while

$$K = \frac{1}{20}\left(\frac{3}{\pi}\right)^{\frac{2}{3}} \frac{h^2}{m_e(\mu_e m_u)^{\frac{5}{3}}}$$

refers to the standard polytropic parameter for $n = 3/2$.

On the other hand, the relativistic polytropic EoS is given simply by inserting $E = pc$ in (26) and using the relation between the pressure and density (recall that for electrons $g = 2$)

$$p = \left(\frac{6\pi^2\hbar^3\rho}{g m_e}\right)^{\frac{1}{3}}, \tag{30}$$

providing

$$P = \frac{\hbar c}{4}\left(\frac{3\pi^2}{m_e^4}\right)^{1/3}\rho^{4/3} + \frac{9}{20} d\sigma\hbar^2 c\left(\frac{3^2\pi^4}{m_e^5}\right)^{1/3}\rho^{5/3}. \tag{31}$$

Note that the phase-space deformation provides a mixture of two polytropes; that is, in the case of non-relativistic case, the modification is in the form of the rigid polytrope, while the relativistic one has a term resembling the non-relativistic polytrope ($n = 3/2$).

5. Bose Statistics

Let us briefly recall the basic equations describing a system of N spinless, non-interacting particles in Palatini gravity [24]. The Hamiltonian is given by $H = \sum_{i=1}^{N} \frac{p_i^2}{2m}$, with $p_i^2 = \mathrm{p}_i \cdot \mathrm{p}_i$ and p_i being the momentum operator of the single-particle with energy $E_p = p^2/2m$. Since the grand partition function of an ideal Bose gas for such a system is given by ($\beta =: (k_B T)^{-1}$)

$$Z = \prod_p \frac{1}{1 - ze^{-\beta E_p}}, \tag{32}$$

an average occupation number for a state p is (the total number of particles $N = \sum_p \langle n_p \rangle$)

$$\langle n_p \rangle = -\frac{1}{\beta}\frac{\partial}{\partial E_p}\ln Z = \frac{ze^{-\beta E_p}}{1 - ze^{-\beta E_p}}, \tag{33}$$

while the equation of state is

$$\beta P V = -\sum_p \ln(1 - ze^{-\beta E_p}). \tag{34}$$

Considering $V \to \infty$ and defining the specific volume $v = V/N$, the above series can be written as

$$\beta P = -\frac{4\pi}{(2\pi\hbar)^3}\int_0^\infty \frac{dp p^2}{1 - \sigma p}\ln\left[1 - ze^{-\beta\frac{p}{2m}}\right] - \frac{\ln(1-z)}{V}, \tag{35}$$

$$\frac{1}{v} = \frac{4\pi}{(2\pi\hbar)^3}\int_0^\infty \frac{dp p^2}{1 - \sigma p}\frac{1}{z^{-1}e^{\beta\frac{p}{2m}} - 1} + \frac{1}{V}\frac{z}{1-z}. \tag{36}$$

Taking into account only the terms linear in the parameter $|\alpha| =: |\sigma|\sqrt{2mk_B T}$, one writes (to see how the series converge, see [24])

$$\beta P = \frac{1}{\lambda^3}\left[g_{5/2}(z) + \frac{2\alpha}{\pi}\mathrm{Li}_3(z)\right] - \frac{\ln(1-z)}{V}, \tag{37}$$

$$\frac{1}{v} = \frac{1}{\lambda^3}\left[g_{3/2}(z) + \frac{2\alpha}{\pi}\mathrm{Li}_2(z)\right] + \frac{1}{V}\frac{z}{1-z}, \tag{38}$$

where $\lambda = \sqrt{\frac{2\pi\hbar^2}{mk_BT}}$ is the thermal wavelength, $g_m(z) = \sum_{n=1}^{\infty} \frac{z^n}{n^m}$, and $\text{Li}_n(z)$ is the polylogarithm function which can be expressed for $|z| < 1$ as $\text{Li}_n(z) = \sum_{k=1}^{\infty} \frac{z^k}{k^n}$. The internal energy is given simply by

$$\frac{U}{V} = -\frac{1}{V}\frac{\partial}{\partial\beta}\ln Z = \frac{3k_BT}{2\lambda^3}\left[g_{5/2}(z) + \frac{2\alpha}{\pi}\text{Li}_3(z)\right]. \tag{39}$$

Comparing (37) with the assumption that its last term can be neglected, we deal with a well known relation between the internal energy and temperature $U = \frac{3}{2}PV$. Notice that in order to study the further properties of an ideal Bose gas, one needs to know the fugacity z dependence on the temperature and specific volume v. We will come back to this issue in Section 5.3.

5.1. Photon Gas

Since photons are massless, their energy will be simply given by $E = cp$. Its wave number and frequency are related by $p = \hbar k = \hbar\omega/c$, and there are two propagating modes, which are taken into account by multiplying the partition function (11) by a factor of 2. The internal energy will be given by

$$U = \frac{V}{\pi^2}\left[\frac{\pi^4(k_BT)^4}{15(\hbar c)^3} + 24\sigma d\frac{(k_BT)^5\zeta(5)}{c^4\hbar^3}\right], \tag{40}$$

where the first term corresponds to Stefan's law. Accordingly, the specific heat will be given by

$$C_V = \frac{V}{\pi^2}\left[\frac{4\pi^4k_B^4T^3}{15(\hbar c)^3} + 120\sigma d\frac{k_B^5T^4\zeta(5)}{c^4\hbar^3}\right], \tag{41}$$

while pressure is simply given by $PV = U/3$.

5.2. Phonon Gas

In spite of not being particles, the vibrations in lattices and in particular its normal modes, phonons, can be mathematically described as bosons. This treatment is specially relevant when considering a solid subject to low temperatures, where phonon interactions are negligible. In this case, we have three propagation modes, in contrast with photons which only have two, and the allowed frequencies are bounded. The energy per atom is

$$\frac{U}{N} = 3\kappa_BTD_3(\beta\hbar\omega) + \frac{9}{4c}\kappa_BTd\sigma\hbar\omega_mD_4(\beta\hbar\omega), \tag{42}$$

where $\omega_m = c\left(\frac{6\pi^2N}{V}\right)^{1/3}$ is the maximum frequency and

$$D_n(x) = \frac{n}{x^n}\int_0^x \frac{t^n}{e^t - 1}dt \tag{43}$$

is the Debye function. By defining the Debye temperature via $\kappa_BT_D = \hbar\omega_m$, and using the solutions for the Debye functions, we can rewrite (42) as

$$\frac{U}{N} = 3\kappa_BT\left(1 - \frac{3}{8}\frac{T_D}{T}\right) + \frac{9}{4c}\kappa_B^2T_DTd\sigma\left(1 - \frac{2}{5}\frac{T_D}{T}\right) \tag{44}$$

for $T >> T_D$ and

$$\frac{U}{N} = 3\kappa_BT\left[\frac{\pi^4}{5}\left(\frac{T}{T_D}\right)^3\right] + 216\zeta(5)\kappa_BTd\sigma(\kappa_BT_D)\left(\frac{T}{T_D}\right)^4 \tag{45}$$

for $T << T_D$.

The specific heat, for high and low temperatures, will be

$$\frac{C_V}{\kappa_B N} = 3\left[1 - \frac{1}{20}\left(\frac{T_D}{T}\right)^2\right] + \frac{9}{4c} d\sigma \kappa_B T_D \left[1 - \frac{1}{18}\left(\frac{T_D}{T}\right)^2\right], T >> T_D \tag{46}$$

and

$$\frac{C_V}{\kappa_B N} = \frac{12\pi^4}{5}\left(\frac{T}{T_D}\right)^3 + 1080\zeta(5)\frac{d\sigma}{c}\kappa_B T_D \left(\frac{T}{T_D}\right)^4, T << T_D. \tag{47}$$

The first term in (46) corresponds to the Dulong–Petit law, namely $C_v = 3N\kappa_B$. However, it should be emphasized that although $T >> T_D$, the temperatures considered here are low enough so that we can consider the phonons to be non-interactive. At very high temperatures, the phonons interaction cannot be overlooked, and the current approach will no longer be valid.

5.3. Bose–Einstein Condensate

The behavior of the fugacity z depends on the properties of the functions $g_{3/2}(z)$ and $\mathrm{Li}_2(z)$. Analyzing Equation (38), one can derive a modified condition which the ideal Bose gas has to satisfy to become the Bose–Einstein condensate ($g_{3/2}(1) \approx 2.612$):

$$\frac{\lambda^3}{v} - \frac{2\alpha}{\pi}\zeta(2) = g_{3/2}(1). \tag{48}$$

It clearly provides the critical value for the specific volume (or critical density $n_c = 1/v_{cr}$):

$$n_c = \left(\frac{1}{4\pi\hbar^2}\right)^{\frac{3}{2}} \left[\zeta(3/2)(2mk_B T)^{\frac{3}{2}} + \sigma\frac{\pi}{3}(2mk_B T)^2\right]. \tag{49}$$

Moreover, the fugacity dependence on the temperature T and specific volume v is

$$z = \begin{cases} 1 \text{ for} & \frac{\lambda^3}{v} - \frac{\pi\alpha}{3} \geq g_{3/2}(1) \\ \text{solution of } \frac{\lambda^3}{v} = \left[g_{3/2}(z) + \frac{2\alpha}{\pi}\mathrm{Li}_2(z)\right] & \text{otherwise.} \end{cases}$$

Hence, the fugacity remains fixed at 1 throughout the Einstein–Bose condensate, indicating a zero chemical potential. In other words, within the region where $\frac{\lambda^3}{v} - \frac{\pi\alpha}{3} \leq g_{3/2}(1)$, we are dealing with the gas phase.

The equation of state and other thermodynamic functions in both regions can be also derived and have the following forms:

$$\beta P = \begin{cases} \frac{1}{\lambda^3}\left[g_{5/2}(z) + \frac{2\alpha}{\pi}\mathrm{Li}_3(z)\right] & \text{if } v > v_{cr}, \\ \frac{1}{\lambda^3}\left[g_{5/2}(1) + \frac{2\alpha}{\pi}\zeta(3)\right] & \text{if } v < v_{cr}, \end{cases}$$

$$\frac{U}{N} = \frac{3}{2}Pv = \begin{cases} \frac{3}{2}\frac{k_B T v}{\lambda^3}\left[g_{5/2}(z) + \frac{2\alpha}{\pi}\mathrm{Li}_3(z)\right] & \text{if } v > v_{cr}, \\ \frac{3}{2}\frac{k_B T v}{\lambda^3}\left[g_{5/2}(1) + \frac{2\alpha}{\pi}\zeta(3)\right] & \text{if } v < v_{cr}, \end{cases}$$

$$\frac{C_V}{Nk_B} = \begin{cases} \frac{15}{4}\frac{v}{\lambda^3}h_1(T) + \frac{3}{2}\frac{Tv}{\lambda^3}h_2(T)\frac{dz}{dT} & \text{if } v > v_{cr}, \\ \frac{15}{4}\frac{v}{\lambda^3}g_{5/2}(1) + \sigma f_1(T) & \text{if } v < v_{cr}, \end{cases}$$

where the functions h_1, h_2, f_1 were defined in [24]. The vapor pressure takes the following form

$$P_0(T) = \frac{k_B T}{\lambda^3}\left[g_{5/2}(1) + \frac{2\sigma\sqrt{2mk_B T}}{\pi}\zeta(3)\right], \tag{50}$$

while a modified latent heat resulting from the Clapeyron equations derived from (50) is (the function f_2 is given by [24])

$$f_2(T) = \frac{48\sqrt{\frac{\hbar^2\pi}{k_B}}}{5g_{5/2}(1)(g_{3/2}(1)v_{cr})^{1/3}}\zeta(3) - \frac{\sqrt{8k_B mT}\zeta(2)}{g_{3/2}(1)}. \tag{51}$$

$$L = \frac{5}{2}\frac{k_B T g_{5/2}(1)}{g_{3/2}(1)}\Big(1 + \frac{\sigma}{\pi}f_2(T)\Big). \tag{52}$$

The Bose–Einstein condensation is then also a first-order phase transition in the considered models of gravity if $L \neq 0$.

6. More Physical Models

Building upon the discoveries outlined in the preceding section of the paper, one can delve into studying toy-model systems. However, it is crucial to handle the conclusions drawn from such investigations with utmost care, as they have the potential to foster misconceptions. Therefore, overly optimistic bounds on a theory parameter, as discussed in [24], or "no-go theorems" for models of gravity with toy-model assumptions [52,60–62] should not be decisive in assessing the plausibility of a theory (see, e.g., [2,7,63,64]).

Therefore, in the following section, we will derive a Fermi EoS with finite temperature corrections and revisit a simple Landau model of liquid helium. This model realistically describes its behavior at low temperatures, in contrast to treating it as an ideal Bose–Einstein condensate.

6.1. Fermi Gas with Finite Temperature Corrections

The aim of the present subsection is to provide a more accurate EoS for a Fermi gas at low temperatures. This approach is particularly significant in the study of substellar objects, such as brown dwarfs, for which electron degeneracy is the main factor driving the stellar evolution and stability. The temperatures considered here are low enough so that $k_B T \ll mc^2$, which means that we are dealing with a non-relativistic gas. In this regime, the energy becomes $E = p^2/2m$, and consequently, the pressure (26) becomes

$$P = a_1\left[\int_0^\infty \frac{E^{3/2}}{1+e^{\beta(E-\mu)}}dE + a_2\int_0^\infty \frac{E^2}{1+e^{\beta(E-\mu)}}dE\right], \tag{53}$$

where $a_1 = \frac{(2m)^{3/2}}{3\pi^2\hbar^3}$ and $a_2 = \frac{3}{4}\sigma d(2m)^{1/2}$. The two integrals above are Fermi integrals with the following solutions (we to refer Appendix A of [65] to the reader interested in the details of this calculations):

$$\begin{aligned}\int_0^\infty \frac{E^{3/2}}{1+e^{\beta(E-\mu)}}dE =& \frac{2}{5}\mu^{5/2} - \frac{1}{8}\beta^{-1}\mu^{3/2}\ln(1+e^{-\beta\mu}) \\ &+ \frac{\pi^2}{4}\beta^{-2}\mu^{1/2} + \frac{3}{4}\beta^{-2}\mu^{1/2}Li_2(-e^{-\beta\mu}) + \ldots \end{aligned} \tag{54}$$

$$\int_0^\infty \frac{E^2}{1+e^{\beta(E-\mu)}}dE = \frac{\mu^3}{3} + \frac{\pi^2}{3}\beta^{-2}\mu^1 + \ldots \tag{55}$$

Then, by setting $\mu = E_F$, we can rewrite the pressure as

$$\begin{aligned} P_F =& a_1\left\{\frac{2A^{5/2}}{5}\left(\frac{\rho}{\mu_e}\right)^{5/3}\left[1 - \frac{5}{16}\psi\ln(1+e^{-1/\psi}) + \frac{5\pi^2}{8}\psi^2 + \frac{15}{8}\psi^2 Li_2(e^{-1/\psi})\right]\right. \\ &+ \left.\frac{a_2}{3}A^3\left(\frac{\rho}{\mu_e}\right)^2(1+\pi^2\psi^2)\right\}, \end{aligned} \tag{56}$$

where $A = \frac{(3\pi^2\hbar^3 N_A)^{2/3}}{2m_e}$ is a constant, and ψ is the degeneracy parameter defined in (28). Notice that the first terms of the EoS (56) corresponds to the pressure for $T = 0$ in a non-deformed phase space, i.e., the polytrope equation for $n = 3/2$, discussed already in Section Non-Relativistic and Relativistic Degenerate Fermi Gas. The first line in the same equation corresponds to the pressure for a finite temperature T in a non-deformed phase space.

In stars with a very low mass, apart from the Fermi pressure (56), we also consider the pressure due to ionized gases. The total pressure will then combine both contributions. If we consider an ionized hydrogen gas, for instance, we have the ideal gas law in the form $P_i = \frac{\kappa\rho T}{\mu_1 m_H}$, where m_H is the mass of the hydrogen atom and $\mu_1^{-1} = ((1 + x_+)X + Y/4)$, with x_+ being the ionization fraction of hydrogen. The equation of state then becomes $P = K'\rho^{\frac{5}{3}} + \sigma K_2'\rho^2$, with

$$K' = \frac{2a_1 A^{5/2}}{5}\mu_e^{-5/3}\left[1 - \frac{5}{16}\psi\ln(1 + e^{-1/\psi}) + \frac{5\pi^2}{8}\psi^2 + \frac{15}{8}\psi^2 Li_2(e^{-1/\psi}) + \alpha\psi\right], \quad (57)$$

and

$$K_2' = \frac{a_1 a_2}{3\sigma}A^3\mu_e^{-2}(1 + \pi^2\psi^2), \quad (58)$$

where $\alpha = 5\mu_e/2\mu_1$. Such an equation of state is used to describe the matter properties in low mass stars and brown dwarfs, in which the finite temperature corrections, providing the time evolution of electron degeneracy, are important, see, e.g., [65–67].

6.2. Liquid Helium He^4

The Landau model [68,69] offers a comprehensive microscopic depiction of a two-fluid model near absolute zero. As T approaches zero, the specific heat of liquid helium behaves as T^3, characteristic of a phonon gas and experimentally confirmed. However, at finite temperatures, an additional term emerges. Thus, the dispersion relation of quasiparticles as a function of wave number k for ^{4}He can be described by

$$\hbar\omega = \begin{cases} \hbar ck & \text{if } k << k_0, \\ \Delta + \frac{\hbar^2(k-k_0)^2}{2\gamma} & \text{if } k \approx k_0, \end{cases}$$

where c denotes the sound velocity, while Δ, k_0, and γ are experimental constants (We will use the data from [70]).

$$c = 239\ \mathrm{m\,s^{-1}},\ \rho = 144\ \mathrm{kg\,m^{-3}},\ \Delta/k_B = 8.65\ \mathrm{K}, k_0 = 1.92 \times 10^{10}\ \mathrm{m^{-1}},\ \gamma = 1.07 \times 10^{-27}\ \mathrm{kg}.$$

In the Landau theory, it is posited that the quantum states of ^{4}He near the ground state can be treated as those of a non-interacting gas with energy levels given by

$$U = E_0 + \sum_k \hbar\omega_k\langle n_k\rangle = E_0 + \frac{V}{2\pi^2}\int_0^\infty \frac{k^2\hbar\omega_k}{e^{\beta\hbar\omega_k} - 1}\frac{dk}{(1 - \sigma\hbar k)}. \quad (59)$$

Here, $\hbar\omega_k$ represents the elementary excitation energy with wave vector k and occupation number $\langle n_k\rangle$. In the second equality, the deformation of the phase space has been taken into account. Now, let us compute the internal energy and its gravity corrections at low temperatures. In this context, only contributions from the phonon and roton components [70,71] affect the energy in Equation (59). They are given, respectively, as follows:

$$E_{\rm phonon} = \frac{V}{2\pi^2}\left(\frac{\pi^4(k_BT)^4}{15\hbar^3c^3} + 24\sigma\frac{(k_BT)^5\zeta(5)}{c^4\hbar^3}\right), \tag{60}$$

$$\frac{E_{\rm roton}}{V} \approx \frac{k_0^2\Delta}{\pi}\sqrt{\frac{\gamma k_BT}{2\pi\hbar^2}}e^{-\frac{\Delta}{k_BT}}(1+\sigma\hbar k_0). \tag{61}$$

Using Expression (15), we derive the specific heat for liquid helium in low temperature

$$C_{\rm He^4} = 20.7T^3 + \frac{387\times10^3}{T^{3/2}}e^{-8.85/T} + \sigma(5.73\times10^{-24}T^4 + \frac{7.83\times10^{-19}}{T^{3/2}}e^{-8.85/T}). \tag{62}$$

7. Discussion and Conclusions

In this paper, our focus was to investigate the impact of Ricci-based gravity theories, such as Palatini $f(R)$ and Eddington-inspired Born–Infeld models, in conjunction with linear Generalized Uncertainty Principle (GUP) models, on systems governed by Bose and Fermi statistics. By leveraging the recently established correspondence between modified gravity and GUP models, we developed a formalism to analyze ideal Bose and Fermi gases. As anticipated, the inclusion of modified gravity or linear GUP resulted in the introduction of additional terms to the familiar expressions. We outlined these modifications for general Fermi gases, degenerate Fermi gases, and Boltzmann gases, as well as Bose gases including phonons, photons, Bose–Einstein condensates, and liquid helium.

It is worth noting that the thermodynamic framework derived from this correspondence is best suited for non-relativistic systems or those characterized by low curvature regimes, such as planets, brown dwarfs, active stars, and white dwarfs. Even white dwarfs, despite their dense nature, can still be considered within a non-relativistic regime due to their size. Consequently, the compactness criterion applicable to our formalism is $\mathcal{C} << 1$.

Let us now briefly summarize the previous and novel results. We start by extending the method found in [7] for Fermi gases to obtain the pressure, number of particles and energy for Bose–Einstein statistics (Equations (18)–(20)). In Section Non-Relativistic and Relativistic Degenerate Fermi Gas, we introduced the corrections to the EoS for a non-relativistic degenerate Fermi gas [25] and included the EoS for the relativistic case. In Section 5, we review some key aspects of the Bose–Einstein statistics and the Bose–Einstein condensate, which can be found in [24]. Additionally, we derive the energy and specific heat for photon and phonon gases. We conclude our analysis in Section 6, with finite temperature corrections to non-relativistic Fermi gases [25] and liquid helium [24].

This intriguing connection between Modified Gravity and GUP models opens up avenues for testing gravitational proposals through various tabletop experiments, some of which have already been explored and documented in the literature. Ongoing research in this direction aims to delve deeper into the implications of Modified Gravity on the microscopic properties of matter, seeking further validation and insights.

Author Contributions: Conceptualization, A.W.; methodology, A.W.; validation, A.W. and D.A.G.; formal analysis, A.W. and D.A.G.; investigation, A.W. and D.A.G.; writing—original draft preparation, A.W. and D.A.G.; writing—review and editing, A.W. and D.A.G.; supervision, A.W.; project administration, A.W. All authors have read and agreed to the published version of the manuscript.

Funding: A.W. acknowledges financial support from MICINN (Spain) Ayuda Juan de la Cierva-incorporación 2020 No. IJC2020-044751-I.

Data Availability Statement: No new data were created or anlyzed in this study. Date sharing is not applicable to this article.

Conflicts of Interest: The authors declare no conflicts of interest.

References

1. Kulikov, I.K.; Pronin, P.I. Low-temperature properties of a quantum Fermi gas in curved space-time. *Int. J. Theor. Phys.* **1995**, *34*, 1843–1854. [CrossRef]
2. Kim, H.C. Physics at the surface of a star in Eddington-inspired Born-Infeld gravity. *Phys. Rev. D* **2014**, *89*, 064001. [CrossRef]
3. Sakstein, J. Testing gravity using dwarf stars. *Phys. Rev. D* **2015**, *92*, 124045. [CrossRef]
4. Wojnar, A.; Velten, H. Equilibrium and stability of relativistic stars in extended theories of gravity. *Eur. Phys. J. C* **2016**, *76*, 697. [CrossRef]
5. Wojnar, A. On stability of a neutron star system in Palatini gravity. *Eur. Phys. J. C* **2018**, *78*, 421. [CrossRef]
6. Sarmah, L.; Kalita, S.; Wojnar, A. Stability criterion for white dwarfs in Palatini f(R) gravity. *Phys. Rev. D* **2022**, *105*, 024028. [CrossRef]
7. Wojnar, A. Fermi gas and modified gravity. *Phys. Rev. D* **2023**, *107*, 044025. [CrossRef]
8. Sakstein, J. Hydrogen burning in low mass stars constrains scalar-tensor theories of gravity. *Phys. Rev. Lett.* **2015**, *115*, 201101. [CrossRef]
9. Olmo, G.J.; Rubiera-Garcia, D.; Wojnar, A. Minimum main sequence mass in quadratic Palatini $f(R)$ gravity. *Phys. Rev. D* **2019**, *100*, 044020. [CrossRef]
10. Crisostomi, M.; Lewandowski, M.; Vernizzi, F. Vainshtein regime in scalar-tensor gravity: Constraints on degenerate higher-order scalar-tensor theories. *Phys. Rev. D* **2019**, *100*, 024025. [CrossRef]
11. Rosyadi, A.; Sulaksono, A.; Kassim, H.A.; Yusof, N. Brown dwarfs in Eddington-inspired Born-Infeld and beyond Horndeski theories. *Eur. Phys. J. C* **2019**, *79*, 1–10. [CrossRef]
12. Wojnar, A. Lithium abundance is a gravitational model dependent quantity. *Phys. Rev. D* **2021**, *103*, 044037. [CrossRef]
13. Delhom-Latorre, A.; Olmo, G.J.; Ronco, M. Observable traces of non-metricity: New constraints on metric-affine gravity. *Phys. Lett. B* **2018**, *780*, 294–299. [CrossRef]
14. Kalita, S.; Sarmah, L.; Wojnar, A. Metric-affine effects in crystallization processes of white dwarfs. *Phys. Rev. D* **2023**, *107*, 044072. [CrossRef]
15. Wojnar, A.; Kalita, S.; Sarmah, L. Effects of modified gravity on microscopic properties and cooling timescale of white dwarfs. *Phys. Lett. B* **2024**, *850*, 138494. [CrossRef]
16. Lecca, P. The effects of gravitational potential on chemical reaction rates. *Proc. J. Physics Conf. Ser.* **2021**, *2090*, 012034. Available online: https://iopscience.iop.org/article/10.1088/1742-6596/2090/1/012034/pdf (accessed on 3 May 2024) [CrossRef]
17. Hossain, G.M.; Mandal, S. Equation of states in the curved spacetime of spherical degenerate stars. *J. Cosmol. Astropart. Phys.* **2021**, *2021*, 026. [CrossRef]
18. Hossain, G.M.; Mandal, S. Higher mass limits of neutron stars from the equation of states in curved spacetime. *Phys. Rev. D* **2021**, *104*, 123005. [CrossRef]
19. Li, J.; Guo, T.; Zhao, J.; He, L. Do we need equation of state in curved spacetime for neutron stars? *arXiv* **2022**, arXiv:2206.02106.
20. Chavanis, P.H. Statistical mechanics and thermodynamic limit of self-gravitating fermions in D dimensions. *Phys. Rev. E* **2004**, *69*, 066126. [CrossRef]
21. Sakstein, J.; Croon, D.; McDermott, S.D. Axion instability supernovae. *Phys. Rev. D* **2022**, *105*, 095038. [CrossRef]
22. Wojnar, A. Unveiling phase space modifications: A clash of modified gravity and the generalized uncertainty principle. *Phys. Rev. D* **2024**, *109*, 024011. [CrossRef]
23. Ali, A.F.; Wojnar, A. A covariant tapestry of linear GUP, metric-affine gravity, their Poincaré algebra and entropy bound. *arXiv* **2024**, arXiv:2401.05941.
24. Wojnar, A. Bose–Einstein Condensate and Liquid Helium He^4: Implications of GUP and Modified Gravity Correspondence. *arXiv* **2024**, arXiv:2401.01159.
25. Pachoł, A.; Wojnar, A. Constraining Snyder and GUP models with low-mass stars. *arXiv* **2023**, arXiv:2307.03520.
26. Pachoł, A.; Wojnar, A. Fermi equation of state with finite temperature corrections in quantum space-times approach: Snyder model vs. GUP case. *Class. Quant. Grav.* **2023**, *40*, 195021. [CrossRef]
27. Kozak, A.; Pachoł, A.; Wojnar, A. Refining Bounds for Snyder and GUP Models through Seismic Wave Analysis. *arXiv* **2023**, arXiv:2310.00913.
28. Kempf, A.; Mangano, G.; Mann, R.B. Hilbert space representation of the minimal length uncertainty relation. *Phys. Rev. D* **1995**, *52*, 1108. [CrossRef] [PubMed]
29. Maggiore, M. A generalized uncertainty principle in quantum gravity. *Phys. Lett. B* **1993**, *304*, 65–69. [CrossRef]
30. Maggiore, M. Quantum groups, gravity, and the generalized uncertainty principle. *Phys. Rev. D* **1994**, *49*, 5182. [CrossRef]
31. Chang, L.N.; Minic, D.; Okamura, N.; Takeuchi, T. Exact solution of the harmonic oscillator in arbitrary dimensions with minimal length uncertainty relations. *Phys. Rev. D* **2002**, *65*, 125027. [CrossRef]
32. Chang, L.N.; Minic, D.; Okamura, N.; Takeuchi, T. Effect of the minimal length uncertainty relation on the density of states and the cosmological constant problem. *Phys. Rev. D* **2002**, *65*, 125028. [CrossRef]
33. Moussa, M. Effect of generalized uncertainty principle on main-sequence stars and white dwarfs. *Adv. High Energy Phys.* **2015**, *2015*, 343284. [CrossRef]
34. Bishop, M.; Lee, J.; Singleton, D. Modified commutators are not sufficient to determine a quantum gravity minimal length scale. *Phys. Lett. B* **2020**, *802*, 135209. [CrossRef]

35. Bishop, M.; Contreras, J.; Singleton, D. A subtle aspect of minimal lengths in the generalized uncertainty principle. *Universe* **2022**, *8*, 192. [CrossRef]
36. Segreto, S.; Montani, G. Extended GUP formulation and the role of momentum cut-off. *Eur. Phys. J. C* **2023**, *83*, 385. [CrossRef]
37. Rashidi, R. Generalized uncertainty principle and the maximum mass of ideal white dwarfs. *Ann. Phys.* **2016**, *374*, 434–443. [CrossRef]
38. Belfaqih, I.H.; Maulana, H.; Sulaksono, A. White dwarfs and generalized uncertainty principle. *Int. J. Mod. Phys. D* **2021**, *30*, 2150064. [CrossRef]
39. Mathew, A.; Nandy, M.K. Existence of Chandrasekhar's limit in generalized uncertainty white dwarfs. *R. Soc. Open Sci.* **2021**, *8*, 210301. [CrossRef]
40. Hamil, B.; Lütfüoğlu, B. New higher-order generalized uncertainty principle: Applications. *Int. J. Theor. Phys.* **2021**, *60*, 2790–2803. [CrossRef]
41. Gregoris, D.; Ong, Y.C. On the Chadrasekhar Limit in Generalized Uncertainty Principles. *arXiv* **2022**, arXiv:2202.13904.
42. Alfonso, V.I.; Bejarano, C.; Jimenez, J.B.; Olmo, G.J.; Orazi, E. The trivial role of torsion in projective invariant theories of gravity with non-minimally coupled matter fields. *Class. Quantum Gravity* **2017**, *34*, 235003. [CrossRef]
43. Borowiec, A.; Ferraris, M.; Francaviglia, M.; Volovich, I. Universality of Einstein equations for the Ricci squared Lagrangians. *Class. Quant. Grav.* **1998**, *15*, 43–55. [CrossRef]
44. Allemandi, G.; Borowiec, A.; Francaviglia, M. Accelerated cosmological models in Ricci squared gravity. *Phys. Rev. D* **2004**, *70*, 103503. [CrossRef]
45. Beltrán Jiménez, J.; Delhom, A. Ghosts in metric-affine higher order curvature gravity. *Eur. Phys. J. C* **2019**, *79*, 656. [CrossRef]
46. Jiménez, J.B.; Delhom, A. Instabilities in metric-affine theories of gravity with higher order curvature terms. *Eur. Phys. J. C* **2020**, *80*, 585. [CrossRef]
47. Vollick, D.N. Palatini approach to Born-Infeld–Einstein theory and a geometric description of electrodynamics. *Phys. Rev. D* **2004**, *69*, 064030. [CrossRef]
48. Jiménez, J.B.; Heisenberg, L.; Olmo, G.J.; Rubiera-Garcia, D. Born–Infeld inspired modifications of gravity. *Phys. Rep.* **2018**, *727*, 1–129. [CrossRef]
49. Toniato, J.D.; Rodrigues, D.C.; Wojnar, A. Palatini $f(R)$ gravity in the solar system: Post–Newtonian equations of motion and complete PPN parameters. *Phys. Rev. D* **2020**, *101*, 064050. [CrossRef]
50. Banados, M.; Ferreira, P.G. Eddington's theory of gravity and its progeny. *Phys. Rev. Lett.* **2010**, *105*, 011101. [CrossRef]
51. Pani, P.; Cardoso, V.; Delsate, T. Compact stars in Eddington inspired gravity. *Phys. Rev. Lett.* **2011**, *107*, 031101. [CrossRef]
52. Pani, P.; Sotiriou, T.P. Surface singularities in Eddington-inspired Born-Infeld gravity. *Phys. Rev. Lett.* **2012**, *109*, 251102. [CrossRef]
53. Cortes, J.L.; Gamboa, J. Deformed classical-quantum mechanics transition. *Phys. Rev. D* **2020**, *102*, 036015. [CrossRef]
54. Ali, A.F.; Das, S.; Vagenas, E.C. Discreteness of space from the generalized uncertainty principle. *Phys. Lett. B* **2009**, *678*, 497–499. [CrossRef]
55. Ali, A.F. Minimal length in quantum gravity, equivalence principle and holographic entropy bound. *Class. Quantum Gravity* **2011**, *28*, 065013. [CrossRef]
56. Abac, A.G.; Esguerra, J.P.H.; Otadoy, R.E.S. Modified structure equations and mass–radius relations of white dwarfs arising from the linear generalized uncertainty principle. *Int. J. Mod. Phys. D* **2021**, *30*, 2150005. [CrossRef]
57. Vagenas, E.C.; Farag Ali, A.; Alshal, H. GUP and the no-cloning theorem. *Eur. Phys. J. C* **2019**, *79*, 276. [CrossRef]
58. Tawfik, A.; Diab, A. Generalized uncertainty principle: Approaches and applications. *Int. J. Mod. Phys. D* **2014**, *23*, 1430025. [CrossRef]
59. Vagenas, E.C.; Ali, A.F.; Hemeda, M.; Alshal, H. Linear and quadratic GUP, Liouville theorem, cosmological constant, and Brick Wall entropy. *Eur. Phys. J. C* **2019**, *79*, 398. [CrossRef]
60. Barausse, E.; Sotiriou, T.P.; Miller, J.C. Curvature singularities, tidal forces and the viability of Palatini f (R) gravity. *Class. Quantum Gravity* **2008**, *25*, 105008. [CrossRef]
61. Barausse, E.; Sotiriou, T.P.; Miller, J.C. A no-go theorem for polytropic spheres in Palatini f (R) gravity. *Class. Quantum Gravity* **2008**, *25*, 062001. [CrossRef]
62. Casado-Turrión, A.; de la Cruz-Dombriz, A.; Dobado, A. Is gravitational collapse possible in f(R) gravity? *Phys. Rev. D* **2022**, *105*, 084060. [CrossRef]
63. Olmo, G.J. Reexamination of polytropic spheres in Palatini f (R) gravity. *Phys. Rev. D* **2008**, *78*, 104026. [CrossRef]
64. Olmo, G.J.; Rubiera-Garcia, D. Junction conditions in Palatini f(R) gravity. *Class. Quantum Gravity* **2020**, *37*, 215002. [CrossRef]
65. Auddy, S.; Basu, S.; Valluri, S. Analytic models of brown dwarfs and the substellar mass limit. *Adv. Astron.* **2016**, *2016*, 5743272. [CrossRef]
66. Benito, M.; Wojnar, A. Cooling process of brown dwarfs in Palatini f(R) gravity. *Phys. Rev. D* **2021**, *103*, 064032. [CrossRef]
67. Kozak, A.; Soieva, K.; Wojnar, A. Cooling process of substellar objects in scalar-tensor gravity. *Phys. Rev. D* **2023**, *108*, 024016. [CrossRef]
68. Landau, L. The theory of superfluidity of helium II. In *An Introduction to the Theory of Superfluidity*; CRC Press: Boca Raton, FL, USA, 2018; pp. 185–204.
69. Tisza, L. The theory of liquid helium. *Phys. Rev.* **1947**, *72*, 838. [CrossRef]

70. Yarnell, J.; Arnold, G.; Bendt, P.; Kerr, E. Excitations in liquid helium: Neutron scattering measurements. *Phys. Rev.* **1959**, *113*, 1379. [CrossRef]
71. Cohen, M.; Feynman, R.P. Theory of inelastic scattering of cold neutrons from liquid helium. *Phys. Rev.* **1957**, *107*, 13. [CrossRef]

universe

MDPI

Article

New Parametrization of the Dark-Energy Equation of State with a Single Parameter

Jainendra Kumar Singh [1,*], Preeti Singh [1], Emmanuel N. Saridakis [2,3,4,*], Shynaray Myrzakul [5] and Harshna Balhara [1]

1 Department of Mathematics, Netaji Subhas University of Technology, New Delhi 110078, India; harshna.ma19@nsut.ac.in (H.B.)
2 National Observatory of Athens, Lofos Nymfon, 11852 Athens, Greece
3 CAS Key Laboratory for Research in Galaxies and Cosmology, University of Science and Technology of China, Hefei 230026, China
4 Departamento de Matemáticas, Universidad Católica del Norte, Avda. Angamos 0610, Casilla, Antofagasta 1280, Chile
5 Ratbay Myrzakulov Eurasian International Centre for Theoretical Physics, Nur-Sultan 010009, Kazakhstan
* Correspondence: jksingh@nsut.ac.in (J.K.S.); msaridak@noa.gr (E.N.S.)

Abstract: We propose a novel dark-energy equation-of-state parametrization, with a single parameter η that quantifies the deviation from ΛCDM cosmology. We first confront the scenario with various datasets, from the Hubble function (OHD), Pantheon, baryon acoustic oscillations (BAO), and their joint observations, and we show that η has a preference for a non-zero value, namely, a deviation from ΛCDM cosmology is favored, although the zero value is marginally inside the 1σ confidence level. However, we find that the present Hubble function value acquires a higher value, namely, $H_0 = 66.624^{+0.011}_{-0.013}$ Km s^{-1} Mpc^{-1}, which implies that the H_0 tension can be partially alleviated. Additionally, we perform a cosmographic analysis, showing that the universe transits from deceleration to acceleration in the recent cosmological past; nevertheless, in the future, it will not result in a de Sitter phase since it exhibits a second transition from acceleration to deceleration. Finally, we perform the statefinder analysis. The scenario behaves similarly to the ΛCDM paradigm at high redshifts, while the deviation becomes significant at late and recent times and especially in the future.

Keywords: dark-energy; equation-of-state parameter; cosmological observations

PACS: 98.80.-k; 95.36.+x; 98.80.Es

1. Introduction

The understanding of fundamental physics is challenged by the universe's rapid expansion, which is the most exciting subject in modern cosmology. This phenomenon indicates that, contrary to popular belief, the cosmos is expanding faster with time. It was initially discovered in the late 20th century through observations of distant supernovae. Under the general relativistic framework, the first general class of explanation involves introducing new and exotic sectors in the universe content, under the umbrella term dark energy [1,2]. Extending the fundamental principles of gravity constitutes the second general class, in which case the cause of acceleration is of gravitational origin [3–7].

Nevertheless, at the phenomenological level, both approaches can be quantified through the (effective) dark-energy equation-of-state parameter w_x. Thus, introducing various parametrizations of w_x allows us to describe the universe's evolution and confront with observational datasets to reveal the required dark-energy features to obtain agreement. Several researchers have embraced a phenomenological strategy for modeling the equation of state (EoS) parameter, representing it as a function of redshift, denoted as $\omega = \omega(z)$ [8], or equivalently as a function of the cosmic scale factor $\omega = \omega(a)$ [9,10]. This methodology streamlines the examination of dark energy (DE) dynamics and enables the exploration of physically significant functions. However, during data fitting, it becomes necessary to simplify the parameterization of the equation of state $\omega(z)$ and subsequently restrict the evolution of $\omega(z)$ based on the parameters we introduced in our parametrization. In particular, starting from the simple cosmological constant, a large number of parametrizations have been introduced in the literature, involving one parameter [11,12] or two parameters, such as the Chevallier–Polarski–Linder (CPL) parametrization [13,14], the linear parametrization [15–17], the logarithmic parametrization [18], the Jassal–Bagla–Padmanabhan parametrization (JBP) [19], the Barboza–Alcaniz (BA) parametrization [20], H_0 problem at low reshift [21–50], etc. Additionally, note that one can impose the parametrization at the deceleration parameter level [51–53], at the equation-of-sate (EoS) parameter level [54], or even at the Hubble parameter level [55–58].

In the present manuscript, we propose a novel dark-energy equation-of-state parametrization, with a single parameter η that quantifies the deviation from ΛCDM cosmology. Additionally, under this scenario, dark energy behaves like a cosmological constant at high redshifts, while the deviation becomes significant at low and recent redshifts, especially in the future. Finally, for $\eta = 0$, we recover the ΛCDM cosmology completely. As we will see, apart from being capable of fitting the data, the new parametrization can partially alleviate the H_0 tension too since it leads to a H_0 value in between the Planck one and the one from direct measurements [59,60]. Recently, several authors have conducted so many remarkable works regarding the measurements of H_0 obtained from cosmological probes for different redshifts, such as a bias-free cosmological analysis with quasars alleviating H_0 tension [61]; new statistical insights and cosmological constraints consisting of gamma-ray bursts, quasars, baryonic acoustic oscillations, and supernovae Ia [62]; reduced uncertainties up to 43% on the Hubble constant and the matter density with the SNIa with a new statistical analysis [63]; on the Hubble constant tension in the SNIa Pantheon sample [64]; on the evolution of the Hubble constant with the SNIa Pantheon sample and baryon acoustic oscillations: a feasibility study for GRB-cosmology in 2030 [65]; and $f(R)$ gravity in the Jordan frame as a paradigm for the Hubble tension, in which the authors provides a subsequent interpretation of the results through an effective Hubble constant that evolves with the redshift in a $f(R)$ modified gravity theory in the Jordan frame [66].

The article is organized as follows. In Section 2, we present the novel dark-energy parametrization. Then, in Section 3, we perform a detailed confrontation with observations, namely, with Hubble function (OHD), Pantheon, and baryon acoustic oscillations (BAO) data. In Section 4, we perform a cosmographic analysis, and we apply the statefinder diagnostic. Finally, Section 5 is devoted to the conclusions. Lastly, the details of the various datasets and the corresponding fitting procedure are given in the Appendix A.

2. New Single-Parameter Equation-of-State Parametrization

In this section, we first briefly review the basic equations of any cosmological scenario, and then we introduce the new parametrization for the dark-energy equation of state, with just a single parameter. We consider the usual homogeneous and isotropic Friedmann–Robertson–Walker (FRW) metric

$$\mathrm{d}s^2 = -\mathrm{d}t^2 + a^2(t)\left[\frac{\mathrm{d}r^2}{1-Kr^2} + r^2\left(d\theta^2 + \sin^2\theta d\phi^2\right)\right], \tag{1}$$

with $a(t)$ the scale factor and K the spatial curvature parameter ($K = 0, -1, +1$ for spatially flat, open, and closed universe, respectively). Furthermore, we consider that the universe is filled with baryonic and dark matter, radiation, and effective dark-energy fluid. Hence, the Friedmann equations that determine the background evolution of the universe are

$$H^2 + \frac{K}{a^2} = \frac{8\pi G}{3}\rho_{tot}, \quad (2)$$

$$2\dot{H} + 3H^2 + \frac{K}{a^2} = -8\pi G\, p_{tot}, \quad (3)$$

where G is the gravitational constant, and $H = \dot{a}/a$ is the Hubble function, with dots marking time derivatives. The total energy density and pressure are thus given as $\rho_{tot} = \rho_r + \rho_b + \rho_c + \rho_x$ and $p_{tot} = p_r + p_b + p_c + p_x$, where the subscripts r, b, c, x stand respectively for radiation, baryon, cold dark matter, and dark energy. As usual, and without loss of generality, we focus on the spatially flat case, and therefore in the following we impose $K = 0$. Finally, assuming that the various sectors do not interact mutually, we deduce that they are separately conserved, following the conservation equations

$$\dot{\rho}_i + 3H(1 + w_i)\rho_i = 0, \quad (4)$$

where $i \in \{r, b, c, x\}$. In the above expression, we have introduced the equation-of-state parameter of each fluid as $p_i \equiv w_i\rho_i$, which yields

$$w_i = -\left(1 + \frac{a}{3\rho_i}\frac{d\rho_i}{da}\right). \quad (5)$$

We proceed by providing, for completeness, the evolution equations of the universe at the perturbation level. In the synchronous gauge, the perturbed metric reads as

$$ds^2 = a^2(\tau)\left[-d\tau^2 + (\delta_{ij} + h_{ij})dx^i dx^j\right], \quad (6)$$

with τ as the conformal time, and where δ_{ij} and h_{ij} denote the unperturbed and the perturbed metric parts (with $h = h^j_j$ the trace). Perturbing additionally the universe fluids and transforming to the Fourier space, we finally extract [67–69]:

$$\delta_i' = -(1 + w_i)\left(\theta_i + \frac{h'}{2}\right) - 3\mathcal{H}\left(\frac{\delta p_i}{\delta \rho_i} - w_i\right)\delta_i - 9\mathcal{H}^2\left(\frac{\delta p_i}{\delta \rho_i} - c^2_{a,i}\right)(1 + w_i)\frac{\theta_i}{k^2}, \quad (7)$$

$$\theta_i' = -\mathcal{H}\left(1 - 3\frac{\delta p_i}{\delta \rho_i}\right)\theta_i + \frac{\delta p_i/\delta \rho_i}{1 + w_i}k^2\,\delta_i - k^2\sigma_i, \quad (8)$$

with primes denoting the conformal-time derivative and with $\mathcal{H} = a'/a$ denoting the conformal Hubble function, and where k is the mode wave number. Moreover, $\delta_i = \delta\rho_i/\rho_i$ stands for the over density of the i-th fluid, $\theta_i \equiv ik^j v_j$ marks the divergence of the i-th fluid velocity, and σ_i is the corresponding anisotropic stress. Lastly, $c^2_{a,i} = \dot{p}_i/\dot{\rho}_i$ is the adiabatic sound speed given as $c^2_{a,i} = w_i - \frac{w_i'}{3\mathcal{H}(1+w_i)}$.

Let us now introduce the new dark-energy parametrization. As usual, knowing the equation of state of a fluid allows us to extract its time evolution by solving Equation (4). For radiation, we have $w_r = 1/3$, and thus we obtain $\rho_r = \rho_{r_0}\, a^{-4}$ (setting the scale factor at present to 1), while for the baryonic and dark matter, we have $w_b = w_c = 0$, which leads to $\rho_b = \rho_{b_0}\, a^{-3}$ and $\rho_c = \rho_{c_0}\, a^{-3}$, where ρ_{i_0} stands for the present density value of the i-th fluid. Concerning the equation-of-state parameter of the dark-energy sector, since it is unknown, as we mentioned in the Introduction, one can consider various parametrizations. Focusing on the barotropic fluid sub-class, we consider that it is a function of time only, or equivalently, of the scale factor a, i.e., $w_x(a)$. Hence, the solution of the EoS Equation (5) leads to

$$\rho_x(a) = \rho_{x_0}\, a^{-3}\exp\left[-3\int_1^a \frac{w_x(a')}{a'}\,da'\right]. \quad (9)$$

In this work, we consider a novel dark-energy equation-of-state parameter (5) as follows:

$$w_x(a) = -1 + \frac{a^{-\eta}\, e^{-2\eta a}\, \eta \arctan a^{-\eta}}{3(1+a^{-2\eta})}, \tag{10}$$

where η is the single parameter. Hence, introducing for convenience the redshift z as the independent variable (where $a^{-1} = 1+z$), the above relation becomes

$$w_x(z) = -1 + \frac{(1+z)^{\eta}\, e^{\frac{-2\eta}{(1+z)}}\, \eta \arctan(1+z)^{\eta}}{3(1+(1+z)^{2\eta})}. \tag{11}$$

Relation (11) is the parametrization that we propose, and in the case $\eta = 0$, we recover the ΛCDM concordance model, where $w_x = -1$ and $\rho_x = \rho_{x_0} = const$, but in the general case, the parameter η quantifies the deviation from the ΛCDM scenario. However, note that for general η, for large redshifts, i.e., for $z \to \infty$, we acquire $w_x \to -1$, which implies that the deviation from the ΛCDM scenario disappears in this regime, and thus, the Big Bang nucleosynthesis bounds are immediately satisfied.

Inserting the above parametrization in the first Friedmann Equation (2), we obtain

$$H = H_0\sqrt{\left[(\Omega_{b_0}+\Omega_{c_0})(1+z)^3 + \Omega_{r_0}(1+z)^4 + \Omega_{x_0} e^{\frac{\eta z}{1+z}}\frac{\arctan(1+z)^{\eta}}{\arctan 1}\right]}, \tag{12}$$

where H_0 is the present value of the Hubble parameter, and where we have introduced the present values of the density parameters $\Omega_{i_0} \equiv \frac{8\pi G}{3H^2}\rho_{i_0}$ (hence, the present value of the total matter density parameter is $\Omega_{m_0} \equiv \Omega_{b_0} + \Omega_{c_0}$). This expression allows us to investigate the cosmological evolution in detail, and confront it with observational datasets. Actually, expression (12), which is a simple deviation from ΛCDM cosmology only at small redshifts, while it recovers ΛCDM scenario at high redshifts, was the motivation behind parametrization (10).

Lastly, from parametrization (11) and the corresponding Hubble function (12), we can straightforwardly calculate various quantities. In particular, the deceleration parameter $q = -1 - \dot{H}H^{-2}$ is given by

$$q(z) = -1 + \frac{\left[3(1+z)^4\Omega_{m_0} + \frac{4\eta\Omega_{x_0}(1+z)^2 e^{\frac{\eta z}{1+z}}\arctan(1+z)^{\eta-1}}{\pi(z^2+2z+2)} + \frac{4\eta\Omega_{x_0} e^{\frac{\eta z}{1+z}}\arctan(1+z)^{\eta}}{\pi}\right]}{2(1+z)\left[(1+z)^3\Omega_{m_0} + \frac{4\Omega_{x_0} e^{\frac{\eta z}{1+z}}\arctan(1+z)^{\eta}}{\pi}\right]}, \tag{13}$$

while the higher-order cosmographic parameters [70] read as

$$j = -q + 2q(1+q) + (1+z)\frac{dq}{dz}, \tag{14}$$

$$s = j - 3j(1+q) - (1+z)\frac{dj}{dz}, \tag{15}$$

$$l = s - 4s(1+q) - (1+z)\frac{ds}{dz}, \tag{16}$$

$$m = l - 5l(1+q) - (1+z)\frac{dl}{dz}. \tag{17}$$

Similarly, for the matter and dark-energy density parameters, we obtain

$$\Omega_m(z) = \frac{1}{1 + \frac{4\Omega_{x_0} e^{\frac{\eta z}{1+z}}\arctan(1+z)^{\eta}}{\pi\Omega_{b_0}(1+z)^3}}, \tag{18}$$

and

$$\Omega_x(z) = \frac{1}{1 + \frac{\pi\,\Omega_{b_0} e^{\frac{-\eta z}{1+z}}\left[(1+z)^3 \arctan(1+z)^{-\eta}\right]}{4\,\Omega_{x_0}}}. \tag{19}$$

3. Observational Constraints

In the previous section, we proposed a new parametrization for the dark-energy equation of state, given by (11), which has a single parameter, namely, η. In this section, we perform a detailed confrontation with various datasets [71–75], focusing on the bounds of η. In particular, we will use data from (i) Hubble function observations (OHD) with 77 data points [76], (ii) Pantheon with 1048 data points [77], and (iii) baryon acoustic oscillations (BAO). The details of the datasets and the corresponding methodology are given in the Appendix A. In our analysis, we use the following priors: $H_0 \in [66, 70]$, $\Omega_{m_0} \in [0.1, 0.4]$, $\Omega_{x_0} \in [0.6, 0.8]$ and $\eta \in [0, 1]$.

Let us now present the constraints we obtain after applying the above formalism and datasets in the Friedmann equations at hand, focusing on the new model parameter η. In Figures 1–4, we present the likelihood contours with 1σ and 2σ confidence levels, around the best-fit values. Additionally, in Table 1, we summarize the obtained results. Finally, in Table 2, we summarize other cosmological parameters, such as the density parameters and the deceleration parameter, the equation-of-state parameters, and the transition redshift.

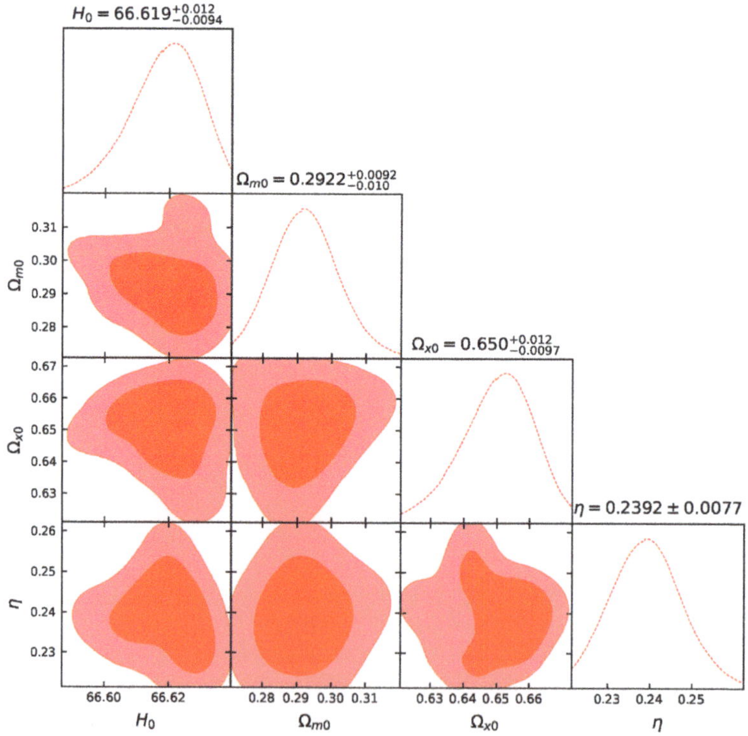

Figure 1. The likelihood contours, with 1σ and 2σ confidence levels, for the $H(z)$ dataset.

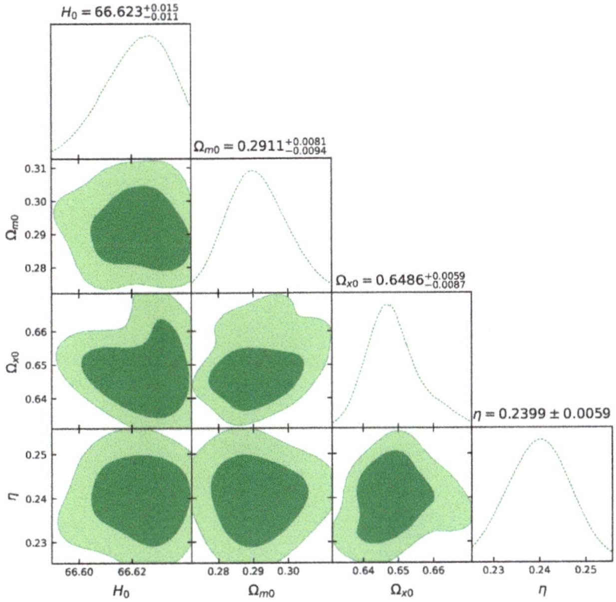

Figure 2. The likelihood contours, with 1σ and 2σ confidence levels, for the *Pantheon* dataset.

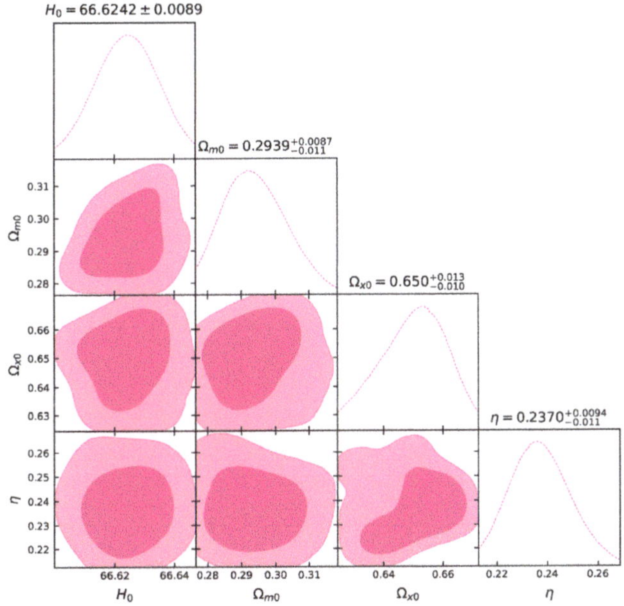

Figure 3. The likelihood contours, with 1σ and 2σ confidence levels, for the *BAO* data.

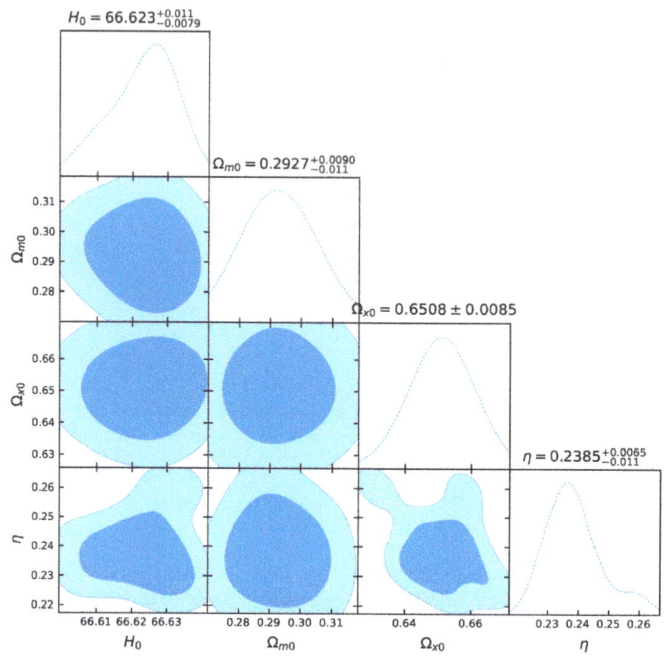

Figure 4. The likelihood contours, with 1σ and 2σ confidence levels, for the *joint* data.

Table 1. Summary of the observational constraints on the model parameters H_0, Ω_{m0}, Ω_{x0} and η from various datasets.

Dataset	H_0 (km/s/Mpc)	Ω_{m0}	Ω_{x0}	η
$H(z)$ (77 points data)	$66.619^{+0.012}_{-0.0094}$	$0.2922^{+0.0092}_{-0.010}$	$0.650^{+0.012}_{-0.0097}$	$0.2392^{+0.0077}_{-0.0077}$
Pantheon	$66.623^{+0.015}_{-0.011}$	$0.2911^{+0.0081}_{-0.0094}$	$0.6486^{+0.0059}_{-0.0087}$	$0.2399^{0.0059}_{0.0059}$
BAO	$66.6242^{+0.0089}_{-0.0089}$	$0.2939^{+0.0087}_{-0.011}$	$0.650^{+0.013}_{-0.010}$	$0.2370^{+0.0094}_{-0.011}$
$H(z)$ + *Pantheon* + *BAO*	$66.623^{+0.011}_{-0.0079}$	$0.2927^{+0.0090}_{-0.011}$	$0.6508^{+0.0085}_{-0.0085}$	$0.2385^{+0.0065}_{-0.011}$

Table 2. Summary of the constraints on the deceleration parameter q, the transition redshift z_{tr} and the current value of the dark-energy equation-of-state parameter w_{x_0}.

Dataset	q_0	z_{tr}	w_{x_0}
$H(z)$ (77 points data)	−0.449196	≃0.7804	−0.704911
Pantheon	−0.449152	≃0.7811	−0.704047
BAO	−0.449141	≃0.7759	−0.707625
$H(z)$ + *Pantheon* + *BAO*	−0.449532	≃0.7804	−0.705775

As we observe, the new model parameter η that quantifies the deviation from ΛCDM cosmology has a preference for a non-zero value, although zero is marginally inside the 1σ confidence level. Concerning H_0, we observe that we obtain a higher value compared to the ΛCDM scenario, although a bit lower than the direct measurements $H_0 = (73.04 \pm 1.04)$ Km s^{-1} Mpc^{-1} at 68% CL, based on the supernovae calibrated by Cepheids [59], which implies that the new dark-energy parametrization at hand can partially alleviate the H_0 tension. This is an additional result of the present work.

Figure 2 is plotted using the Pantheon dataset comprising 1048 data points, and a full covariance matrix is used. In the case of Figure 2, we notice that the shape of the contour is not oval, i.e., the posterior distribution for η behaves as a bimodal distribution, and therefore we need to test the convergence for the Pantheon dataset. The Gelman–Rubin convergence test [78] is one of the statistical tools in Bayesian inference that is widely used to access the convergence of the chains. The test is based on the idea that multiple MCMC chains with different starting points should converge to the same posterior distribution if they have been run for long enough, that is, after several steps. Essentially, this test evaluates, for each parameter of the discussed model, the term called potential scale reduction $\hat{R}$, which is the ratio between the variance W within a chain and the variance $\mathrm{Var}(\theta)$ among the chains:

$$\hat{R} = \sqrt{\frac{\mathrm{Var}(\theta)}{W}}. \tag{20}$$

Also, in the likelihood contour of the Pantheon dataset, there is a possibility of ± 1.2 for the H_0 value and ± 0.29 for the α value within 100 steps of burn-in, which were used to run the MCMC chains.

The maximum Gelman–Rubin diagnostic across the model parameters is labeled as Max Gelman–Rubin $\hat{R}$ in the header and is less than 1.2. Refs. [78,79] suggest that diagnostic $\hat{R}$ values greater than 1.2 for any of the model parameters should indicate non-convergence. The contour plots in the plane $\eta - H_0$ with 1σ and 2σ errors are given in Figure 2, and the corresponding best-fit values of H_0, and η for different datasets are given in Table 1.

4. Cosmographic Analysis and Statefinder Diagnostic

In this section, for completeness, we perform the cosmographic analysis of the cosmological scenario with the new dark-energy parametrization, and we apply the statefinder diagnostic. For simplicity, we neglect the radiation sector.

Let us start with the deceleration parameter given in (13). Using the best-fit values of the model parameters given in Tables 1 and 2, we plot $q(z)$ in Figure 5. As we can see, we obtain the transition from deceleration to acceleration at the transition redshift z_{tr} in agreement with observations. However, it is interesting to note that the novel parametrization at hand will lead to a second transition in the future, at redshift z_{tr_2}, from acceleration to deceleration (at around $z_{tr_2} \approx -0.9$).

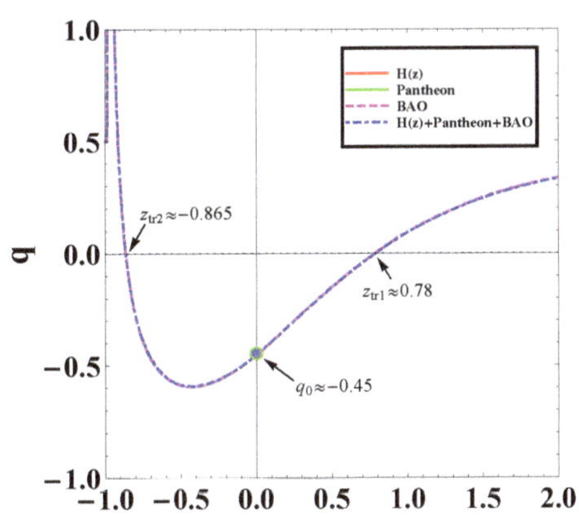

Figure 5. The evolution of the deceleration parameter q in terms of the redshift z, using the best-fit values of the model parameters given in Tables 1 and 2. The red dots mark the present values.

We proceed to the examination of the other cosmographic parameters given in Equations (14)–(17). In particular, we use the best-fit values of the model parameters given in Tables 1 and 2, and in Figure 6, we present their evolution. Additionally, in Table 3 we summarize their values at present. Since in the ΛCDM paradigm, the value of the jerk parameter is equal to unity ($j = 1$), the deviation from $j = 1$ quantifies the deviation of a dark-energy scenario from the concordance model. Again, we find that the new proposed dark-energy parametrization behaves similarly to ΛCDM scenario at high redshifts, while the deviation becomes more significant at late and recent times, and especially in the future. Finally, the same features can be obtained from the evolution of the snap s, lerk l, and m parameters.

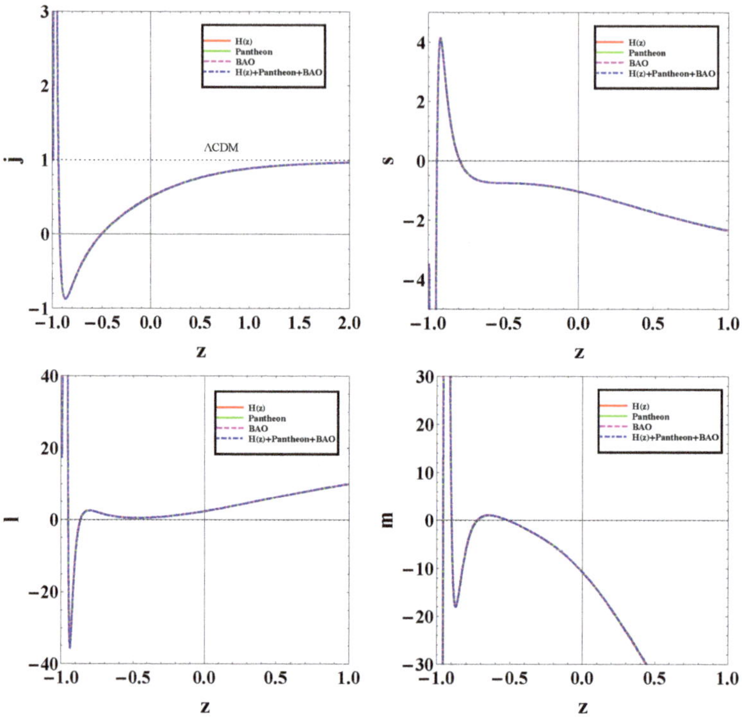

Figure 6. The evolution of the cosmographic parameters j, s, l, m given in (14)–(17), in terms of the redshift z, using the best-fit values of the model parameters given in Tables 1 and 2.

Table 3. Summary of the constraints on the present values of the cosmographic parameters, namely jerk j, snap s, lerk l and m parameters, as well as on the present values of the statefinder diagnostic parameters r and s^*.

Dataset	j	s	l	m	r	s^*
$H(z)$ (77 points data)	0.506092	−1.02864	2.37997	−10.7706	$0.506^{+0.341}_{-0.269}$	$0.174^{+0.124}_{-0.123}$
Pantheon	0.504616	−1.02915	2.37783	−10.7529	$0.505^{+0.238}_{-0.200}$	$0.174^{+0.074}_{-0.076}$
BAO	0.51083	−1.02745	2.38876	−10.8356	$0.511^{+0.227}_{-0.201}$	$0.172^{+0.074}_{-0.073}$
$H(z)$ + *Pantheon* + *BAO*	0.507417	−1.02752	2.37906	−10.7734	$0.508^{+0.225}_{-0.198}$	$0.1729^{+0.073}_{-0.074}$

Let us now come to the statefinder diagnostic, which is based on higher derivatives of the scale factor [80–82]. In particular, one introduces a pair of geometrical parameters $\{r, s^*\}$ in order to examine the dynamics of different dark-energy models [83,84]. The pair of parameters $\{r, s^*\}$ are defined as:

$$r = \frac{\dddot{a}}{aH^3}, \quad s^* = \frac{r-1}{3(q-\frac{1}{2})}, \tag{21}$$

with $q \neq \frac{1}{2}$. For our parametrization (9), the expression for r is found to be

$$r = \frac{r_1 + r_2 + r_3}{(1+z)^2[2+z(2+z)]^2 \tan^{-1}(1+z)^2 \left[\pi(1+z)^3\Omega_{b_0} + 4e^{\frac{\eta z}{1+z}}\Omega_{x_0} \tan^{-1}(1+z)^\eta\right]}, \quad (22)$$

where

$$\begin{aligned} r_1 &= \pi(1+z)^5[2+z(2+z)]^2\Omega_{m_0} \tan^{-1}(1+z)^2 + 2e^{\frac{\eta z}{1+z}}(1+z)^4(\eta-1)\eta\Omega_{x_0} \tan^{-1}(1+z)^\eta, \\ r_2 &= 4e^{\frac{\eta z}{1+z}}\eta\{2\eta - 3 + z[2\eta - 7 + z(\eta - 2z - 6)]\}\Omega_{x_0} \tan^{-1}(1+z)^{1+\eta}, \\ r_3 &= 2e^{\frac{\eta z}{1+z}}[2+z(2+z)]^2[2(1+z)^2 - 4(1+z)\eta + \eta^2]\Omega_{x_0} \tan^{-1}(1+z)^{2+\eta}. \end{aligned} \quad (23)$$

Finally, the expression for s^* is obtained using (13) and (22).

In Figure 7, we present trajectories for different observational datasets in the $q - r$ plane. As we can see, all trajectories start from the decelerating zone, enter into the accelerating zone behaving close to ΛCDM at present, and in the far future, they converge to the CDM model without a cosmological constant (namely, the $SCDM$ model) without resulting to the de Sitter (dS) phase. The present values of parameters $\{r, s^*\}$ of the statefinder diagnostic are also given in Table 3. As we can see, the new parametrization at hand at high redshifts behaves like ΛCDM, while the deviation appears at small redshifts and present time.

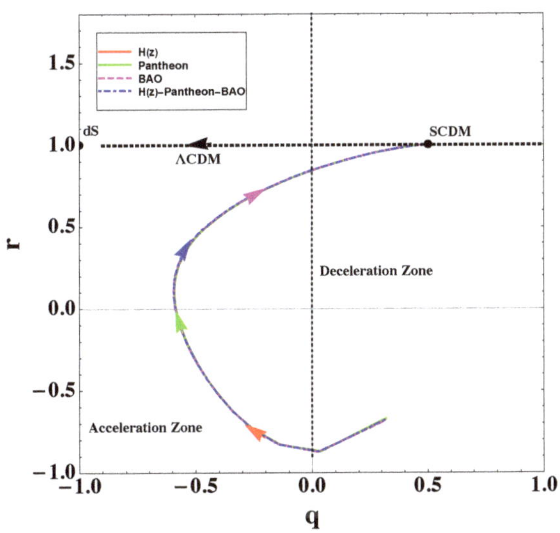

Figure 7. Statefinder diagnostic trajectories in the $q - r$ plane, using the best-fit values of the model parameters given in Tables 1 and 2.

5. Conclusions

In this work, we proposed a novel dark-energy equation-of-state parametrization, with a single parameter η that quantifies the deviation from ΛCDM cosmology. Firstly, we confronted the scenario with various datasets, from the Hubble function (OHD), Pantheon dataset, baryon acoustic oscillations (BAO) observations, and their joint dataset, and we presented the corresponding likelihood contours. As we saw, the new model parameter η has a preference for a non-zero value, namely, a deviation from ΛCDM cosmology is favored, although the zero value is marginally inside the 1σ confidence level. However, interestingly enough, we found that H_0 acquires a higher value compared to the ΛCDM scenario, which implies that the new dark-energy parametrization at hand can partially alleviate the H_0 tension.

Additionally, we performed a cosmographic analysis, examining the cosmographic parameters, namely, deceleration q, jerk j, snap s, lerk l, and m parameters. As we showed, in the scenario at hand, the universe transits from deceleration to acceleration in the recent cosmological past; however, in the future it will not result in a de Sitter phase since a second transition (at $z_{tr_2} \approx -0.9$) will lead from acceleration to deceleration. Additionally, we found that the scenario behaves similarly to the ΛCDM paradigm at high redshifts, while the deviation becomes significant at late and recent times (and thus, the H_0 tension is alleviated), and especially in the future.

Finally, we performed a statefinder diagnostic analysis. As we saw, all trajectories start from the decelerating zone, enter into the accelerating zone behaving close to ΛCDM at the present time, while in the far future, they converge to deceleration without resulting in the de Sitter phase.

In summary, the new parametrization of the dark-energy equation of state with a single parameter is efficient in describing the data, as well as being able to alleviate the H_0 tension. Hence, it would be worthy to proceed to more detailed investigations, such as examining it at the perturbation level, and in particular, relating to the σ_8 tension. Such an analysis will be performed in a separate project.

Author Contributions: Conceptualization, J.K.S.; methodology, J.K.S. and H.B.; software, P.S. and H.B.; validation, P.S., H.B. and S.M.; formal analysis, J.K.S., H.B. and E.N.S.; investigation, J.K.S. and E.N.S.; data curation, J.K.S. and H.B.; writing—original draft preparation, J.K.S. and E.N.S.; writing—review and editing, J.K.S. and E.N.S.; visualization, P.S. and H.B.; supervision, J.K.S. and E.N.S.; project administration, J.K.S. and E.N.S. All authors have read and agreed to the published version of the manuscript.

Funding: This research received no external funding.

Data Availability Statement: All data used in this work were taken from published papers of other groups, publicly available, and were given proper credits.

Acknowledgments: J. K. Singh wishes to thank M. Sami and S. G. Ghosh for the fruitful discussions. The authors also express their thanks to the referees for their valuable comments and suggestions.

Conflicts of Interest: The authors declare no conflicts of interest.

Appendix A. Observational Data

In this appendix, we present the observational datasets we use in our analysis, and we provide the relevant methodology and the corresponding χ^2.

Appendix A.1. H(z) Data

In the case of observational Hubble data (OHD), the corresponding χ^2 of the maximum likelihood analysis is given by

$$\chi^2_{OHD}(\eta, H_0) = \sum_{i=1}^{77} \left[\frac{H_{th}(\Omega_{m0}, \Omega_{x0}, H_0, \eta, z_i) - H_{obs}(z_i)}{\sigma_{H(z_i)}} \right]^2, \tag{A1}$$

where $H(z_i)$ is evaluated at redshift z_i, while H_{th} and H_{obs} represent the theoretical and observed values, and $\sigma_{H(z_i)}$ is the standard deviation. The detailed $H(z)$ data, namely, the 77 points, are given in Table A1 below.

Table A1. The 77 Hubble parameter data from $H(z)$ measurements used in the present analysis in units of $\mathrm{km\,s^{-1}Mpc^{-1}}$. Method (a) corresponds to the cosmic chronometric method, method (b) to the BAO signal in the galaxy distribution, and method (c) to the BAO signal in the $Ly\alpha$ forest distribution alone, or cross-correlated with $QSOs$.

z	$H(z)$ **(km/s/Mpc)**	**Method**	**Reference**
0.00	69.1 ± 1.3	a	[85]
0.07	70.4 ± 20	a	[86]
0.07	69.0 ± 19.6	a	[86]
0.09	70.4 ± 12.2	a	[87]
0.10	70.4 ± 12.2	a	[86]
0.120	68.6 ± 26.2	a	[86]
0.12	70.0 ± 26.7	a	[85]
0.170	83.0 ± 8	a	[87]
0.170	84.7 ± 8.2	a	[87]
0.179	76.5 ± 4	a	[88]
0.1791	75.0 ± 4	a	[88]
0.199	76.5 ± 5.1	a	[88]
0.1993	75.0 ± 5	a	[88]
0.200	72.9 ± 29.6	a	[86]
0.20	74.4 ± 30.2	a	[86]
0.24	81.5 ± 2.7	b	[89]
0.27	78.6 ± 14.3	a	[87]
0.280	88.8 ± 36.3	a	[86]
0.28	90.6 ± 37.3	a	[85]
0.35	84.4 ± 8.6	b	[85]
0.3519	83.0 ± 14	a	[88]
0.352	84.7 ± 14.3	a	[88]
0.38	81.5 ± 1.9	b	[90]
0.3802	83.0 ± 13.5	a	[91]
0.3802	84.7 ± 14.1	a	[91]
0.40	95.0 ± 17	a	[87]
0.40	96.9 ± 17.3	a	[87]
0.4004	77.0 ± 10.2	a	[91]
0.4004	78.6 ± 10.4	a	[91]
0.4247	87.1 ± 11.2	a	[91]
0.4247	88.9 ± 11.4	a	[91]
0.43	88.3 ± 3.8	a	[85]
0.44	84.3 ± 7.9	a	[92]
0.4497	92.8 ± 12.9	a	[91]
0.4497	94.7 ± 13.1	a	[91]
0.470	89.0 ± 34.0	a	[93]
0.47	90.8 ± 50.6	a	[93]
0.4783	80.0 ± 99.0	a	[91]
0.4783	82.5 ± 9.2	a	[91]
0.48	99.0 ± 63.2	a	[93]
0.51	90.8 ± 1.9	b	[90]
0.57	98.8 ± 3.4	b	[86]
0.593	104.0 ± 13.0	a	[88]
0.593	106.1 ± 13.3	a	[88]
0.60	89.7 ± 6.2	a	[86]
0.61	97.8 ± 2.1	b	[90]
0.64	98.82 ± 2.98	b	[94]
0.6797	92.0 ± 8	a	[88]
0.68	93.9 ± 8.1	a	[88]
0.73	99.3 ± 7.1	a	[92]
0.7812	105.0 ± 12	a	[88]
0.781	107.1 ± 12.2	a	[88]
0.875	127.6 ± 17.3	a	[88]
0.8754	125.0 ± 17	a	[88]
0.88	91.8 ± 40.8	a	[95]
0.880	90.0 ± 40	a	[93]
0.90	69.0 ± 12	a	[87]
0.90	119.4 ± 23.4	a	[87]
0.900	117.0 ± 23	a	[87]
1.037	157.2 ± 20.4	a	[88]
1.037	154.0 ± 20	a	[88]
1.30	171.4 ± 17.3	a	[87]
1.300	168.0 ± 17	a	[87]
1.363	160.0 ± 33.6	a	[96]

Table A1. *Cont.*

z	$H(z)$ (km/s/Mpc)	Method	Reference
1.363	163.3 ± 34.3	a	[96]
1.430	177.0 ± 18	a	[87]
1.43	180.6 ± 18.3	a	[87]
1.530	140.0 ± 14	a	[87]
1.53	142.9 ± 14.2	a	[87]
1.750	202.0 ± 40	a	[87]
1.75	206.1 ± 40.8	a	[87]
1.965	186.5 ± 50.4	a	[96]
1.965	190.3 ± 51.4	a	[96]
2.30	228.0 ± 8.1	c	[97]
2.34	226.5 ± 7.1	c	[97]
2.36	230.6 ± 8.2	c	[98]

Appendix A.2. Pantheon Data

We use the latest published dataset for the supernovae type *Ia* Pantheon sample, consisting of 1048 data points. In our analysis, we utilize these data points, which have been confirmed spectroscopically by SNe*Ia* and cover the redshift range of $0.01 < z < 2.26$. The $\chi^2_{Pantheon}$ function for the Pantheon dataset is taken as:

$$\chi^2_{Pan} = \sum_{i=1}^{1048} \left[\frac{\mu_{th}(\Omega_{m0}, \Omega_{x0}, H_0, \eta, z_i) - \mu_{obs}(z_i)}{\sigma_{\mu(z_i)}} \right]^2, \tag{A2}$$

where μ_{th} and μ_{obs} are the theoretical and observed distance modulus, and $\sigma_{\mu(z_i)}$ the standard deviation. The distance modulus $\mu(z)$ is defined as

$$\mu(z) = m - M = 5LogD_l(z) + \mu_0, \tag{A3}$$

where m and M denote the apparent and absolute magnitudes. Additionally, the luminosity distance $D_l(z)$ for the flat universe and the nuisance parameter μ_0 are given by

$$D_l(z) = (1+z)H_0 \int_0^z \frac{1}{H(z^*)} dz^*, \tag{A4}$$

and

$$\mu_0 = 5Log\left(\frac{H_0^{-1}}{1Mpc}\right) + 25, \tag{A5}$$

respectively.

Appendix A.3. Baryon Acoustic Oscillations (BAO)

Concerning baryon acoustic oscillations (BAO), we use the data from the Sloan Digital Sky Survey ($SDSS$) [99], $6dF$ Galaxy survey ($6dFGS$) [100], $BOSS\ CMASS$ [101] and three parallel measurements from the $WiggleZ$ survey [92]. In the BAO observations, the distance redshift ratio d_z is

$$d_z = \frac{r_s(z^*)}{D_v(z)}, \tag{A6}$$

where $z^* = 1090$ is the redshift at the time of photon decoupling [102], and $r_s(z^*)$ is the corresponding comoving sound horizon [103]. The dilation scale defined by Eisenstein et al. [104] is

$$D_v(z) = \left[(1+z)^2 \frac{d_A^2(z) z}{H(z)} \right]^{\frac{1}{3}}, \tag{A7}$$

where $d_A(z)$ is the angular diameter distance, which essentially is a geometric mean of two transverse and one radial direction. The value of χ^2_{BAO} is given by [105]

$$\chi^2_{BAO} = A^T C^{-1} A, \tag{A8}$$

where

$$A = \begin{bmatrix} \frac{d_A(z^*)}{D_v(0.106)} - 30.95 \\ \frac{d_A(z^*)}{D_v(0.2)} - 17.55 \\ \frac{d_A(z^*)}{D_v(0.35)} - 10.11 \\ \frac{d_A(z^*)}{D_v(0.44)} - 8.44 \\ \frac{d_A(z^*)}{D_v(0.6)} - 6.69 \\ \frac{d_A(z^*)}{D_v(0.73)} - 5.45 \end{bmatrix},$$

and the inverse covariance matrix C^{-1} is

$$C^{-1} = \begin{bmatrix} 0.48435 & -0.101383 & -0.164945 & -0.0305703 & -0.097874 & -0.106738 \\ -0.101383 & 3.2882 & -2.45497 & -0.0787898 & -0.252254 & -0.2751 \\ -0.164945 & -2.454987 & 9.55916 & -0.128187 & -0.410404 & -0.447574 \\ -0.0305703 & -0.0787898 & -0.128187 & 2.78728 & -2.75632 & 1.16437 \\ -0.097874 & -0.252254 & -0.410404 & -2.75632 & 14.9245 & -7.32441 \\ -0.106738 & -0.2751 & -0.447574 & 1.16437 & -7.32441 & 14.5022 \end{bmatrix},$$

approaching the correlation coefficients available in [105,106].

Appendix A.4. Joint Analysis

In the case where some of the above datasets are used simultaneously, the corresponding χ^2 arises from the sum of the separate ones. In particular, we will use the following combinations:

$$\chi^2_{HPB} = \chi^2_{OHD} + \chi^2_{Pan} + \chi^2_{BAO}, \tag{A9}$$

References

1. Copeland, E.J.; Sami, M.; Tsujikawa, S. Dynamics of dark energy. *Int. J. Mod. Phys. D* **2006**, *15*, 1753–1935. [CrossRef]
2. Cai, Y.F.; Saridakis, E.N.; Setare, M.R.; Xia, J.Q. Quintom Cosmology: Theoretical implications and observations. *Phys. Rep.* **2010**, *493*, 1–60. [CrossRef]
3. Felice, A.D.; Tsujikawa, S. $f(R)$ theories. *Living Rev. Relativ.* **2010**, *13*, 3. [CrossRef]
4. Capozziello, S.; Laurentis, M.D. Extended Theories of Gravity. *Phys. Rep.* **2011**, *509*, 167–321. [CrossRef]
5. Cai, Y.F.; Capozziello, S.; Laurentis, M.D.; Saridakis, E.N. $f(T)$ teleparallel gravity and cosmology. *Rep. Prog. Phys.* **2016**, *79*, 106901. [CrossRef]
6. Nojiri, S.; Odintsov, S.D.; Oikonomou, V.K. Modified Gravity Theories on a Nutshell: Inflation, Bounce and Late-Time Evolution. *Phys. Rep.* **2017**, *692*, 1–104. [CrossRef]
7. Saridakis, E.N.; Lazkoz, R.; Salzano, V.; Moniz, P.V.; Capozziello, S.; Jiménez, J.B. *Modified Gravity and Cosmology: An Update by the CANTATA Network*; Springer: Cham, Switzerland, 2021.
8. Castillo-Santos, M.N.; Hernández-Almada, A.; García-Aspeitia, M.A.; Magaña, J. An exponential equation of state of dark energy in the light of 2018 CMB Planck data. *Phys. Dark Univ.* **2023**, *40*, 101225. [CrossRef]
9. Liu, D.J.; Li, X.Z.; Hao, J.; Jin, X.H. Revisiting the parametrization of Equation of State of Dark Energy via SNIa Data. *Mon. Not. R. Astron. Soc.* **2008**, *388*, 275–281. [CrossRef]
10. Perkovic, D.; Stefancic, H. Barotropic fluid compatible parametrizations of dark energy. *Eur. Phys. J. C* **2020**, *80*, 629. [CrossRef]
11. Gong, Y.G.; Zhang, Y.Z. Probing the curvature and dark energy. *Phys. Rev. D* **2005**, *72*, 043518. [CrossRef]
12. Yang, W.; Pan, S.; Valentino, E.D.; Saridakis, E.N.; Chakraborty, S. Observational constraints on one-parameter dynamical dark-energy parametrizations and the H_0 tension. *Phys. Rev. D* **2019**, *99*, 043543. [CrossRef]
13. Chevallier, M.; Polarski, D. Accelerating universes with scaling dark matter. *Int. J. Mod. Phys. D* **2001**, *10*, 213–223. [CrossRef]
14. Linder, E.V. Exploring the expansion history of the universe. *Phys. Rev. Lett.* **2003**, *90*, 091301. [CrossRef]
15. Cooray, A.R.; Huterer, D. Gravitational lensing as a probe of quintessence. *Astrophys. J.* **1999**, *513*, L95. [CrossRef]
16. Astier, P. Can luminosity distance measurements probe the equation of state of dark energy. *Phys. Lett. B* **2001**, *500*, 8–15. [CrossRef]

17. Weller, J.; Albrecht, A. Future supernovae observations as a probe of dark energy. *Phys. Rev. D* **2002**, *65*, 103512. [CrossRef]
18. Efstathiou, G. Constraining the equation of state of the universe from distant type Ia supernovae and cosmic microwave background anisotropies. *Mon. Not. R. Astron. Soc.* **1999**, *310*, 842–850. [CrossRef]
19. Jassal, H.K.; Bagla, J.S.; Padmanabhan, T. Observational constraints on low redshift evolution of dark energy: How consistent are different observations? *Phys. Rev. D* **2005**, *72*, 103503. [CrossRef]
20. Barboza, E.M., Jr.; Alcaniz, J.S. A parametric model for dark energy. *Phys. Lett. B* **2008**, *666*, 415–419. [CrossRef]
21. Banerjee, A.; Cai, H.; Heisenberg, L.; Colgáin, E.; Sheikh-Jabbari, M.M.; Yang, T. Hubble sinks in the low-redshift swampland. *Phys. Rev. D* **2021**, *103*, L081305. [CrossRef]
22. Lee, B.H.; Lee, W.; Colgáin, E.; Sheikh-Jabbari, M.M.; Thakur, S. Is local H_0 at odds with dark energy EFT? *J. Cosmol. Astropart. Phys.* **2022**, *2022*, 004.
23. Ma, J.Z.; Zhang, X. Probing the dynamics of dark energy with novel parametrizations. *Phys. Lett. B* **2011**, *699*, 233–238. [CrossRef]
24. Nesseris, S.; Perivolaropoulos, L. A Comparison of cosmological models using recent supernova data. *Phys. Rev. D* **2004**, *70*, 043531. [CrossRef]
25. Linder, E.V.; Huterer, D. How many dark energy parameters? *Phys. Rev. D* **2005**, *72*, 043509. [CrossRef]
26. Feng, B.; Li, M.; Piao, Y.S.; Zhang, X. Oscillating quintom and the recurrent universe. *Phys. Lett. B* **2006**, *634*, 101–105. [CrossRef]
27. Zhao, G.B.; Xia, J.Q.; Li, H.; Tao, C.; Virey, J.M.; Zhu, Z.H.; Zhang, X. Probing for dynamics of dark energy and curvature of universe with latest cosmological observations. *Phys. Lett. B* **2007**, *648*, 8–13. [CrossRef]
28. Nojiri, S.; Odintsov, S.D. The Oscillating dark energy: Future singularity and coincidence problem. *Phys. Lett. B* **2006**, *637*, 139–148. [CrossRef]
29. Saridakis, E.N. Theoretical Limits on the Equation-of-State Parameter of Phantom Cosmology. *Phys. Lett. B* **2009**, *676*, 7–11. [CrossRef]
30. Dutta, S.; Saridakis, E.N.; Scherrer, R.J. Dark energy from a quintessence (phantom) field rolling near potential minimum (maximum). *Phys. Rev. D* **2009**, *79*, 103005. [CrossRef]
31. Lazkoz, R.; Salzano, V.; Sendra, I. Oscillations in the dark energy EoS: New MCMC lessons. *Phys. Lett. B* **2011**, *694*, 198. [CrossRef]
32. Feng, L.; Lu, T. A new equation of state for dark energy model. *J. Cosmol. Astropart. Phys.* **2011**, *2011*, 034. [CrossRef]
33. Saridakis, E.N. Phantom evolution in power-law potentials. *Nucl. Phys. B* **2009**, *819*, 116–126. [CrossRef]
34. Felice, A.D.; Nesseris, S.; Tsujikawa, S. Observational constraints on dark energy with a fast varying equation of state. *J. Cosmol. Astropart. Phys.* **2012**, *2012*, 029. [CrossRef]
35. Saridakis, E.N. Quintom evolution in power-law potentials. *Nucl. Phys. B* **2010**, *830*, 374–389. [CrossRef]
36. Feng, C.J.; Shen, X.Y.; Li, P.; Li, X.Z. A New Class of Parametrization for Dark Energy without Divergence. *J. Cosmol. Astropart. Phys.* **2012**, *2012*, 023. [CrossRef]
37. Basilakos, S.; Solá, J. Effective equation of state for running vacuum: 'mirage' quintessence and phantom dark energy. *Mon. Not. R. Astron. Soc.* **2014**, *437*, 3331–3342. [CrossRef]
38. Pantazis, G.; Nesseris, S.; Perivolaropoulos, L. Comparison of thawing and freezing dark energy parametrizations. *Phys. Rev. D* **2016**, *93*, 103503. [CrossRef]
39. Valentino, E.D.; Melchiorri, A.; Silk, J. Reconciling Planck with the local value of H_0 in extended parameter space. *Phys. Lett. B* **2016**, *761*, 242–246. [CrossRef]
40. Chávez, R.; Plionis, M.; Basilakos, S.; Terlevich, R.; Terlevich, E.; Melnick, J.; Bresolin, F.; González-Morán, A.L. Constraining the dark energy equation of state with H_{II} galaxies. *Mon. Not. R. Astron. Soc.* **2016**, *462*, 2431–2439. [CrossRef]
41. Zhao, G.B.; Raveri, M.; Pogosian, L.; Wang, Y.; Crittenden, R.G.; Handley, W.J.; Percival, W.J.; Beutler, F.; Brinkmann, J.; Zhang, H.; et al. Dynamical dark energy in light of the latest observations. *Nat. Astron.* **2017**, *1*, 627–632. [CrossRef]
42. Yang, W.; Nunes, R.C.; Pan, S.; Mota, D.F. Effects of neutrino mass hierarchies on dynamical dark energy models. *Phys. Rev. D* **2017**, *95*, 103522. [CrossRef]
43. Valentino, E.D.; Melchiorri, A.; Linder, E.V.; Silk, J. Constraining Dark Energy Dynamics in Extended Parameter Space. *Phys. Rev. D* **2017**, *96*, 023523. [CrossRef]
44. Di Valentino, E. Crack in the cosmological paradigm. *Nat. Astron.* **2017**, *1*, 569–570. [CrossRef]
45. Yang, W.; Pan, S.; Paliathanasis, A. Latest astronomical constraints on some nonlinear parametric dark energy models. *Mon. Not. R. Astron. Soc.* **2018**, *475*, 2605–2613. [CrossRef]
46. Pan, S.; Saridakis, E.N.; Yang, W. Observational Constraints on Oscillating Dark-Energy Parametrizations. *Phys. Rev. D* **2018**, *98*, 063510. [CrossRef]
47. Pan, S.; Yang, W.; Singha, C.; Saridakis, E.N. Observational constraints on sign-changeable interaction models and alleviation of the H_0 tension. *Phys. Rev. D* **2019**, *100*, 083539. [CrossRef]
48. Pan, S.; Yang, W.; Valentino, E.D.; Saridakis, E.N.; Chakraborty, S. Interacting scenarios with dynamical dark energy: Observational constraints and alleviation of the H_0 tension. *Phys. Rev. D* **2019**, *100*, 103520. [CrossRef]
49. Singh, J.K.; Balhara, H.; Bamba, K.; Jena, J. Bouncing cosmology in modified gravity with higher-order curvature terms. *J. High Energy Phys.***2023**, *3*, 191. [CrossRef]
50. Singh, J.K.; Shaily; Ram, S.; Santos, J.R.L.; Fortunato, J.A.S. The constrained cosmological model in Lyra geometry. *Int. J. Mod. Phys. D* **2023**, *32*, 2350040. [CrossRef]

51. Cunha, J.V.; Lima, J.A.S. Transition Redshift: New Kinematic Constraints from Supernovae. *Mon. Not. R. Astron. Soc.* **2008**, *390*, 210–217. [CrossRef]
52. Akarsu, O.; Dereli, T.; Kumar, S.; Xu, L. Probing kinematics and fate of the Universe with linearly time-varying deceleration parameter. *Eur. Phys. J. Plus* **2014**, *129*, 22. [CrossRef]
53. Xu, L.; Lu, J. Cosmic constraints on deceleration parameter with Sne Ia and CMB. *Mod. Phys. Lett. A* **2009**, *24*, 369–376. [CrossRef]
54. Colgáin, E.Ó.; Sheikh-Jabbari, M.M.; Yin, L. Can dark energy be dynamical? *Phys. Rev. D* **2021**, *104*, 023510. [CrossRef]
55. Singh, J.K.; Bamba, K.; Nagpal, R.; Pacif, S.K.J. Bouncing cosmology in $f(R,T)$ gravity. *Phys. Rev. D* **2018**, *97*, 123536. [CrossRef]
56. Singh, J.K.; Nagpal, R. FLRW cosmology with EDSFD parametrization. *Eur. Phys. J. C* **2020**, *80*, 295. [CrossRef]
57. Nojiri, S.; Odintsov, S.D.; Oikonomou, V.K. Singular inflation from generalized equation of state fluids. *Phys. Lett. B* **2015**, *747*, 310–320. [CrossRef]
58. Nagpal, R.; Singh, J.K.; Beesham, A.; Shabani, H. Cosmological aspects of a hyperbolic solution in $f(R,T)$ gravity. *Ann. Phys.* **2019**, *405*, 234–255. [CrossRef]
59. Abdalla, E.; Abellán, G.F.; Aboubrahim, A.; Agnello, A.; Akarsu, Ö.; Akrami, Y.; Alestas, G.; Aloni, D.; Amendola, L.; Pettorino, V.; et al. Cosmology intertwined: A review of the particle physics, astrophysics, and cosmology associated with the cosmological tensions and anomalies. *J. High Energy Astrophys.* **2022**, *34*, 49–211.
60. Valentino, E.D.; Saridakis, E.; Riess, A. Cosmological tensions in the birthplace of the heliocentric model. *Nat. Astron.* **2022**, *6*, 1353–1355. [CrossRef]
61. Lenart, A.Ł; Bargiacchi, G.; Dainotti, M.G.; Nagataki, S.; Capozziello, S. A Bias-free Cosmological Analysis with Quasars Alleviating H_0 Tension. *Astrophys. J. Suppl. Ser.* **2023**, *264*, 46.
62. Dainotti, M.G.; Bargiacchi, G.; Bogdan, M.; Capozziello, S.; Nagataki, S. Reduced uncertainties up to 43% on the Hubble constant and the matter density with the SNe Ia with a new statistical analysis. *arXiv* **2023**, arXiv:2303.06974.
63. Dainotti, M.G.; Simone, B.D.; Schiavone, T.; Montani, G.; Rinaldi, E.; Lambiase, G. On the Hubble constant tension in the SNe Ia Pantheon sample. *Astrophys. J.* **2021**, *912*, 150. [CrossRef]
64. Schiavone, T.; Montani, G.; Dainotti, M.G.; Simone, B.D.; Rinaldi, E.; Lambiase, G. Running Hubble constant from the SNe Ia Pantheon sample? *arXiv* **2022**, arXiv:2205.07033.
65. Dainotti, M.G.; Simone, B.D.; Schiavone, T.; Montani, G.; Rinaldi, E.; Lambiase, G.; Bogdan, M.; Ugale, S. On the Evolution of the Hubble Constant with the SNe Ia Pantheon Sample and Baryon Acoustic Oscillations: A Feasibility Study for GRB-Cosmology in 2030. *Galaxies* **2022**, *10*, 24. [CrossRef]
66. Schiavone, T.; Montani, G.; Bombacigno, F. $f(R)$ gravity in the Jordan frame as a paradigm for the Hubble tension. *Mon. Not. R. Astron. Soc.* **2023**, *522*, L72–L77. [CrossRef]
67. Mukhanov, V.F.; Feldman, H.A.; Brandenberger, R.H. Theory of cosmological perturbations. *Phys. Rep.* **1992**, *215*, 203–333. [CrossRef]
68. Ma, C.P.; Bertschinger, E. Cosmological perturbation theory in the synchronous and conformal Newtonian gauges. *Astrophys. J.* **1995**, *455*, 7. [CrossRef]
69. Malik, K.A.; Wands, D. Cosmological perturbations. *Phys. Rep.* **2009**, **475**, 1–51. [CrossRef]
70. Bolotin, Y.L.; Cherkaskiy, V.A.; Ivashtenko, O.Y.; Konchatnyi, M.I.; Zazunov, L.G. Applied Cosmography: A Pedagogical Review. *arXiv* **2018**, arXiv:1812.02394.
71. Shaily; Singh, A.; Singh, J.K.; Ray, S. Late time phantom characteristic of the model in $f(R,T)$ gravity with quadratic curvature term. *arXiv* **2024**, arXiv:2402.01780.
72. Singh, J.K.; Shaily; Balhara, H.; Bamba, K.; Jena, J. Cosmic analysis of a model in higher-order gravity theory. *Astron. Comput.* **2024**, *46*, 100790. [CrossRef]
73. Singh, J.K.; Balhara, H.; Shaily; Singh, P. The constrained accelerating universe in $f(R,T)$ gravity. *Astron. Comput.* **2024**, *46*, 100795.
74. Balhara, H.; Singh, J.K.; Saridakis, E.N. Observational constraints and cosmographic analysis of $f(T,T_G)$ gravity and cosmology. *arXiv* **2024**, arXiv:2312.17277.
75. Riess, A.G.; Kirshner, R.P.; Schmidt, B.P.; Jha, S.; Challis, P.; Garnavich, P.M.; Esin, A.A.; Carpenter, C.; Grashius, R.; Schild, R.E.; et al. BV RI light curves for 22 type Ia supernovae. *Astron. J.* **1999**, *117*, 707–724. [CrossRef]
76. Shaily; Zeyauddin, M.; Singh, J.K.; Santos, J.R.L. Dynamics of a hyperbolic solution in Scale-covariant theory. *arXiv* **2022**, arXiv:2207.05076.
77. Scolnic, D.M.; Jones, D.O.; Rest, A.; Pan, Y.C.; Chornock, R.; Foley, R.J.; Huber, M.E.; Kessler, R.; Narayan, G.; Riess, A.G.; et al. [Pan-STARRS1]. The Complete Light-curve Sample of Spectroscopically Confirmed SNe Ia from Pan-STARRS1 and Cosmological Constraints from the Combined Pantheon Sample. *Astrophys. J.* **2018**, *859*, 101. [CrossRef]
78. Gelman, A.; Rubin, D.B. Inference from Iterative Simulation Using Multiple Sequences. *Stat. Sci.* **1992**, *7*, 457–472. [CrossRef]
79. Brooks, P.S.; Gelman, A. General methods for monitoring convergence of iterative simulations. *J. Comput. Graph. Stat.* **1998**, *7*, 434–455. [CrossRef]
80. Sami, M.; Shahalam, M.; Skugoreva, M.; Toporensky, A. Cosmological dynamics of a nonminimally coupled scalar field system and its late time cosmic relevance. *Phys. Rev. D* **2012**, *86* 103532. [CrossRef]

81. Myrzakulov, R.; Shahalam, M. Statefinder hierarchy of bimetric and galileon models for concordance cosmology. *J. Cosmol. Astropart. Phys.* **2013**, *2013*, 047. [CrossRef]
82. Rani, S.; Altaibayeva, A.; Shahalam, M.; Singh, J.K.; Myrzakulov, R. Constraints on cosmological parameters in power-law cosmology. *J. Cosmol. Astropart. Phys.* **2015**, *2015*, 031. [CrossRef]
83. Sahni, V.; Saini, T.D.; Starobinsky, A.A.; Alam, U. JStatefinder—A new geometrical diagnostic of dark energy. *J. Exp. Theor. Phys. Lett.* **2003**, *77*, 201–206. [CrossRef]
84. Alam, U.; Sahni, V.; Saini, T.D.; Starobinsky, A.A. Exploring the expanding universe and dark energy using the statefinder diagnostic. *Mon. Not. R. Astron. Soc.* **2003**, *344*, 1057–1074. [CrossRef]
85. Farooq, O.; Madiyar, F.R.; Crandall, S.; Ratra, B. Hubble parameter measurement constraints on the redshift of the deceleration–acceleration transition, dynamical dark energy, and space curvature. *Astrophys. J.* **2017**, *835*, 26. [CrossRef]
86. Zhang, M.-J.; Xia, J.-Q. Test of the cosmic evolution using Gaussian processes. *J. Cosmol. Astropart. Phys.* **2016**, *2016*, 005. [CrossRef]
87. Simon, J.; Verde, L.; Jimenez, R. Constraints on the redshift dependence of the dark energy potential. *Phys. Rev. D* **2005**, *71*, 123001. [CrossRef]
88. Moresco, M.; Cimatti, A.; Jimenez, R.; Pozzetti, L.; Zamorani, G.; Bolzonella, M.; Dunlop, J.; Lamareille, F.; Mignoli, M.; Welikala, N.; et al. Improved constraints on the expansion rate of the Universe up to z~1.1 from the spectroscopic evolution of cosmic chronometers. *J. Cosmol. Astropart. Phys.* **2012**, *2012*, 006. [CrossRef]
89. Gaztanaga, E.; Cabre, A.; Hui, L. Clustering of luminous red galaxies—IV. Baryon acoustic peak in the line-of-sight direction and a direct measurement of $H(z)$. *Mon. Not. R. Astron. Soc.* **2009**, *399*, 1663–1680. [CrossRef]
90. Alam, S.; Ata, M.; Bailey, S.; Beutler, F.; Bizyaev, D.; Blazek, J.A.; Bolton, A.S.; Brownstein, J.R.; Burden, A.; Zhao, G.B.; et al. The clustering of galaxies in the completed SDSS-III Baryon Oscillation Spectroscopic Survey: Cosmological analysis of the DR12 galaxy sample. *Mon. Not. R. Astron. Soc.* **2017**, *470*, 2617–2652. [CrossRef]
91. Moresco, M.; Pozzetti, L.; Cimatti, A.; Jimenez, R.; Maraston, C.; Verde, L.; Thomas, D.; Citro, A.; Tojeiro, R.; Wilkinson, D. A 6% measurement of the Hubble parameter at z~0.45: Direct evidence of the epoch of cosmic re-acceleration. *J. Cosmol. Astropart. Phys.* 2016, *2016*, 014. [CrossRef]
92. Blake, C.; Brough, S.; Colless, M.; Contreras, C.; Couch, W.; Croom, S.; Croton, D.; Davis, T.M.; Drinkwater, M.J.; Yee, H.K.; et al. Energy Survey: Joint measurements of the expansion and growth history at z < 1. *Mon. Not. R. Astron. Soc.* **2012**, *425*, 405–414.
93. Ratsimbazafy, A.L.; Loubser, S.I.; Crawford, S.M.; Cress, C.M.; Bassett, B.A.; Nichol, R.C.; Väisänen, P. Age-dating luminous red galaxies observed with the Southern African Large Telescope. *Mon. Not. R. Astron. Soc.* **2017**, *467*, 3239–3254. [CrossRef]
94. Wang, Y.; Zhao, G.B.; Chuang, C.H.; Ross, A.J.; Percival, W.J.; Gil-Marín, H.; Cuesta, A.J.; Kitaura, F.S.; Rodriguez-Torres, S.; Zhu, F.; et al. The clustering of galaxies in the completed SDSS-III Baryon Oscillation Spectroscopic Survey: Tomographic BAO analysis of DR12 combined sample in configuration space. *Mon. Not. R. Astron. Soc.* **2017**, *469*, 3762–3774. [CrossRef]
95. Stern, D.; Jimenez, R.; Verde, L.; Kamionkowski, M.; Stanford, S.A. Cosmic chronometers: Constraining the equation of state of dark energy. I: $H(z)$ measurements. *J. Cosmol. Astropart. Phys.* **2010**, *2010*, 008. [CrossRef]
96. Moresco, M. Raising the bar: New constraints on the Hubble parameter with cosmic chronometers at z~2. *Mon. Not. R. Astron. Soc. Lett.* **2015**, *450*, L16–L20. [CrossRef]
97. Delubac, T.; Bautista, J.E.; Rich, J.; Kirkby, D.; Bailey, S.; Font-Ribera, A.; Slosar, A.; Lee, K.; Pieri, M.M.; York, D.G.; et al. Baryon acoustic oscillations in the Lyα forest of BOSS DR11 quasars. *Astron. Astrophys.* **2015**, *574*, A59. [CrossRef]
98. Font-Ribera, A.; Kirkby, D.; Miralda-Escudé, J.; Ross, N.P.; Slosar, A.; Rich, J.; Aubourg, É.; Bailey, S.; Bhardwaj, V.; York, D.G.; et al. Quasar-Lyman α forest cross-correlation from BOSS DR11: Baryon Acoustic Oscillations. *J. Cosmol. Astropart. Phys.* **2014**, *2014*, 027. [CrossRef]
99. Padmanabhan, N.; Xu, X.; Eisenstein, D.J.; Scalzo, R.; Cuesta, A.J.; Mehta, K.T.; Kazin, E. A 2 per cent distance to z = 0.35 by reconstructing baryon acoustic oscillations—I. Methods and application to the Sloan Digital Sky Survey. *Mon. Not. R. Astron. Soc.* **2012**, *427*, 2132–2145. [CrossRef]
100. Beutler, F.; Blake, C.; Colless, M.; Jones, D.H.; Staveley-Smith, L.; Campbell, L.; Parker, Q.; Saunders, W.; Watson, F. The 6dF Galaxy Survey: Baryon acoustic oscillations and the local Hubble constant. *Mon. Not. R. Astron. Soc.* **2011**, *416*, 3017–3032. [CrossRef]
101. Anderson, L.; Aubourg, E.; Bailey, S.; Beutler, F.; Bhardwaj, V.; Blanton, M.; Zhao, G.-B.; BOSS Collaboration. The clustering of galaxies in the SDSS-III Baryon Oscillation Spectroscopic Survey: Baryon acoustic oscillations in the Data Releases 10 and 11 Galaxy samples. *Mon. Not. R. Astron. Soc.* **2014**, *441*, 24–62. [CrossRef]
102. Ade, P.A.R.; Aghanim, N.; Arnaud, M.; Ashdown, M.; Aumont, J.; Zonca, A. Planck 2015 results. XIII. Cosmological parameters. *Astron. Astrophys.* **2016**, *594*, A13.
103. dos Santos, M.V.; Reis, R.R.; Waga, I. Constraining the cosmic deceleration-acceleration transition with type Ia supernova, BAO/CMB and $H(z)$ data. *J. Cosmol. Astropart. Phys.* **2016**, *2016*, 066. [CrossRef]
104. Eisenstein, D.J.; Zehavi, I.; Hogg, D.W.; Scoccimarro, R.; Blanton, M.R.; Nichol, R.C.; Scranton, R.; Seo, H.; Tegmark, M.; York, D.G. Detection of the baryon acoustic peak in the large-scale correlation function of SDSS luminous red galaxies. *Astrophys. J.* **2005**, *633*, 560. [CrossRef]

105. Giostri, R.; dos Santos, M.V.; Waga, I.; Reis, R.R.R.; Calvao, M.O.; Lago, B.L. From cosmic deceleration to acceleration: New constraints from SN Ia and BAO/CMB. *J. Cosmol. Astropart. Phys.* **2012**, *2012*, 027. [CrossRef]
106. Hinshaw, G.; Larson, D.; Komatsu, E.; Spergel, D.N.; Bennett, C.; Dunkley, J.; Nolta, M.R.; Halpern, M.; Hill, R.S.; Wright, E.L.; et al. Nine-year Wilkinson Microwave Anisotropy Probe (WMAP) observations: Cosmological parameter results. *Astrophys. J. Suppl. Ser.* **2013**, *208*, 19. [CrossRef]

universe

MDPI

Article

On the Effects of Non-Metricity in an Averaged Universe

Anish Agashe [1,2,*] and Sai Madhav Modumudi [2]

1 Department of Physics, St. Mary's College of Maryland, 47645 College Dr., St. Mary's City, MD 20686, USA
2 Department of Physics, University of Texas at Dallas, 800 W Campbell Rd., Richardson, TX 75080, USA; saimadhav.modumudi@utdallas.edu
* Correspondence: anagashe@smcm.edu

Abstract: In the covariant averaging scheme of macroscopic gravity, the process of averaging breaks the metricity of geometry. We reinterpret the back-reaction within macroscopic gravity in terms of the non-metricity of averaged geometry. This interpretation extends the effect of back-reaction beyond mere dynamics to the kinematics of geodesic bundles. With a 1 + 3 decomposition of the spacetime, we analyse how geometric flows are modified by deriving the Raychaudhuri and Sachs equations. We also present the modified forms of Gauss and Codazzi equations. Finally, we derive an expression for the angular diameter distance in the Friedmann Lemaître Robertson Walker universe and show that non-metricity modifies it only through the Hubble parameter. Thus, we caution against overestimating the influence of back-reaction on the distances.

Keywords: non-metricity; back-reaction; averaging problem; 1 + 3 formalism

PACS: 98.80.-k; 04.50.Kd

MSC: 83D05; 83F05

1. Introduction

The universe is traditionally modelled by employing cosmological solutions to the Einstein field equations. In these models, the matter distribution is taken to be averaged over cosmological scales and hence, the Einstein tensor needs to be averaged as well [1,2]. However, averaging tensor fields on a differentiable manifold is not straightforward, and this is a well-known problem in relativistic cosmology [3,4]. The difficulty arises because of the lack of a mathematically rigorous integration operation on non-Euclidean geometries. The non-linearity of the Einstein field equations exacerbates the problem since the averaged dynamics of a given metric are different from the dynamics of an averaged metric, i.e., $\langle \boldsymbol{G}[\boldsymbol{g}]\rangle \neq \boldsymbol{G}[\langle \boldsymbol{g}\rangle]$. This is referred to as the averaging problem in the literature. There is a long history of approaches to this problem (see [3,5,6] and the references within). In all the approaches, however, the problem is effectively addressed by modifying the dynamics of the averaged spacetime. This modification may take different forms in various approaches and is generically referred to as cosmological back-reaction [5,6]. One such approach is macroscopic gravity [7–10], which offers a non-perturbative and fully covariant solution to the averaging problem.

Macroscopic gravity calculates the spacetime averages of arbitrary tensor fields through averaging bi-vectors and Lie dragging of the averaging regions. When this averaging scheme is applied to general relativity (GR), the Riemannian[1] structure of the geometry is broken. The averaged geometry (the geometry at the macroscopic scales) is equipped with a metric-incompatible but symmetric connection that arises from averaging. This leads to a non-vanishing non-metricity but a vanishing torsion. The dynamics of macroscopic gravity are given by taking a spacetime average of the Einstein field equations.

The back-reaction in these dynamics is also given in terms of the non-metricity, as has been noted in [11,12].

The aim of this work is to analyse the effects of the non-metricity in macroscopic gravity on various aspects of relativistic cosmology. We achieve this by treating macroscopic gravity as a metric-affine theory. In such a theory, the geometry is endowed with a metric and an independent connection, which is precisely how macroscopic gravity proposes the averaged geometry to be. This means that the back-reaction (arising due to non-metricity) modifies the kinematics in the averaged universe. This work emphasises this by demonstrating its effects on geometric flows, which in turn affect the distance measures in cosmology.

We perform a 1 + 3 decomposition of the macroscopic geometry. This enables us to derive the kinematics of geodesic flows. We show that both the Raychaudhuri and Sachs optical equations have contributions from the non-metricity. In other words, back-reaction in macroscopic gravity modifies not only the dynamics but also the kinematics. However, it turns out that the non-metricity has no effect on the cross-sectional area of geodesic bundles. We use the Sachs equation to derive an expression for the angular distances in an arbitrary spacetime. As a quick application of this geometric setup, we present the case of a spatially flat Friedman–Lemaître-Robertson–Walker (FLRW) universe and discuss how the non-metricity affects the distance measurements. We also show that the spatial hypersurfaces have a non-Riemannian geometry and derive the modified Gauss and the Codazzi equations.

The organisation of the paper is as follows: in Section 2, we present the basic formalism of macroscopic gravity and reinterpret it as a metric-affine theory. In Section 3, we derive the structure equations for geometric flows. Then, in Section 4, we discuss the geometry of spatial hypersurfaces and derive the Gauss–Codazzi equations. In Section 5, we derive an expression for the angular distances. We show that the back-reaction affects the distances only through the dynamics (Hubble expansion). Finally, in Section 6, we discuss the results in this paper and provide a few concluding remarks.

The notation used is as follows: Greek indices run from 0 to 3 while Latin indices run from 1 to 3. The partial derivative is denoted by ∂, while ∇ and $\overline{\nabla}$ denote the covariant derivatives with respect to the non-Riemannian and the Levi–Civita connections, respectively. Averaged quantities are denoted using angular brackets, $\langle\ldots\rangle$. The indices within $(\ldots)$ and $[\ldots]$ are symmetrised/anti-symmetrised and underlined indices are not included in (anti-)symmetrisation. The sign convention followed is the 'Landau-Lifshitz Space-like Convention (LLSC)' [13].

2. Averaging the (Pseudo-)Riemannian Geometry

2.1. Basics of Macroscopic Gravity Formalism

In this section, we briefly present the basic formalism of macroscopic gravity. We use the same terminology as used in [7–9,11]. We will reinterpret certain quantities within the macroscopic gravity formalism in terms of more familiar geometric objects in a metric-affine gravity. Macroscopic gravity employs a covariant averaging procedure to average out the (pseudo-)Riemannian geometry and the field equations of general relativity [7–11,14,15]. Given a geometric object, $p^{\alpha}_{\beta}(x)$, defined on an n-dimensional differentiable metric manifold $(\mathcal{M}, g)$, the spacetime averaged value of this object is defined as:

$$\left\langle p^{\alpha}_{\beta}(x)\right\rangle = \frac{\int_{\Sigma} \mathcal{A}^{\alpha}_{\mu'}(x,x')p^{\mu'}_{\nu'}(x')\mathcal{A}^{\nu'}_{\beta}(x',x)\sqrt{-g'}\,\mathrm{d}^n x'}{\int_{\Sigma}\sqrt{-g'}\,\mathrm{d}^n x'} \tag{1}$$

where the average has been taken over a compact region $\Sigma \subset \mathcal{M}$ around a supporting point $x \in \Sigma$; $\int_{\Sigma}\sqrt{-g'}\,\mathrm{d}^n x'$ is the volume (V_{Σ}) of the region Σ and $g' = \det\left[g_{\alpha\beta}(x')\right]$. The integration is carried out for all the points $x' \in \Sigma$. The integrand $\mathcal{A}^{\alpha}_{\mu'}(x,x')p^{\mu'}_{\nu'}(x')\mathcal{A}^{\nu'}_{\beta}(x',x)$ is called the bilocal extension of the object $p^{\alpha}_{\beta}(x)$ and the objects $\mathcal{A}^{\alpha}_{\mu'}(x,x')$ and $\mathcal{A}^{\nu'}_{\beta}(x',x)$ are called the bilocal averaging operators.

The average is taken over finite volume regions in the manifold (i.e., V_Σ is finite). Hence, well defined microscopic objects average out without the integral diverging in Equation (1). In the limit of vanishing averaging volume, the average of the object is equal to the object itself [7–9]. Furthermore, there is freedom to choose these averaging operators according to the problem at hand. However, all these operators must have the following properties: coincidence limit, idempotency, the existence of an inverse, and factorisation (see [7,8] for details). Moreover, the averaging procedure is constructed in such a way that the average of a quantity is uniquely defined and does not depend on the choice of the averaging operators [7–9].

The average of the microscopic Riemann curvature tensor is written as $\langle R^\alpha{}_{\beta\rho\sigma}\rangle$ and behaves like a curvature tensor [7–9,11]. Further, bilocal objects defined as $\mathcal{F}^\alpha{}_{\beta\rho}(x,x') \equiv \mathcal{A}^\alpha_{\epsilon'}\left(\partial_\rho \mathcal{A}^{\epsilon'}_\beta + \nabla_{\sigma'}\mathcal{A}^{\epsilon'}_\beta \mathcal{A}^{\sigma'}_\rho\right)$ behave as connection coefficients at x, and hence can be considered the bilocal extension of the microscopic connection coefficients, $\Gamma^\alpha{}_{\beta\rho}$. They have the following coincidence limit:

$$\lim_{x'\to x} \mathcal{F}^\alpha{}_{\beta\rho} = \Gamma^\alpha{}_{\beta\rho} \tag{2}$$

The average of the connection coefficients, $\langle\Gamma^\alpha{}_{\beta\rho}\rangle = \frac{\int \mathcal{F}^\alpha{}_{\beta\rho}\sqrt{-g'}\,\mathrm{d}^n x'}{\int \sqrt{-g'}\,\mathrm{d}^n x'}$, are the macroscopic affine connection coefficients [7,8]. Then, one can construct a curvature tensor, $M^\alpha{}_{\beta\rho\sigma}$, with these macroscopic connection coefficients. This will not be the same as the average of the microscopic curvature tensor, but rather is related to it through

$$M^\alpha{}_{\beta\rho\sigma} = \langle R^\alpha{}_{\beta\rho\sigma}\rangle + 2\left\langle \mathcal{F}^\delta{}_{\beta[\rho}\mathcal{F}^\alpha{}_{\underline{\delta}\sigma]}\right\rangle - 2\left\langle \mathcal{F}^\delta{}_{\beta[\rho}\right\rangle\left\langle \mathcal{F}^\alpha{}_{\underline{\delta}\sigma]}\right\rangle \tag{3}$$

The additional terms on the right hand side arise from the fact that the curvature is a non-linear function of the connection coefficients.

One can calculate another connection, $\Pi^\alpha{}_{\beta\rho}$, associated with the curvature tensor $\langle R^\alpha{}_{\beta\rho\sigma}\rangle$. Therefore, in the macroscopic spacetime, there are two connections associated with two curvature tensors, which are related to each other via Equation (3). An affine deformation tensor, $A^\alpha{}_{\beta\rho} = \langle\Gamma^\alpha{}_{\beta\rho}\rangle - \Pi^\alpha{}_{\beta\rho}$, can be defined, which measures the difference between the two connections. This deformation tensor exists due to the metric incompatibility of the connection $\Pi^\alpha{}_{\beta\rho}$. That is, for macroscopic metric, $g_{\alpha\beta}$, we have:

$$g_{\alpha\beta|\rho} \neq 0 \tag{4}$$

where $|$ denotes the covariant derivative with respect to the connection $\Pi^\alpha{}_{\beta\rho}$.

Further, looking at the construction of both the curvature tensors in terms of their respective connections, Equation (3) takes the form:

$$A^\alpha{}_{\beta[\rho\|\sigma]} - A^\epsilon{}_{\beta[\rho}A^\alpha{}_{\underline{\epsilon}\sigma]} = -\frac{1}{2}P^\alpha{}_{\beta\rho\sigma} \tag{5}$$

where $\|$ denotes the covariant derivative with respect to the metric-compatible (Levi–Civita) connection, $\langle\boldsymbol{\Gamma}\rangle$, of the macroscopic manifold. The tensor, $\boldsymbol{P}$, is dubbed the gravitational polarisation tensor [7], and is related to the the bilocal objects $\boldsymbol{\mathcal{F}}$, as follows:

$$P^\alpha{}_{\beta\rho\sigma} = 2\left\langle \mathcal{F}^\delta{}_{\beta[\rho}\mathcal{F}^\alpha{}_{\underline{\delta}\sigma]}\right\rangle - 2\left\langle \mathcal{F}^\delta{}_{\beta[\rho}\right\rangle\left\langle \mathcal{F}^\alpha{}_{\underline{\delta}\sigma]}\right\rangle \tag{6}$$

The tensor, $\boldsymbol{P}$, although not a curvature tensor, follows the algebraic and differential properties of one. From Equations (3) and (6), we can see the that it measures the difference between the two curvature tensors—a curvature deformation tensor:

$$P^\alpha{}_{\beta\rho\sigma} = M^\alpha{}_{\beta\rho\sigma} - \langle R^\alpha{}_{\beta\rho\sigma}\rangle \tag{7}$$

Both the curvature tensors, $\boldsymbol{M}$ and $\langle \boldsymbol{R} \rangle$, follow the same algebraic identities and the differential Bianchi identities with respect to their corresponding connections. The former ensures that the connection, Π, although metric-incompatible, is symmetric. Moreover, the averaging of the differential Bianchi identities for the microscopic curvature tensor leads to the following relations:

$$\left\langle R^{\alpha}{}_{\beta[\rho\sigma} \right\rangle_{,\epsilon]} + \left\langle R^{\gamma}{}_{\beta[\rho\sigma} \mathcal{F}^{\alpha}{}_{\underline{\gamma}\epsilon]} \right\rangle - \left\langle R^{\alpha}{}_{\gamma[\rho\sigma} \mathcal{F}^{\gamma}{}_{\underline{\beta}\epsilon]} \right\rangle = 0 \tag{8}$$

In order to evaluate the above equations, a splitting rule is required for the products in the second and third terms. This is constructed using an object called the connection correlation, which is defined as [7,8]:

$$Z^{\alpha}{}_{\beta[\gamma}{}^{\mu}{}_{\underline{\nu}\sigma]} = \left\langle \mathcal{F}^{\alpha}{}_{\beta[\gamma} \mathcal{F}^{\mu}{}_{\underline{\nu}\sigma]} \right\rangle - \left\langle \mathcal{F}^{\alpha}{}_{\beta[\gamma} \right\rangle \left\langle \mathcal{F}^{\mu}{}_{\underline{\nu}\sigma]} \right\rangle \tag{9}$$

Using Equations (6) and (9), we see that,

$$P^{\alpha}{}_{\beta\rho\sigma} = 2Z^{\epsilon}{}_{\beta[\rho}{}^{\alpha}{}_{\underline{\epsilon}\sigma]} \tag{10}$$

In summary, when the averaging procedure of macroscopic gravity (1) is applied to the four-dimensional (pseudo-)Riemannian manifold of GR, one finds that, at macroscopic scales, the manifold is characterised by a metric, two symmetric affine connections[2], two curvature tensors associated with these connections, and a correlation tensor that needs to be constructed in order to solve the differential Bianchi identities for these curvature tensors.

The Field Equations of Macroscopic Gravity

The average of Einstein field equations with a cosmological constant:

$$\langle R^{\alpha}{}_{\beta} \rangle - \frac{1}{2}\delta^{\alpha}_{\beta} \langle R \rangle = \kappa \langle T^{\alpha}{}_{\beta} \rangle - \Lambda \delta^{\alpha}_{\beta} \tag{11}$$

yields the field equations of macroscopic gravity, which are given by: [7,8],

$$\overline{R}^{\alpha}{}_{\beta} - \frac{1}{2}\delta^{\alpha}_{\beta}\overline{R} = \kappa \langle T^{\alpha}{}_{\beta} \rangle - \Lambda \delta^{\alpha}_{\beta} + \left(Z^{\alpha}{}_{\beta} - \frac{1}{2}\delta^{\alpha}_{\beta} P \right) \tag{12}$$

where $Z^{\epsilon}{}_{\gamma} = Z^{\epsilon\mu}{}_{\mu\gamma}$, $Z^{\epsilon}{}_{\mu\nu\gamma} = 2Z^{\epsilon}{}_{\mu[\alpha}{}^{\alpha}{}_{\underline{\nu}\gamma]}$, $P = P^{\mu}{}_{\mu}$, $P_{\mu\nu} = P^{\epsilon}{}_{\mu\epsilon\nu}$, and $\langle \boldsymbol{T} \rangle$ is the averaged energy-momentum tensor (usually taken to be that of a perfect fluid in cosmology). Therefore, upon averaging, the field equations are modified and now include additional terms, given (partially) in terms of the non-metricity of the spacetime.

2.2. Non-Metricity of the Averaged Geometry

As discussed above, applying the averaging procedure of macroscopic gravity to GR leads to the interesting result that, at macroscopic scales, the manifold is non-Riemannian. That is, the manifold has a Riemannian structure only at small (microscopic) scales. At such scales, GR is a valid theory of gravity. However, when one is concerned with large scales (like in cosmology), an averaging has to be performed on the geometry, and in the process, the Riemannian structure of the manifold is broken. In other words, for cosmological analyses, one must work with a theory of gravity that employs a non-Riemannian geometrical framework. The theory of macroscopic gravity is precisely such a theory.

According to the macroscopic gravity formalism, the non-Riemannian (macroscopic) manifold is characterised by a symmetric but metric-incompatible connection, given by $\boldsymbol{\Pi}$. The covariant derivative of the metric (of the macroscopic manifold) with respect to this connection is not zero. The non-metricity tensor is then defined as:

$$Q_{\rho\alpha\beta} = \frac{1}{2}\nabla_{\rho} g_{\alpha\beta} \tag{13}$$

Using this, the general connection, $\mathbf{\Pi}$, can be decomposed as:

$$\Pi^{\rho}{}_{\alpha\beta} = \overline{\Gamma}^{\rho}{}_{\alpha\beta} + D^{\rho}{}_{\alpha\beta} \tag{14}$$

where $\overline{\mathbf{\Gamma}}$ is the Riemannian part (Levi–Civita connection)[3], and $\boldsymbol{D}$ is the so-called disformation tensor that encapsulates the deviation from the Riemannian nature of geometry. The disformation is given in terms of the non-metricity as:

$$D^{\rho}{}_{\alpha\beta} := Q^{\rho}{}_{\alpha\beta} - Q_{\alpha}{}^{\rho}{}_{\beta} - Q_{\beta}{}^{\rho}{}_{\alpha} \tag{16}$$

Further, the Ricci identity is given by:

$$(\nabla_{\alpha}\nabla_{\beta} - \nabla_{\beta}\nabla_{\alpha})V^{\rho} = R^{\rho}{}_{\sigma\alpha\beta}V^{\sigma} \tag{17}$$

where $R^{\rho}{}_{\sigma\alpha\beta}$, is the Riemann curvature tensor defined in terms of the general connection of the non-Riemannian geometry:

$$R^{\rho}{}_{\sigma\alpha\beta} := \partial_{\alpha}\Pi^{\rho}{}_{\sigma\beta} - \partial_{\beta}\Pi^{\rho}{}_{\sigma\alpha} + \Pi^{\rho}{}_{\mu\alpha}\Pi^{\mu}{}_{\sigma\beta} - \Pi^{\rho}{}_{\mu\beta}\Pi^{\mu}{}_{\sigma\alpha} \tag{18}$$

Using Equation (14), it can be shown that:

$$R^{\rho}{}_{\sigma\alpha\beta} = \overline{R}^{\rho}{}_{\sigma\alpha\beta} + \overline{\nabla}_{\alpha}D^{\rho}{}_{\sigma\beta} - \overline{\nabla}_{\beta}D^{\rho}{}_{\sigma\alpha} + D^{\rho}{}_{\mu\alpha}D^{\mu}{}_{\sigma\beta} - D^{\rho}{}_{\mu\beta}D^{\mu}{}_{\sigma\alpha} \tag{19}$$

where $\overline{R}^{\rho}{}_{\sigma\alpha\beta}$ is the Riemannian part of the curvature tensor, which is the same as $\boldsymbol{M}$ of the previous section, and $\overline{\nabla}$ is the covariant derivative both defined with the Levi–Civita connection. One can define a new quantity that captures the deviation of the curvature tensor from its Riemannian counterpart:

$$N^{\rho}{}_{\sigma\alpha\beta} := \overline{\nabla}_{\alpha}D^{\rho}{}_{\sigma\beta} - \overline{\nabla}_{\beta}D^{\rho}{}_{\sigma\alpha} + D^{\rho}{}_{\mu\alpha}D^{\mu}{}_{\sigma\beta} - D^{\rho}{}_{\mu\beta}D^{\mu}{}_{\sigma\alpha} \tag{20}$$

It is not difficult to see that the so-called affine deformation tensor, $\boldsymbol{A}$, and the polarisation tensor, $\boldsymbol{P}$, of macroscopic gravity are, respectively, the disformation tensor, $\boldsymbol{D}$, and the object, $\boldsymbol{N}$, of the metric–affine gravity literature. Henceforth, we will use these symbols to denote the affine deformation tensor and polarisation tensor since we are treating the theory of macroscopic gravity as a metric–affine theory.

3. Structure Equations for Geometric Flows

Since the geometry at the macroscopic scales is no longer Riemannian, determining the geodesic curves (curves with extremum length) is not trivial. In Riemannian geometry, the autoparallel curves with respect to the Levi-Civita connection are also the geodesic curves. Then, the geodesic equation is given by:

$$u^{\beta}\overline{\nabla}_{\beta}u^{\alpha} = 0 \tag{21}$$

However, in general, autoparallel curves (of a non-Riemannian geometry) are not necessarily geodesics. In fact, since, in non-Riemannian geometry, there exist additional geometric fields (like non-metricity), the paths traced by freely moving physical particles (the geodesics) will depend on whether these particles interact with these 'new' geometric fields [18]. In cosmology, one takes the content of the universe to be a perfect fluid or dust. Since such matter does not interact with non-metricity, despite the averaged geometry being non-Riemannian, the path that is traced is taken to be the Riemannian geodesics given by Equation (21).

We will now derive equations governing the evolution of the congruences of geodesics on the averaged (non-Riemannian) geometry. These are sometimes called kinematic equations. Given a congruence, one can characterise its flow in terms of the evolution of the irreducible components of its velocity gradient. Physically, these equations characterise the flow of the fluid content of the Universe (for the timelike curves) and bundle of photons (for the null curves). We will derive (in detail) the equations for the trace part of the velocity gradient (also called expansion) for both the timelike and null curves. This will provide us with the macroscopic gravity analogue of the Raychaudhuri and Sachs equations [19,20].

3.1. The Raychaudhuri Equation

To derive the Raychaudhuri equation, we begin with considering a congruence of timelike geodesics. Let u^α be the vector field tangent to the congruence (the velocity vector). In cosmology, this is simply the comoving velocity of the (fluid) content of the Universe, $u^\alpha = (1,0,0,0)$. The norm of such a velocity is normalised as $u^\alpha u_\alpha = -1$ [21]. We are interested in the transverse evolution of the congruence, and hence we define a transverse metric in the following manner [21]:

$$h_{\alpha\beta} = g_{\alpha\beta} + u_\alpha u_\beta \tag{22}$$

The 3-metric $h_{\alpha\beta}$ is completely transverse to the congruence, i.e., we have $u^\alpha h_{\alpha\beta} = 0 = h_{\alpha\beta}u^\beta$. It is easy to see that $h^\alpha{}_\alpha = 3$. This splits the spacetime into temporal and spatial parts, referred to as $1+3$ decomposition. All the quantities transverse to the congruence reside on a spatial hypersurface orthogonal to it. Here, we will investigate the transverse properties of the congruence while, in the next section, we present the properties of spatial hypersurfaces.

Using the 3-metric, one can separate the transverse and longitudinal parts of the velocity gradient, as follows:

$$^{(\mathrm{T})}\nabla_\alpha u^\beta = h^\rho{}_\alpha h^\beta{}_\epsilon \nabla_\rho u^\epsilon \tag{23}$$

$$^{(\mathrm{L})}\nabla_\alpha u^\beta = \nabla_\alpha u^\beta -{}^{(\mathrm{T})}\nabla_\alpha u^\beta \tag{24}$$

where ∇ represents the covariant derivative with respect to the non-Riemannian connection, $\mathbf{\Pi}$. It is easy to observe that $u^\alpha\, {}^{(\mathrm{T})}\nabla_\alpha u^\beta = 0 ={}^{(\mathrm{T})}\nabla_\alpha u^\beta u_\beta$. The transverse velocity gradient can be decomposed into its irreducible components as follows:

$$\theta := \delta^\alpha_\beta\, {}^{(\mathrm{T})}\nabla_\alpha u^\beta = h^\alpha{}_\beta \nabla_\alpha u^\beta \tag{25}$$

$$\omega_\alpha{}^\beta :={}^{(\mathrm{T})}\nabla_{[\alpha}u^{\beta]} = h^\rho{}_{[\alpha}h^{\beta]}{}_\epsilon \nabla_\rho u^\epsilon \tag{26}$$

$$\sigma_\alpha{}^\beta :={}^{(\mathrm{T})}\nabla_{(\alpha}u^{\beta)} - \frac{1}{3}\theta h_\alpha{}^\beta = h^\rho{}_{(\alpha}h^{\beta)}{}_\epsilon \nabla_\rho u^\epsilon - \frac{1}{3}\theta h_\alpha{}^\beta \tag{27}$$

where θ is the trace part, known as expansion, ω is the anti-symmetric part, known as rotation or vorticity, and σ is the symmetric traceless part, known as shear.

Using the expression for the expansion scalar, we can write:

$$\theta = \nabla_\alpha u^\alpha + D_{\alpha\beta\rho}u^\alpha u^\beta u^\rho \tag{28}$$

where we have used $u^\alpha \nabla_\alpha u^\beta = D^\beta{}_{\alpha\rho}u^\alpha u^\rho$, which follows from Equation (21). One can decompose the first term on the right hand side of the above equation into a Riemannian and a non-Riemannian part. The Riemannian part would be the usual expansion scalar: $\overline{\theta} = \overline{\nabla}_\alpha u^\alpha$. Then, we have:

$$\theta = \overline{\theta} + D^\alpha{}_{\alpha\beta}u^\beta + D_{\alpha\beta\rho}u^\alpha u^\beta u^\rho \tag{29}$$

Taking a derivative on both sides gives us the Raychaudhuri equation in the averaged geometry:

$$\frac{\mathrm{d}\theta}{\mathrm{d}\lambda} = -\overline{R}_{\alpha\beta}u^\alpha u^\beta - \frac{1}{3}\overline{\theta}^2 + \overline{\omega}^\alpha{}_\beta \overline{\omega}_\alpha{}^\beta - \overline{\sigma}^\alpha{}_\beta \overline{\sigma}_\alpha{}^\beta + \frac{\mathrm{d}}{\mathrm{d}\lambda}\left(D^\alpha{}_{\alpha\beta}u^\beta + D_{\alpha\beta\rho}u^\alpha u^\beta u^\rho\right) \tag{30}$$

where the terms with a bar over them are Riemannian quantities. The last term on the right hand side is the contribution from averaging (the back-reaction), provided solely on terms of the non-metricity of the averaged spacetime.

3.2. The Sachs Optical Equation

The Sachs optical equation is the null counterpart of the Raychaudhuri equation. To derive this, we begin by considering a congruence of null geodesics with tangent vector k^α. The norm of the tangent is, $k^\alpha k_\alpha = 0$. In this case, the transverse metric is given by:

$$H_{\alpha\beta} = g_{\alpha\beta} + k_\alpha n_\beta + n_\alpha k_\beta \tag{31}$$

where n^α is an auxiliary null vector satisfying the conditions $n^\alpha n_\alpha = 0$ and $n^\alpha k_\alpha = -1$. Once again, one can check, $k^\alpha H_{\alpha\beta} = 0 = H_{\alpha\beta}k^\beta$, and $h^\alpha{}_\alpha = 2$.

The steps involved in deriving the Sachs equation are analogous to the previous section. We define the transverse part of the velocity gradient as:

$${}^{(\mathrm{T})}\nabla_\alpha k^\beta = H^\rho{}_\alpha H^\beta{}_\epsilon \nabla_\rho k^\epsilon \tag{32}$$

This can be can be decomposed into irreducible components, as follows:

$$\Theta := \delta^\alpha_\beta\,{}^{(\mathrm{T})}\nabla_\alpha k^\beta = H^\alpha{}_\beta \nabla_\alpha k^\beta, \tag{33}$$

$$\Omega_\alpha{}^\beta :={}^{(\mathrm{T})} \nabla_{[\alpha} k^{\beta]} = H^\rho{}_{[\alpha} H^{\beta]}{}_\epsilon \nabla_\rho k^\epsilon, \tag{34}$$

$$\Sigma_\alpha{}^\beta :={}^{(\mathrm{T})} \nabla_{(\alpha} k^{\beta)} - \frac{1}{3}\Theta H_\alpha{}^\beta = H^\rho{}_{(\alpha} H^{\beta)}{}_\epsilon \nabla_\rho k^\epsilon - \frac{1}{3}\Theta H_\alpha{}^\beta, \tag{35}$$

Using the definition of the expansion scalar, we can obtain:

$$\Theta = \nabla_\alpha k^\alpha + D_{\alpha\beta\rho} n^\alpha k^\beta k^\rho + D_{\alpha\beta\rho} k^\alpha k^\beta n^\rho \tag{36}$$

where we used $k_\beta \nabla_\alpha k^\beta = D_{\alpha\beta\rho} k^\alpha k^\beta$. The first term on the right hand side can again be decomposed into Riemannian and non-Riemannian parts. The Riemannian part is the usual null expansion scalar, $\overline{\Theta} = \overline{\nabla}_\alpha k^\alpha$. This gives us:

$$\Theta = \overline{\Theta} + D^\alpha{}_{\alpha\beta} k^\beta + D_{\alpha\beta\rho} n^\alpha k^\beta k^\rho + D_{\alpha\beta\rho} k^\alpha k^\beta n^\rho \tag{37}$$

Taking a derivative on both sides leads to the Sachs optical equation in the averaged geometry:

$$\begin{aligned}\frac{\mathrm{d}\Theta}{\mathrm{d}\lambda} = -\overline{R}_{\alpha\beta} k^\alpha k^\beta - \frac{1}{2}\overline{\Theta}^2 + \overline{\Omega}^\alpha{}_\beta \overline{\Omega}_\alpha{}^\beta - \overline{\Sigma}^\alpha{}_\beta \overline{\Sigma}_\alpha{}^\beta \\ + \frac{\mathrm{d}}{\mathrm{d}\lambda}\left(D^\alpha{}_{\alpha\beta} k^\beta + D_{\alpha\beta\rho} n^\alpha k^\beta k^\rho + D_{\alpha\beta\rho} k^\alpha k^\beta n^\rho\right)\end{aligned} \tag{38}$$

Therefore, the expansion of both the timelike and null geodesics is affected by the non-metricity that arises due to averaging. The non-metricity in macroscopic gravity partially constitutes the cosmological back-reaction. Hence, we have shown here that the back-reaction modifies not only the dynamics but also the kinematics in the averaged Universe. This is expected since, in addition to the field equations, the process of averaging in macroscopic gravity also modifies the underlying geometric structure.

Expansion Scalar and Cross-Section of Null Bundles

It is useful (especially in cosmology) to show how the expansion scalar is related to the cross-sectional area of the congruence. A formal way of achieving this can be found in [22]. We will follow an analogous treatment here. Consider a cross-section around a point P on a curve in the congruence. This can be constructed by taking a set of points, P', on the neighbouring curves that correspond to the same parameter value as that of P. Let this set of points be called S_P. To see the change in the cross-section, we need to compare

S_P to a similar set of neighbouring points at some other point, Q, along the same curve. Let us call this S_Q. Since the points on S_P (and S_Q) are also points on neighbouring curves; they can be used to label the curves, thus providing us with a consistent coordinate system, y^A $(A = 2, 3)$, on the cross-section. Then, one can construct a 2-metric on the cross-section, as follows:

$$ds^2 = H_{AB}\, dy^A dy^B \tag{39}$$

Using this, it can be shown that (see [22]) the fractional rate of the change in the cross-sectional area is:

$$\frac{1}{A}\frac{dA}{d\lambda} = \frac{1}{2}H^{AB}\frac{dH_{AB}}{d\lambda} \tag{40}$$

Calculating the right hand side of the above equation, we can obtain:

$$\frac{1}{2}H^{AB}\frac{dH_{AB}}{d\lambda} = \nabla_\alpha k^\alpha - D^\alpha{}_{\alpha\beta}k^\beta \tag{41}$$

Using Equation (37), this becomes:

$$\frac{1}{2}H^{AB}\frac{dH_{AB}}{d\lambda} = \overline{\Theta} \tag{42}$$

Therefore, the fractional rate of change of the cross-sectional area of the null geodesic congruence in the averaged geometry is equal to the Riemannian part of the expansion scalar:

$$\overline{\Theta} = \frac{1}{A}\frac{dA}{d\lambda} \tag{43}$$

An important point to note here is that this is turning out to be the case since we are dealing with geodesic curves ($k^\alpha \overline{\nabla}_\alpha k^\beta = 0$). For non-geodesic curves, one should expect contributions from non-metricity[4]. The fractional rate of change in the cross-sectional area is closely related to the angular diameter distances in cosmology. The consequence of Equation (42) is that the non-metricity will contribute to the distances only through the dynamics.

4. The Geometry of Hypersurfaces

In this section, we will continue with the $1+3$ decomposition of the spacetime presented in the last section. However, here we will present the equations governing the geometric properties of the spatial hypersurfaces. This approach is commonly employed in GR to split the curvature and the field equations into temporal and spatial parts. This leads to the well-known Gauss-Codazzi equations and the definitions of the shift and lapse functions. Such a splitting is particularly useful in probing various physical situations in numerical relativity.

Let us consider a spacelike hypersurface normal to a timelike geodesic. The 3-metric, $h_{\alpha\beta}$, in Equation (22) serves as the metric for this hypersurface. This is sometimes referred to as the induced metric. In the signature adapted in this paper, this is a positive, definite, non-Riemannian metric. To analyse the geometric properties of the hypersurface, one needs to take the spatial projection of the spacetime objects. This is achieved using the induced metric in a similar manner as Equation (23).

4.1. The Non-Metricity of the Hypersurface

The spatial projection of the covariant derivative is defined as follows:

$${}^{(3)}\nabla_\mu T^{\alpha_1\alpha_2\ldots}{}_{\beta_1\beta_2\ldots} := h^{\alpha_1}{}_{\rho_1}h^{\alpha_2}{}_{\rho_2}\ldots h^{\epsilon_1}{}_{\beta_1}h^{\epsilon_2}{}_{\beta_2}\ldots h^\nu{}_\mu \nabla_\nu T^{\rho_1\rho_2\ldots}{}_{\epsilon_1\epsilon_2\ldots} \tag{44}$$

This serves as the covariant derivative on the hypersurface. Using this definition, we can write:

$${}^{(3)}\nabla_\rho h_{\alpha\beta} = h^\epsilon{}_\rho h^\mu{}_\alpha h^\nu{}_\beta \nabla_\epsilon h_{\mu\nu} = 2h^\epsilon{}_\rho h^\mu{}_\alpha h^\nu{}_\beta Q_{\epsilon\mu\nu} \tag{45}$$

Therefore, we see that the spatial covariant derivative of the three-metric is not zero. In other words, the spatial connection is not metric-compatible[5]. This means that the hypersurface retains the non-Riemannian nature. We define the the non-metricity of the hypersurface as follows:

$$^{(3)}Q_{\rho\alpha\beta} := \frac{1}{2}\,^{(3)}\nabla_\rho h_{\alpha\beta} = h^\epsilon{}_\rho h^\mu{}_\alpha h^\nu{}_\beta Q_{\epsilon\mu\nu} \tag{46}$$

Similarly, one can evaluate that $\frac{1}{2}\,^{(3)}\nabla_\rho h^{\alpha\beta} = -\,^{(3)}Q_\rho{}^{\alpha\beta}$. Note, however, that $^{(3)}\nabla_\rho h^\alpha{}_\beta = 0$.

The extrinsic curvature of the hypersurface is defined as follows:

$$K_{\alpha\beta} := \,^{(3)}\nabla_\alpha u_\beta \tag{47}$$

where $\boldsymbol{u}$ is a timelike vector normal to the hypersurface, i.e., $u^\alpha K_{\alpha\beta} = 0 = K_{\alpha\beta}u^\beta$. Due to the non-metricity of the hypersurface, we have:

$$^{(3)}\nabla_\alpha u^\beta = K_\alpha{}^\beta - 2h^\rho{}_\alpha h^\beta{}_\epsilon Q_\rho{}^{\epsilon\sigma}u_\sigma \tag{48}$$

where $K_\alpha{}^\beta = g^{\rho\beta}K_{\alpha\rho}$. Since the connection is torsion-less (and, hence, so is the connection on the hypersurface), we have $K_{\alpha\beta} = K_{\beta\alpha}$.

4.2. The Gauss Equation

The Gauss equations arises from evaluations of the Ricci identity on the hypersurface. This provides an expression for the Riemannian curvature of the full spacetime in terms of the components projected onto its hypersurface. To derive the Gauss equation, we begin by defining the Riemann tensor on the hypersurface in the following manner:

$$\left[^{(3)}\nabla_\alpha\,^{(3)}\nabla_\beta - ^{(3)}\nabla_\beta\,^{(3)}\nabla_\alpha\right]n^\rho = \,^{(3)}R^\rho{}_{\sigma\alpha\beta}n^\sigma \tag{49}$$

where $\boldsymbol{n}$ is some vector tangential to the hypersurface, i.e., $n^\alpha u_\alpha = 0$. Evaluating the left side of the above equation, we can obtain:

$$\begin{aligned}^{(3)}R^\rho{}_{\sigma\alpha\beta}n^\sigma = &\left[^{(3)}\nabla_\alpha u_\sigma\,^{(3)}\nabla_\beta u^\rho - \,^{(3)}\nabla_\beta u_\sigma\,^{(3)}\nabla_\alpha u^\rho\right]n^\sigma \\ &+ h^\rho{}_\epsilon h^\mu{}_\alpha h^\nu{}_\beta(\nabla_\mu\nabla_\nu - \nabla_\nu\nabla_\mu)n^\epsilon\end{aligned} \tag{50}$$

Using Equations (17), (47), (48), and recognising, $n^\phi = h^\phi{}_\sigma n^\sigma$, we finally obtain the Gauss equation:

$$^{(3)}R^\rho{}_{\sigma\alpha\beta} = h^\rho{}_\epsilon h^\phi{}_\sigma h^\mu{}_\alpha h^\nu{}_\beta R^\epsilon{}_{\phi\mu\nu} + K_{\alpha\sigma}K^\rho{}_\beta - K_{\beta\sigma}K^\rho{}_\alpha - 2h^\rho{}_\epsilon Q_\nu{}^\epsilon{}_\mu u^\mu\left(K_{\alpha\sigma}h^\nu{}_\beta - K_{\beta\sigma}h^\nu{}_\alpha\right) \tag{51}$$

Contracting the appropriate indices, we can write similar equations for the Ricci tensor and Ricci scalar:

$$^{(3)}R_{\sigma\beta} = h^\phi{}_\sigma h^\nu{}_\beta R_{\phi\nu} + K_{\alpha\sigma}K^\alpha{}_\beta - K_{\beta\sigma}K^\alpha{}_\alpha - 2Q_\nu{}^\epsilon{}_\mu u^\mu\left(K_{\epsilon\sigma}h^\nu{}_\beta - K_{\beta\sigma}h^\nu{}_\epsilon\right) \tag{52}$$

$$^{(3)}R = R + R_{\alpha\beta}u^\alpha u^\beta + K_{\alpha\beta}K^{\alpha\beta} - (K^\alpha{}_\alpha)^2 - 2Q_\alpha{}^\beta{}_\rho u^\rho\left(K^\alpha{}_\beta - K^\epsilon{}_\epsilon h^\alpha{}_\beta\right) \tag{53}$$

The above identities are useful in performing a $1+3$ splitting of the field equations.

4.3. The Codazzi Equation

The Gauss equation relates the Riemann curvature tensor with all components projected onto the hypersurface to the extrinsic curvature. Similar to this, one can project only three components of the Riemann tensor onto the hypersurface and one component along a timelike vector: $\boldsymbol{u}$. This quantity can also be written in terms of the extrinsic curvature. This is achieved through the Codazzi equation.

To derive the Codazzi equation, we start by evaluating the following:

$$ {}^{(3)}\nabla_\rho K_{\alpha\beta} = -\left(K_{\rho\alpha}h^\epsilon{}_\beta + K_{\rho\beta}h^\epsilon{}_\alpha\right)D_{\mu\nu\epsilon}u^\mu u^\nu + h^\epsilon{}_\rho h^\mu{}_\alpha h^\nu{}_\beta \nabla_\epsilon \nabla_\mu u_\nu \tag{54} $$

By anti-symmetrising ρ and α, the first term in the brackets on the right hand side vanishes. The second term turns into the Ricci identity (17). Taking care of raising and lowering of the indices introduces a few non-metricity terms, finally leading to the following:

$$ \begin{aligned} {}^{(3)}\nabla_\rho K_{\alpha\beta} - {}^{(3)}\nabla_\alpha K_{\rho\beta} = h^\epsilon{}_\rho h^\mu{}_\alpha h^\nu{}_\beta \Big(R_{\epsilon\sigma\mu\nu} + 4\nabla_{[\epsilon}Q_{\mu]\nu\sigma}\Big)u^\sigma \\ + \left(K_{\alpha\beta}h^\epsilon{}_\rho - K_{\rho\beta}h^\epsilon{}_\alpha\right)D_{\mu\nu\epsilon}u^\mu u^\nu \end{aligned} \tag{55} $$

This is the Codazzi equation. Contracting the indices ρ and β in the above equation, we can obtain:

$$ \begin{aligned} {}^{(3)}\nabla_\rho K^\rho{}_\alpha - {}^{(3)}\nabla_\alpha K^\beta{}_\beta = -h^\mu{}_\alpha R_{\sigma\mu}u^\sigma + h^\mu{}_\alpha h^\epsilon{}_\nu \Big(\nabla_{[\epsilon}Q_{\mu]}{}^\nu{}_\sigma + Q_{[\epsilon}{}^{\rho\nu}Q_{\mu]\rho\sigma}\Big)u^\sigma \\ + \Big(K^\epsilon{}_\alpha - K^\beta{}_\beta h^\epsilon{}_\alpha\Big)D_{\mu\nu\epsilon}u^\mu u^\nu - 4\,{}^{(3)}Q_{(\rho}{}^{\rho\beta}K_{\alpha)\beta} \end{aligned} \tag{56} $$

Using the $1+3$ covariant formalism presented here, it will be straightforward to decompose the four dimensional averaged geometry into foliations and fibrations, as is frequently done within GR. The definition of the lapse function and the shift vector remain the same and the $1+3$ decomposition of the metric retains the same form [24].

5. FLRW Cosmology in Macroscopic Gravity

The field equations of macroscopic gravity, obtained by averaging the Einstein field equations, contain different traces of the connection correlation and the non-Riemannian part of the curvature tensor. These additional terms in Equation (12) are the result of a change in the geometric structure of the spacetime due to averaging. These terms constitute precisely what is referred to as the cosmological back-reaction in the literature. Considering a particular averaged geometry (i.e., taking an ansatz on the metric of the averaged spacetime), one can find the back-reaction in terms of the metric coefficients. This was previously done for the FLRW cosmological model [25–27]. Here, we will build upon the analyses in [25–27] and discuss how the non-metricity affects the distance measurements in cosmology.

5.1. Dynamics and Expansion History

Let us begin by considering the metric at the macroscopic scales, which is the spatially flat FLRW metric:

$$ \mathrm{d}s^2 = -\mathrm{d}t^2 + a^2(t)\Big(\mathrm{d}r^2 + r^2\mathrm{d}\theta^2 + r^2\sin^2\theta\mathrm{d}\phi^2\Big) \tag{57} $$

Considering the averaged matter distribution to be a perfect fluid, we can obtain the following:

$$ \langle T^\epsilon{}_\gamma\rangle = (\rho+p)u^\epsilon u_\gamma + p\delta^\epsilon_\gamma \tag{58} $$

where $\rho \equiv \rho(t)$ and $p \equiv p(t)$ are the energy density and pressure for the averaged matter–energy content, respectively, and $u^\alpha = (1,0,0,0)$ is the average four-velocity of the fluid. Then, writing Equation (12) explicitly provides us with the modified Friedmann equations [25,26]:

$$ 3\left(\frac{\dot a}{a}\right)^2 = 8\pi\rho + \Lambda - \frac{\mathcal{Q}}{a^2} \tag{59} $$

$$ 2\frac{a\ddot a}{a^2} + \left(\frac{\dot a}{a}\right)^2 = -8\pi p + \Lambda - \frac{\mathcal{Q}}{3a^2} \tag{60} $$

where $\mathcal{Q}$ is a constant coming from the back-reaction term and 'dot' represents the derivative with respect to the cosmological time, t. One can immediately notice that the back-reaction term evolves as a curvature.

The first equation above can be written in the form:

$$H^2 = \frac{8\pi}{3}(\rho_m + \rho_\Lambda + \rho_\mathcal{Q}) \tag{61}$$

where $H(t) = \frac{\dot{a}}{a}$ is the Hubble parameter; we defined, $\rho_\Lambda = \frac{\Lambda}{8\pi}$ and $\rho_\mathcal{Q} = -\frac{\mathcal{Q}}{8\pi a^2}$, and we added the subscript m to differentiate the matter density from other energy densities. Further, by defining the associated density parameters as, $\Omega_X = \frac{8\pi\rho_X}{3H^2}$, we can write,

$$1 = \Omega_m + \Omega_\Lambda + \Omega_\mathcal{Q} \tag{62}$$

Multiplying both sides by a factor of $\frac{H^2}{H_0^2}$, we can obtain:

$$H^2 = H_0^2\left(\Omega_{m_0}a^{-3} + \Omega_{\Lambda_0} + \Omega_{\mathcal{Q}_0}a^{-2}\right) = H_0^2\left[\Omega_{m_0}(1+z)^3 + \Omega_{\Lambda_0} + \Omega_{\mathcal{Q}_0}(1+z)^2\right] \tag{63}$$

where $\Omega_{X_0} = \frac{8\pi\rho_{X_0}}{3H_0^2}$, $\rho_{\mathcal{Q}0} = -\frac{\mathcal{Q}}{8\pi}$ is the 'non-metricity density' at present, and we have used $a = \frac{1}{1+z}$. The above expression provides the expansion history of the universe, taking into account the effect due to non-metricity at the macroscopic scales. One can see that, depending on the sign of the constant, $\mathcal{Q}$, a spatially flat FLRW universe may evolve identically to a spatially curved one.

5.2. Angular Diameter Distance

In observational cosmology, one deals with large amounts of photons, and therefore, it is the averaged photon paths that should be used to calculate distances. In the context of macroscopic gravity, this means that the paths of photon bundles (null geodesics) on the macroscopic geometry should be used to determine the distance measures in cosmology. The angular diameter distance, D_A, is related to the cross-sectional area, A, of null congruence, as [28]:

$$A \propto D_A^2 \tag{64}$$

Using this, one can write:

$$\frac{1}{A}\frac{\mathrm{d}A}{\mathrm{d}\lambda} = \frac{2}{D_A}\frac{\mathrm{d}D_A}{\mathrm{d}\lambda} \tag{65}$$

Using Equation (43), this becomes:

$$\overline{\Theta} = \frac{2}{D_A}\frac{\mathrm{d}D_A}{\mathrm{d}\lambda} \tag{66}$$

Taking a derivative on both sides, one can obtain:

$$\frac{2}{D_A}\frac{\mathrm{d}^2D_A}{\mathrm{d}\lambda^2} = \frac{\mathrm{d}\overline{\Theta}}{\mathrm{d}\lambda} + \frac{1}{2}\overline{\Theta}^2 = -\overline{R}_{\alpha\beta}k^\alpha k^\beta + \overline{\Omega}^\alpha{}_\beta\overline{\Omega}_\alpha{}^\beta - \overline{\Sigma}^\alpha{}_\beta\overline{\Sigma}_\alpha{}^\beta \tag{67}$$

where, in the second equality, we have used the Sachs Equation (38).

The above equation holds for arbitrary spacetimes. For a spatially flat FLRW spacetime, we have, $\overline{\mathbf{\Omega}} = 0 = \overline{\mathbf{\Sigma}}$, and the four-velocity of null geodesics is given by [29],

$$k^\alpha = \left(-a^{-1}, a^{-2}, 0, 0\right) \tag{68}$$

Using this, one can calculate:

$$-\overline{R}_{\alpha\beta}k^\alpha k^\beta = \frac{2\dot{H}}{a} \tag{69}$$

Using this in Equation (67), we can obtain:

$$\frac{d^2 D_A}{d\lambda^2} - \frac{1}{a^2}\frac{dH}{dt} D_A = 0 \tag{70}$$

Using $k^0 = \frac{dt}{d\lambda} = -\frac{1}{a}$, the above equation becomes:

$$\frac{d^2 D_A}{dt^2} - H\frac{dD_A}{dt} - \frac{dH}{dt} D_A = 0 \tag{71}$$

Writing in terms of the scale factor, a, this becomes:

$$\frac{d^2 D_A}{da^2} + \frac{1}{H}\frac{dH}{da}\frac{dD_A}{da} - \frac{1}{aH}\frac{dH}{da} D_A = 0 \tag{72}$$

Now, making a variable change, $\chi(a) = \int \frac{da}{a^2 H}$, we can obtain:

$$\frac{d^2 D_A}{d\chi^2} - 2aH\frac{dD_A}{d\chi} - \frac{dH}{d\chi} D_A = 0 \tag{73}$$

which can be reduced to:

$$\frac{d^2}{d\chi^2}\left(\frac{D_A}{a}\right) = 0 \tag{74}$$

Solving the above differential equation leads to an expression for the angular diameter distance in the FLRW geometry:

$$D_A(a) = a\chi \tag{75}$$

Therefore, we see that the non-metricity affects the distance measurements only through its presence in the Hubble parameter. That is, even though the cross-sectional area of null congruences only depends on the Riemannian part of the expansion scalar, since photon trajectories are also affected by the Hubble parameter (H), the distance measurements are modified by the non-metricity. In Figure 1, we plot the expansion history and the angular distance to show the modifications caused by the small values of the non-metricity parameter.

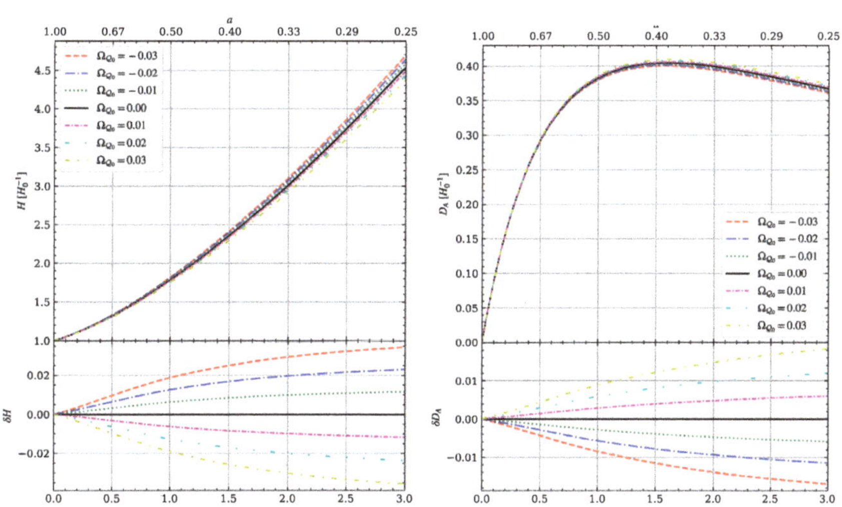

Figure 1. Hubble parameter (**left**) and angular diameter distance (**right**) at later times ($0 < z < 3$) for small values of the non-metricity parameter ($-0.03 \leq \Omega_{\mathcal{Q}_0} \leq 0.03$). The lower panel shows the relative difference between ΛCDM values and the $\mathcal{Q}$-ΛCDM values.

The effects of the non-metricity (back-reaction) become more pronounced for large redshifts. For small, negative values of the non-metricity parameter, the Hubble parameter becomes larger than its ΛCDM value. This could prove useful in the context of Hubble tension [30], in which the cosmic microwave data suggest a smaller value of the Hubble constant than the local observations. However, it is worth noting that the back-reaction parameter was previously constrained to have small positive values [31,32].

6. Summary and Discussion

In macroscopic gravity, the averaging procedure breaks the metricity of geometry, and the cosmological back-reaction can be provided in terms of the non-metricity of the spacetime. Since the non-metricity results from a change in the geometric structure of the spacetime, the effects of the back-reaction are not limited to dynamics. In this paper, we analysed the effects of non-metricity on the geometric properties of the averaged universe. The primary goal of this paper was to demonstrate how non-metricity modifies kinematics by deriving the Raychaudhuri and Sachs equations. To be able to do so, we first dealt with the questions the the non-Riemannian nature of spacetime raises about the trajectories of physical particles. It has been argued that only matter that possesses a 'microstructure' (properties like hypercharge and dilation) couples with the non-Riemannian geometric fields and thus will be affected by them [18]. In the metric-affine gravity literature, such matter has been termed a 'hyperfluid' [33,34]. Ordinary matter (e.g., the perfect fluid or dust used in cosmology) does not possess this microstructure. Hence, despite the averaged geometry being non-Riemannian, we took the paths traced by physical particles to be simply Riemannian geodesics.

Having clarified this issue, we presented a $1+3$ decomposition of the averaged spacetime geometry within macroscopic gravity. We started by deriving the Raychaudhuri and Sachs equations which govern the expansion of the geodesic flows. Physically, these equations characterise the flow of the cosmic fluid and photon bundles, respectively. Both of these equations were modified due to non-metricity. However, we found that only the Riemannian part of the expansion scalar was related to the fractional rate of change in the cross-sectional area of the geodesic bundles. Then, we presented the geometric properties of the spatial hypersurfaces orthogonal to these geodesics. Using a definition of the spatial covariant derivative, we showed that the spatial hypersurfaces are also non-Riemannian in nature. This became even more apparent in the expressions for the extrinsic curvature. Using the Ricci identity for hypersurfaces, we derived the Gauss–Codazzi equations. As expected, they showed us that non-metricity modifies the Riemann curvature projected onto the hypersurfaces.

Finally, we applied the results derived in this paper to observational cosmology. Using the Sachs optical equation, we derived the angular diameter distance formula. We showed that, despite the existence of a non-metricity, the angular diameter distances are not explicitly modified by it. This is a consequence of the cross-sectional area depending only on the Riemannian part of the expansion scalar. Therefore, one needs to be careful when using distances for observational cosmology within macroscopic gravity. Including the averaged Ricci tensor directly in the Sachs equation might result in an incorrect evaluation of the effects of back-reaction on distances in cosmology. However, it is important to note that the non-metricity does modify the distances implicitly through the dynamics (Hubble parameter). We plotted the expansion history and the angular distance formula to demonstrate the deviation from their ΛCDM counterparts for small back-reaction values. We noted that a small negative value of the back-reaction parameter could help alleviate the Hubble tension. This warrants further investigations into the averaging problem.

Author Contributions: Conceptualization, A.A.; methodology, A.A. and S.M.M.; formal analysis, A.A. and S.M.M.; writing—original draft preparation, A.A.; writing—review and editing, A.A. and S.M.M. All authors have read and agreed to the published version of the manuscript.

Funding: This research received no external funding.

Data Availability Statement: Data are contained within the article.

Acknowledgments: We thank Mustapha Ishak for the useful comments. A.A. would also like to thank the anonymous reviewer of one of our previous manuscripts, whose comments were crucial in the conception of this paper.

Conflicts of Interest: The authors declare no conflicts of interest.

Appendix A. Constraint on the Connection Correlation

To derive a solution to the averaged field Equation (12), we need to determine the form of connection correlation for a given macroscopic geometry. This is achieved by solving the following algebraic and differential constraints on the connection correlation [7,8],

$$Z^{\alpha}{}_{\beta\gamma}{}^{\mu}{}_{\nu\sigma} = -Z^{\alpha}{}_{\beta\sigma}{}^{\mu}{}_{\nu\gamma} \tag{A1}$$

$$Z^{\alpha}{}_{\beta\gamma}{}^{\mu}{}_{\nu\sigma} = -Z^{\mu}{}_{\nu\gamma}{}^{\alpha}{}_{\beta\sigma} \tag{A2}$$

$$Z^{\alpha}{}_{\beta[\gamma}{}^{\mu}{}_{\nu\sigma]} = 0 \tag{A3}$$

$$Z^{\epsilon}{}_{\epsilon\gamma}{}^{\mu}{}_{\nu\sigma} = 0 \tag{A4}$$

$$Z^{\alpha}{}_{\beta\mu}{}^{\gamma}{}_{\delta\nu}u^{\nu} = 0 \tag{A5}$$

$$Z^{\alpha}{}_{\beta[\gamma}{}^{\mu}{}_{\underline{\nu}\sigma||\lambda]} = 0 \tag{A6}$$

$$Z^{\epsilon}{}_{\beta[\gamma}{}^{\mu}{}_{\underline{\nu}\sigma}\overline{R}^{\alpha}{}_{\underline{\epsilon}\lambda\rho]} - Z^{\alpha}{}_{\epsilon[\gamma}{}^{\mu}{}_{\underline{\nu}\sigma}\overline{R}^{\epsilon}{}_{\underline{\beta}\lambda\rho]} + Z^{\alpha}{}_{\beta[\gamma}{}^{\epsilon}{}_{\underline{\nu}\sigma}\overline{R}^{\mu}{}_{\underline{\epsilon}\lambda\rho]} - Z^{\alpha}{}_{\beta[\gamma}{}^{\mu}{}_{\underline{\epsilon}\sigma}\overline{R}^{\epsilon}{}_{\underline{\nu}\lambda\rho]} = 0 \tag{A7}$$

where, $\boldsymbol{u}$ is the unit time-like four-vector field and $\overline{\boldsymbol{R}}$ is the Riemannian part of the curvature tensor for the (assumed) macroscopic metric.

Equations (A1)–(A5) in Appendix A are algebraic constraints and do not depend on the macroscopic geometry. Solving these leaves one with 121 independent components in the connection correlation. Equations (A5)–(A7) ensure that the higher-order correlations are zero. One also assumes that the connection correlation is invariant under the same group of isometries as the macroscopic metric [35]. Given this, the connection correlation can be determined completely. A systematic way of deriving the above macroscopic gravity identities is presented in [26,27,35]. The disformation tensor can also be solved in a similar manner.

Notes

1 The signature is implicitly assumed to be Lorentzian. Therefore, the term 'Riemannian' here is used to mean *pseudo*-Riemannian.

2 Gravity theories with multiple connections both in the Riemannian and non-Riemannian framework are already present in the literature [16,17].

3 The Levi–Civita connection can be written in terms of metric as:

$$\overline{\Gamma}^{\rho}{}_{\alpha\beta} = \frac{1}{2}g^{\rho\sigma}\left(\partial_{\alpha}g_{\beta\sigma} + \partial_{\beta}g_{\sigma\alpha} - \partial_{\sigma}g_{\alpha\beta}\right) \tag{15}$$

4 In an even more general, metric-affine geometry, where the connection is not symmetric, one would have contributions from torsion even in the case of geodesic curves [23].

5 This could mean that a covariant averaging, albeit over three-space, necessarily breaks the Riemannian nature of the space. One needs to be careful about this when comparing macroscopic gravity to spatial averaging schemes that deal only with scalars.

References

1. Ellis, G. Relativistic Cosmology: Its Nature, Aims and Problems. *Fundam. Theor. Phys.* **1984**, *9*, 215–288. [CrossRef]
2. Ellis, G.; Stoeger, W. The 'fitting problem' in cosmology. *Class. Quantum Gravity* **1987**, *4*, 1697–1729. [CrossRef]
3. Ellis, G.F.; Buchert, T. The Universe seen at different scales. *Phys. Lett. A* **2005**, *347*, 38–46. [CrossRef]
4. van den Hoogen, R. Averaging Spacetime: Where do we go from here? In Proceedings of the 12th Marcel Grossmann Meeting on General Relativity, Paris, France, 12–18 July 2009; Volume 3, pp. 578–588. [CrossRef]
5. Clifton, T. Back-Reaction in Relativistic Cosmology. *Int. J. Mod. Phys. D* **2013**, *22*, 1330004. [CrossRef]
6. Buchert, T.; Räsänen, S. Backreaction in Late-Time Cosmology. *Annu. Rev. Nucl. Part. Sci.* **2012**, *62*, 57–79. [CrossRef]
7. Zalaletdinov, R.M. Averaging out the Einstein equations. *Gen. Relativ. Gravit.* **1992**, *24*, 1015–1031. [CrossRef]
8. Zalaletdinov, R. Towards a theory of macroscopic gravity. *Gen. Relativ. Gravit.* **1993**, *25*, 673–695. [CrossRef]
9. Mars, M.; Zalaletdinov, R.M. Space–time averages in macroscopic gravity and volume-preserving coordinates. *J. Math. Phys.* **1997**, *38*, 4741–4757. [CrossRef]
10. Zalaletdinov, R.M. Averaged Lagrangians and MacCallum–Taub's Limit in Macroscopic Gravity. *Gen. Relativ. Gravit.* **1996**, *28*, 953–979. [CrossRef]
11. Zalaletdinov, R.M. Space-time Averages of Classical Physical Fields. *Ann. Eur. Acad. Sci.* **2004**, 344–371. [CrossRef]
12. Van Den Hoogen, R.J. Towards a covariant smoothing procedure for gravitational theories. *J. Math. Phys.* **2017**, *58*, 122501. [CrossRef]
13. Misner, C.W.; Thorne, K.; Wheeler, J. *Gravitation*; W. H. Freeman: San Francisco, CA, USA, 1973.
14. Zalaletdinov, R. The averaging problem in cosmology and macroscopic gravity. *Int. J. Mod. Phys. A* **2008**, *23*, 1173–1181. [CrossRef]
15. Zalaletdinov, R.M. Averaging problem in general relativity, macroscopic gravity and using Einstein's equations in cosmology. *Bull. Astron. Soc. India* **1997**, *25*, 401–416.
16. Khosravi, N. Geometric massive gravity in multiconnection framework. *Phys. Rev. D* **2014**, *89*, 024004. [CrossRef]
17. Iosifidis, D.; Pallikaris, K. Biconnection gravity as a statistical manifold. *Phys. Rev. D* **2023**, *108*, 044026. [CrossRef]
18. Obukhov, Y.N.; Puetzfeld, D. Demystifying autoparallels in alternative gravity. *Phys. Rev. D* **2021**, *104*, 044031. [CrossRef]
19. Raychaudhuri, A. Relativistic cosmology. 1. *Phys. Rev.* **1955**, *98*, 1123–1126. [CrossRef]
20. Sachs, R.; Bondi, H. Gravitational waves in general relativity. VI. The outgoing radiation condition. *Proc. R. Soc. Lond. Ser. A Math. Phys. Sci.* **1961**, *264*, 309–338. [CrossRef]
21. Iosifidis, D.; Vogiatzis, I.G.; Tsagas, C.G. Friedmann-like universes with non-metricity. *Eur. Phys. J. C* **2023**, *83*, 216. [CrossRef]
22. Poisson, E. *A Relativist's Toolkit: The Mathematics of Black-Hole Mechanics*; Cambridge University Press: Cambridge, UK, 2009. [CrossRef]
23. Agashe, A. Kinematics in metric-affine geometry. *Phys. Scr.* **2023**, *98*, 105210. [CrossRef]
24. Capozziello, S.; Finch, A.; Said, J.L.; Magro, A. The 3 + 1 formalism in teleparallel and symmetric teleparallel gravity. *Eur. Phys. J. C* **2021**, *81*, 1141. [CrossRef]
25. Coley, A.A.; Pelavas, N.; Zalaletdinov, R.M. Cosmological Solutions in Macroscopic Gravity. *Phys. Rev. Lett.* **2005**, *95*, 151102. [CrossRef]
26. van den Hoogen, R.J. A complete cosmological solution to the averaged Einstein field equations as found in macroscopic gravity. *J. Math. Phys.* **2009**, *50*, 082503. [CrossRef]
27. Wijenayake, T.; Ishak, M. Expansion and Growth of Structure Observables in a Macroscopic Gravity Averaged Universe. *Phys. Rev. D* **2015**, *91*, 063534. [CrossRef]
28. Peebles, P. *Principles of Physical Cosmology*; Princeton University Press: Princeton, NJ, USA, 2019; Volume 99.
29. Plebanski, J.; Krasinski, A. *An Introduction to General Relativity and Cosmology*; Cambridge University Press: Cambridge, UK, 2012.
30. Di Valentino, E.; Mena, O.; Pan, S.; Visinelli, L.; Yang, W.; Melchiorri, A.; Mota, D.F.; Riess, A.G.; Silk, J. In the realm of the Hubble tension—A review of solutions. *Class. Quantum Gravity* **2021**, *38*, 153001. [CrossRef]
31. Clarkson, C.; Clifton, T.; Coley, A.; Sung, R. Observational Constraints on the Averaged Universe. *Phys. Rev. D* **2012**, *85*, 043506. [CrossRef]
32. Wijenayake, T.; Lin, W.; Ishak, M. Averaged universe confronted with cosmological observations: A fully covariant approach. *Phys. Rev. D* **2016**, *94*, 083501. [CrossRef]
33. Obukhov, Y.N.; Tresguerres, R. Hyperfluid: A Model of classical matter with hypermomentum. *Phys. Lett. A* **1993**, *184*, 17–22. [CrossRef]
34. Iosifidis, D. The Perfect Hyperfluid of Metric-Affine Gravity: The Foundation. *J. Cosmol. Astropart. Phys.* **2021**, *4*, 72. [CrossRef]
35. Van Den Hoogen, R. Spherically Symmetric Solutions in Macroscopic Gravity. *Gen. Relativ. Gravit.* **2008**, *40*, 2213–2227. [CrossRef]

universe

MDPI

Review

Energy-Momentum Squared Gravity: A Brief Overview

Ricardo A. C. Cipriano [1], Nailya Ganiyeva [1], Tiberiu Harko [2], Francisco S. N. Lobo [1,3], Miguel A. S. Pinto [1,3,*] and João Luís Rosa [4,5]

1 Instituto de Astrofísica e Ciências do Espaço, Faculdade de Ciências da Universidade de Lisboa, Campo Grande, Edifício C8, 1749-016 Lisbon, Portugal; fc51825@alunos.fc.ul.pt (R.A.C.C.); fc57452@alunos.fc.ul.pt (N.G.); fslobo@fc.ul.pt (F.S.N.L.)
2 Department of Physics, Babes-Bolyai University, Kogalniceanu Street, 400084 Cluj-Napoca, Romania; tiberiu.harko@aira.astro.ro
3 Departamento de Física, Faculdade de Ciências da Universidade de Lisboa, Campo Grande, Edifício C8, 1749-016 Lisbon, Portugal
4 Institute of Theoretical Physics and Astrophysics, University of Gdańsk, Jana Bażyńskiego 8, 80-309 Gdańsk, Poland; joao.rosa@ug.edu.pl
5 Institute of Physics, University of Tartu, W. Ostwaldi 1, 50411 Tartu, Estonia
* Correspondence: mapinto@fc.ul.pt

Abstract: In this work, we present a review of Energy-Momentum Squared Gravity (EMSG)—more specifically, $f(R, T_{\mu\nu}T^{\mu\nu})$ gravity, where R represents the Ricci scalar and $T_{\mu\nu}$ denotes the energy-momentum tensor. The inclusion of quadratic contributions from the energy-momentum components has intriguing cosmological implications, particularly during the Universe's early epochs. These effects dominate under high-energy conditions, enabling EMSG to potentially address unresolved issues in General Relativity (GR), such as the initial singularity and aspects of big-bang nucleosynthesis in certain models. The theory's explicit non-minimal coupling between matter and geometry leads to the non-conservation of the energy-momentum tensor, which prompts the investigation of cosmological scenarios through the framework of irreversible thermodynamics of open systems. By employing this formalism, we interpret the energy-balance equations within EMSG from a thermodynamic perspective, viewing them as descriptions of irreversible matter creation processes. Since EMSG converges to GR in a vacuum and differences emerge only in the presence of an energy-momentum distribution, these distinctions become significant in high-curvature regions. Therefore, deviations from GR are expected to be pronounced in the dense cores of compact objects. This review delves into these facets of EMSG, highlighting its potential to shed light on some of the fundamental questions in modern cosmology and gravitational theory.

Keywords: modified gravity; energy-momentum squared gravity; non-conservation of the energy-momentum tensor; irreversible thermodynamics of open systems; cosmology; compact objects; black holes; wormholes

1. Introduction

The discovery of late-time cosmic acceleration [1,2] has spurred extensive research into modified theories of gravity [3–12] as an alternative to dark energy [13]. Among these, several theories incorporate non-minimal couplings between geometry and matter [14–18], leading to intriguing modifications in the gravitational dynamics. A distinctive feature of these theories is the non-conservation of matter's energy-momentum tensor. The explicit coupling between geometry and matter results in a non-zero covariant derivative of the energy-momentum tensor, leading to non-geodesic motion and the emergence of an extra force. This additional force modifies the standard motion of test particles, influencing both local and cosmological scales. Notable examples include $f(R, \mathcal{L}_m)$ [19], $f(R, T)$ [20,21], $f(R, T_{\mu\nu}T^{\mu\nu})$ [22,23], and $f(R, T, R_{\mu\nu}T^{\mu\nu})$ [24,25] theories of gravity. Here, R and $R_{\mu\nu}$ represent the Ricci scalar and tensor, respectively; $\mathcal{L}_m$ denotes the matter

Lagrangian density; and $T = T^{\mu}{}_{\mu} = g^{\mu\nu}T_{\mu\nu}$ is the trace of the energy-momentum tensor ($T_{\mu\nu}$). In a cosmological context, the geometry–matter couplings introduce compelling phenomenological effects, enabling a unified description of the different cosmological epochs. For instance, these modified gravity models prove invaluable for describing interactions between dark energy and dark matter, providing insights into the mechanisms driving the late-time cosmic speed-up by modifying the gravitational field equations. Moreover, they offer potential explanations for the observed behavior of the galactic flat rotation curves. In this framework, the extra terms in the gravitational field equations alter the equations of motion for test particles, introducing a supplementary gravitational interaction that can account for the rotation curves without invoking dark matter.

A novel gravitational framework known as energy-momentum squared gravity (EMSG) or $f(R, T_{\mu\nu}T^{\mu\nu})$ gravity [22] has garnered significant attention in the scientific community since its introduction. The pioneering paper laid the groundwork for this theory, which was later refined [23]. The action was originally expressed as

$$S = \frac{1}{2\kappa^2}\int \sqrt{-g}f(R, T_{\mu\nu}T^{\mu\nu})d^4x + S_m\,, \tag{1}$$

where $\kappa^2 = 8\pi G/c^4$ is a constant; S_m is the action of the matter Lagrangian, i.e., $S_m = \int \sqrt{-g}L_m d^4x$; R is the Ricci scalar; and, finally, $T_{\mu\nu}T^{\mu\nu}$ is the self-contraction of the stress energy tensor that characterizes the aforementioned theory. In [23], the action functional of the theory was simplified to the specific case of $f(R, T_{\mu\nu}T^{\mu\nu}) = R - 2\Lambda - \eta T_{\mu\nu}T^{\mu\nu}$, where η is a coupling constant and Λ is the cosmological constant, which was subsequently dubbed EMSG. This abbreviation has since become the standard term for the theory. However, some authors, such as the authors of [26], refer to the original form (as seen in Equation (1)) as generalized energy-momentum squared gravity (gEMSG), to differentiate it from other formulations, such as the power law described in the functional form of $f = f\left(R, \left(T_{\mu\nu}T^{\mu\nu}\right)^n\right)$ (see Ref. [27]) or the inclusion of the matter Lagrangian in the functional (see Ref. [28]). In fact, establishing a consistent notation has been crucial for maintaining coherence and aligning with established conventions in the literature. Thus, in this work, in order to avoid confusion and to simplify definitions, we generically consider the theories described by the action in Equation (1) as energy-momentum squared gravity (EMSG).

Contrary to $f(R, T)$ gravity, EMSG induces quadratic contributions of the energy-momentum components to the right-hand side of the field equations. This self-contraction of the energy-momentum tensor has particularly interesting cosmological consequences during early epochs of the Universe due to its effects being predominant under high-energy regimes, which then allows EMSG to address some of the open questions left in general relativity (GR), such as the initial singularity, dense compact astrophysical objects, or even the big bang nucleosynthesis under some particular models. In fact, an extensive body of literature surrounding the general formulation of this theory includes numerous papers on a plethora of topics, for instance, in a cosmological setting [26,29–35] and compact objects [36–56], among other issues. In fact, energy-momentum squared symmetric teleparallel gravity has also been explored to explain the cosmological dynamics of both the early and the late Universe without resorting to the invocation of dark energy [57]. Tight constraints have also been obtained on specific parameters of EMSG by binary pulsar observations [58].

Indeed, these higher-order quadratic terms lead the field equations to resemble other deeply interesting theories, such as loop quantum gravity [59] or brane-world cosmologies [60–62]. In particular, as we will see, EMSG replaces terms such the energy density (ρ) with $\rho(1 \pm \mathcal{O}(\rho^2))$, where the negative contribution is associated with loop quantum gravity [59] and the positive contribution is associated with the brane-world cosmologies [62] while having an entirely different foundation. It is worth noting that EMSG accomplishes all this without introducing novel forms of fluid stresses (such as extra scalar fields or bulk viscosity) [36]. In regions of high curvature, such as the dense cores of neutron stars or black holes, the energy-momentum tensor becomes significant, causing EMSG corrections to deviate from GR. These deviations manifest in the following several key ways: (i) EMSG changes the internal structure equations of compact objects, affecting density profiles, pressure distributions, and mass–radius relationships; (ii) gravitational waves from compact object mergers may differ under EMSG and be potentially detectable by observatories; (iii) EMSG could influence the formation, evolution, and stability of black holes, neutron stars, and wormholes, as well as the end states of stellar collapse.

The resurgence in popularity of this theory can be attributed to several key developments. Its momentum grew rapidly with the publication of [23], which demonstrated that under a particular yet simple Lagrangian ($f = R + \gamma T_{\mu\nu}T^{\mu\nu}$), the early epochs of the cosmos had a minimum length and a finite maximum energy density. This finding resolves the primordial singularity without resorting to quantum gravity by generating a bouncing cosmology while maintaining a consistent sequence of cosmic epochs and adequately explaining cosmic behavior. Consequently, EMSG can be considered an emergent universe scenario while still being a subset of the k-essence models [63]. It is worth noting that not all EMSG solutions yield these results. However, from a theoretical perspective, it is still a promising indication that EMSG might be an interesting theory to explore further. Although it has been argued that a cosmological evolution that can regularly connect the early-universe bounce to a viable de Sitter late-time universe should not generally exist [30], this issue is addressed by the existence of a vacuum energy density in EMSG [64].

Finally, some theoretical remarks regarding EMSG are warranted. Similar to $f(R,T)$ gravity, EMSG does not satisfy the conservation of the energy-momentum tensor. This implies that the paths of test particles may exhibit non-geodesic motion. However, in the special case where $T_{\mu\nu}T^{\mu\nu} = 0$, the standard geodesic path is recovered [65]. Moreover, EMSG may not be equivalent to $f(R)$ gravity when the trace of the energy-momentum tensor ($T = g^{\mu\nu}T_{\mu\nu}$) is zero. For instance, in the case of radiation, although $T = 0$, EMSG still contains non-vanishing terms [31,32]. These differences are explored below.

This review paper is organized in the following manner: In Section 2, we briefly present the formalism of EMSGin the geometric and scalar-tensor representations and consider an extended theory of gravity with an n-th-order invariant constructed of contractions of the energy-momentum tensor. In Section 3, we explore the thermodynamics of open systems with the possibility of irreversible matter creation processes due to the non-conservation of the energy-momentum tensor in EMSG. In Section 4, we briefly review the fundamentals of cosmology in EMSG and explore some general cosmological models. In Section 5, we explore compact objects in EMSG—in particular, the possibility of black hole solutions—and expand on wormhole geometries. Finally, in Section 6, we summarize and conclude this research.

Henceforth, to simplify the notation, we denote the self-contraction of the energy-momentum tensor as the following scalar: $\mathcal{T} = T_{\mu\nu}T^{\mu\nu}$. Although this notation is used for brevity, there are instances where displaying the full contraction provides a deeper understanding of the calculations. Whenever $\mathcal{T}$ is mentioned, it refers to this specific contraction. Additionally, it should be noted that most of the analysis presented here is based on the authors' original research. However, readers have access to a substantial body of literature for further exploration and context.

2. $f(R,\mathcal{T})$ Gravity: Formalism

2.1. Action and Field Equations

In this section, we provide a brief overview of the geometric formalism of the theory. The action (S) that describes $f(R,\mathcal{T})$ gravity is given by

$$S = \frac{1}{2\kappa^2}\int_\Omega \sqrt{-g} f(R,\mathcal{T}) d^4x + \int_\Omega \sqrt{-g}\mathcal{L}_m \, d^4x, \tag{2}$$

Varying the action (2) with respect to the metric provides the following modified gravitational field equations:

$$f_R R_{\mu\nu} - \frac{1}{2} g_{\mu\nu} f - (\nabla_\mu \nabla_\nu - g_{\mu\nu}\Box) f_R = 2\kappa^2 T_{\mu\nu} - f_{\mathcal{T}} \Theta_{\mu\nu}, \tag{3}$$

where $\Box = \nabla^\mu \nabla_\mu$ is the usual D'Alembert operator, $f_R \equiv \partial f / \partial R$ and $f_{\mathcal{T}} \equiv \partial f / \partial \mathcal{T}$, and the auxiliary tensor ($\Theta_{\mu\nu}$) is defined as

$$\Theta_{\mu\nu} \equiv \frac{\delta \mathcal{T}}{\delta g^{\mu\nu}}. \tag{4}$$

Taking into account the following explicit variation:

$$\begin{aligned}
\delta\left(T_{\alpha\beta} T^{\alpha\beta}\right) &= \delta\left(g^{\alpha\mu} g^{\beta\nu} T_{\alpha\beta} T_{\mu\nu}\right) = 2\left(T^{\alpha\beta} \frac{\delta T_{\alpha\beta}}{\delta g_{\mu\nu}} + T^\sigma_\mu T_{\nu\sigma}\right) \delta g^{\mu\nu} \\
&= 2\left[T^{\mu\beta}\left(g_{\alpha\beta} \frac{\partial \mathcal{L}_m}{\partial g^{\mu\nu}} + \frac{\delta g_{\alpha\beta}}{\delta g^{\mu\nu}} \mathcal{L}_m - 2 \frac{\partial^2 \mathcal{L}_m}{\partial g^{\mu\nu} \partial g^{\alpha\beta}}\right) + T^\nu_\mu T_{\nu\sigma}\right] \delta g^{\mu\nu} \\
&= 2\left[T^{\alpha\beta}\left\{-\mathcal{L}_m\left(g_{\mu\alpha} g_{\beta\nu} - \frac{1}{2} g_{\mu\nu} g_{\alpha\beta}\right) - \frac{1}{2} g_{\alpha\beta} T_{\mu\nu} - 2 \frac{\partial^2 \mathcal{L}_m}{\partial g^{\mu\nu} \partial g^{\alpha\beta}}\right\} + T^\sigma_\mu T_{\nu\sigma}\right] \delta g^{\mu\nu},
\end{aligned} \tag{5}$$

the auxiliary tensor $\Theta_{\mu\nu}$ can be written as

$$\Theta_{\mu\nu} = -2\mathcal{L}_m\left(T_{\mu\nu} - \frac{1}{2} g_{\mu\nu} T\right) - T T_{\mu\nu} + 2 T_\mu{}^\alpha T_{\alpha\nu} - 4 T^{\alpha\beta} \frac{\partial^2 \mathcal{L}_m}{\partial g^{\mu\nu} \partial g^{\alpha\beta}}. \tag{6}$$

The energy-momentum tensor for EMSG is generically not conserved and is transparent according to the following relation:

$$\nabla^\mu T_{\mu\nu} = -f_{\mathcal{T}} g_{\mu\nu} \nabla^\mu (T_{\alpha\beta} T^{\alpha\beta}) + 2\nabla^\mu (f_{\mathcal{T}} \theta_{\mu\nu}), \tag{7}$$

which has implications that are further explored below.

2.2. Scalar-Tensor Representation

In addition to the familiar geometrical representation, EMSG can also be described using the scalar-tensor representation. This approach, which has gained popularity recently, separates the scalar and tensor parts of the variational action. It results in second-order equations in the metric, simplifying calculations and potentially addressing issues like Ostrogradsky instabilities [66,67]. For a detailed derivation of EMSG in this representation, we refer the reader to [32].

Let us start by defining the action in the scalar-tensor representation as

$$S = \frac{1}{2\kappa^2} \int_\Omega \sqrt{-g}[\phi R + \psi \mathcal{T} - V(\phi, \psi)] d^4x + \int_\Omega \sqrt{-g} \mathcal{L}_m d^4x, \tag{8}$$

where ϕ and ψ are the characteristic scalar fields of this representation, which are non-minimally coupled to the Ricci scalar (R) and the contraction of the energy-momentum tensor ($\mathcal{T}$), respectively. Moreover, $V(\phi, \psi)$ represents the potential of the two scalar fields. Varying the action (8) with respect to the metric leads to the following field equations:

$$\phi\left(R_{\mu\nu} - \frac{1}{2} R g_{\mu\nu}\right) + \frac{1}{2} g_{\mu\nu} V - (\nabla_\mu \nabla_\nu - g_{\mu\nu}\Box)\phi = 8\pi T_{\mu\nu} - \psi\left(\Theta_{\mu\nu} - \frac{1}{2} g_{\mu\nu} \mathcal{T}\right), \tag{9}$$

which are equivalent to Equation (3). Notably, we have, yet again, introduced the auxiliary tensor (still defined by Equation (6)), with f_R and $f_{\mathcal{T}}$ being replaced by the scalar fields of ϕ and ψ, respectively. Furthermore, the extra terms in Equation (9) result from the proper definition of V.

For particular cosmological models, in order to obtain an expression of the potential, we vary the action with respect to both scalar fields (ϕ and ψ), which yields

$$V_\phi = R, \tag{10}$$

$$V_\psi = \mathcal{T}, \tag{11}$$

respectively, where $V_\phi \equiv \partial V/\partial\phi$ and $V_\psi \equiv \partial V/\partial\psi$. Indeed, to find the analytical expression for $V(\phi,\psi)$, one can integrate Equations (10) and (11). Lastly, by taking the covariant derivative of (9), we obtain the following corresponding conservation equation:

$$8\pi\nabla_\nu T^{\mu\nu} = \nabla_\nu(\psi\Theta^{\mu\nu}) - \frac{1}{2}g^{\mu\nu}[R\nabla_\nu\phi + \nabla_\nu(\psi\mathcal{T} - V)]. \tag{12}$$

2.3. Geometrical and Scalar-Tensor $f(R,\mathcal{T}^{(n)})$ Gravity

It is possible to generalize the concept of EMSG presented above to an arbitrary finite power of the scalar ($\mathcal{T}$), i.e., $\mathcal{T}^{(n)}$. This work was developed in [68] by considering an extension of the action (2) that includes a function of the scalar curvature and the higher-order invariant, defined as

$$\begin{aligned} \mathcal{T}^{\mu\ (n)}_{\ \nu} &\equiv T^{\mu}_{\ \alpha_1} T^{\alpha_1}_{\ \alpha_2} \dots T^{\alpha_{n-1}}_{\ \ \nu}, \\ \mathcal{T}^{(n)} &\equiv \mathcal{T}^{\mu\ (n)}_{\ \mu}. \end{aligned} \tag{13}$$

The potential physical significance of this expansion is still under investigation as research continues to progress. However, for the present, it serves as a valuable mathematical generalization. With this in mind, to derive the field equations for the model, we consider a general function ($f(R,\mathcal{T}^{(n)})$) that depends on the exponent n. The starting point for the geometrical representation is now the following higher-order gravitational action:

$$S = \frac{1}{2\kappa^2}\int \mathrm{d}^4x\sqrt{-g}\,f\left(R,\mathcal{T}^{(n)}\right) + \int_\Omega \sqrt{-g}\mathcal{L}_m d^4x. \tag{14}$$

We assume that the matter Lagrangian ($\mathcal{L}_m$) only depends on the metric tensor and not on its derivatives. By varying Equation (14) with respect to the metric, we obtain the following field equations:

$$f_R R_{\mu\nu} - \frac{1}{2}g_{\mu\nu}f + (g_{\mu\nu}\Box - \nabla_\mu\nabla_\nu)f_R = \kappa^2 T_{\mu\nu} - f_{\mathcal{T}^{(n)}}\Theta^{(n)}_{\mu\nu}, \tag{15}$$

where $f_{\mathcal{T}^{(n)}} \equiv \partial f/\partial\mathcal{T}^{(n)}$ and $f_R \equiv \partial f/\partial R$. The auxiliary tensor of the n-th order ($\Theta^{(n)}_{\mu\nu} \equiv \delta\mathcal{T}^{(n)}/\delta g^{\mu\nu}$), which can be explicitly obtained, similarly to (6), takes the following form:

$$\Theta^{(n)}_{\mu\nu} = n\mathcal{T}^{(n)}_{\mu\nu} - n\mathcal{L}_m\mathcal{T}^{(n-1)}_{\mu\nu} + n\left(\frac{1}{2}g_{\mu\nu}\mathcal{L}_m - T_{\mu\nu}\right)\mathcal{T}^{(n-1)} - 2ng^{\alpha\gamma}\frac{\partial^2\mathcal{L}_m}{\partial g^{\mu\nu}\partial g^{\beta\gamma}}\mathcal{T}^{\beta(n-1)}_{\alpha}, \tag{16}$$

which allows us to describe the geometrical field equations for $f(R,\mathcal{T}^{(n)})$ gravity.

On the other hand, in the scalar-tensor representation, in a similar manner as above, one can start from a generalization of Equation (8) as follows:

$$S = \frac{1}{2\kappa^2}\int_\Omega \sqrt{-g}[\phi R + \psi\mathcal{T}^{(n)} - V(\phi,\psi)]d^4x + \int_\Omega \sqrt{-g}\mathcal{L}_m d^4x, \tag{17}$$

which yields a similar set of field equations given by

$$\phi\left(R_{\mu\nu} - \frac{1}{2}Rg_{\mu\nu}\right) + \frac{1}{2}g_{\mu\nu}V - (\nabla_\mu\nabla_\nu - g_{\mu\nu}\Box)\phi = 8\pi T_{\mu\nu} - \psi\left(\Theta^{(n)}_{\mu\nu} - \frac{1}{2}g_{\mu\nu}\mathcal{T}^{(n)}\right), \tag{18}$$

which results in the now more general conservation equation given by

$$\left(\kappa^2 - \psi\right)\nabla^\mu T_{\mu\nu} = \left(T_{\mu\nu} + \Theta^{(n)}_{\mu\nu}\right)\nabla^\mu\psi + \psi\nabla^\mu\Theta^{(n)}_{\mu\nu} - \frac{1}{2}g_{\mu\nu}\left[R\nabla^\mu\varphi + \nabla^\mu\left(\psi\mathcal{T}^{(n)} - V\right)\right]. \tag{19}$$

Taking into account the definition of Equation (16), we have our field equations of a scalar-tensor gravity ($f(R, \mathcal{T}^{(n)})$). In particular, for the case of $n = 2$, the general $2n$-th auxiliary tensor takes the following form:

$$\Theta^{(2)}_{\mu\nu} = -2\mathcal{L}_m\left(T_{\mu\nu} - \frac{1}{2}g_{\mu\nu}T\right) - TT_{\mu\nu} + 2T_\mu{}^\alpha T_{\alpha\nu} - 4T^{\alpha\beta}\frac{\partial^2\mathcal{L}_m}{\partial g^{\mu\nu}\partial g^{\alpha\beta}}, \tag{20}$$

recovering the expression for the conservation equation previously derived in Equation (9).

$$\left(\kappa^2 - \psi\right)\nabla^\mu T_{\mu\nu} = \left(T_{\mu\nu} + \Theta^{(2)}_{\mu\nu}\right)\nabla^\mu\psi + \psi\nabla^\mu\Theta^{(2)}_{\mu\nu} - \frac{1}{2}g_{\mu\nu}\left[R\nabla^\mu\varphi + \nabla^\mu\left(\psi T^{(2)} - V\right)\right]. \tag{21}$$

As mentioned in [68], one may determine the extra force that appears from this non-conservation of the energy-momentum tensor in a manner analogous to that reported in [14], and it was shown to be related to MOND theory and other anomalies in test particle motions. Indeed, this fact may generically constitute a reliable test bed for these theories at local scales and at the Galactic level.

3. Thermodynamics of Open Systems

Large-scale matter creation is of the utmost importance in cosmology, as it can provide a description of the origin of the large-scale structure, as well as for the evolution of the various components that constitute the cosmological fluid. Motivated by the relevance of this topic, in this section, we summarize some of the existent literature regarding matter creation in the cosmological context, with an emphasis on the irreversible thermodynamics of open systems. Afterward, we apply the aforementioned framework to a spatially flat, homogeneous, and isotropic universe and briefly discuss its consequences.

3.1. Particle Production in Cosmology

The investigation of particle production processes in expanding universes started with the pioneering works of Erwin Schrödinger in 1939 and 1940 [69,70]. While studying a quantum wave packet propagating in an expanding Friedmann–Lemaitre–Robertson–Walker (FLRW) universe, Schrödinger reached a peculiar conclusion, namely that a scalar particle could have a non-zero probability of stimulating the creation of a pair of scalar particles spontaneously. As pointed out by Leonard Parker in Ref. [71], instead of having scalar particles, if one considers photons, then a single one could induce the creation of a pair of them. Indeed, Schrödinger referred to this occurrence as an "alarming" phenomenon. Unfortunately, since quantum field theory in curved spacetime was in its early stages at the time, some technical details regarding Schrödinger's approach were not sufficiently robust. Thus, the idea of exploring particle creation in expanding universes was abandoned for a while. In addition, it is important to mention that in curved, expanding spacetime, the question "what is matter?" is more difficult to answer because of the lack of a global time symmetry, a problem intrinsically related to the definition of energy in gravitation.

It was only in the late 1960s that the same Leonard Parker brought this research program back (although unaware of Schrödinger's works at the beginning), formulating the first successful mechanism for the production of particles by gravity [72–75]. Using the more developed mathematical apparatus of quantum field theory in curved spacetime, Parker demonstrated that there is a deep relationship between an expanding FLRW geometry and particle production. By considering a quantized massive scalar field in an expanding FLRW geometry, whose initial vacuum state (with no particles) is the Minkowski vacuum, Parker verified that the particle number associated with the final vacuum state of such an evolution was not zero. The physical process responsible for this peculiar result was attributed to the expansion of the Universe itself, meaning that a massive scalar field evolving in an FLRW geometry can give rise to particles spontaneously. Furthermore, it is also worth mentioning that this matter creation can be seen as the process responsible for originating the power spectrum of the inflation field perturbations during inflation [71], demonstrating possible imprints on physical observations.

Later, in the 1980s, Nobel Prize winner Ilya Prigogine and collaborators [76–78] searched for an alternative framework to accommodate cosmological particle production with the entropy evolution of the Universe. Since the semiclassical Einstein field equations used by Parker (and subsequently by others [79–81]) are both adiabatic and reversible, Prigogine et al. argued that these could not provide a natural explanation for the increase in cosmological entropy accompanied by the creation of particles due to the non-reversible nature of this process. As such, they proposed an alternative cosmology based on the irreversible thermodynamics of open systems, in which the explanation for macroscopic matter and entropy production relies on a reinterpretation of the matter energy-momentum tensor that includes an irreversible matter creation term.

Although this framework was successful at the time, because it was employed in GR, a physical interpretation of this irreversible process was missing. The main problem was, again, the adiabatic and reversibility character of the Einstein field equations or, equivalently, the covariant conservation of the energy-momentum tensor. However, in the last two decades, the appearance of classical modified gravity theories that contain previously discussed non-minimal geometry–matter couplings gave a new meaning to Prigogine's approach to cosmology. Since in all these theories, the matter energy-momentum tensor is not conserved, this feature, in the context of the thermodynamics of open systems, allows one to physically interpret such a non-conservation as an irreversible energy flow from the gravitational field to the matter sector that could result in particle creation [82]. The effects and implications of the irreversible matter creation processes on late cosmological evolution have been studied in some of these modified gravity theories by the authors of the present paper [32,82–84] (see [85] for a review). In addition, particle creation may also result from the following different processes:

- Vacuum instability in the presence of both gravitational and gauge fields, possibly resulting from the conformal trace anomaly, as shown in [86];
- The existence of quadratic curvature terms in the action of Weyl gravitational theory and the direct interaction of the perfect fluid particles. Hence, in such models, particles may also be created directly from the vacuum [87];
- Cosmological models such as the one presented in [88], in which there is an interaction between dark energy and massive particle pairs that can produce both stable and unstable particle pairs.

In summary, many different physical mechanisms exist that may generate particles. Nonetheless, we emphasize that since in EMSG, the energy-momentum tensor is not conserved, and the natural framework to study particle production is through the lens of the thermodynamics of open systems [82].

3.2. Thermodynamic Interpretation of Irreversible Matter Creation

In this subsection, we introduce the basics of this formalism by considering a spatially flat FLRW universe. Generally speaking, the application of the irreversible thermodynamics of open systems in cosmology relies on the basic assumption that the Universe can be seen as an effective thermodynamic open system [85]. From this premise, it follows that the bulk of the system corresponds to the observable Universe and the boundary to the apparent horizon (which, in a flat FLRW geometry, coincides with the Hubble radius), and the surroundings correspond to the unobservable universe [85].

This framework consists of two main equations, namely the first and second laws of thermodynamics.

3.2.1. First Law of Thermodynamics: Temperature Evolution

In particular, for a homogeneous universe, the heat (dQ) is negligible, so the first law can be expressed as

$$d(\rho V) + pdV = \frac{h}{n} d(nV), \tag{22}$$

where ρ is the energy density of the cosmological fluid, V the volume, p is the usual thermodynamic pressure of the fluid, $h = \rho + p$ is the enthalpy per unit volume, and $n = N/V$ is the particle number density. In addition, it is assumed that p and ρ describe non-exotic matter contents. Therefore, these must satisfy the following various energy conditions:

$$\rho \geq 0, \qquad p \geq 0, \qquad \rho + p \geq 0, \qquad \rho + 3p \geq 0, \qquad \rho \geq |p|. \tag{23}$$

To further develop Equation (22), one can take the time derivative and describe the volume of the Universe in terms of the scale factor ($a(t)$) as $V(t) = a^3(t)$, which yields

$$\dot{\rho} + 3H(\rho + p) = \frac{\rho + p}{n}(\dot{n} + 3Hn), \tag{24}$$

with $H = \dot{a}/a$ being the Hubble function. In order to further develop the above equation, it is convenient to introduce two quantities, namely the number current, defined as

$$N^{\mu} \equiv n u^{\mu}, \tag{25}$$

and the particle production rate (Γ). This quantity expresses the number of particles that are being created in comparison with the total number of particles at a specific instant, i.e.,

$$\Gamma = \frac{1}{N}\frac{dN}{dt} = \frac{1}{nV}\frac{d}{dt}(nV). \tag{26}$$

From Equations (25) and (26), it follows that the covariant derivative of the number current in the flat FLRW geometry has the following form:

$$\nabla_{\mu} N^{\mu} = \dot{n} + 3Hn \equiv n\Gamma, \tag{27}$$

By inserting Equation (27) into Equation (24), we obtain the particle production rate in terms of the Hubble function, pressure, and energy density.

$$\Gamma = 3H + \frac{\dot{\rho}}{\rho + p}. \tag{28}$$

Additionally, in the context of the thermodynamics of open systems, the total pressure makes two contributions, namely the usual thermodynamic pressure (p) and the creation pressure (p_c), i.e., $\tilde{p} = p + p_c$. The creation pressure can be viewed as the pressure responsible for describing, in an effective way, the irreversible matter creation processes that occur within the bulk of the open system, which, in this case, is the observable Universe. Hence, Equation (22) can be expressed as

$$d(\rho V) + (p + p_c)dV = 0. \tag{29}$$

With all the previous considerations in mind, one can find a relation between the creation pressure and the particle production rate.

$$p_c = -\left(\frac{\rho + p}{3H}\right)\Gamma. \tag{30}$$

With this framework established, it is feasible to obtain a cosmological temperature evolution. To do so, we start by assuming that the energy density (ρ) and the pressure (p) are both functions of the particle number density (n) and of the temperature ($\mathbb{T}$), that is,

$$\rho = \rho(n, \mathbb{T}), \qquad p = p(n, \mathbb{T}). \tag{31}$$

This allows us to express the first law as

$$\left(\frac{\partial \rho}{\partial n}\right)_{\mathbb{T}} \dot{n} + \left(\frac{\partial \rho}{\partial \mathbb{T}}\right)_{n} \dot{\mathbb{T}} + 3(\rho + p)H = (\rho + p)\Gamma. \tag{32}$$

From the equation above and using some useful thermodynamic relations, one can arrive at the following equation for the temperature evolution:

$$\frac{\dot{\mathbb{T}}}{\mathbb{T}} = c_s^2(\Gamma - 3H), \tag{33}$$

where we naturally define the speed of sound as $c_s = \sqrt{(\partial p/\partial \rho)_n}$ (for a more in depth derivation, we refer the reader to [85]). Then , the general solution of Equation (33) can be expressed as

$$\mathbb{T}(t) = \mathbb{T}_0 \exp\left\{c_s^2 \int_0^{t'} \left[\Gamma(t') - 3H\right] dt'\right\}, \tag{34}$$

where $\mathbb{T}_0 = \mathbb{T}(0)$ is the constant initial temperature.

3.2.2. Second Law of Thermodynamics: Entropy Evolution

Moreover, the second law of thermodynamics for an open system is usually stated as

$$d\mathcal{S} = d_e\mathcal{S} + d_i\mathcal{S} \geq 0, \tag{35}$$

where $d_e\mathcal{S}$ is called the entropy flow and $d_i\mathcal{S}$ represents entropy creation. It was demonstrated in [83] that these two entropy terms can be respectively written as

$$d_e\mathcal{S} = \frac{dQ}{\mathbb{T}}, \tag{36a}$$

$$d_i\mathcal{S} = \frac{s}{n} d(nV). \tag{36b}$$

As stated above, we are considering a homogeneous universe ($dQ = 0$); therefore, the entropy flow vanishes ($d_e\mathcal{S} = 0$). This implies that the variation of entropy is only dependent on the entropy of creation. Taking into account these considerations, the second law of thermodynamics is reduced to

$$d\mathcal{S} = \frac{s}{n} d(nV) \geq 0. \tag{37}$$

By recalling the definition of the creation rate in Equation (28), we can rewrite Equation (37) as

$$\dot{\mathcal{S}} = \mathcal{S}\Gamma \geq 0, \tag{38}$$

which has the following solution:

$$\mathcal{S}(t) = \mathcal{S}_0 \exp\left[\int_0^t \Gamma(t') dt'\right], \tag{39}$$

where $\mathcal{S}_0 = \mathcal{S}(0)$ is the constant initial entropy. It is important to note that the derivations presented above are not specific to EMSG but general solutions for modified theories of gravity working under the following assumptions:

1. Universe locally considered as an open system;

2. Geometry being flat FLRW.

Furthermore it is noteworthy that for theories where the energy-momentum tensor is conserved, the particle production rate also vanishes.

4. Cosmology of $f(R,\mathcal{T})$ Gravity

In this section, we briefly review the fundamentals of the cosmology of $f(R,\mathcal{T})$ gravity. We introduce the generalized Friedmann equations and discuss some of their implications. We examine the behavior of an important observational quantity, the deceleration parameter, in detail. Additionally, we cover the de Sitter evolutionary phases and explore some general cosmological models.

4.1. The Generalized Friedmann Equations

The first step in investigating the cosmological implications of $f(R,\mathcal{T})$ gravity is to consider the dynamical evolution within the flat, homogeneous, and isotropic Friedmann–Lemaître–Robertson–Walker metric given by

$$ds^2 = -dt^2 + a^2(t)\left[dr^2 + r^2\left(d\theta^2 + \sin^2\theta d\varphi^2\right)\right], \tag{40}$$

written in spherical coordinates as (t,r,θ,φ), where $a(t)$ is the scale factor, the homogeneity of cosmological spacetime requires that the scalar fields $\phi \equiv \phi(t)$ and $\psi \equiv \psi(t)$, and all thermodynamic quantities describing the matter content of the Universe depend only on the time coordinate (t). Additionally, we introduce the Hubble parameter ($H \equiv \dot{a}/a$, where $(\dot{} \equiv d/dt)$ denotes the derivative with respect to time).

We assume that the matter content of the Universe consists of an isotropic perfect fluid, with the energy-momentum tensor given by

$$T_{\mu\nu} = (\rho + p)u_\mu u_\nu + p g_{\mu\nu}, \tag{41}$$

where $\rho(t)$ is the energy density of the Universe and $p(t)$ is the isotropic pressure. Moreover, u^μ is the velocity four-vector of the fluid, satisfying the normalization condition of $u_\mu u^\mu = -1$.

The matter energy-momentum tensor (41) can be obtained from two different forms of the matter Lagrangian after variation with respect to the tensor component metric [89,90]. First, we note that the variation of the Lagrangian of a perfect fluid must be constrained by the requirements of the conservation of the rates of the entropy (s) and of particle production. The baryon number flux vector density is defined according to $n^\mu = nu^\mu\sqrt{-g}$, where the baryon number density n is given by

$$n = \sqrt{\frac{1}{g}n^\mu n^\nu g_{\mu\nu}}. \tag{42}$$

We require that the variation of the thermodynamic variables satisfies the constraints of $\delta s = 0$ and $\delta n^\mu = 0$ [89,90]. For the variation of the particle flux (nu^μ), we obtain the following relationship:

$$\nabla_\mu(nu^\mu) = \frac{1}{\sqrt{-g}}\frac{\partial}{\partial x^\mu}n^\mu = \frac{1}{\sqrt{-g}}\partial_\mu n^\mu. \tag{43}$$

Now, we take the variation of the gradient of δn^μ, and we take into account the relation of $\partial_\mu\delta = \delta\partial_\mu$. Thus, we obtain

$$\delta(\partial_\mu n^\mu) = \sqrt{-g}\left[\delta(\nabla_\mu(nu^\mu)) + \frac{1}{2}\nabla_\mu(nu^\mu)g^{\alpha\beta}\delta g_{\alpha\beta}\right] = 0 \tag{44}$$

This relation guarantees that the rate of particle production in the fluid is preserved under the variation (δ). This is a weaker demand than imposing the condition that the number density be conserved ($\nabla_\mu(nu^\mu) = 0$) from the beginning. A similar relation for the entropy ($\delta(n^\mu\partial_\mu s) = 0$) can also be easily obtained [89,90].

Now, we consider the assumption that the fluid matter Lagrangian is $\mathcal{L}_m = -\rho$. We also assume that the fluid obeys an equation of state ($\rho = \rho(n,s)$). From the thermodynamic equality ($(\partial\rho/\partial n)_s = w$, where w is the specific enthalpy $w = (\rho + p)n$), we obtain the relation of $\delta\rho = w\delta n$ [89,90].

With the above choice of the Lagrangian, the matter action is

$$S_m = -\int \rho\sqrt{-g}d^4x. \tag{45}$$

For the variation of the action (45), we first obtain

$$\delta S_m = -\int \left[\delta\rho\sqrt{-g} + \rho\delta\sqrt{-g}\right]d^4x. \tag{46}$$

By taking the variation of Equation (42), we find

$$\delta n = \frac{n}{2}(-g)u^\mu u^\nu\left(\frac{\delta g_{\mu\nu}}{g} - \frac{g_{\mu\nu}}{g^2}\delta g\right) = -\frac{n}{2}(u^\mu u^\nu + g^{\mu\nu})\delta g_{\mu\nu}. \tag{47}$$

Hence, Equation (46) becomes

$$\begin{aligned} \delta S_m &= -\frac{1}{2}\int [wn(u^\mu u^\nu + g^{\mu\nu}) - \rho g^{\mu\nu}]\sqrt{-g}d^4x \\ &= \frac{1}{2}\int [(\rho + p)u^\mu u^\nu + pg^{\mu\nu}]d^4x. \end{aligned} \tag{48}$$

This immediately yields the relation expressed by (41) for the matter energy-momentum tensor.

The same expression for the matter energy-momentum tensor is obtained by assuming that $\mathcal{L}_m = p$. We now express the equation of state in the form of $p = p(w', s)$, where $w' = (\rho + p)/\rho_0$, with ρ_0 being the rest mass density. In this case, we have the following thermodynamic relation [89,90]:

$$dp = \rho_0 dw' - \rho_0\mathbb{T}ds. \tag{49}$$

Generally, the four-velocity of the fluid can be represented as [89,90]

$$u^\mu = \frac{1}{w'}(\partial_\nu\phi + \alpha\partial_\nu\beta + \theta\partial_\nu s)g^{\mu\nu}, \tag{50}$$

where ϕ, α, β, θ, and s are the velocity potentials (scalar fields), which, together with w' and $g_{\mu\nu}$, comprise the dynamical variable of the fluid to be varied freely. Varying with respect to the metric and using the normalization condition of u^μ, we obtain

$$\delta w' = -\frac{1}{2w'}(\partial_\nu\phi + \alpha\partial_\nu\beta + \theta\partial_\nu s)(\partial_\mu\phi + \alpha\partial_\mu\beta + \theta\partial_\mu s)\delta g^{\mu\nu}. \tag{51}$$

By taking into account that $\delta g^{\mu\nu} = -g^{\mu\alpha}g^{\nu\beta}g_{\alpha\beta}$, we obtain

$$\delta w' = \frac{w'}{2}u^\mu u^\nu\delta g_{\mu\nu}. \tag{52}$$

By adopting the action expressed as

$$S_m = \int p\sqrt{-g}d^4x, \tag{53}$$

for the fluid and by assuming that the fluid is isothermal ($\mathbb{T} = cons\tan t$) and that $\delta s = 0$, the variation of Equation (53) yields

$$\delta S_m = \frac{1}{2}\int \left(\rho_0 w' u^\mu u^\nu + p g^{\mu\nu}\right)\sqrt{-g}\delta g_{\mu\nu} d^4x, \tag{54}$$

a relation from which we immediately reobtain Equation (41).

In the following, for the matter Lagrangian, we adopt the expression of $\mathcal{L}_m = p$ (see Ref. [91]), which, for the auxiliary tensor, yields $\Theta_{\mu\nu}$ the following expression:

$$\Theta_{\mu\nu} = -\left(\rho^2 + 4\rho p + 3p^2\right)u_\mu u_\nu. \tag{55}$$

To investigate the cosmological implications of the EMSG theory, we adopt the comoving reference frame, with $u_\mu = (1,0,0,0)$. Then, we obtain the system of two generalized Friedmann equations in $f(R,\mathcal{T})$ gravity given by

$$3H^2 = \frac{1}{\phi}\left[\kappa^2\rho + \frac{1}{2}V + \frac{1}{2}\psi\left(\rho^2 + 8\rho p + 3p^2\right) - 3H\dot{\phi}\right] = \kappa^2_{\text{eff}}\,\rho + \rho_{\text{eff}}, \tag{56a}$$

$$2\dot{H} = \frac{1}{\phi}\left[-\kappa^2(\rho+p) - \psi\left(\rho^2 + 4\rho p + 3p^2\right) + H\dot{\phi} - \ddot{\phi}\right] = -\kappa^2_{\text{eff}}\,(\rho+p) - (\rho_{\text{eff}} + p_{\text{eff}}), \tag{56b}$$

where $\kappa^2_{\text{eff}} = \kappa^2/\phi$, and we denote

$$\rho_{\text{eff}} = \frac{1}{\phi}\left[\frac{1}{2}V + \frac{1}{2}\psi\left(\rho^2 + 8\rho p + 3p^2\right) - 3H\dot{\phi}\right], \tag{57}$$

and

$$p_{\text{eff}} = \frac{1}{\phi}\left[-\frac{1}{2}V + \frac{1}{2}\psi\left(\rho^2 + 8\rho p + 3p^2\right) - 3H\dot{\phi} + \ddot{\phi}\right], \tag{58}$$

where ρ_{eff} and p_{eff} represent the effective scalar-tensor contributions to the cosmological evolution equations resulting from the coupling between matter and geometry via the square of the matter energy-momentum tensor. Moreover, the gravitational coupling constant becomes time-dependent, indicating the time variation of the gravitational constant in this model. The equations of motion for the scalar fields of ϕ and ψ are given by

$$V_\phi = R = 6\left(\dot{H} + 2H^2\right), \tag{59a}$$

$$V_\psi = \mathcal{T} = \rho^2 + 3p^2. \tag{59b}$$

4.1.1. The Energy Balance Equation

From Equations (56a) and (56b), one can obtain the generalized energy-conservation equation in the following form:

$$\frac{d}{dt}\left(\kappa^2_{\text{eff}}\,a^3\rho\right) + p\frac{d}{dt}\left(\kappa^2_{\text{eff}}\,a^3\right) + \frac{d}{dt}\left(a^3\rho_{\text{eff}}\right) + p_{\text{eff}}\frac{d}{dt}a^3 = 0. \tag{60}$$

Explicitly, we obtain

$$\dot{\rho} + 3H(\rho+p) = -\frac{1}{\kappa^2_{\text{eff}}}(\rho+p)\frac{d}{dt}\kappa^2_{\text{eff}} - \frac{1}{\kappa^2_{\text{eff}}}[\dot{\rho}_{\text{eff}} + 3H(\rho_{\text{eff}} + p_{\text{eff}})] \tag{61}$$

$$= (\rho+p)\Gamma, \tag{62}$$

which yields the following expression for the particle creation rate:

$$\Gamma = -\frac{1}{\kappa^2_{\text{eff}}}\frac{d}{dt}\kappa^2_{\text{eff}} - \frac{1}{(\rho+p)\kappa^2_{\text{eff}}}[\dot{\rho}_{\text{eff}} + 3H(\rho_{\text{eff}} + p_{\text{eff}})]. \tag{63}$$

4.1.2. The Deceleration Parameter

An important cosmological parameter, the deceleration parameter, is defined as

$$q = \frac{d}{dt}\frac{1}{H} - 1 = -\frac{\dot{H}}{H^2} - 1. \tag{64}$$

With the use of the gravitational field equations, the deceleration parameter is obtained in the following form:

$$q = \frac{3}{2}\frac{\kappa_{\rm eff}^2(\rho + p) + (\rho_{\rm eff} + p_{\rm eff})}{\kappa_{\rm eff}^2\rho + \rho_{\rm eff}} - 1, \tag{65}$$

or explicitly,

$$q = \frac{3}{2}\frac{\kappa^2(\rho + p) + \psi(\mathcal{T} + 4\rho p) - H\dot{\phi} + \ddot{\phi}}{\kappa^2\rho + \frac{1}{2}V + \frac{1}{2}\psi(\mathcal{T} + 8\rho p) - 3H\dot{\phi}} - 1. \tag{66}$$

Thus, the condition for accelerating expansion ($q < 0$) yields

$$\kappa_{\rm eff}^2\left(\frac{\rho}{3} + p\right) < -\left(\frac{\rho_{\rm eff}}{3} + p_{\rm eff}\right). \tag{67}$$

Explicitly, we obtain the condition for the accelerated expansion as

$$\kappa_{\rm eff}^2\left(\frac{\rho}{3} + p\right) < \frac{V}{3} - \frac{2}{3}\psi(\mathcal{T} - 2\rho p) + H\dot{\phi} - \ddot{\phi}. \tag{68}$$

For a dust Universe, the late time-accelerated phase occurs for

$$\kappa_{\rm eff}^2\rho < \frac{V}{3} - 2\psi\rho^2 + 3H\dot{\phi} - 3\ddot{\phi}. \tag{69}$$

4.1.3. Dark Matter and Dark Energy

The EMSG theory generates, in its two scalar field representations, an effective energy and an effective pressure of an essentially geometric nature, as expressed by Equations (57) and (58), respectively. These effective geometric quantities can be interpreted as describing both dark matter and dark energy at the same time. The presence of an effective gravitational constant (κ^2/ϕ) could already have some important implications for the behavior of massive particles gravitating around galaxies—behavior usually explained by assuming the presence of dark matter. However, an effective gravitational constant appearing in Newton's law ($F_{grav} = -G_{eff}(r)mM/r^2$) can already provide some insights into the dark matter problem, since, for the rotational velocity, the modified Newtonian–Keplerian law ($v^2/r = G_{eff}M/r^2$) yields $v \approx \sqrt{GM/\phi r}$, which could explain the flatness of the rotation curves once $\phi \sim 1/r$. The effective energy density (ρ_{DM}) and pressure (p_{DM}) of dark matter can be inferred from Equations (57) and (58) by assuming that in the case of pressureless matter with $p = 0$, $\rho_{DM} = p_{DM} = (\psi/2\phi)\rho^2$. Hence, in this model, there is a close relationship between (effective) dark matter and ordinary matter. Interestingly enough, in this model, dark matter satisfies the stiff (causal) equation of state, with $\rho = p$.

From a cosmological point of view, $\rho_{\rm eff}$ and $p_{\rm eff}$ can be interpreted as globally describing both dark components of the Universe, and in our analysis we do not make an explicit distinction between dark matter and dark energy, treating the dark components of the Universe as a single physical entity. However, by considering the above interpretation of dark matter, we may assume that the dark energy term is described by time-dependent energy density and pressure expressed as $\rho_{DE} = (V/2 - 3H\dot{\phi})/\phi$ and $p_{DE} = (-V/2 - 3H\dot{\phi} + \ddot{\phi})/\phi$, respectively. The equation of state of the dark energy depends on the explicit choice of V and on the dynamical evolution of the two scalar fields. Then, a necessary condition for the late accelerated expansion of the Universe is that the parameter of the dark energy equation of state ($w = p_{DE}/\rho_{DE}$) must be negative, i.e., $w < 0$.

4.2. de Sitter Expansion

The fact that the Universe experiences an accelerated expansion, which may end in a de Sitter-type phase of the Universe, is an important result of observational cosmology. Hence, viable theoretical cosmological models must have an explicitly exponentially expanding solution. We now consider the de Sitter-type expansion of the Universe with $H = H_0 =$ constant in $f(R, \mathcal{T})$ gravity. The theory admits exponential expansion for both vacuum and dust universes.

4.2.1. Self-Interacting Potential and Constant-Density de Sitter Expansion

We first consider the self-interacting potential during de Sitter evolution in the constant-density dust case, which is obtained in the de Sitter phase with $H = H_0 =$ constant, $p = 0$, and $\rho = \rho_0 =$ constant. Then, by integrating (59a) and (59b), we immediately find

$$V(\phi, \psi) = 12H_0^2\phi + \rho_0^2\psi + \Lambda_0, \tag{70}$$

where Λ_0 is the resulting constant of integration. For this specific form of the potential, the energy balance equation becomes

$$\kappa^2[\dot{\rho} + 3H(\rho + p)] + (\rho + p)(\rho + 3p)(\dot{\psi} + 3H\psi) + \psi[\dot{\rho}(\rho + 4p) + \dot{p}(4\rho + 3p)] = 0. \tag{71}$$

The matter creation rate for the constant-density de Sitter phase is given by

$$\Gamma = -\frac{1}{\kappa^2}\left[(\dot{\psi} + 3H\psi)(\rho + 3p) + \psi\frac{\dot{\rho}(\rho + 4p) + \dot{p}(4\rho + 3p)}{(\rho + p)}\right]. \tag{72}$$

The creation pressure can be obtained as

$$p_c = \frac{\rho + p}{3H\kappa^2}\left[(\dot{\psi} + 3H\psi)(\rho + 3p) + \psi\frac{\dot{\rho}(\rho + 4p) + \dot{p}(4\rho + 3p)}{(\rho + p)}\right]. \tag{73}$$

4.2.2. The Vacuum de Sitter Solution

For the vacuum case, with $\rho = p = 0$, for the de Sitter expansion, the potential (V) becomes $V(\phi, \psi) = 12H_0^2\phi$. The evolution equation of the scalar field (ϕ) that follows from Equation (56b) is

$$H_0\dot{\phi} - \ddot{\phi} = 0, \tag{74}$$

with the general solution of

$$\phi(t) = C_1\frac{e^{H_0 t}}{H_0} + C_2, \tag{75}$$

where C_1 and C_2 are arbitrary constants of integration. For this solution, Equation (56a) gives $3C_2H_0^2 = 0$, a condition that is satisfied by $C_2 = 0$. Hence, the exact vacuum de Sitter solution corresponds to an exponentially increasing $\phi = C_1 e^{H_0 t}/H_0$ in the presence of an arbitrary field (ψ). The particle creation rate (Γ) is zero ($\Gamma \equiv 0$) for the vacuum Universe, and the energy balance equation (Equation (71)) is identically satisfied. It is important to note that in the present model, matter creation takes place only if matter already exists.

4.2.3. de Sitter Solution with Constant Matter Density

We now assume that the Universe consists of pressureless dust, with $p = 0$, and that the matter density is constant ($\rho = \rho_0$). From Equation (71), we obtain the following differential equation for ψ:

$$\dot{\psi}(t) + 3H_0\psi + \frac{3\kappa^2 H_0}{\rho_0} = 0, \tag{76}$$

with the general solution of

$$\psi(t) = e^{-3H_0 t}\left[\psi_0 + \frac{\kappa^2}{\rho_0}\left(1 - e^{3H_0 t}\right)\right], \tag{77}$$

where $\psi_0 = \psi(0)$. From Equation (56a), it follows that the differential equation that describes the scalar field (ϕ) is given by

$$\dot{\phi}(t) - H_0\phi - \frac{1}{6H_0}\left[\psi\left(\rho_0^2 - \rho_0\right) + 2\kappa^2\rho_0 + \Lambda_0\right] = 0, \tag{78}$$

with the general solution of

$$\begin{aligned}\phi(t) \quad &= \quad \frac{1}{12H_0^2}\left\{e^{-3H_0 t}\left[-2e^{3H_0 t}\Lambda_0 - \rho\left(k^2 + \rho\psi_0\right)\right.\right.\\ &\left.\left. + e^{4H_0 t}\left(2\Lambda_0 + \kappa^2\rho + 12H_0^2\phi_0 + \rho^2\psi_0\right)\right]\right\},\end{aligned} \tag{79}$$

where $\phi_0 = \phi(0)$. The creation pressure and the creation rate can now be obtained as

$$\Gamma = 3H_0, \quad p_c = -\rho_0. \tag{80}$$

4.2.4. de Sitter Solution with Arbitrary Matter Density

For a de Sitter Universe with time-varying, non-zero matter density, the generalized Friedmann equations (Equations (56a) and (56b)) take the following form:

$$3H_0^2\phi = \kappa^2\rho + \frac{1}{2}V + \frac{1}{2}\psi\rho^2 - 3H_0\dot{\phi}, \tag{81}$$

and

$$-\kappa^2\rho - \psi\rho^2 + H_0\dot{\phi} - \ddot{\phi} = 0, \tag{82}$$

The solutions of the generalized Friedmann equations, as well as the cosmological dynamics, depend on the form of the potential (V). In the following, we assume that V can be represented as

$$V(\phi,\psi) = 12H_0^2\phi - \frac{\alpha^2}{\psi}. \tag{83}$$

Then from Equation (59b), we first obtain $V_\psi = \alpha^2/\psi^2 = \rho^2$, $\psi = \alpha/\rho$, and $\dot{\psi} = -\alpha\dot{\rho}/\rho^2$.

Equation (71) takes the following form:

$$\dot{\rho}\left(\kappa^2 + \psi\rho\right) + \rho^2(\dot{\psi} + 3H_0\psi) + 3\kappa^2 H_0\rho = 0, \tag{84}$$

and

$$\dot{\rho} + 3H_0\left(1 + \frac{\alpha}{\kappa^2}\right)\rho = 0. \tag{85}$$

Hence, the matter density is obtained as

$$\rho(t) = \rho_0 e^{-3H_0\left(1+\frac{\alpha}{\kappa^2}\right)t}, \tag{86}$$

where $\rho_0 = \rho(0)$. From Equation (82), for ϕ, we obtain the following equation:

$$\ddot{\phi} - H_0\dot{\phi} + \left(\kappa^2 + \alpha\right)\rho_0 e^{-3H_0\left(1+\frac{\alpha}{\kappa^2}\right)t} = 0, \tag{87}$$

the general solution of which is expressed as

$$\phi(t) = \frac{c_1 e^{H_0 t}}{H_0} + c_2 - \frac{\kappa^4 \rho_0 e^{-\frac{3H_0 t\left(\alpha+\kappa^2\right)}{\kappa^2}}}{3H_0^2(3\alpha + 4k^2)}, \tag{88}$$

where c_1 and c_2 are arbitrary constants of integration. The first Friedmann Equation (56a) yields $3H_0^2 c_2 = 0$, or $c_2 = 0$. Thus, this solution satisfies all cosmological evolution equations.

In the present model, the particle creation rate, given by

$$\Gamma = -\frac{1}{\kappa^2}[(\dot{\psi} + 3H\psi)\rho + \psi\dot{\rho}], \tag{89}$$

becomes

$$\Gamma = -\frac{3\alpha}{\kappa^2} H_0. \tag{90}$$

Thus, for the de Sitter-type solution corresponding to the potential in Equation (83), the thermodynamics of open systems require a negative particle creation rate, corresponding to a decrease in the density of ordinary matter.

4.3. Matter and Radiation Domination Phases

When the Universe was about 47,000 years old, it entered into the matter-dominated phase, in which the energy density of matter largely exceeded the energy densities of radiation and dark energy (cosmological constant). The matter-dominated phase lasted until the Universe was about 9.8 billion years old [92]. During this phase, the standard Big Bang cosmological scenario predicts that the scale factor of the Universe evolved as $t \propto t^{2/3}$, while the deceleration parameter took a constant value of $q = 1/2$. The matter density varied during this period as $\rho(t) \propto 1/a^3(t) = 1/t^2$. In the following, we investigate the conditions under which a similar matter-dominated era could exist in the energy-momentum squared gravity theory, assuming a more general form for the scale factor than that of standard general relativity ($a = t^n$, $n > 0$).

4.3.1. Models with a Quadratic Additive Potential

We consider a case in which during the matter-dominated phase, the potential ($V(\phi, \psi)$) has the following quadratic form:

$$V(\phi, \psi) = \frac{\alpha}{2}\phi^2 + \frac{\beta}{2}\psi^2, \tag{91}$$

where α and β are constants. Then, from Equation (59a), we obtain

$$\alpha\phi = \frac{6n(2-n)}{t^2}, \tag{92}$$

while Equation (59b) yields

$$\beta\psi = \rho^2. \tag{93}$$

With these choices for V, ϕ, and ψ, the generalized Friedmann equations of the EMSG theory take the following forms:

$$27\beta n^2(2n-1) + \alpha\beta\kappa^2 t^4 \rho(t) + \alpha t^4 \rho^4(t) = 0 \tag{94}$$

and

$$36\beta n(n-1) + \alpha\beta\kappa^2 t^4 \rho(t) + \alpha t^4 \rho^4(t) = 0. \tag{95}$$

For n, the required equivalence of Equations (94) and (95) yields the following algebraic equation:

$$-54\beta n^3 + 99\beta n^2 - 36\beta n = 0, \tag{96}$$

which fixes n as

$$n = 0, n = \frac{1}{2}, n = \frac{4}{3}. \tag{97}$$

The matter density can be obtained by solving any of the algebraic equations (Equation (94) or (95)).

For $n = 1/2$, the density takes a constant value, i.e., $\rho(t) = \rho_0 = (-\beta)^{1/3}$. For $n = 4/3$, the density can be obtained as a solution of the following algebraic equation:

$$\alpha t^4 \rho^4(t) + \kappa^2 \alpha \beta t^4 \rho(t) + 80\beta = 0, n = \frac{4}{3}. \tag{98}$$

In the first order of approximation, we neglect the term containing the fourth power of the density, obtaining

$$\kappa^2 \rho(t) \approx -\frac{80}{\alpha}\frac{1}{t^4}, n = \frac{4}{3}. \tag{99}$$

In order to assure the physical viability of the density, the ratio of $1/\alpha$ must be negative, i.e., $1/\alpha < 0$.

Hence, we have obtained two distinct solutions of the EMSG theory in the matter-dominated phase of cosmological evolution in the presence of an additive quadratic potential (V). In the first solution, $a(t) = t^{1/2}$, $H(t) = 1/2t$, $\rho(t) =$ constant, and $q = 1$. The density of the Universe is constant, and the expansion is decelerating. The constancy of density is maintained by the particle creation rate, given by $\Gamma = 3H = 3/2t$, which decreases over time. But, even so, it can maintain the constancy of the matter density. In the second solution, $a = t^{4/3}$, $H(t) = 4/37$, $\rho(t) \propto 1/t^4$, and $q = -1/4$. In this model, the Universe is already accelerating in the matter-dominated phase but with a relatively low value of the deceleration parameter. By taking the derivative of the density equation (Equation (98)), we obtain

$$\frac{\dot{\rho}}{\rho} = \frac{320\beta}{\alpha\rho(\rho^3 + \kappa^2\beta)}, \tag{100}$$

a relation that allows for the expression of the particle creation rate as

$$\Gamma = 3H + \frac{320\beta}{\alpha\rho(\rho^3 + \kappa^2\beta)}, n = \frac{4}{3}. \tag{101}$$

In the approximation of $\rho \propto t^{-4}$, the particle creation rate is $\Gamma = 3H + \dot{\rho}/\rho \approx 0$. However, when using the exact solution of the algebraic density equation, the particle creation rate is non-zero, indicating the presence of matter creation during an accelerating expansionary phase.

4.3.2. Radiation-Dominated Models

In the very early stages of cosmological evolution, most of the total energy density was in the form of radiation, which was the major constituent of the Universe. With the expansion of the Universe and the cooling down of radiation, matter became the dominant component. Although we are presently in a phase dominated by dark energy, an understanding of the roles of radiation and matter is fundamental for the description of the evolution of the early Universe.

We now consider the effects of the EMSG theory on radiation-dominated cosmological models. The generalized Friedmann equations in the presence of radiation with energy density (ρ_r) and the equation of state ($p_r = \rho_r/3$) take the following forms:

$$3H^2 = \frac{1}{\phi}\left[\kappa^2\rho_r + \frac{1}{2}V + 2\psi\rho_r^2 - 3H\dot{\phi}\right] \tag{102}$$

and

$$2\dot{H} = \frac{1}{\phi}\left[-\kappa^2\frac{4}{3}\rho_r - \frac{8}{3}\psi\rho_r^2 + H\dot{\phi} - \ddot{\phi}\right]. \tag{103}$$

For the potential (V), we assume a simple, multiplicative form so that

$$V(\phi,\psi) = \alpha\phi\psi, \tag{104}$$

where α is a constant. Hence, we immediately obtain

$$V_\phi = \alpha\psi = 6\left(\dot{H} + 2H^2\right), \alpha\phi = \rho_r^2. \tag{105}$$

Then, the generalized Friedmann equations of the EMSG theory become

$$15\dot{H} - \frac{6H\dot{\rho}}{\rho} + 27H^2 + \frac{\alpha\kappa^2}{\rho} = 0, \tag{106}$$

$$18\dot{H} + \frac{-6H\dot{\rho} + 4\alpha\kappa^2 + 6\ddot{\rho}}{3\rho} + 32H^2 + \frac{2\dot{\rho}^2}{\rho^2} = 0. \tag{107}$$

We again look for solutions of the above generalized Friedmann equations with $a(t) = t^n$, where n is a constant. With this choice of the scale factor, Equations (106) and (107) take the following forms

$$-\alpha k^2 t^2 - 27n^2\rho + 15n\rho + 6nt\dot{\rho} = 0, \tag{108}$$

and

$$\frac{\ddot{\rho}}{\rho} + \frac{\dot{\rho}^2}{\rho^2} - \frac{n\dot{\rho}}{t\rho} + \frac{2\alpha k^2}{3\rho} + \frac{n(16n-9)}{t^2} = 0, \tag{109}$$

respectively. From Equation (108), we obtain $\dot{\rho}$ as

$$\dot{\rho} = \frac{\alpha\kappa^2 t}{6n} + \frac{(9n-5)\rho}{2t}. \tag{110}$$

By taking the time derivative of the above equation and substituting the result into Equation (109), we obtain the following for the time variation of the density the first-order differential equation:

$$\dot{\rho} = \frac{3n\rho\{8\alpha k^2(1-3n)t^2 - 3n[n(127n-134)+35]\rho\} - \alpha^2k^4t^4}{18n^2(9n-5)t\rho}. \tag{111}$$

Equation (111) cannot be generally solved for arbitrary values of n. However, some simple particular solutions can be obtained for some particular values of n. Thus, for $n = 1/3$, we obtain the following expression for the variation of matter density in the radiation-dominated phase of EMSG theory cosmology:

$$\rho(t) = \sqrt{c_1 t^{20/9} + \frac{9}{32}\alpha^2 k^4 t^4}, \tag{112}$$

where c_1 is an arbitrary constant of integration. $\rho(t)$ has the property of $\rho(0) = 0$. The density of the Universe is increasing during the expansionary phase, and a large amount of radiation is produced. This qualitative behavior is similar to that considered in the warm inflationary scenario [93–98], which represents an interesting and successful alternative to standard inflation and reheating. In warm inflation, not only the dynamics of the scalar field but also the creation and effects of radiation during the process of accelerated expansion are considered. Radiation is produced via the decay of the scalar field as a result of a dissipative process. A similar result can be obtained in the framework of the EMSG theory. The non-conservation of the matter energy-momentum tensor, which is associated with particle creation, allows for the production of photons during the early expansion of the

Universe. Hence, radiation creation could take place as triggered by the presence of the two essentially geometric scalar fields (ϕ and ψ), and the interplay between these two fields and gravity may be responsible for the creation of the matter content of the early Universe. It is also important to note that in the present model, matter creation takes place during a decelerating phase of evolution, with $q = 2$. Of course, other values of n could lead to power law accelerating models, which may also explain other observational features of the early Universe.

4.4. General Cosmological Models

We now investigate cosmological models with the Hubble function (H), an arbitrary function of time whose behavior must be obtained from the generalized Friedmann equations. We also require that H tend towards a constant value in the final stages of cosmological evolution, that is, that the Universe end in a de Sitter-type phase.

4.4.1. The Dimensionless Representation

A significant simplification of the mathematical formalism can be obtained by introducing a set of the dimensionless variables (h, τ, r, U, Ψ), defined as

$$H = H_0 h, \qquad \tau = H_0 t, \qquad \rho = \frac{3H_0^2}{\kappa^2} r, \qquad V = 3H_0^2 U, \qquad \psi = \frac{\kappa^4}{9H_0^2}\Psi. \tag{113}$$

Then, the system of differential equations describing cosmological evolution in $f(R, \mathcal{T})$ gravity takes the following form:

$$h^2 = \frac{1}{\phi}\left(r + \frac{1}{2}U + \frac{1}{6}\Psi r^2 - h\frac{d\phi}{d\tau}\right), \tag{114}$$

$$\frac{dh}{d\tau} = \frac{1}{\phi}\left(-\frac{3}{2}r - \frac{1}{2}\Psi r^2 + \frac{1}{2}h\frac{d\phi}{d\tau} - \frac{1}{2}\frac{d^2\phi}{d\tau^2}\right), \tag{115}$$

$$\left(1 + \frac{1}{3}r\Psi\right)\frac{dr}{d\tau} + 3hr + \frac{1}{3}r^2\left(\frac{d\Psi}{d\tau} + 3h\Psi\right) = 0. \tag{116}$$

The system of Equations (114)–(116) contains two independent dynamical equations but four unknown functions (h, r, ϕ, Ψ). To close the system, one must first specify the potential (U); then, using Equations (59a) and (59b), one can obtain two relations for ϕ and ψ, given by

$$U_\phi = 2\left(\frac{dh}{d\tau} + 2h^2\right), \qquad U_\Psi = \frac{1}{3}r^2. \tag{117}$$

The potential can now be represented as

$$U_\phi = \frac{2}{\phi}\left(\frac{1}{2}r - \frac{1}{6}\Psi r^2 - \frac{3}{2}h\frac{d\phi}{d\tau} - \frac{1}{2}\frac{d^2\phi}{d\tau^2}\right). \tag{118}$$

4.4.2. The Redshift Representation

A comparison with the observational data of cosmological models can be performed in the easiest way by considering the evolution of the models and adopting redshift (z) as an independent variable. Redshift is defined as

$$1 + z = \frac{1}{a}, \tag{119}$$

yielding $\frac{d}{d\tau} = -(1+z)h\frac{d}{dz}$.

After introducing the new variable ($d\phi/d\tau = \theta$), the cosmological evolution equations in the two scalar field representations of $f(R,\mathcal{T})$ gravity can be formulated in a redshift space as a first-order dynamical system given by

$$-(1+z)h\frac{d\phi}{dz} = \theta, \tag{120}$$

$$-(1+z)h\frac{dh}{dz} = \frac{1}{\phi}\left[-\frac{3}{2}r - \frac{1}{2}\Psi r^2 + \frac{1}{2}h\theta + \frac{1}{2}(1+z)h\frac{d\theta}{dz}\right], \tag{121}$$

$$-(1+z)\left(1+\frac{1}{3}r\Psi\right)h\frac{dr}{dz} + 3hr + \frac{1}{3}r^2\left[-(1+z)h\frac{d\Psi}{dz} + 3h\Psi\right] = 0, \tag{122}$$

$$(1+z)h\frac{d\theta}{dz} = \phi U_\phi - r + \frac{1}{3}\Psi r^2 + 3h\theta, \tag{123}$$

$$(1+z)h\frac{dh}{dz} = 2h^2 - \frac{1}{2}U_\phi, \tag{124}$$

$$U_\Psi = \frac{1}{3}r^2. \tag{125}$$

The system of Equations (120)–(125) must be solved together with the initial conditions of $h(0) = 1$, $r(0) = r_0$, $\phi(0) = \phi_0$, and $\theta(0) = \theta_0$. No dynamical evolution equation for ψ exists, but the evolution of h is described by two equivalent equations. Thus, one can consider only one of Equations (121) and (124) when solving the cosmological field equations.

To describe the accelerating/decelerating phases of the evolution of the Universe, we use the deceleration parameter (q), which takes the following form in the redshift space:

$$q = \frac{d}{d\tau}\frac{1}{h} - 1 = (1+z)\frac{1}{h}\frac{dh}{dz} - 1. \tag{126}$$

The negative sign of q indicates an accelerated evolution of the Universe, whereas a positive sign describes the decelerating stages of evolution.

The particle creation rate is obtained from the following relation:

$$\Gamma = H_0\left[3h - (1+z)\frac{1}{r}\frac{dr}{dz}\right] = H_0\Sigma, \tag{127}$$

while the creation pressure is given by

$$P_c = \frac{p_c\kappa^2}{H_0^2} = -\frac{r}{h}\Sigma. \tag{128}$$

4.4.3. Specific Cosmological Models: A Qualitative Discussion

In the following, we briefly discuss, , the general properties of two distinct cosmological models corresponding to two independent functional forms of the potential (U) from a qualitative point of view.

Additive power law potential: $U(\phi,\Psi) = 2\alpha\phi^{n+1} + \frac{1}{3}\beta\Psi^{m+1}$. The first case corresponds to a potential ($U(\phi,\psi)$) with an additive algebraic structure of the following form:

$$U(\phi,\Psi) = 2\alpha\phi^{n+1} + \frac{1}{3}\beta\Psi^{m+1}, \tag{129}$$

where α, β, n, and m are constants. The system of differential equations describing cosmological evolution must be solved with initial conditions or $h(0) = 1$, $r(0) = r_0$, $\phi(0) = \phi_0$, and $\theta(0) = \theta_0$. In this model, cosmological evolution depends on the model parameters, as well as on the initial conditions for the scalar field (ϕ) and its derivative. The variations of the dimensionless Hubble function ($h(z)$) and of the deceleration parameter of this model are represented in Figure 1.

For $n \approx 1$, $m = 2$, $\beta = 1$, and α in the range of $\alpha \in (0.38, 0.42)$, this model provides a very good description of the observational data for the Hubble function up to a redshift ($z \approx 1$) [32]. However, at higher redshifts ($z > 1$), important differences appear with respect to the observational data and the ΛCDM predictions, at least for the considered range of model parameters. At redshifts greater than $z = 1$, the Hubble function increases faster as compared to the ΛCDM model. The evolution of the deceleration parameters of the two models also shows important differences.

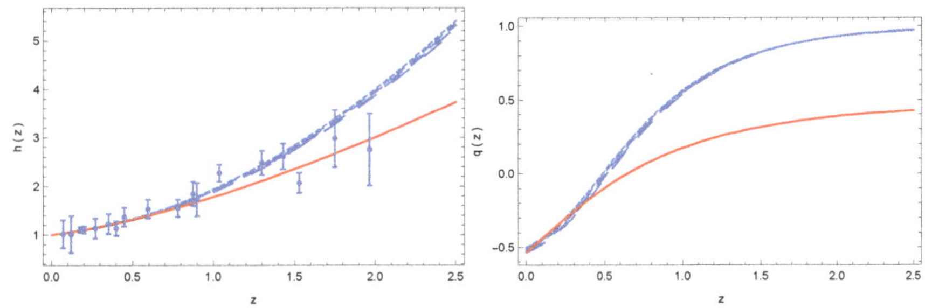

Figure 1. The evolution of the dimensionless Hubble function ($h(z)$) (**left panel**) and of the deceleration parameter ($q(z)$) (**right panel**) as a function of the redshift (z) in the EMSG cosmological model with $U(\phi, \Psi) = 2\alpha\phi^{n+1} + \frac{1}{3}\beta\Psi^{m+1}$ for $n = 0.975$, $m = 2$, $\beta = 1$, $r(0) = 0.30$, and different values of α ($\alpha = 0.368$, solid curve; $\alpha = 0.370$, dotted curve; $\alpha = 0.372$, short dashed curve; $\alpha = 0.374$, dashed curve; $\alpha = 0.376$, long dashed curve). Predictions of the standard ΛCDM model are represented by the solid red curve, while the observational data are shown with their error bars.

At redshifts of $z > 1.5$, the deceleration parameter in $f(R, \mathcal{T})$ gravity takes higher values close to $q \approx 1$. The transition to the accelerated expansion phase occurs at redshifts of $z < 0.5$. On the other hand, the present-day value of the deceleration parameter is dependent on the model parameters.

Additive–multiplicative potential: $U(\phi, \psi) = \alpha\phi + (1/3)\beta\Psi + (\gamma/3)\phi^n\Psi^{m+1}$. As a second cosmological model, one can consider the model corresponding to potential (U) with a non-additive structure, given by

$$U(\phi, \psi) = \alpha\phi + \frac{1}{3}\beta\Psi + \frac{\gamma}{3}\phi^n\Psi^{m+1}, \tag{130}$$

where α, β, γ, n, and m are constants. The variations of the dimensionless Hubble function ($h(z)$) and of the deceleration parameter for this model are represented in Figure 2.

For $n = 10^{-8}$, $m = 1.05$, $\beta = 0.02$, $\gamma = 0.04$, and $\alpha \in (3.50, 3.6)$, the model provides a good description of the observational data and of the Hubble function up to a redshift of $z \approx 2$ [32]. Differences in the behavior of the deceleration parameter appear at both low and high redshifts. The model also predicts a lower matter density as compared to ΛCDM [32].

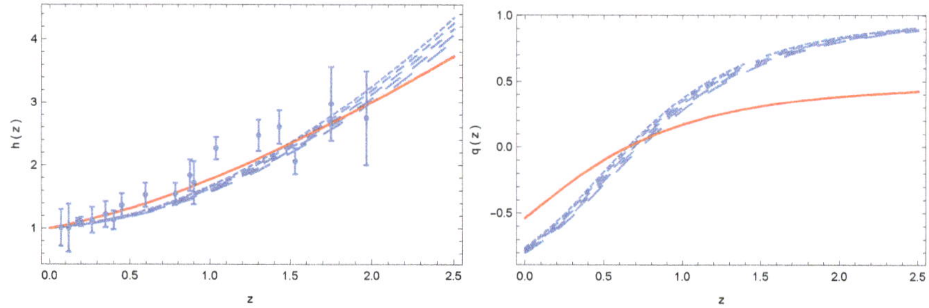

Figure 2. The evolution of the dimensionless Hubble function ($h(z)$) (**left panel**) and of the deceleration parameter ($q(z)$) (**right panel**) as a function of the redshift (z) in the EMSG cosmological model with $U(\phi,\Psi)=U(\phi,\psi)=\alpha\phi+\frac{1}{3}\beta\Psi+\frac{\gamma}{3}\phi^{n}\Psi^{m+1}$ for $n=10^{-8}$, $m=1.05$, $\beta=0.02$, $\gamma=0.04$, $r(0)=0.30$, and different values of α ($\alpha=0.352$, solid curve; $\alpha=0.354$, dotted curve; $\alpha=0.356$, short dashed curve; $\alpha=0.358$, dashed curve; $\alpha=0.360$, long dashed curve). Predictions of the standard ΛCDM model are represented by the red solid curve, while the observational data are shown with their error bars. The initial conditions used to integrate the generalized Friedmann equations are $\phi(0)=0.10$, and $\theta(0)=-0.01$.

4.5. Summary and Discussion

Our preliminary results on the cosmology of $f(R,\mathcal{T})$ gravity point towards the possibility that the considered theory and the cosmological models that can be built using it can provide qualitative and quantitative descriptions of the observational data once the optimal values of the model parameters have been obtained either by a trial-and-error method or by systematic fitting of the observational data. The models we discussed above can provide both a qualitative and quantitative descriptions of the observational Hubble data for a limited range of redshifts, and they can also reproduce the predictions of the ΛCDM model.

An interesting issue related to $f(R,\mathcal{T})$ gravity and to the cosmological models based upon it is the possibility of at least alleviating, if not solving, the Hubble tension. One can see from the first Friedmann equation (Equation (56a)) that the present-day value of the Hubble function is fully determined by the present-day values of the energy density of matter, potential, and the two scalar fields. These quantities satisfy the constraint of the (closure relation).

$$H_0^2\frac{\dot{\phi}(0)}{\phi(0)}H_0=\kappa^2\frac{\rho(0)}{3\phi(0)}+\frac{1}{6}\frac{V(0)}{\phi(0)}+\frac{1}{6}\frac{\psi(0)}{\phi(0)}\rho^2(0). \tag{131}$$

By appropriately fixing the initial values of the matter energy density, the potential (V), the ϕ and ψ fields, and their derivatives, the present-day value of the Hubble function can be obtained in concordance with supernova data.

Moreover, the function $H(z)$ can generally be represented as

$$H(z)=H_0\Phi(z), \tag{132}$$

where

$$\Phi(z)=\sqrt{\frac{r(z)+U(z)/2+\Psi(z)r^2(z)}{1-(1+z)\frac{d}{dz}\ln\phi(z)}}. \tag{133}$$

If the ΛCDM model is correct for all redshifts, then $H(0)=H_0\Phi(0)$, $\Phi(0)=1$, and H_0 can be determined from the value obtained from supernova observations ($H_0=H_0^{sup}=H(0)/\Phi(0)$). But if the ΛCDM model is not correct for all z, then $\Phi(0)$ may be different from one or have values of the order of one that minimally deviate from the ΛCDM value. On the other hand, as estimated from the early Universe, $\Phi(z)$ yields a different value for H_0. Let us assume that at a redshift of z_r, the function $\Phi(z)$ takes a value of $\Phi(z_r)=0.92$. Then, the present-day value of the Hubble function is $H_0 = H(z_r)/\Phi(z_r) = H_0^{Planck}$. Therefore, we have

$$\frac{H_0^{Planck}}{H_0^{sup}} = \frac{H(z_r)}{H(0)} \times \frac{\Phi(0)}{\Phi(z_r)} = \zeta. \tag{134}$$

For $\zeta = 0.92$, $H_0^{Planck} = \zeta \times H_0^{sup} = 0.92 = 67.16$ km/s/Mpc [32]. Numerical values of this order of magnitude for ζ can be obtained in the $f(R, T_{\mu\nu}T^{\mu\nu})$ gravity theory for appropriate initial conditions of the two scalar fields. Of course, the form of the potential (U) must also be known. As an additional constraint on the theoretical models, one can also use the condition of the existence of small deviations only from the ΛCDM model, a condition that must be valid, especially in the period of formation of galaxies and stars, that is, during the reionization phase, and the cosmic dawn era, corresponding to $z < 11$.

The behavior of the cosmological models of $f(R, \mathcal{T})$ gravity theory is strongly influenced by the choice of the potential (V) of the two scalar fields. This choice is relatively arbitrary; however, the form of the potential can be determined by comparison of the theoretical predictions with observational data. Constraints on the potential may be found from other astrophysical and cosmological observations, including the study of gravitational waves, black holes, gravitational lensing, or structure formation.

$f(R, \mathcal{T})$ gravity provides an interesting approach to gravitational force, and it certainly has the potential to explain the yet-unknown physical aspects of cosmology. In this brief presentation, we have outlined the basic theoretical concepts and mathematical tools necessary to further develop the theory.

5. Compact Objects in $f(R, \mathcal{T})$ Gravity

It is interesting to note that EMSG reduces to GR in vacuum, where differences only arise in the presence of an energy-momentum distribution. These effects become significant in regions of high curvature. Thus, deviations from GR are expected within compact objects. In this context, $f(R, \mathcal{T})$ gravity has been studied in [36,99–101], including black holes in [23,37,102,103] and wormholes in [44,65,104–106]. More specifically, the deviations from the predictions of GR due to EMSG are expected to become pronounced in the high-density cores of neutron stars [36]. Here, the hydrostatic equilibrium equations in EMSG were derived and solved numerically to obtain the neutron star mass–radius relations for several realistic equations of state. The existing observational measurements of the masses and radii of neutron stars were then used to constrain the free parameter that characterizes the coupling between matter and spacetime in EMSG.

In the following sections, we outline several strategies to obtain black hole solutions and briefly review specific wormhole geometries obtained by several of the authors.

5.1. Black Holes in EMSG Coupled with Electrodynamics

In this section, we present the formalism of EMSG coupled with electrodynamics and consider a strategy to obtain black hole solutions.

Consider and action (2) expressed in the following form:

$$S = \frac{1}{2\kappa^2}\int_\Omega \sqrt{-g} f(R, \mathcal{T}) d^4x + \int_\Omega \sqrt{-g}\mathcal{L}_m^{(em)}\, d^4x, \tag{135}$$

where $\mathcal{L}_m^{(em)}$ is the Maxwell Lagrangian given by

$$\mathcal{L}_m^{(em)} = -\frac{1}{4}F_{\mu\nu}F^{\mu\nu} = -\frac{1}{4}\mathcal{F}, \tag{136}$$

where $\mathcal{F} = F_{\mu\nu}F^{\mu\nu}$, and the electromagnetic tensor is defined as $F_{\mu\nu} = \partial_\mu A_\nu - \partial_\nu A_\mu$. In the case of linear electrodynamics (Maxwell), the energy-momentum tensor is given by

$$T_{\mu\nu}^{(em)} = F_{\mu\sigma}F_\nu{}^\sigma - \frac{1}{4}g_{\mu\nu}\mathcal{F}. \tag{137}$$

To obtain the gravitational field equation in the presence of the electromagnetic field. we vary Equation (135) with respect to the $g_{\mu\nu}$ metric, which yields

$$f_R R_{ab} - \frac{1}{2} g_{ab} f - (\nabla_a \nabla_b - g_{ab}\Box) f_R = 2\kappa^2 T^{(em)}_{ab} - f_{\mathcal{T}} \Theta^{(em)}_{ab}, \tag{138}$$

where $\Theta^{(em)}_{\mu\nu}$ is given by

$$\Theta^{(em)}_{\mu\nu} = \frac{1}{16\pi^2} \left[4F_\gamma^{\ \beta} F^{\gamma\alpha} F_{\alpha\mu} F_{\beta\nu} - F^\rho_{\ \mu} F_{\rho\nu} \mathcal{F} \right]. \tag{139}$$

Finally incorporating both Equations (137) and (139) into Equation (138) yields the following:

$$\begin{aligned} f_R R_{ab} - \frac{1}{2} g_{ab} f - (\nabla_a \nabla_b - g_{ab}\Box) f_R = \\ 2\kappa^2 \left(F_{\mu\sigma} F_\nu^{\ \sigma} - \frac{1}{4} g_{\mu\nu} \mathcal{F} \right) - \frac{1}{16\pi^2} f_{\mathcal{T}} \left[4F_\gamma^{\ \beta} F^{\gamma\alpha} F_{\alpha\mu} F_{\beta\nu} - F^\rho_{\ \mu} F_{\rho\nu} \mathcal{F} \right], \end{aligned} \tag{140}$$

which are the gravitational field equations assuming that the only matter in the Universe is described by $T^{(em)}_{\mu\nu}$. Specific solutions for $f(R, \mathcal{T})$ were obtained in [37,40–42,54], and we refer the reader to these articles for more details.

In principle, one can also consider baryonic matter contributions ($\mathcal{L}^{(B)}_m$) to the matter total Lagrangian density so that $\mathcal{L}_m = \mathcal{L}^{(B)}_m + \mathcal{L}^{(em)}_m$, and the gravitational field equations generalize to

$$\begin{aligned} f_R R_{ab} - \frac{1}{2} g_{ab} f - (\nabla_a \nabla_b - g_{ab}\Box) f_R = 2\kappa^2 \left\{ T^{(B)}_{ab} + F_{\mu\sigma} F_\nu^{\ \sigma} - \frac{1}{4} g_{\mu\nu} \mathcal{F} \right\} \\ - f_{\mathcal{T}} \left\{ \Theta^{(B)}_{ab} + \frac{1}{16\pi^2} \left[4F_\gamma^{\ \beta} F^{\gamma\alpha} F_{\alpha\mu} F_{\beta\nu} - F^\rho_{\ \mu} F_{\rho\nu} \mathcal{F} \right] \right\}. \end{aligned} \tag{141}$$

However, we do not explore this topic further at this stage, as it is still a work in progress, and developments are ongoing. We look forward to sharing more detailed insights once the work has reached a more finalized stage.

As one may expect, the standard Maxwell equations for EMSG are also different. It is more useful to use the Euler–Lagrange equations applied to the vector potential (A_μ) given by

$$\nabla_\mu \left[\frac{\partial \mathcal{L}}{\partial(\nabla_\mu A_\nu)} \right] - \frac{\partial \mathcal{L}}{\partial A_\nu} = 0, \tag{142}$$

which yield

$$\begin{aligned} \nabla_\mu \left[\frac{\partial}{\partial(\nabla_\mu A_\nu)} \left\{ \frac{1}{2\kappa^2} \sqrt{-g} f(R, \mathcal{T}) + \sqrt{-g} \mathcal{L}^{(em)}_m \right\} \right] = 0, \\ \nabla_\mu \left[\frac{\partial}{\partial(\nabla_\mu A_\nu)} \left\{ f(R, \mathcal{T}) + 2\kappa^2 \mathcal{L}^{(em)}_m \right\} \right] = 0. \end{aligned} \tag{143}$$

Using Equation (136), these take the following form:

$$\begin{aligned} \nabla_\mu \left[\frac{\partial f}{\partial(\nabla_\mu A_\nu)} \right] - \frac{\kappa^2}{2} \nabla_\mu \left[\frac{\partial \mathcal{F}}{\partial(\nabla_\mu A_\nu)} \right] = 0, \\ \nabla_\mu \left[f_{\mathcal{T}} \left\{ \frac{1}{2\pi^2} \left(F_\gamma^{\ \nu} F^{\mu\rho} F^\gamma_{\ \rho} - \frac{1}{4} F^{\mu\nu} \mathcal{F} \right) \right\} \right] - 2\kappa^2 \nabla_\mu F^{\mu\nu} = 0, \end{aligned} \tag{144}$$

which yield the modified Maxwell equations for EMSG as follows:

$$\nabla_\mu F^{\mu\nu} = \frac{1}{4\kappa^2\pi^2}\nabla_\mu\left[f_{\mathcal{T}}\left(F_\gamma{}^\nu F^{\mu\rho}F^\gamma{}_\rho - \frac{1}{4}F^{\mu\nu}\mathcal{F}\right)\right]. \tag{145}$$

One may also generalize the action (135) to EMSG coupled with nonlinear electrodynamics (NLED), where the Maxwell Lagrangian ($\mathcal{L}_m^{(em)}$) is substituted with an NLED Lagrangian density ($\mathcal{L}_m^{(\mathrm{N})}(\mathcal{F})$). Varying the action (135) with respect to the vector potential, one obtains the following equation of motion: $\nabla_\mu(\mathcal{L}_\mathcal{F}\, F^{\mu\nu}) = \sqrt{-g}\,\partial_\mu(\sqrt{-g}\,\mathcal{L}_\mathcal{F}\, F^{\mu\nu}) = 0$, where we denote $\mathcal{L}_\mathcal{F} = \partial\mathcal{L}_m^{(\mathrm{N})}(\mathcal{F})/\partial\mathcal{F}$. Now, we can explore novel black hole solutions (including regular black holes and black bounces) with static and spherical symmetry by coupling EMSG with NLED, as carried out extensively in the recent literature [107–113]. This is work in progress and will shortly be submitted for publication.

5.2. Wormhole Geometries

In this section, we focus on wormholes, which are structures that connect two different spacetime manifolds or two distinct regions within the same spacetime manifold. Wormholes have been extensively studied within the framework of general relativity (GR) [114–119]. A crucial requirement for a wormhole to be traversable is the flaring-out condition [114], which, when applied through the Einstein field equations, leads to the violation of the null energy condition (NEC) and, consequently, all other energy conditions [116,120,121]. Matter that violates the NEC is referred to as *exotic* matter, and it lacks physical relevance due to the scarcity of experimental evidence supporting its existence.

One way to avoid the use of exotic matter to sustain wormhole geometries is to consider these objects within the context of modified theories of gravity [44,65,105,122–133]. In these theories, the additional curvature components of the gravitational sector preserve the geometry of the wormhole throat, making it traversable while keeping the matter components non-exotic. This outcome can be achieved in various modifications of GR, from $f(R)$ gravity and its extensions [134–140] to couplings between curvature and matter [141,142], theories with additional fundamental fields [143,144], Einstein–Cartan gravity [145], Gauss–Bonnet gravity [146–148], and brane-world scenarios [149,150]. Regarding wormhole physics in the context of $f(R,\mathcal{T})$ gravity, specific solutions were initially found using a Noether symmetry approach; however, these lacked physical relevance, as they violated the NEC [44,65]. More recently, physically relevant traversable wormhole spacetimes that satisfy all energy conditions have been found [105], and these are the solutions presented in this section.

5.2.1. Metric and Field Equations

We start by considering a static and spherically symmetric traversable wormhole metric, which can be written in spherical coordinates (t,r,θ,φ) as

$$ds^2 = -e^{\zeta(r)}dt^2 + \left[1-\frac{b(r)}{r}\right]^{-1}dr^2 + r^2d\Omega^2, \tag{146}$$

where $\zeta(r)$ is the redshift function, $b(r)$ is the shape function, and $d\Omega^2 = d\theta^2 + \sin^2\theta d\varphi^2$ is the solid-angle surface element. The necessary conditions to ensure the traversability of the wormhole are the finiteness of the redshift function throughout the entire spacetime, i.e., $|\zeta(r)| < \infty$, to avoid event horizons, therefore allowing an observer to cross the wormhole's interior without being trapped, and the flaring-out condition at the wormhole throat $(r=r_0)$[1], which is given by the following boundary conditions:

$$b(r_0) = r_0, \qquad b'(r_0) < 1. \tag{147}$$

Given these requirements, we consider the following two broad families of solutions for the functions of $\zeta(r)$ and $b(r)$:

$$\zeta(r) = \zeta_0 \left(\frac{r_0}{r}\right)^{\alpha}, \qquad b(r) = r_0 \left(\frac{r_0}{r}\right)^{\beta}, \tag{148}$$

where ζ_0 is an arbitrary constant to be specified and α and β are arbitrary positive exponents.

Regarding the matter sector, we assume that the matter distribution is well described by an anisotropic perfect fluid. Thus, the energy-momentum tensor ($T_{\mu\nu}$) is given by

$$T_{\mu}^{\nu} = \mathrm{diag}(-\rho, p_r, p_t, p_t), \tag{149}$$

where $\rho \equiv \rho(r)$ represents the energy density, $p_r \equiv p_r(r)$ denotes the radial pressure, and $p_t \equiv p_t(r)$ is the tangential pressure, which depend only on the radial coordinate (r) to maintain the spherical symmetry of the wormhole. Under these assumptions, the matter Lagrangian is given by $\mathcal{L}_m = \frac{1}{3}(p_r + 2p_t)$, and consequently, the auxiliary tensor ($\Theta_{\mu\nu}$) results in

$$\Theta_{\mu\nu} = -\frac{2}{3}(p_r + 2p_t)\left(T_{\mu\nu} - \frac{1}{2}g_{\mu\nu}T\right) - TT_{\mu\nu} + 2T_{\mu}^{\sigma}T_{\sigma\nu}. \tag{150}$$

We developed the methodology to obtain wormhole solutions by considering that the function of $f(R, \mathcal{T})$ is separable and linear in both R and $\mathcal{T}$, that is,

$$f(R, \mathcal{T}) = R + \gamma\mathcal{T}, \tag{151}$$

and briefly generalized it to cases where the function includes highers powers of $\mathcal{T}$ along the way, which can be seen in detail in [105].

Under this assumption, the field equations in Equation (3) and the conservation equation in Equation (7) become

$$G_{\mu\nu} = 8\pi T_{\mu\nu} - \gamma\left(\Theta_{\mu\nu} - \frac{1}{2}g_{\mu\nu}\mathcal{T}\right), \tag{152}$$

$$8\pi\nabla_b T^{\mu\nu} = \gamma\nabla_b\left(\Theta^{\mu\nu} - \frac{1}{2}g^{\mu\nu}\mathcal{T}\right), \tag{153}$$

respectively. Now, with the assumptions from Section 5.2, the field equations in Equation (152) yield three independent components, which are expressed as follows:

$$8\pi\rho = \frac{\gamma}{6}\left(p_r^2 - 2p_t^2 - 3\rho^2 - 8p_r p_t - 8p_r\rho - 16p_t\rho\right) - \frac{\beta}{r^2}\left(\frac{r_0}{r}\right)^{\beta+1}, \tag{154}$$

$$\begin{aligned} 8\pi p_r = {} & \frac{\gamma}{6}\left(p_r^2 + 2p_t^2 - 3\rho^2 - 12p_r p_t + 4p_r\rho - 4p_t\rho\right) \\ & -\frac{1}{r^2}\left(\frac{r_0}{r}\right)^{\beta+1} - \frac{\alpha\zeta_0}{r^2}\left(\frac{r_0}{r}\right)^{\alpha}\left[1 - \left(\frac{r_0}{r}\right)^{\beta+1}\right], \end{aligned} \tag{155}$$

$$\begin{aligned} 8\pi p_t = {} & -\frac{\gamma}{6}\left(p_r^2 + 6p_t^2 + 3\rho^2 + 2p_r p_t + 2p_r\rho - 2p_t\rho\right) \\ & + \frac{1+\beta}{2r^2}\left(\frac{r_0}{r}\right)^{\beta+1} + \frac{\alpha^2\zeta_0^2}{4r^2}\left(\frac{r_0}{r}\right)^{2\alpha}\left[1 - \left(\frac{r_0}{r}\right)^{\beta+1}\right] \\ & + \frac{\alpha\zeta_0}{4r^2}\left(\frac{r_0}{r}\right)^{\alpha}\left[2\alpha - (1 + 2\alpha + \beta)\left(\frac{r_0}{r}\right)^{\beta+1}\right]. \end{aligned} \tag{156}$$

Equations (154)–(156) form a system of three equations with three unknowns, namely ρ, p_r, and p_t. Each of these equations is quadratic in its respective unknown, implying that the system can potentially yield up to eight independent solutions. However, the nature of these solutions may vary, including the possibility of some solutions being complex, depending on the specific values of the parameters involved.

5.2.2. Wormhole Solutions

Due to the complexity of the system of Equations (154)–(156), analytical solutions for ρ, p_r, and p_t cannot be obtained explicitly, even when considering specific choices for the free parameters of r_0, α, β, γ, and ζ_0. However, analytical solutions for these quantities can still be obtained using a recursive approach. We begin by choosing specific values for the free parameters and solve the system to find $\rho(r_0)$, $p_r(r_0)$, and $p_t(r_0)$ at an initial radius of $r = r_0$. This provides an initial set of values $\{\rho_0^i, p_{r0}^i, p_{t0}^i\}$ for $i \in \{1, \ldots, 8\}$ corresponding to the eight independent solutions of the system. Then, for each solution, the radius (r) is incremented in small steps, such as $r_{n+1} = r_n + \epsilon$, where ϵ is a small increment, to compute $\rho(r_{n+1})$, $p_r(r_{n+1})$, and $p_t(r_{n+1})$. By recursively applying this process up to a sufficiently large radius (r), one can analytically determine the behavior of the solutions. The same method is applicable to any form of the function of $f(R, \mathcal{T})$ containing higher powers of $\mathcal{T}$ and no crossed $R\mathcal{T}$ terms, given that the relation between the matter fields remains algebraic, although the set of solutions is larger.

Among the obtained solutions, we focus only on those that are astrophysically relevant, meaning that their matter components satisfy the energy conditions (see Equation (23)). For a diagonal energy-momentum tensor ($T_{\mu\nu}$) as given in Equation (149), these energy conditions are expressed as follows:

$$\rho + p_r \geq 0, \quad \rho + p_t \geq 0, \quad \rho \geq 0, \quad \rho + p_r + 2p_t \geq 0, \quad \rho \geq |p_r|, \quad \rho \geq |p_t|. \tag{157}$$

From the set of eight solutions for the matter quantities within the theory, solutions that violate any of these energy conditions are discarded, and only those that satisfy all of the energy conditions are considered.

As a specific example, let us consider the parameter combination of $\alpha = \beta = -\gamma = 1$, $r_0 = 3M$, and $\zeta_0 = -\frac{6}{5}$.[2] The matter density (ρ), as well as the combinations of $\rho + p_r$, $\rho + p_t$, $\rho + p_r + 2p_t$, $\rho - |p_r|$, and $\rho - |p_t|$, are plotted in Figure 3 for the solution that satisfies all previously mentioned energy conditions.

Similar to findings in linear $f(R, T)$ gravity [140], from the example above and from all the solutions satisfying the energy conditions that we have found for the linear case of the function of $f(R, \mathcal{T})$, we concluded that solutions satisfying the energy conditions throughout the entire spacetime could only be obtained by taking negative values of γ, as shown in Figure 3. However, when considering higher powers of $\mathcal{T}$, i.e., $f(R, \mathcal{T}) = R + \gamma\mathcal{T} + \sigma\mathcal{T}^n$, we found that solutions satisfying all the energy conditions can be achieved even for positive values of γ, provided that σ remains negative. Furthermore, for both cases, a notable characteristic stood out, which is the fact that despite the spacetime being asymptotically flat, the matter components do not vanish across the entire radial coordinate range, indicating non-localized behavior, as seen in Figure 3. Therefore, to improve the physical relevance of these solutions, we must match them with an exterior vacuum spacetime with a finite radius. We address this issue in the next section.

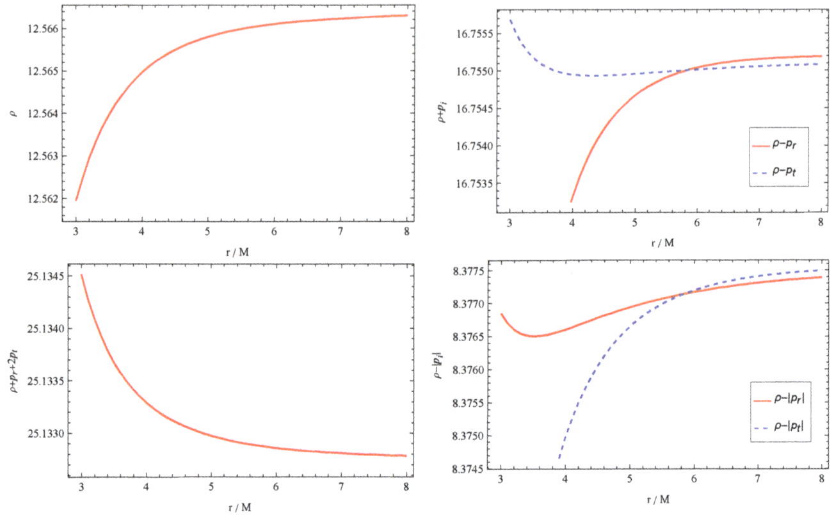

Figure 3. Considering the parameters $\alpha = \beta = -\gamma = 1$, $r_0 = 3M$, and $\zeta_0 = -\frac{6}{5}$, the matter density ρ and the combinations of $\rho + p_r$, $\rho + p_t$, $\rho + p_r + 2p_t$, $\rho - |p_r|$, and $\rho - |p_t|$ are presented.

5.2.3. Junction Conditions and Matching

To obtain physically relevant spacetime solutions that describe localized objects, one has to employ the junction conditions to perform matching between the interior and exterior spacetimes at a finite radius. In the context of GR, these equations were derived long ago [151] and have been applied in various astrophysical scenarios, such as in the analysis of traversable wormholes [152–157], fluid stars [158–160], and gravitational collapse [161,162]. Since these conditions are theory-dependent, several studies have analyzed them in the context of different modified theories of gravity (we refer to [106] for a review), from $f(R)$ gravity and its extensions [163–169] to theories with additional fundamental fields [170–173]; metric-affine gravity [174–176]; and, more recently, in the context of $f(R, \mathcal{T})$ gravity [105]. In this section, we present the latter work, including both the derivation of the junction conditions and the matching performed between the interior wormhole spacetime and an exterior vacuum spacetime.

Junction Conditions

We begin the derivation of the junction conditions by first considering the linear case of the function, that is, $f(R, \mathcal{T}) = R + \gamma\mathcal{T}$, and later briefly extend the analysis to cases where the function includes higher powers of $\mathcal{T}$, namely $f(R, \mathcal{T}) = R + \gamma\mathcal{T} + \sigma\mathcal{T}^n$.

Let us consider a spacetime manifold (Ω) consisting of two distinct and complementary regions ($\Omega^\pm$), each described by metric tensors ($g^\pm_{\alpha\beta}$) in their respective coordinate systems ($x^\alpha_\pm$). We define Ω^+ as the exterior spacetime and Ω^- as the interior spacetime. The interface between $\Omega^\pm$ is a three-dimensional hypersurface (Σ) with a metric (h_{ab}) expressed in terms of a coordinate system (y^a), where Latin indices exclude the direction orthogonal to Σ. The projection tensors from the four-dimensional spacetime (Ω) onto the hypersurface (Σ) are expressed as $e^\alpha_a = \partial x^\alpha / \partial y^a$. The normal vector on Σ is defined as $n_\alpha = \epsilon \partial_\alpha l$, where l is the affine parameter along geodesics orthogonal to Σ, and ϵ takes values 1, -1, and 0 for space-like, time-like, and null geodesic congruences, respectively. By construction, note that $n^\alpha e^a_\alpha = 0$ holds true. Using this notation, the induced metric (h_{ab}) and the extrinsic curvature (K_{ab}) of the hypersurface (Σ) are expressed as

$$h_{ab} = e^\alpha_a e^\beta_b g_{\alpha\beta}, \qquad K_{ab} = e^\alpha_a e^\beta_b \nabla_\alpha n_\beta. \tag{158}$$

To derive the junction conditions, we employ the distribution formalism. In this approach, any quantity (X) and its derivative ($\nabla_\alpha X$) are expressed in terms of distribution functions as follows:

$$X = X^+\theta(l) + X^-\theta(-l), \tag{159}$$

$$\nabla_\alpha X = \nabla_\alpha X^+\theta(l) + \nabla_\alpha X^-\theta(-l) + \epsilon n_\alpha [X]\delta(l), \tag{160}$$

where $X^\pm$ represents the quantity (X) in the spacetimes ($\Omega^\pm$); $\theta(l)$ is the Heaviside distribution function defined as $\theta(l) = 0$ for $l < 0$, $\theta(l) = 1$ for $l > 0$, and $\theta(l) = \frac{1}{2}$ for $l = 0$; $\delta(l) = \partial_l\theta(l)$ is the Dirac delta distribution; and we have introduce the definition of $[X]$ to represent the jump of X across Σ, that is,

$$[X] = X^+|_\Sigma - X^-|_\Sigma. \tag{161}$$

If the quantity (X) is continuous across Σ, then $[X] = 0$. Additionally, notice that by definition, we have $[n^\alpha] = [e^\alpha_a] = 0$.

To obtain the junction conditions, we start by writing every quantity appearing in the field equations (see Equation (152)) in the distribution formalism. Let us begin with the metric of $g_{\alpha\beta}$, which, in the distributional formalism, becomes

$$g_{\alpha\beta} = g^+_{\alpha\beta}\theta(l) + g^-_{\alpha\beta}\theta(-l). \tag{162}$$

With the metric expression in Equation (162), we proceed to compute the Christoffel symbols ($\Gamma^\gamma_{\alpha\beta}$) associated with the metric of $g_{\alpha\beta}$, which involves calculating the derivatives ($\partial_\lambda g_{\alpha\beta}$). Using Equation (160), these derivatives are expressed as $\partial_\lambda g_{\alpha\beta} = \partial_\lambda g^+_{\alpha\beta}\theta(l) + \partial_\lambda g^-_{\alpha\beta}\theta(-l) + \epsilon n_\lambda [g_{\alpha\beta}]\delta(l)$. However, the presence of the term proportional to $\delta(l)$ poses a challenge when defining the Riemann tensor ($R^\alpha_{\beta\lambda\gamma}$) in the distributional formalism. Since the Riemann tensor ($R^\alpha_{\beta\lambda\gamma}$) depends on products between the Christoffel symbols ($\Gamma^\gamma_{\alpha\beta}$), it results in terms proportional to $\delta^2(l)$, which are singular in the distributional formalism. The junction conditions are introduced precisely to eliminate such singular terms from the field equations when expressed in the distribution formalism. Thus, to remove the singular term arising in the Riemann tensor ($R^\alpha_{\beta\lambda\gamma}$), we impose that the metric of $g_{\alpha\beta}$ must be continuous across Σ, i.e., $[g_{\alpha\beta}] = 0$. Considering $[e^\alpha_a] = 0$, this condition can be expressed in a coordinate-independent manner by projecting both indices onto the hypersurface (Σ), resulting in

$$[h_{ab}] = 0. \tag{163}$$

Equation (163) represents the first junction condition, stipulating that the induced metric at Σ must be continuous. Using this result, the derivatives of $g_{\alpha\beta}$ become

$$\partial_\lambda g_{\alpha\beta} = \partial_\lambda g^+_{\alpha\beta}\theta(l) + \partial_\lambda g^-_{\alpha\beta}\theta(-l). \tag{164}$$

We can now compute the Christoffel symbols in the distributional formalism and subsequently derive the Riemann tensor, along with its contractions, namely the Ricci tensor (R_{ab}) and the Ricci scalar (R), which are well-defined. These quantities are expressed as follows:

$$R_{\alpha\beta} = R^+_{\alpha\beta}\theta(l) + R^-_{\alpha\beta}\theta(-l) - \left(\epsilon[K_{ab}]e^a_\alpha e^b_\beta + n_\alpha n_\beta [K]\right)\delta(l), \tag{165}$$

$$R = R^+\theta(l) + R^-\theta(-l) - 2\epsilon[K]\delta(l), \tag{166}$$

where $K = h^{ab}K_{ab}$ represents the trace of the extrinsic curvature.

Considering the matter sector, it is useful to link the existence of a thin shell at the hypersurface (Σ) with any terms proportional to $\delta(l)$ in the gravitational sector of the modified field equations. Therefore, in the distribution formalism, we express the energy-momentum tensor as follows:

$$T_{\alpha\beta} = T^{+}_{\alpha\beta}\theta(l) + T^{-}_{\alpha\beta}\theta(-l) + S_{\alpha\beta}\delta(l), \tag{167}$$

where $S_{\alpha\beta} = S_{ab}e^{a}_{\alpha}e^{b}_{\beta}$, and S_{ab} denotes the three-dimensional energy-momentum tensor of the thin shell. To obtain the scalar ($\mathcal{T}$) in the distributional formalism, one can simply contract $T_{\alpha\beta}$ with itself using the previous expression (167), resulting in

$$\mathcal{T} = \mathcal{T}^{+}\theta(l) + \mathcal{T}^{-}\theta(-l) + \left(T^{+}_{\alpha\beta} + T^{-}_{\alpha\beta}\right)S^{\alpha\beta}\delta(l) + S_{\alpha\beta}S^{\alpha\beta}\delta^{2}(l), \tag{168}$$

Note that the term proportional to $\delta^2(l)$ in Equation (168) is singular in the distributional formalism and needs to be removed. The only possible approach to eliminate the term proportional to $\delta^2(l)$ in $\mathcal{T}$ is to require the energy-momentum tensor of the thin shell to vanish, that is,

$$S_{\alpha\beta} = 0. \tag{169}$$

When Equation (169) holds true, the matching is referred to as a smooth matching. Unlike in several other theories of gravity, where this kind of matching is considered a specific case of a broader thin-shell matching, in $f(R,\mathcal{T})$ gravity, smooth matching between two spacetimes is the only method allowed to maintain the regularity of the action.

Based on the definitions provided above and under the restriction in Equation (169), we project the field equations for the function of $f(R,\mathcal{T}) = R + \gamma\mathcal{T}$ that are presented in Equation (152) onto the hypersurface (Σ) with $e^{\alpha}_{a}e^{\beta}_{b}$, for which they take the form of $[K_{ab}] - [K]h_{ab} = 0$. Taking the trace of this result with h^{ab} yields $[K] = 0$. Substituting this back into the original equation results in

$$[K_{ab}] = 0. \tag{170}$$

Therefore, the second junction condition implies that the extrinsic curvature (K_{ab}) must be continuous across Σ.

In summary, the matching between two spacetimes in linear $f(R,\mathcal{T})$ gravity must always be smooth, meaning it must occur without a thin shell. The two junction conditions that must be satisfied are the same as in GR—the induced metric (h_{ab}) and the extrinsic curvature K_{ab} must be continuous across the hypersurface (Σ) as follows:

$$[h_{ab}] = 0, \qquad [K_{ab}] = 0. \tag{171}$$

When considering a higher-order power of $\mathcal{T}$ in the function, the field equations include products in the form of $\mathcal{T}^{n-1}\Theta_{ab}$ and powers of $\mathcal{T}^{n}$. For the linear case, we have shown that for $\mathcal{T}$ to be well-defined in the distributional formalism, the matching must be smooth, meaning there is no thin shell, i.e., $S_{\alpha\beta} = 0$. This requirement ensures that $\mathcal{T}$ and the auxiliary tensor ($\Theta_{\alpha\beta}$), as defined in Equation (150), are completely regular, featuring only terms proportional to $\theta(l)$ and no terms proportional to $\delta(l)$. As a result, the products of $\mathcal{T}$ and $\Theta_{\alpha\beta}$, as well as the powers of $\mathcal{T}^{n}$, maintain this regularity. Therefore, introducing a higher-order power law of $\mathcal{T}$ in the function of $f(R,\mathcal{T})$ does not lead to additional junction conditions, as long as crossed $R\mathcal{T}$ terms are absent.

Matching with an Exterior Vacuum

Let us now apply the junction conditions derived in the previous Section Junction Conditions to match the interior wormhole spacetime with an exterior spherically symmetric and static vacuum solution. The metrics for the interior and exterior spacetimes are given by

$$ds^2_{-} = -Ce^{\zeta_0\left(\frac{r_0}{r}\right)^{\alpha}}dt^2 + \left[1 - \left(\frac{r_0}{r}\right)^{\beta+1}\right]^{-1}dr^2 + r^2d\Omega^2, \tag{172}$$

$$ds_+^2 = -\left(1 - \frac{2M}{r}\right)dt^2 + \left(1 - \frac{2M}{r}\right)^{-1} dr^2 + r^2 d\Omega^2, \tag{173}$$

respectively. The metric in Equation (172) is derived from Equation (146) using the suggested ansatz for the redshift and shape functions, as given in Equation (148), and C is introduced for convenience to ensure that the time coordinates in both the interior and exterior metrics coincide. The metric in Equation (173) corresponds to the Schwarzschild solution with a mass of M [177].

The analysis becomes more convenient if we begin with the second junction condition (Equation (171)). Due to the spherical symmetry of the metrics under consideration, the extrinsic curvatures ($K_{ab}^{\pm}$) have only two independent components, namely K_{00} and $K_{\theta\theta} = K_{\phi\phi} \sin^2 \theta$. Thus, we derive two independent constraints on the matching, that is, $[K_{00}] = 0$ and $[K_{\theta\theta}] = 0$, which are given by

$$\frac{\alpha \zeta_0}{2} \left(\frac{r_0}{r}\right)^{\alpha} \sqrt{1 - \left(\frac{r_0}{r}\right)^{\beta+1}} + \frac{M}{r} \sqrt{\frac{r}{r - 2M}} = 0, \tag{174}$$

$$\sqrt{1 - \left(\frac{r_0}{r}\right)^{\beta+1}} = \sqrt{1 - \frac{2M}{r}}, \tag{175}$$

respectively. Solving the second of these conditions for the radius (r) using Equation (175), we obtain unique real solutions for $M > 0$ and $r_0 > 0$, given by

$$r_\Sigma = (2M)^{-\frac{1}{\beta}} (r_0)^{1+\frac{1}{\beta}}, \tag{176}$$

which corresponds to the radius (r_Σ) at which the matching must be performed. The radius (r_Σ) must satisfy $r_\Sigma > 2M$ to prevent event horizons in the complete wormhole spacetime, which implies that the throat radius (r_0) should also satisfy $r_0 > 2M$ for any $\beta \geq 1$. Introducing the obtained solutions for r_Σ back into the first condition (Equation (174)), we can solve it with respect to the value of ζ_0, for which matching at the radius ($r = r_\Sigma$) is possible. Doing so, the expression for ζ_0 becomes

$$\zeta_0 = \frac{(2M)^{\frac{1-\alpha+\beta}{\beta}} (r_0)^{\frac{\alpha}{\beta}}}{\alpha \left[(2M)^{1+\frac{1}{\beta}} - (r_0)^{1+\frac{1}{\beta}} \right]}. \tag{177}$$

Given that $r_0 > 2M$, as derived from the previous constraint, it follows that for any $\alpha \geq 1$ and $\beta \geq 1$, we have $\zeta_0 < 0$. This is expected because negative values of ζ_0 ensure that the derivative of g_{00} maintains consistent signs in both the interior and exterior metrics, which is necessary for a smooth matching.

Let us now consider the first junction condition given in Equation (171). Since the angular components of the metrics in Equations (172) and (173) coincide, the angular parts of the induced metric ($h_{\alpha\beta}$) are straightforwardly continuous. Therefore, the continuity condition ($[h_{00}] = 0$) is analyzed independently and takes the following form:

$$C e^{\zeta_0 \left(\frac{r_0}{r}\right)^{\alpha}} = \left(1 - \frac{2M}{r}\right). \tag{178}$$

We take the results derived from the second junction condition, namely the radius (r_Σ) from Equation (176), where the matching occurs, and the corresponding value of ζ_0 from Equation (177), and substitute them into Equation (178) and solve for the constant (C), obtaining

$$C = \left[1 - \left(\frac{2M}{r_0}\right)^{1+\frac{1}{\beta}}\right] e^{-\alpha \left[\left(\frac{r_0}{2M}\right)^{1+\frac{1}{\beta}} - 1\right]}. \tag{179}$$

Since $r_0 > 2M$, the constant (C) remains strictly positive, regardless of the values of $\alpha \geq 1$ and $\beta \geq 1$, therefore preserving the correct metric signature.

To summarize, given $r_0 > 2M$, $\alpha \geq 1$, and $\beta \geq 1$, the second junction condition ($[K_{\alpha\beta}] = 0$) determines the radius (r_Σ) for the matching (see Equation (176)) and the corresponding value of ζ_0 (see Equation (177)). Meanwhile, the first junction condition ($[h_{\alpha\beta}]$) sets the value of the constant (C), ensuring the continuity of the complete spacetime metric (see Equation (179)).

Let us provide the following example. Consider the specific case where $r_0 = 3M$, $\alpha = 1$, and $\beta = 1$. With these parameters, Equation (176) gives the matching radius at $r_\Sigma = \frac{9}{2}M$, Equation (177) gives $\zeta_0 = -\frac{6}{5}$, and Equation (179) results in $C = \frac{5}{9}e^{\frac{4}{5}}$. The g_{00} components of the interior, exterior, and matched metrics are plotted in the left panel of Figure 4. Notice that the g_{00} component of the metric transitions smoothly from the interior to the exterior metric, ensuring the continuity of both the induced metric and the extrinsic curvature.

Analyzing the radial component of the g_{rr} metric, as shown in the right panel of Figure 4, we observe that even though the radial component is not directly constrained by the junction conditions, as both the induced metric (h_{ab}) and the extrinsic curvature (K_{ab}) are three-dimensional tensors on the hypersurface (Σ), we find that g_{rr} remains continuous, although not differentiable at $r = r_\Sigma$. This continuity of g_{rr} is anticipated when considering its dependence on the mass function within a spherical hypersurface of radius r ($m(r)$), as given by $g_{rr} = \left(1 - \frac{2m(r)}{r}\right)^{-1}$, from which we obtain $m(r) = \frac{r_0}{2}\left(\frac{r_0}{r}\right)^\beta$ (see Equations (148) and (172)). Indeed, since the matching between the interior and exterior spacetimes are smooth, i.e., lacking a thin shell, it follows that the mass function ($m(r)$) is continuous at r_Σ, which, in turn, ensures the continuity of the g_{rr} component of the metric.

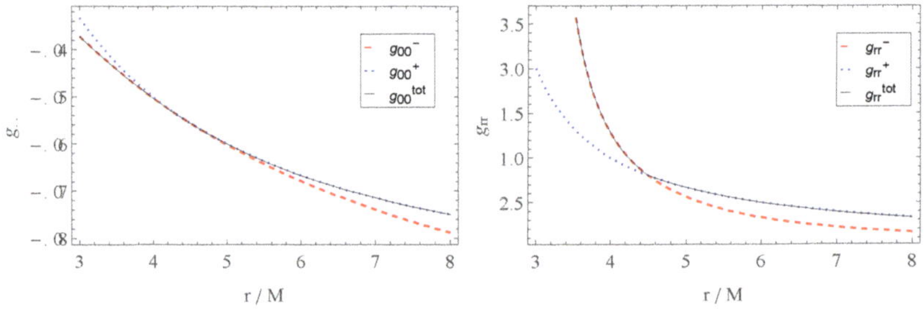

Figure 4. The interior wormhole spacetime in Equation (172) (red dashed curve) and the exterior Schwarzschild spacetime in Equation (173) (blue dotted curve) for $\beta = 1$, $r_0 = 3M$, and $\alpha = 1$. The thin black line represents the solution obtained by matching the interior and exterior solutions at $r = r_\Sigma$ for the g_{00}^{tot} component (**left panel**) and g_{rr}^{tot} component (**right panel**).

5.2.4. Summary

In conclusion, this analysis of traversable wormhole spacetimes within the framework of $f(R, \mathcal{T})$ gravity for a linear model on both R and $\mathcal{T}$ demonstrates the existence of numerous traversable wormhole solutions that satisfy all energy conditions, therefore possessing significant physical relevance. An intriguing aspect of the obtained solutions is that, although the spacetime metrics are asymptotically flat, the $f(R, \mathcal{T})$ theory permits matter distributions that are not asymptotically vacuum and, therefore, not localized. However, the localization of the solutions is achievable via the use of the theory's junction conditions. These conditions were derived, revealing that only smooth matching is allowed under this theory, as the scalar ($\mathcal{T}$) becomes singular in the presence of a thin shell. Consequently, the junction conditions for a smooth matching reduce to those of GR, i.e., the continuity of the induced metric and extrinsic curvature at the hypersurface separating the interior and exterior spacetime regions. Upon performing this matching, localized wormhole solutions satisfying all energy conditions throughout the entire spacetime were obtained, highlighting their particular astrophysical relevance. Moreover, the methods introduced in this work can be straightforwardly generalized to more complex dependencies of the function of $f(R, \mathcal{T})$ in $\mathcal{T}$, provided that crossed terms between R and $\mathcal{T}$ are absent. Furthermore, due to the requirement that the matching in this theory be smooth, the absence of these crossed terms also ensures that no additional junction conditions arise, allowing for effective localization of the solutions obtained in the same manner as in the linear ($\mathcal{T}$) counterpart.

6. Conclusions

The exploration of EMSG presents a novel framework for extending our understanding of gravitational phenomena beyond the limits of GR. This comprehensive study encompasses the theoretical formalism, thermodynamic implications, cosmological models, and the nature of compact objects within EMSG, providing a robust foundation for future research and potential observational confirmations. In this work, we delved into the mathematical formulation of EMSG by presenting the action and field equations in the geometrical and scalar-tensor representations, laying the groundwork for the theory. The thermodynamic aspects of EMSG are crucial in understanding the Universe's evolution. We also examined particle production in cosmology, highlighting how EMSG influences matter creation. The thermodynamic interpretation of irreversible matter creation provides insights into a deeper connection between thermodynamics and cosmological expansion. This motivated the analysis of the cosmological implications of EMSG. The generalized Friedmann equations were derived, incorporating the energy balance equation and the deceleration parameter, and various de Sitter expansion scenarios were analyzed, including self-interacting potentials and constant-density solutions. The exploration of general cosmological models through dimensionless and redshift representations offered a qualitative discussion of specific models and their observational prospects.

In fact, matter creation processes are believed to play a fundamental role in quantum field theoretical approaches to gravity, where these processes emerge naturally within the framework. A well-established result in quantum field theory, particularly in curved spacetimes, is the creation of quanta from minimally coupled scalar fields in an expanding Friedmann–Robertson–Walker (FRW) universe. This phenomenon, as rigorously demonstrated by Leonard Parker in his pioneering research, has become a cornerstone of our understanding of particle creation in cosmological settings [71–75]. Parker's work, beginning in the late 1960s, showed that the expansion of the Universe leads to the spontaneous creation of particles—an insight that has profound implications for both quantum mechanics and cosmology. His findings revealed that the dynamics of an expanding universe could result in the production of both fermions and bosons, establishing a deep connection between the geometry of spacetime and quantum field behavior.

Given the significance of these results, finding an equivalent microscopic quantum description of the matter creation processes discussed in the present work could provide critical insights into the physical mechanisms that govern particle generation through the coupling of gravity and matter geometry. Although this analysis lies outside the scope of the present work, by developing such a microscopic perspective, we could further illuminate the fundamental processes by which particles are created in the Universe, potentially bridging the gap between quantum field theory and gravitational phenomena. These models, which offer a synthesis of classical and quantum descriptions of the Universe, provide a robust platform for exploring how particles(both fermions and bosons) are generated in an expanding cosmological background. Parker's extensive body of work serves as a foundational reference for these explorations, offering both theoretical insights and analytical techniques necessary to delve into the complexities of particle creation [71–75]. These seminal works provide the theoretical framework and foundational principles that continue to guide research on how the expanding Universe influences quantum fields, leading to the creation of matter as we observe it. Through these studies, the intricate relationship between gravity, quantum fields, and matter creation is progressively being unraveled, offering new perspectives on the nature of the Universe. Work along these lines is currently underway.

Furthermore, we investigated the impact of EMSG on compact objects. Indeed, EMSG reduces to GR in vacuum, making them indistinguishable without matter or energy. However, in the presence of an energy-momentum distribution, differences arise because EMSG includes higher-order corrections dependent on the square of the energy-momentum tensor. These corrections become significant in high-curvature regions, such as the dense cores of neutron stars, black holes, or wormholes, leading to noticeable deviations from GR. In this work, electrodynamics within the EMSG framework were explored, suggesting potential deviations in electromagnetic phenomena near massive objects. Wormhole geometries were also studied, with a focus on the metric and field equations, and specific wormhole solutions were obtained. Junction conditions and matching provide the criteria for viable physical models. Therefore, the dense cores of compact objects serve as ideal laboratories to test the predictions of EMSG against those of GR. By studying high-curvature regions, we can gain insights into the validity of EMSG and potentially identify observable phenomena that distinguish it from GR. As observational technology advances, particularly in the realms of gravitational wave astronomy and high-energy astrophysics, these deviations could provide crucial evidence supporting or refuting the modifications introduced by EMSG.

An interesting avenue for further exploration would be to consider the effects of non-spherical symmetry, particularly those induced by the distribution of angular momentum. When a compact object is set into slow rotation, as described in the works of Hartle [178,179], both the geometry of the surrounding spacetime and the interior distribution of stress energy experience significant alterations. These modifications become even more pronounced in the presence of EMSG. In the context of EMSG, the slow rotation of a compact object could lead to complex changes in the spacetime geometry, potentially affecting the stability, structure, and observable properties of these objects. The inclusion of angular momentum distribution in such models is likely to yield new insights into the behavior of rotating compact objects, particularly in how they deviate from the predictions of GR. Given the complexity and importance of these considerations, a detailed analysis of the impact of angular momentum on non-spherical compact objects within the framework of EMSG would be a valuable contribution to the field. However, this investigation involves substantial theoretical and computational challenges, which we intend to address in future work. This future research will aim to provide a more comprehensive understanding of how rotation and EMSG interact to shape the properties of compact astrophysical objects.

In this context of compact objects, it is also of interest to define the mass of these geometries as viewed by a distant observer, which is, indeed, not a trivial matter. Here, one can, indeed, enumerate several concepts, such as the Komar mass [180,181], ADM mass (Arnowitt-Deser-Misner Mass) [182,183], and Bondi mass [184,185], which are not equivalent in GR but are related concepts that apply in different contexts. The Komar mass is the mass associated with stationary spacetimes, specifically those with a time-like Killing vector field. The Komar mass is a measure of the total mass (including contributions from both matter and gravitational energy) within a certain region of spacetime. It is applicable primarily in spacetimes with certain symmetries (e.g., stationary or axisymmetric spacetimes) and is not defined in non-stationary situations. The ADM mass is a measure of the total mass (including gravitational energy) of an isolated system as seen from spatial infinity. It is defined in asymptotically flat spacetimes and represents the conserved quantity associated with the asymptotic time translation symmetry. The Bondi mass is defined at null infinity and measures the total mass of an isolated system as seen by a distant observer, accounting for any energy radiated away as gravitational waves. Unlike the ADM mass, which is defined in a spatial context, the Bondi mass is relevant in dynamical spacetimes where gravitational radiation is present. In GR, these three concepts are not directly equivalent but can be related under specific conditions. For instance, in a stationary, asymptotically flat spacetime, the Komar mass and ADM mass can be shown to be equivalent. This is because both are designed to measure the total mass of a static or stationary system, although using different methods. The ADM mass and Bondi mass differ because the Bondi mass decreases as gravitational waves carry energy away from the system. In a stationary situation with no gravitational radiation, they would be equivalent, but in general, the Bondi mass is less than the ADM mass for a radiating system.

In modified theories of gravity, such as in EMSG, the relationship between these masses can differ significantly from GR due to changes in the underlying field equations and the nature of gravitational interaction. Here, the Komar mass typically relies on the Einstein field equations and the presence of a time-like Killing vector. In modified gravity theories, the field equations are altered, and the existence and properties of Killing vectors may also change, leading to different expressions or interpretations for Komar mass. The ADM mass is defined based on the Hamiltonian formulation of GR, which may be altered in modified theories of gravity. As a result, the ADM mass could take a different form or might not even be well-defined, depending on the asymptotic structure of the modified theory. Relative to the Bondi mass, in theories with different gravitational radiation properties or different treatments of null infinity, the definition of Bondi mass could also be modified. The relationship between energy carried away by gravitational waves and the reduction in Bondi mass could differ from those in GR.

In conclusion, the study of EMSG provides a compelling extension to existing gravitational theories, offering potential solutions to longstanding problems in cosmology and astrophysics. The formalism developed here lays a solid foundation for further theoretical exploration and potential empirical verification. By bridging the gap between gravity, thermodynamics, and quantum effects, EMSG emerges as a promising candidate for a more comprehensive theory of gravity, potentially furthering our understanding of the Universe.

Author Contributions: Conceptualization, R.A.C.C. and F.S.N.L.; methodology, R.A.C.C., N.G., T.H., F.S.N.L., M.A.S.P. and J.L.R.; software, R.A.C.C., N.G., T.H., F.S.N.L., M.A.S.P. and J.L.R.; validation, R.A.C.C., N.G., T.H., F.S.N.L., M.A.S.P. and J.L.R.; formal analysis, R.A.C.C., N.G., T.H., F.S.N.L., M.A.S.P. and J.L.R.; investigation, R.A.C.C., N.G., T.H., F.S.N.L., M.A.S.P. and J.L.R.; resources, R.A.C.C., N.G., T.H., F.S.N.L., M.A.S.P. and J.L.R.; writing—original draft preparation, R.A.C.C. and F.S.N.L.; writing—review and editing, R.A.C.C., N.G., T.H., F.S.N.L., M.A.S.P. and J.L.R.; visualization, R.A.C.C., N.G., T.H., F.S.N.L., M.A.S.P. and J.L.R.; supervision, F.S.N.L.; project administration, F.S.N.L.; funding acquisition, R.A.C.C., N.G., F.S.N.L. and M.A.S.P. All authors have read and agreed to the published version of the manuscript.

Funding: This research was funded by the Fundação para a Ciência e a Tecnologia (FCT) under research grants UIDB/04434/2020, UIDP/04434/2020, and PTDC/FIS-AST/0054/2021.

Data Availability Statement: Data are contained within the article.

Acknowledgments: F.S.N.L. acknowledges support from the Fundação para a Ciência e a Tecnologia (FCT) Scientific Employment Stimulus contract with reference CEECINST/00032/2018.

Conflicts of Interest: The authors declare no conflicts of interest.

Notes

1. The flaring-out condition in the neighborhood of the throat takes the form of $(b - b'r)/b^2 > 0$ [114].
2. While the choice of ζ_0 at this point is somewhat arbitrary, we chose this particular value for reasons that we clarify in the subsequent section. Various other values of ζ_0, including positive values, would yield qualitatively similar solutions.

References

1. Riess, A.G. et al. [Supernova Search Team]. Observational evidence from supernovae for an accelerating universe and a cosmological constant. *Astron. J.* **1998**, *116*, 1009–1038. [CrossRef]
2. Perlmutter, S. et al. [Supernova Cosmology Project]. Measurements of Ω and Λ from 42 High Redshift Supernovae. *Astrophys. J.* **1999**, *517*, 565–586. [CrossRef]
3. Capozziello, S. Curvature quintessence. *Int. J. Mod. Phys. D* **2002**, *11*, 483–492. [CrossRef]
4. Nojiri, S.; Odintsov, S.D. Introduction to modified gravity and gravitational alternative for dark energy. *Int. J. Geom. Methods Mod. Phys.* **2007** *4*, 115–145. [CrossRef]

5. Lobo, F.S.N. The Dark side of gravity: Modified theories of gravity. *arXiv* **2008**, arXiv:0807.1640.
6. Sotiriou, T.P.; Faraoni, V. $f(R)$ Theories of Gravity. *Rev. Mod. Phys.* **2010**, *82*, 451–497. [CrossRef]
7. Nojiri, S.; Odintsov, S.D. Unified cosmic history in modified gravity: From F(R) theory to Lorentz non-invariant models. *Phys. Rep.* **2011**, *505*, 59–144. [CrossRef]
8. Olmo, G.J. Palatini Approach to Modified Gravity: F(R) Theories and Beyond. *Int. J. Mod. Phys. D* **2011**, *20*, 413–462. [CrossRef]
9. Capozziello, S.; De Laurentis, M. Extended Theories of Gravity. *Phys. Rep.* **2011**, *509*, 167–321. [CrossRef]
10. Clifton, T.; Ferreira, P.G.; Padilla, A.; Skordis, C. Modified Gravity and Cosmology. *Phys. Rep.* **2012**, *513*, 1–189.
11. Harko, T.; Lobo, F.S.N. *Extensions of $f(R)$ Gravity: Curvature-Matter Couplings and Hybrid Metric-Palatini Theory;* Cambridge University Press: Cambridge, UK, 2018; ISBN 978-1-108-42874-3; 978-1-108-58457-9
12. Saridakis, E.N. et al. [CANTATA]. *Modified Gravity and Cosmology: An Update by the CANTATA Network;* Springer: Cham, Switzerland, 2021; ISBN 978-3-030-83714-3; 978-3-030-83717-4; 978-3-030-83715-0
13. Copeland, E.J.; Sami, M.; Tsujikawa, S. Dynamics of dark energy. *Int. J. Mod. Phys. D* **2006**, *15*, 1753–1936. [CrossRef]
14. Bertolami, O.; Boehmer, C.G.; Harko, T.; Lobo, F.S.N. Extra force in f(R) modified theories of gravity. *Phys. Rev. D* **2007**, *75*, 104016. [CrossRef]
15. Harko, T.; Lobo, F.S.N.; Otalora, G.; Saridakis, E.N. Nonminimal torsion-matter coupling extension of f(T) gravity. *Phys. Rev. D* **2014**, *89*, 124036. [CrossRef]
16. Harko, T.; Lobo, F.S.N.; Otalora, G.; Saridakis, E.N. $f(T,\mathcal{T})$ gravity and cosmology. *J. Cosmol. Astropart. Phys.* **2014**, *12*, 021. [CrossRef]
17. Harko, T.; Lobo, F.S.N. Generalized curvature-matter couplings in modified gravity. *Galaxies* **2014**, *2*, 410–465. [CrossRef]
18. Harko, T.; Koivisto, T.S.; Lobo, F.S.N.; Olmo, G.J.; Rubiera-Garcia, D. Coupling matter in modified Q gravity. *Phys. Rev. D* **2018**, *98*, 084043. [CrossRef]
19. Harko, T.; Lobo, F.S.N. $f(R,L_m)$ gravity. *Eur. Phys. J. C* **2010**, *70*, 373–379. [CrossRef]
20. Harko, T.; Lobo, F.S.N.; Nojiri, S.; Odintsov, S.D. $f(R,T)$ gravity. *Phys. Rev. D* **2011**, *84*, 024020. [CrossRef]
21. Barrientos, E.; Lobo, F.S.N.; Mendoza, S.; Olmo, G.J.; Rubiera-Garcia, D. Metric-affine f(R,T) theories of gravity and their applications. *Phys. Rev. D* **2018**, *97*, 104041. [CrossRef]
22. Katırcı, N.; Kavuk, M. $f(R,T_{\mu\nu}T^{\mu\nu})$ gravity and Cardassian-like expansion as one of its consequences. *Eur. Phys. J. Plus* **2014**, *129*, 163. [CrossRef]
23. Roshan, M.; Shojai, F. Energy-Momentum Squared Gravity. *Phys. Rev. D* **2016**, *94*, 044002. [CrossRef]
24. Haghani, Z.; Harko, T.; Lobo, F.S.N.; Sepangi, H.R.; Shahidi, S. Further matters in space-time geometry: $f(R,T,R_{\mu\nu}T^{\mu\nu})$ gravity. *Phys. Rev. D***2013**, *88*, 044023. [CrossRef]
25. Odintsov, S.D.; Sáez-Gómez, D. $f(R,T,R_{\mu\nu}T^{\mu\nu})$ gravity phenomenology and ΛCDM universe. *Phys. Lett. B* **2013**, *725*, 437–444. [CrossRef]
26. Bahamonde, S.; Marciu, M.; Rudra, P. Dynamical system analysis of generalized energy-momentum-squared gravity. *Phys. Rev. D* **2019**, *100*, 083511. [CrossRef]
27. Akarsu, Ö.; Katirci, N.; Kumar, S. Energy-momentum powered gravity and cosmic acceleration. In Proceedings of the PoS CORFU2017, Corfu, Greece, 2–28 September 2017; p. 105.
28. Akarsu, Ö.; Barrow, J.D.; Uzun, N.M. Screening anisotropy via energy-momentum squared gravity: ΛCDM model with hidden anisotropy. *Phys. Rev. D* **2020**, *102*, 124059. [CrossRef]
29. Board, C.V.R.; Barrow, J.D. Cosmological Models in Energy-Momentum-Squared Gravity. *Phys. Rev. D* **2017**, *96*, 123517; Erratum: *Phys. Rev. D* **2018**, *98*, 129902. [CrossRef]
30. Barbar, A.H.; Awad, A.M.; AlFiky, M.T. Viability of bouncing cosmology in energy-momentum-squared gravity. *Phys. Rev. D* **2020**, *101*, 044058. [CrossRef]
31. Akarsu, Ö.; Katırcı, N.; Kumar, S. Cosmic acceleration in a dust only universe via energy-momentum powered gravity. *Phys. Rev. D* **2018**, *97*, 024011. [CrossRef]
32. Cipriano, R.A.C.; Harko, T.; Lobo, F.S.N.; Pinto, M.A.S.; Rosa, J.L. Gravitationally induced matter creation in scalar-tensor $f(R,T_{\mu\nu}T^{\mu\nu})$ gravity. *Phys. Dark Universe* **2024**, *44*, 101463 [CrossRef]
33. Sharif, M.; Zeeshan Gul, M. Stability Analysis of the Inhomogeneous Perturbed Einstein Universe in Energy–Momentum Squared Gravity. *Universe* **2023**, *9*, 145. [CrossRef]
34. Canuto, Á.J.C.; Santos, A.F. Gödel-type universes in energy–momentum-squared gravity. *Eur. Phys. J. C* **2023**, *83*, 404. [CrossRef]
35. Shahidi, S. Non-minimal energy–momentum squared gravity. *Eur. Phys. J. C* **2021**, *81*, 274. [CrossRef]
36. Akarsu, Ö.; Barrow, J.D.; Çıkıntoğlu, S.; Ekşi, K.Y.; Katırcı, N. Constraint on energy-momentum squared gravity from neutron stars and its cosmological implications. *Phys. Rev. D* **2018**, *97*, 124017. [CrossRef]
37. Chen, C.Y.; Chen, P. Eikonal black hole ringings in generalized energy-momentum squared gravity. *Phys. Rev. D* **2020**, *101*, 064021. [CrossRef]
38. Sharif, M.; Zeeshan Gul, M. Dynamics of charged anisotropic spherical collapse in energy-momentum squared gravity. *Chin. J. Phys.* **2021**, *71*, 365–374. [CrossRef]
39. Yousaf, Z.; Bhatti, M.Z.; Farwa, U. Evolution of axially and reflection symmetric source in energy–momentum squared gravity. *Eur. Phys. J. Plus* **2022**, *137*, 49. [CrossRef]

40. Sharif, M.; Naz, S. Stable charged gravastar model in $f(\Re, \mathbf{T}^2)$ gravity with conformal motion. *Eur. Phys. J. Plus* **2022**, *137*, 421. [CrossRef]
41. Sharif, M.; Naz, S. Impact of charge on gravastars in $f(\Re, \mathbf{T}^2)$ gravity. *Mod. Phys. Lett. A* **2022**, *37*, 2250065. [CrossRef]
42. Sharif, M.; Anjum, A. Impact of charge on the complexity of static sphere in $f(R, \mathbf{T}^2)$ gravity. *Eur. Phys. J. Plus* **2022**, *137*, 602. [CrossRef]
43. Sharif, M.; Gul, M.Z. Anisotropic compact stars with Karmarkar condition in energy-momentum squared gravity. *Gen. Relativ. Gravit.* **2023**, *55*, 10. [CrossRef]
44. Zeeshan Gul, M.; Sharif, M. Traversable Wormhole Solutions Admitting Noether Symmetry in Theory. *Symmetry* **2023**, *15*, 684.
45. Sharif, M.; Naz, S. Study of decoupled gravastars in energy–momentum squared gravity. *Ann. Phys.* **2023**, *451*, 169240. [CrossRef]
46. Sharif, M.; Naz, S. Viable decoupled solutions in energy–momentum squared gravity. *Pramana* **2023**, *97*, 116. [CrossRef]
47. Sharif, M.; Naz, S. Stable gravastars with Krori–Barua metric in f(R,T2) gravity. *Ann. Phys.* **2023**, *457*, 169426. [CrossRef]
48. Sharif, M.; Gul, M.Z. Compact stellar objects in $f(\Re, \mathbf{T}^2)$ gravity. *Pramana* **2023**, *97*, 122. [CrossRef]
49. Kazemi, A.; Roshan, M.; De Martino, I.; De Laurentis, M. Jeans analysis in energy–momentum-squared gravity. *Eur. Phys. J. C* **2020**, *80*, 150. [CrossRef]
50. Sharif, M.; Manzoor, S. Compact objects admitting Finch–Skea symmetry in f(R,T2) gravity. *Ann. Phys.* **2023**, *454*, 169337. [CrossRef]
51. Hosseini Mansoori, S.A.; Felegray, F.; Talebian, A.; Sami, M. PBHs and GWs from $\mathbb{T}^2$-inflation and NANOGrav 15-year data. *J. Cosmol. Astropart. Phys.* **2023**, *08*, 067. [CrossRef]
52. Sharif, M.; Zeeshan Gul, M. Noether Symmetries and Some Exact Solutions in f(R, T ^{2}) Theory. *J. Exp. Theor. Phys.* **2023**, *136*, 436–445. [CrossRef]
53. Nasir, M.M.M.; Bhatti, M.Z.; Yousaf, Z. Influence of EMSG on complex systems: Spherically symmetric, static case. *Int. J. Mod. Phys. D* **2023**, *32*, 2350009. [CrossRef]
54. Pretel, J.M.Z.; Tangphati, T.; Banerjee, A. Relativistic structure of charged quark stars in energy–momentum squared gravity. *Ann. Phys.* **2023**, *458*, 169440. [CrossRef]
55. Naz, S.; Sharif, M. Gravastars with Kuchowicz Metric in Energy-Momentum Squared Gravity. *Universe* **2022**, *8*, 142. [CrossRef]
56. Sharif, M.; Naz, S. Gravastars with Karmarkar condition in $f(R, T^2)$ gravity. *Int. J. Mod. Phys. D* **2022**, *31*, 2240008. [CrossRef]
57. Rudra, P. Energy–momentum squared symmetric Teleparallel gravity: $f(Q, T_{\mu\nu}T^{\mu\nu})$ gravity. *Phys. Dark Universe* **2022**, *37*, 101071. [CrossRef]
58. Nazari, E.; Roshan, M.; De Martino, I. Constraining energy-momentum-squared gravity by binary pulsar observations. *Phys. Rev. D* **2022**, *105*, 044014. [CrossRef]
59. Ashtekar, A.; Singh, P. Loop Quantum Cosmology: A Status Report. *Class. Quantum Gravity* **2011**, *28*, 213001. [CrossRef]
60. Maartens, R. Brane world gravity. *Living Rev. Relativ.* **2004**, *7*, 7. [CrossRef]
61. Maartens, R.; Koyama, K. Brane-World Gravity. *Living Rev. Relativ.* **2010**, *13*, 5. [CrossRef]
62. Brax, P.; van de Bruck, C. Cosmology and brane worlds: A Review. *Class. Quantum Gravity* **2003**, *20*, R201–R232. [CrossRef]
63. Armendariz-Picon, C.; Mukhanov, V.F.; Steinhardt, P.J. Essentials of k essence. *Phys. Rev. D* **2001**, *63*, 103510. [CrossRef]
64. Nazari, E.; Sarvi, F.; Roshan, M. Generalized Energy-Momentum-Squared Gravity in the Palatini Formalism. *Phys. Rev. D* **2020**, *102*, 064016. [CrossRef]
65. Sharif, M.; Zeeshan Gul, M. Viable wormhole solutions in energy–momentum squared gravity. *Eur. Phys. J. Plus* **2021**, *136*, 503. [CrossRef]
66. Woodard, R.P. Avoiding dark energy with 1/r modifications of gravity. *Lect. Notes Phys.* **2007**, *720*, 403–433.
67. Ostrogradski, M. *Mémoires sur les Équations Différentielles, Relatives au Problème des Isopérimètres*; Mémoires de l'Académie Impériale des Sciences: St-Pétersbourg, Russia, 1850; Volume 4, p. 385.
68. Abedi, H.; Bajardi, F.; Capozziello, S. Linearized field equations and extra force in f(R,T(n)) extended gravity. *Int. J. Mod. Phys. D* **2022**, *31*, 2240015. [CrossRef]
69. Schrödinger, E. The proper vibrations of the expanding universe. *Physica* **1939**, *6*, 899. [CrossRef]
70. Schrödinger, E. The General Theory of Relativity and Wave Mechanics. *Physica* **1940**, *46*, 25.
71. Parker, L. Particle creation and particle number in an expanding universe. *J. Phys. A* **2012**, *45*, 374023. [CrossRef]
72. Parker, L. Particle creation in expanding universes. *Phys. Rev. Lett.* **1968**, *21*, 562–564. [CrossRef]
73. Parker, L. Quantized fields and particle creation in expanding universes. 1. *Phys. Rev.* **1969**, *183*, 1057–1068. [CrossRef]
74. Parker, L. Quantized fields and particle creation in expanding universes. 2. *Phys. Rev. D* **1971**, *3*, 346–356; Erratum in *Phys. Rev. D* **1971**, *3*, 2546–2546. [CrossRef]
75. Parker, L. Particle creation in isotropic cosmologies. *Phys. Rev. Lett.* **1972**, *28*, 705–708; Erratum in *Phys. Rev. Lett.* **1972**, *28*, 1497. [CrossRef]
76. Prigogine, I.; Geheniau, J.; Gunzig, E.; Nardone, P. Thermodynamics and cosmology. *Gen. Relativ. Gravit.* **1989**, *21*, 767–776. [CrossRef]
77. Prigogine, I.; Géhéniau, J. Entropy, matter, and cosmology. *Proc. Nat. Acad. Sci. USA* **1986**, *83*, 6245–6249. [CrossRef] [PubMed]
78. Prigogine, I.; Geheniau, J.; Gunzig, E.; Nardone, P. Thermodynamics of cosmological matter creation. *Proc. Nat. Acad. Sci. USA* **1988**, *85*, 7428. [CrossRef]
79. Ford, L.H. Gravitational Particle Creation and Inflation. *Phys. Rev. D* **1987**, *35*, 2955. [CrossRef]

80. Brout, R.; Englert, F.; Spindel, P. Cosmological Origin of the Grand Unification Mass Scale. *Phys. Rev. Lett.* **1979**, *43*, 417; Erratum in *Phys. Rev. Lett.* **1979**, *43*, 890. [CrossRef]
81. Grib, A.A.; Mamaev, S.G.; Mostepanenko, V.M. Particle Creation from Vacuum in Homogeneous Isotropic Models of the Universe. *Gen. Relativ. Gravit.* **1976**, *7*, 535–547 [CrossRef]
82. Harko, T. Thermodynamic interpretation of the generalized gravity models with geometry—Matter coupling. *Phys. Rev. D* **2014**, *90*, 044067. [CrossRef]
83. Pinto, M.A.S.; Harko, T.; Lobo, F.S.N. Gravitationally induced particle production in scalar-tensor f(R,T) gravity. *Phys. Rev. D* **2022**, *106*, 044043. [CrossRef]
84. Harko, T.; Lobo, F.S.N.; Mimoso, J.P.; Pavón, D. Gravitational induced particle production through a nonminimal curvature–matter coupling. *Eur. Phys. J. C* **2015**, *75*, 386. [CrossRef]
85. Pinto, M.A.S.; Harko, T.; Lobo, F.S.N. Irreversible Geometrothermodynamics of Open Systems in Modified Gravity. *Entropy* **2023**, *25*, 944. [CrossRef] [PubMed]
86. Chernodub, M.N. Conformal anomaly and gravitational pair production. *arXiv* **2023**, arXiv:2306.03892.
87. Berezin, V.A.; Dokuchaev, V.I. Cosmological particle creation in Weyl geometry. *Class. Quantum Gravity* **2023**, *40*, 015006. [CrossRef]
88. Xue, S.S. Massive particle pair production and oscillation in Friedman Universe: Reheating energy and entropy, and cold dark matter. *Eur. Phys. J. C* **2023**, *83*, 355. [CrossRef]
89. Schutz, F. Perfect Fluids in General Relativity: Velocity Potentials and a Variational Principle. *Phys. Rev. D* **1970**, *2*, 2762. [CrossRef]
90. Brown, J.D. Action functionals for relativistic perfect fluids. *Class. Quantum Gravity* **1993**, *10*, 1579. [CrossRef]
91. Bertolami, O.; Lobo, F.S.N.; Paramos, J. Non-minimum coupling of perfect fluids to curvature. *Phys. Rev. D* **2008**, *78*, 064036. [CrossRef]
92. Dodelson, S. *Modern Cosmology*; Academic Press: San Diego, CA, USA, 2003.
93. Berera, A.; Fang, L.Z. Thermally Induced Density Perturbations in the Inflation Era. *Phys. Rev. Lett.* **1995**, *74*, 1912. [CrossRef]
94. Berera, A. Warm Inflation. *Phys. Rev. Lett.* **1995**, *75*, 3218. [CrossRef]
95. Harko, T.; Sheikhahmadi, H. Irreversible thermodynamical description of warm inflationary cosmological models. *Phys. Dark Universe* **2020**, *28*, 100521. [CrossRef]
96. Berera, A. The Warm Inflation Story. *Universe* **2023**, *9*, 272. [CrossRef]
97. Kamali, V.; Motaharfar, M.; Ramos, R.O. Recent Developments in Warm Inflation. *Universe* **2023**, *9*, 124. [CrossRef]
98. Matei, T.M.; Harko, T. Warm inflation in a Universe with a Weylian boundary. *Phys. Dark Universe* **2024**, *46*, 101578. [CrossRef]
99. Nari, N.; Roshan, M. Compact stars in Energy-Momentum Squared Gravity. *Phys. Rev. D* **2018**, *98*, 024031. [CrossRef]
100. Singh, K.N.; Banerjee, A.; Maurya, S.K.; Rahaman, F.; Pradhan, A. Color-flavor locked quark stars in energy–momentum squared gravity. *Phys. Dark Universe* **2021**, *31*, 100774. [CrossRef]
101. Sharif, M.; Zeeshan Gul, M. Dynamics of spherical collapse in energy–momentum squared gravity. *Int. J. Mod. Phys. A* **2021**, *36*, 2150004. [CrossRef]
102. Rudra, P.; Pourhassan, B. Thermodynamics of the apparent horizon in the generalized energy–momentum-squared cosmology. *Phys. Dark Universe* **2021**, *33*, 100849. [CrossRef]
103. Khodadi, M.; Firouzjaee, J.T. A survey of strong cosmic censorship conjecture beyond Einstein's gravity. *Phys. Dark Universe* **2022**, *37*, 101084. [CrossRef]
104. Moraes, P.H.R.S.; Sahoo, P.K. Nonexotic matter wormholes in a trace of the energy-momentum tensor squared gravity. *Phys. Rev. D* **2018**, *97*, 024007. [CrossRef]
105. Rosa, J.L.; Ganiyeva, N.; Lobo, F.S.N. Non-exotic traversable wormholes in $f\left(R, T_{ab}T^{ab}\right)$ gravity. *Eur. Phys. J. C* **2023**, *83*, 1040.
106. Rosa, J.L. Junction conditions in gravity theories with extra scalar degrees of freedom. *Phys. Rev. D* **2024**, *109*, 064018. [CrossRef]
107. Lobo, F.S.N.; Rodrigues, M.E.; de Sousa Silva, M.V.; Simpson, A.; Visser, M. Novel black-bounce spacetimes: Wormholes, regularity, energy conditions, and causal structure. *Phys. Rev. D* **2021**, *103*, 084052. [CrossRef]
108. Rodrigues, M.E.; Silva, M.V.d.S. Source of black bounces in general relativity. *Phys. Rev. D* **2023**, *107*, 044064. [CrossRef]
109. Junior, J.T.S.S.; Rodrigues, M.E. Coincident $f(\mathbb{Q})$ gravity: Black holes, regular black holes, and black bounces. *Eur. Phys. J. C* **2023**, *83*, 475. [CrossRef]
110. Fabris, J.C.; Junior, E.L.B.; Rodrigues, M.E. Generalized models for black-bounce solutions in f(R) gravity. *Eur. Phys. J. C* **2023**, *83*, 884. [CrossRef]
111. Junior, J.T.S.S.; Lobo, F.S.N.; Rodrigues, M.E. (Regular) Black holes in conformal Killing gravity coupled to nonlinear electrodynamics and scalar fields. *Class. Quantum Gravity* **2024**, *41*, 055012. [CrossRef]
112. Junior, J.T.S.S.; Lobo, F.S.N.; Rodrigues, M.E. Black holes and regular black holes in coincident $f(\mathbb{Q}, \mathbb{B}_Q)$ gravity coupled to nonlinear electrodynamics. *Eur. Phys. J. C* **2024**, *84*, 332. [CrossRef]
113. Junior, J.T.S.S.; Lobo, F.S.N.; Rodrigues, M.E. Black bounces in conformal Killing gravity. *Eur. Phys. J. C* **2024**, *84*, 557. [CrossRef]
114. Morris, M.S.; Thorne, K.S. Wormholes in spacetime and their use for interstellar travel: A tool for teaching general relativity. *Am. J. Phys.* **1988**, *56*, 395. [CrossRef]
115. Morris, M.S.; Thorne, K.S.; Yurtsever, U. Wormholes, Time Machines, and the Weak Energy Condition. *Phys. Rev. Lett.* **1988**, *61*, 1446–1449. [CrossRef]
116. Visser, M. *Lorentzian Wormholes: From Einstein to Hawking*; Springer: New York, NY, USA, 1996.

117. Lemos, J.P.S.; Lobo, F.S.N.; Quinet de Oliveira, S. Morris-Thorne wormholes with a cosmological constant. *Phys. Rev. D* **2003**, *68*, 064004. [CrossRef]
118. Visser, M.; Kar, S.; Dadhich, N. Traversable wormholes with arbitrarily small energy condition violations. *Phys. Rev. Lett.* **2003**, *90*, 201102. [CrossRef]
119. Kar, S.; Sahdev, D. Evolving Lorentzian wormholes. *Phys. Rev. D* **1996**, *53*, 722–730. [CrossRef] [PubMed]
120. Hawking, S.W.; Ellis, G.F.R. *The Large Scale Structure of Space-Time*; Cambridge University Press: Cambridge, UK, 1973. ISBN 978-1-00-925316-1.
121. Sajadi, S.N.; Riazi, N. Gravitational lensing by multi-polytropic wormholes. *Can. J. Phys.* **2020**, *98*, 1046–1054. [CrossRef]
122. Agnese, A.G.; La Camera, M. Wormholes in the Brans-Dicke theory of gravitation. *Phys. Rev. D* **1995**, *51*, 2011–2013. [CrossRef]
123. Nandi, K.K.; Bhattacharjee, B.; Alam, S.M.K.; Evans, J. Brans-Dicke wormholes in the Jordan and Einstein frames. *Phys. Rev. D* **1998**, *57*, 823–828. [CrossRef]
124. La Camera, M. Wormhole solutions in the Randall-Sundrum scenario. *Phys. Lett. B* **2003**, *573*, 27–32. [CrossRef]
125. Lobo, F.S.N. Exotic solutions in general relativity: Traversable wormholes and 'warp drive' spacetimes. In *Classical and Quantum Gravity Research*; Christiansen, M.N., Rasmussen, T.K., Eds.; Nova Science Publishers: Hauppauge, NY, USA, 2008; p. 1.
126. Garattini, R.; Lobo, F.S.N. Self sustained phantom wormholes in semi-classical gravity. *Class. Quantum Gravity* **2007**, *24*, 2401–2413. [CrossRef]
127. Lobo, F.S.N. General class of wormhole geometries in conformal Weyl gravity. *Class. Quantum Gravity* **2008**, *25*, 175006. [CrossRef]
128. Garattini, R.; Lobo, F.S.N. Self-sustained traversable wormholes in noncommutative geometry. *Phys. Lett. B* **2009**, *671*, 146–152. [CrossRef]
129. Lobo, F.S.N.; Oliveira, M.A. General class of vacuum Brans-Dicke wormholes. *Phys. Rev. D* **2010**, *81*, 067501. [CrossRef]
130. Montelongo Garcia, N.; Lobo, F.S.N. Exact solutions of Brans-Dicke wormholes in the presence of matter. *Mod. Phys. Lett. A* **2011**, *40*, 3067–3076. [CrossRef]
131. Garattini, R.; Lobo, F.S.N. Self-sustained wormholes in modified dispersion relations. *Phys. Rev. D* **2012**, *85*, 024043. [CrossRef]
132. Myrzakulov, R.; Sebastiani, L.; Vagnozzi, S.; Zerbini, S. Static spherically symmetric solutions in mimetic gravity: Rotation curves and wormholes. *Class. Quantum Gravity* **2016**, *33*, 125005. [CrossRef]
133. Lobo, F.S.N. (Ed.) *Wormholes, Warp Drives and Energy Conditions*; Fundamental Theories of Physics; Springer: New York, NY, USA, 2017; Volume 189.
134. Lobo, F.S.N.; Oliveira, M.A. Wormhole geometries in f(R) modified theories of gravity. *Phys. Rev. D* **2009**, *80*, 104012. [CrossRef]
135. Capozziello, S.; Harko, T.; Koivisto, T.S.; Lobo, F.S.N.; Olmo, G.J. Wormholes supported by hybrid metric-Palatini gravity. *Phys. Rev. D* **2012**, *86*, 127504. [CrossRef]
136. Rosa, J.L.; Lemos, J.P.S.; Lobo, F.S.N. Wormholes in generalized hybrid metric-Palatini gravity obeying the matter null energy condition everywhere. *Phys. Rev. D* **2018**, *98*, 064054. [CrossRef]
137. Rosa, J.L. Double gravitational layer traversable wormholes in hybrid metric-Palatini gravity. *Phys. Rev. D* **2021**, *104*, 064002. [CrossRef]
138. Rosa, J.L.; Lemos, J.P.S. Junction conditions for generalized hybrid metric-Palatini gravity with applications. *Phys. Rev. D* **2021**, *104*, 124076. [CrossRef]
139. Rosa, J.L.; André, R.; Lemos, J.P.S. Traversable wormholes with double layer thin shells in quadratic gravity. *Gen. Relativ. Gravit.* **2023**, *55*, 65. [CrossRef]
140. Rosa, J.L.; Kull, P.M. Non-exotic traversable wormhole solutions in linear $f(R,T)$ gravity. *Eur. Phys. J. C* **2022**, *82*, 1154. [CrossRef]
141. Garcia, N.M.; Lobo, F.S.N. Wormhole geometries supported by a nonminimal curvature-matter coupling. *Phys. Rev. D* **2010**, *82*, 104018. [CrossRef]
142. Montelongo Garcia, N.; Lobo, F.S.N. Nonminimal curvature-matter coupled wormholes with matter satisfying the null energy condition. *Class. Quantum Gravity* **2011**, *28*, 085018. [CrossRef]
143. Harko, T.; Lobo, S.N.; Mak, M.K.; Sushkov, S.V. Modified-gravity wormholes without exotic matter. *Phys. Rev. D* **2013**, *87*, 067504. [CrossRef]
144. Anchordoqui, L.A.; Perez Bergliaffa, S.E.; Torres, D.F. Brans-Dicke wormholes in nonvacuum space-time. *Phys. Rev. D* **1997**, *55*, 5226–5229. [CrossRef]
145. Di Grezia, E.; Battista, E.; Manfredonia, M.; Miele, G. Spin, torsion and violation of null energy condition in traversable wormholes. *Eur. Phys. J. Plus* **2017**, *132*, 537. [CrossRef]
146. Bhawal, B.; Kar, S. Lorentzian wormholes in Einstein—Gauss-Bonnet theory. *Phys. Rev. D* **1992**, *46*, 2464. [CrossRef]
147. Dotti, G.; Oliva, J.; Troncoso, R. Exact solutions for the Einstein-Gauss-Bonnet theory in five dimensions: Black holes, wormholes and spacetime horns. *Phys. Rev. D* **2007**, *76*, 064038. [CrossRef]
148. Mehdizadeh, M.R.; Kord Zangeneh, M.; Lobo, F.S.N. Einstein-Gauss-Bonnet traversable wormholes satisfying the weak energy condition. *Phys. Rev. D* **2015**, *91*,084004. [CrossRef]
149. Bronnikov, K.A.; Kim, S.W. Possible wormholes in a brane world. *Phys. Rev. D* **2003**, *67*, 064027. [CrossRef]
150. Lobo, F.S.N. A General class of braneworld wormholes. *Phys. Rev. D* **2007**, *75*, 064027. [CrossRef]
151. Israel, W. Singular hypersurfaces and thin shells in general relativity. *Nuovo Cimento B* **1966**, *44*, 1–14. [CrossRef]
152. Visser, M. Traversable wormholes: Some simple examples. *Phys. Rev. D* **1989**, *39*, 3182. [CrossRef] [PubMed]
153. Visser, M. Traversable wormholes from surgically modified Schwarzschild spacetimes. *Nucl. Phys. B* **1989**, *328*, 203. [CrossRef]

154. Lobo, F.S.N. Energy conditions, traversable wormholes and dust shells. *Gen. Relativ. Gravit.* **2005**, *37*, 2023–2038. [CrossRef]
155. Lobo, F.S.N. Surface stresses on a thin shell surrounding a traversable wormhole. *Gen. Relativ. Gravit.* **2004**, *21*, 4811–4832. [CrossRef]
156. Lobo, F.S.N. Phantom energy traversable wormholes. *Phys. Rev. D* **2005**, *71*, 084011. [CrossRef]
157. Lobo, F.S.N. Stability of phantom wormholes. *Phys. Rev. D* **2005**, *71*, 124022. [CrossRef]
158. Schwarzschild, K. *Über das Gravitationsfeld Einer Kugel aus Inkompressibler Flüssigkeit Nach der Einsteinschen Theorie*; Sitzungsberichte der Königlich-Preussischen Akademie der Wissenschaften Berlin: Berlin, Germany, 1916; pp. 424–434.
159. Rosa, J.L.; Piçarra, P. Existence and stability of relativistic fluid spheres supported by thin-shells. *Phys. Rev. D* **2020**, *102*, 6. [CrossRef]
160. Rosa, J.L. Observational properties of relativistic fluid spheres with thin accretion disks. *arXiv* **2023**, arXiv:2302.11915.
161. Oppenheimer, J.R.; Snyder, H. On Continued Gravitational Contraction. *Phys. Rev.* **1939**, *56*, 455. [CrossRef]
162. Rosa, J.L.; Carloni, S. Junction conditions for general LRS spacetimes in the $1+1+2$ covariant formalism. *arXiv* **2023**, arXiv:2303.12457.
163. Senovilla, J.M.M. Junction conditions for F(R)-gravity and their consequences. *Phys. Rev. D* **2013**, *88*, 064015. [CrossRef]
164. Vignolo, S.; Cianci, R.; Carloni, S. On the junction conditions in $f(R)$-gravity with torsion. *Class. Quantum Gravity* **2018**, *35*, 095014. [CrossRef]
165. Reina, B.; Senovilla, J.M.M.; Vera, R. Junction conditions in quadratic gravity: Thin shells and double layers. *Class. Quantum Gravity* **2016**, *33*, 105008. [CrossRef]
166. Deruelle, N.; Sasaki, M.; Sendouda, Y. Junction conditions in f(R) theories of gravity. *Prog. Theor. Exp. Phys.* **2008**, *119*, 237–251. [CrossRef]
167. Olmo, G.J.; Rubiera-Garcia, D. Junction conditions in Palatini $f(R)$ gravity. *Class. Quantum Gravity* **2020**, *37*, 215002. [CrossRef]
168. Rosa, J.L. Junction conditions and thin shells in perfect-fluid $f(R,T)$ gravity. *Phys. Rev. D* **2021**, *103*, 104069. [CrossRef]
169. Rosa, J.L.; Rubiera-Garcia, D. Junction conditions of Palatini f(R,T) gravity. *Phys. Rev. D* **2022**, *106*, 064007. [CrossRef]
170. Suffern, K.G. Singular hypersurfaces in the Brans-Dicke theory of gravity. *J. Phys. A Math. Gen.* **1982** *15*, 1599. [CrossRef]
171. Barrabes, C.; Bressange, G.F. Singular hypersurfaces in scalar—Tensor theories of gravity. *Class. Quantum Gravity* **1997**, *14*, 805–824. [CrossRef]
172. Padilla, A.; Sivanesan, V. Boundary Terms and Junction Conditions for Generalized Scalar-Tensor Theories. *J. High Energy Phys.* **2012**, *08*, 122. [CrossRef]
173. Casado-Turrión, A.; de la Cruz-Dombriz, Á.; Jiménez-Cano, A.; Maldonado Torralba, F.J. Junction conditions in bi-scalar Poincaré gauge gravity. *J. Cosmol. Astropart. Phys.* **2023**, *07*, 023. [CrossRef]
174. de la Cruz-Dombriz, Á.; Dunsby, P.K.S.; Saez-Gomez, D. Junction conditions in extended Teleparallel gravities. *J. Cosmol. Astropart. Phys.* **2014**, *12*, 048. [CrossRef]
175. Arkuszewski, W.; Kopczynski, W.; Ponomarev, V.N. Matching Conditions in the Einstein-Cartan Theory of Gravitation. *Commun. Math. Phys.* **1975**, *45*, 183–190. [CrossRef]
176. Macias, A.; Lammerzahl, C.; Pimentel, L.O. Matching conditions in metric affine gravity. *Phys. Rev. D* **2002**, *66*, 104013. [CrossRef]
177. Schwarzschild, K. On the gravitational field of a mass point according to Einstein's theory. *Sitzungsber. Preuss. Akad. Wiss. Berlin Math. Phys.* **1916**, *1916*, 189–196.
178. Hartle, J.B. Slowly rotating relativistic stars. 1. Equations of structure. *Astrophys. J.* **1967**, *150*, 1005–1029. [CrossRef]
179. Hartle, J.B.; Thorne, K.S. Slowly Rotating Relativistic Stars. II. Models for Neutron Stars and Supermassive Stars. *Astrophys. J.* **1968**, *153*, 807. [CrossRef]
180. Komar, A. Covariant conservation laws in general relativity. *Phys. Rev.* **1959**, *113*, 934–936. [CrossRef]
181. Wald, R.M. *General Relativity*; The University of Chicago Press: Chicago, IL, USA, 1984.
182. Arnowitt, R.L.; Deser, S.; Misner, C.W. The Dynamics of general relativity. *Gen. Relativ. Gravit.* **2008**, *40*, 1997–2027. [CrossRef]
183. Misner, C.W.; Thorne, K.S.; Wheeler, J.A. *Gravitation*; W. H. Freeman: New York, NY, USA, 1973; ISBN 978-0-7167-0344-0; 978-0-691-17779-3.
184. Bondi, H.; van der Burg, M.G.J.; Metzner, A.W.K. Gravitational waves in general relativity. 7. Waves from axisymmetric isolated systems. *Proc. R. Soc. Lond. A* **1962**, *269*, 21–52.
185. Sachs, R.K. Gravitational waves in general relativity. 8. Waves in asymptotically flat space-times. *Proc. R. Soc. Lond. A* **1962**, *270*, 103–126.

MDPI AG
Grosspeteranlage 5
4052 Basel
Switzerland
Tel.: +41 61 683 77 34

Universe Editorial Office
E-mail: universe@mdpi.com
www.mdpi.com/journal/universe

www.ingramcontent.com/pod-product-compliance
Lightning Source LLC
Chambersburg PA
CBHW061507220926
51850CB00004B/189